International Conference on Computer Applications in Shipbuilding (ICCAS 2022)

Yokohama, Japan
13 - 15 September 2022

ISBN: 978-1-7138-8296-1

Printed from e-media with permission by:

Curran Associates, Inc.
57 Morehouse Lane
Red Hook, NY 12571

Some format issues inherent in the e-media version may also appear in this print version.

The Royal Institution of Naval Architects
8-9 Northumberland Street
London, WC2N 5DA
United Kingdom

Phone: 020 7235 4622
Fax: 020 7259 5912

publications@rina.org.uk

Additional copies of this publication are available from:

Curran Associates, Inc.
57 Morehouse Lane
Red Hook, NY 12571 USA
Phone: 845-758-0400
Fax: 845-758-2633
Email: curran@proceedings.com
Web: www.proceedings.com

160 YEARS OF HISTORY, KNOWLEDGE AND EXPERIENCE

THE ROYAL INSTITUTION OF NAVAL ARCHITECTS

International Conference on Computer Applications in Shipbuilding ICCAS 2022

13th-15th September 2022
Yokohama, Japan

Sponsored by:

BETA SIMULATION SOLUTIONS
CADMATIC
ClassNK
PROSTEP integrate the future
SSI
SIEMENS

TOGETHER WE WILL SHAPE OUR FUTURE

THE ROYAL INSTITUTION
OF NAVAL ARCHITECTS

International Conference on Computer Applications in Shipbuilding

ICCAS 2022

13th -15th of September 2022
Yokohama, Japan

Papers published in these conference Proceedings are peer reviewed.

THE ROYAL INSTITUTION OF NAVAL ARCHITECTS
8-9 Northumberland Street
London WC2N 5DA

Telephone: 020 7235 4622
Fax: 020 7259 5912

ISSN No: 2752-6569

CONTENTS

STREAMLINING THE SHIP STRUCTURAL OPTIMIZATION PROCESS BY USING AN EARLY 3D PRODUCT MODEL 1
M Amano, Y Tanaka, Sumitomo Heavy Industries Marine & Engineering Co, Japan
T Masui, T Hulkkonen NAPA, Japan/Finland

SUSTAINABLE SHIP DESIGN WITH MODULAR SHIPBUIDLING APPROACH 15
R Audoire, Dassault Systèmes, France

WIND ASSISTED PROPULSION: INVESTIGATIONS ON THE USE OF SHIP HOLISTIC MODELS FOR PERFORMANCE ANALYSIS 27
A Bellot[1], A I Ölçer, R Baumler, World Maritime University, Sweden
C Bouallou, M Nemer, Mines Paris[1], France

DIGITAL ENTERPRISE PLATFORM - ENABLING EFFICIENT CAD, PLM, ERP INTEGRATION ON SHIPYARDS 49
J Bitomsky, M Grau, J Lützenberger and C Zerbst, PROSTEP AG, Germany

SMART DIGITAL SHIPYARDS WITH MODEL-BASED MANUFACTURING 59
O Chouche, Dassault Systèmes, France

COUPLING A HIGH FIDELITY NEAR SURFACE EFFECTS MODEL INTO PARAMARINE S4 65
P Crossland and C J M Forrest, QinetiQ Ltd, UK

AUTONOMOUS MOBILE ROBOTS' INTRODUCTION IN SHIPYARDS' MANUFACTURING PROCESS 77
C Dentesano, Fincantieri S.p.A., Italy

AUTOMATED GENERATION AND LOW-EFFORT AUTHORING OF COMMISSIONING CONTENT IN THE MARITIME INDUSTRY 85
A Elzalabany, V Settler, T Jansen, R Rost and H Lödding Hamburg University of Technology, Germany

SIMULATION TECHNIQUES IN THE DEVELOPMENT OF THE QUIESCENT PERIOD PREDICTION (QPP) FLIGHT DECK MOTION FORECASTING TOOL 97
B Ferrier, Hoffman Engineering, USA
R J Taylor, Royal Navy, UK
M R Belmont, J Christmas, University of Exeter, UK
L Fedrick, MOD, UK

VIRTUAL REALITY BEYOND DESIGN REVIEWS IN SHIPBUILDING: THE NEED FOR INDUSTRY-TAILORED IMMERSIVE DATA INTERACTION 111
N Fourrier1, M Benaouicha, SEGULA Technologies, France
G Moreau, IMT Atlantique, France
JM Normand, Nantes Université1, France

DESIGN OF A MOBILE APPLICATION TO ASSESS THE STABILITY OF SMALL FISHING BOATS 121
A Grech La Rosa, C Ryan, G Thomas, University College London, UK
L Huang, Cranfield University, UK
W Hetharia, University of Pattimura, Indonesia
S Riyadi, D Setyawan, I Utama, Institut Teknologi Sepuluh Nopember, Indonesia

VEHICLE HOT SPOT DETECTOR AND DANGEROUS GOODS DETECTOR TO FIRE IGNITION PREVENTION IN RO-RO SHIPS 137
M Haas, H Brumm, L Wallimannand and T Oe, SICK AG, Germany
A Marrero, CENIT Group for Innovation in Multimodal Transport at CIMNE (UPC), Spain

IoS-OP: INITIATIVES FOR SHIP OPERATION DATA COLLECTION, DISTRIBUTION, AND UTILIZATION 147
Y Ikeda and V Sharma, Ship Data Center, Japan

A ROUTE PLANNING METHOD FOR COASTAL NAVIGATION OF SMALL SHIPS 153
D G Jeong, M I Roh, Seoul National University, Republic of Korea
K S Kim, University of Ulsan, Republic of Korea
J S Lee, D H Kim, and W S Jang, Avikus Co.,Ltd, Republic of Korea

INTERFACE DEVELOPMENT BETWEEN THE 3D CAD SOFTWARE AND THE STRUCTURAL STRENGTH ASSESSMENT SOFTWARE FOR EFFICIENT CLASSIFICATION APPROVAL 157
J O Kim, J H Lee and H G Park, Korean Register, Republic of Korea
T G Jang, NAPA Korea, Republic of Korea

A METHOD OF VARIABLE RECOGNITION AND CONNECTION FOR REVIEWING SHIP REGULATIONS 163
M C Kong, M I Roh, K S Kim, Seoul National University, Republic of Korea
J Kim, H Park, and J Lee, Korea Register, Republic of Korea

ACCELERATING JAPANESE SHIPBUILDING DX ADOPTION 169
D Larkins, S Crook, and G Goulanian, SSI, Canada

FROM POINT CLOUD TO CAD-MODEL BASED ON AI 179
J Luetzenberger, M Grau, J Stjepandic and C Zerbst, PROSTEP AG, Germany

DIGITAL TRANSFORMATION THROUGH DATA-DRIVEN AND DATA-CENTRIC APPROACHES WITH ARTIFICIAL INTELLIGENCE 189
A Maimun, C L Siow, and J Khairuddin, Universiti Teknologi Malaysia, Malaysia
K Hiekata, University of Tokyo, Japan

PERFORMANCE OPTIMISATION OF SOLID OXIDE FUEL CELLS THROUGH WASTE HEAT RECOVERY SYSTEMS FOR MARINE APPLICATIONS 199
P Manias, C McKinlay, D A Hudson and S R Turnock, University of Southampton, UK

DEVELOPMENT OF PLM SYSTEM FOR PRODUCTION PLANNING AND PRODUCTION CONTROL IN SHIPBUILDING 213
K Matsuo, M Morishita and T Taniguchi, National Maritime Research Institute, Japan

DYNAMIC MODELLING OF AMMONIA CRACKERS AND HYDROGEN PEM FUEL CELLS FOR SHIPPING APPLICATIONS 219
CJ McKinlay, P Manias, SR Turnock and DA Hudson, University of Southampton, UK

DEVELOPMENT OF A PIPE FABRICATION PROCESS DETERMINATION SYSTEM USING GRAPH DATABASE AND PROCESS SIMULATION 233
Y Mihara and T Mitsuyuki, Yokohama National University, Japan

APPLICATION OF MIXED REALITY TECHNOLOGY TO SHIP MANUFACTURING PROCESS 241
Y Mimori and T Hiraki, Mitsubishi Shipbuilding Co., Ltd., Japan

ZERO INFRASTRUCTURE GEOLOCATION OF NEARBY FIRST RESPONDERS ON RO-RO VESSELS 249
P Mpeis and D Zeinalipour-Yazti, University of Cyprus, Cyprus
J Bleye Vicario, Centro Jovellanos, Spain

SEMI-AUTOMATIC DISTRIBUTED SHIP SERVICE SYSTEMS ROUTING FRAMEWORK FOR SUBMARINE EARLY-STAGE DESIGN 265
M H Mukti, R J Pawling, and D J Andrews, University College London, UK

AN EXPANDED APPLICATION OF BASIC SHIP-PLANNING SUPPORT SYSTEM USING BIG DATA IN MARITIME LOGISTICS FOR PANAMAX AND CAPESIZE DRY BULK CARRIERS 287
D A F Muzhoffar, K Hamada, T Saito, Hiroshima University, Japan
Y Wada, National Maritime Research Institute, Japan
H Harada, Tsuneishi Shipbuilding Co., Ltd., Japan

VOYAGE OPTIMIZATION WITH WIND PROPULSION 301
V Paakkari, Norsepower Ltd, Finland
H Wang and C Stigler Napa Ltd, Finland

DEVELOPMENT OF CAD BASED STABILITY CALCULATION SOFTWARE FOR PURE-LOSS OF STABILITY AND PARAMETRIC ROLLING FAILURE MODE OF 2ND GENERATION INTACT STABILITY CRITERIA 305
J Y Park, M S Jang and H G Park, Korean Register, Republic of Korea

A NEW APPROACH FOR USING CAD AND PLM INTEGRATION 313
R Perez Fernandez, Siemens Digital Industry Software, Spain

THE BENEFITS OF HAVING A BEST-IN-CLASS SHIPBUILDING TOOL 321
R Perez Fernandez, Siemens Digital Industry Software, Spain

INTERCEPTOR EFFECTS ON A 3D RECTANGULAR PLATES IN A CALM WATER BY USING COMPUTATIONAL FLUID DYNAMICS (CFD) 327
AMF Putra and H Suzuki, Osaka University, Japan

STOWAGE PLANNING TOOL SUPPORTING FIRE RISK MANAGEMENT: A NEW WAY FOR CARGO DISTRIBUTION 337
F Rodero and A Marrero, CENIT Group for Innovation in Multimodal Transport, Spain
E S Petersen, Norwegian University of Science and Technology, Norway

MAKING 3D MODEL-BASED APPROVAL A REALITY 349
M Sakagami and T Otaguro, ClassNK, Japan
T Masui and P Lin, NAPA Japan, Japan
Y Shimakawa and S Hisano, Nihon Shipyard, Japan

DIGITAL PLATFORM ENABLING ROBOTIC SURVEY, REPAIR & AGILE MANUFACTURING OF SHIPS AND WATERCRAFT 363
F T Santo and D Williams, Lancaster University, UK

CONCEPTION AND IMPLEMENTATION OF A HETEROGENEOUS SHIP-SPECIFIC SENSOR SYSTEM FOR THE ACQUISITION OF RELEVANT DATA FOR AUTONOMOUS OPERATION 367
E Schütz, T Holzki, C Genschorek and A Loewe TU Berlin, Germany

INFORMATION MANAGEMENT IN SHIPBUILDING PROJECTS - INFORMATION FLOW FROM 3D DESIGN TO PRODUCTION DATA 377
L Seppälä, CADMATIC, Finland

ABOUT CREATING OPEN-SOURCE 3D MODULAR VIEWER IN PYTHON 385
G Sikic, LINA et al, Croatia
P Prebeg, UNIZAG FSB, Croatia

ABOUT USING THE GAME ENGINES IN SHIPBUILDING, SOFTWARE DEVELOPER'S PERSPECTIVE 393
G Sikic, LINA et al, Croatia

DYNAMIC OPTIMIZATION OF PORT OPERATIONS ONBOARD A TYPICAL RO-RO VESSEL, AIDED BY A SMART DECISION SUPPORT SYSTEM 401
C Singh Uppal, Indian Navy, India
S Babu PK, CUSAT, India

DEVELOPMENT OF A COMMUNICATION PLATFORM TO SUPPORT CONSENSUS-BUILDING AMONG STAKEHOLDERS IN SHIPPING DECARBONIZATION 411
K Takahashi and Y Ushijima, Mitsubishi Research Institute, Japan
Y Fujita and T Saito, Koga Software Company, Japan
H Yamato, Japan Agency for Marine-Earth Science and Technology, Japan
R Kakiuchi, Planning and Design Center for Greener Ships, Japan
S Wanaka, National Maritime Research Institute, Japan

A STUDY OF PROCESS SIMULATION BASED ON A MULTI-AGENT SYSTEM FOR SHIPBUILDING 421
T Taniguchi and K Matsuo, National Institute of Maritime, Port and Aviation Technology, Japan

DEVELOPMENT OF A SHIPPING MARKET FORECASTING SYSTEM USING VESSEL MOVEMENT DATA AND ITS PRACTICAL APPLICATION 427
Y Wada and T Majima, National Maritime Research Institute, Japan
K Hamada and H Hirayama, Hiroshima University, Japan
D Watanabe, Marubeni Corporation, Japan
Y Igarashi, Kawasaki Kisen Kaisha, Ltd., Japan

STUDY ON SIMULATION BASED EVALUATION OF ROUTE AND CARGO SPECIFIC PROJECT FOR INTRODUCING DECARBONIZED SHIPS 437
S Wanaka, National Maritime Research Institute, Japan
R Kakiuchi, Planning and Design Center for Greener Ships, Japan
N Fujita, Mitsubishi Research Institute, Japan
K Hiekata, University of Tokyo, Japan
H Yamato, Japan Agency for Marin-Earth Science and Technology, Japan

STUDY ON SIMULATION BASED EVALUATION OF ROUTE AND CARGO SPECIFIC PROJECT FOR INTRODUCING DECARBONIZED SHIPS 451
S Wanaka, National Maritime Research Institute, Japan
R Kakiuchi, Planning and Design Center for Greener Ships, Japan
N Fujita, Mitsubishi Research Institute, Japan
K Hiekata, University of Tokyo, Japan
H Yamato, Japan Agency for Marin-Earth Science and Technology, Japan

AI/ML APPLICATIONS FOR SHIP DESIGN 463
J Hodges, M P Wheeler, and M Belhocine Siemens Digital Industry Software, USA
J Henry, Monolith AI, UK

AUTOMATIC DATA EXTRACTION AND UNFOLDING FOR SHIP HULL PLATES ON CATIA 475
Yu Fu, Wei Wei, Wuyi Xie, Yong Hu1 and Chaoyan Huang

STREAMLINING THE SHIP STRUCTURAL OPTIMIZATION PROCESS BY USING AN EARLY 3D PRODUCT MODEL

M Amano, **Y Tanaka**, Sumitomo Heavy Industries Marine & Engineering Co, Japan
T Masui, **T Hulkkonen** NAPA, Japan/Finland

SUMMARY

Optimization of ship structure has been a target for ship designers ever since the beginning of modern shipbuilding. At the same time, class society rules and regulations have been continuously evolving and guiding shipbuilders in their use of more advanced methods and tools, which will further facilitate the optimization of structures against the latest knowledge. This paper presents the results of a project in which the whole rule-based structural optimization process is built inside one technology platform.

This platform combines the necessary modelling work, prescriptive rule check, FEM-based rule check and further weight and other calculations in one automated and streamlined process. This will have a tremendous effect on the throughput time of iteration cycles. The paper will present the process and technologies used for integrating the different components into the common design platform. It will also show some results achieved in a commercial cargo ship design.

NOMENCLATURE

.NET	A cross-platform implementation for running websites, services, and console apps on Windows.
k	Material factor.
λ_{yperm}	Permissible yield utilisation factor.
σ_{VM}	Von Mises stress.
API	Application Programming Interface is a set of NAPA libraries facilitating the interfacing of programs.
CFD	Computational Fluid Dynamics.
CSR	Common Structural rules by IACS [1].
DSA	Direct Strength Assessment.
GUI	Graphical User Interface
FEM	Finite Element Method.
IACS	International Association of Classification Societies.
MGR	NAPA Manager Application, an application development technology by NAPA.
MOGA	Multi-Objective Genetic Algorithm.
NAPA	NAPA Ltd, a software development company and its NAPA software.
Plugin	An extension to NAPA API in .NET environment in form of a dynamic link library (DLL).
PSM	Primary Supporting Members
R_Y	Nominal yield stress of the material, taken equal to 235/k.
SHI-ME	Sumitomo Heavy Industries Marine & Engineering Co, a company operating in marine industries.
STYPE	Structure Type, representing a typical structure of the ship such as the main deck, transverse bulkheads
Windows	Microsoft Windows is a graphical operating system developed and published by Microsoft.

1. INTRODUCTION

Ship structural optimization is an important and complex topic, however, new rules and regulations make it increasingly time-consuming. In addition to this, innovative technologies and manufacturing processes create new possibilities, as well as challenges, for finding the optimum solution. Repeating the old design methods and processes is no longer an optimum solution, and there is a need for more precise design checks.

This paper presents the implementation and results of a project aimed at creating an automatic design process for optimizing a tanker's structures with CSR rules by using its proposed DSA concept. The need for this kind of integrated design system has been previously highlighted by Shimakawa in [2], for example.

The project has been executed in collaboration with SHI-ME and NAPA. SHI-ME has helped with the specification of the desired process and the development work for the tools used providing the initial scantlings for typical constructions. NAPA's contribution has been in software development for structural modelling, FEM meshing and DSA rule check, in addition to providing a platform for the developed and integrated solution.

Chapter 2 presents a conventional design process, which is used typically at shipyards building cargo vessels, with a particular focus on highlighting the most time-consuming tasks in this process, which often limits possibilities for finding optimum solutions within the available time. Also, the role of weight estimation in structural design is emphasized and its relationship with the strength assessment process.

Chapter 3 presents the technology and procedures used in the streamlined process. The streamlined process has similar steps as used in the conventional design process, but the process is built into an integrated 3D design environment. The modelling part covers the parametrized ship geometry, including main structures like bulkheads and decks and also structural details like stiffeners and seams. Chapter 3 also shows how the shipyard's practices for initial scantling estimations can be implemented and how the rule check based on DSA can be streamlined. By including the geometry into the 3D product model of NAPA, we can ensure that all the necessary information needed for structural optimization is also compatible with naval architecture calculations and further downstream design.

Following on from this, Chapters 4 and 5 present some of the practical results and conclusions achieved with this streamlined process. In summary, the design iteration cycle has a remarkable improvement in throughput time, which allows more iteration cycles during the available design period, further reduces the possible weight margins and brings new cost savings. Quick design cycles also make it possible to test unconventional, yet optimum, design solutions, beyond the realms of traditional practices, while ensuring they are backed by solid data on design and weight estimation already in early design phases. This makes it possible to start innovating and producing new vessel designs, helping meet rapidly evolving market demands and improving the competitiveness of shipyards.

2. CONVENTIONAL PROCESS

In this chapter, the design process of conventional initial planning and structural design will be reviewed and the issues that this approach is trying to solve will be clarified.

2.1 INITIAL PLANNING

In the cargo ship design, as shown in Figure 1, the main dimensions and general arrangements are studied in the initial design phase based on the shipowner's requirements, while the hull structure, hull form, main equipment and other outfitting designs are initiated using the basic information agreed at the initial design phase. Changes to such basic information can cause a lot of backtracking work in each design, not only in the hull structure. There is also a time limit in the initial planning phase, where basic specifications have to be presented to the shipowner within a short period of time, say a few weeks. Therefore, a design procedure is traditionally adopted in which the basic information that has been settled is taken over and each design is continued.

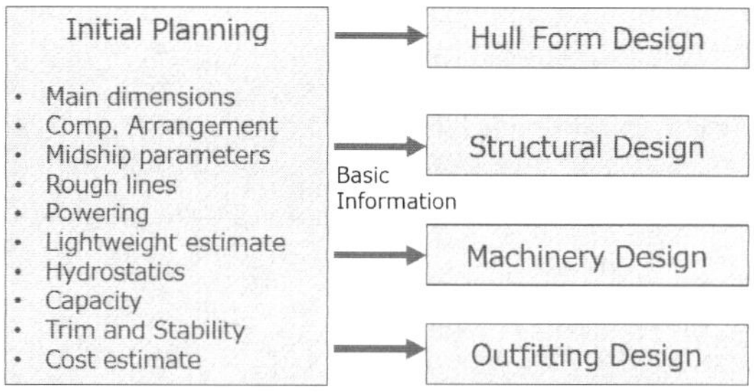

Figure 1. Conventional Early-Stage Design Process

2.2 IMPORTANCE OF HULL WEIGHT

However, initial planning requires information that is uncertain and unclear at that stage, such as lightweight, which is needed for payload and stability assessments. While hull weight accounts for the majority of lightweight in cargo ships, the fact is that the exact hull weight cannot be settled until the structural design has been carried out. Therefore, shipyards have generally adopted a method of estimating hull weights based on statistical data of actual vessels built in the past and using them to estimate hull weights. A certain margin is applied for the weight estimation, and it is expected that the weight will be kept within that range in the later design stages. However, in recent years, it has become increasingly difficult to use past construction experience as it is, due to the class societies' continuously evolving structural rules and the development of new, innovative ships using new energy sources to enhance competitiveness. A rational weight estimation method that does not rely solely on past experience is desired. If the margin is found to be exceeded in the later stages of design, the initially planned deadweight and performance cannot be ensured, which may lead to a significant backward step in the design. Such a situation must be avoided. On the other hand, an excessive weight margin will result in a redundant design in all aspects, including ship performance and cost, and will make the ship less competitive and lose profitability.

In addition to this, hull weight reduction has become increasingly important in recent years. Since steel costs account for roughly 30 to 40% of the cost of building a cargo ship, the shipyard's competitiveness is directly connected to its ability to lighten the hull weight. For example, the recent rise in steel prices has become more pronounced, and this has had a significant impact on shipyards' profitability. Furthermore, a lighter hull also contributes to the economics of operation and environmental performance. This is because the transportation efficiency and fuel efficiency can be improved if the hull structure can be made lighter, as a ship with a lighter hull weight can carry more cargo, or a ship with a lighter hull weight can have better propulsive performance for the same deadweight.

Thus, improving the accuracy of hull weight estimation and lightening the hull weight are regarded as critical business issues for cargo vessel builders.

2.3 STRUCTURAL DESIGN

To determine the hull weight, it is necessary to determine the scantling of each structural member in the structural design. However, the time required to finalize the scantling of structural members tends to increase as the complexity of class societies' rules, such as the CSR, becomes more extensive to comply with safety demands.

Recently, 3D design, in which structural design is performed directly on a 3D model, has become popular. However, speaking of only the design of a midship structure in the early stages of structural design, 2D drawing is often used as a starting point because of its handiness. The following is a typical design workflow for the midship structure of tankers, aiming at the point at which scantlings of structural members can be fixed and hull weights can be obtained.

I. Prepare a 2D drawing of the midship structure.
II. Input the data into the class society's prescriptive rule calculation software. Iterate the process manually until the rule requirements are met, e.g., by changing the scantlings or stiffener positions.
III. Create a FEM model and perform DSA. There, the results of the prescriptive rule calculations are reflected in the FEM model. Increase the plate thickness or add reinforcements until the rule requirements are met.
IV. Reflect the results in the 2D drawing.
V. Summarize the hull weight of the midship structure from the 2D drawing.

Table 1 shows the time required to design the midship structure of a medium-sized tanker. The time required for each step depends on various conditions and is only an approximation. Anyway, it is clear from the table that the works required for the prescriptive rule calculations and DSA are the bottleneck of the entire process.

Table 1. Approximate Time Required to Design a Midship Structure of a Medium-Sized Tanker

I. Creation of a drawing of the midship structure	15-20H
II. Performing prescriptive rule calculations	50-100H
III. Creation of FEM model and performing DSA	200-300H
IV. Updating the drawing	5-10H
V. Summarize the hull weight	15-20H

In other words, it takes one to two months to determine reasonable scantlings of structural members and obtain hull weights for the midship structures only. This is the reason why, in the initial planning stage, where there is a time limitation of a few weeks, we have no choice but to estimate hull weights by relying on previous statistical data.

However, it may be possible to further reduce the hull weight by placing the compartment arrangements and principal parameters, which are usually fixed in the initial plans, as design variables for structural design. In the development of a standard ship design, structural studies are sometimes conducted with these as design variables to obtain the optimal compartment arrangement and principal dimensions [3]. However, even in such cases, the study is usually limited to performing rule calculations for a specified set of design patterns because of time constraints. Though there must be many structural members for which the results of DSA were dominant, thus presenting a challenge in terms of validity of the result as an optimized design.

2.4 KEY ISSUES AND GOALS

Table 2 summarizes the key issues and the goals to be accomplished for the initial planning and structural design phases of the tanker design, which are to be tackled in this project.

Table 2. Key Issues and Goals for the Initial Planning and Structural Design of the Tanker Design

Current Situation	Ideal Situation
• Time limitations during initial planning make it difficult to optimize hull weight by taking compartments and principal parameters as design variables. • Estimation method of hull weights that relies solely on past statistical data • In structural design, a large amount of time is required to determine reasonable scantlings of structural members.	• Optimization study of hull weight can be performed using the compartment and principal parameters as design variables. • Highly accurate hull weight estimation method (appropriate margin) that does not solely rely on past statistical data • In structural design, reasonable scantlings of structural members can be obtained in a short time.
↓	↓
Unoptimized design with redundant hull weight, losing competitiveness and profitability	Optimized design with lighter hull weight, gaining competitiveness and profitability

3. STREAMLINED PROCESS

As a solution to the issues described in the previous section, we aimed to build an integrated system that can be used in the initial planning phase by fully automating the complicated processes of structural strength calculations, which takes a great amount of time. By incorporating SHI-ME's design standards into the 3D modelling logic, the system also aimed to eliminate the need for drawing preparation.

3.1 NAPA AS A PLATFORM FOR DESIGN OPTIMIZATION

The NAPA system and its 3D product model were adopted as the design platform for automating the structural strength calculation process. The reasons for this are:

One Integrated Database

- Single database containing hull form, compartment and structure.
- Any information concerning the hull of the vessel is available on a single database. For example, naval architectural data such as compartment volume, loading conditions and still water bending moment and shear force, and ship structural data such as weight, painting area, welding length, etc.
- The capability of high-quality auto FEM meshing, complying with CSR.

Flexible 3D Model

- Easy modification of design parameters, hull form, frame spacing, compartments and primary members.
- A quick review of design alternatives.
- Possibility to enhance NAPA model along with design progress for later stage use.

Customization and Automation

- Automation, personalization and interfacing using NAPA-specific API and macro language.
- Custom process management tool named NAPA Manager.
- Optimization methods such as MOGA, systematic variation, etc.

3.2 OUTLINE OF THE OPTIMIZATION SYSTEM

The primary goal of this project is to be able to grasp the effects of design parameters related to compartment arrangement and principal parameters as "primary independent variables" for hull weights systematically by automatically performing parametric studies of prescriptive rule calculations and DSA for the midship structure of a medium-sized tanker. The geometry of PSM, longitudinal stiffener and seam arrangements are also treated as "secondary independent variables" so that the optimal structural arrangement can be studied after the primary independent variables have been selected. Initial input data such as hull form data studied in the NAPA system by the initial planning engineer can be reused as is, but the system is also designed to allow direct input of such data. All in all, the highest priority was to ensure that the system can be effectively utilized in real work and that the entire process can be fully automated.

- Supported ship type: Medium-sized tanker (Figure 3)
- Design variables: Table 4 and Figure 5.

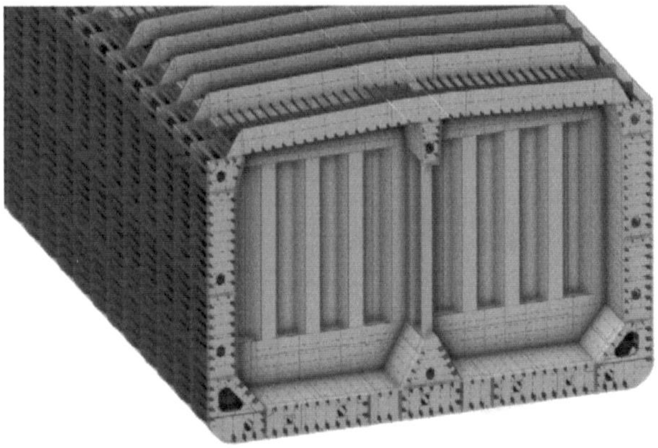

Figure 3. Example of a medium-sized tanker

Table 4: Design Variables and Initial Information in the Optimization System

Initial Information	Primary Independent Variables	Secondary Independent Variables	Dependent Variables
Information required to define design loads. For example, • Main dimensions • Hull form, • Transverse bulkhead positions arrangement • Permissible longitudinal bending moment and shear force, etc.	Variables representing the compartment arrangement and the principal parameters in the midship structure. This variable is used for the parametric study.	Variables related to the PSM geometry and longitudinal stiffener/seam arrangement. In the parametric studies, the comparative evaluation is performed for the same secondary independent variable values.	Variables determined following the primary and secondary independent variables. PSM geometries and longitudinal stiffener arrangement are used as a reference, and the stiffener and seam arrangement on the transverse members are determined with certain standardization logic.

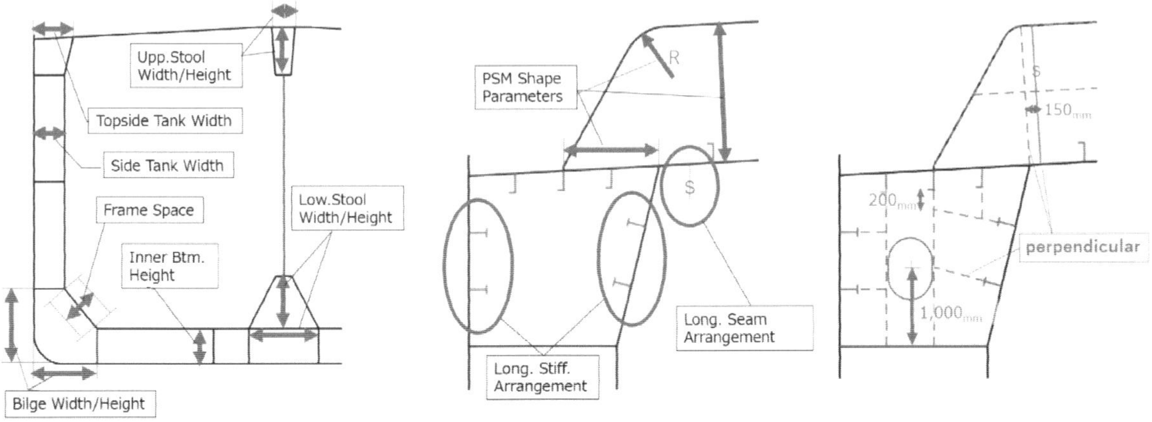

Figure 5: Primary Independent Variables (left), Secondary Independent Variables (centre), Dependent Variables (right)

3.3 OVERALL FLOW

The overall flow of the developed system is described below (Figure 6). Each of the following steps can be executed manually or automatically. It is also possible to execute all the steps fully automatically.

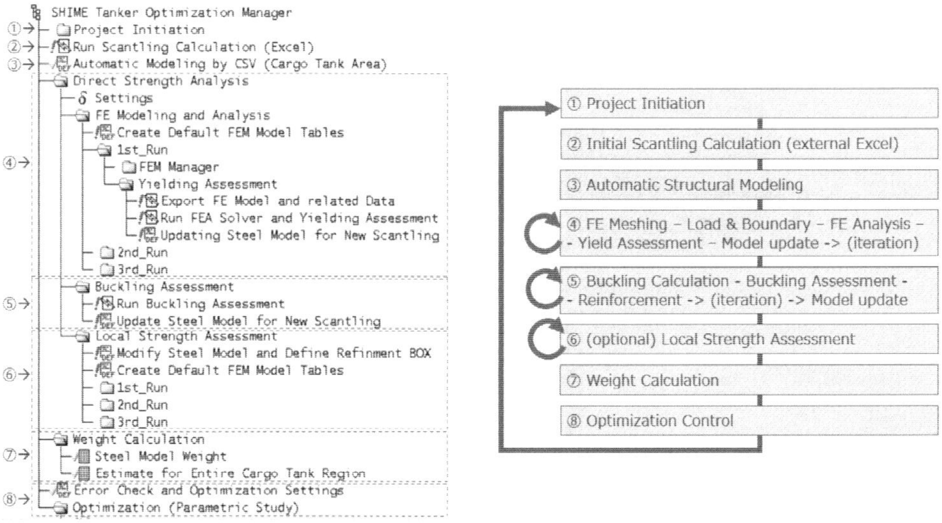

Figure 6. Tree structure of the optimization MGR (left) and overall flow (right)

The features of this system will be introduced in the following sections. However, some features will be limited to a brief description only.

3.4 PROJECT INITIATION

This step is to define the initial information and the primary and secondary independent variables.

3.5 INITIAL SCANTLING CALCULATION

Perform prescriptive rule calculations for hull girder strength, local strength, and buckling per CSR Chapters 5, 6, and 8. This calculation tool was developed by SHI-ME using Microsoft Excel.

As soon as the optimization MGR outputs the initial information and the primary and secondary independent variable values in a CSV file format, it executes a Windows Batch file and launches the rule calculation tool. This tool automatically reads the CSV file and starts the rule calculations. After the calculations are finished, it provides information on the stiffener profiles and plate thicknesses that satisfy the rule requirement values for longitudinal and transverse members. This is exported in a specified CSV file format. The information in this file, together with the primary and secondary independent variable values, serve as the basis for creating the 3D product model.

3.6 AUTOMATIC STRUCTURAL MODELLING

As soon as the optimization MGR imports the CSV file generated after the Initial Scantling Calculation, 3D modelling is automatically started. The content of the created 3D product model is described below and in Figure 7.

Longitudinal structures:
- Main Objects, Plates*, Stiffeners*, Seams*, Openings* Note: (*) is only for the central 3 holds

Transverse structures:
- Main Objects, Plates*, Stiffeners*, Brackets*, Openings*

Compartment models

Among these, the logic for the arrangement of stiffener, seams, brackets and openings on the transverse structures was based on the SHI-ME design standard. In other words, these arrangements are treated as dependent variables whose positions are determined based on the PSM geometry and longitudinal stiffener arrangement information and so on specified in the CSV file and the logic based on SHI-ME's design standards.

Traditionally, it would take several weeks to create a 3D model, even for the midship structures; however, highly sophisticated standardization based on the shipyard's design standards enable full automation, making it possible to create such a model within a few seconds. In addition to the structural model of the central 3 cargo holds, the surface-based main object model is automatically created for the other cargo holds based on the information in the CSV file, as well as its compartment model. The main object model is used as fundamental data for weight estimation, and the latter is used to automatically calculate the corrosion reduction values and the load conditions later used in the DSA process.

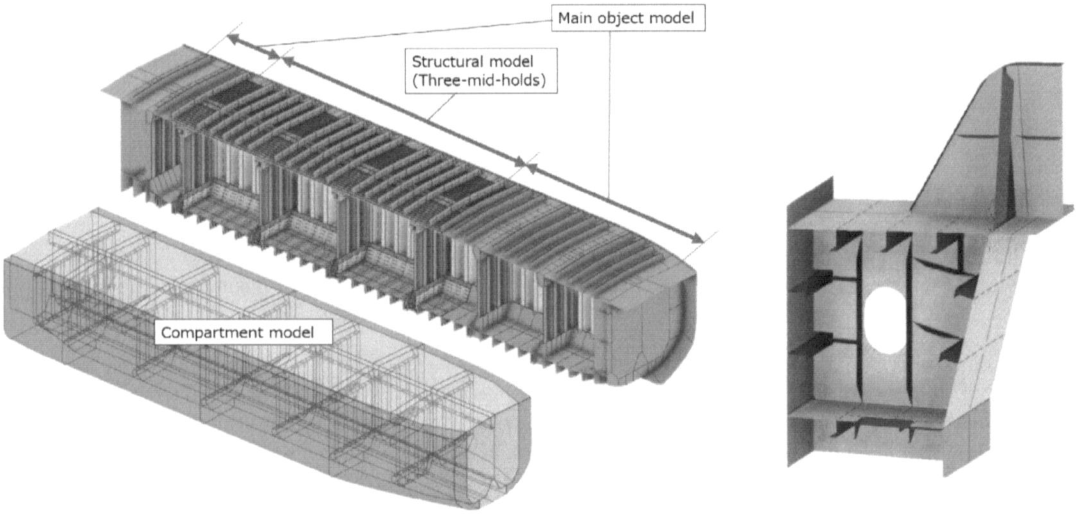

Figure 7. Automatically Created 3D Product Model

3.7 FEM MESHING

NAPA has a sophisticated auto-meshing feature that converts the structural model into a high-quality FEM model. The relevant functionalities are incorporated into the optimization MGR. The following features of the NAPA FEM meshing feature made it possible to generate high-quality FEM models fully automatically without the need for manual mesh modifications in this system.

- Create a high-quality FEM model automatically based on an as-built 3D geometry of the structures.
- Structural members such as stiffeners, openings, seams, and brackets can be idealized automatically based on several idealization options. In this manager application, these meshes are idealized following CSR Chapter 7.
- FEM model can be converted to net scantling model using CSR compatible corrosion table when exporting in BDF format.
- Fine mesh model can be created automatically to the specified locations using local refinement functions

In general, FEM model creation is also a very time-consuming process. However, the coarse mesh model of the central 3 cargo holds and specific fine mesh models can be created automatically in a few seconds using this system. As Figure 8 shows, the FEM model is created automatically.

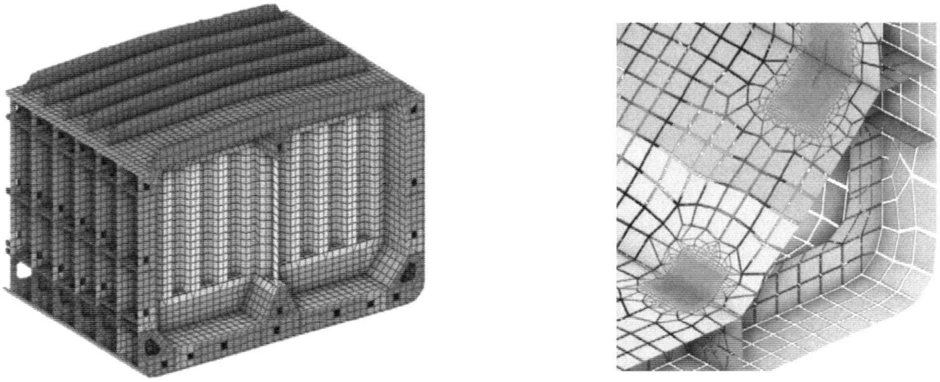

Figure 8. Automatically generated FEM models (left: Coarse mesh model, right: Fine mesh models)

3.8 RULE CHECK WITH DSA RULES

The structural strength evaluation covers the stress assessment and buckling. Together these requirements cover the needs of the structural optimization process for weight and production cost estimations.

The rule check stress assessment part is based on requirements of CSR Chapter 7 for DSA of cargo hold structures and local structures [1]. The buckling analysis is based on CSR Chapter 8 by using the buckling requirements of DSA. All the analyses are based on the net scantling approach, where the FEM analyses and structural capacity estimation concern net and gross scantlings case by case according to the CSR rules.

In this project, the rule calculation engine is built into the NAPA system by using the plugin technology available in NAPA .NET API. The benefit of using a plugin is the direct access to information without any intermediate files. However, the FEM solver used here is an external one and the input data for the solver and the calculation results are handled in the Windows file system level. The rough description of the rule check process is shown in Figure 9.

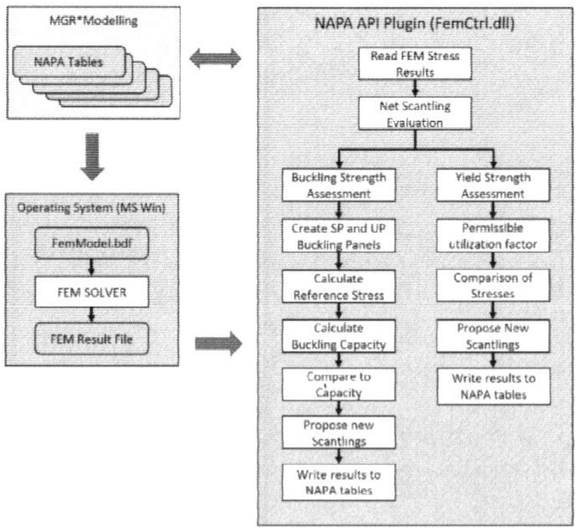

Figure 9: The Rule Check Process on DSA

3.8 (a) Yield strength assessment

The CSR yield strength assessment is based on the von Mises stress σ_{VM} calculated based on the membrane normal and shear stresses of the shell elements' centroid of the mid-plane. These stresses are directly available in most of the commercial FEM solver result files for each element. The plugin reads directly these results from the FEM solver for further processing.

The yield criteria are formulated as:

$$\frac{\sigma_{VM}}{R_Y} \leq \lambda_{yperm}$$

σ_{VM} is the von Mises stress
λ_{yperm} is the permissible utilisation factor
R_Y is the nominal yield stress of the material.

The permissible utilisation factor is defined in the CSR rules and it depends on several factors like:

- Static or dynamic load
- Corrugated or plain panel
- Coarse mesh, fine mesh, very fine mesh.

The automated optimization process needs information from the structural model and the applied FEM analyses. By comparing each element's stress utilization to the permissible one, it is possible to decide the acceptance for each plate and stiffener in the 3D model.

The plugin is proposing also new structural scantling if the utilization of stresses exceeds the permissible values. New scantling proposals and the status of each structural element are returned to the NAPA database by using the API functionality.

3.8 (b) Buckling strength assessment

For buckling strength assessment one of the key parts is to create buckling panels. CSR has capacity calculation guidelines for stiffened panels (SP) and unstiffened panels (UP). SP panel calculation includes the stiffener buckling together with plate panel buckling. Unstiffened panel buckling is concerning most of all only the elementary plate panels (EPP) between the stiffening elements.

The buckling capacity calculation in CSR formulas is based on the assumption of regular rectangular panels. In the case of irregular unstiffened panels, their shape is idealized to a corresponding regular shape with CSR guidelines. The plugin is creating the buckling panels automatically from the FEM mesh for each optimization's iteration cycle. The buckling panels are created from shell elements surrounded by the beam elements and other crossing shell elements as indicated in Figure 10.

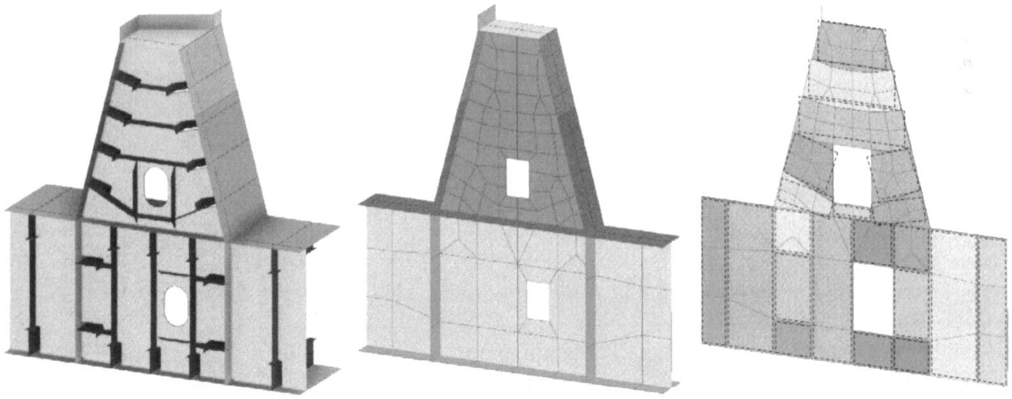

Figure 10: Structure model of a web frame, shell elements and the buckling panels with idealised shape.

For both SP and UP there are two assessment methods, A and B, the selection of which depends on the flexibility of the surrounding boundaries. CSR is guiding the usage of A and B assessments directly depending on the typical structure types like in Figure 11.

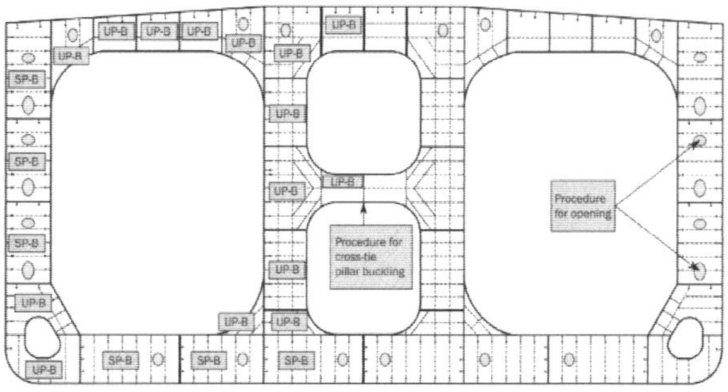

Figure 11: Assessment methods in tanker web frame structures [1].

The CSR buckling capacity calculation for DSA does not provide any direct closed form solution for buckling capacity, but the capacity is searched from the interaction formulas, which have the stress multiplier factors as unknown parameters. The interaction formulas are non-linear having many solution branches and the proper solution is found by applying several validation criteria. Optional solution strategies in CSR January 2018 version are iterative, or it is still possible to work out the equations in an open form. The current version of the plugin is now using the open-form equations strategy.

The buckling reference stress is calculated from the local element stresses by using the stress-based method introduced in CSR. The local stresses are transformed first to panel local directions and then the method determines stress distribution along the edges of the considered buckling panel by 2nd order polynomial curve or by linear distribution using least square method and by weighted average approach. The solving of the panel longitudinal reference stress requires forming and solving a set of 3 linear equations for each buckling panel. For transverse reference stress, the stress distribution is assumed to be a linear producing set of 2 linear equations.

The buckling criteria is defined as:

$$\eta_{act} = \frac{W_{act}}{W_u} < \eta_{all}$$

W_{act} is the applied equivalent stress
W_u is the equivalent buckling capacity
η is the buckling utilization factor
η_{all} is the allowed buckling utilization factor

The allowed buckling utilization factor varies between 0.65 – 1.00 and it depends on different parameters like:

- Structural component
- Panel type
- Load combinations
- Corrugation parameters.

For the automated optimization process, the plugin extracts the needed information from the 3D product model and the applied FEM analysis results. By comparing each element's stress utilization to the permissible one, it is possible to decide the acceptance for each plate and stiffener in the 3D model.

The plugin also introduces new structural scantling if the utilisation of stresses exceeds the permissible values. New scantling proposals and the status of each structural element are sent to the NAPA database by using the API functionality.

3.9 WEIGHT CALCULATION

Among the central 3 cargo holds for which a structural model was created, only the CSR evaluation area (the central 1 cargo hold and its aft and front frame locations) was subjected to yield and buckling evaluation and reinforcements. Using the NAPA function, the weight of the structural model of this evaluated cargo hold is calculated.

In addition, a table summarizing the weight per unit area is automatically generated for every STYPE and member type (plate, stiffener, and bracket). The purpose of this unit weight table is to estimate the weight for the entire cargo holds by multiplying the weight per unit area and the area of the main object model in each part. In outside midship structures, however, the unit weight changes due to changes in the design load, etc. Therefore, this system allows any correction coefficients for blocks divided into an arbitrary longitudinal range of the entire cargo hold. This can be expressed in a simple formula as follows.

$$W_{total} = \sum W_{block}$$
$$W_{block} = \sum A \times \left(U_p + U_s + U_b\right) + w_o$$

W_{total} is the estimated weight of entire cargo holds
W_{block} is the estimated weight for a block divided into an arbitrary longitudinal range of entire cargo holds
A is the total area of main objects belonging to certain STYPE group in a block
U_p is the weight per unit area for plates belonging to certain STYPE group in a block
U_s is the weight per unit area for stiffeners belonging to certain STYPE group in a block
U_b is the weight per unit area for brackets belonging to certain STYPE group in a block
w_o is the weight of other structural members

This correction coefficient is intended to be applied with reference to past ship experience or simple prescriptive rule-based calculations.
The aforementioned procedures related to weight estimation are also designed to support automatic processing.

3.10 RUNNING THE OPTIMIZATION PROCESS

The entire process can be automatically executed using NAPA's optimization MGR functionality. It has several optimization algorithms such as MOGA, but to grasp the overall weight effects of different primary independent variables (i.e., to perform sensitivity analysis using parametric studies), the Systematic Variation method was used as the default in this manager. This means that hundreds of cases can be executed automatically, with variables such as bilge size, inner bottom height, side tank width, and frame space. A similar study can be performed under new conditions by resetting the initial information on main dimensions and hull form.

In addition, this system has a sort of local optimization mechanism to obtain reasonable scantling, as shown in Figure 6, which automatically iterates through each of the yields and buckling assessment processes. In other words, reinforcements are added, or plate thicknesses are increased based on the designated strengthening procedure until the required values of the rules are satisfied.

To improve practicality, the system can also allow batch execution of selected processes. For cases where errors occur, such as FEM analysis being stopped due to a failure of the mesh shape, the system can automatically detect and report the error so that the problematic case and the relevant process in which the error occurred can be identified quickly.

4. RESULTS ACHIEVED

SHI-ME applied the new tool in the development of a new vessel and verified its effectiveness. As a result, midship structure evaluation time during the new ship development phase was dramatically reduced by streamlining the process and using new tools. The conventional process of the structural design required about 300 hours per case, but the new process takes only 1 to 3 hours per case. (Note: The study time with this tool depends on the number of iterations of FE analyses.) The reduction in man-hours for creating FEM models for DSA, using direct calculation software provided by class society, and summarizing the analysis results, is noteworthy.

Using the developed tool, the following comparisons can be made in a short time and accurate weights can be obtained.

- Weight comparison by bilge size
- Weight comparison by frame space
- Weight comparison by PSM size (web depth, edge bracket evaluation)
- Weight comparison by height position of horizontal girders on the transverse bulkhead
- Weight comparison by longitudinal stiffener spaces
- Combination of the above, etc.

Since the study time was significantly reduced, it became possible to compare cargo hold weights for many cases. As a result, the possibility of hull steel weight reduction was expanded.

As an example, a comparison of the impacts of bilge size and frame space on hull steel weights is shown in Figure 12.

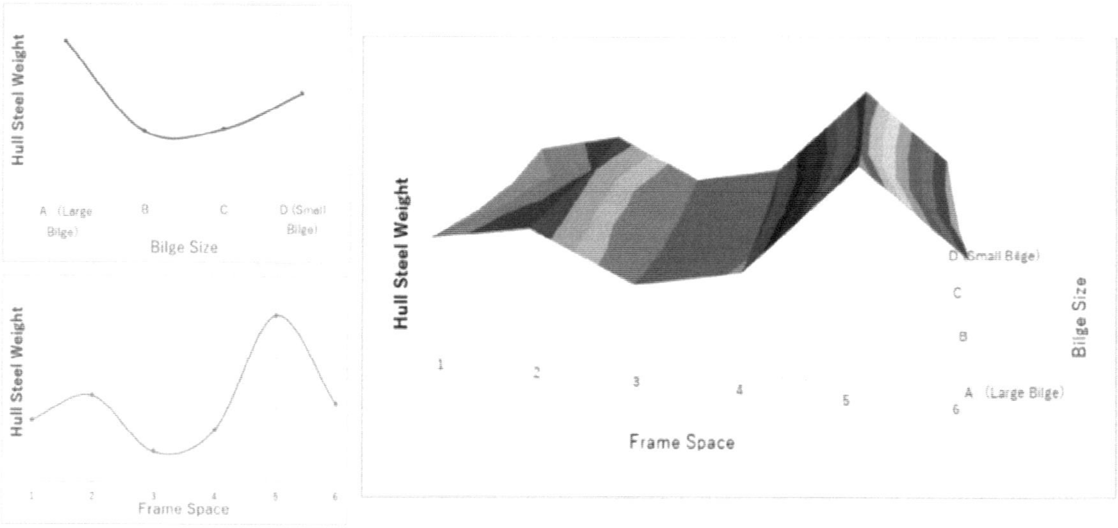

Figure 12: Impacts of bilge size and frame space on hull steel weights

5. CONCLUSIONS AND FUTURE WORK

5.1 CONCLUSION

We streamlined the design process for the initial planning phase and developed a tool for this purpose. As a result, the lead time for the strength evaluation and weight estimation of the midship structure of tankers can be significantly reduced.

The outcomes and findings obtained through the project are described below:

- The initial planning process could be streamlined by using a single 3D product model and by standardizing and automating time-consuming tasks.
- NAPA is a useful platform for studying optimal ship designs because it can handle the processes of naval architectural calculations, hull structure, structural analysis, and rule calculations and integrate them into a single solution.
- The optimization system developed significantly improved the initial planning process and expanded the possibilities for hull steel weight reduction, which is extremely valuable for SHI-ME.
- The 3D product models automatically created through the optimization system can be adapted and utilized in the next design process. A similar kind of improvement is possible also for other ship types.

Through this experience, we can see the immense potential for automation through the advanced use of appropriate design systems. From a different perspective, this may indicate that the roles and competencies required of structural designers may change in the future. In other words, the future role of the designers would be to standardize the company's design know-how and turn this into a system. The authors expect that, in addition to existing engineering skills and experience, software skills, such as design standardization, and system design will be necessary to fulfil these roles in the future.

5.2 FUTURE WORK

We have several ideas for making the newly developed system more practical and effective. These include:

- Inclusion of trim and stability calculation process; currently, longitudinal bending moment and shear force distribution for a set of loading conditions are separately performed based on a compartment arrangement. A further decrease in design lead time can be expected by including these functions in the system.
- Improvement of FEM automatic mesh quality; mesh quality is sometimes difficult to ensure in some complicated structural arrangements, and this affects stress evaluation results to some extent. Further improvement of automatic mesh quality is expected.
- Inclusion of a class society's rule calculation system; the rule calculations in the developed system refer to the CSR rules, but the system itself is not authorized by the class society. To be able to use the system as it is for class approval and maintenance due to rule updates, etc., it is ideal to incorporate the program provided by the class society in the optimization system.

- Evaluation of construction cost; although this system targets the minimization of hull steel weight, this system should be able to include a construction cost perspective by using numerical information such as welding lengths and paint areas obtained from the same 3D product model.
- Low-code automatic modelling methodology; in this system, the arrangement of stiffeners and seams on transverse members, which is treated as a dependent variable, is reproduced by a program based on standardized logic. Design automation will become even more prominent if a generic tool can be developed that allows automatic modelling based on such standard logic to be easily performed through GUI and setting information, instead of hand-written programs.

We will consider the implementation of these improvements in the future as we continue to apply them in practice.

6. REFERENCES

1. IACS, 'Common Structural Rules for Bulk Carriers and Oil Tankers', 01 Jan 2018.

2. SHIMAKAWA, Process Integration of Structural Design Promoted by 3D Model Based Approval, *International Conference on Computer Applications in Shipbuilding 2019*, 24-26 September 2019

3. WINTER, Optimizing Ships Using the Holistic Accelerated Concept Design Methodology, *Practical Design of Ships and Other Floating Structures 2019*, 22-26 September 2019

7. AUTHORS BIOGRAPHY

Mr. Masayoshi Amano holds the current position of Senior Engineer in Advanced Development Group at Sumitomo Heavy Industries Marine & Engineering Co., Ltd. (SHI-ME) He is currently responsible for the initial structural design for new ship development projects. He joined Sumitomo Heavy Industries, Ltd. in 1999 and has 21 years of experience in structural design.

Mr. Yosuke Tanaka holds the current position of Senior Engineer in Structure Design Group at Sumitomo Heavy Industries Marine & Engineering Co., Ltd. (SHI-ME) He has been responsible for the structural design since joining SHI-ME in 2006.

Mr. Takayoshi Masui holds the current position of Deputy Managing Director in NAPA Japan. He is currently responsible for the NAPA Steel software and business development in NAPA Group. His previous experience includes 4 years working at shipyards in Japan and 12 years in NAPA Group.

Mr. Tapio Hulkkonen holds the position of Director, Product Management in NAPA Design Solutions. He is responsible for NAPA software-related product management tasks. His previous experience includes 23 years working at shipyards in Finland and 17 years in NAPA Group.

SUSTAINABLE SHIP DESIGN WITH MODULAR SHIPBUIDLING APPROACH

R Audoire, Dassault Systèmes, France

SUMMARY

The present paper outlines the approach used for the European project called NAVAIS (New, Advanced and Value-Added Innovative Ships) to reconsider the ship initial design process flow that contributes to minimize the impact on the environment. The main challenge focusses on modular design at the initial design stage, to foster re-use of standard logical modules, and digital collaboration platform technics between all actors. The core process relies on Model Based System Engineering (MBSE), moving from an Engineering To Order (ETO) conventional ship design method, to a Configuration To Order (CTO) method. Another benefit of this new approach is the ability for the class society to pre-approve the various pre-designed systems modules in advance, thus saving time during the final approval process

The overall initial design approach includes a specific attention on how a single collaborative platform can orchestrate several processes coming from different stakeholders having specific expertise. The domains covered are parametric hull form design, logical modular module design, ship resistance curve calculation, propulsion equipment prediction, ship energy efficiency and underwater radiation noise analysis.

The paper presents the research work that has been achieved on that topic so far and will benefit from demos and videos that have been produced by the members of the project.

NOMENCLATURE

CATIA: is a product line of Computer-Aided Three-dimensional Interactive Applications from Dassault Systèmes [1]
CFD: Computational Fluid Dynamics
CO2: Carbone dioxide gas found in the atmosphere
CtO: Configuration to Order
EtO: Engineering to Order
IMO: International Maritime Organization
MARPOL: The International Convention for the Prevention of Pollution from Ships
MBSE: Model-Based System Engineering
Modelica: is an object-oriented language for modeling and simulating complex multi-domain systems.
NAVAIS: New, Advanced and Value-Added Innovative Ship
NOx: Oxides of nitrogen emissions (atmospheric pollutants)
P&ID: Piping and Instrumentation Diagram
R&D: Research & Development
RANS: The Reynolds-Averaged Navier-Stokes equations
RFLP: Requirements, Functional, Logical and Physical view (description) of the core elements of Systems Engineering
SEECAT: Ship Energy Efficiency Calculation and Analysis Tool from Bureau Véritas [2]
SIMULIA: is a product line of computer simulation applications from Dassault Systèmes
SOx: Sulphur Oxides emissions (atmospheric pollutants)
URN: Underwater Radiation Noise
3DEXPERIENCE: is a platform used as a collaborative environment

1. INTRODUCTION

This paper is the logical continuation of the one presented at ICCAS 2019 [3] in Rotterdam which presented the major challenges of the European project "New Advanced Value Added Innovation in Shipbuilding" (NAVAIS) [4] [5]. It aims to share some of the results of the new approaches defined and tested during the project that ended on May 31, 2022. In particular, those directly related to the initial design modular that minimizes the impact on the environment. The following, has been developed within the NAVAIS project funded by the Europeans' Horizon 2020 research and innovation programme (contract No.: 769416) [6] [7]

2. THE MODEL BASED SYSTEMS ENGINEERING APPROACH

In order to increase their market competitiveness various ship designers and builders have to challenge themselves in finding the right balance between customer requirements for engineered-to-order designs (EtO) and cost and lead time efficiency.

For serving individual customer needs, with minimum development, manufacturing and service efforts, the manufacturing industry common solution is the adoption of the "Configure to Order" (CtO) business model. To successfully apply the CtO business model, the appropriate platform-based product family should be established together with a number of possible variants enabling a decent level of customization of the selected platform. It follows that within the CtO business model the pre-contract and contract steps are carried out by the ship designer, prior the engagement with a specific ship-owner, based on extensive market analysis. As matter of fact, for such markets the vast majority of requirements are generally accepted by the ship-owners since they could be considered industry standards (e.g., bollard pull of a harbour tug). The following described business model is relatively new to ship designing and building industries, whereas is well used in automotive and aerospace ones. Where applicable the CtO business model will imply the need of defining the modular architecture for the platform-based product family relying on the Model based Systems Engineering (MBSE) approach.

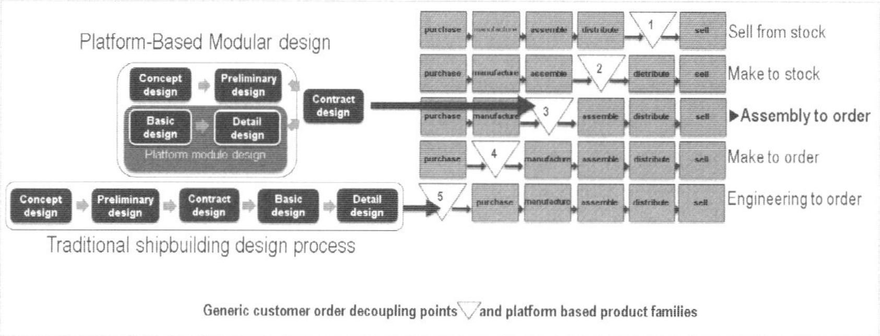

Fig.1: From "Engineering to Order", design process to "Configure to Order" approach.

3. THE MBSE GENERAL PRINCIPLE APPLIED TO A MODULAR PRODUCT ARCHITECTURE

The modular architecture products have a one-to-one correspondence between a component and its function, so that the different functions of the product are, to the extent possible, allocated to separate product modules. In other words, the modular architecture allows decomposing a product into smaller blocks, i.e., modules, with specified interfaces. Modular architectures are flexible in structure, with highly standardized interoperability and standard connections for subsystems. Modular design identifies particular functions necessary to achieve the overall product purpose. Standard assemblies are then used to undertake these individual functions. Subsequently, the assemblies or modules are brought together to form the complete product, which can then perform its complete function. In a glance, the objective of modularization is to define groups of related products that share common features, components, subsystems, interfaces and manufacturing processes that satisfy a wide range of customer requirements. Then, to ensure that each requirement is taken into account from the initial design phase, the MBSE method becomes an obvious solution to use.

3.1 THE "REQUIREMENT-FUNCTIONAL-LOGICAL-PHYSICAL" (RFLP) AS A MBSE FRAMEWORK

The MBSE method is an approach based on the visual representation (model) of the system module to be developed, rather than using a text document whose interpretation usually leads to errors, whose consequences often have repercussions on the entire design and production chain. As of today, product development systems are often complex because they include mechanical components as well as fluidic, electrical, electronic components and incorporating embedded software. Therefore, in order to achieve an optimal result, it is essential that all these components work in full coherence. That is where system-engineering processes come into play because they allow system engineers to share information seamlessly between disciplines. Indeed, the interdependence between the resources and the components mentioned above requires a perfect understanding of the requirements, so that the updating is carried out automatically at the level of all the departments involved in the company.

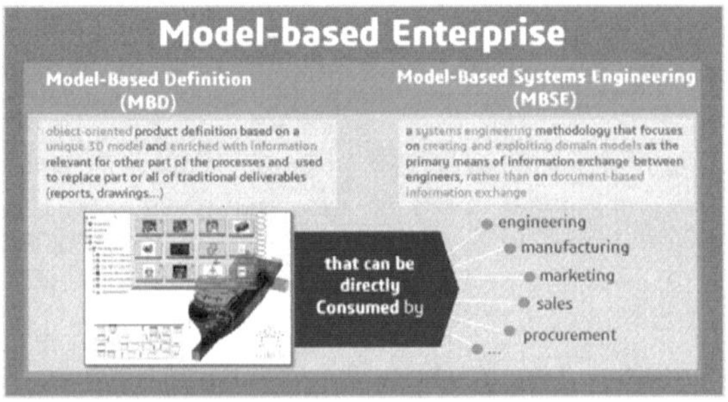

Fig.2: The Model-Based method: a unique object-oriented product definition

The RFLP approach is a framework to engage on the MBSE approach within the **3D**EXPERIENCE platform. The RFLP decomposition of systems, which initially grew popular among embedded software development teams, has now become more widely adopted across manufacturing industries. The benefit of the RFLP approach is that it enables product teams to analyse design elements independently, opening the possibility for reuse, and providing a logical path for integration to gain a holistic view of the product definition.

The RFLP framework supports a MBSE process. It is a unified system definition with four fundamentals facets.

- Requirements (R): "What the system module should satisfy?"
- Functional (F): "What the system module should do?"
- Logical (L): "What technology should be used (how the system module will be constituted)?"
- Physical (P): "How the real system module will be realized?"

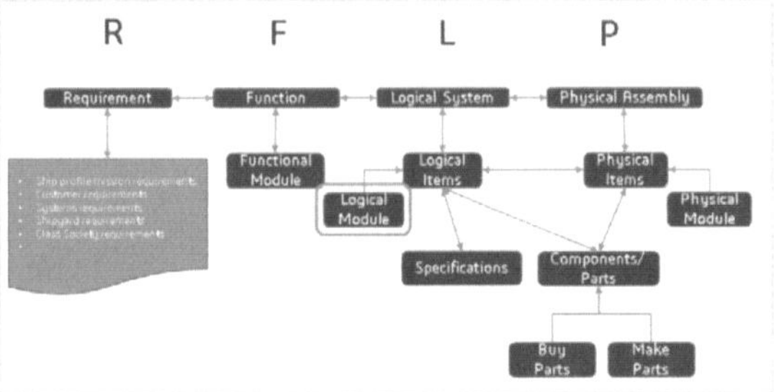

Fig.3: Module structure mapped with RFLP framework

4. THE MBSE APPLIED TO MODULAR SHIP INITIAL DESIGN

The project focussed on the definition and development of the ship functional systems with their respective scalable and interconnected system modules. For a handier approach to the definition and development of the functional modules of the platform, it was decided to split the modules into two main clusters: hull and ship systems.

The requirements, functions, logical architecture, and main components have been collected for each considered system of the different vessels in order to develop a single system architecture that can be applied to all ship family members (i.e.: workboat). The obtained system architecture fulfils a set of requirements and functionalities that will span the entirety of the RFLP facets. For each system, a finite number of module sizes have been defined in order to cover the entire range of the vessel requirements. The following systems have been considered:

- Propulsion system,
- Bilge system (including internal firefighting and deck-wash functions),
- Fuel oil system (including transfer and supply functions),
- Cooling system,
- Fresh and black water systems,
- Lubrication oil system (including clean and dirty oil functions),
- Hydraulic oil system,
- External fire-fighting system.

Note: In the current paper we will focus only on the RFLP approach carried out for the bilge system including internal firefighting and deck-wash functions. This system is on board of each vessel, and it is required by the class society. Then in addition, of the MBSE approach developed for the systems, we included additional heterogeneous tools developed by other partners to complete the initial design process, such as a resistance prediction tool, a propulsion prediction tool, a ship energy efficiency & emission calculation tool and an Underwater Radiation Noise (URN) prediction tool. This was done in order to know from the initial design phase, what are the possible impacts on the environment of such or such technological or design choices and to perform, if necessary, some design changes accordingly.

4.1 THE MBSE APPLIED TO THE BILGE SYSTEM

Following the RFLP approach, the four facets of the bilge system have been defined and implemented in the platform as described below.

4.1 (a) A complete list of Requirements (R) that the system has to fulfil

The system requirements (pump capacity, diameters, pressure, …) are defined, implemented and stored within the data base in addition to the vessel requirements (vessel length, speed, bollard pool, …) based on market analysis or customer expectations.

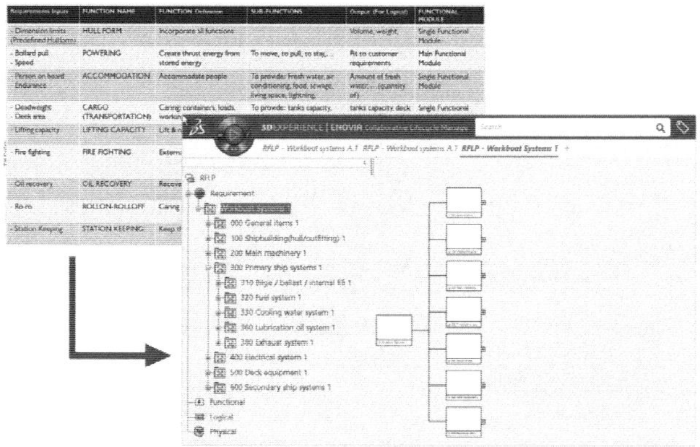

Fig.4: Ship systems requirements structure tree

4.1 (b) A functions (F) list that the system has to fulfil

The functional modules are resulting from the system requirements. They are aggregating systems and sub-systems modules that are defined based on some kind of master Piping and Instrumentation Diagram (P&ID) document managing all interfaces (mechanical, physical, electrical, fluidic,…). For sure, the functional modules are linked to a dedicated set of requirements, and this allows to have the traceability all along the RFLP facets definition. A separate functional module is considered for the hull form, which is viewed as a particular function. However, this function that we called "floating" also relies on a set of dedicated requirements which defined the main hull form characteristics (length, beam, draught, depth, entrance length, mid-ship body length).

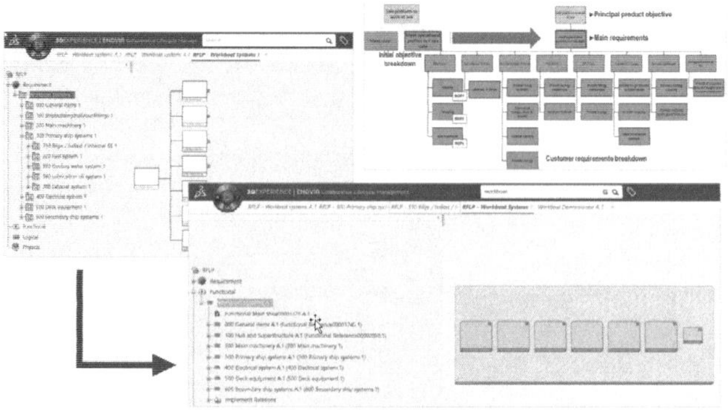

Fig.5: Ship systems functions structure three built from the requirements

4.1 (c) A library of system re-usable logical (L) modules:

The logical system modules are defined as the smallest logical component, hence a group of physical components, presented in the detailed P&ID of a given system, that is inter-connectable and can be plugged in with other modules and/or systems and can be considered as stand-alone production process.

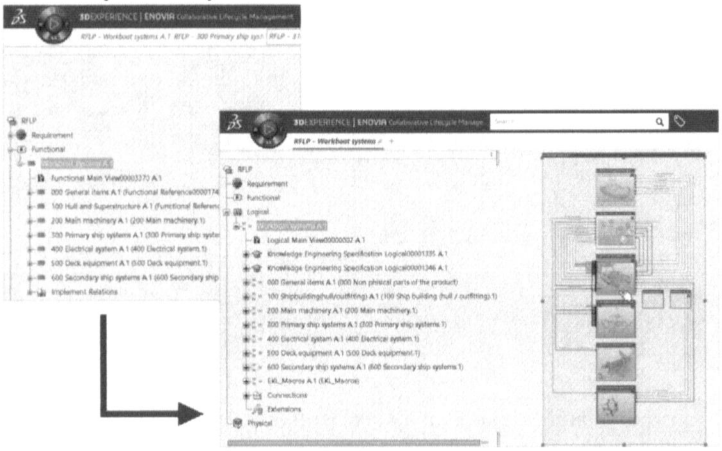

Fig.6: Ship system logical structure tree built from the functional tree

The necessary components that composed the logical system are coming from catalogues and the resulting standard logical system modules are stored in library as a template.

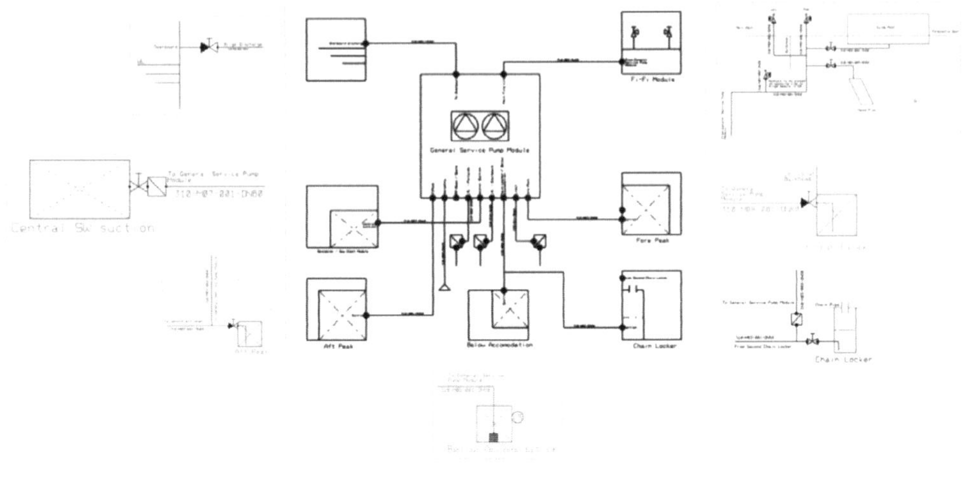

Fig.7: Master Bilge P&ID diagram composed of standard logical system modules from library

For each system, a flow chart with the sizing calculation/simulation has been defined to make sure that the customers, the classification society, and the shipyard high requirements and other system requirements are taken into account to dimension the system elements and the equipment as well. This includes the necessary sizing calculation check rules developed within the platform using the native Engineering Knowledge language (EKL) to verify that the requirement values are respected.

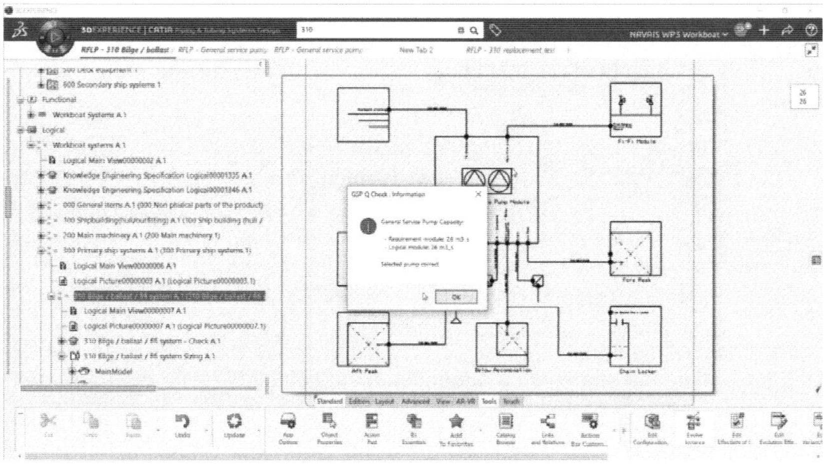

Fig.8: General service pump capacity check rule using EKL

Note: the electrical system has been handled as a separate module since its support all other on board systems, taking input from all the mechanical systems that have to be dimensioned first.

4.1 (d) The physical (P) 3D modules layout

Based on the logical modules definition, the 3D instantiation is performed inside the digital twin of the ship. The 3D components and equipment instantiation are driven by the logical specifications, making sure that the 3D objects will respect the specifications from the 2D logical module definition.

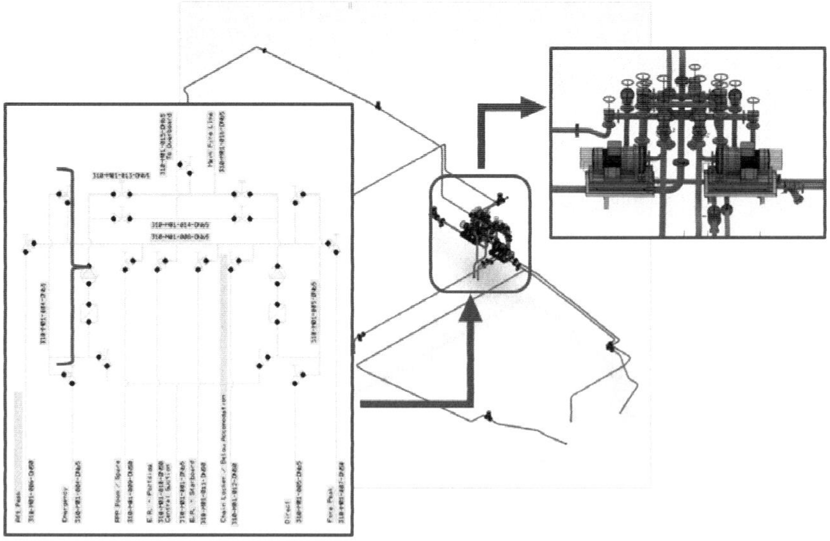

Fig.9: 3D Bilge system instantiation based on the logical module specifications

4.1 (e) The physical (P) 3D parametric hull form

An initial parametrized hull form was created based on six parameters directly linked to the main vessel requirements. These parameters can be adjusted to drive the parametric shape of the hull designed, using the Aesthetical Shape Modeler (FFS) application. Three parameters are defining the main dimensions of the platform and the other three are influencing the vessel resistance. The obtained hull shapes have than been used for resistance simulations using RANS equations primarily used to describe turbulent flow (RANS CFD solvers) or others regression methods. This allowed us to adjust the ship hull form to the customer needs in term of vessel dimensions and have impact on the ship resistance curve calculation.

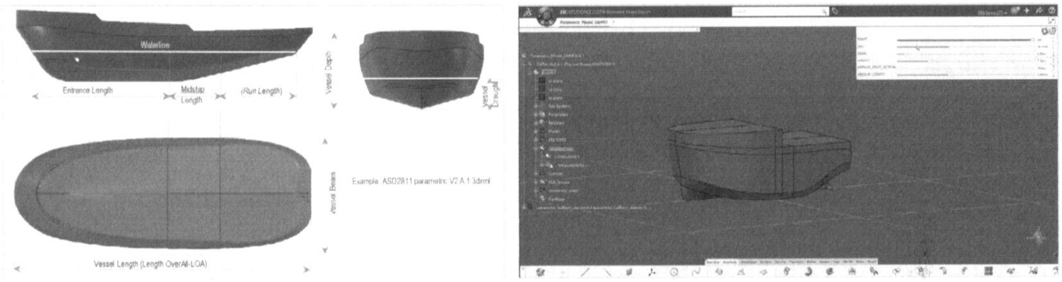

Fig.10: The six requirements parameters that drive the ship hull form

5. ADVANCED SHIP INITIAL DESIGN PROCESS FLOW INCLUDING THE MBSE APPROACH AND EXTERNAL TOOLS

5.2 HETEROGENEOUS PARTNERS TOOLS

Ultimately, when the ship digital twin starts to be designed at the RFL levels, we can start to make sure that the vessel requirements are fulfilled in addition to the ones dedicated to the systems. To achieve this objective, specific "niche" tools were developed by partners of the NAVAIS consortium, who have in-depth expertise in their specific fields. These tools use the digital twin data already available in the platform, created during the initial design phase. Their successive sequences make it possible to validate certain environmental KPIs (Key Performance Indicators) such as:
• Low design impact,
• Low energy and low emissions impact (IMO/MARPOL standards),
• Ship energy efficiency calculation,
• Underwater Radiation Noise estimation,

This set of tools is giving additional inputs to the engineers, during the initial design stage, to understand the impacts of their technical choices on the environment and to revise them according to requirements, rules and standards from the International Maritime Organization (IMO) for instance. Following IMO rules made the NAVAIS project one with low design impact and low operational environment footprint.

The following chapter describes the purpose of each tool.

5.3 HETEROGENEOUS PARTNERS TOOLS SPECIFICATIONS

Each partner developed their own tool using different technologies such as web application, script, and executable or system modelling model based on the Modelica language. The next part is describing the purpose of each tool and their specific input-output flow relying on the MBSE data from the platform.

5.3 (a) Ship resistance curve calculation tool (from MARIN [8]):

It is a web application that calculates a final resistance curve based on hull frictional resistance, wave making resistance, eddy resistance and air resistance. The input data are the six mains parameters that drive the 3D parametric hull form dimensions created using CATIA modeler with the **3D**EXPERIENCE platform apps. Such parameters are defined as ship geometrical Requirements in the platform and also as logical parameters which compose the hull logical (L facet) module.

5.3 (b) Propulsion prediction tool (from DAMEN [9]):

The tool is a python script developed by the DAMEN R&D department. It relies on a combination of main equipment such as engine, gearbox, propeller and ship loading conditions. This tool allows to predict the fuel consumption, the bollard pull and speeding taking into account some abatement options at the initial design stage. The input data is the speed/resistance table issued from the ship resistance curve calculation, and the propulsion equipment specifications defined in the platform at the logical stage of the module definition (engine, gearbox, propeller…).

5.3 (c) Ship Energy Efficiency Calculation and Analysis tool (from Bureau Veritas [8]):

The SEECAT Version 2 app is new and 100% developed on the **3D**EXPERIENCE platform using standard Modelica libraries. It provides a holistic approach to simulate, understand and then optimize the ship design and operations for energy efficiency and air pollution prevention. It simulates the different energy systems (diesel engines, propulsion system, steam production, cooling system, waste heat recovery etc.) and the energy transformations shared between them (fuel, heat, steam, mechanical power). The simulation is performed in time domain according to a pre-selected operational profile allowing for real time monitoring of fuel consumption and of CO2, NOx and SOx emissions. The input data is the ship displacement (i.e.: the weight of water a ship displaces of when fully loaded and submerged to her load lines). This information is computed based on the six mains parameters that drive the 3D parametric hull form dimensions created using CATIA modeler from the **3D**EXPERIENCE platform.

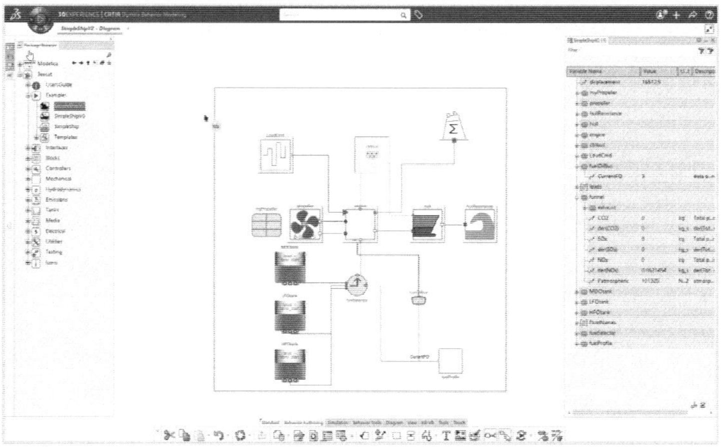

Fig.11: SEECAT Version 2 - interface in the **3D**EXPERIENCE platform

5.3 (d) Under Water Radiation tool (from MARIN):

A regression model is created using results from many propeller cavitation noise computations. Then machinery noise contribution is added. Finally, a noise spectrum in several operation conditions is proposed to the engineer to figure out what the noise levels of the current design will be. It is helpful to compare the predicted URN levels against the set limits and to make eventual decision of design change. The input data is the propulsion equipment specifications defined in the **3D**EXPERIENCE platform at the logical stage of the module definition (engine, gearbox, propeller…) and some operating conditions from the SEECAT tool integrated in the **3D**EXPERIENCE platform (propeller rotation rate, propeller thrust, ship speed…).

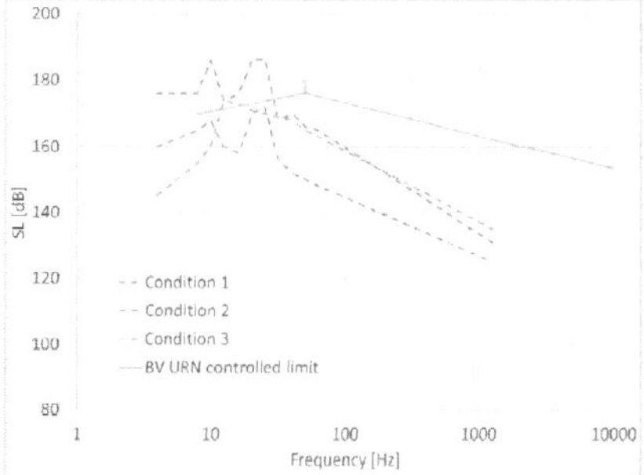

Fig.12: Noise spectrum example based on frequency band

The first difficulty encountered was to identify in an exhaustive manner the parameters expected by each of the tools and in what format. After successive adaptations, each tool was tested individually and then manually tested in sequence to ensure the continuity of the process and the generation of the expected final outputs. These were the ship energy efficiency KPIs and the noise spectra based on some requirements and pre-selected logical modules specifications from the **3D**EXPERIENCE platform. Thus, the main phases of the initial design process were linked as described in the figure below that shows: the ship resistance curve generation, the prediction propulsion hypothesis, the ship energy efficiency calculation and the URN computation process.

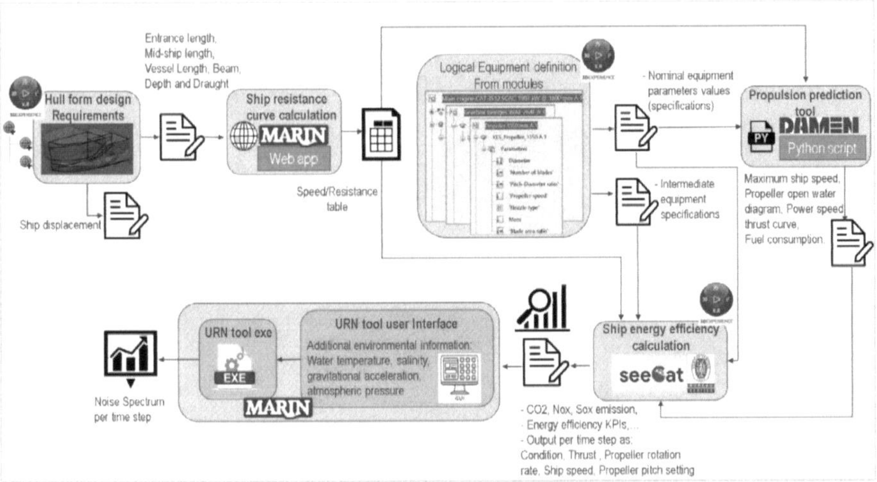

Fig.13: Heterogeneous Partner's tools and **3D**EXPERIENCE platform - initial design process flow input/output

6. THE SHIP INITIAL DESIGN PROCESS FLOW ORCHESTRATION

The process described above really has a high benefit, as successive iterations are necessary to respect the vessel specifications. Indeed, the verification and validation of the ship's requirements and the environmental rules from the IMO and from the MARPOL convention require the analysis of several module configurations. In order to have a quick help in the final decision, it is necessary to think about automating the whole process.
The challenge was then to be able to orchestrate the sequence of these different external tools while relying on MBSE data from the platform. This proved possible thanks to the openness capabilities of the **3D**EXPERIENCE platform and the use of the Process Composer application from the Dassault Systèmes' SIMULIA brand. Indeed, this application is providing capabilities to automate simulation processes by capturing and automating the execution of multiple tasks necessary in the innovation, verification & validation cycle. In addition, it allows to capture complex integration workflows that take into account variability of a high number of design parameters and multiple simulation tools for use in initial design exploration.

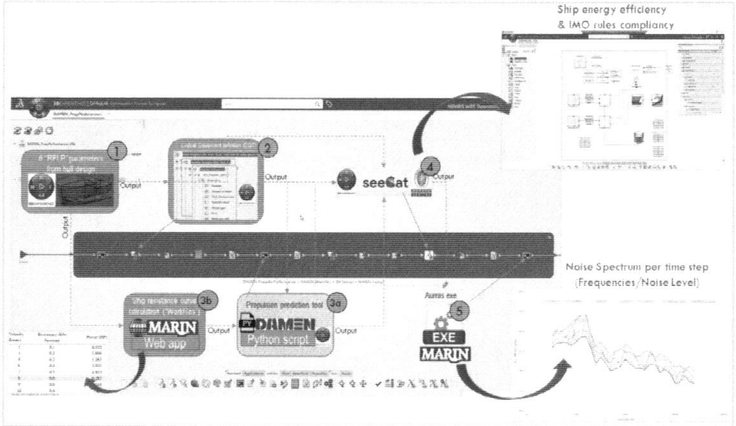

Fig.14: the Process Composer application applied used for initial design on the workboat demonstrator.

7. CONCLUSION

The MBSE approach is for sure the best way to manage complex systems. Especially when they have multiple connections to each other and when there are so many project vessel requirements (R facet of the product) as ship mission and profile, customer needs, systems specifications, class rules…
The idea to address the CtO design focusing on pre-defined logical modules (L facet of the product) with the traceability to primary product definition (R facet of the product) was a clever one.
Indeed, including the notion of modularity from the initial design phase without waiting for the 3D definition of the digital twin allows greater flexibility to answer quickly to the call for tenders. The logical modules associated with all the primary and secondary functions (F facet of the product) are already defined in the form of the standard modules in the library. They can then be easily resized to meet the product requirements changes and assembled together to verify and validate the final modules size and the final configuration of the vessel. Also, because the technological choices have already been implemented in the standard logic modules, a pre-approval of these modules by the Class allows the system engineer to work in compliance with the rules and to confirm his choices while minimizing the risks during the final approval of the ship.
With logical modular design, assemblies are treated like individual components. The platform-based product family designed with a proper modular architecture makes it possible to select and purchase assemblies (i.e., modules) in the same way as components are procured. Thanks to the "Configure to Order" business model, the cost or lead-time of the vessel will decrease since:

- The pre-contractual and contractual steps are completed prior the engagement with the ship-owner,
- The classification review process is limited to the review of the variants of the products selected by the ship-owner and not of the entire vessel,
- A large number of components and submodules can be purchased and assembled independently from the vessel to be built at any moment.

At a glance, the main advantages of model-based systems engineering are threefold: rigorous design based on requirements, management of product lines and reuse and thus simplification of manufacturing processes.

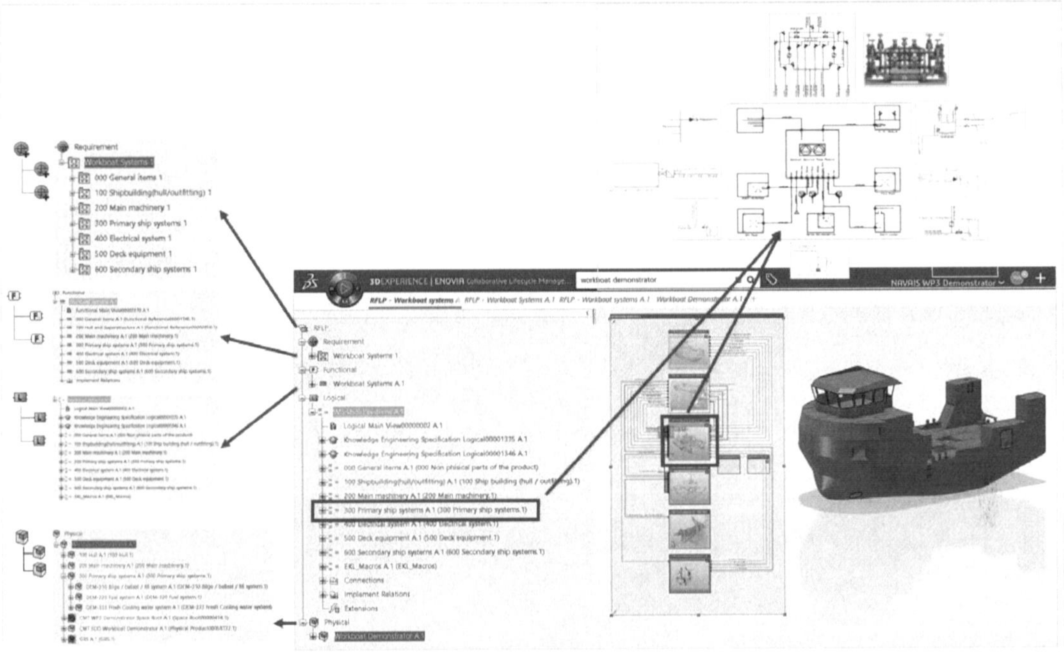

Fig.15: one of the final NAVAIS demonstrator using the MBSE approach of the **3D**EXPERIENCE platform

Finally, the main challenge for the NAVAIS project was not only related to the questioning of the EtO design approach versus the CtO modular one. Nor was it about implementing the MBSE approach. It was rather to understand in its entirety the digital transformation of the study phase of a ship, with the final objective of validating a sustainable by design product, thanks to a single collaborative openness platform.

8. ACKNOWLEDGEMENTS

The project has received funding from the European's Horizon 2020 research and innovation program, under contract number 769419. Horizon 2020 is the financial instrument implementing the Innovation Union, a Europe 2020 flagship initiative aimed at securing Europe's global competitiveness. Horizon 2020 [6] is the biggest EU Research and Innovation program ever with nearly €80 billion of funding available over 7 years (2014 to 2020) – in addition to the private investment that this money will attract. It promises more breakthroughs, discoveries and world-firsts by taking great ideas from the lab to the market.

9. REFERENCES

1. Dassault Systèmes, the 3DEXPERIENCE Company (https://www.3ds.com/)

2. BUREAU VERITAS Marine and Offshore site (https://marine-offshore.bureauveritas.com/)

3. ICCAS 2019 International Conference on Computer Applications in Shipbuilding 2019, 24-26 September 2019, Rotterdam "Platform-based modular product family design" (https://www.rina.org.uk/ICCAS_2019.html)

4. NAVAIS, "Challenge New Advanced Value Added Innovation in Shipbuilding", 2019 (https://www.navais.eu/)

5. MG-2.4-2017-Complex and value-added specialized vessels, 2017 (https://cordis.europa.eu/programme/rcn/701935/en)

6. EUROPEAN COMMISION, "Horizon 2020-The EU framework program for Research and Innovation", 2019 (https://ec.europa.eu/programmes/horizon2020/en)

7. EC-The Community Research and Development Information Service (CORDIS), "NAVAIS Fact Sheet", 2018 (https://cordis.europa.eu/project/rcn/216018/factsheet/en)

8. MARIN Research Institute Netherlands (https://www.marin.nl/en)

9. DAMEN Shipyards Gorinchem site (https://www.damen.com/companies/damen-shipyards-gorinchem)

9. AUTHORS BIOGRAPHY

Richard Audoire holds the current position of Marine & Offshore Industry Solution Technical Director at Dassault Systèmes. He is responsible for managing innovative projects to reinforce the business value proposition delivered to Marine & Offshore companies. His previous experience includes working closely with leading Dassault Systèmes' customers such as Damen, Bureau Veritas, Naval Group, Heesen Yachts and Meyer Werft.

WIND ASSISTED PROPULSION: INVESTIGATIONS ON THE USE OF SHIP HOLISTIC MODELS FOR PERFORMANCE ANALYSIS

A Bellot[1], **A I Ölçer**, **R Baumler**, World Maritime University, Sweden
C Bouallou, **M Nemer,** Mines Paris[1], France

SUMMARY

Wind propulsion is one of the most promising technologies to reduce carbon intensity of ships. However, with variable thrust production and induced drift action, sail's contribution to propulsion is to be evaluated in different ways from traditional propulsors. For sail retrofit projects, where vessels are initially designed to be optimal at single transit speeds with propulsion engine, wind assistance may also induce a degradation of operational efficiency and limit the extend of energy savings. For these reasons, performance analysis should evolve to account for complex energy transfers between propulsion domain and energy systems. A novel Modelica based model is presented in this article, where manoeuvres and energy modelling are used jointly to predict overall efficiency of wind assisted ship concepts. The model is used to evaluate the impact of retrofitting 4 Flettner rotors on the Duisburg test case (DTC) equipped with diesel engines mechanical propulsion system, with oil fired and exhaust gases steam plants and electricity generation from gensets and turbogenerator. Finally, from the analysis of DTC performance in a sweep of wind conditions, potential solutions to increase sailing efficiency are presented, offering new research directions in vessel design adaptation to natural propulsion.

NOMENCLATURE

$\square_{RB}$ Rigid body inertia matrix
$\square_{\square}$ Added mass matrix (kg)
$V = (u, v, w, p, q, r)$ Velocity matrix 6DOFS (m/s) (rad/s)
F Resultant force and moment matrix (N) (N.m)
m Ship displacement (kg)
(x_G, y_G, z_G) Centre of gravity position (m)
$F_{x\,rotor}$ Flettner rotor longitudinal force (N)
$F_{y\,rotor}$ Flettner rotor lateral force (N)
C_L Rotor lift coefficient
C_D Rotor drag coefficient
α Spin ratio coefficient
ρ_{air} Air density (kg/m^3)
V_{aw} Apparent wind velocity vector (m/s)
A Rotor wind exposed surface (m^2)
T Transformation matrix from world to local frame
$\overrightarrow{V_{tw}}$ True wind velocity vector (m/s)
$\overrightarrow{V_{ship}}$ Ship velocity vector (world frame) (m/s)
n_{rotor} Revolution per second rotor (Hz)
d Rotor's diameter (m)
C_{Thrust} Rotor thrust coefficient
C_{drift} Rotor drift coefficient
$w_{water\,feed}$ Feed water flow to boiler (kg/s)
$w_{evaporation}$ Evaporation flow in water drum (kg/s)
$w_{to\,EGB}, w_{from\,EGB}$ Flow from/to exhaust gas boiler (kg/s)
m_{water} Mass of water in water drum (kg)
$h_{water\,feed}$ Specific enthalpy of feed water (J/ K.kg)
h_{dew} Specific enthalpy of water at dew point (J/ K.kg)
$h_{from\,EGB}$ Specific enthalpy of water from EGB (J/ K.kg)
h_{water} Specific enthalpy of water in water drum (J/ K.kg)
Q_{burner} Burner heat flow to water drum (W)
h_{bubble} Specific enthalpy at bubble point (J/ K.kg)
P_{boiler} Boiler pressure (Pa)
η_{burner} Burner efficiency ratio
w_{fuel} Fuel flow to burner (kg/s)
LHV_{fuel} Fuel lower heating value (J/kg)
$w_{steam\,in}, w_{steam\,out}$ Steam flow in/out of steam drum (kg/s)
m_{steam} Mass of steam in steam drum (kg)
h_{steam} Specific enthalpy of steam in steam drum (J/ K.kg)
V_{drum} Volume of steam drum (m^3)
S Available surface in heat exchanger (m^2)
S_{di} Surface required for diphasic transformation (m^2)
S_{mon} Surface left for monophasic transformation (m^2)
K Heat exchange global coefficient (W/m^2.K)
$T_{A1}, \ldots, T_{B2}$ Temperature of fluids across exchanger (K)
$h_{A1}, \ldots h_{B2}$ Specific enthalpy of fluids across exchanger (J/ K.kg)
w_A, w_B Flow of fluids in heat exchanger (kg/s)

C_{pA}, C_{pB} Heat capacity of fluids in exchanger (J/ K.kg)
$P_{A1}, \dots, P_{B2}$ Pressure of fluids in exchanger (Pa)
R Ratio of heat capacity rates
NUT Number of transfer units in exchanger
ε Efficiency of exchanger
V_s Stator input voltage (V)
i_s Stator current (A)
E Electromotive force in stator (V)
i_R Rotor current (A)
ω_{rotor} Rotational velocity of rotor (rad/s)
τ_{rotor} Torque at rotor (N.m)
f Frequency of electrical signal at stator (Hz)
ω_{elec} Rotational velocity of electrical signal (Hz)
p Number of pairs of poles in generator
φ Phase angle between voltage and intensity (rad)
THD Total harmonic distortion of signal at rectifier
S_{AC} Apparent power input to rectifier bridge (VA)
V_{DC} Power output of rectifier bridge (V)
i_{DC} Output direct current of rectifier bridge (A)
V_{AC} Alternative current voltage input to rectifier (V)
α Firing angle of thyristors in rectifier bridge (rad)
M Matrix of possible combination of genset use
$M_{optim\ power}$ Matrix of optimal loads at genset (VA)
$S_{switchboard}$ Apparent power at switchboard (VA)
$S_{difference}$ Difference between produced power at power at switchboard (VA)
$S_{optim\ genset1}, \dots, S_{optim\ genset4}$ Optimal power load of gensets (VA)
$S_{rated\ genset1}, \dots, S_{rated\ genset4}$ Rated power of gensets (VA)
$S_{genset1}, \dots, S_{genset4}$ Power delivered by gensets (VA)
$V_{genset1}, \dots, V_{genset4}$ Gensets voltage condition (V)
$f_{switchboard}$ Frequency at switchboard (Hz)
$f_{genset1}, \dots, f_{genset4}$ Frequency at gensets (Hz)
$i_{genset1}, \dots i_{genset4}$ Currents produced by gensets (A)

1. INTRODUCTION

From an historical perspective, wind has always been the most common and convenient way to propel vessels across waters. However, for at least 100 years, and even more importantly since the deployment of container transportation, shipping and ship design was drove by economic optimisation in a context of low energy costs, increasing freight rates and intensification of international logistics. As a consequence, modern vessels were shaped to maximise cargo capacity over hydrodynamic efficiency leading to the development of the emergence of propulsion systems leaning on massive fossil fuel supplies and powerful internal combustion engines. However, in the context of emission reduction objectives (IMO, 2018) and constraining environmental regulations (IMO, 2011, 2014, 2016, 2021a, 2021b) new methodologies in ship building might appear. In a transition era, wind assisted ships are therefore attracting a lot of attention. However, fuel and wind ships cannot be built according to similar design principles, and in particular optimisation research cannot be achieved in the same way for two completely different propulsion technologies (Kralj & Klarin, 2016). Also, for hybrid vessels with sails and propeller, energy flows on board should be analysed in a holistic manner to evaluate energy saving potentialities. Holistic ship design (Papanikolaou, 2010), supported by advanced energy modelling tools (Dimopoulos et al., 2014; Mermiris et al., 2011; Tillig et al., 2017), is the most viable way to overcome technical barriers in the uptake of wind propulsion. As a result, new digital applications are needed to support designers in the evaluation of concept model and drive future research towards more efficient ship concepts.

In this article, an analysis of the impacts of sails on a fuel propulsion optimised vessel is evaluated with a multi-physic model. In the first section, a literature review in holistic ship modelling is presented. In a second part, more information can be found on the theorical framework of the newly developed model. Using the presented model, case study results on sails impact on performance further described and commented, while the final chapter intends to comment on design options to increase efficiency of sail assisted vessels.

2. LITERATURE REVIEW

The ship simulation problem has been extensively studied in the past decade aiming at several objectives such as manoeuvers, course keeping ability or energy consumption predictions.

Starting from 2011, D. Mermeris et al. analysed the potential of digital models to "*understand[...] how [ship] systems interact with each other's from a design and operational point of view, and which are their effect on fuel consumption*" as, "*contrary of an offshore rig, ships are travelling for a substantial proportion of their lives and the encountered environmental conditions cannot be taken into consideration in generalized format of average values*" (Mermiris et al., 2011). The simulation object focused on propulsion using sets of basic component models for dynamic prediction of consumption (referred as Dynamic Energy Modelling) (DEM). DEM uses sequential and causal modelling, balancing input/output signals with transfer functions, and sequential calculations time steps. The model was then developped in (Cichowicz et al., 2015) with system ugrades, and achieving complete system modelling for ship's propulsion with auxiliary systems. Same modelling approach can be found in (Dimopoulos et al., 2014) with a special attention given on generic and modular modelling of marine system. The developped simulation tool also aimed at supporting concept evaluation in naval architecture with model-based system engineering (MBSE) in a commercial service to shipowners: the tool was finally supported by DNV-GL in COSSMOS (Compplex Ship Systems Modelling and Simulation) with g-PROMS as simulation platform. More specifically, MBSE for ship design aims at:

(a) evaluate performance of a technologic set under real service condition (dynamic or steady state modelling) and,
(b) provide optimising tools for simulation parameters (such as battery capacity of an hybrid propulsion plant as an exemple) One of the main achievment of COSSMOS in marine engineering remains the integration of a configurable library of marine block components under a time domain dynamic energy modelling approach.

In (Tillig et al., 2017) a holistic ship model accounting for interactions between hull efficiency and energy systems with power transfers between mechanical, thermal and electric domains. The simulation tool (for propulsion system from hull to fuel consumption) is then declined in two versions, one targeting static power prediction (steady state calculations) and the other to analyse dynamic conditions (transient states calculations).

Yet improvement of models can be achieved with the addition of manoeuvring and course keeping simulations. Ship trajectory predictions, modelled in 2 (planar) or 3 (spatial) dimensions, are necessary to evaluate the true performance of ships equipped with sails. The manoeuvring modelling question was addressed in 1980 with the seminal work of M. Abkowits (Abkowits, 1980). In (Abkowits, 1980) vessel's planar motions is modelled from resultant hydrodynamic forces and propeller thrust on the vessel. While propeller thrust can be expressed with generic models such as B-screw polynomial series from (Oosterveld & van Oossanen, 1975), hydrodynamic forces and moments needs to be approximated with Taylor's decomposition functions of independent system variables. On the other hand, calm water resistance models can be approximated from (Holtrop & Mennen, 1982) and (Holtrop & Mennen, 1984) or extracted from model tests experiments. Also, aerodynamic loads can be modelled in a generic way using inputs from (Isherwood, 1973), where apparent wind load is calculated from wind inputs (velocity, relative direction) and geometric parameters for hull and superstructures. In (Yasukawa & Yoshimura, 2015; Yoshimura, 2005), rudder action is separately modelled and added to the general model. Additionnal models are frequently developped with upgrades in accuracy, or with additionnal interaction models such as shallow water effect.

So far, ship models for manoeuvring or energy consumption have followed separate development paths and different research objectives. While manoeuvering models intended to evaluate ship's trajectory as a function of propeller and rudder actions, energy models were targeting fuel consumption prediction as a function of thrust power or ship longitudinal velocity. However, with wind propelled vessels, where propeller thrust requirement results from course keeping strategy and sail drift and drag forces, these approaches must be combined together in further holistic energy optimisation research.

3. MODELLING STRATEGY

In the continuation of past developments, to integrate sail power in energy prediction tools, a holistic ship model with manoevring and energy models was developped on Dymola (Dassault, 2022) and encodded with modelica language (Modelica, n.d.). The ship modelling philosophy is thus applied to:

- Manoeuvring, where resultant forces and moments acting on hull is converted to planar acceleration and rotation.
- Propulsion system including thrust production with sails and propeller, in addition to rudder action.
- Electrical plant, where energy is processed from fuel to electrical domain and fed to electrical consumers
- Thermal plant, where energy is processed from fuel or exhaust gases to steam and fed to consumers and/or condenser.

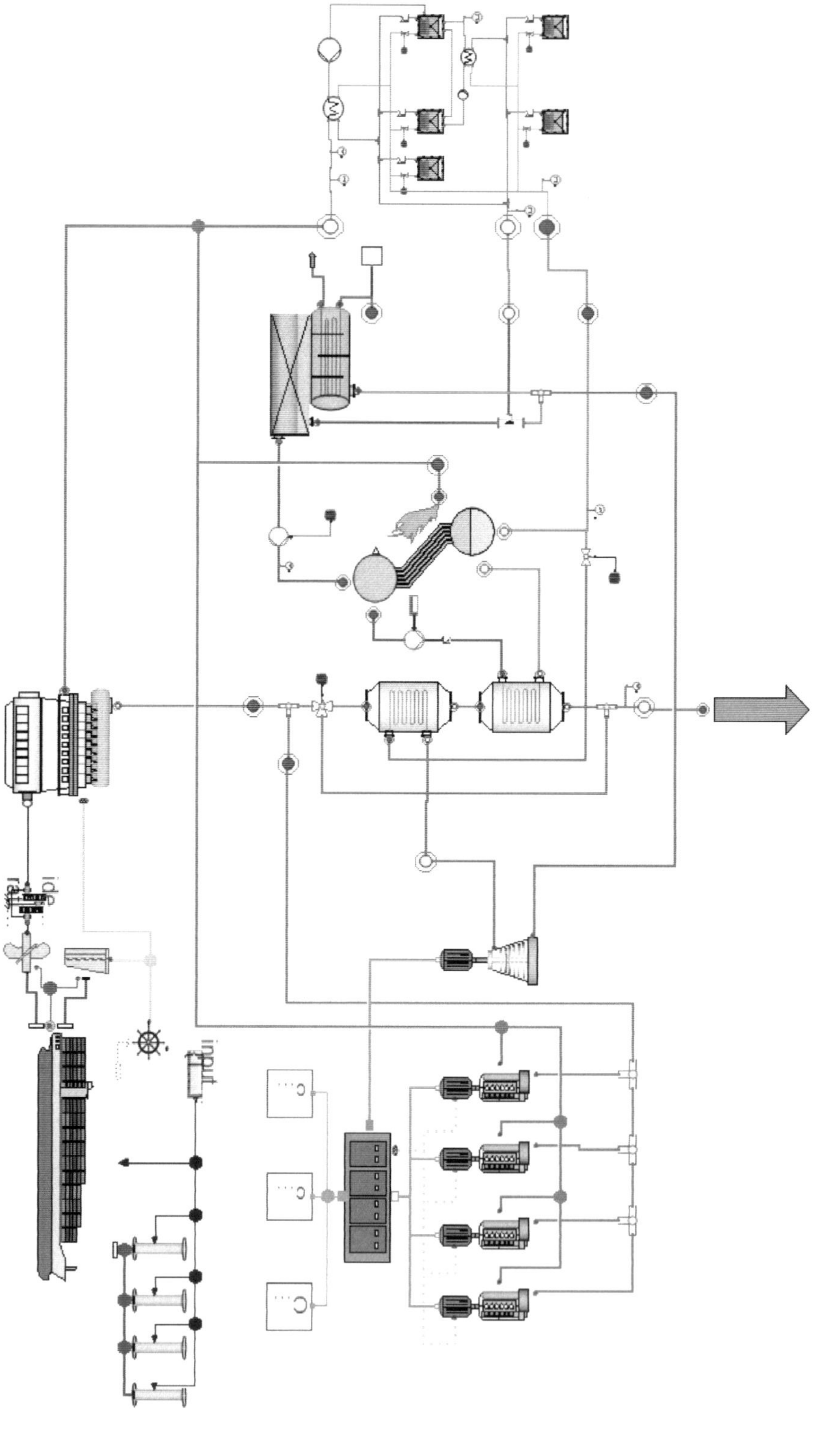

Figure 1. Modelling interface for the simulation of a vessel on mechanical propulsion system (ICE+ST) with Flettner rotors

3.1 MANOEUVRING MODELS

The modelica model calculates the trajectory of the vessel from the expression of mechanical interactions applied on hull in a set local frame. A mechanical solver relates resultant forces and moments at midship to acceleration vector in local frame; Acceleration vector can be converted in world frame using translation along a position vector $\vec{r}$ (position vector) and a rotation according to a transformation matrix T calculated from cardan angles.

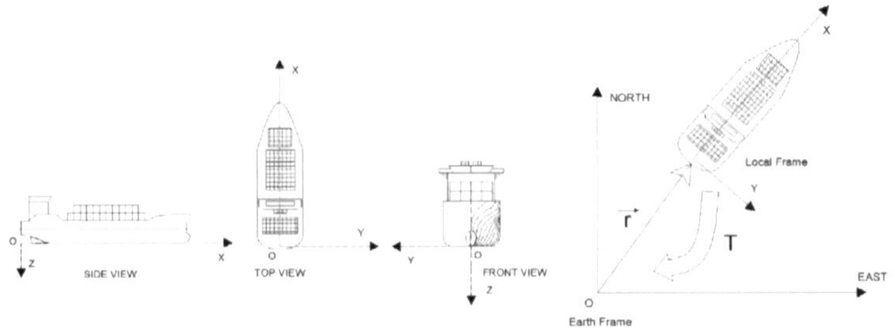

Figure 1. Local frame definition for the manoeuvring model

Using fondamental equations of motion, the trajectory of the vessel is calculated in a matricial form with intertia, coriolis and added mass matrix in the same standards as (Fossen & Fjellstad, 1995):

$$(M_{RB} + M_A).\dot{V} + C_{RB}.V = F$$

The rest of the manoeuvring model follows same logic as (Yasukawa & Yoshimura, 2015) with a separate rudder model and accounts for aerodynamics loads, hydrodynamic performances, calm water resistance and hull to propeller interactions using experimental data (CFD, towing tank, open water propeller or wind tunnel analysis) or alternatively generic models or such as (Isherwood, 1973), (Oosterveld & van Oossanen, 1975) and (Holtrop & Mennen, 1982, 1984). Added wave resistance can also be calculated in a generic way with sea margin or models, (Liu et al., 2016).

In addition to the propeller Flettner rotor can be added to the propulsion mix and modelled with generic models where lift and drag forces are expressed from coefficients:

$$F_{x\,rotor} = \frac{1}{2} \,.\, C_L(\alpha) \,.\, \rho \,.\, V_{aw}^2 \,.\, A$$
$$F_{y\,rotor} = \frac{1}{2} \,.\, C_D(\alpha) \,.\, \rho \,.\, V_{aw}^2 \,.\, A$$

Apparant wind V_{aw}vector is calculated from ship word speed vector $\overrightarrow{V_{ship}}$ and word true wind vector $\overrightarrow{V_{tw}}$ as such:

$$\overrightarrow{V_{aw}} = T.(\overrightarrow{V_{tw}} - V_{ship})$$

C_L and $C_D$are expressed as a function of spin ratio coefficient α from generic relationship (Sun et al., 2010), where α is a control variable, function of rotor's velocity of rotation:

$$\alpha = \frac{n.\pi.d}{V_{aw}}$$

Instantaneous spin factor α is controlled in the simulation using apparent off-bow wind angle AWA in order to maximise longitudinal force, which is calculated from thrust coefficient C_{Thrust} (projection of drag and lift coefficient in the direction of ship longitudinal axis).

$$C_{Thrust}(\alpha) = C_L(\alpha).\sin(|AWA|) - C_D(\alpha).\cos(AWA)$$
$$C_{Drift}(\alpha) = -sign(AWA) * C_L(\alpha).\cos(AWA) - C_D(\alpha).\sin(AWA)$$

Consequently, in the simulation, thrust and draft coefficient are calculated with the following relationship:

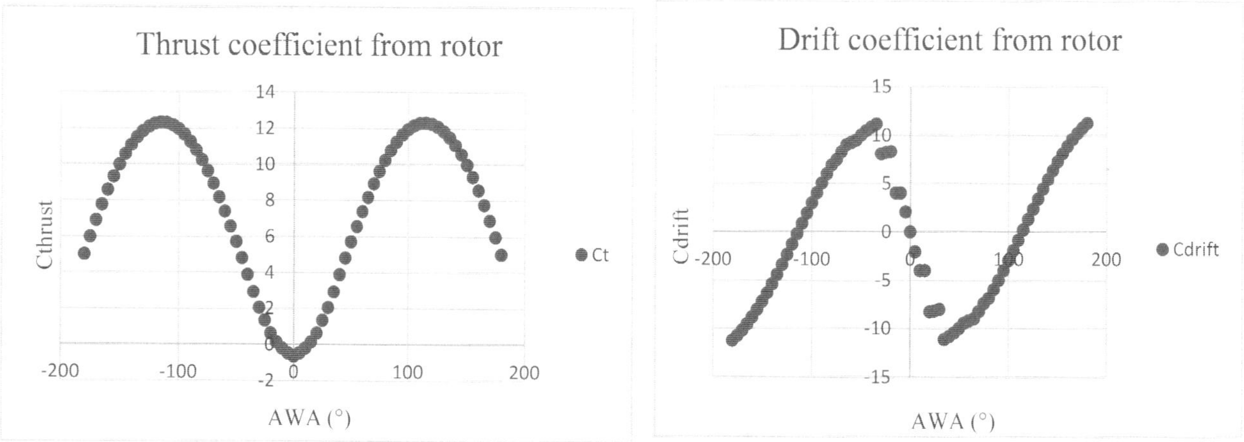

Figure 2: Thrust and drift coefficient calculation as a function of apparent wind angle

3.2 THERMAL SYSTEM MODELS

Steam system is modelled from Modelica library of medium models, based on IF97 standards for steam and water (IAPWS, 2007). In the heating network, fluid composition is given by pressure and specific enthalpy state variables and mass flow rate and enthalpy flow variables. Related state functions such as temperature, density or entropy can therefore be calculated from initial inputs. In the simulation platform, steam heating and power generation system includes the following sub-models:

- An oil-fired boiler model with water and steam drums
- Several diphasic fluid-fluid exchangers with variations including a meta-model of atmospheric condenser, exhaust gas evaporator and superheater and tank heating system
- A generic steam turbine model

3.2 (a) Oil fired boiler model

The boiler converts liquid water in saturated steam using heating power from oil fired burner. Boiler's drum pressure and enthalpy are directly influenced by the balance of input/output energy (oil burner, freshwater feed, exhaust gas boiler circulation and steam consumption flow). The boiler model behaves in a dynamic way with a real time monitoring of mass and specific energy contained in water and steam drums.

First, the mass of water in water drum is calculated from inflow and outflow mass equation:

$$w_{water\ feed} + w_{evaporation} + w_{to\ EBG} + w_{from\ EGB} = \frac{dm_{water}}{dt}$$

Evaporation flow in water drum is calculated from inflow and outflow energy equation in water drum :

$$w_{water\ feed}.h_{water\ feed} + w_{evaporation}.h_{dew} + w_{from\ EGB}.h_{from\ EGB} + w_{to\ EGB}.h_{water}(t) + Q_{burner} = \frac{d(m_{water}.h_{water})}{dt}$$

With:

$$h_{water}(t) = h_{bubble}(P(t))$$

Heat flow produced by the oil-fired burner, is roughly calculated from fuel lower heating value and consumption flow, with a parameter efficiency coefficient:

$$\eta = 0.8.$$

$Q_{burner} = \eta * LHV * w_{fuel}$

In steam drum, water evaporation and energy increase are accumulated and pressures builds-up along with density increase:

$w_{steam\ in} + w_{steam\ out} = \frac{dm_{steam}}{dt}$ (1)

$w_{steam\ in}.h_{steam} + w_{steam\ out}.h_{steam} = \frac{dh_{steam}}{dt}$

With:

$h_{steam}(t) = h_{dew}(P(t))$

And:

$\rho = \frac{m_{steam}}{V_{drum}}$ (2)

Therefore, by combining (1) and (2):

$w_{steam\ in} + w_{steam\ out} = \frac{d\rho(h,P)}{dt}.V$ *(steam drum volume is supposed constant)*

Which is equivalent to:

$w_{steam\ in} + w_{steam\ out} = (\frac{\partial\rho}{\partial P}.\frac{dP}{dt} + \frac{\partial\rho}{\partial h}.\frac{dh_{steam}}{dt}).V$

With:

$h_{steam} = h_{dew}(P(t))$

$\frac{\partial\rho}{\partial P} = f_{IF97}(P(t), h_{steam})$

$\frac{\partial\rho}{\partial h} = g_{IF97}(P(t), h_{steam})$

Density derivative on pressure at constant enthalpy can be calculated according to IF97 standard from instantaneous state variable)

Which allows the calculation of boiler pressure variable P.

3.2 (b) Exhaust gas boiler, condenser, and tank heating model

The exhaust gas boiler, just as the condenser or the tank heating model are based on generic diphasic exchanger equations between two fluids at different temperature levels. The model evaluates the surface required for diphasic exchange S_{di} (transformation of steam into water or water into steam) and surface left for subcooling or superheating S_{mon} with regards to the total available surface of exchange:

$S = S_{di} + S_{mon}$

For the heat transfers in diphasic mode, the logarithmic mean temperature difference methodology (LMTD) is used to predicts output temperatures of fluids:

$K.S_{di}.\frac{|T_{Bx}-T_{A1}|-|T_{B2}-T_{A1}|}{\ln(|T_{Bx}-T_{A1}|)-\ln(|T_{B2}-T_{A1}|)} = w_A.|h_{buble\ or\ dew} - h_{A\,1}|$

and for heat transfers during monophasic exchanges:

$T_{B\,x} = T_{B\,1}(1-\varepsilon) + T_{A1}(1+\varepsilon)$

$w_B.Cp_B.|T_{B\,1} - T_{B\,x}| = w_A.|h_{A\,2} - h_{dew}|$

Where heat transfer efficiency ε is calculated from number of transfer unit (NUT) in the exchanger.

$$R = \frac{\min(w_B.Cp_B, w_A.Cp_A)}{\max(w_B.Cp_B, w_A.Cp_A)}$$

$$NUT = \frac{K_{cond} \cdot S_{mon}}{\min(w_B.Cp_B, w_A.Cp_A)}$$

$\varepsilon = f(R, NUT)$ *(Epsilon is calculated according to efficiency functions for the selected exchanger type)*

And finally, output specific enthalpy with subcooling or superheating is calculated from the overall energy conservation equation:

$$w_A.|h_{A\,2} - h_{A\,1}| = w_B.Cp_B.|T_{B\,1} - T_{B\,2}|$$

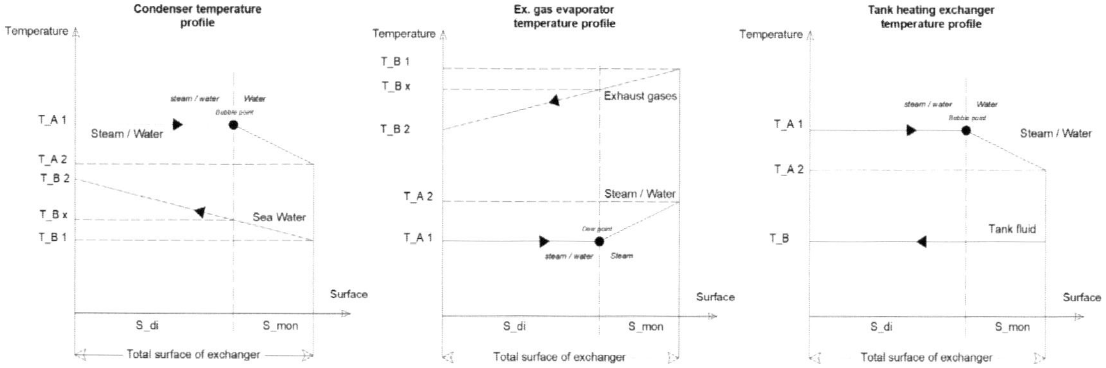

Figure 3. Temperature profiles across the diphasic exchangers (condenser, exhaust gas evaporator, tank heater)

3.2 (c) Steam turbine

The steam turbine is modelled as an irreversible adiabatic transformation with variable isentropic efficiency ratio function of input pressure ratio and steam mass flow rate. Isentropic efficiency is calculated from 3rd degree polynomials with coefficient identification from input turbine's parameters:

- Isentropic efficiency equals 80% at nominal pressure ratio
- Isentropic efficiency equals 50% at 120% of nominal pressure ratio
- Isentropic efficiency equals 50% at 60% of nominal pressure ratio
- Isentropic efficiency equals 20% at 40% of nominal pressure ratio

And steam flow accros the turbine is calculated from input pressure ratio and nominal pressure ratio:

- Steam flow equals nominal steam flow at 120% nominal pressure ratio
- Steam flow equals nominal steam flow at nominal pressure ratio
- Steam flows equals 95% of steam flow at 60% of nominal pressure ration
- Steam flows equals 0% of steam flow at 0% of nominal pressure ration

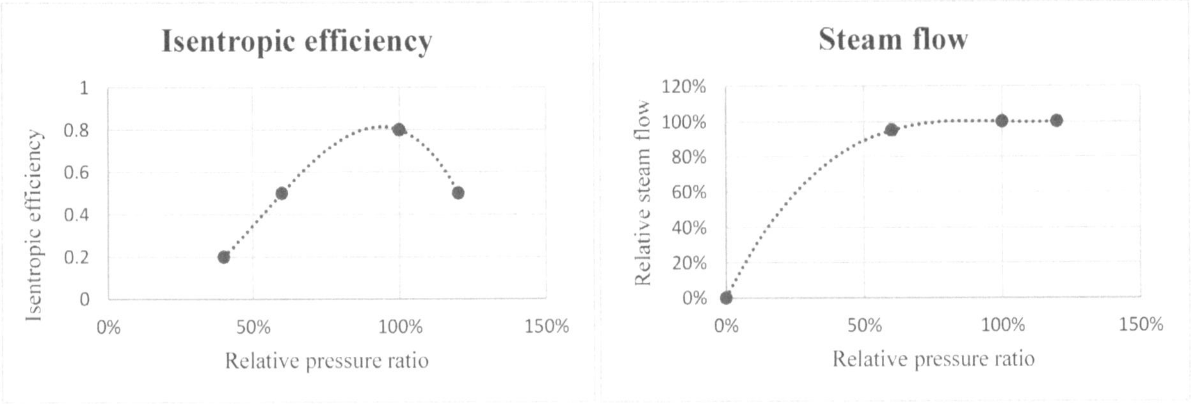

Figure 4. Temperature profile across the exhaust gas boiler exchanger

Mechanical losses in turbine are also modelled with a constant efficiency ratio of 95%. Also the turbine velocity is supposed constant as generator regulates torque on shaft to maintain a 60Hz alternative current production.

3.3 ELECTRICAL SYSTEM MODELS

Alternative electrical currents are modelled with complex phasors in quasi-stationary modes for potential and current. Potential phase is constrained to zero while current can have a positive or negative phase angle (power factor).

$\underline{v} = V.e^{j.0} = V$

$\underline{i} = I.e^{j.\theta} = I.(\cos(\theta) + \text{j}.\sin(\theta))$

Electrical lines are modelled in equivalent 1 phase system, assuming that in any conditions, multiphases lines are balanced with a constant phase angle of $\pm\frac{2\pi}{3}$. Electrical power through balanced 3-multiphase is then 3 times higher than electrical power through the modelled electrical line:

$\underline{S} = 3.\underline{v}.\underline{i}$

Electrical system accounts for the production, distribution, and consumption of electrical energy on board. The electrical simulation platform involves 4 synchronous generators, a main switchboard with power management system and consumers modelled in a generic form.

3.3 (a) Synchronous generator model

The synchronous generator is modelled according to Behn Eschenburg standards with equivalent vector format:

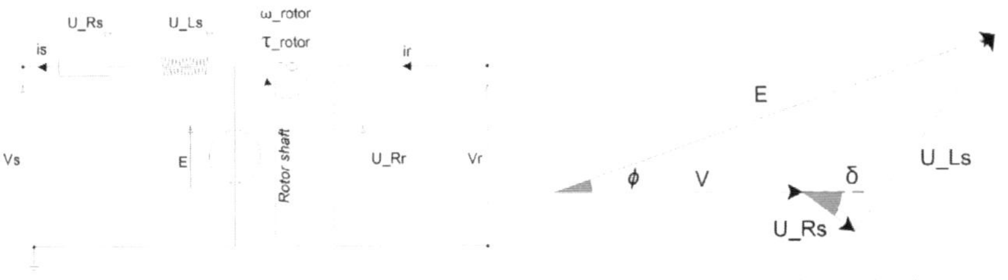

Figure 5: Equivalent 1 phase model for synchronous electrical generators with Behn Eschenburg vectors correspondance

Electromotive force's norm $|\underline{E}|$is calculated as a function of rotor current i_r while argument arg $(\underline{\mathrm{E}}) = \delta$ is a floating variable calculated from shaft torque and stator current. Electrical torque applied on mechanical shaft (resistive torque) is modelled without losses of energy (magnetic losses are neglected):

$$|\underline{E}| = f(i_R)$$
$$\tau.\omega_{rotor} = 3.Real(\underline{E}.\underline{i_{stator}})$$

On the other side, rotor velocity is defined as a function of electrical stator electrical frequency f and generator's number of pairs of magnetic poles p:

$$f = 2\pi.\omega_{elec} = p.\omega_{rotor}$$

Current $i_r$is collected from generator power output with a rectifier bridge of 3 thyristors modelled in a simplified way from energy losses in total harmonic distortion and heat rejection:

$$3.S_{AC}.(\frac{cos\varphi}{\sqrt{1+THD^2}} - \frac{Heat_{nom}}{S_{nom}}). = v_{DC}.i_{DC}$$

And voltage relationship between AC and DC lines is modelled in a static way with average equation:

$$v_{DC} = 3.\frac{\sqrt{6}}{\pi}.V_{ac}.\cos(\alpha)$$

3.3 (b) Main switchboard and power management system

Switchboard model aims at splitting the consumed load and in an optimal way and command gensets loads accordingly while optimising system efficiency. Gensets instantaneous load is imposed at switchboard with an encoded algorithm :

$$\text{With } M = \begin{pmatrix} 1 & 0 & 0 & 0 \\ 0 & 1 & 0 & 0 \\ 0 & 0 & 1 & 0 \\ 0 & 0 & 0 & 1 \\ 1 & 1 & 0 & 0 \\ 1 & 0 & 1 & 0 \\ 1 & 0 & 0 & 1 \\ 0 & 1 & 1 & 0 \\ 0 & 1 & 0 & 1 \\ 0 & 0 & 1 & 1 \\ 1 & 1 & 1 & 0 \\ 1 & 1 & 0 & 1 \\ 1 & 0 & 1 & 1 \\ 0 & 1 & 1 & 1 \\ 1 & 1 & 1 & 1 \end{pmatrix} \quad \text{and} \quad M_{optim.power} = \begin{pmatrix} S_{optim\ power\ genset\ 1} \\ S_{optim\ power\ genset\ 2} \\ S_{optim\ power\ genset\ 3} \\ S_{optim\ power\ genset\ 4} \end{pmatrix}$$

Matrix of possible combination of gensets in operations

Matrix of gensets optimal power configurations

Total consumed power $S = |\underline{S_{switchboard}}|$ is compared to $M \times M_{\_optim\ power}\ [i]$ to find a minimal deviation from optimal working conditions:

Algorithm
$S_{difference} = S$
$k = 0$
for i in 1 → 15
if $|S - M \times P_{optim.power}\ [i]| < S_{difference}$ **then**
$S_{difference} = |S - M \times S_{optim.power}\ [i]|$
$k = i$
else $S_{difference} = S_{difference}$
$k = k$
end if

Finally, the remaining power difference $S_{difference}$ from optimal working condition is divided and imposed proportionally to each operating genset (each operating gensets increase or decrease their load from optimal position to reach total consumed power with the same percentage of rated power).

$$S_{gensets\ 1} = M[1,k].S_{optim\ genset\ 1} + S_{difference}.\frac{S_{rated\ genset\ 1}}{\sum S_{rated\ gensets}}$$

$$S_{gensets\ 2} = M[2,k].S_{optim\ genset\ 2} + S_{difference}.\frac{P_{rated\ genset\ 2}}{\sum S_{rated\ gensets}}$$

$$S_{gensets\ 3} = M[3,k].S_{optim\ genset\ 3} + S_{difference}.\frac{S_{rated\ genset\ 3}}{\sum S_{rated\ gensets}}$$

$$S_{gensets\ 4} = M[4,k].S_{optim\ genset\ 4} + S_{difference}.\frac{S_{rated\ genset\ 4}}{\sum S_{rated\ gensets}}$$

Power is converted to intensity with the hypothesis of a constant voltage and frequency level at switchboard and by applying the same power factor to each synchronous generator:

$$V_{switchboard} = V_{genset\ 1} = V_{genset\ 2} = V_{genset\ 3} = V_{genset\ 4}$$

$$f_{switchboard} = f_{genset\ 1} = f_{genset\ 2} = f_{genset\ 3} = f_{genset\ 4}$$

And for i in 1 to 4:

$$\left| \underline{V_{genset\ i}}.\underline{i_{genset\ i}} \right| = S_{genset\ i}$$

$$\arg\left(\underline{i_{genset\ i}} \right) = \arg\left(\underline{S} \right)$$

3.4 PRIMARY POWER CONVERTERS MODELS

At the other end of the energy simulation chain, propulsion and auxiliary engines are modelled using industrial data. The propulsion engine model relates input shaft velocity and output produced power according to manufacturer optimised torque control strategy. The engine chart, from input shaft velocity provides therefore:

- Instantaneous fuel consumption
- Output exhaust gas flow
- Output exhaust gas temperature
- Output mechanical power

Auxiliary engine model, on the other hand, works at constant rotational velocity. Therefore, from input power command, the auxiliary engine model calculates:

- Instantaneous fuel consumption
- Output exhaust gas flow
- Output exhaust gas temperature
- Output mechanical torque

4. CASE STUDY

The ship simulation platform was configured for the simulation of the Duisburg test case hull (Post Panamax container ship with 14.000 TEU capacity) with hydrodynamic and propulsion data coming from towing tank and propulsion test performed in (El Moctar et al., 2012) and from manoeuvring model (Taimuri et al., 2020). Propulsion systems were sized to optimise consumption of the vessel at a transit speed of 22 knots in calm waters (in this article, added wave resistance is ignored in order to illustrate correlations between wind conditions and propulsion and auxiliary systems behaviours):

Main engine was chosen to run in optimal condition for a contracted maximum continuous rating (CMCR) of 24 kn and continuous service rating (CSR) of 22 kn using technical data for 9X92-B 58050kW engine, extracted from WinGD™ engine sizing software.

The thermal system was designed to maximise steam production at CSR with a minimal outlet flow of exhaust gas at 190°C, producing therefore 2.7 kg/s of saturated steam (8.5 bars) used either for fuel heating or superheated at 260°C and used in turbogenerator (producing approximatively 300kW at CSR). Tank heating was sized according to the following parameters:

		Mass / Temperature (t / °C)	Fuel flow (kg/s)	Consumption (kW)
Tank heating	Fuel storage tank	*18,000 t @ 45°C*	-	*1500*
	Fuel settling tank	*2,500 t @ 60°C*	-	*00*
	Fuel daily tank	*300 t @ 85 °C*	-	*200*
	FO separator	*85°C out*	*3 kg/s*	-
	FO line heater	*150°C out*	*Total consumption*	-
	Sludge settling tank	*3 t @ 80°C*	-	*50*
	Sludge tank	*1 t @ 80°C*	-	*20*
Support function	HVAC heater	-	-	*100*
	Sanitary water heater	-	-	*50*

The electric production plant was sized according to the following parameters:

		Power (kVA)	Power factor $cos\varphi$
Cargo load	2,000 x TEU Reefers	*9 500*	*0.7*
Support function	Various load - machinery	*1 000*	*0.8*
	Various load - hotel	*500*	*0.9*

Finally, 4 Flettner rotor (4m diameter and 8 m height) were fitted on ship sides above cargo (at 260 and 200 meters from aft perpendicular, at port and starboard of cargo central area). The vessel was thus analysed for potential energy savings and efficiency on hybrid propeller and sailing modes.

4.1 VALIDATION OF RESULTS FOR MANOEUVERS AND PROPULSION LOAD

The model was evaluated in turning and zig zag tests and compared with similar dynamic simulations such as (Taimuri et al., 2020), (Chillcce & el Moctar, 2018). The predictions obtained for the simulation of turning manoeuvre seems to corroborate previous results with equivalent advance (approx. $3.5 \times L_{pp}$),), transfer (approx. $1.5 \times L_{pp}$) and tactical diameter (approx. $3.2 \times L_{pp}$) (see annex 1: Results of manoeuvring test simulations). In addition, propulsion tests for the vessel operation at transit speed are equivalent to the results from model test, with equivalents value for hull, propeller and propulsive efficiency (see annex 2. Propulsive efficiency value from loading test at transit speed - 22kn). As the model provides satisfactory results for vessel dynamics and propulsive power, the model could be used to assess the impact of added wind power on performances.

4.2 EVALUATION OF THE REDUCTION/INCREASE OF THRUST REQUIREMENT AT TRANSIT SPEED

To evaluate the propeller thrust reduction with 4 Flettner rotors, DTC was set at different speed on a set route (North 00°) at different speed and submitted to a sweep of input wind conditions, with wind direction from -180° to 180° and velocity from 0kn to 30 kn and compared to reference vessel without sails.

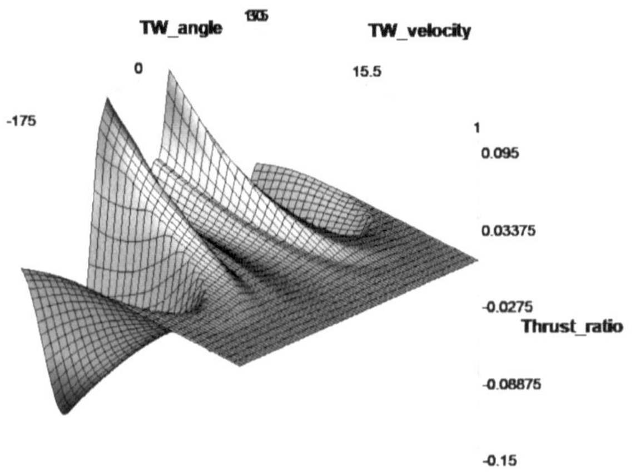

Thrust saving ratio chart at 22 kn

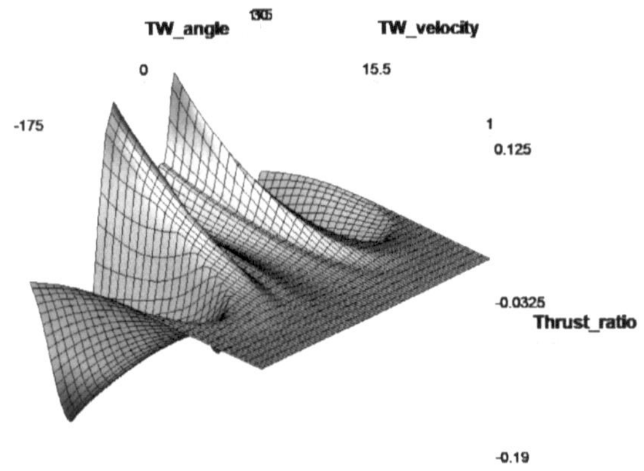

Thrust saving ratio chart at 20 kn

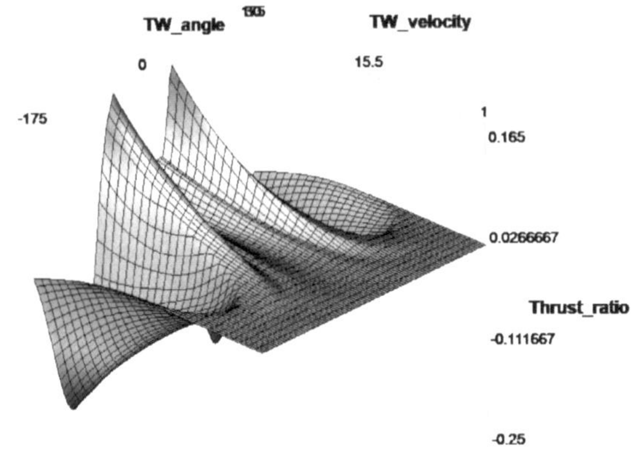

Thrust saving ratio chart at 18 kn

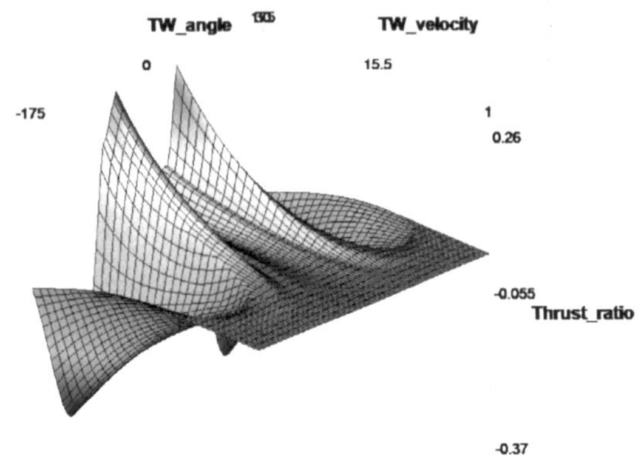

Thrust saving ratio chart at 16 kn

Figure 6. Thrust ratio of wind assisted DTC for wind between 0 and 30 kn at different transit speed

As a result, for DTC with 4 Flettner rotors and low transit speed, the thrust at propeller can be reduced as much as 37% compared to a reference vessel maintaining same speed with propeller. At 22kn, however, rotors cannot reduce thrust power above 15%. Speed reduction, in addition to reductions of overall hydrodynamic resistance, can also be a way to maximise savings from sails and limit further fuel consumption of vessels. In a sail retrofit case study, therefore, shipowners might consider speed reduction as a way to increase energy savings and reduce return on investments on sails. In the following, 2 transit velocities will therefore be investigated, a reference speed at 22kn and an alternative case with speed reduction at 18kn. In addition to thrust reduction, further sail propulsion implications on energy efficiency shall be evaluated at maximal wind velocity and compared to reference vessel at 22 kn.

In addition, an exclusion area was used for rotor control avoid propulsion power increase in high drift conditions (exclusion area regroup wind parameter inducing positive thrust ratio):

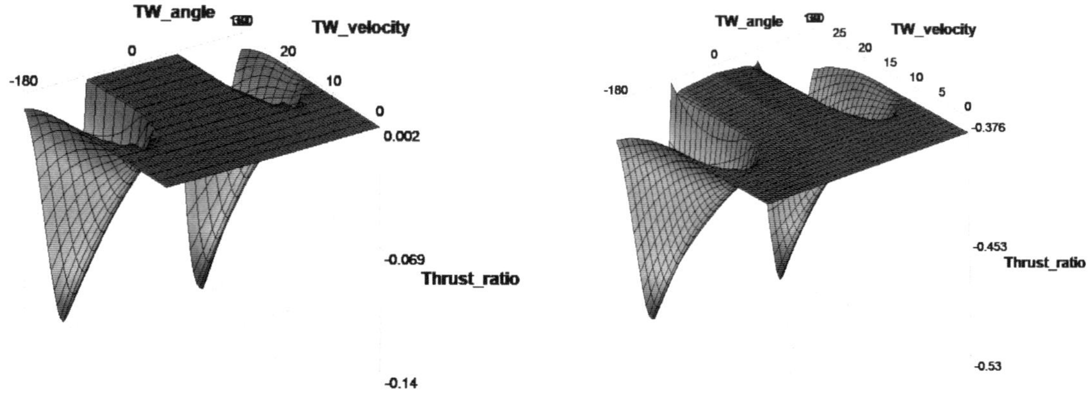

Thrust savings ratio with rotors at 22 kn *Thrust savings ratio with rotors and 18kn speed reduction*

Figure 7. Thrust ratio chart with optimised sail control at 22 kn and 18 kn

4.3 EVALUATION OF IMPACTS ON PROPULSION EFFICIENCY

With sails, propeller power requirements cannot be solely calculated from a deduction of longitudinal resistances at set speed and sail thrust. The reduction of thrust at propeller results from sail thrust but also from route equilibrium conditions defined by drift and rudder angle. Also, compared to traditional vessels, propeller thrust is to be considered variable as a function of wind conditions. With additional propulsion power, and water inflow velocity at propeller, mechanical propulsion efficiency may vary in favourable or unfavourable way. Indeed, fixed pitch propellers are usually designed to provide optimal power at a constant propeller advance ratio $J = \frac{V_a}{n.D}$. With sail power, and a significant reduction of propeller velocity at transit speed, the overall propulsion ratio may shift from design condition.

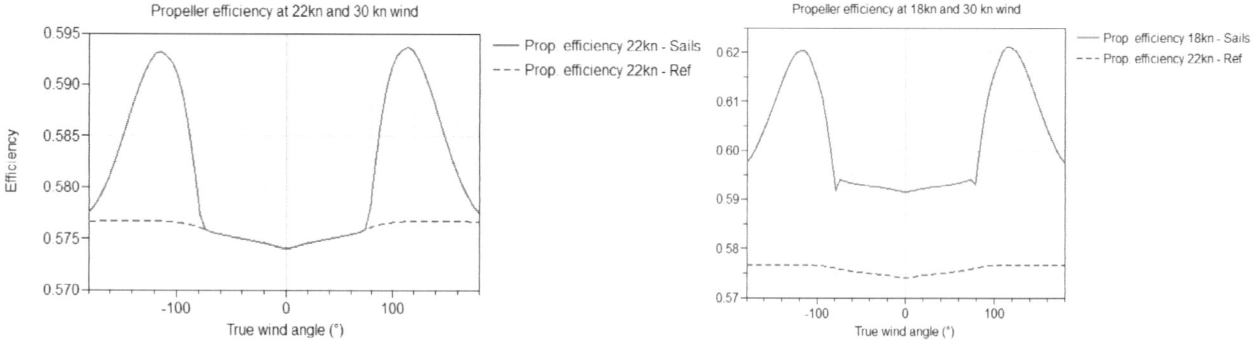

Figure 8. Influence of wind direction (constant velocity) over propeller efficiency for DTC with Flettner rotors

For DTC, however, propeller efficiency seems not to be a critically impacted by sail power. This can be explained by the propeller design characteristics in (El Moctar et al., 2012) that were not exactly optimal at 22kn transit speed. In alternative cases, when transit speed matches exactly with optimal advance ratio for the propeller, efficiency reductions are to be expected. Nonetheless, main propulsion engine power reduction is to be calculated with dynamic models able to predict variation of propeller operating condition as a function of wind direction and speed.

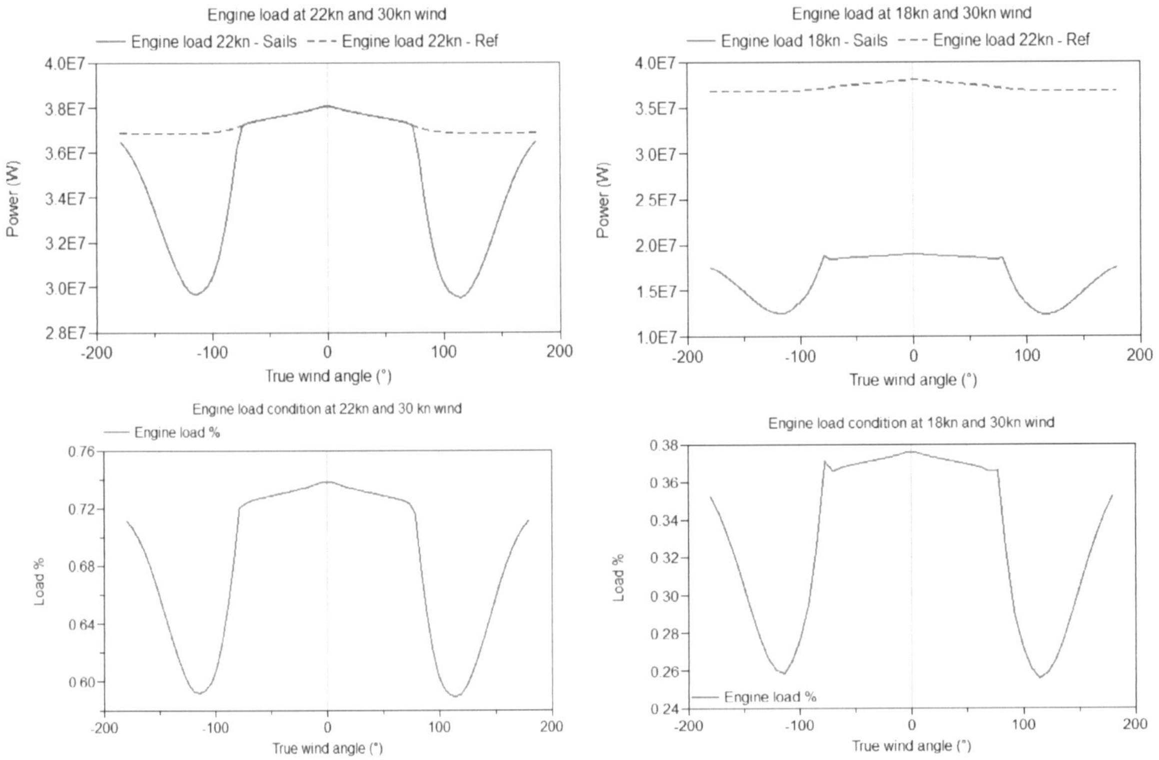

Figure 9. Influence of wind direction (constant velocity) over propulsion engine load for DTC with Flettner rotors

Propulsion engines, just like propellers, are designed to be optimal for a set transit speed. If engine's operating conditions changes from optimal, consequently, engine efficiency shall be reduced. Mismatch between engine rating, propeller design and additional sail power is therefore an element of design subject to be optimised in further works. In practice, new methodologies for propeller and propulsion engine sizing are required to account for an increase variability in thrust production and operating conditions.

4.4 EVALUATION OF IMPACTS ON AUXILIARY SYSTEMS

Further consequences can also be analysed in auxiliary systems. For vessels equipped with heat recovery systems on exhaust gases, such as evaporators for heating and superheaters for turbogenerators, changes in engine exhaust gases flow or temperature would affect electricity and heat power stations. On one hand, diesel generators would increase load to compensate for the decrease power production in turbo generator. Also, reduction in saturated steam production could create shortages of heating power, which would be compensated by the oil-fired boiler. Finally, changes in raw fuel consumption at propulsion engine, diesel gensets or boiler would also alter the heating power required by fuel heating and conditioning system.

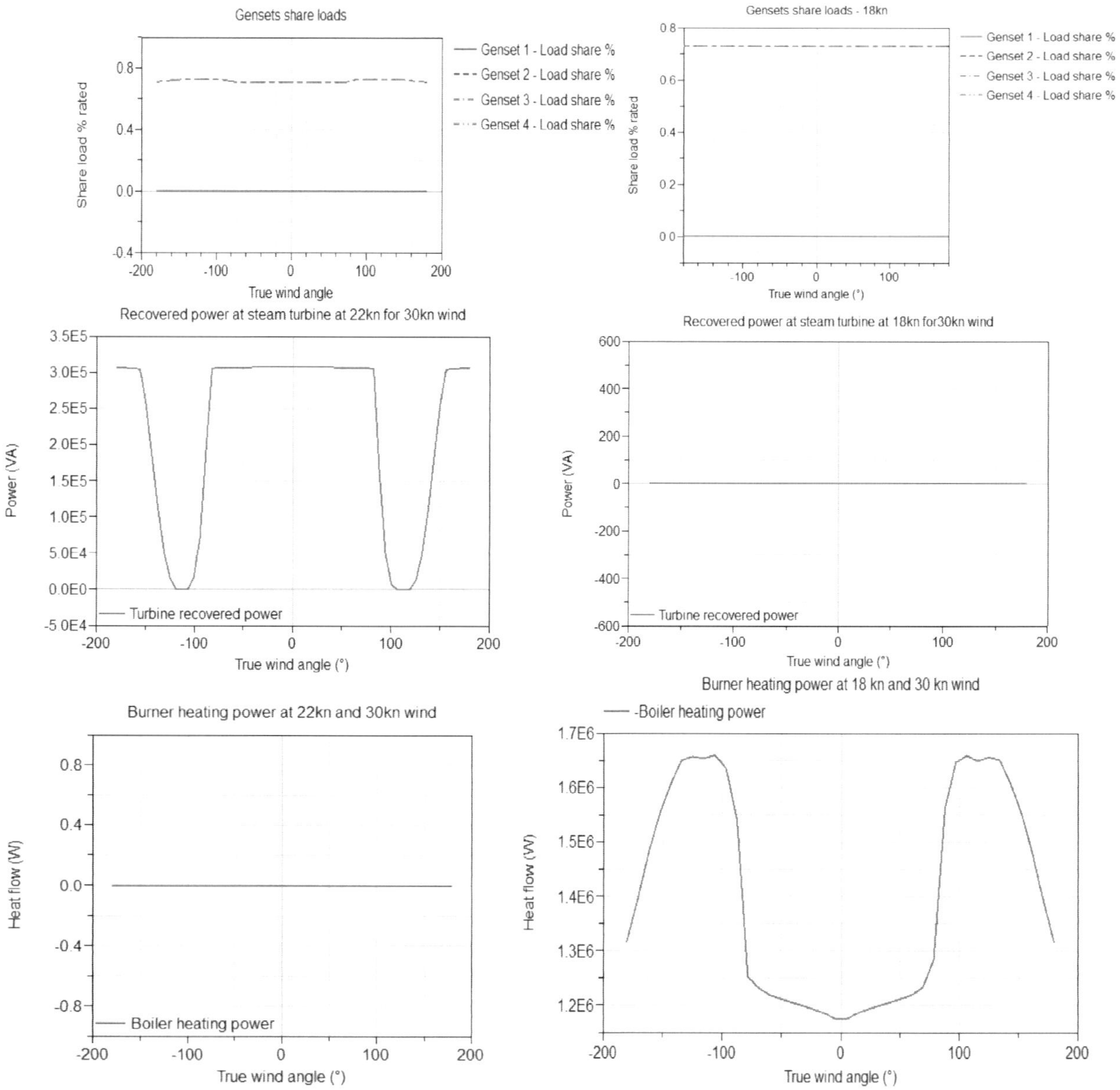

Figure 10. Influence of wind direction (constant velocity) over auxiliary system behaviours for DTC with Flettner rotors

Auxiliary systems operating conditions can also be impacted quite significatively. On one hand, shortage of exhaust gases may limit the extend of available superheated steam for turbogenerators and induce load increases over gensets. Also, while enough steam is produced at 22kn to maintain heating system with exhaust boiler, the rest of engine gases calories is therefore to be by-passed to the atmosphere. In more critical situations, at 18kn, steam production shortage affects both steam turbine and heating system. As a consequence, the oil-fired boiler compensates the loss with extra fuel consumption. Finally, shut down of turbogenerator do affects genset load condition. In DTC however, due to the difference of ratings between gensets (5MW) and turbine (300kW), load conditions for the engines changes by 2% (3 engines) or 3% (2 engines) and 6% (1 engine).

Accounting for the influence of sails over hydrodynamic, propulsion and auxiliary systems efficiency, with a dynamic ship simulator it is therefore possible to extract general function of fuel consumption from wind conditions.

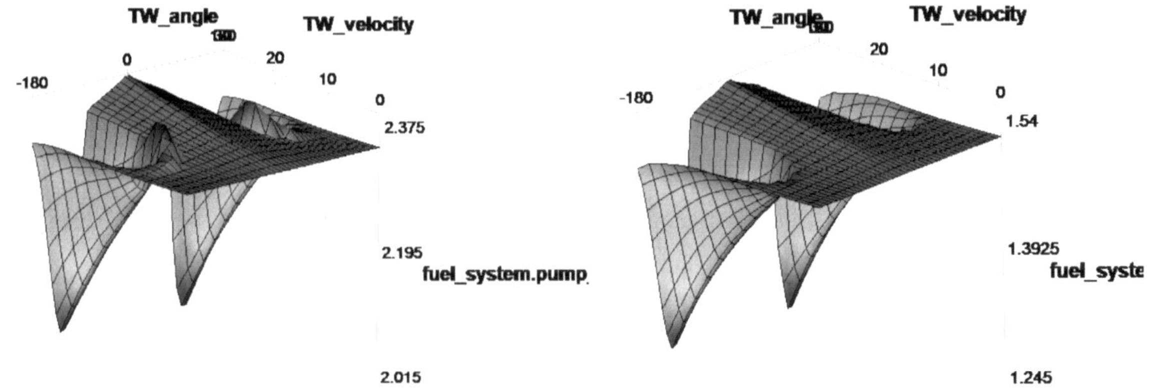

Figure 11. Fuel consumption chart of DTC with Flettner rotors at different transit speed and for a range of wind conditions

5. COMMENTS ON WAYS TO IMPROVE EFFICIENCY OF WIND ASSISTED VESSELS

Efficiency and fuel consumption of vessels with sails are complex functions to be modelled with many variables from the hydrodynamic, propulsion, and auxiliary systems domains. Combining dynamic models for manoeuvring and energy transfers seems to be the most effective way to account for entire systems interactions and tank to wake energy performance. The simulation here performed accounts for motion with 3 degrees of freedoms and provide satisfactory results for the simulation of large vessels equipped with small sail devices. However, additional modifications to the model are required to evaluate the impact of trim and list on sailing performance. Developing such a modelling tool with 5 degrees of freedom, might be necessary for the modelling of sailing ship (major part of propulsive power being supported by sail system) or vessels where list and trim might have an important impact on performance and stability. In addition, wave resistance was excluded of the scope of this analysis to highlight single correlations between sails and wind with energy production for propulsion and auxiliary systems. While wave influences should not be ignored in realistic simulation cases, investigating energy management in calm water provides interesting insights on cargo ship design efficiency in sailing modes. In the previous section, at least 3 sail-ship influences were measured:

On the hydrodynamic level, developing hulls with increased counter resistance to lateral forces, with lower B/L ratios or block coefficients C_B, could be a relevant lead of investigations. In addition, appendices could be used to enhance course stability at the cost of low added resistance. Yet further studies are required to optimise such appendices and reduce overall course keeping added resistances from a set of design parameters and rated sail power.

On the propulsion and engine level, new sizing methodologies are required to account for variability in operating conditions. Instead of fixed-point optimisation, stochastic methodologies combining ship modelling and statistical evaluation of sailing route could improve design efficiency and minimise emissions. Finally, parameter sweeps, or deeper optimisation technics based on genetic algorithm could be used to size propellers and engines according to a global optimal in a set of continuous operating conditions. Such optimality research could also be supported by propellers and engines resilient to variations in load conditions. On one side, controllable pitch propellers, while having lower maximal efficiency than fixed type, could globally perform better in part loads while offering a control option on engine velocity for different ship velocity. Indeed, engines with controllable fixed propellers, usually shows better resilience to output power variations with an adapted torque and velocity control strategy. Another solution would be to use electrical powertrains and propulsion engines with stable efficiency ratios between 95% and 99%. At final, variations of propulsive power would be absorbed by gensets with lesser impact on fuel efficiency.

Finally, thermal, and electrical systems are impacted by the variations of load at propulsion engine. Heat recovery systems are designed to maximise efficiency at transit speed. In partial loads, shortage of power is compensated by fuel fired boiler and auxiliary gensets. On the contrary, for higher loads, excess flow of gas is simply by-passed and lost in atmosphere or excess steam is dumped to condenser. One of the best ways to rationalise energy use and maintain a high level of heat recovery rate could be to use energy storage devices in the form of thermos tanks or battery system. Such capacities could therefore off-peak energy delivery and store excess energy production otherwise.

Further work on wind assisted profusion modelling would benefit to the uptake low energy intensive shipping. Notably, model evaluation with different powertrains in case studies are necessary to define new standards in the development of wind propelled vessels. This research approach is to be supported by novel simulation tools, including Modelica based models but alternatively with any other dynamic multi-physics simulation software.

6. CONCLUSION

The model presented in this article was developed in continuation of the previous works on the simulation of energy flows and manoeuvring of vessels from physical and semi-empirical models. Multi-physic simulation platform such as Dymola, or Modelica based programming tools show encouraging results both in the accuracy of predictions and in computational performance.

At least, for traditional cargo vessels (containers, dry bulk, or general cargo) propulsion, heat and electricity power production is responsible for most of the fuel consumption. Therefore, simulation models, encoded to perform dynamic manoeuvring simulation along with system modelling demonstrates a great potential as design support tool. Applied to the investigation of energy conversation between fuel to vessel motion, models are capable of classifying elements of design per efficiency losses and thus directs optimisation efforts in relevant directions. On the other hand, using proper parameter sweep, the tool can be used to create datasets of vessel general consumptions, modelled in a high-fidelity manner, but simple enough to perform efficiently in complex optimisation algorithms such as voyage optimisation. Combined in one single effort, design and voyage optimisation can also be performed by combining dynamic programming and genetic algorithm, to outputs both optimal ship designs and optimal control strategy.

As a result of the analysis of the Duisburg Test Case (DTC), fitting wind propulsion assistance on post panamax container vessels, while reducing the consumption of fuel, alters overall ship efficiency at hydrodynamic level, but also the propeller, the main engine, and support systems. Consequently, fuel savings from sails cannot be evaluated in a simple and generic way for all sorts of vessels. Results obtained from the analysis also show that optimisation margin for vessels adaptation to natural propulsion is rather important. For each case study, depending on the relative share of wind and mechanical propulsion power, hull and systems optimal design may differ greatly. For this reason, a wide variety of design standards might emerge out of the research of wind ship building applied to different category of vessels (capacity, transit speed, navigation area).

7. ANNEXES: Validation of model's predictions

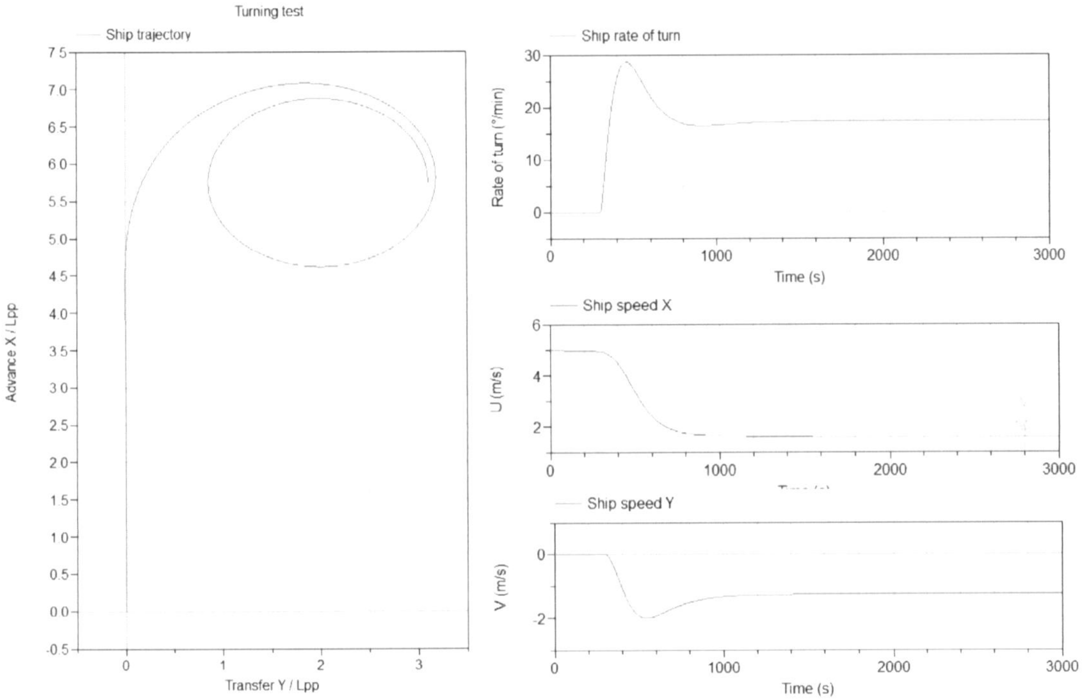

Turning test was performed with constant propeller revolution of 40 rpm and with rudder angle $\delta = 35°$

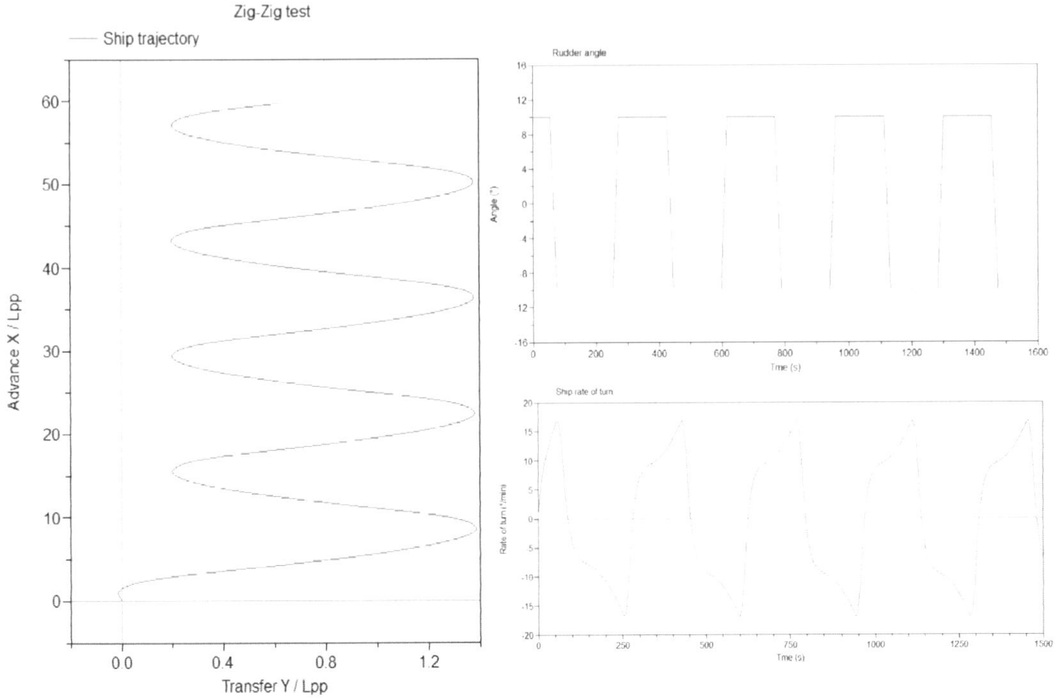

Zig-Zag test was performed at constant velocity of 7 m/s

Annex 1: Results of manoeuvring test simulations

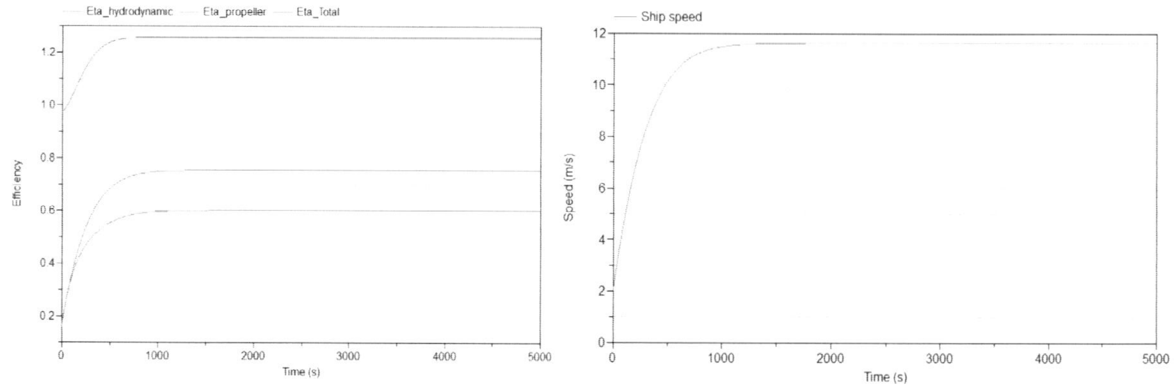

Annex 2. Propulsive efficiency value from loading test at transit speed - 22kn

8. REFERENCES

1. Abkowits, M. (1980). Measurment of hydrodynamic characteristics from ship manoeuvring trials by system identification. *Transactions on SNAME*, *No. 88*, 283–318.

2. Chillcce, G., & el Moctar, O. (2018). A numerical method for manoeuvring simulation in regular waves. *Ocean Engineering*, *170*(December 2017), 434–444. https://doi.org/10.1016/j.oceaneng.2018.09.047

3. Cichowicz, J., Theotokatos, G., & Vassalos, D. (2015). Dynamic energy modelling for ship life-cycle performance assessment. *Ocean Engineering*, *110*, 49–61. https://doi.org/10.1016/j.oceaneng.2015.05.041

4. Dassault. (2022). *DYMOLA 2022X*.

5. Dimopoulos, G. G., Georgopoulou, C. A., Stefanatos, I. C., Zymaris, A. S., & Kakalis, N. M. P. (2014). A general-purpose process modelling framework for marine energy systems. *Energy Conversion and Management*, *86*, 325–339. https://doi.org/10.1016/j.enconman.2014.04.046

6. El Moctar, O., Shigunov, V., & Zorn, T. (2012). Duisburg test case: Post-panamax container ship for benchmarking. *Ship Technology Research*, *59*(3), 50–64. https://doi.org/10.1179/str.2012.59.3.004

7. Fossen, T. I., & Fjellstad, O.-E. (1995). Nonlinear modelling of marine vehicles in 6 degrees of freedom. *Mathematical Modelling of Systems*, *1*(1), 17–27. https://doi.org/10.1080/13873959508837004

8. Holtrop, J., & Mennen, G. G. . (1982). An approximate power prediction method. *International Shipbuilding Progress*, *29*(No. 335).

9. Holtrop, J., & Mennen, G. G. J. (1984). A statistical re-analysis of resistance and propulsion data. *International Shipbuilding Progress*, *No. 31*(No. 335).

10. IAPWS (2007). *Revised release on the IAPWS industrial formulation 1997 for the thermodynamic properties of water and steam.*

11. IMO (International Maritime Organization). (2011). *Amendments to the annex of the protocol of 1997 to amend the international convention for the prevention of pollution from ships (Inclusion of regulations on energy efficiency for ships)*.

12. IMO (International Maritime Organization). (2014). *Guidelines on survey and certification of the energy efficiency design index*.

13. IMO (International Maritime Organization). (2016). *Amendments to the annex of the protocol of 1997 to amend the international convention for the prevention of pollution from ships (Data collection system for fuel oil consumption of ships)*.

14. IMO (International Maritime Organization). (2018). *Initial IMO strategy on reduction of GHG emissions from ships*.

15. IMO (International Maritime Organization). (2021a). *Amendments to the annex of the protocol of 1997 to amend the international convention for the prevention of pollution from ships (Revised MARPOL annex VI)*.

16. IMO (International Maritime Organization). (2021b). *Guidelines on operational carbon intensity indicators and calculation methods*.

17. Isherwood, R. M. (1973). Wind resistane of merchant ships. *The Royal Institution of Naval Architects*.

18. Kralj, D. M., & Klarin, B. (2016). Wing sails for hybrid propulsion of a ship. *Journal of Sustainable Development of Energy, Water and Environment Systems*. https://doi.org/10.13044/j.sdewes.2016.04.0001

19. Lee, T. I., Ahn, H. S., Lee, H. S., & Yum, O. D. J. (2003). An empirical prediction of hydrodynamic coefficients for modern ship hulls. *MARSIM 2003*.

20. Liu, Shukui & Shang, Baoguo & Papanikolaou, Apostolos & Bolbot, Victor. (2016). An Improved Formula for Estimating the Added Resistance of Ships in Engineering Applications. Journal of Marine Science and Application. 15. 10.1007/s11804-016-1377-3.

21. Mermiris, D., Vassalos, D., Dodworth, K., Bell, B., & Sfakianakis, D. (2011). *Dynamic energy modeling - a new approach to energy efficiency modeling. June 2014.*

22. *Modelica.* (n.d.). Modelica Association,. https://modelica.org/

23. Oosterveld, M. W. C., & van Oossanen, P. (1975). Further Computer-Analyzed Data of the Wageningen B-Screw Series. *International Shipbuilding Progress*, *22*(251), 251–262. https://doi.org/10.3233/isp-1975-2225102

24. Papanikolaou, A. (2010). Holistic ship design optimization. *CAD Computer Aided Design*, *42*(11), 1028–1044. https://doi.org/10.1016/j.cad.2009.07.002

25. Taimuri, G., Matusiak, J., Mikkola, T., Kujala, P., & Hirdaris, S. (2020). A 6-DoF maneuvering model for the rapid estimation of hydrodynamic actions in deep and shallow waters. *Ocean Engineering*, *218*, 108103. https://doi.org/10.1016/j.oceaneng.2020.108103

26. Tillig, F., Ringsberg, J. W., Mao, W., & Ramne, B. (2017). A generic energy systems model for efficient ship design and operation. *Proceedings of the Institution of Mechanical Engineers Part M: Journal of Engineering for the Maritime Environment*, *231*(2), 649–666. https://doi.org/10.1177/1475090216680672

27. Sun, C., Mullin, T., Van Wijngaarden, L., & Lohse, D. (2010). Drag and lift forces on a counter-rotating cylinder in rotating flow. *Journal of Fluid Mechanics, 664*, 150-173. doi:10.1017/S0022112010003666

28. Yasukawa, H., & Yoshimura, Y. (2015). Introduction of MMG standard method for ship maneuvering predictions. *Journal of Marine Science and Technology (Japan)*, *20*(1), 37–52. https://doi.org/10.1007/s00773-014-0293-y

29. Yoshimura, Y. (2005). Mathematical Model for Manoeuvring Ship Motion (MMG Model). *Workshop on Mathematical Models for Operations Involving Ship-Ship Interaction*, *August*, 1–6.

DIGITAL ENTERPRISE PLATFORM - ENABLING EFFICIENT CAD, PLM, ERP INTEGRATION ON SHIPYARDS

J Bitomsky, M Grau, J Lützenberger and **C Zerbst,** PROSTEP AG, Germany

SUMMARY

The design and production of vessels is nowadays supported by a plethora of IT tools. This comprises both tools which are very specific to shipbuilding, like hydrostatic and -dynamic design tools, as well as generic tools, like generic CAD or ERP systems, used also by other industries. Most of these tools started as island solutions, resulting in difficulties to manage the complete vessel design or even all shipyard processes as a holistic approach. Such a holistic approach allows data reuse wherever needed, without manual transfer causing additional effort and potential errors. This paper presents several use-cases for such an approach, which are beneficial for shipyards. Some of them are driven by step-by-step improvements of a shipyard's operation, whereas others are needed to capitalize on the benefits of a merger of two shipyards.

NOMENCLATURE

BOM	Bill of Materials
CAD	Computer-Aided Design
DMS	Document Management System
DMU	Digital Mockup
EBOM	Engineering Bill of Material
ERP	Enterprise Resource Planning
MBOM	Manufacturing Bill of Material
PDM	Product Data Management
PLM	Product Lifecycle Management
PMI	Post Merger Integration
WBS	Work Breakdown Structure

1. INTRODUCTION

The first steps in the digitalisation process were taken by shipyards decades ago. In the 1970s, companies started to develop in-house IT solutions to support their drawing-based ship development or to provide data for NC-controlled machinery. Over the years a growing number of vendors entered the market providing CAD tools, ERP systems, document management (DMS), and later product data management (PDM) as standard software to support product development and manufacturing. Many of today's standard PLM software vendor's history traces back to IT departments of manufacturing companies – across all industries and including shipbuilding. By changing from in-house developed applications to standard software applications, the manufacturing companies were able to focus on their core products, strengthen the robustness of processes and increase their efficiency significantly.

Twenty years later, a paradigm shifts from drawing-based to model-based product creation processes started to evolve, significantly picking up speed during the last decade. This means drawings, BOM lists, and reports more and more become derived information, created only as needed from model-type artefacts – such as 3D CAD, simulation, and product structures – as the primary information being subject to revision, configuration, and change management. It started in industries like automotive and aerospace and is nowadays also identified by leading shipyards as the future way of product development. Meanwhile, model-based product development turns out to be the major precondition to establishing product information models for manufacturing and operational purposes, commonly known as digital twins. These digital representations of the shipyard´s products allow for a completely new level of product lifecycle support including manufacturing methods, field data collection, and related all-new business concepts.

To enable this, internal and cross-company business processes need to be shaped accordingly. PLM integration is becoming a crucial part here. However, so far integration is often realized by in-house developments of the shipyard`s IT departments. These integration interfaces reflect today's workflows but are typically point-to-point connections lacking flexibility and adaptability due to their hardwired architecture. Maintenance is expensive and time-consuming. Software updates to the connected systems must be addressed and the integration solutions need to be adopted respectively by the shipyard`s IT experts each time. In addition, increased complexity due to model-based product development, specialized applications, and the need to include new types of platforms, such as digital twins of product and manufacturing into the integration architecture challenge in-house solutions to their limits. A sufficient and economical operation can no longer be guaranteed.

This paper illustrates how we propose to solve these integration problems. Chapter 2 introduces the term of semantic integration categorized into different integration levels. Chapter 3 illustrates scenarios showing how semantic integration is used in real-world shipyards. The final chapter 4 contains our conclusions on what is achievable with today's technology and what benefits are possible to achieve by the companies.

2. SEMANTIC: SEAMLESS PLM INTEGRATION

In the past, integration was often done only on BOM or CAD level. It was sufficient to transfer e.g. plate parts drawing to be used as input for plate manufacturing. But to address the increased integration requirements, systems need to be connected at a different – higher – semantic level. Approaches inherently capable of dealing with the data semantics of product development, or the product lifecycle in general provide the capabilities for seamless integration of the various enterprise systems and to cope with the new complexity. This puts integration at the centre of a company's enterprise application architecture and on par with existing enterprise systems such as CAD, PDM, and ERP. Here too, solutions based on standard products are more cost-effective and lower risk than in-house developments. Standard integration products ensure changes in processes and updates to connected software systems can be handled without the need for an internal software development project. Standard integration products are updated by their supplier just like other commercial software applications and provide maturity, robustness, and reliability. This allows IT to focus on the operation of software instead of developing it thereby significantly increasing their efficiency as an additional benefit.

2.1 SEMANTICAL PLM INTEGRATION LEVELS IN SHIPBUILDING

Semantic integration is the resolution of semantic conflicts between heterogeneous data. Simplified, these conflicts deal with different meanings of the terms and concepts used. A very generic example can be illustrated with the word "Rock". Rock describes a music genre and a stone. Only the context of the whole sentence provides the required information for the recipient to interpret it correctly. Otherwise, the information "Rock" is only transferred to the recipient, but the recipient can do nothing meaningful with it.

With PLM integration, the semantics are defined by the respective project and the IT applications used. For example, a shipbuilding CAD system has a defined pipeline object type. Within this IT application, certain functions and processes can be applied to this object type because the IT application interprets this object type correctly. With a CAD to PDM integration, this object type must be interpreted correctly in the target system. It must be adapted to the data model in the target system during the transfer. Furthermore, the transfer must also contain further information on the project context. This is the only way for the target system to interpret the received pipeline correctly and initiate the next defined steps.

We propose to differentiate between four levels of PLM integration, as can be seen in Figure 1. In general, the semantics but also the implementation effort increases from level 0 to level 3. It's important to understand that each level has its justification. This justification depends on the relevant use case.

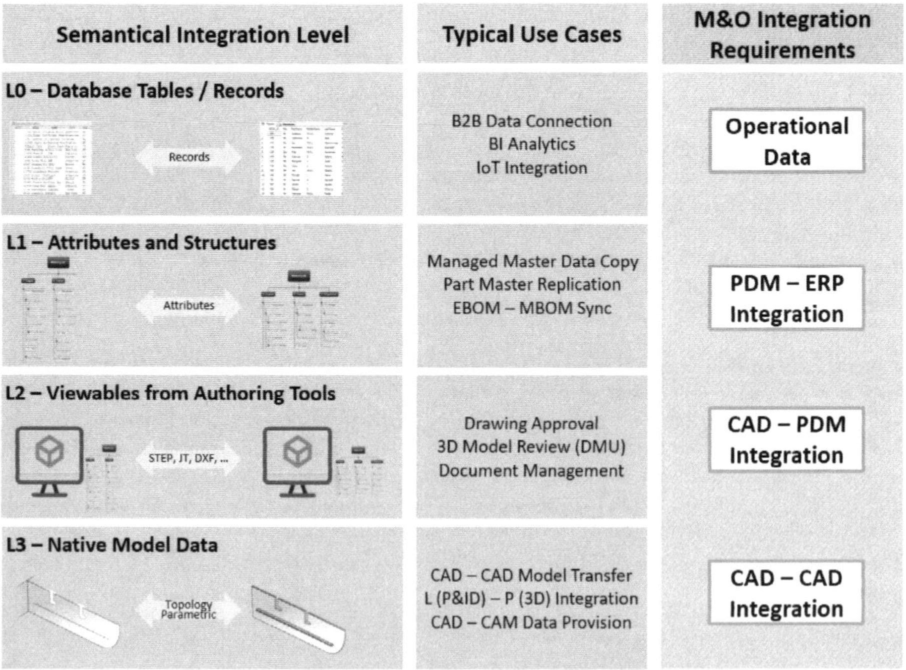

Figure 1: Semantic PLM Integration Levels in Shipbuilding

2.1 (a) Level 0 - Database Tables & Records

Level 0 describes, for example, a simple copy of database entries from a source to a target application. In general, this level contains a small degree of semantics, but typically has a rather large number of data records instead. One use case could be the transfer of operating data, e.g., from a forklift control system or another track and trace system to an application for evaluation.

2.1 (b) Level 1 – Attributes and Structures

Level 1 describes the transfer of product structures with parts and selected metadata. In contrast to Level 0, Level 1 already has a higher degree of semantics. Often, this Level 1 takes place between PDM and ERP systems. In shipbuilding, very typical use cases are the transfer of part master data or EBOM - MBOM synchronization. Both use cases require a modification or transformation of the structure so that the transferred information fits the structure and data model in the target application.

2.1 (c) Level 2 – Geometry Representation from Authoring Tools

Level 2 integration is based on the Level 1 integration and adds geometry representations to the transfer. These geometry representations do not contain the full set of 3D information, e.g. the type and parameters of cut-outs applied on a plate, but only its final geometrical shape. Typically, this Level 2 integration is relevant for many of the use cases between CAD and PLM systems. The 3D and 2D information from a CAD system is often a piece of required information in a PLM system. In shipbuilding, the most common use cases describe the transfer of drawings, diagrams, and 3D model parts. The shipyard's EBOM is typically defined in the PLM system and provides a 3D definition of the particular part from the various CAD systems. A DMU process displays all 3D parts in the right context. The PLM system can only display the 3D parts when the 3D format was converted accordingly to the viewing capabilities of the PLM system. Otherwise, the PLM system contains all required information but cannot interpret it correctly to provide the required added value to the users.

Figure 2: AVEVA Marine MTP Project in Siemens Teamcenter [2].

2.1 (d) Level 3 – Native Model Data

Level 3 integration is the most complicated stage of PLM integration in shipbuilding. For Level 3 it is required that the transferred CAD model is transferred in such a way, that the user can continue working with it in the target application. The semantic information is transferred when needed and applicable, rebuilding the equivalent 3D CAD model in the target system. For shipbuilding steel model semantic transfer a level 3 integration handles the topological boundaries of the panels, the type and parameters of features like holes, cut-outs, and notches. For an outfitting model semantic transfer it handles system, pipelines and -spools, and connection points of pipes and equipment.

The resulting models have a native quality, allowing users to seamlessly modify and adapt converted vessels with the same functionality as if the model would have been designed directly in the target CAD application. Some examples of a converted ship model are contained in Figure 3. In comparison, in level 2 the finished 3D component is transferred, and in level 3 the recipe to generate the 3D model is transferred into the language of the target system. This implies that the source and target application should have similar capabilities or concepts to describe topological relationships or parametric features, like in a transfer from CAD-to-CAD application. The CAD capability question has been investigated in our paper at ICCAS 2017 [1], containing our analysis of differences between dedicated 'intent-driven' shipbuilding systems and more general mechanical CAD-based systems.

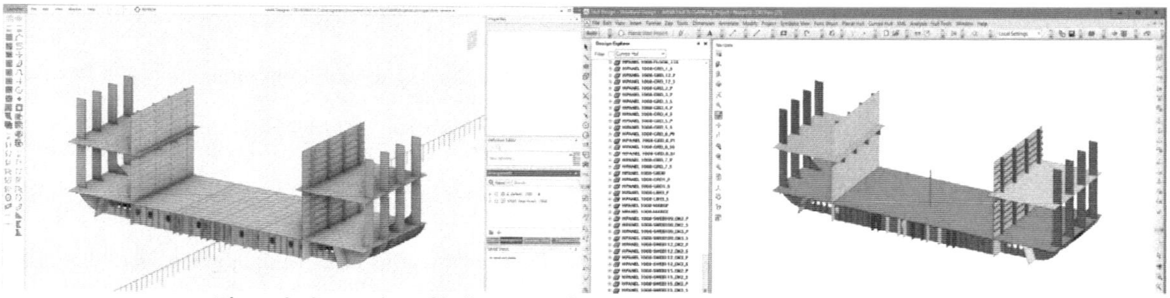

Figure 3: Conversion of hull structure from Napa Model in Aveva Marine [1]

2.1 (e) Automation of PLM integrations

In summary, it becomes clear that each of these four integration levels has its justification. Even if many things are technically possible, the technical solution must fit the problem. There is no once-fits-all integration. However, PLM integration is much more than selecting the right level of integration. The frequency of transfer and the amount of data to be transferred drive the effort needed to automate the interfaces. If we have a daily transfer of one million parts, a pure manual solution will become too costly and unreliable in the long run. Fully automating a transfer of 100 metadata attributes happening once a month could be an exaggeration as well [3].

3. PLM INTEGRATION SCENARIOS FROM THE SHIPBUILDING INDUSTRY

Chapter 2 describes a possible categorization of PLM integration levels for shipbuilding. Just because there is a technical possibility, it is not a reason to use it. Only a valid business requirement can guide the selection of the technical solution. This chapter is presenting example scenarios from the worldwide shipbuilding industry. The authors of this paper are in contact with many shipyards around the globe and provide vendor-independent support to improve their shipbuilding PLM capabilities. The following scenarios do not claim to enumerate all valid ones that business may require, but some of the most important scenarios that we encountered during the recent years.

3.1 SCENARIO 1: POST MERGER INTEGRATION (PMI)

In the last two decades, international shipbuilding has been characterized by many mergers of shipyards. For example, just recently (2020) in Japan, Japan Marine United Corporation and Imabari Shipbuilding initially announced their intention to combine efforts in a bid to enhance Japan's competitive position in the shipbuilding industry. The Japanese industry has been eclipsed by the efforts in China and South Korea. The plan is to use the resources of the two shipbuilders to maximize the benefits and increase their overall competitiveness. Nihon Shipyard will focus on a common and future-proof design [4]. About three years ago (20219), China completed the merger of CSIC and CSSC to create a shipbuilding company that can compete with global heavyweights in the maritime industry [5]. Last but not least, MEYER Werft acquired STX Finland in 2014 [6]. Today, the MEYER shipyards build cruise liners on a common platform approach to be more efficient than other shipyards in that segment [9]. Not all mergers and acquisitions are as big as the mentioned examples. Last year (2021), the superyacht builder Nobiskrug was acquired by Flensburger Schiffbau-Gesellschaft (FSG). FSG expects that the Nobiskrug acquisition enables better access to the attractive market of superyacht construction [7].

These references already show several motivations for mergers in shipbuilding. Mergers take place because the shipyards want to expand their market share in the existing segment. Mergers also take place because shipyards want to enter new segments. These are only two out of several more motivations for mergers. For this reason, there are also different types of mergers as illustrated in Figure 4. On one side of the spectrum, previously independent parts aim to become one integrated enterprise. This necessitates unifying the processes and IT infrastructures for the newly built shipyard organization. On the other side of the spectrum, the merger focus is more on collaboration to allow acting as one unit on the market, e.g. bidding for a specific project, than to completely harmonising internally. This type is better served by setting up well-defined interfaces and centralizing e.g. marketing and material procurement but then leaving the rest of the shipyards as is. In each case, the IT solution will look differently on those two sides of the spectrum. And as always you find a lot of situations in the spectrum between these two.

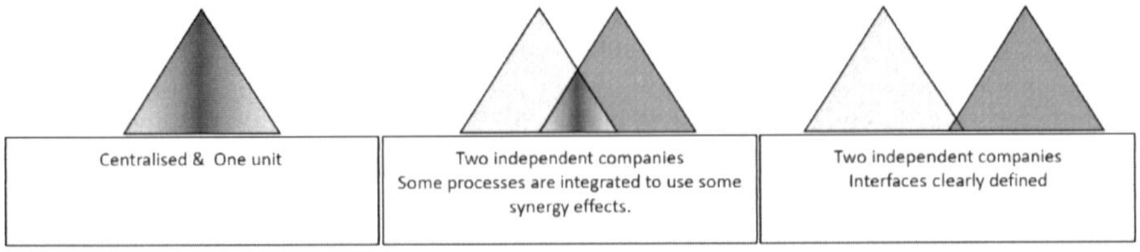

Figure 4: Three simplified Models of Mergers

In recent years, we gained a lot of experience by actively supporting shipyards in such post-merger integration situations. Choosing the right merger model is a fundamental decision and must fit the anticipated business. In addition to the overall merger strategy, we noticed three essential components of a successful merger:

- Communication of changes into the new organisation.
- Collaboration across the different IT landscapes and business processes of the shipyard locations.
- Project Management of the Post Merger Integration (takes several years).

Each of the three components has two important aspects. First, it is the political process that must provide governance and direction for the PMI process. Second, it is the technical PMI process that ensures the development of the wanted IT implementation and environment. Figure 5 shows perfectly how these two aspects must be synchronised for a successful merger. It also illustrates that Business is the No. 1 driver for the synergy effects of any merger. For example, supply chain synergies are dependent on a common view of suppliers and commodities which require: (1)a combination of processes and systems and (2)IT integration.

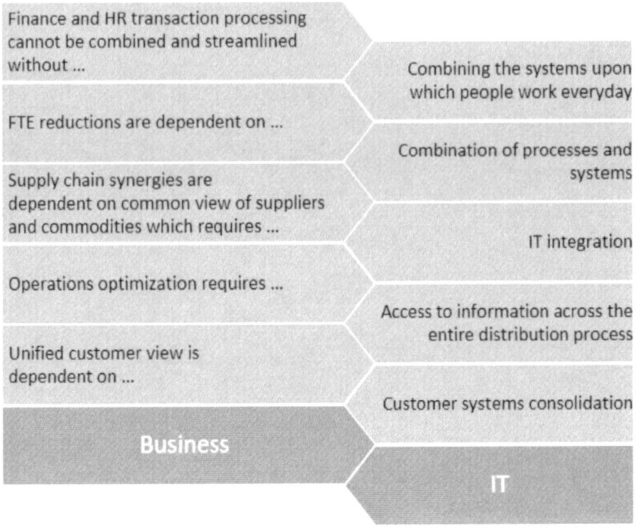

Figure 5: The secret to Success for Mergers [8]

From our experience, the two most common requirements for PLM integrations in mergers are:

- Early in the PMI process: Integration of the different CAD applications into one harmonized PLM system (PLM Integration Level 2). Often this is the defined technical way to bring the bigger user group together so that they have the same set of information available.
- Later in the PMI process: Replacement of legacy systems, like redundant CAD applications. This step can be motivated by several reasons. Often, the new shipyard organization wants to have the CAD users working closely together in one single leading CAD application. This also allows the shipyard to optimize the license handling from a commercial point of view. From a technical point of view, the PLM integration Levels 2 and 3 can apply here. Level 2 is used, when the CAD model should only be transferred as a "reference project" to a target application. Level 3, when the shipyard plans to continue this project (or a sister ship, or a series of ships) in the future in the new CAD application.

3.2 SCENARIO 2: THE INTEGRATED SHIPYARD

Today, this scenario is very common in the global shipbuilding industry. In the past, a shipyard often had a shipbuilding CAD system for the design and an ERP system for the material handling and manufacturing work package definition. For a few years we can see the trend is to introduce PLM systems to enable configuration management for engineering and design data thus enabling traceability in engineering change processes and providing a status - containing single source of truth towards downstream processes like work preparation, planning and procurement. The key driver is the need to manage complex vessels or those with a lot of information required by the customer. Typically, these are governmental orders such as NAVY vessels or complex commercial vessels like cruise ships or FPSO.

From our point of view, the introductions of a PLM system is largely due to two reasons:

- Amount of project data: All information (e.g. changes, communications, decisions, releases, status, ...) that are not managed in CAD or ERP increases massively the amount and complexity of the projects. Other industries have been using successfully PLM systems for this purpose for years.
- Model-based: The paradigm shifts from drawing-based to model-based product creation processes. Traditionally, drawings have been approved and revised. Shipyards use document management systems for this purpose. If the entire model is revised in the future, then a shipyard needs a system for managing the product model data: a PLM System. Mechanical CAD and shipbuilding CAD systems do not have this capability.

PLM integration is closely referenced to the respective IT applications. Before a shipyard can say unequivocally which information is to be transferred from which source system to which target system, the shipyard must specify exactly which information is to be generated and required in which system (CAD, PDM, PLM, ERP, etc.). Each shipbuilding project executes the process shown in Figure 6. This figure does not show any IT systems, it is primarily about the interaction of the individual product views in the project and this figure looks a little different for each shipyard.

A representative example from the shipbuilding industry might look like this:

- Parts & Material master is defined in ERP System.
- Physical (3D) Model is done in a classic shipbuilding CAD application.
- Functional Breakdown is realised in a PLM system.

Consequently, the following PLM Integrations are required in that simplified example:

- Catalogue Parts from ERP to CAD
- Unmodeled Parts from ERP to PLM
- Physical (3D) Model information from CAD to PLM

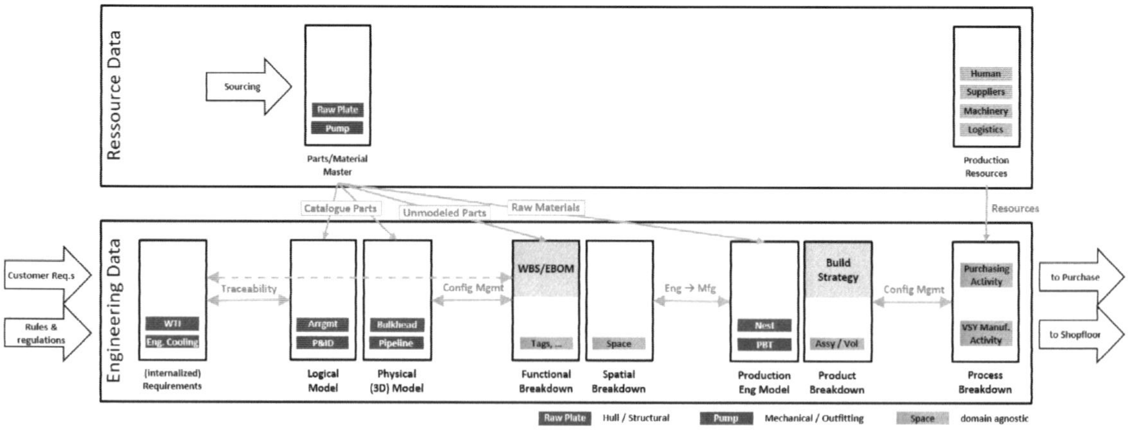

Figure 6: The different Model Views in a Shipbuilding Project

As a rule, shipyards do not start from scratch but have an existing IT landscape. A classic shipbuilding CAD system is often used. The ERP system is either a medium-sized commercial tool or sometimes even a homegrown system. These existing systems of the shipyard already cover a large part in Figure 6. The resulting white spots must be filled with new IT systems. The final PLM architecture then reflects the high-level requirements of the PLM integration. From our experience, the most common requirement for PLM integration at shipyards deals with shipbuilding CAD to PLM nowadays.

PLM integration between shipbuilding CAD and a PLM system requires some defined business rules before a technical integration can be implemented. Below, three relevant areas for PLM integration are listed:

Synchronized EBOM and CAD breakdown (PLM Integration Level 1):

- Typically, the shipyards EBOM or WBS is managed in the PLM system. This structure contains information from different CAD systems but also a lot of non-CAD information.
- In CAD, the breakdown of hull and outfitting parts evolves and needs to be merged into the higher-ranking EBOM/WBS in the PLM system.

Creation of parts in defined IT system (PLM Integration Level 1 & 2):

- A part, e.g. main engine, is often defined before the 3D model starts. Typically, this information flows from PLM (EBOM) to CAD.
- A part, e.g. regular pump, exists as a CAD catalogue part master and is placed as an instance in the 3D CAD model. The integration is from CAD to PLM (EBOM).

Configuration Management (PLM Integration Level 1)

- Defined concept of how Shipbuilding CAD parts are revised in PLM (which transfer is an update and which transfer is a higher revision). A key challenge is often, that a shipbuilding CAD application does not support configuration management, but the EBOM/WBS requires it. Configuration Management is the enabler for Change Management, Sistership Handling, drawing-free processes, and many more.

3.3 SCENARIO 3: ENABLING BEST TOOL IN CLASS FOR SHIPBUILDING

Another benefit of integrating the tools in shipyards is, that it allows companies to pick the best tool in class to perform a specific task. As described in Figure 7, the shipbuilding process is typically subdivided into multiple phases and covers multiple domains like ship steel, piping, and HVAC. The complete digital toolset needs to cover use-cases from all phases and domains. But this does not imply, that there is exactly *one* tool to cover all phases and domains. Quite often there is a combination of multi-phase and multi-domain tools integrated into the toolset used in the shipyard. Without the necessary integration in place, you need to ponder the gain you get from using that best tool with the effort it takes to take over the generated information. If the effort is too high, you better stay with a mediocre, but well-integrated solution. But then you are not able to improve your overall efficiency or reduce throughput time and the use of resources.

From our experience, the two most common requirements for PLM integrations of this described use case are:

- Native design transfer from a Basic Design application to Detailed Design application (PLM Integration Level 3). Often, we hear, that such a design transfer saves weeks to months of work in design. Most often, it is mentioned by the shipyards that they spend these gained weeks to generate a more mature basic design. This reduces many downstream design changes and by keeping the overall ship development process short.
- Native design transfer from Detailed Design to Manufacturing applications (PLM Integration Level 3). For a couple of years, more and more shipyards are interested in a native transfer from their detailed CAD model to the IT systems, which control the automated manufacturing processes of assembly panels, micro-panels, built T-bars etc.. This is already state of the art for plate and profile production, but still sees a lot of potentials when it comes to panel-lines and such.

To enable shipyards to use this best tool-in-class approach, we are permanently working to develop out-of-the-box integrations which could be easily applied. We presented our first implementation for a high-level, cross-tool interface at ICCAS 2017 in Singapore [1]. It covers the integration from Catia to AVEVA Marine Hull for steel design and CATIA to CADMATIC for outfitting design. Based on this, we started developing an industrialized solution with NAPA OY [3], presented in 2019 at the ICCAS in the Netherlands. This interface offers a high-level transfer of the steel design performed in NAPA Designer to AVEVA Marine Hull. It is implemented as export connector for NAPA Designer and respectively import connector for AVEVA Marine. The resulting quality is so good, that its now used by several shipyards to leverage the possibilities of NAPA Steel for very fast steel design and then continue in AVEVA Marine to create the final production level design.

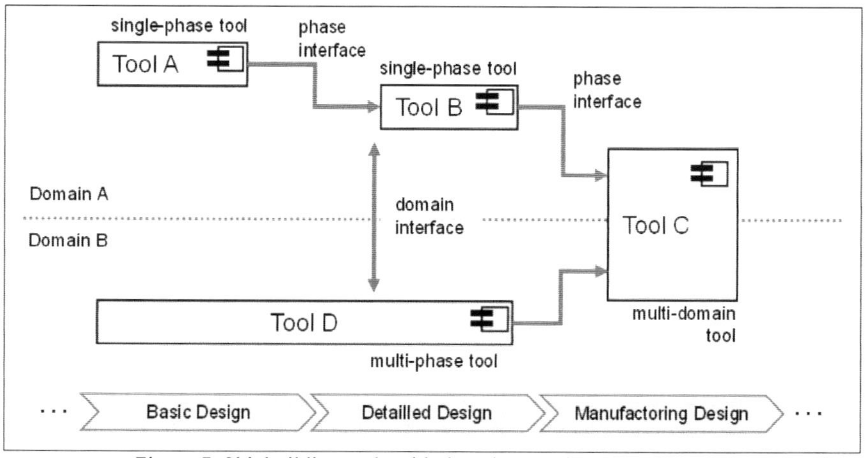

Figure 7: Shipbuilding tools with domain and phase interfaces [3]

4. CONCLUSION

As we discussed in chapters 2 and 3, integrating existing or additional tools could become a business enabler for shipyards. There are several reasons driving those ideas, most notably these are the need to improve the overall efficiency, requirements driven from the customer side and last, but not least, factors like mergers and acquisitions driven by management or even government. When faced with such a task, the integration levels described in chapter 2 may help you, to better decide what type of integration you are striving for. In our opinion, there is no one-size-fits-all solution, so it is quite useful to balance the users' demand (make everything possible) with the effort to setup and maintain such an integration.

We are permanently working on reducing this effort. That's why we not only proving bespoke solutions for the shipbuilding industry, but we are also increasing our range of tailored solutions. PROSTEP's digital enterprise platform OpenPDM Ship provides integration made from standard products. We are able to address both vertical integrations of shipbuilding CAD applications with common enterprise systems to manage configuration and change management or feed material sourcing and production planning as well as horizontal integration between authoring tools and systems. For the vertical integrations we support a whole range of PLM (e.g. ARAS, Enovia, Teamcenter, etc.) or ERP systems (e.g. IFS, SAP, etc.) in additional to shipbuilding specific CAD systems. Those are also bases for native level integration for systems like AVEVA Marine, CADMATIC, or NAPA. To connect the various systems to OpenPDM, connectors are available as maintained standard product components. Each connector provides read and write access to the connected system using interfaces provided by the system vendors. On their backside, all connectors share a common interface to provide PLM data into the OpenPDM platform in a normalized format while maintaining the proprietary data semantics.

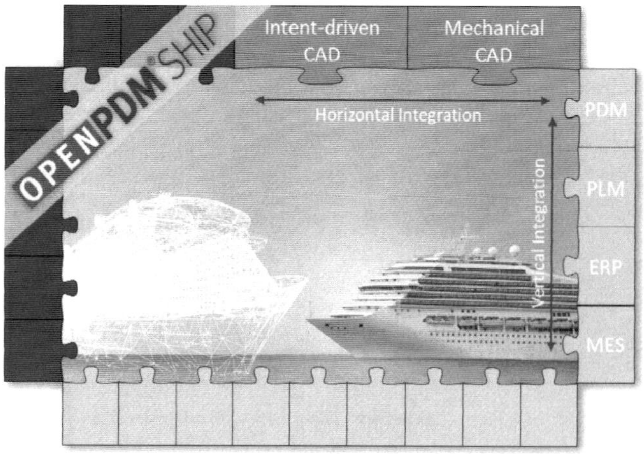

Figure 8: Vertical and horizontal integration offered by OpenPDM Ship [3]

5. REFERENCES

1. DUSCH, T., FRANKE, B., GRAU, M., ZERBST, C. 'Intent-driven CAD vs. Mechanical CAD in Shipbuilding', ICCAS, Singapore, 2017

2. PROSTEP AG, 'AVEVA to TEAMCENTER'. 2021 https://vimeo.com/506992397. Accessed 2022

3. GRAU, M., GRESENS, C., ZERBST, C. 'Enabling best Tool in Class for Shipbuilding', ICCAS, Netherlands, 2019

4. The Maritime Executive, 'Japan Launches New Nihon Shipyard in JV Between JMU and Imabari'. 2021 https://maritime-executive.com/article/japan-launches-new-nihon-shipyard-in-jv-between-jmu-and-imabari. Accessed 2022

5. Ship Technology, 'China completes merger of CSIC and CSSC to create shipbuilding firm'. 2019 https://www.ship-technology.com/news/china-csic-cssc-merger/. Accessed 2022

6. Meyer Werft, 'MEYER WERFT joins forces with STX Finland'. 2014 https://www.meyerwerft.de/en/press/press_detail/meyer_werft_joins_forces_with_stx_finland.jsp Accessed 2022

7. Marinelog, 'FSG acquires superyacht shipyard Nobiskrug'. 2021 https://www.marinelog.com/news/fsg-acquires-superyacht-shipyard-nobiskrug/. Accessed 2022

8. Forrester Consulting. 'The secret to Success for Mergers'. 2008

9. Wikipedia, 'Excellence-class cruise ship'. 2022. https://en.wikipedia.org/wiki/Excellence-class_cruise_ship

6. AUTHORS BIOGRAPHY

Jan Bitomsky, MBA holds the position of PLM Shipbuilding Senior Consultant at PROSTEP. His professional background is Naval Architect with a strong focus on PLM in shipbuilding. Previous experience includes projects in the fields of CAD, CAM, and PLM for various shipyards around the world.

Dr.-Ing. Matthias Grau is currently Account Manager Shipbuilding Industry. As a Naval Architect, he has been involved in various shipyards as well as in CAD/PLM system development on the vendor side. In his role as a PLM Strategy Consultant, he also gained experience in other industries such as aerospace and automotive.

Johannes Lützenberger, Dipl.-Ing. holds the current position of Technical Account Manager Plant at PROSTEP. As a mechanical engineer, he has been involved in several PLM-projects in various industries. In his role as Product Manager 3DigitalTwin he is responsible for the development of the service for an AI-based object recognition in point clouds.

Dr.-Ing. Carsten Zerbst PMP holds the current position of Manager System Integration at PROSTEP. He is responsible to define and implement solutions for maritime customers. He obtained a doctorate from the Technical University Hamburg-Harburg in the area of onboard data integration.

SMART DIGITAL SHIPYARDS WITH MODEL-BASED MANUFACTURING

O Chouche, Dassault Systèmes, France

SUMMARY

The biggest challenge for shipyards is to deliver on time and with the expected quality. Each day of delay can cost shipyards up to one million dollars. With small margins, less than 2% on average for the 50 biggest Shipyards WW, shipyards can easily end up losing money. In addition, ships are becoming a lot more complex. This is due to the technological advancements imposed by mandated customer experiences and environmental requirements. By 2050, there are lots of new regulations for emissions shipping and neutral carbon policies. This creates challenges for Korea to produce more complex ships and for China to move from volume production of low value ships to volume production of complex ships. This is good news for Smart Digital Shipyards transformation, this is because Smart Digital Shipyard Transformation is essential for managing complexity and large volumes.

Smart Digital Shipyards covers a huge portion of the life cycle of shipbuilding. From requirements to digital manufacturing, to manufacturing execution. Model-based manufacturing will ensure all the information required for shipbuilding will be available in a unified model within a collaborative platform. This will seamlessly support all the manufacturing processes without having to go through non-value add activities.

Until today, many shipbuilders still use disjointed primitive tools to manage sophisticated processes. Tools such as Excel spreadsheets as well as lists and 2D drawings printed on paper. Production teams manage warehouses with tools developed over a decade ago. Shipbuilders also address production issues inefficiently – they often rack up a significant amount of additional work hours, going beyond the agreed project timeline and incurring massive extra costs. In the past, this approach may have been acceptable, but today's environment is less forgiving.

This paper introduces the major transformation undergoing at some leading shipyards, by replacing the traditional document-based processes with a model-based manufacturing approach, leveraging ship virtual twins from design to production, achieving a strategic move towards "Smart Digital Shipyards".

1. INTRODUCTION

Products are no longer enough for today's consumers. True value is linked to the multitude of customized services associated with a product, as well as the experience the user derives from its use. Consumers value experience over all else and demand that their lives be enriched in meaningful ways. Experience Thinking is a framework for innovation, focusing on engaging consumers with breath taking product experiences that are smarter, intuitive and sustainable. This evolution in consumer's expectations is putting a lot of pressure on innovators to produce complex products. Manufacturing these complex products and delivering them on time, with the expected quality while staying profitable becomes very hard to manage without a well-oiled and integrated Smart Digital Shipyard.

Why Smart Digital Shipyards and what this notion brings to the table. The following are high level values that can be harness by transforming to a Smart Digital Shipyard:

- Be able to efficiently and quickly commit to a new ship order. This means the ability to commit to a start and delivery date for this new ship order before signing the contract.
- Deliver ships on time and with the expected quality.
- Improve shipyard productivity and efficiency enabling higher capacity.
- Have the ability to quickly monitor and assess the status and health of shipyard resources.
- Improve worker safety and reduce liabilities.
- Improve business image, quality, and reliability. Of course, this will improve business order book and profitability.

Smart Digital Shipyards is a transformative vision that is very challenging to realize. This is because it embodies so many different elements that must be connected in harmony to produce the desired results. The objective of this transformation is to achieve significant improvement in productivity, efficiency, repeatability, reliability, quality, traceability, and health and safety over the lifecycle of vessels manufacturing and operation.

What's essential here is to make sure this transformation can be performed while staying profitable, competitive, and relevant and at the same time ensure on time introduction of new experiences. To achieve this, stakeholders around the world are forced to rethink the way they design, manufacture, and operate. The ability to truly transform processes and take on a holistic approach is critical. Therefore, an integrated business platform that provides end-to-end digital continuity from ideation to ship in service by integrating design, engineering, simulation, manufacturing, and business processes in a single collaborative environment is an essential enabler in carrying out this profound transformation.

What makes this venture extremely attractive is the tremendous, estimated growth of such transformation. The Smart Digital Shipbuilding sector is poised for explosive growth – from $591m in 2019 to $2.7b by 2027, with an impressive 21.1% compound annual growth rate.

2. ELEMENTS OF A SMART DIGITAL SHIPYARD

Smart Digital Shipyards expresses the idea that the digital world of product engineering, digital manufacturing, the real world of manufacturing operation management and operations can share a single digital source for comprehensive management. All Marine and Offshore companies can monitor every last detail of their industrial operations for every product and process by adopting a single digital model approach throughout the development of concept design, detail design, digital manufacturing, manufacturing operations management, and operations to optimize overall company performance. This fosters continuous improvement, innovative design, quality and on time manufacturing, and precise traceability. This also promotes collaboration, operational excellence and agility.

Transforming to a Smart Digital Shipyard, is definitely not a small feat. It will require investments in mindset, machines, connectivity, and data mining. This is a cultural change in all aspect of the shipyard and manufacturing processes. To make this transformation plausible and business friendly, it is critical to adopt these technologies incrementally and wherever makes sense to the business. Forcing an entire transformation all at the same time is not viable or acceptable to most businesses.

The following is a brief description of the elements that must be considered during the journey towards achieving a Smart Digital Shipyard.

2.1 COLLABORATIVE PLATFORM

To achieve a Smart Digital Shipyard, manufacturing silos must be eliminated. There are three aspects to breaking the silos: having a collaborative platform that will host all the information; adopting a data driven approach; and the use of integrated software tools and machines to create the data and the relationships between the data to ensure traceability. It's clear that a collaborative business platform is the cornerstone that is required to eliminate silos, moving from a static, filed-based world to a digitally connected world, where live data drives innovation, processes, and business-decisions. Having all the data in a single collaborative platform will ensure traceability, alleviate the risk of data inconsistencies and potential errors due to working with out-of-date and/or incorrect data.

Connected technologies and services allow people, machines, and communities to communicate in a secure and controlled environment. It enables a new era of innovation on extended ecosystems and fosters a truly open platform innovation. It also enables improved manufacturing planning and execution, conformity to standards and regulations, certification and validation of as-design to as-built, and value chain relationships via an interactive 360° program management approach and full traceability.

2.2 DATA DRIVEN

Data is at the centre of innovation. Having all the data in one collaborative platform will foster a data centric and data driven environment. This centricity will promote data connectivity across all Smart Digital Shipyard lifecycle phases. This includes engineering, digital manufacturing, planning and scheduling, and manufacturing execution information. Relationships between data objects are defined which allow users to more effectively leverage information to perform their daily jobs. An example of leveraging this data includes assessing the manufacturing objects that will be impacted by a design change. Another example is validating the as-design to the as-built to ensure manufacturing is performed within the allowable tolerances of the approved design. With one unified 3D digital model to represent the entire lifecycle of the shipyard and the ship, all stakeholders are always certain that they are working on the latest data without worrying about version mismatch or data duplication.

2.3 MODEL-BASED-MANUFACTURING

The model-based approach is a core enabler of the smart digital shipyard transformation and is valuable to ensure manufacturing effectiveness. This approach provides the ability for the information to flow seamlessly from one project phase to the next. No need for translation or migration of data between the different phases. It is one single model that supports all the phases of the entire lifecycle: conception, design, digital manufacturing, planning, manufacturing execution, and operation. Specifically, during each phase, the model is enriched with additional information which is then used by other enterprise users and downstream processes.

In the design phase for instance, all engineering disciplines whether they're responsible for structure, piping, safety or structural stability can access and work on the same model to do their part. With Model-based approach, all the information is in the model and can be used to enable the next step in the lifecycle. Once the design is stable and ready for manufacturing, the Engineering Bill Of Material (EBOM) can be used to generate the Manufacturing Bill Of Material (MBOM). The MBOM can then be used to generate the Process Bill Of Material (PBOM). The PBOM will include the material, the operations, the sequence of the operations, and the required resources (man or machine). The PBOM can then be used for manufacturing planning and manufacturing execution. With Model-based approach, any potential issue can therefore be immediately detected and corrected. Also, with model-based-manufacturing, all the deliverables (1D, 2D drawings, 3D models, videos) that are needed during any manufacturing process can be immediately derived from information included in the 3D master model. In the manufacturing planning phase, as an example, a well-known Italian ship design company [1], took advantage of a single 3D master model to streamline its processes. The company had to provide shipyards with high quality manufacturing documentation for hull structure scantling and fabrication, as well as for systems design and installation. This documentation indicated where to cut the plates and instructed on how the ship should be assembled block by block. These manufacturing instructions, comprising 3D assembly documentation and simulations as well as detailed instructions for each task, were automatically generated from the 3D master model. In the past, without the single 3D master model, every design unit had to be created from scratch.

2.4 INTEGRATED SHIPYARD LAYOUT, BUILD STRATEGY, AND PLANNING TOOLS

Shipbuilders need to ensure they are able to commit to a new ship order. This means the ability to commit to a start and delivery date for this new ship order before signing the contract. This must be done in the context of the entire shipyards for all the ships that are currently In-Work and the new ship that needs to be delivered at a specific future date. The integration between digital shipyard layout, a build strategy definition, and an advanced planning and scheduling tool will enable shipyards to do just that. For shipbuilders to assess this ability, the main questions/statements to consider are the following:

- Can we quickly and easily define different build strategies for a ship and then converge towards the most efficient strategy?
- Can we have a real time understanding of resource availability for the new ship in the context of the shipyard and all the In-Work ships?
- Can we protect existing scheduled ship commitments?
- Can we avoid penalties due to late deliveries?
- Are we able to protect shipyard credibility by delivering on time?

If you find yourself in doubt on any of the questions above, you should seriously consider exploring this element of Smart Digital Shipyards.

2.5 INTEGRATED MANUFACTURING ENGINEERING, PLANNING AND SCHEDULING, AND MANUFACTURING EXECUTION TOOLS

Many companies use multiple disjointed tools which leads to inconsistencies and inefficiencies in their general business processes. This lack of integration forces conversion of data and manual interfacing between tools, which results in lost time, potential loss of accuracy, and inefficient feedback loop. These inefficiencies will tremendously impact the decision process and ultimately business activities.
An integrated solution [2] will greatly enable and promote the following high-level scenarios:

- Design change propagation to digital manufacturing and then assess the impact on the physical manufacturing.
- Automatic generation and synchronization of Manufacturing Bill Of Material (MBOM) from the Engineering Bill Of Material (EBOM).
- Automatic generation and synchronization of Process Bill Of Material (PBOM) from the MBOM.
- Use the PBOM to populate an Advanced Planning and Scheduling tool.
- Use the PBOM to generate deliverables (1D, 2D drawings, 3D models, videos) that are needed during any manufacturing process.
- Use the planning tool to drive the Material Requirement Planning (MRP) system.
- Use the planning tool and the PBOM to drive the manufacturing execution system.
- Continuously update, in real time, the planning tool based on the manufacturing execution status.

2.6 CONNECTED SYSTEMS AND SHIPYARD MONITORNG

When talking about Smart Digital Shipyards, this topic of connected digital systems is the prime topic that is always front and center. This is of course at the core of Industry 4.0 [4] which includes: internet of things (IoT), internet of services (IoS), master data management (MDM), high performance computing (HPC), supply chain tower (SCT), augmented reality and virtual reality (AR/VR), artificial intelligence (AI), machine learning (ML) and collaboration platforms (CP).

The fundamental premise of Smart Digital Shipyard is the vertical and horizontal integration of the shipyard infrastructure and processes. This connectivity provides a platform for the development of technical or digital functionalities that enable smart production, for example analytics, process automation and simulation and real time monitoring through extensive use of sensors enabling control, condition monitoring and predictive maintenance.

3. CONCLUSIONS

As an integrated business platform, Dassault Systèmes' **3D**EXPERIENCE® platform is an essential enabler of Smart Digital Shipyards. Experiential platforms are fundamental because they bring together talents, ideas, solutions and data. They enable the implementation of innovative supply strategies (interconnected modular manufacturing systems) and demand strategies (customized and connected products and services).

The **3D**EXPERIENCE platform provides a single, real-time source of data/information across the entire ship/platform lifecycle, from conception to design, engineering, manufacturing, operation and maintenance. It uses a single, intelligent 3D master representation for all processes which helps break silos, avoid duplication and inconsistency, reduce costly errors and delays, support decision making, and improve global efficiency/productivity.

Leading companies across a number of industries see the strong value proposition of the **3D**EXPERIENCE business platform to bring their different disciplines together and power processes across different disciplines, from ideation, design, scientific simulation and manufacturing to marketing and sales, using meaningful data analytics, thus enabling end-to-end digital continuity. This results in significant improvement in their innovation capacity and operational efficiency. The move to the **3D**EXPERIENCE platform is accelerated by increasing product complexity, need for connectivity and collaboration, and the necessity to make sense of an exponentially larger amount of data.

The **3D**EXPERIENCE platform provides social and collaborative applications to:

- Connect people to imagine sustainable innovations together.
- Help all disciplines involved in a marine and offshore project to collaborate more effectively.
- Facilitate planning and management of the complete ecosystem including the supply chain.

3D Modelling applications to:

- Virtually design smart ship or smart platform (high quality, energy efficient, safe, and regulation compliant) with world class 3D modelling capabilities from requirements to manufacturing, to operation and maintenance.

Content and simulation applications to:

- Virtually experience a ship or a platform before it actually exists with realistic simulation to explore its real-world behaviour.
- Deliver a unique, stunning immersive buying experience customers can trust.
- Maximize design and manufacturing efficiency by re-imagining planning, management, and optimization of industrial operations.
- Ensure compliance with regulations and owners' requirements.
- Optimize on-board safety, operations, and maintenance.

Information intelligence applications to:

- Transform a massive amount of data coming from different sources into tangible insights and business action with real time, dynamic dashboards.
- Reveal the hidden information intelligence needed to understand, participate, and act.
- Help track and monitor all project-related information from a single-entry point.

4. REFERENCES

1. Dassault Systèmes, 'NAOS Ship and Boat Design case study', *3ds.com*, 2017. https://www.naos-design.com/en/

2. Dassault Systemes Marine & Offshore Industry https://www.3ds.com/industries/marine-offshore

3. Dassault Systemes 3DEXPERIENCE Platform https://www.3ds.com/3dexperience

4. Manufacturing Industry 4.0

5. NAVAL TECHNOLOGY – Maritime 4.0 shapes the digital shipyard of the future. https://www.naval-technology.com/homepage/maritime-4-0-shapes-the-digital-shipyard-of-the-future/

6. The Digital Shipyard – Flinders University – Australian Industry Transformation Institute https://www.flinders.edu.au/content/dam/documents/research/aiti/The_Digital_Shipyard_Opportunities_and_Challenges.pdf

5. AUTHORS BIOGRAPHY

Oussama Chouche is an Industry Solution Director at Dassault Systèmes. Main responsibility includes the development of Industry Solution Experiences in the Manufacturing domain for Marine & Offshore companies. Previous experience includes working for EPC firms to design and simulate Fluids, Electrical, and Steel Structure Systems. This includes initial, basic, and detail design.

COUPLING A HIGH FIDELITY NEAR SURFACE EFFECTS MODEL INTO PARAMARINE S4

P Crossland and **C J M Forrest,** QinetiQ Ltd, UK

SUMMARY

When a submarine is operating at depths typical of those required when using the periscope, snorting during the period of recharging the batteries or, in some cases, at slow to zero speed whilst hovering, the submarine can experience complex oscillatory motion due to higher order sea loads. The second order vertical plane loads can manifest themselves as a mean and time varying suction force effectively causing the submarine to be drawn to the surface risking broaching and increasing the possibility of detection. Current methodologies for modelling near surface effects tend to utilize coefficient-based methods to represent the first and second order wave forces on a submarine body. These empirical methods may be suitable for some applications but are not able to capture all the effects of the submarine operating close to the free surface. As the body approaches the free surface, the flow field around the submarine is modified by the gravitational wave effects, changing the pressure distribution across the body and hence the forces and moments. This change in force is dependent upon the shape of the body and the environment and, therefore, it is impractical that a coefficients-based approach could cover all these scenarios to any great level of fidelity. This paper presents a potential flow method for quantifying the hydrodynamic loads on a fully appended submarine when in proximity and under the influence of surface waves. The paper presents some validation of the method and describes its integration with the Paramarine S4 submarine manoeuvring software.

1. INTRODUCTION

Whilst most of the research has been targeted towards understanding the performance of a submarine operating in deep water (away from the effects of the free surface), see [1], submarines do sometimes operate at shallow submergence depths, under the influence of surface waves. In this case, they can experience complex oscillatory motions due to the sea load effects and whilst experimental and numerical techniques exist to help understand the particular aspect of a submarine operating close to the free surface, [2] for example, it has not been the focus of research to the same extent of the manoeuvring and control performance of a deeply submerged submarine. However, these sea load effects, particularly for depthkeeping, manifest themselves as a time varying suction force that can cause the submarine to be drawn to the surface risking broaching; or the submarine, after compensating for the suction effects, coupled with the effects of hull compressibility from a change to a deeper depth, may be significantly heavy unless rapid de-ballasting, through pumping, can be achieved to regain a safe trim state.

The effect of the waves on a submarine are related to geometrical characteristics such as hull length and hull diameter and to physical characteristics such as depth of submergence, speed, sea state and heading to the waves; physical characteristics, like depth of submergence for operations such as snorting, are driven by the design and limitations of such systems. Speed and heading can be, to some extent, directed by operating procedures that aim to maximise the mission effectiveness. Environmental physical characteristics such as sea states are not controllable and the design requirement would be to, ostensibly, maximise the sea state at which periscope depthkeeping or snorting operations can take place. Consideration of the control surface design, including the size of appendages, their means of actuation and the control algorithms, provides the most immediate solution to the challenge of operating a submarine close to the free surface. However, this control system design requires a detailed understanding of the magnitude of the wave induced forces early in the design of the submarine to ensure that any control surfaces are sufficiently large, that their means of actuation, such as servos, provide the appropriate level of control authority and that any control algorithms are sufficient to meet the operational requirements of the submarine.

This paper presents a numerical method for quantifying the hydrodynamic loads on a fully appended submarine when in proximity and under the influence of surface waves based on potential flow theory. The paper presents some validation of the method and describes its integration with the Paramarine S4 submarine manoeuvring software.

2. MODELLING SUBMARINES UNDER WAVES

2.1 EXPLANATION OF THE FREE SURFACE PHENOMENA

When a submarine is operating under an incoming wave, it can experience wave-induced forces that create a complex oscillatory motion [3]. The resultant complex motion can include oscillatory components at frequencies that are typical of those contained within a wave spectrum (so called first order responses that are linearly proportional to the individual wave amplitudes, [4]. The motion responses may also contain sum (high) and difference (low) frequencies that are beyond the dominant frequency region of the wave spectrum (called second order responses that are proportional to the square of the wave amplitudes, [4]. There are some problems, like hull structure vibrations, where the high frequency components are of interest. However, for the motion of a submarine near the surface and its control, the second order effects of the low frequency and non-zero mean components are the major concerns. To explain the phenomena further, Figure 1 shows a submarine at forward speed in a regular wave of wavelength approximately equal to the length of the submarine and wave amplitude approximately equal to a quarter of the hull diameter. The figure includes an illustration of the flow regime around the submarine when under the influence of the free surface.

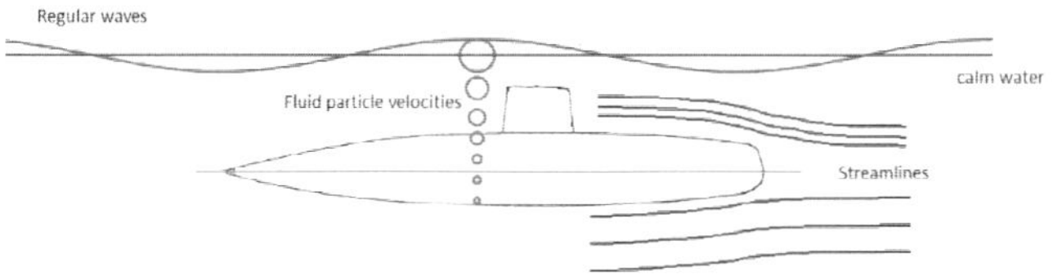

Figure 1: Flow features around a submarine under the influence of surface waves

In the first instance of steady flow, when the submarine is moving through calm water (no incoming waves) at constant speed, the presence of the free surface will mean that the streamlines (shown in Figure 1) over the upper surface of the submarine are closer together when compared to its lower surface thus creating a pressure differential between these two surfaces. This difference in pressure would create a force that will draw the submarine to the surface, a suction force. Actually, [5] noted that the suction force is upwards for low Froude number (based on submarine length) but becomes downwards at higher Froude numbers. This change in direction of the suction force is attributed to the complex wave pattern forming on the free surface above the submarine, [5]. However, it should be noted that typical maximum speeds, when a submarine is close to the free surface, correspond to Froude number around 0.25, [6]. Therefore, at these low Froude numbers, the vertical force is almost always upwards.

When a submarine is close to a free surface, in the presence of waves, the flow effects are modified by the incident wave; for linear Stokes waves, the fluid particle trajectory is circular in nature (in deep water) and the amplitude (the radius of the circle) decreases exponentially with depth; this rate of reduction is related to wavelength. This means that, in the first instance, the effect on the flow due to the wave is smaller when the submergence of the submarine becomes larger, and it also means that the orbital velocities of the incoming wave on the upper surface of the submarine are greater than those on the lower surface. Furthermore, the calm water streamlines due to forward speed (shown in Figure 1) will be influenced by the wave diffraction due to the presence of the body and wave radiation effects as the body oscillates in response to the waves; based on linear theory the combined effects are obtained through superposition. If the slowly varying force in waves (which can vary in sign) becomes a sufficiently large suction force that cannot be countered in a timely way by the control system in the submarine, the force can draw the submarine to the surface causing it to broach and thus increasing the possibility of detection. Moreover, submarine operators tend to take on additional ballast to reduce the possibility of broaching that can leave the submarine in a "heavy trim" state, which means, if the slowly varying force becomes positive (downwards), it could lead to excessive depth excursions before recovering depth control.

2.2 AXES SYSTEM AND EQUATIONS OF MOTION

Use is made of two coordinate axes systems; define as a body-fixed coordinate system Oxyz (with origin at x_0, y_0, z_0) and an earth-fixed coordinate system O'x'y'z'; z' is vertically downwards and z'=0 is the mean horizontal surface at the water surface. The motion of the body is described by Newton's Law in terms of the conservation of both linear and angular momentum, where the rate of change of momentum (or angular momentum) of a rigid body is equal to the total force (or moment) acting on the body.

The mass of the body is denoted as m, with rotational inertial matrix I_o given as:

$$I_o = \begin{bmatrix} I_{xx} & -I_{xy} & -I_{xz} \\ -I_{yx} & I_{yy} & -I_{yz} \\ -I_{zx} & -I_{zy} & I_{zz} \end{bmatrix}$$

The forces and moments on the body are defined.

$F = [X, Y, Z]^T$ is the external force, $M = [K, M, N]^T$ is the external moment, where the components of the forces and moments are in the earth fixed system (x'y'z' directions respectively) with $v = [u, v, w]^T$ as the linear velocity and $\omega = [p, q, r]^T$ as the angular velocity where the components of these variables are in the body-fixed system (xyz directions respectively). The sign convention is given in Figure 2.

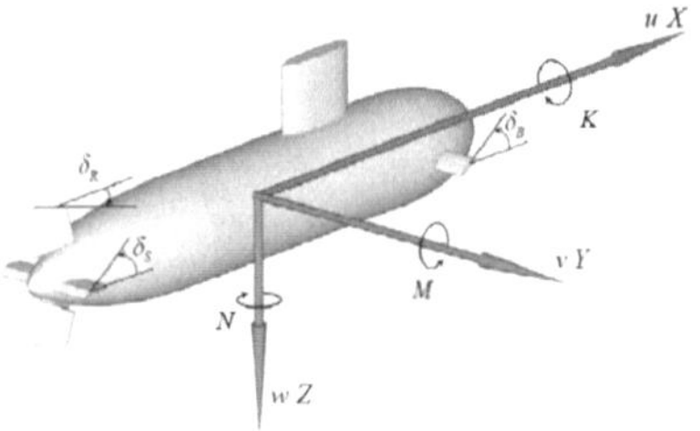

Figure 2: Submarine axis sign convention

The equations of motions are taken from [7], and represent a set of fully coupled, non-linear, equations in which there are no assumptions other than the body is rigid and the mass is constant.

$$m[\dot{u} - vr + wq - x_g(q^2 + r^2) + y_g(pq - \dot{r}) + z_g(pr + \dot{q})] = X$$

$$m[\dot{v} - wp + ur - y_g(r^2 + p^2) + z_g(qr - \dot{p}) + x_g(qp + \dot{r})] = Y$$

$$m[\dot{w} - uq + vp - z_g(p^2 + q^2) + x_g(rp - \dot{q}) + y_g(rq + \dot{p})] = Z$$

$$I_{xx}\dot{p} + (I_{zz} - I_{yy})qr - (\dot{r} + pq)I_{xz} + (r^2 - q^2)I_{yz} + (pr - \dot{q})I_{xy}$$

$$+m[y_g(\dot{w} - uq + vp) - z_g(\dot{r} - wp + ur)] = K$$

$$I_{yy}\dot{q} + (I_{xx} - I_{zz})rp - (\dot{p} + qr)I_{xy} + (p^2 - r^2)I_{xz} + (qp - \dot{r})I_{yz}$$

$$+m[z_g(\dot{u} - vr + wq) - x_g(\dot{w} - uq + vp)] = M$$

$$I_{zz}\dot{r} + (I_{yy} - I_{xx})pq - (\dot{q} + rp)I_{yz} + (q^2 - p^2)I_{xy} + (rq - \dot{p})I_{zx}$$

$$+m[x_g(\dot{v} - wp + ur) - y_g(\dot{u} - vr + wq)] = N$$

$[X\ Y\ Z\ K\ M\ N]^T$ represents all the forces on the submarine including hydrodynamic forces, restoring forces, forces due to control surfaces and external forces due to the waves.

As computational capabilities have increased over the years, Computational Fluid Dynamics (CFD) has been used to obtain predictions through free-running simulations where the hydrodynamic forces and moments on the manoeuvring surface ship or submarine are used directly in solving the equations of motion. These approaches can provide more reliable predictions of manoeuvring motions than simplified theoretical methods because of the ability to include both viscous and rotational effects in the solution to the problem. One of the problems with these approaches still lies in the computational cost in directly calculating the free running trajectory of a manoeuvring ship or submarine, which is considerably high for many applications. This computational cost increases significantly when considered the additional problem of manoeuvring in waves, largely because of the long durations required for the computation run to obtain meaningful statistics. Other less computationally intensive approaches will have simplifying assumptions, whilst requiring sufficient fidelity to be representative of the problem. These simplified approaches generally require a means of solving the manoeuvring and seakeeping problems separately, see [7], for example. There are two main approaches for dealing with the coupled manoeuvring and seakeeping, manoeuvring in waves, problem: the first method, called the two-time scales method, is based on a coupled but independent solution of the manoeuvring and seakeeping problems, in the low and high frequency range respectively. This method, introduced by [8], assumes that the motions from the manoeuvring and seakeeping problems can be linearly superimposed on each other. The second method is based on a unified description of the hydrodynamic problem, where the forces due to the manoeuvring and seakeeping behaviour are linearly superimposed in the same equations of motion, [9], for example.

Reference [10] considered the hydrodynamic manoeuvring forces and the wave induced, or seakeeping, forces by developing a simplified theoretical model to obtain the forces on a vessel manoeuvring in waves. However, they accounted for the limitations of this simplified theoretical model for the manoeuvring problem and replaced those predicted forces with the manoeuvring forces predicted by an experiment based manoeuvring model. For the mathematical model described in this current paper, the hydrodynamic forces on the vessel were decomposed, as follows, in the same way as [10]:

$$F(W, M) = F_{WM}(W, M) - F_{WM}(O, M) + F_M(M)$$

Where

F_{WM} are the forces predicted by the simplified model

W is a generalised set of parameters defining the waves; $W = 0$ means the still water condition

M is a generalised set of parameters defining the manoeuvre; $M = 0$ corresponds to the more typical seakeeping condition and not associated with any manoeuvring.

Reference [10] considered that $F_{WM}(O, M)$ would be less accurate than those manoeuvring forces predicted by an experiment-based mathematical model, represented by $F_M(M)$. This decomposition of the hydrodynamic forces and moments was adopted here because of the known limitations of predicting the forces and moments on a manoeuvring submarine using inviscid, potential flow, theory (due to inability to predict flow separation on the body). Also, this decomposition has been considered because a full set of hydrodynamic derivatives is often available for a submarine early in its design. This decomposition implies that the manoeuvring forces and wave forces do not interact with each other. When the forces and moments on the body are known, the accelerations can be found and through time integrations the translational and rotational velocities and, subsequently, the translational and rotational displacements can be found; the key challenge is the determination of the total forces and moments on the submarine for each time step.

2.3 NUMERICAL APPROACH

The numerical approach, described here, modelled the motion of a submarine under the free surface as a set of equations based on the quasi-steady forces due to manoeuvring coupled with a potential flow approach to model the force components due to the waves. The quasi-steady components are those obtained from the same process of Taylor series expansion that results in the standard derivative-based model given by Gertler and Hagen [1967]. The effects of the waves are considered as being high frequency responses that include the first and second order 6 degrees of freedom body forces and moments on a submarine in an irregular wave. In this particular problem of the numerical solution to the unsteady problem of a fully submerged submarine under the influence of surface waves, panel methods are widely used, utilising Boundary Element Methods (BEM) to provide a solution to the continuity equation by using source and doublet distributions on the body boundary, [11].

The theory is developed based on the assumption that the fluid surrounding the totally submerged body is inviscid, incompressible and constant density and the flow around the body is irrotational. It is assumed that the body is moving with constant forward speed at a mean depth beneath the free surface. The fluid velocity components, in Earth axes, can then be described by a scalar velocity potential ϕ such that:

$$\begin{bmatrix} u'_f \\ v'_f \\ w'_f \end{bmatrix} = \begin{bmatrix} -\partial\phi/\partial x' \\ -\partial\phi/\partial y' \\ -\partial\phi/\partial z' \end{bmatrix}$$

The problem is defined with respect to Earth axes by equations associated with mass continuity, body kinematics, free surface dynamics and free surface kinematics

Mass continuity equation: $\nabla^2\phi(x', y', z', t) = 0$ everywhere in the fluid

Body impermeable condition means that at any point on the body-fluid boundary there is no relative normal fluid velocity component. Hence, the fluid velocity component in the direction of the local normal vector must equal the corresponding velocity component associated with the same point on the body given by:

Kinematic body boundary condition: $\frac{\partial\phi}{\partial n} = -\bar{V}'.\bar{n}$ on the surface of the body

where $\bar{n} = (n_x, n_y, n_z)$ is the local normal vector of the body surface and $\bar{V}' = (u', v', w')$ is its corresponding velocity vector of the body. The convention adopted for $\bar{n}$ is that it should be in a direction pointing into the fluid, away from the interior of the body.

The far field condition is given by:

$$\nabla\phi \to 0 \text{ as } r \to \infty$$

with $r^2 = x^2 + y^2 + z^2$

The elevation of the free surface η in the vertical direction in functional form referred to the O'x'y' plane is $\eta(x', y', t)$. The free surface is assumed to be a single value function of the two horizontal coordinates x' and y' in the earth fixed coordinate system.

Kinematic free surface condition:

$$u_f' \frac{\partial \eta}{\partial x'} + v_f' \frac{\partial \eta}{\partial y'} = w_f' \; on \; z' = 0$$

The dynamic free surface condition follows from the application of the Bernoulli equation with the pressure being assumed to be equal to the constant atmospheric pressure.

Dynamic free surface condition:

$$\frac{\partial \phi}{\partial t} - \frac{1}{2} \nabla \phi . \nabla \phi = 0 \; on \; z' = 0$$

The pressure on the body can be obtained from the unsteady form of the Bernoulli pressure equation:

$$\frac{p}{\rho} = \frac{\partial \phi}{\partial t} + g z' - \frac{1}{2} \nabla \phi . \nabla \phi$$

The steady state problem (forward speed in calm water) can be solved based on the exact or simplified free surface boundary condition. For the limiting case of low Froude number and submergence depths typical of periscope depthkeeping PD operations, treating the free surface as rigid is considered sufficient for understanding the surface suction on a submarine and its impact on submarine control, [3].

For the linear or the first order unsteady problem, the velocity potential ϕ can be defined in terms of a known incident wave potential ϕ_I and an unknown body potential ϕ_b due to wave diffraction and radiation, such that

$$\phi = \phi_I + \phi_b$$

Consider an incident wave of amplitude a and a e eng h o aga ing i h a heading ang e μ with respect to the x'-axis of the Earth-fixed coordinate system. The incident velocity potential ϕ_I of a regular wave, first order, in deep water takes the form:

$$\phi_I(x', y', z', t) = \frac{ag}{\omega} \cos(kx' \cos\psi_w + ky' \sin\psi_w - \omega t) e^{-kz'}$$

To extend the time domain computation to irregular, short crested, waves, the assumption of the linear superposition of sinewaves is used to determine the wave induced flow velocities on the submarine hull (hence solve for the body boundary condition) and over the appendages. In this case, the total potential of an irregular short, crested wave will be;

$$\phi_I(x', y', z', t) = \sum_{j=1}^{N_\mu} \sum_{i=1}^{N_W} \sqrt{G(\mu_j, \omega_i)} \frac{a_i g}{\omega_i} \cos\left(k_i x' \cos\psi_{w_j} + k_i y' \sin\psi_{w_j} - \omega_i t + \varepsilon_i + \varepsilon_j\right) e^{-k_i z'}$$

Where, N_W is the number of wave components

N_μ is the number of spreading angles

$G(\mu_j, \omega_i)$ is the spreading function

μ_j is the spreading angle

ω_i is the frequency

k_i is the wave number; $k = \frac{\omega^2}{g}$

$\psi_{w_j} = \psi_w + \mu_j$

ψ_w is the submarine heading

ε are the random phases.

To determine the body potential, ϕ_b, a singularity distribution over the body surface may be used, each of which is associated with a panel on the body surface. In this case at a point on the body surface a line normal to the local surface of the body point is constructed. The strengths of these singularities are chosen such that the total potential satisfies the impermeable kinematic boundary condition. The flat and rigid free surface condition is satisfied using an image of the body above the free surface, or the double body theory, which leaves the mass continuity equation satisfied automatically by the nature of the singularities used.

For the 3D case, [12] represented the potential for a body S in a steady uniform flow with source distribution. This is an example of the constant source distribution procedure. Through a source distribution σ over the body surface, the potential ϕ_b due to the body (without the free surface effect) is written as:

$$\phi_b(x', y', z') = \oiint_S \frac{\sigma(q)}{r(p,q)} ds$$

where $r(p,q)$ is the distance from the source point q on the surface to the field point p (x', y', z') and $\frac{-1}{4\pi r}$ is the velocity potential due to a source of unit strength.

The form of ϕ_b in the above equation implicitly satisfies both the mass continuity in the fluid domain, or the Laplace equation, and the far field condition; the function σ, which is the source strength distribution, must be determined to satisfy the boundary condition on the body, that requires the normal derivative of ϕ_b to satisfy the impermeable condition. The source distribution approach used here is that first developed in [12] in which the body surface (the hull and the bridge fin) is discretized. The other appendages are treated differently.

For a lifting, thin, finite aerofoil moving at a constant velocity with the free stream velocity vector $\bar{V}_\infty$ at a small angle of attack relative to a coordinate system attached to the aerofoil, the velocity field for the potential flow problem is obtained by solving the Laplace equation subject to the boundary condition on the appendage. The simplest means of solving this 3D problem is to assume that the chordwise circulation, at any spanwise station, is replaced by a single concentrated vortex with circulation.

$$\Gamma = \Gamma(y)$$

placed along a single spanwise line, [11] which is usually along the quarter chord and solved using the lifting line theory developed in [13]. A solution to the spanwise circulation $\Gamma = \Gamma(y)$ can be obtained by describing the unknown distribution in terms of a trigonometric expansion. In the context of appendages on a submarine, [14] used a spanwise coordinate, θ_A, from which the following Fourier expansion was used.

$$\Gamma(\theta_A) = 4b_e U_\infty \sum_{n=1}^{\infty} A_n \sin(n\theta_A)$$

where

θ_A is the spanwise coordinate given by:

$$\varsigma = \frac{y_A - (b - b_e)}{b_e} = -cos\theta_A$$

y_A is the distance from the hull surface to the geometric centre of the appendage.

For appendages that are typical of those used for submarines, [14] used an approximation $c(\theta)$ such that:

$$c(\theta) = c_r(1 - cos\theta_A(1 - e))$$

with

$e = \frac{c_r}{c_t}$ as the taper ratio, c_r the root chord, c_t the tip chord

By re-arranging the above equations, [14] reduced the problem to determining the unknown coefficients A_n by solving the simulation equations.

$$\sum_{n=1}^{\infty} A_n \sin(n\theta_A)(n\mu + sin\theta_A) = \mu\beta_L sin\theta_A$$

where

$$\mu = \frac{\pi}{4b_e} c_r(1 - cos\theta_A(1 - e))$$

The lift generated by the whole appendage is the spanwise integration of the section lift. To account for the effect of the hull on the appendage, [14] used an empirical function, k_{BL}, that reduced the circulation close to the hull itself. The expression used by [14], was modified in the present study to take account of the presence of the hull in the following way:

$$L_c = b_e\rho \int_{\theta_m}^{\pi} k_{BL}\Gamma(\theta_A) U_A sin\theta_A d\theta_A$$

U_A is the spanwise velocity (in the axial direction) obtained from the off-body velocities derived from solving the body boundary problem.

3. WAVE INDUCED MOTION ON A LIGHTLY RESTRAINED SUBMARINE MODEL AT PERISCOPE DEPTH

In order to provide a dataset that can be used for validation, an experiment using a lightly restrained submarine model was performed in the Ship Tank at QinetiQ Haslar using the Submarine Research Model (SRM), see Figure 3, configured as the EUCLID 10.17 geometry. The SRM, a QinetiQ submarine model capability described by [6], was lightly restrained using two towing swords that attached the model to the Ship Tank carriage; each tow sword was attached to the model via a pivot assembly that allowed the model to heave and pitch.

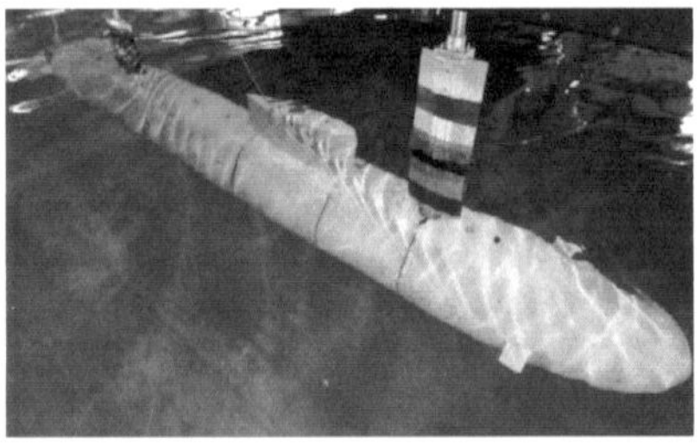

Figure 3: SRM on the carriage in the Ship Tank at QinetiQ Haslar

The towing arrangement that attached the swords to the Ship Tank carriage included two sets of compression springs, one set for each tow post, to provide the stiffness in the vertical plane. Each tow post had two compression springs that had one end rigidly fixed to a load cell to measure the vertical force directly. In this configuration, only one spring on each tow post provided the restoring force (when that spring was under compression) – when one spring was under compression the other one would provide no force at all. Figure 4 shows the aft tow post spring assembly with the principal components marked that includes a load cell mounted to the end of the spring as the spring compressed the load cell would measure the force and simple geometric calculations could then be performed using the load cell readings from the forward and aft spring assemblies to derive the mean heave force and pitch moment. In this example, as the stern of the model rises in waves, the spring, shown in the figure, would compress to provide the restoring force; the compressed spring would impart a force on the load cell from which this component of the overall second order force can be determined. A shaft collar was included to prevent the movement of the tow post to an extent that the spring gets compressed beyond its elastic limit, potentially leading to overloading and damaging the load cell. A displacement transducer was fitted to each tow post to measure its vertical motion from which the vertical plane rigid body motions of the model were derived.

In such tests, the towing rig should be designed so that the natural frequency of the restraining system is significantly less than the lowest frequency to which the linear response of the model may be still noticeable; the lower the spring stiffness, the lower the natural frequency of this spring/mass system would be. So, the spring constants are chosen so that the natural frequencies of the vessel motions, as a consequence of the presence of the springs in the restraining system, are sufficiently far removed from the frequency range of the regular waves. However, if the spring constants were too low then they would not have the capability to withstand the expected vertical plane forces due to the second order wave effects, and thus not being able to prevent either the model from broaching or the springs reaching their compression limit. For the purpose of these tests, a spring stiffness was chosen to reduce the natural frequency as much as practicable whilst limiting the vertical displacement due to springs responding to the second order wave force to prevent the bridge fin from emerging from the water. The minimum wave encounter frequency that was tested was approximately 2 rads/s (or 0.65 Hz) and the natural frequency of the system was found to be 0.08 Hz using springs with a stiffness of 0.238 N/mm.

Model tests were carried out in regular, bichromatic and irregular waves in head seas at zero and at forward speed (Fr = 0.14); the case of forward speed also included tests in calm water. Each of the single wave frequencies were paired to form the two frequency wave systems that were generated for the bichromatic wave tests (for the bichromatic tests, where two waves are paired, the amplitude of each wave is half of the amplitude in regular waves) and, finally, a series of tests were performed in long crested irregular head waves, using an ITTC two-parameter spectrum formulation. The wave amplitudes were chosen to ensure that the first order motions remained sufficiently low to prevent the springs going beyond their elastic limit and to prevent the tow rig from experiencing excessive second order loads. The full analysis of these data is presented in [15]. Due to limitations on the size of this paper, only a small sample of the results are presented here.

Figure 4: Aft tow post spring assembly

For this experiment, the lightly restrained model was able to respond to first order forces, where the restraining system is designed to have an insignificant influence on the first order motion, but the mean forces are countered by the spring system in the towing assembly. Therefore, for regular waves (single frequency), the heave and pitch motion time histories were analysed according to the regression analysis procedure developed by [16]. Figure 6 shows the results of the computations and measurements with the model travelling with forward speed ($F_r = 0.14$). Presented are the first order heave and pitch responses as functions of the non-dimensional wavelength; the heave motion is non-dimensionalised with respect to the wave amplitude and pitch with respect to the wave slope. Comparisons show that the predicted first order motions agree with the experimental data. Figure 7 shows the comparison of the mean heave forces obtained from the lightly restrained case compared with a set of fully restrained model tests with the same hull form, [17]; it is clear that the fully restrained case gives larger suction forces than the lightly restrained case. Comparing the computations with the experimental data shows that these second order forces are under-predicted by the time domain method; the overall predicted magnitudes are smaller than that measured in experiments. In all cases, the component of vertical force due to the forward speed of the body in calm water has been removed from the presentation of the data; the mean heave force (Z') is the mean force due entirely to the presence of regular waves.

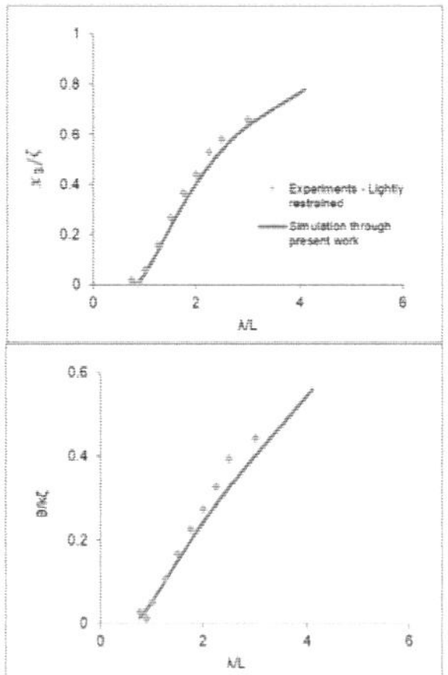

Figure 6: First order heave and pitch responses of the lightly restrained model in regular head waves ($F_r = 0.14$)

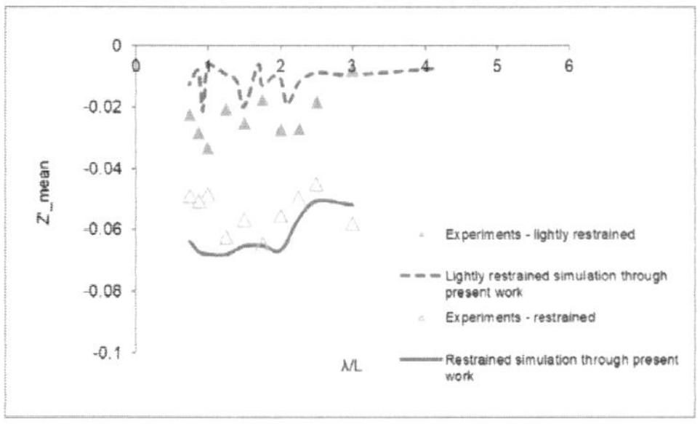

Figure 7: Mean heave force in regular head waves ($F_r = 0.14$)

In summary, the predictions of the first order motions compared very well with those measured during the experiments, which suggests that the experimental rig design is appropriate for measuring first order motion and second order forces. The simulations confirm the results from the experiments in that the mean second order heave force for a lightly restrained body is less than that for a fully restrained body.

4. INTERFACING WITH Paramarine S4

Paramarine is a ship and submarine design package running on Microsoft Windows. It contains elements to assist with the synthesis of a new marine vehicle, together with various Naval Architectural analyses as shown in Figure 8.

Surface vessel	***Submarine***
Statical stability including Probabilistic Damage	Surfaced and submerged stability
Ultimate strength, critical section and plate/grillage design	Pressure hull strength analysis
Manoeuvring via BCTQ Mathman and Triman codes	Manoeuvring, S4
Strip theory seakeeping	
Resistance, propulsion and endurance using regression series	

Figure 8: Types of analysis provided by the Paramarine software

Within Paramarine, S4 ("Submarine State Space Simulation") is a general-purpose time-domain submarine simulation package which has been developed over 20 years to meet the needs of the UK Ministry of Defence. It provides various capabilities summarized in Figure 9. One notable feature of S4 is that it is designed to be extended via external software in the form of dynamic-link libraries (DLLs). This enables a customer organisation to use S4 alongside their own software, for which no sharing of IPR is required.

Acceptance criteria	Built-in and external autopilots	MBT blow modelling
Hull, bow and stern coefficient models	Cruciform, X and inverted-Y stern types	Built-in and external modelling of appendage dynamics
Modelling of fixed and azimuthing thrusters	Detailed pumping and flooding models	Built-in and external trim control systems
Triggering of commands within a manoeuvre via numerous methods	Various wave spectra for shallow-water simulation	Haslar and Wageningen propulsion models
Safe Manoeuvring Envelope generation	Time histories and graphs of output quantities	Comparison with model-scale and full-scale trials data
Compressibility modelling	Adaptive or fixed solver time step	3D visualisation of run trajectory

Figure 9: Summary of S4 capabilities

The calculation elements of the Paramarine software, including S4, are written in C++ which provides good performance for what are often heavily computational analyses. An implementation of the code to predict forces and moments at periscope depth based upon the methodology outlined in Section 2.3 (hereafter referred to as "Perisim") was available only in "breadboard" form in the Matlab system provided by The Mathworks Inc. So the first challenge with integrating the Perisim approach into S4 lay in finding a reliable method of turning the Matlab code into C++ to dovetail with the rest of the system. The solution to this issue was found to lie with an open-source linear algebra library called Armadillo ([18], [19]). Armadillo provides equivalent approaches and syntax to many of the matrix definition and manipulation features of Matlab, whilst enabling compilation to C++ resulting in a huge speed improvement. One of the only downsides with Armadillo is the need to adjust hard-wired array indices by 1, since the Matlab code uses 1-based arrays whereas the Armadillo code works with 0-based indices as is normal in C++.

Every S4 simulation consists of an initialisation phase, during which the submarine's control surface and propulsion parameters are automatically adjusted according to the user's requirements for initial depth, pitch and speed. Following this there is a simulation loop which consists of calculation of forces and moments and then incorporating their effect on the submarine's accelerations in six degrees of freedom. Having achieved a translation into C++ of the Perisim force and moment calculation code, it was possible to call the code from within the force and moment estimation part of the S4 simulation loop.

In order for the calculations to be meaningful, it is necessary to ensure that each piece of data provided from S4 to Perisim (and vice-versa) is correctly converted to allow for any differences in axes systems, datums, and sign conventions. One particular area needing attention lay in the area of faceting. The Perisim calculations are critically dependent upon a good triangular mesh representing the submarine's combined hull and fin. (As described above, the other appendages are dealt with using lifting line theory and their outer geometry does not need to be defined in facetised form.) It soon became clear that the default meshing of the submarine geometry provided by the Siemens Parasolid modeller used within Paramarine was inadequate, and a new bespoke meshing algorithm was developed to suit the Perisim requirements, providing a controlled number of triangles of non-slender aspect ratio without large size variation, whilst providing a good rendition of the submarine's shape. Figure 10 provides an example.

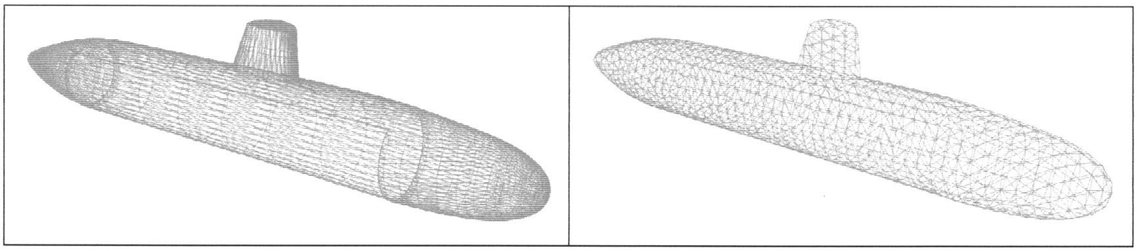

Figure 10: (left) default Parasolid faceting (9700 facets); (right) bespoke Perisim faceting (2000 facets)

There are two aspects to the Perisim approach: forces and moments due to a submerged vessel operating in proximity to the calm air/water interface ("near surface"), and forces and moments due to the effect of waves ("wave induced"). During development it became clear that the first of these would need to be pre-calculated before a simulation run, owing to the need to invert a large square matrix at each depth considered. "Near surface" forces and moments would then be interpolated within the simulation from a depth-based lookup table. The nature of the interpolation is logarithmic with depth as this matches the nature of the effects being modelled.

The "wave induced" calculations simply require matrix multiplication and can therefore be performed within the simulation loop. There is however also in this case a need to pre-calculate a variety of parameters based upon a calm water surface in advance of the calculations performed at each timestep.

5. CONCLUSIONS

This paper outlines a method for estimating the forces and moments on a submerged vehicle operating in proximity with a calm water surface and a wave spectrum suitable for use during design. It improves upon existing coefficient-based approaches by considering the physics of the problem in sufficient detail, whilst providing a more rapid analysis than can be obtained by Computational Fluid Dynamics (CFD). The approach has been cross-checked with a number of physical model-scale trials conducted using QinetiQ's Ship Tank and agreement has been found to be good.

Using this approach it will be possible for submarine designers to estimate the amount of Out-Of-Trim compensating water to take on prior to operating at Periscope Depth, and it will be possible to assess the performance of candidate autopilot algorithms for the problem of operating at near-surface depths in a seaway.

6. ACKNOWLEDGEMENTS

The authors gratefully acknowledge the support of the UK Ministry of Defence in undertaking the work described here.

7. REFERENCES

1. RAY, A., SESHADRI V., SINGH SN., SEN D., 'Manoeuvring studies of underwater vehicles – a review'. *International Journal of Maritime Engineering*, 2008.

2. VEILLON A., AILLARD JM., BRUNET, P., 'Submarine depth control under waves: an experimental approach'. *RINA Warship '96: naval Submarines 5 – the total weapon system, London*, 1996.

3. MUSKER, AJ., LOADER, PR., BUTCHER, MC., Simulation of a submarine under waves. *International Shipbuilding Progress, 35, No. 404, pp 389-410,* 1988.

4. PINKSTER, JA., Mean and low frequency wave drifting forces on floating structures. *Ocean Engineering, Vol 6, pp593-615*, 1979.

5. RENILSON, M. Submarine Hydrodynamics. *Springer International Publishing, ISBN 978-3-319-79056-5*, 2018.

6. BAYLISS, JA, KIMBER, NI, MARCHANT, P, BRYARS J. Submarine trials and experimentation – dealing with real-life data. *RINA Warship 2005: Naval Submarines 8, London*, 2005.

7. FOSSEN, TI., Guidance and control of ocean vehicles. *John Wiley and Sons Ltd, ISBN 0 471 94113 1*, 1994.

8. HIRANO, M., TAKASHINA, J., TAKESHI, K., SARUTA, T., Ship turning trajectory in regular waves. *Published in the West Japan Society of Naval Architecture*, 1980.

9. BAILEY, PA., PRICE, WG., TEMAREL P., A unified mathematical model describing the manoeuvring of a ship travelling in a seaway. *Trans RINA, W6*, 1997.

10. SUTULO, S., GUEDES SOARES, G., A unified nonlinear mathematical model for simulating ship manoeuvring and seakeeping in regular waves. *International Conference On Marine Simulation And Ship Manoeuvrability, Terschelling, The Netherlands*, 2006.

11. KATZ, J., PLOTKIN, A., Low speed aerodynamics – second edition. *Cambridge University Press, ISBN 0 521 66219 2*, 2001.

12. HESS JL., SMITH AMO., Calculation of non-lifting potential flow about arbitrary three-dimensional bodies. *Douglas Aircraft Division Report No. E.S. 40622*, 1962.

13. PRANDTL, L., Über Flüssigkeitsbewegung bei sehr kleiner Reibung. *Verhandlungen des III. Internationalen Mathematiker Kongresses, Heidelberg, pp. 484–491, Teubner, Leizig*, 1904.

14. LLOYD, ARJM., Progress towards a rational method of predicting submarine manoeuvres, *RINA symposium on Naval Submarines, London*, 1983.

15. CROSSLAND, P., Hydrodynamic control of a submarine close to the sea surface, *PhD Thesis, University College London*, 2022.

16. 23RD INTERNATIONAL TOWING TANK CONFERENCE OCEAN ENGINEERING COMMITTEE. Analysis procedure for model tests in regular waves. *No, 7.5 – 02 07 - 03.2*, 2002.

17. MARTINUSSEN K., Submarine motions in confined waters – tests in the Ocean Basin Laboratory close to the surface in oblique waves. *EUCLID/RTP 10.17/MARINTEK/WE-2.5/TR/1/1, Marintek Report 810009.00.01*, 2006.

18. SANDERSON, C., CURTIN, R., Armadillo: a template-based C++ library for linear algebra. *Journal of Open Source Software, Vol. 1, pp. 26, 2016.*

19. SANDERSON, C., CURTIN, R., Practical sparse matrices in C++ with hybrid storage and template-based expression optimisation. *Mathematical and Computational Applications, Vol. 24, No. 3, 2019.*

9. AUTHOR'S BIOGRAPHY

Paul Crossland currently leads the Submarine and Propulsion Team at QinetiQ Haslar. He is responsible for the delivery of the submarine hydrodynamics programme for the UK MoD, developing experimental and computational techniques for assessing the performance of submarines. Recently, he has been awarded a PhD by UCL for his research in increasing the understanding of the behaviour of submarine close to the free surface.

Charles Forrest is a Principal Software Architect also based at QinetiQ Haslar. He has been involved in the development of the Paramarine software, including S4, for many years.

AUTONOMOUS MOBILE ROBOTS' INTRODUCTION IN SHIPYARDS' MANUFACTURING PROCESS

C Dentesano, Fincantieri S.p.A., Italy

SUMMARY

In the context of an exponential growth of digitalization in the naval production process, the introduction of emerging technologies becomes a key point for the future competitiveness of companies. As a matter of fact, it is fundamental to optimize the process, exploiting newest solutions to improve production tracking, quality control and timesaving.
Specifically, autonomous mobile robot technologies (AMR) are becoming increasingly relevant in recent years, thanks to the advances in the scientific community and to the growing number of market options.
In this scenario, it was decided to use AMRs to face different problems in the shipyard, such as autonomous inspections and material transport, to provide our employees with a concrete and automated support during the tasks of every day. As a result, our work led to a cloud ecosystem where our workers have the possibility to assign specific tasks to robots, like setting up real stocktaking missions or making requests for missing material, while putting their effort in more worthwhile activities. Humans and different kinds of mobile robots can operate within the same environment without hindering each other, and the autonomous mode can be easily turned into a controlled one.
In the following, we are going to present the research and development journey which took us to the deployment of a concrete solution for improving both efficiency and quality of our shipyards, which fits perfectly into the framework of digital innovation of our company.

NOMENCLATURE

AMR	Autonomous Mobile Robot
ROS	Robot Operating System
IoT	Internet of Things
CV	Computer Vision

1. INTRODUCTION

1.1 CONTEXT

The study and decision to introduce self-driving robots in shipbuilding manufacturing process is part of the digital transformation process ongoing in the company shipyards. The goal is to optimise, track, and have greater control over the many processes and activities of shipbuilding using information systems, cloud platforms, and edge technologies.
As the goal of this work is a production ready system, our efforts are not only oriented on the Autonomous Mobile Robots (AMRs) platforms itself, but also on the control and monitoring system they will require to be used by the business key users.
With these premises the application quickly become a large and complex system and his implementation requires the use of multiple cross-sector technologies. In fact, rather than the robot operating system (ROS) for the devices low level control was fundamental the use of the IoT (Internet of Things) cloud services fleet management and message dispatching, the artificial intelligence algorithms for an augmented perception and multiple cloud server-less functions to process the IoT telemetry. The Artificial Intelligence application is divided into two main streams, the computer vision to improve the robot's perception of the world around it and the machine learning techniques to assist it in decision making.
The study presented was carried out in one of the areas where this type of technology is most widespread globally, namely the warehouse. The goal of our activities will be to progressively automate the entire area with a tailor-made solution, while keeping intact the already tested internal operating procedures and without altering the structure of the warehouse.
In this area, the activities that can be automated are many, such as inventory inspections, internal movement tracking, material transport, shelf picking and refilling, and quality control. Our research will focus on inventory inspections (through photographic inspection and material counting) and asset tracking.

1.2 THE PROJECT

From the outset, the project design encompasses an ecosystem, capable of connecting and managing multiple robots simultaneously in the performance of various missions. The solution is technologically divided into three major areas, Figure 1, the IoT cloud architecture for fleet management, the development of the control and perception system on the device via the ROS operating system, and the development and training of neural networks for the accomplishment of the more complex or generic tasks.

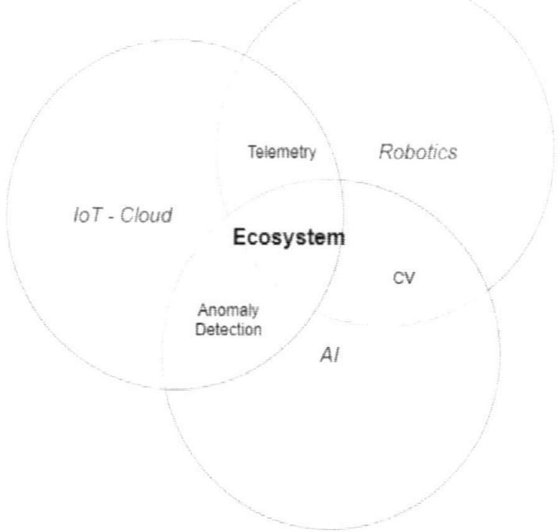

Figure 1: AMR Application Domain

In the following, we first describe how the overall architecture of the solution was composed, then describe in depth the framework developed for the robotic platforms and how we applied computer vision and artificial intelligence to accomplish the proposed tasks. Finally, we will summarise the results obtained from devices, in production, during autonomous inspection tasks.

1.3 DEVICES AND EXPERIMENTS

Given the wide variety of devices on the market, not only by size or range but also by kinematic system, it was decided to study very different devices to compare their capabilities and usability in the production environment. For this purpose, the platforms chosen for the experimentation were a quadruped and a differential track rover. The quadruped was chosen for its ability to move with agility, omnidirectionally even in confined spaces; in contrast, the tracked rover has less freedom to manoeuvre but greater autonomy and robustness.
The comparison between the two devices was made using the same framework for control and navigation in both systems, and their performance was measured with different metrics to calculate time, efficiency, and accuracy in navigation and positioning.

2. CLOUD ARCHITECTURE

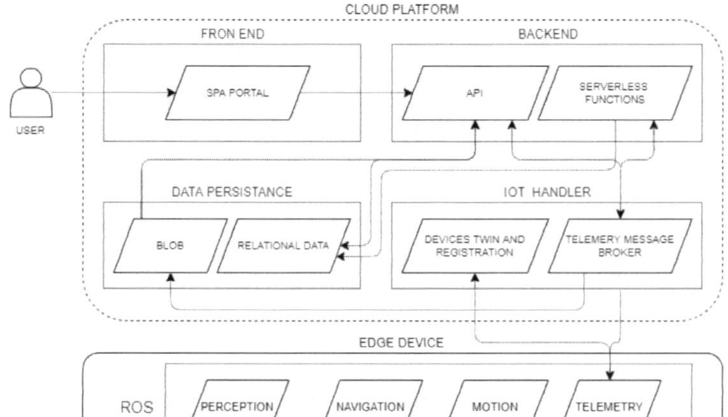

Figure 2: Solution high level architecture

As introduced above, the ecosystem is based on a cloud environment of SaaS (Software as a Service) services whose central point is the services for managing IoT devices. Each autonomous platform will behave as an IoT device, which during the performance of its activities will produce the telemetry and reporting messages, via the MQTT protocol, necessary to monitor its progress and collect the required data from the field.
The activities to be performed by the robot are communicated to the device via cloud-to-device messages according to a generic template representing the various instructions that the robot will carry out in sequence.
The main components of the architecture represented in Figure 2 are therefore:

- IoT handling services

  The IoT handling services include the ability to register and define the devices in the cloud infrastructure, receiving telegrams and redirecting them in real time to the back-end service or to a generic telemetry storage space.
- Back-end services

  A web API to provide data to the user interface and execute business logic on the received configurations of devices, operating areas, and missions to accomplish. Some server-less functions for processing robot telemetry and other external events are also included.
- Front-end services

  To allow users to easily manage the fleet, a web application was built with three main functions: mission creation, mission planning and report analysis.

The architecture is based on an event-driven design, where the triggering event is the queuing of a mission to be accomplished (through the scheduling functionality) by the user. The flow then proceeds from the field devices to the telemetry message persistence services.

3. ROBOTICS

The device control software is based on the ROS framework and consists of a collection of independent nodes that exchange information through topics, starting from the data collected by the sensors up to higher-level elements. This system allows a very layered and decoupled software architecture, making it possible to transfer software through different platforms by replacing only the lowest-level node that manages the hardware actuators of the current platform. Thus, defining the format of the messages exchanged within the framework consists of publishing the data interfaces between the nodes.

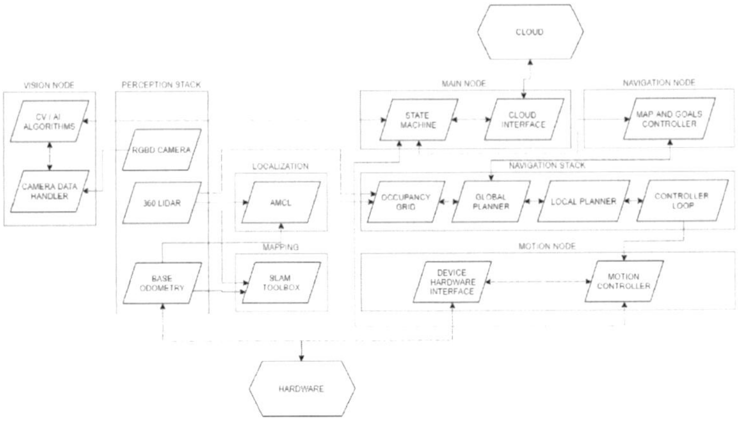

Figure 3: ROS nodes architecture

Within this framework, our integration acts as a glue between several state-of-the-art ROS stacks such as ROS-Navigation [1] for handling the navigation algorithms, SLAM-Toolbox [2] for mapping through pose-graphs, AMCL [3] for correction and relocation on the map. In addition, several internally developed nodes take care of the reception and execution of missions through a state machine and the processing of the collected images through AI algorithms.
The robots used were, however, always equipped with similar sensors to enable them to be used reliably. The sensors used were therefore a frontal depth camera (RGBD) and a 360° LIDAR which, depending on the platform used, could be either 2d or 3d (multi-planar LIDAR).
Navigation is based, as mentioned, on the ROS Navigation stack, and the algorithm chosen to create the local path is Teb Local planner [4], acronym for Time-elastic-band, an optimiser of the trajectory according to the time required to complete it.

3.1 NAVIGATION IMPLEMENTATION CHALLENGES

- Obstacles close to the device and out of field of views

  One of the most obvious problems with using 360 LIDAR placed on top of devices is the detection of obstacles witch height is up to about ¾ the height between the centre of the LIDAR and the floor. Even with a multi-planar LIDAR, the problem still arises when these very low obstacles leave the sensor frustum and approach the device. In detail, the challenge is to find an algorithm that allows to retain the memory of obstacle presence of the recorded obstacles now under the frustum while at the same time freeing up the viewable space permitting the robot's manoeuvring. The solution adopted consists in the use of the spatio-temporal-voxels algorithm from Steve Macenski [5], which allows, by means of a temporal decay of the obstacles present in the frustum, to free up the space covered by the sensor's field of vision and at the same time maintain the memory of obstacles no longer present in the same place, thus exponentially lengthening the persistence time.

- Odometry accuracy and re-localisation

  A precise robot positioning system is a key point to achieve reliable results across all the operation time. During our tests we noticed how the accuracy of the calculated odometry can drastically differ between platforms. The observed error could imply a consistent deviation during rotation, like in the case of the differential tracked platform, or a slow deviation in all directions while walking, like in the case of the quadruped platform. To solve these problems a fine-tuned configuration [6] of the navigation stack was essential. The key point was tuning the noise coefficients of AMCL over the reported odometry.

3.2 COMPARISON EXPERIMENTS BETWEEN THE DIFFERENT PLATFORMS

In the following presentation of the experiments purposes, we will describe the methodologies and the results of the comparison between a quadruped device and a differential tracked rover.

3.2 (a) Premises and purposes

Once the architecture of the ROS nodes was defined and the core software developed, the behaviour of the different platforms in the production environment was tested to identify their potential. The objectives of the tests to which the platforms were subjected were to identify through quantifiable parameters the ability to move through space in the shortest possible time and with the highest final accuracy.

3.2 (b) Methods

The two platforms were loaded with the same software and planning algorithms, differences in configuration concern only the conformation of the platform itself, the arrangement of the sensor space and the device odometry deviation. The starting point of the mission and the goals to be achieved are shared, while the done beforehand environment mapping was performed separately by each device based on its own space perception.

- Type of missions

The missions are divided into three kinds, navigation in a static and obstacle-free environment as baseline for both platforms, navigation on the same path in the presence of dynamic obstacles (moving people), and as the last scenario, navigation in the presence of detectable and undetectable static obstacles.

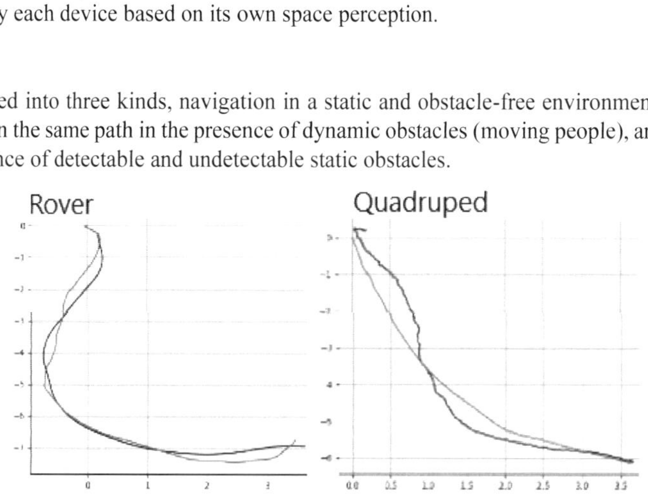

Figure 4: Planned and travelled paths from the same instructions by the two devices

- Metrics

To evaluate the effectiveness of the devices, the navigation time, the deviation between the planned and executed optimal route, and the smoothness of the route itself were measured. In addition, based on the intermediate positions obtained, also the positioning error at waypoints was calculated.
Since the path is re-planned during execution, the calculation of the distance between planned and executed path is composed by two errors Eq 1, the error with respect to the starting plan and the error with respect to the recalculated plan at the instant of execution.

$$\frac{\sum_{i=1}^{n} \left(d(r_i, p_{0,i}) + d(r_i, p_{rt_i,i}) \right) / 2}{n}$$

Eq 1: Distance Metric from planned path

The path smoothness Eq 2, is calculated only on the executed path as the result of the curvature coefficient between the first and the second derivative of two sequential path points.

$$f_i'(x) = \frac{p_{i+1,y} - p_{i-1,y}}{2 \cdot |p_{i+1,x} - p_{i-1,x}|}$$

$$f_i''(x) = \frac{(p_{i+1,y} - p_{i-1,y}) + p_{i-1,y}}{|p_{i+1,x} - p_{i-1,x}|2}$$

$$\text{smoothness} = \frac{\sum_{i=0,n} \left(\frac{f''_i(x)}{\left(1 * (f'_i(x)^2)\right)^{(3/2)}} \right)^2}{n}$$

Eq 2: Planned path smoothness metric

3.2 (c) Results

The results of the experiments show that although the paths planned by the quadruped device were shorter Table 1, due to the smaller size of the device, the continuous corrections necessary to maintain the gait, caused a greater walked distance.

Table 1. Planned vs Travelled Distances.

Device -Plan	Baseline	Dynamic	Rough floor
Rover planned	94.57	94.20	100.25
Rover executed	92.55	97.34	91.96
Quadruped planned	86.65	86.31	85.52
Quadruped executed	117.64	118.19	112.87

The observed difference in the walked distances is also reflected in the trajectory smoothness index, Table 2, which is far higher on the quadruped trajectories.

Table 2. Smoothness metric of executed paths.

Device	Baseline	Dynamic	Rough floor
Rover	51.7	51.790	194.43
Quadruped	105.36	50.1	5214.96

Regarding the distances from the planned paths and the reported positions we can observe how the rover remains close to the initial planned path, Table3, compared to the quadruped. Accordingly with these results, considering our warehouse spaces we decided to adopt the tracked rover in production for the autonomous inspections use case.

Table 3. Distance between executed path and planned path.

Device	Baseline	Dynamic	Rough floor
Rover	1.07	2.95	2.35
Quadruped	3.20	3.16	3.31

4. ARTIFICIAL INTELLIGENCE

The application of artificial intelligence in our study is geared towards improving the perception of the surrounding environment so that devices can perform the required inspections while also drastically reducing the client-side configurations required for a proper functioning.
Reducing these configurations in fact allows a marked improvement in the maintainability of the entire system over the long term. The first algorithm used for this purpose is based on OCR technology and its application is aimed at reading the material codes on signs placed near each location in the warehouse. The ability to read these codes allows the system to autonomously understand the type of material, which is analysed, remaining consistent even when the arrangement of the materials changes, avoiding the user having to continually update the layout of the warehouse.

4.1 TEXT DETECTION AND OCR

The pipeline implemented for text detection consists of several sequential algorithms, the first of which is oriented towards the detection of the position of the text within the image. This allows to determine the bounding box containing the material code even in the case of imperfect positioning of the robot. The algorithm chosen in our implementation is craft-text-detector [7] which relies on a pre-trained convolutional network as an extractor of the basic image information and use them to produce via a multilayer neural network the bounding boxes and the corresponding confidence.
These extracted image portions are to be transformed into text using OCR.
In the current implementation, the algorithm chosen for this processing is Google's Tesseract-OCR [8]. Tesseract is a well-known and widely used algorithm, the crucial points in its use, based on our implementation experience are the resolution of the extracted patch and a correct pre-processing. The image patch pre-processing is oriented to reduce the noise and in a clear binarization of it with the goal of obtaining more sharp text edges possible.

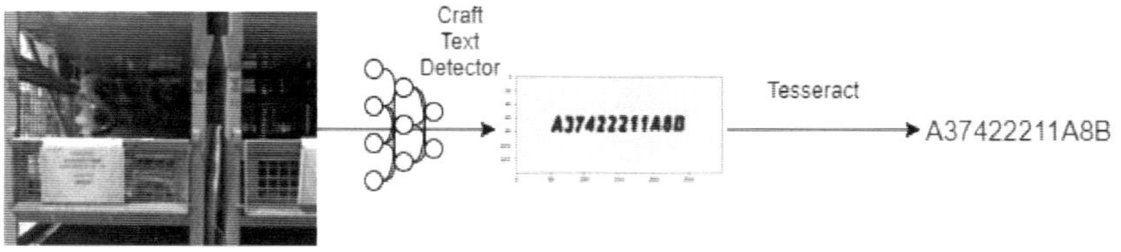

Figure 5: Text Detection Pipeline

The last step of our pipeline is matching the extracted text with the material codes stored in the warehouse, a crucial step as the OCR algorithm could misunderstand some characters or just ignore them resulting in a shorter code. The string-matching method called SequenceMatcher, based on gestalt pattern matching [9] algorithm, tries to compute a "human-friendly diff" between two sequences. Unlike e.g. UNIX (tm) diff, the fundamental notion is the longest contiguous and junk-free matching sub-sequence, where for *junk* are intended all the characters matching with similar shapes.

4.2 OBJECT COUNTING

The second algorithm adopted in our solution aims to make the robot able to count the number of materials stored in a certain location, this ability allows the system to produce, daily, a punctual report of the warehouse inventory status. The produced report will warn the operator in case of a misalignment with the inventory tracked by the business ERP (Enterprise Research Planning) systems.

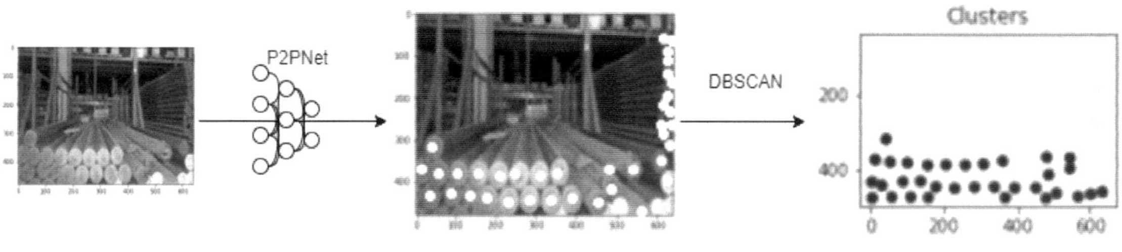

Figure 6: Object Counting Pipeline

The solution adopted is an algorithm designed to solve the crowd counting problem, which presents many similarities with our case as the high and varying density of the objects to count inside the frame, the morphological similarities between

them, and being a supervised technique, it requires a manual tagging of the objects, but contrary to what is required in the case of object detection which needs precise boxes, it just requires a single central point per object.
Specifically, the chosen algorithm was the Point-Based Framework presented in [10] by Song, Qingyu and Wang, an alternative solution from the classical density-based methods used in this field, which giving a more precise prediction of the objects' locations enable us to filter the detected items in the scene. These needs become relevant when the target location is not the only one present in the scene, as in Figure 6.
After the inference on the frame trough the Point-based network an unsupervised clustering (DBSCAN [11]) is performed to get the target cluster. The DBSCAN algorithm, being a density-based clustering technique, finds core samples of high density and expands clusters from them. The metric used to evaluate the density is the Euclidean distance between the points.

5. CONCLUSIONS

The operative procedure followed in the production environment consisted in the following steps:

- The warehouse manager schedules the robot activities by the provided web portal
- The warehouse operator turns on the device at the end of the work shift, so that the device will operate during warehouse closure
- The robot queries the cloud platform requesting the tasks to be accomplished and accordingly with the scheduled trigger starts autonomously to perform it

In a two-week period, the rover, the best IoT platform under the circumstances of our tests, performed 211 checks by detecting and counting 80 varied materials. The reading of identification labels was effective in 70% of cases, a loss of 30% due to environmental conditions that made it impossible to read them.
The next steps to improve performance are to research a better label affixing technique.

5.1 NEXT STEPS AND CURRENT DEVELOPMENTS

We already focused our current developments on optimise localisation, made the robot able to Geo-localise and track active Bluetooth tags inside of the warehouse to track various assets. What is also in plan is the development of a path planning algorithm that will made the rover capable to move with a trailer attached to achieve the goal of autonomous material transportation.

6. REFERENCES

1. E. Marder-Eppstein, et al., The Office Marathon: Robust Navigation in an Indoor Office Environment, In International Conference on Robotics and Automation (2010).

2. S. Macenski and I. Jambrecic, SLAM Toolbox: SLAM for the dynamic world, Journal of Open-Source Software 6 (2021) 2783, 10.21105/joss.02783

3. P. Pfaff, W. Burgard, and D. Fox, Robust Monte-Carlo Localization Using Adaptive Likelihood Models, vol. 22 (2006) pp. 181–194, 10.1007/11681120_15.

4. F. H. C. Rosmann and T. Bertram, Integrated online trajectory planning and optimization in distinctive topologies, Robotics and Autonomous Systems 88 (2017) 142–153

5. S. Macenski, D. Tsai, and M. Feinberg, Spatio-temporal voxel layer: A view on robot perception for the dynamic world, International Journal of Advanced Robotic Systems 17 (2020),10.1177/1729881420910530

6. K. Zheng, ROS Navigation Tuning Guide (2017), 10.48550/ARXIV.1706.09068, URL https://arxiv.org/abs/1706.09068.

7. Y. Baek, et al., Character Region Awareness for Text Detection (2019), 10.48550/ARXIV.1904.01941, URL https://arxiv.org/abs/1904.01941.

8. R. Smith, An Overview of the Tesseract OCR Engine, In Proc. Ninth Int. Conference on Document Analysis and Recognition (ICDAR) (2007) pp. 629–633.

9. J. W. Ratcliff and D. Metzener, Pattern Matching: The Gestalt Approach (1988)

10. Q. Song, et al., Rethinking Counting and Localization in Crowds:A Purely Point-Based Framework (2021), 10.48550/ARXIV.2107.12746, URL https://arxiv.org/abs/2107.12746.Rev. 1

11. M. Ester, et al., A Density-Based Algorithm for Discovering Clusters in Large Spatial Databases with Noise, In KDD (1996)

AUTOMATED GENERATION AND LOW-EFFORT AUTHORING OF COMMISSIONING CONTENT IN THE MARITIME INDUSTRY

A Elzalabany, V Settler, T Jansen, R Rost and **H Lödding** Hamburg University of Technology, Germany

SUMMARY

Commissioning in the maritime industry is a long crucial phase in which all essential components in the ship are tested for faults and defects before a final acceptance form is signed by the client. Due to the variety of components in a ship and the complex hierarchical structure of such components, it is typical to have thousands of tests that should be performed throughout the commissioning phase. For every new ship, commissioning planners initially spend hours preparing the commissioning tests by reading technical documents, extracting relevant information, and applying the gathered knowledge to the unique equipment configuration in the ship. This leads to high preparation efforts and planning mistakes, e.g., unconsidered dependencies, and as a result, commissioning, and the shipbuilding process as a whole may be delayed.

This paper lays the foundation for automating the process of authoring commissioning content for maritime productions. The paper presents a digital modeling approach for ship components, which facilitates the development of commissioning-specialized software. The model shown in this paper takes textual and geometrical information of ship components as input data and produces generalized commissioning templates that can be used to generate commissioning tests automatically.

NOMENCLATURE

BOM	Bill of materials
HOG	Histogram of oriented gradients
IoT	Internet of things

1. INTRODUCTION

A newly produced ship must undergo a process called *ship commissioning* to ensure that the requirements set by the client and the classification society are fulfilled [1]. Throughout the process, commissioning personnel perform several inspections to verify that the right components are installed correctly and that all components are fully functional. Such inspections are referred to as *commissioning tests*. A commissioning test document provides information about the component to be inspected, how to inspect the component, and which resources are needed to fulfil the task. That being said, every shipyard structures the information in the commissioning test document differently. Moreover, a commissioning test document serves as an acceptance protocol that has to be signed by the client, shipbuilder, and classification society at the end of the commissioning process.

The commissioning process consists of three primary phases: preparation, execution, and documentation. During the preparation phase, commissioning planners first view the BOM and then start with authoring commissioning test documents based on the components in the BOM. Then, in the execution phase, commissioning engineers perform all the inspections defined in the commissioning test documents (with and without the presence of the client and the classification society members). Finally, the results of the inspections are signed by all parties, and a protocol is generated and archived.

It takes longer than one year to commission a ship successfully [2] because of the high number of components in the ship; moreover, a ship is a one-of-a-kind production with a unique configuration of components, which causes the commissioning process to vary from one ship to another. Due to the high workload in the process, it is evident that the commissioning process significantly impacts the business and that enhancing the process would positively affect the shipbuilding business as a whole.

One of the main hurdles to enhancing the commissioning process is the reliance on paper documents. Since a ship contains thousands of components, it is expected to have thousands of physical sheets of paper representing all commissioning tests and protocols. Moreover, technical drawings and data sheets, used as supplementary material for executing commissioning tests, also take the form of paper documents. Authoring, editing, and archiving such a significant volume of paper-based commissioning documents is an extensive process that requires high effort and time. Moreover, paper documents are limited in presenting information; for example, dependencies between commissioning tests are not easily identified when viewing static paper documents. Nevertheless, the shortcomings of using paper-based documents in one-of-a-kind productions are known, and research continues to present promising digital solutions to reduce the use of paper documents in shipbuilding [3, 4, 5].

As a continuation of work presented in a previously published paper [6], this paper proposes a digital system that enables commissioning planners to author maritime commissioning tests systematically and with low effort. The proposed authoring system provides a clustering approach to arrange similar components in the BOM in meaningful groups so that commissioning tests can be automatically generated in bulk, saving the effort and time of creating each commissioning test from scratch.

The paper starts with an overview of the commissioning process and discusses the deficits of workflow methods in the commissioning preparation and execution phases. Then, chapter 3 presents a digital authoring system developed to overcome the deficits of the conventional workflow methods. Finally, chapter 4 proposes a machine learning approach for automatic clustering of ship components to enhance the digital authoring system and reduce the time and effort needed for authoring commissioning content.

2. OVERVIEW OF THE COMMISSIONING PROCESS

Before a ship can be successfully commissioned, commissioning engineers must inspect all components in the ship. In the case of any defects or abnormalities, a solution has to be provided. Therefore, to save time, shipbuilders do not wait for the completion of ship construction to perform inspections. Instead, the standard approach is to execute commissioning tests parallel to the production process [1]. However, production activities often interfere with commissioning. Such interferences occur whenever a production activity occupies the resources needed to perform a commissioning test or when the vicinity is blocked due to critical production activities. For example, welding near a component to be inspected prevents the commissioning engineer from accessing the location where the component resides. Commissioning planners need to consider such interferences when authoring and scheduling commissioning tests. However, keeping track of such interferences is not always feasible, especially because one-of-a-kind ship production is not as predictable as serial production.

Interferences and dependencies between commissioning tests show how challenging it is to maintain a productive workflow of preparing and executing commissioning tests using the traditional paper-based methods. The following sections provide more information regarding the conventional workflow of preparing and executing commissioning tests and derive the requirements and motivation for developing a digital authoring system.

2.1 COMMISSIONING PREPARATION PHASE AND WORKFLOW DEFICITS

To prepare commissioning content, commissioning planners require technical and organizational information. Technical information exists as documents that contain technical specifications of the components in the BOM. Such specifications are relevant for the inspection tasks to be performed during commissioning. For example, to author a commissioning test for a pressure sensor, the commissioning planner needs to know the location of the sensor to be inspected, the sensor's measuring range, and the signals emitted by the sensor. Such information is structured in dedicated documents referred to as measurement lists. In addition, subcontractors provide other technical documents and drawings that are also relevant for commissioning preparation.

Furthermore, organizational information is relevant for scheduling commissioning tests and booking the resources needed to perform the commissioning tests. For example, production milestones and deadlines are important for commissioning planners to schedule the commissioning tests as efficiently as possible.

When the required information is available, commissioning planners view the BOM and proceed with authoring commissioning content for every component in the BOM. It is worth noting that there are two main types of commissioning content to be authored: internal commissioning test documents and final acceptance checklists. Internal commissioning tests describe inspections performed mainly by commissioning engineers without external participants, i.e., classification society members and clients. On the other hand, acceptance tests are performed in the presence of the client and classification society members. When all commissioning tests are authored, a schedule is set for performing the inspections. Finally, the authored commissioning content is handed over to commissioning engineers allowing the commissioning execution phase to begin. Figure 1 shows a simplified process model for authoring commissioning content.

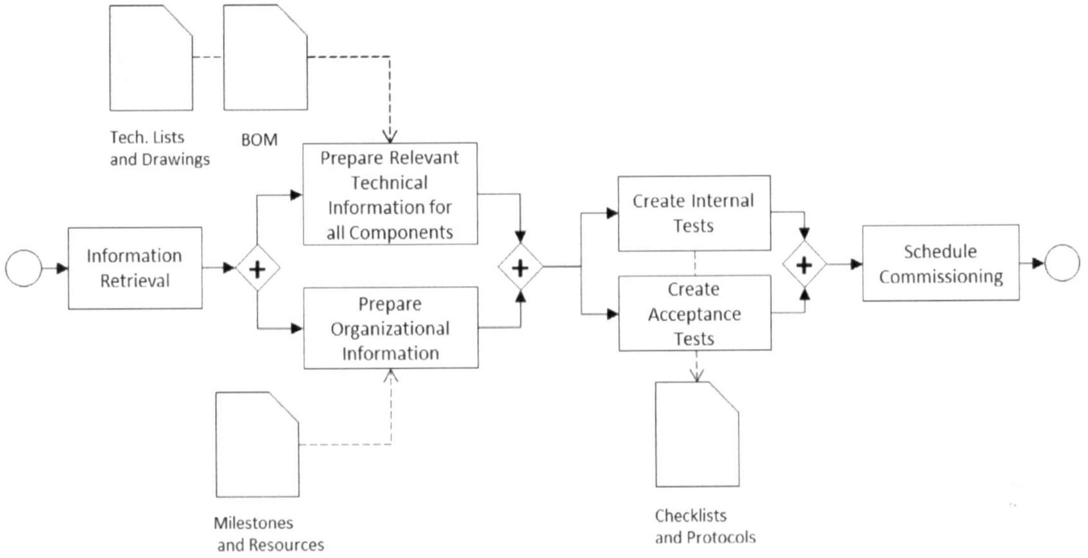

Figure 1: Simplified process model of authoring commissioning content.

The authoring of commissioning tests requires a high volume of manual work since every ship has a different configuration of components. As a result, commissioning planners spend a long time reading technical documents for each component in the BOM and listing relevant parameters in paper-based documents. This process is repeated for thousands of components in the ship, which results in unnecessarily high overhead costs. To improve the process, integrating the BOM in a digital twin offers the possibility of realizing an interactive representation of all components in the ship. The commissioning planner can choose and assign commissioning-relevant documents and parameters to each component in the BOM. Maintaining a link to the digital twin can be beneficial in the long term, where machine learning algorithms can be applied to automatically predict classes of components and assign commissioning-relevant parameters.

Furthermore, due to the uncertainties in one-of-a-kind production, the production schedule is not usually reliable, causing dependent activities to suffer delays and interruptions. As shown in Figure 1, commissioning planners need organizational information (highly dependent on the production schedule) for commissioning preparation. The result is an unrealistic commissioning schedule and sub-optimal allocation of resources. For example, it is not uncommon that a commissioning engineer gathers the documents and tools needed to inspect a machine and goes to the inspection location only to find that the machine is not yet installed or that electricity is not available. Relying on static information sources for commissioning planning is a leading cause of such problems. A solution would be to access real-time information on a digital twin for dynamic planning of the commissioning process.

2.2 COMMISSIONING EXECUTION PHASE AND WORKFLOW DEFICITS

Content authored in the preparation phase is delivered to commissioning engineers, who perform all the commissioning tests in two main stages: *internal* and *acceptance testing*. Faults that may appear in the internal stage must be rectified before moving to the final stage of acceptance testing. Therefore, commissioning engineers follow a process for submitting detailed feedback in case of any deviation from the requirements, hence the use of checklists and protocols throughout the internal testing stage. Moreover, internal testing can occur parallel to production, while acceptance testing takes place after the ship is entirely constructed.

Commissioning engineers perform two main types of tests: *installation* and *behavioural tests*. An installation test is primarily concerned with verifying that a component is installed. Once confirmed that the indicated component is properly installed, the engineer monitors the behaviour of the component by placing the component in various test scenarios that can indicate if the component is functioning as intended. For example, a test scenario can verify if an alarm will be triggered when a temperature reading (recorded by a temperature sensor) reaches a specific threshold. The alarm example shows that commissioning tests should be executed in a meaningful sequence. For instance, if the alarm test scenario is executed before a faulty temperature sensor is inspected, the alarm will not trigger, giving a false indication that the alarm unit might be defective. Therefore, the commissioning test execution sequence is restricted by a set of technical dependencies. Commissioning engineers rely on experience to execute the tests in the right sequence that preserves such dependencies.

Due to the reliance on paper-based protocols and checklists, commissioning engineers write their remarks and feedback on site primarily using a pen and paper. In a test scenario requiring the commissioning engineer to read discrete values from multiple similar components, the engineer would locate every component and write down the values one after another. Such an approach is not optimal if the values are supposed to be recorded simultaneously. It is even more complicated if the values to be recorded are continuous and have been observed over a period of time. As mentioned earlier, if the BOM components are modelled in a digital twin environment, it will be possible to utilize IoT for automatic reporting of discrete and continuous readings obtained from components on the ship. Such an approach is only possible if the commissioning tests are digitally authored and maintained in the same digital twin environment.

As stated before, the commissioning test execution sequence is restricted by technical dependencies due to the complex behaviour of components in the ship. Even though experienced commissioning engineers can detect such dependencies and perform the commissioning tests accordingly, it is desirable to adopt a more systematic approach for determining the optimal execution sequence early in the preparation phase. Currently, it is not feasible to arrange over ten thousand commissioning tests in an optimal sequence due to the use of paper documents and the lack of a data model that enables visualization of such complex meshes of dependencies.

3. AUTOMATED AUTHORING OF COMMISSIONING CONTENT

Conventional methods used in authoring commissioning content result in the deficits discussed in chapter 2. The leading cause of such deficits is the lack of a holistic data model that would enable a digital-twin-based authoring workflow. This chapter proposes a data model for commissioning content and presents a digital authoring system developed based on the proposed model. The digital authoring system utilizes a template-based authoring workflow that allows commissioning planners to automatically generate commissioning tests with low effort.

3.1 DATA MODEL FOR MARITIME COMMISSIONING CONTENT

Since the commissioning process revolves around validating the technical specifications of physical components in the ship, it is clear that the central entity in the data model for commissioning should be the *component*. To realize a digital twin infrastructure, physical components in the ship must have a digital representation that assists users with retrieving and processing relevant data for commissioning preparation. Basic component details can be retrieved from the BOM and 3D models, while technical information, e.g., dependencies (between components) and technical parameters, can be obtained from technical lists and data sheets. The developed data model supports storing component technical information in separate entities such as *technical parameters* and *dependencies*, see Figure 2.

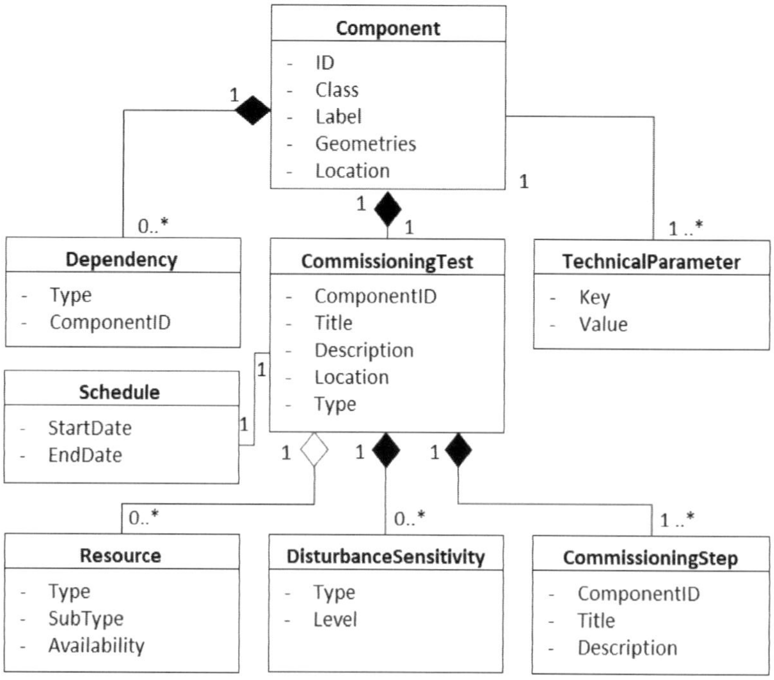

Figure 2: Simplified maritime commissioning data model.

Commissioning tests were modelled based on analyses of the commissioning preparation phase. The *commissioning test* model contains basic information about which component to inspect, where and how to perform the test, and the type of test to be performed. Moreover, organizational information relevant to production and commissioning, such as *schedules*, *resources*, and *disturbance sensitivities*, have dedicated independent models, as shown in Figure 2. The analyses disclosed that a single commissioning test incorporates several inspections to be performed in a specific sequence. Therefore, an additional model, referred to as *commissioning steps,* was developed to provide commissioning planners with the ability to create several inspection steps within each commissioning test.

3.2 TEMPLATE-BASED AUTHORING WORKFLOW

The data model shown in Figure 2 achieves a digital representation of the ship's physical components and their technical characteristics. In other words, a physical component evolves from being a textual item in the BOM to being a *smart component* aware of its role in the commissioning process. This means that commissioning planners can request information from the component directly. For example, Figure 3 shows a screenshot of the authoring system where a temperature sensor is selected from a list of components. The sensor contains attributes such as THRESHOLD_TEMP_MIN and THRESHOLD_TEMP_MAX, which are technical attributes relevant to the commissioning process.

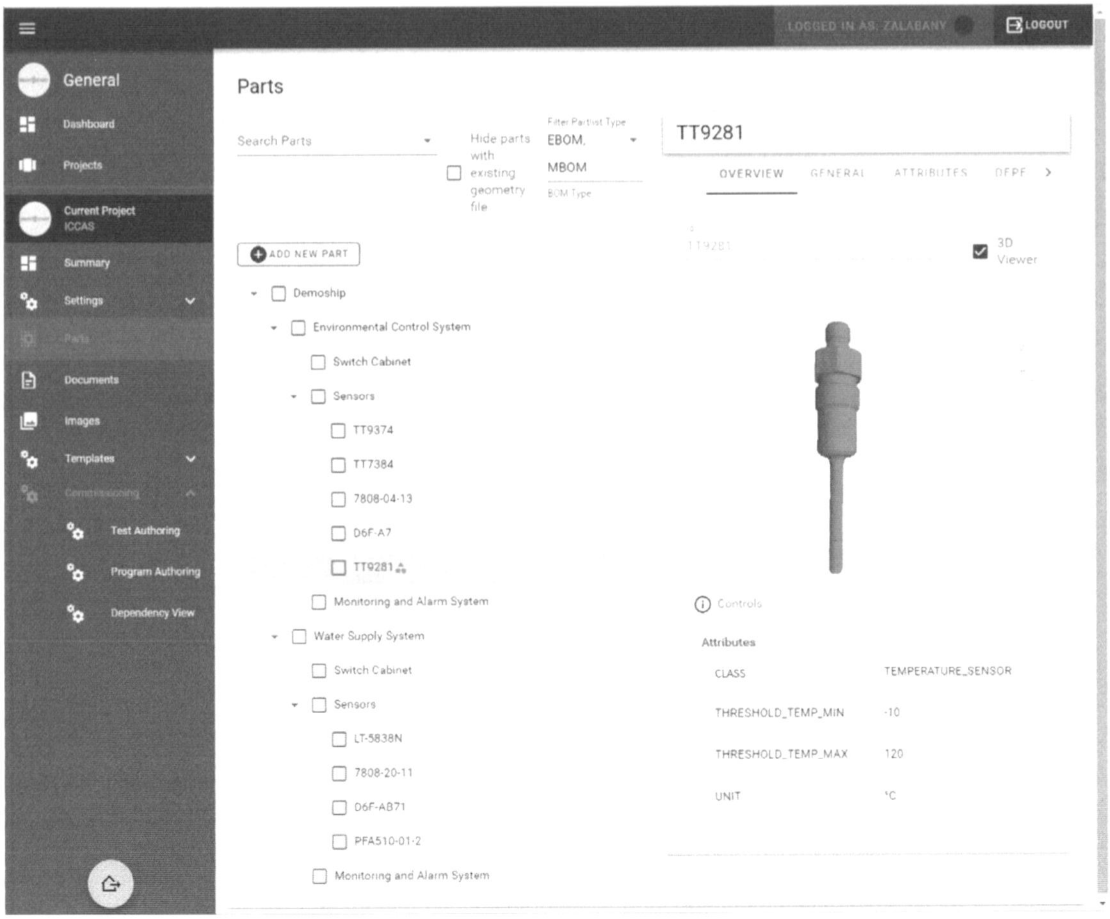

Figure 3: Temperature sensor as a smart component.

The sensor object also contains an attribute referred to as CLASS that carries the value TEMPERATURE_SENSOR. The idea of using the CLASS attribute is to group the components in classes, where every class contains the same set of technical attributes. For example, every component object that belongs to the class of TEMPERATURE_SENSOR has the attributes THRESHOLD_TEMP_MIN and THRESHOLD_TEMP_MAX. However, the values assigned to these attributes differ from one component to another.

Manual authoring of commissioning tests for every component in the ship is time-consuming. It is more efficient to adopt an approach for authoring a commissioning test blueprint once per component class and then reuse the data of the blueprint across all components of the same class to generate commissioning tests automatically. Commissioning test blueprints are referred to in this paper as *commissioning templates*. Commissioning templates can have the same data model as commissioning tests; however, commissioning templates do not contain concrete values since a commissioning template is only a blueprint that includes the rules for generating commissioning tests automatically.

Figure 4 shows an abstract example of using a commissioning template for the class TEMPERATURE_SENSOR to automatically generate commissioning tests. The template contains concrete information such as a description, a title, and a type. Additionally, the template includes parametrized information such as the location of the commissioning test to be generated and the technical attributes needed for executing the commissioning steps. The commissioning planner first creates the template on the digital authoring system, and then the system can search the BOM for all components that contain the class attribute: TEMPERATURE_SENSOR.

In Figure 4, it is shown that the system matches the commissioning template with three components that belong to the class of TEMPERATURE_SENSOR. Consequently, the system will generate three commissioning tests and replace parametrized information in the template with concrete values obtained from each component object. For example, when the system generates a commissioning test for component TT9281, it replaces the attribute THRESHOLD_TEMP_MIN in the first commissioning step by the value -10.

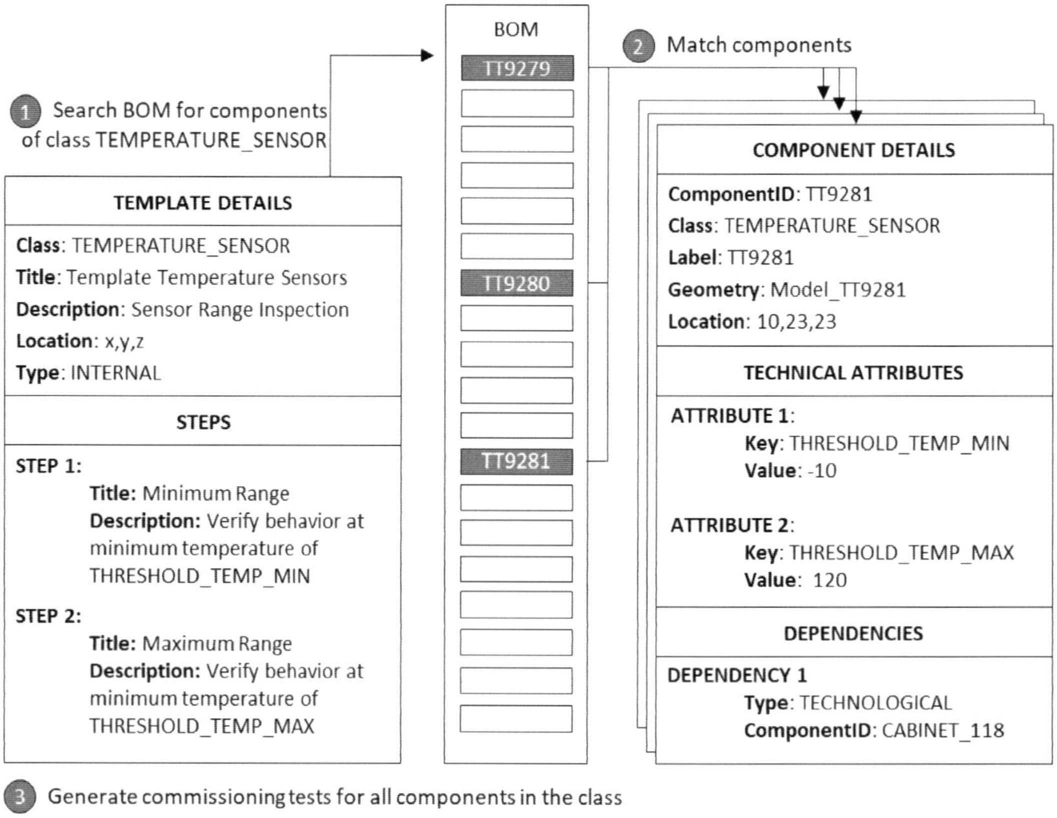

Figure 4: Template-based authoring.

The template-based authoring workflow also allows commissioning planners to create *resource sub-templates* that define the types of resources needed to perform a commissioning test. When authoring a commissioning template for class TEMPERATURE_SENSOR, commissioning planners can indicate that the commissioning engineer would need a tool of type SET_POINT_GENERATOR to perform the test. When the authoring system generates concrete tests using the commissioning template, it searches a resource pool for tools of type SET_POINT_GENERATOR that will be available at the time of test execution and reserves the tool for the commissioning engineer.

Moreover, to resolve the interferences between commissioning and production activities, commissioning templates contain a dedicated sub-template for *disturbance sensitivities*. For example, it is not possible to perform a commissioning test of type TEMPERATURE_SENSOR when there is a welding activity near the location of the test. Therefore, when authoring the commissioning template of class TEMPERATURE_SENSOR, commissioning planners can mark the commissioning template as sensitive to welding. When scheduling the generated commissioning tests, it will be visible which tests are sensitive to welding so that no interferences would occur between production and commissioning. Figure 5 shows a screenshot from the digital authoring system where a commissioning template for the class of TEMPERATURE_SENSOR is being created.

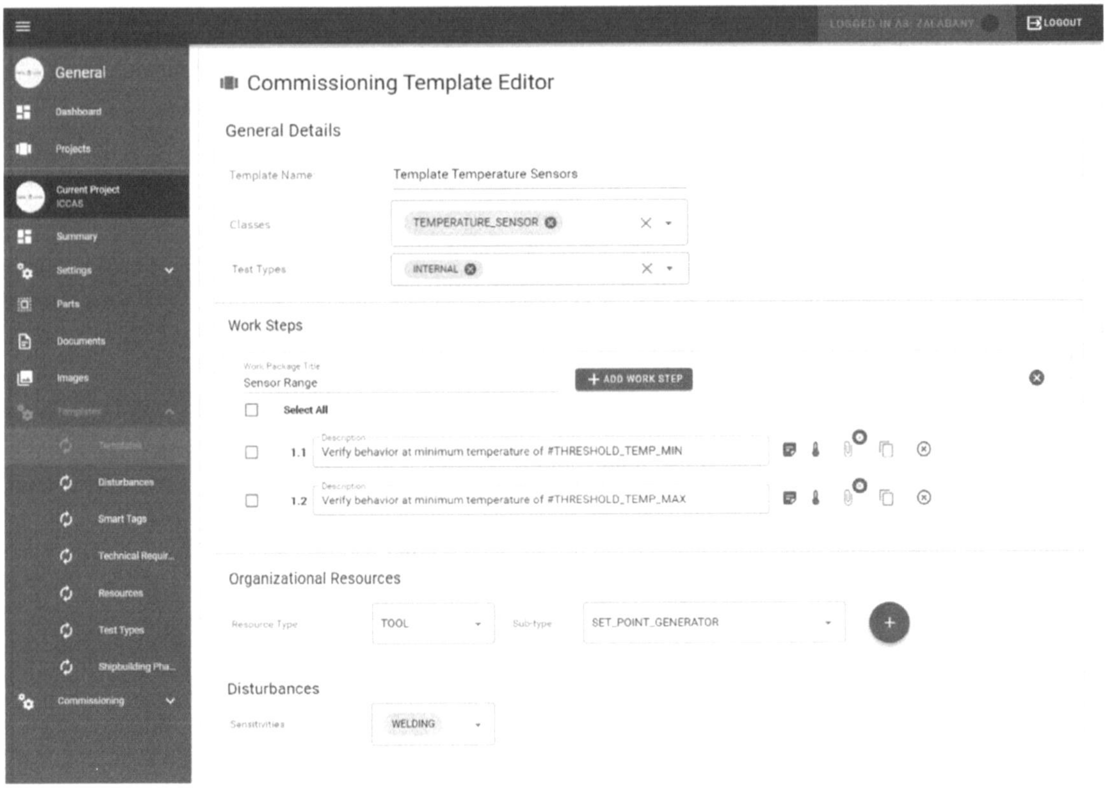

Figure 5: Commissioning template creation in the digital authoring system.

3.3 AUTHORING AND VISUALIZATION OF DEPENDENCIES BETWEEN COMPONENTS

Maintaining an overview of dependent components in the ship is unachievable using conventional authoring methods. Therefore, the data model introduced in Figure 2 takes account of this problem and includes an entity for storing information about component dependencies. In addition, the digital authoring system offers topological visualization for components in the BOM. This visualization scheme allows commissioning planners to create, edit, and view dependencies between all components in a user-friendly manner.

When a commissioning project is initialized, the components in the BOM will be unaware of their dependencies and will be visualized as floating nodes in the digital authoring system. Commissioning planners can create directional links between single component nodes to define a dependency. However, such an approach is still not efficient due to the high number of components in the BOM. Therefore, a mechanism was developed to group component nodes into classes allowing the user to create dependency links between classes of components instead of single component nodes, to save time and effort.

For example, all sensor components in the ship that can trigger an alarm are dependent on switch cabinets. So to create a dependency link between all sensor components and switch cabinets, it would be more efficient to link the component classes TEMPERATURE_SENSOR, PRESSURE_SENSOR, and FLOW_SENSOR directly to the components SWITCH_CABINET_1 and SWITCH_CABINET_2 as in Figure 6.

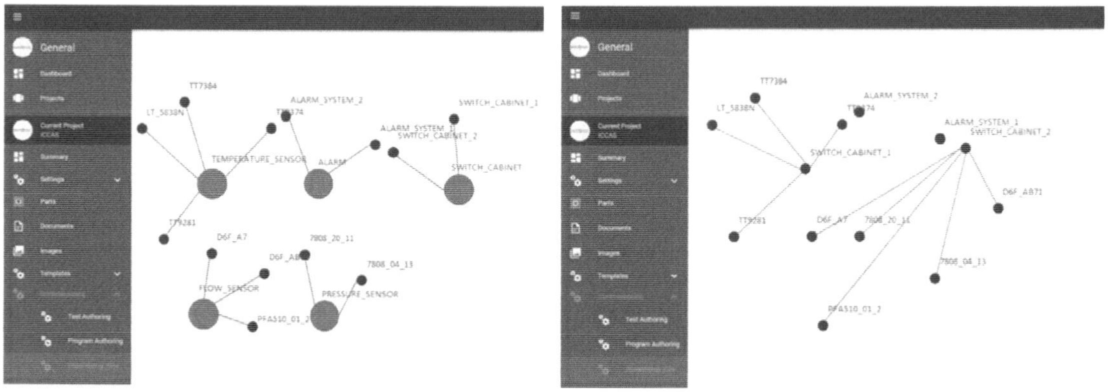

(a) Components grouped in classes (b) Dependencies

Figure 6: Topological visualization of components and their dependencies.

Figure 6 (a) shows all the component nodes (small nodes) and their respective class nodes (large nodes). When the user creates a link from a large node to a small node, the system understands that every component that belongs to the large node (all sensors) depends on a single small node (the switch cabinet). Accordingly, the system creates dependency links from every sensor component to the switch cabinet components, see Figure 6 (b).

4. AUTOMATIC GENERATION OF COMPONENT CLASSES USING MACHINE LEARNING

The digital authoring system presented in Chapter 3 utilizes a workflow that requires ship components to be arranged in classes. Therefore, a class attribute is added to the component data model, as shown in Figure 2. However, due to the large number of ship components, it is not practical to manually assign a class for each component in the BOM. Accordingly, this chapter introduces an approach that relies on image analysis and machine learning algorithms for component clustering.

4.1 OVERVIEW OF THE CLUSTERING APPROACH

Even though components in the BOM are arranged in assembly groups, clustering components based on their technical behaviour in the ship is more beneficial for authoring commissioning tests. To achieve automatic clustering of components, the most intuitive approach is to adopt a naming scheme that identifies components with the same technical behaviour. However, a consistent naming scheme is difficult to achieve since the components are supplied to the shipyard by various sub-contractors that follow different guidelines for naming. Moreover, changing the name of every component in the assembly structure induces high effort and time. Therefore, clustering components solely based on their names is unreliable, and an additional clustering method would be more advantageous.

The clustering method presented in this paper utilizes 3D models of components in the BOM to extract visual features of components and place components with identical or similar visual features in the same class. Such a method helps recognize identical components that exist in different assembly groups. For example, identical temperature sensors in different locations in the ship will have the same 3D model. Still, their names in the BOM will differ according to their position in the assembly hierarchy.

Due to the complexity and the large size of 3D models used in ship design, it is not feasible to work directly on the 3D model to perform component clustering. A typical solution to such a problem is to apply view-based methods for 3D object recognition [7, 8, 9]. View-based object recognition methods work on two-dimensional images constructed from the 3D object to be recognized [9]. The solution presented in the following sections consists of three main steps:

- 2D image construction from 3D models
- Dataset preparation and feature extraction
- Unsupervised learning for component clustering

4.2 2D IMAGE CONSTRUCTION FROM 3D MODELS

In order to apply view-based object recognition, 2D images must be constructed from 3D models in the ship. Therefore, an application was developed that creates minimum bounding boxes for each 3D object and aligns the object so that the longest side of the bounding box is parallel to the global X-axis, and the second longest side is parallel to the global Y-axis. When the object is properly aligned, images are automatically captured from different angles. Figure 7 (a) shows a 3D model of a pressure sensor for which a minimum bounding box is created and aligned. Figure 7 (b) shows the result of image capturing for different ship components.

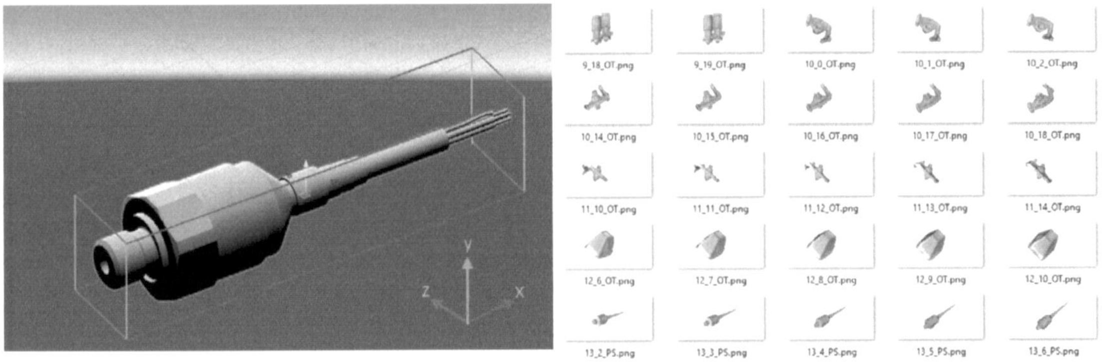

(a) Pressure sensor minimum bounding box (b) 2D images generated from 3D objects in the BOM

Figure 7: 2D image construction from 3D models.

4.3 DATASET PREPARATION AND FEATURE EXTRACTION

The second step of the solution is to prepare the images generated in step 1 by applying pixel value normalization and size reduction to ensure that the images are uniform in terms of colour, size, and pixel value range. When the images are prepared, the stage of feature extraction can start.

Because of the large number of component images, it is important to apply an efficient feature extraction method such as the Histogram of Oriented Gradients (HOG) developed by Dalal and Triggs [10]. The HOG technique is widely used in various computer vision applications due to its high speed and efficient CPU utilization. Briefly summarized, the HOG technique works by computing first-order gradients of the image to capture texture information and contour, which provides resilience to illumination changes. Then the image is divided into regions or cells, and the local gradient is calculated per cell. It is possible that the effects of illumination are still present, and therefore, to counter such a problem, the cells are arranged in groups, and the gradient values of each group of cells are normalized. Finally, normalized groups are concatenated into one vector forming the HOG *feature vector*. The feature vector can then be used later to apply machine learning algorithms. Figure 8 shows an image of a sensor before and after applying the HOG technique to extract features.

(a) Input image of pressure sensor (b) Histogram of Oriented Gradients image

Figure 8: Image of pressure sensor before and after applying HOG.

4.4 UNSUPERVISED LEARNING FOR COMPONENT CLUSTERING

The last step of the provided solution is to cluster all images by applying unsupervised clustering to the HOG representations of the images. Unsupervised clustering is useful when there is a lack of 3D models that can form a training set. Moreover, unsupervised clustering is sufficient in this case since the clustering happens once when a commissioning project is initialized on the digital authoring system. The idea is to provide the user with different clusters of components and ask the user to provide a meaningful name for each cluster. The clustering algorithm employed in this stage is DBSCAN [11]. The primary reason for using DBSCAN, instead of other algorithms such as K-Means, is that DBSCAN does not require specifying the number of clusters.

A test scenario has been prepared where arbitrary 3D models of different ship components are collected and used as input data for clustering—for example, multiple models of sensors and ship engine parts such as cylinder heads and filters. Then the previously discussed three steps are applied to the models, and a plot for the clustering algorithm result is generated as in Figure 9.

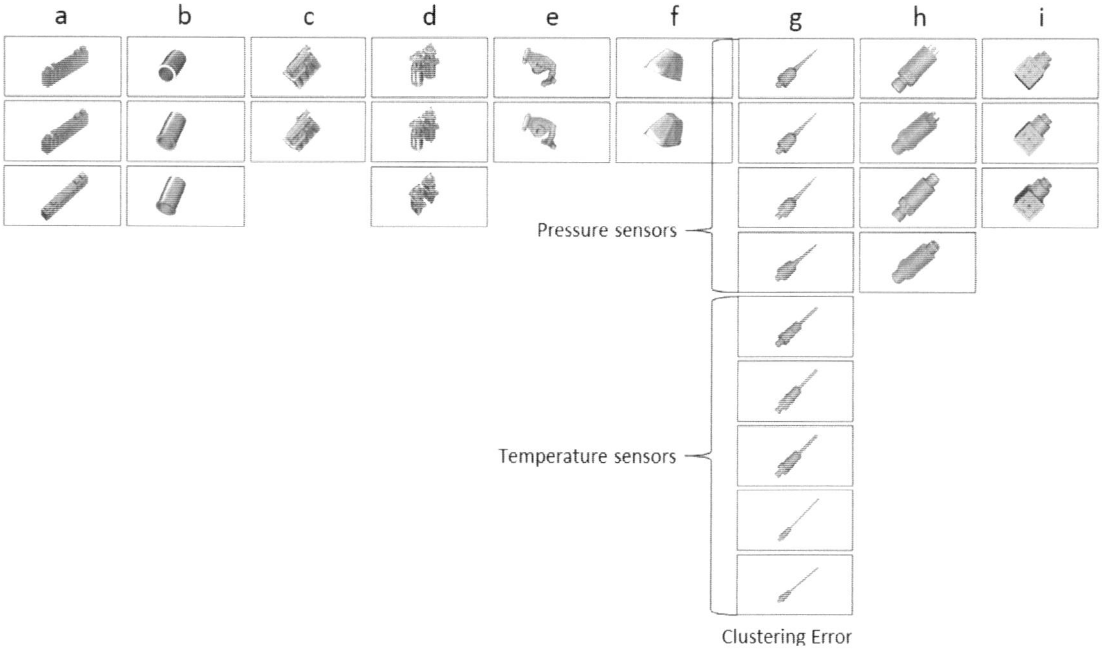

Figure 9: Component clusters generated by the DBSCAN algorithm.

Figure 9 shows that the algorithm has detected 9 clusters labelled alphabetically; the user can then assign a name for each cluster. The algorithm succeeds in detecting identical and similar objects even when the components have different orientations. However, the figure shows that cluster g contains two types of sensors (temperature and pressure sensors), a clustering error due to visual similarities between both sensor types. To overcome such an error and improve the clustering performance, the feature vector may be extended in future research, e.g., by incorporating additional data from the BOM, such as component dimensions.

5. CONCLUSIONS

To enhance the process of authoring maritime commissioning content, this paper presents a software application that allows commissioning planners to generate the required content automatically and with low effort. The developed application is built on a digital twin infrastructure incorporating a comprehensive data model for maritime commissioning information. The paper introduces the data model and demonstrates how utilizing such a model can link physical components of the ship to a digital twin that can store and process information relevant to the commissioning process, which makes automatic generation of commissioning content possible.

The application works in two phases: preparation and automatic generation. In the preparation phase, ship components must be clustered so that components with the same behaviour and technical parameters would belong to the same *class*. Commissioning planners can then create *commissioning templates* that define parametrized test specifications per class of component. In the automatic generation phase, the application iterates through all components in the ship and applies respective commissioning templates to automatically generate test specifications for each component according to the rules defined in the template. By using the application, commissioning planners avoid creating test specifications for each individual component in the ship from scratch, which saves time and effort.

Due to the high number of ship components, manual clustering of components is not feasible; therefore, the paper presents an approach that uses machine learning for the automatic clustering of ship components. The idea is to scan 3D models of components in the ship, apply view-based object recognition to each component, and extract feature vectors. Then, components can be automatically clustered by applying the unsupervised clustering algorithm (DBSCAN) to the feature vectors. The proposed automatic clustering approach is still developing to employ a higher dimensional feature space that includes additional features obtained from the bill of materials. Using additional features can improve the clustering accuracy.

6. ACKNOWLEDGEMENTS

The presented work is funded by the German Federal Ministry of Economic Affairs and Climate Action (Bundesministerium für Wirtschaft und Klimaschutz – BMWK) due to a decision of the German Bundestag.

7. REFERENCES

1. WEBER, K. H., 'Inbetriebnahme Verfahrenstechnischer Anlagen', 5th ed., Springer, Berlin / Heidelberg, 2019.

2. EBERTSHÄUSER, S. et al., 'BIM-basierte Integrale Planung: Gemeinsamer Schlussbericht', *Karlsruher Institut für Technologie (KIT)*, 2020.

3. HALATA, P. S., 'Augmented-Reality-gestützte Informationsbereitstellung für die Unikatproduktion', *Dissertation, Institut für Produktionsmanagement und -technik, Technische Universität Hamburg*, 2018.

4. FRIEDEWALD, A. et al., 'Die Produktivitätswirkung von Augmented Reality in der Unikatfertigung', *Megatrend Digitalisierung - Potenziale der Arbeits- und Betriebsorganisation, GITO mbH Verlag*, 2016, 141–162.

5. MELUZOV, N. et al., 'Holistic Maintenance Support with a Digital Assistance System', *Bertram, V. (ed.), 19th International Conference on Computer and IT Applications in the Maritime Industries (COMPIT '20)*, 2020, 48–62.

6. ELZALABANY, A. et al., 'Digitalization of Maritime Commissioning: A System for Automatic Content Creation and Augmented Reality-based Commissioning Execution', *Bertram, V. (ed.), 21st International Conference on Computer and IT Applications in the Maritime Industries (COMPIT '22)*, 2022, 323–335.

7. SHI, B. et al., 'DeepPano: Deep Panoramic Representation for 3-D Shape Recognition', *IEEE Signal Processing Letters*, 22 (12), 2015, 2339–2343.

8. SINHA, A. et al., 'Deep Learning 3D Shape Surfaces Using Geometry Images', *Leibe, B. et al. (eds.), Computer Vision – ECCV 2016, Springer International Publishing, Cham*, 2016, 223–240.

9. SU, H. et al., 'Multi-view Convolutional Neural Networks for 3D Shape Recognition', *Bajcsy, R. et al. (eds.), 2015 IEEE International Conference on Computer Vision [ICCV 2015], IEEE Computer Society, Los Alamitos, CA*, 2015, 945–953.

10. DALAL, N., TRIGGS, B., 'Histograms of Oriented Gradients for Human Detection', *Schmid, C. et al. (eds.), 2005 IEEE Computer Society Conference on Computer Vision and Pattern Recognition (CVPR'05), IEEE Computer Society, Los Alamitos, CA*, 2005, 886-893.

11. ESTER, M. et al., 'A Density-Based Algorithm for Discovering Clusters in Large Spatial Databases with Noise', *Simoudis, E. et al. (eds.), Proceedings of the Second International Conference on Knowledge Discovery and Data Mining, AAAI Press, Menlo Park, CA,* 1996, 226–231.

8. AUTHORS BIOGRAPHY

Ahmed Elzalabany holds the current position of Research Associate and doctoral candidate at the Institute of Production Management and Technology at Hamburg University of Technology in Germany. His research focus is on applying IoT and machine learning technologies to one-of-a-kind production.

Vincent Settler holds the current position of Research Associate at the Institute of Production Management and Technology at Hamburg University of Technology in Germany. His research focus is on augmented reality as well as 3D data acquisition and processing.

Tim Jansen holds the current position of Research Associate and doctoral candidate at the Institute of Production Management and Technology at Hamburg University of Technology in Germany. His research focus is on developing situation-aware digital assistance systems for one-of-a-kind production.

Robert Rost holds the current position of Chief Engineer at the Institute of Production Management and Technology at Hamburg University of Technology in Germany. His research focus is on applying augmented reality and machine learning technologies to one-of-a-kind production.

Hermann Lödding holds the current position of Professor of Production Management and Director of the Institute of Production Management and Technology at Hamburg University of Technology in Germany. His research focus is on production planning and control.

SIMULATION TECHNIQUES IN THE DEVELOPMENT OF THE QUIESCENT PERIOD PREDICTION (QPP) FLIGHT DECK MOTION FORECASTING TOOL

B Ferrier, Hoffman Engineering, USA
R J Taylor, Royal Navy, UK
M R Belmont, J Christmas, University of Exeter, UK
L Fedrick, MOD, UK

SUMMARY

Quiescent Period Prediction (QPP) purpose is to provide ship motion conditions and motion predictions with sufficient forecasted time to launch, recover, and complete other motion sensitive tasks regardless of the seaway. QPP operates within a federated architecture system containing measurement instruments such as a Wave Radar system used to map remote sea surfaces several hundred meters in advance of the ship. This enables the computation of future wave forces acting on the virtual vessel. Quiescent Period Prediction (QPP) system achieves this from the measured sea surface, a short-term deterministic wave model can be constructed allowing the wave system to be propagated to the ship's location. The ability to fully define a ship's motion as a function of a recovery device, greatly enhances the overall system operational capability whilst fundamentally reducing the inherent risks associated with platform manipulation in high sea conditions. This article describes launch and recovery simulation model and the corresponding verification and validation of the QPP data program testing the accuracy of the simulation modeling. Verification information is discussed regarding the sensor operation and interfaces collected during this trial; a QPP simulation design specification is being produced based on High Level Architecture (HLA) technics. From initial simulation and trial results, QPP has been shown to offer the ability for operators to identify the most favourable deck motion periods to recover an approaching vehicle. The next challenge being studied from the data results is a notional navigational guidance system tool and visual landing aid to be trialled by simulation.

1. INTRODUCTION

The Simulation Based Acquisition (SBA) Group is the technical authority across the UK Ministry of Defence's Defence Equipment & Support Organisation, Maritime Combat Systems for Maritime Simulation development. The Simulation Based Acquisition Programme on behalf of the Royal Navy (RN) is administrative set up to support MOD use of simulation in its role as intelligent customer. To deliver its remit, the programme relies on a corporate strategy of engineering simulation technology ownership underpinned by external sourcing of world class subject matter expertise. An overarching technical infrastructure and discrete technology components are readily available for RN projects to use; the lead contractor acting as a co-ordinating prime and the primary representative for other SME supplier organisations.

SBA has been used effectively and extensively by MOD since 2000 to support the management and reduction of key risks across a wide range of projects. MOD policy places clear emphasis on managing and reducing project risks particularly during the earlier stages to minimise the likelihood of cost and time overruns. Consideration of the use of simulation is often the preferred alternative to traditional approaches, such as sea trials or scale model testing, to predict and verify the safety and performance of new systems. Simulation technology is being used to contribute to safety case assessment, performance prediction and project assurance during the major acquisition decision milestones.

Since 2000, SBA has been responsible for providing the UK technical support package to NATO on Virtual Ships. Future simulation project arrangements under consideration by various groups of nations that can leverage the outputs above include: -

- Extending High Sea State Operations using Quiescent Period Prediction.
- Launching and Recovering Small Boats and UAVs in Higher Sea States,

Both scenarios are discussed later in this article.

2. SIMULATION AHEAD OF ACQUISITION OR OPERATION

One of the most challenging tasks in the deployment of ship-based manned and unmanned VTOL aviation is the launch and recovery and deck handling operational performance on smaller ship platforms. The same may be said of ship Well-deck-based surface boats, like Rhib and LCACs. Awareness and prediction of the ship's motion characteristics, position and movement in the ocean is key to ship-based air operations by targeting the right time to deploy and recover. The ships above waterline design coupled by the propulsion systems, affects ship's motion and ship motion induced air wake behaviour.

Advances in computer science, telecommunications and naval engineering analysis, have enabled the 'virtual prototyping' of warship systems and processes with high-fidelity performance prediction, visualization, interference detection, automated generation of production data and virtual manufacturing.

Evolving simulation techniques, begun during the NATO/PfP Interoperability and Re-Use Study (NIREUS), are offering a substantial contribution to meeting desired risk, cost and time reductions within ship programs **[1]**. SBA processes and technologies are further improving engineering integrity, whole system integration and super-system optimization. Simulation is permitting testing and evaluation of designs in advance of the production of the real system by integrating:

- 3D walk-through;
- Equipment interference checking and operating limits;
- Human factors aspects such as fields of view and task performance;
- Digital demonstrators and mock-ups;
- Assessment of design alternatives;
- Crew traffic flow studies;
- Deployment & recovery of equipment;
- Manoeuvring, sea keeping and hydrodynamics;
- Performance and interoperability of sensors, weapons and command & control systems
- Survivability (susceptibility, vulnerability and recoverability);
- Simulation of the performance of mechanical systems.

For use in design validation and assessing safety, typical maritime simulations may require the modeling of:

- ship motion;
- hydrodynamic and aerodynamic effects;
- system dynamic interfaces;
- propulsion and helm control systems;
- mechanical systems used in vehicle deployment and recovery, ship-to-ship transfer of solids and liquids, etc;
- sensors such as optical, radar, sonar.

3. QUIESCENT PERIOD PREDICTION (QPP)

A key user requirement of any naval aviation or small surface embarked boat is the ability to launch and recover the vehicle in high sea states. One of the means to improve the safety of high sea state recoveries is to predict when such quiescent periods of ship motion will occur and their duration. However, proving this capability at sea carries a significant risk. A pragmatic alternative is to simulate the recovery process, and to this end MOD has developed a modular simulation of vehicle recoveries.

In high sea states it is known that there are periods of time when a ship's motion alternates between large and more modest excursions. One means of improving the safety of high sea state recovery is to predict when such quiescent periods of ship motion will occur and their duration.

Quiescent Period Prediction (QPP) is a class of technologies that use remote sea surface profile measurements to predict the future wave forces acting upon the vessel. Such technologies have time horizons measured in minutes and offer predictions of the actual motions of the vessel. This type of technology is designed primarily to identify so called "Quiescent Periods" in the next few minutes, which make it possible to safely execute an operation when the mean prevailing conditions exceed the traditionally considered safe limits

A Quiescent Period Prediction (QPP) system achieves this by using a wave sensor system to measure the sea surface several hundred meters in advance of the ship **[2]**. From the measured sea surface, a short-term deterministic wave model can be constructed allowing the wave system to be propagated to the ship's location. The ship's response to the wave spectrum is calculated to determine the level of quiescence at the time of encounter (figure 1).

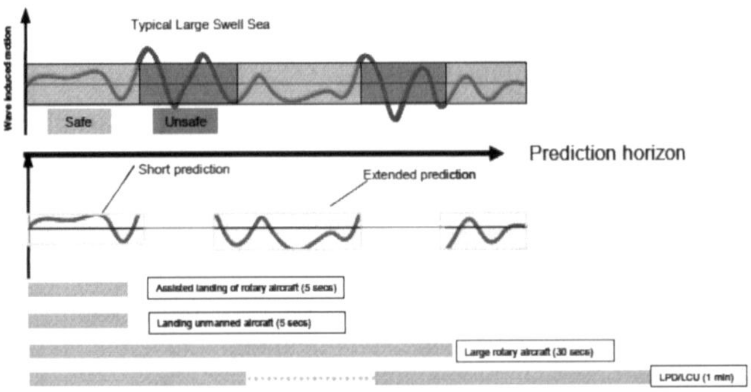

Figure 1: Relationship between Components of Wave induced motion and prediction

This is a continuous process and can be self-monitored. The components that make up a QPP system are shown in figure 2.

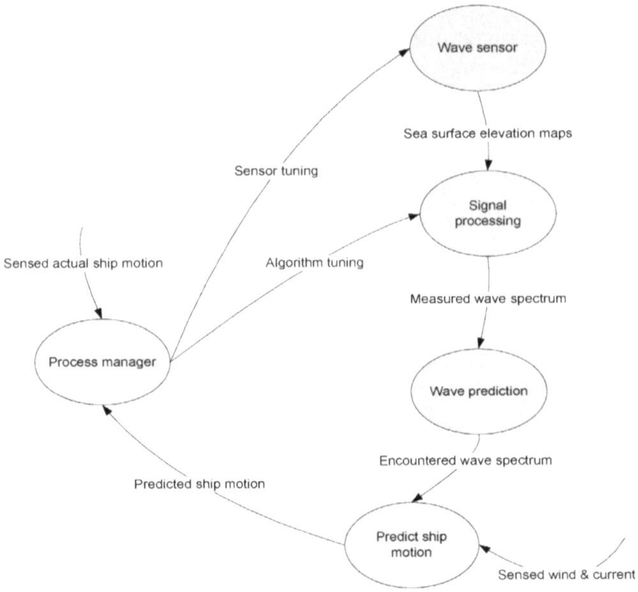

Figure 2: Components of a QPP system

A key part of the system is the wave sensor device, which can be composed of Light Detection and Ranging (LIDAR) or Radio Detection and Ranging (RADAR). A LIDAR system will require a mechanism to scan a laser beam over the sea surface, whereas a radar system can cover a larger area with a single radar pulse. The performance of a candidate sensor will need to be evaluated in a trial, and if proven, the data collected can then be used to validate a simulation of a complete QPP system.

QPP DESIGN BY SIMULATION

The QPP simulation will be an HLA federation. The design will be developed in accordance with the draft NATO STANAG 4684 and associated draft Allied Naval Engineering Publication ANEP 84 for Virtual Ships. The first part of the design process includes the creation of a concept model. The purpose of the QPP is to provide ship motion conditions and motion predictions with sufficient time to launch, recover, and complete other motion sensitive tasks regardless of the seaway. The objective is to expand ship operating deck limits to approximately Sea State 6+. Based on lessons learned from the use of the QPP it is conceivable that, once integrated, shipboard communities could operate in much higher sea conditions recovering the air/sea vehicles with elevated confidence and reliability. The computational flow diagram corresponding to QPP system components is displayed in Figure 3.

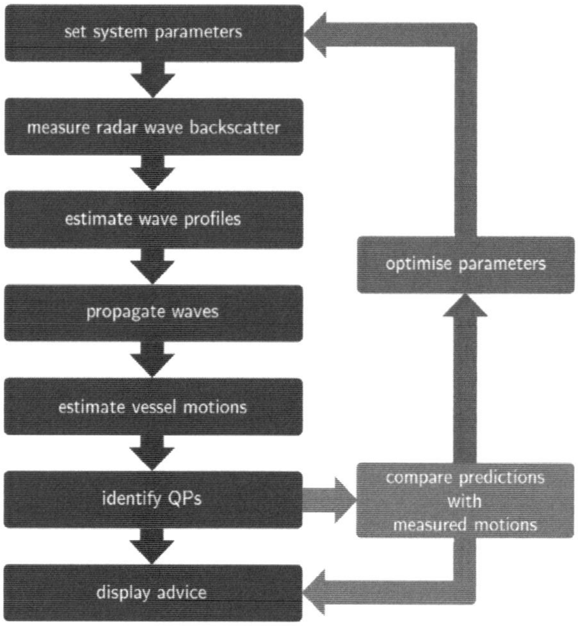

Figure 3: QPP-type Computational Flow Diagram

The simulation components can be described as those representing:

- The real world or ground truth
- Sensors
- QPP processes
- External interface

In any simulation of this nature, it is important to recognise that simulation of the real world must be different to the way that real world is sensed. Thus, simulation of the real world or so-called "ground truth" should be undertaken using the highest fidelity, practicable, techniques. Each of these systems and related air vehicle operational components are federates, along with the measured environment, which is represented as federates, too. In an initial stage, the measured federates would be simulated. A typical High-Level Architecture (HLA) design defines 6 plug and play federates (Figure 4), connected via the HLA Run-time Infrastructure (RTI) software. The structure is applicable to either a crewed or uncrewed scenario.

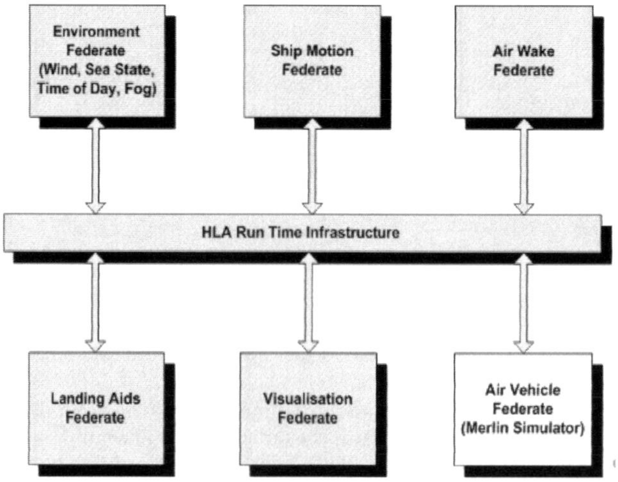

Figure 4: Federation Architecture

The QPP software package is designed to be integrated into a full motion simulator (in the crewed case) and a Tactical Control Station (TCS) (in the UAV case). This is a means to estimate system effectiveness as a function of simulated ship motion, visual environment, and synthetic operational systems, and to compare the results to related recorded at-sea data (Ferrier et al. 2019).

QPP systems require: a remote wave measuring system, typically a wave profiling RADAR, a powerful intelligent decision-making computer system, and advice aids for human operators, or contained within the flight control system of unmanned systems. They typically also include a Landing Period Designator (LPD) sub-system to provide an emergency last minute wave-off capability.

Application Program Objective
The long horizon forecast program objective is to produce a design for a sea going QPP system based around an intelligent decision-making artificial intelligence wrapper (the AI shell) that exploits data from existing proven hardware and can learn from its previous activity. It applies verified short horizon prediction LPD indices in mapping operationally acceptable seaway regions. Short horizon prediction LPD is described later in the paper.

The QPP system delivers three main benefits:

i. It provides a speed-polar representation of the seaway around the vessel permitting ship's crew to modify ship's heading and speed to ensure the best ship's conditions to complete motion sensitive tasks at hand.
ii. It identifies the safest, most quiescent, period in the vessel's wave driven motion over the next minute which corresponds to the safest instant to execute launch/recovery.
iii. The QPP system will incorporate an operator display of the existing confidence level in the predictions. This will allow the operator to appreciate when the system is at its most reliable and avoids the problem of "it works sometimes but we are not sure when".

The QPP system is composed of an AI Shell and LPD measured motion information that accepts data from existing deck interface measurement devices, validated in sea trials and/or deployment, thus reducing the need for extensive certification and testing formalities.

In almost all fully integrated technology transition cases, the integration to various independent components presents a high-risk factor. The QPP Phase 1 work used existing data recordings and representative ship transfer functions along with existing measurement modular devices. The technical risk will be mitigated through simulation and verification lab-based operation Qualification and Systems Integration Tests as well as a shore based QPP device test build-up before proceeding to underway qualification. Central in QPP instrumental components involves the integration and application of wave measured radar system.

4. HELICOPTER (UAV) – SHIP INTEFACE

Experimental / Simulated Dynamic Interface
It is important to develop options to reduce the cost and cycle time associated with the test and evaluation process, to enhance the productivity of flight test team members, and to improve the safety of flight test operations, especially in adverse environments.

Simulated dynamic interface strategies have been developed over several simulation projects. The earliest High-Level Architecture (HLA) simulation was called NIREUS (NATO Interoperability and RE-Use Study) which was followed by the SAIF (Ship/Air Interface Framework) programs. The purpose of these complex programs is to use the HLA standards to integrate air vehicle simulations, ship simulations and environment models to aid assessment of the dynamic interface for a range of helicopter / ship and UAV / ship combinations. The initial phase of the SAIF program focused -on SHOL prediction where operations may involve recovery in other environmental factors including, high levels of turbulence to new naval vessels. The primary future SAIF objective is to use the simulation to minimize the time and cost required for first of class sea trials.

Participating in Envelope Expansion /SHOL trials provides the means to verify/validate the simulation modules. It also ensures the forecasting algorithms are accurate, particularly in face of new non-traditional ship profiles/platforms. A brief example, follows. Recall, the objective is to recover the aircraft aboard a moving vessel within reasonable safety margins regardless of the seaway.

DYNAMIC INTERFACE
Dynamic Interface (DI) is defined as the study of the relationships between an air vehicle and a moving platform. It is performed to reduce risks and maximize operational flexibility **[3]**. Globally, DI is concerned with the effects that one (1) free body has in respect to another free body. Historically, this means the effects that a ship may have on a recovering or launching of air vehicles. However, more recent studies have concluded that the same principles apply to other motion related activities, such as, the boarding of Landing Crafts and/or the Landing Craft Air Cushion (LCAC) vessels into the wells of Amphibious Warfare Ships, the docking of submarines or the launching of unsophisticated missiles.

DI is divided into two broad categories: analytical and at-sea experimental analysis. The analytical provides for a mathematical analysis and solution to flight operations. The methods are not mutually exclusive. Neither method alone can produce a comprehensive and timely solution of the deck interface problem **[4].**

The traditional approach for DI is the experimental approach, which investigates operational launch and recovery of vehicles, engaging and disengaging of rotors, vertical replenishment and helicopter in-flight refuelling envelopes. However, the experimental approach primary focus is on launch and recovery envelope development and expansion. "Shipboard suitability testing" assesses the adequacy, effectiveness, and safety of shipboard aviation. Testing methodologies and procedures have been standardized by such departments at NAVAIR (Patuxent River).

Dynamic Interface UAV Envelope Expansion

DI testing of unmanned vehicles is not straight forward. The methodology of replacing piloted evaluations with operator estimates and the corresponding test criteria, is established prior to actual testing. However, it is often found that assumptions applied early in testing may need real-time adjustments to adapt to circumstances found during the test.

Naval air training and operating procedures standardization (NATOPS) or ship-helicopter operational limit (SHOL) limits are determined based on wind and ship motion data as assembled by conducting various launches, approaches, and recovery events. Central in the SHOL determination is the pilot evaluation. The unmanned configuration during testing normally is instrumented to provide, amongst other parameters, oleo compression and deflection, torque monitoring along with indications of precise weight on wheels or skids. The number of landings per condition becomes almost academic since there is no piloted variation or technique since the piloting is the same as originally programmed. The settled approach is to assess system performance using multiple launch and recovery cycles but only one recovery is required to justify an envelope expansion.

The behaviour of an air vehicle to the deck environment of a given ship platform, is the primary objective of the DI envelope expansion tests. In the piloted case, handling qualities, pilot workload, flight control positions and aircraft limitations are used to define the progress of shipboard launch and recovery envelope development (control margins, torque, rotor speed, power margins, excessive pilot workload, inadequate visual cues, inadequate clearance between aircraft and ship structure). Pilots assign ratings to both the approach and landing, and the takeoff and departure sequences for a specific wind over deck (WOD) condition, in accordance with guidelines set forth in the Deck Interface Pilot Effort Scale (DIPES). Data is evaluated after each flight and plotted on a polar plot. Data points are interpolated to fill in the missing data. The missing data is primarily due to weather conditions which prevents the ship from obtaining the various wind speeds.

Impacts of Deck motion on deck limits

The simulation of ship platforms and systems to assess their interfaces to other platforms and systems need to be made more available. Early use of trusted simulation can allow major ship projects to be de-risked during design to a greater degree than has been possible in the past. During ship design phases, the performance of platform operations and platform interactions such as helicopter recovery, replenishment at sea, vehicle docking, and vehicle deployment and recovery can now be assessed in all required sea and weather conditions **[6].** Establishing the accuracy of simulations developed for supporting design is vitally important perhaps more so than for simulations that have been developed specifically for training. Numerical inaccuracies in training systems can normally be identified early by experienced operators (not necessarily my experience, particularly with respect to ship air wake).

The calculated system limitations provide experimental DI with the necessary data to more effectively set testing strategy to probe the limiting conditions **[7].** Continuing the theme of ship's motion and having created a large library of data on the target ship, it is possible to derive tendencies of ship's characteristics to the most basic of parameters, such as, ship's speed. Figure 5 displays the effect that ship's speed has on roll motion (here assessed at 10 feet significant wave height).

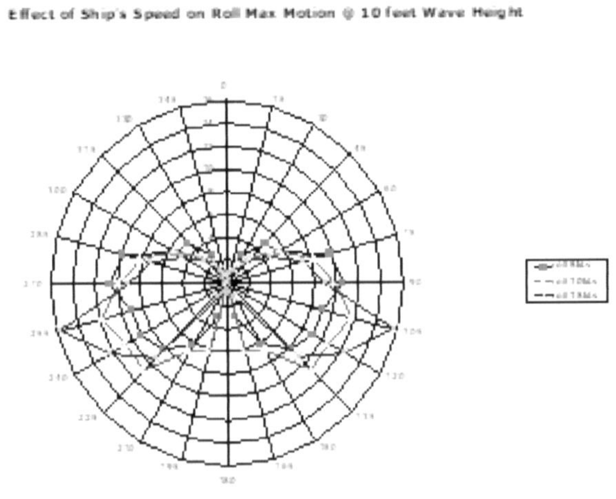

Figure 5: Effect Ship's speed on Roll Motion

Roll appears to improve slightly with increasing ship's speed in ahead seas through 90° relative wave angles. From midship to 135° relative wave angles, roll angle becomes larger with increasing ship's speed **[8].** It then falls away at angles greater than 150° to 180° relative wave angles. Studying pitch (see figure 6), the tendencies are opposite those of roll.

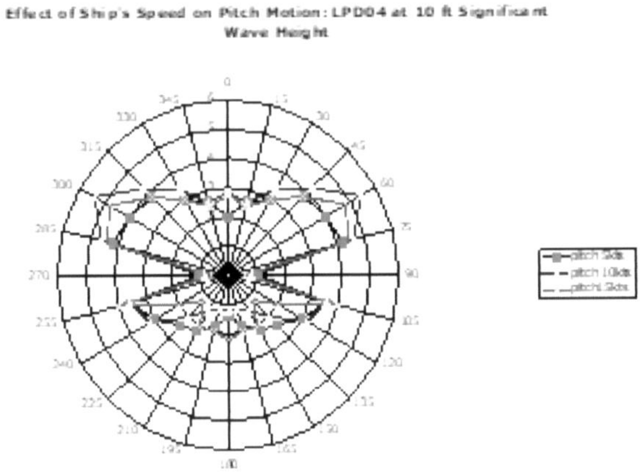

Figure 6: Effects of Ship's Speed on Pitch Motion

Much as expected, in the ahead seas condition pitch motion increases with increasing ship's speed. The following seas cases are more complicated, but overall follow the same theme.

Assessment was made concerning landing gear responses to encountered deck loads from several ship models. Identification of incidents of gear slide (coefficient of friction = 0) and gear dampener stroke are produced. This is done for each motion scenario. A sample analysis is given in Figure 7. The example was randomly taken from data file 10301511, which translated means, 10 knots ship's velocity, and 30 degrees relative wave angle 15 feet significant wave height and 11 seconds wave period. Calculated encountered axial forces are shown over the course of the time history. Fz (vertical forces) is the largest contributor to the resultant force vector (Fr). For this scenario, Fx (longitudinal force vector) is statistically slightly greater than those of Fy (lateral force vector). At the point where the coefficient of friction measures zero (or the index of slide is greater than one), an aircraft displacement is indicated. The sum of the force distribution studies produces a helicopter encountered force deck limit diagram (rondelle, speed polar or O'Reilly plot) **[9].**

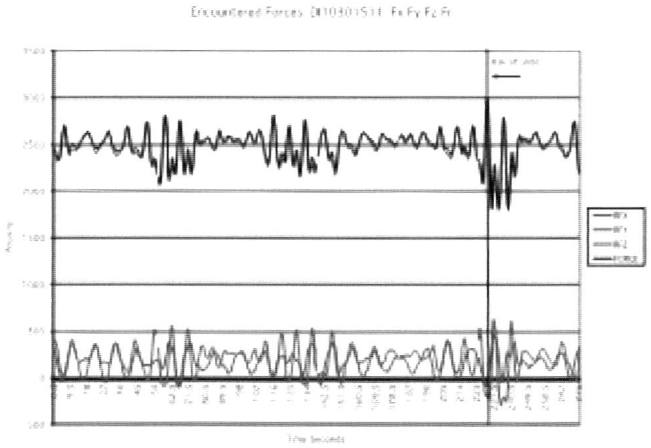

Figure 7: Landing Gear Encountered Deck Forces

As developed earlier for vehicle interface simulation, Figure 8 represents deck limits as a function of deck encountered forces, in this case onboard the LPD-4 running at 5 to 15 knots of ship's speed.

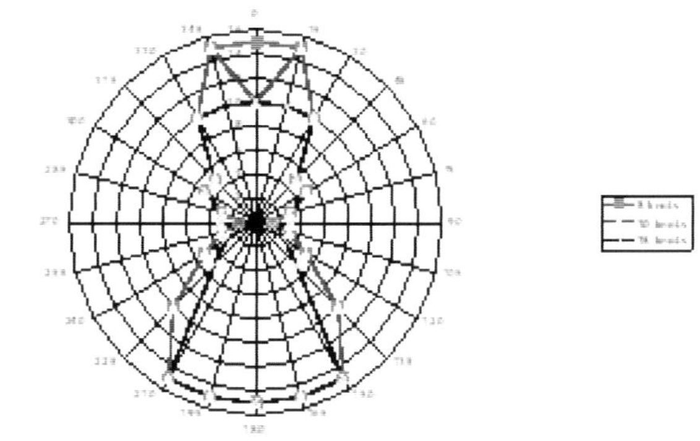

Figure 8: Encountered Deck Forces

Reading the diagram is like that for aircraft stability. Anything inside the hourglass figure is within limits where there no measurable risk of aircraft incident recorded. Anywhere outside the hourglass, an incident of landing gear faults was recorded and is, therefore, out of limit. As an example, at 60 degrees relative wave angle, seaway significant wave heights lower than 5 feet are within limits. Ahead seas show good stability between 330 – 360/0 – 30 degrees relative wave angles and in following seas 150 – 180 – 210 degrees of relative wave angles. Helicopter displacement risk appears to be focused on the beams. Areas without risk on the beams occur with less than 3 feet significant wave height (or sea state three). The most heavily constrained areas are formed from 60 – 120 degrees of relative wave angles and from 240 – 300 degrees of relative wave angles.

The effects of ship's speed on the deck limits are displayed in Figure 8. The interface between the LPD-4 response characteristics combined with the Fire Scout aircraft stability parameters, appears to be particularly sensitive, with minor adjustments in following seas, to ahead seas at higher ship's speeds. From the diagram, ahead seas (330 – 30 degrees relative wave angle) is most stable at 9 feet significant wave height. The stable condition quickly decays at relative wave angles greater than 030 – 135 and 225 – 330 degrees of relative wave angles.

Flight Deck Limits and Best Course to Steer

As described earlier, the Energy Index is an empirical formulation used to identify quiescent deck periods that is computed based on measured (or computed) ship deck motions against operational launch and recovery deck limits.

The time required to raise the deck from minimal motion to unacceptable motion is called the risetime which may be analytically or experimentally determined. The risetime is a thumb print characteristic of the ship's response and rarely changes. In terms of the energy index scale, it is defined as the period that is measured from the end of a green signal to the positive side of the red line. The risetime is mirrored by a fall-time, which is the time-period measured from the negative side of the red line to the negative side of the green line. A typical recorded trace is displayed in Figure 9.

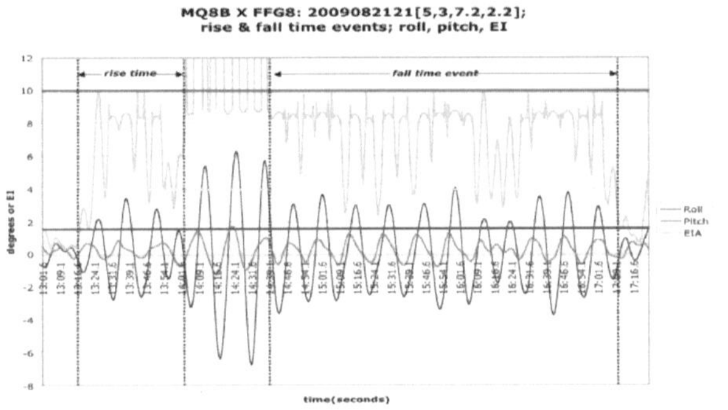

Figure 9: Typical Rise and Fall time Event

The very safe deck is a special condition in which there is insufficient energy in the aircraft-ship system to raise the deck out of limit for some defined time-period or risetime, and it is this concept that was used in the simulators to indicate when it was safe to initiate landing. Deck Status is transmitted to the pilot in flight using an external light indicator or to the Tactical Control System using a graphic user interface (GUI) coupled with data transmitted to an unmanned vehicle.

Assuming the minimum average rise time for a given vessel remains constant, the aircraft descends within the rise time value, and it is assured a recovery within the deck motion limits. Essentially, the methodology summarized above is a formulation to quantify operating beyond the static deck limits as defined in NATOPS or SHOL. Figure 10 displays graphically the safe operating envelope beyond static limit operations. The base envelope is taken from 10 knots of ship speed. As before, any points within the hour-glass structure are conditions within limits and contain no appreciable probability of out-of-limit deck motions. Outside of that structure contains motions which are considered by static reference as out-of-limit **[10]**.

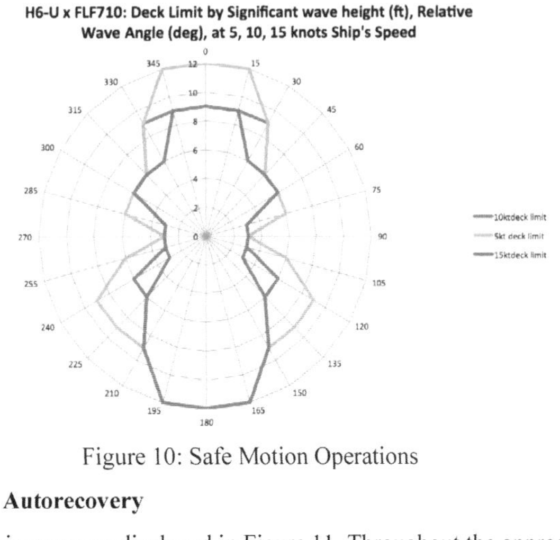

Figure 10: Safe Motion Operations

LPD integration in generic UAS Autorecovery

LPD indicates the onset of deck quiescence as displayed in Figure 11. Throughout the approach and initial hover, LPD is monitoring the deck. It is only at the low hover that the UAV Autoland system would accept a Green Deck signal from the LPD. It is when the EI signals the onset of quiescence sending its signal through the uplink, which sets the air vehicle on its descent to the deck.

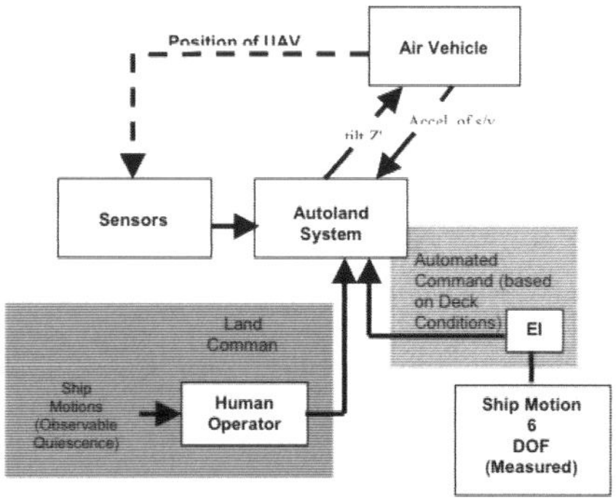

Figure 11: Notional Autoland System

A similar EI computation is used in QPP to determine zones of acceptable flight operations.

QPP AOG Best Course to Steer

To best describe spatially the measured seas acceptable for flight operations, the Energy Index is used. As described earlier, one of the computational approaches is to use the QPP algorithms to measure in real-time the seaway around the mother ship. The converted data are coupled with stochastic ship transfer function information to forecast the ship's motions for a given ship's heading and speed. It is estimated the data will be computed for a diagram representing the ship's motions up to 2 km from the initial measurement position.

A speed-polar representation of the seaway to guide ship's command and deck hand activity, appears as a synoptic map created by the lay-over of various measured and computed parameters. Each parameter speed-polar lay-over is application specific, computed as a function of the acceptability of ship's conditions to complete a specific motion sensitive task (Figure 12). This is based on NAVSEA SEA 851, formally known as the Seakeeping Department, Active Operator Guidance System's post-processor **[11]**.

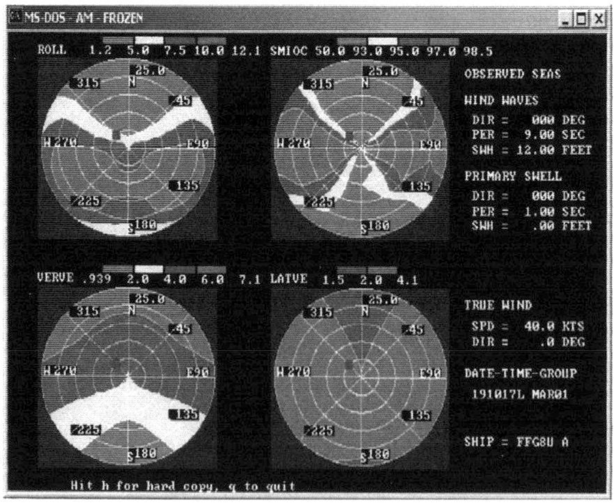

Figure 12: Parameter Map Speed Polar

A limit is defined by the impact that a certain ship motion condition may impose on the air/ship vehicle's structural integrity or that may result in undesired dynamic response. The sum of these lay overs produces an active operator's guidance map (Figure 13). All the various QPP components either exist or require limited developmental modification.

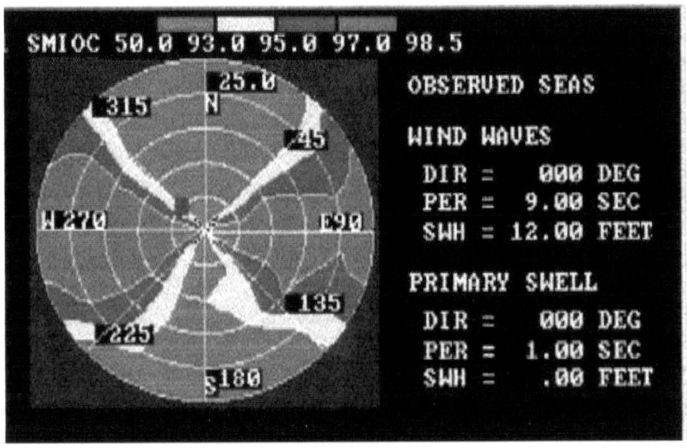

Figure 13: Best Course to Steer

Multiple solution convergence methodologies may be offered using, for example, real-time long horizon forecasting alone or coupled with seaway real-time mapping providing stochastic (ship transfer function) based distance motion forecasts updated several times per second.

QPP can be applied to all ship motion sensitive activities. Throughout air vehicle approach and initial hover, QPP would be producing operator guidance maps allowing the ship to adopt a heading and speed acceptable for air vehicle recovery. It is only at the low hover position that the QPP short-horizon component would be used to verify that the deck was measured in real-time as being in a quiescent period condition.

5. QPP SUMMARY AND CONCLUSIONS

Parts of the MQ-8 future improvements involve the enhancement of the associated system components, such as that being done for the L&R segment with an eye on devices like QPP. The primary goal for conducting DI analysis is to expand existing operating envelopes and increase AV's availability thereby improving overall naval effectiveness. The objective of DI is to determine the maximum safe air vehicle/ship platform operational limitations. Given an air/ship system and inherent operational limitations, DI strives to increase tactical flexibility for any set of environmental conditions. Modeling and Simulation complements DI testing to delineate system limitations. The calculated system limitations provide experimental DI with the necessary data to more effectively set testing strategy to probe expected limiting conditions.

In the future QPP system, a short horizon algorithm is used as verification that the capture or recovery is occurring within specified limits. The QPP algorithms uses real-time information coupled with stochastic ship transfer function information to compute, for a given ship's heading and speed, what the ship's motions will be up to 2 km from the initial measurement position.

The QPP algorithms are programmed by simulation into a high-level architecture system permitting plug and play federates representing various motion measurement devices being applied to the QPP system. Verification and validation of QPP functionality takes the form of breadboard software integration. This is an iterative process measuring the fidelity of the actual system against the software model, and the efficiency of the developing software wrapper which is specified by various motion-based applications.

Multiple solution convergence methodologies may be offered using, for example, real-time long horizon forecasting alone or coupled with seaway real-time mapping providing stochastic (ship transfer function) based distance motion forecasts updated several times per second.

QPP can be applied to all ship motion sensitive activities. In the NSRS case, the NATO Submarine Rescue Ship uses an A-frame mounted to a vessel of opportunity (VOO) to deploy and recover the Submersible Rescue Vessel (SRV) which is Unmanned. The greatest risk is during the recovery operation where there is a risk of collision between the SRV and the mothership. The SRV is towed at a safe distance from the ship until it is determined that the conditions are safe to perform a recovery. At this point the SRV is vulnerable to collisions with the ship, so the final recovery sequence needs to be performed quickly. QPP offers the ability to bring the vehicle into the danger zone only when the ship motions are most favourable.

Another motion sensitive operation is the case of the Unmanned (or Manned) air vehicle recovery. Throughout air vehicle approach and initial hover, QPP would be producing operator guidance maps allowing the ship to adopt a heading and speed acceptable for air vehicle recovery. It is only at the low hover position that the QPP short-horizon component would be used to verify that the deck was measured in real-time as being in a quiescent period condition.

In both the NSRS and UAV cases, the motion forecasting and measurement components have been extensively tested by the Navy and has been successfully demonstrated individually. Assembling the QPP components by simulation sets the baseline of verification of system of systems. The next step requires the components to be integrated into a Navy QPP Test bed. In all cases, the QPP components have demonstrated accuracy through sea state 6. By themselves, the QPP components are fully developed with a TRL rating of 8; simulation integration as federates than transition breadboard test bed,

In a related but future QPP study activity, it was observed that certain ship deck condition tendencies were apparently related. Operating over deck whilst the ship is experiencing a quiescent ship motion period normally has the secondary effect of minimized air wake confusion. This is owing to fewer ship structure excursions into and out of the air flow. To better define deck airflow around the ship it is proposed to integrate a Doppler LIDAR instrumented federate. The purpose of this is to predict the future vessel air wake and look for quiescent periods in this paralleling the vessel motion QPP technique. Preliminary test results from a proof-of-concept trial, are briefly discussed. The ability to fully define a ship's motion and air wake condition as a function of a recovery device greatly enhances the overall system operational capability whilst fundamentally reducing the inherent risks associated with platform manipulation in high sea conditions.

6. ACKNOWLEDGEMENTS

The authors wish to pay tribute to the technical competence and outstanding professionalism of retired Dr. John Duncan's former head of SBA, his staff and associated team members in the early stages of the Quiescent Period Prediction project. This report is dedicated to the memory of NAWCAD's late Chief Engineer Robert Ernst, and to the late Dr. Peter J F O'Reilly and Dean Carico, fathers of the Dynamic Interface discipline.

7. ABBREVIATIONS

DI	Dynamic Interface Study
DOF	Degree-of-freedom
EI	Energy Index (calculated by LPD)
HLA	High Level Architecture
LPD	Landing Period Designator
LPDLOOP	Landing Period Designator software
MRU	Motion Reference Unit
NATOPS	Naval Air Training and Operating Procedures Standardization
NIREUS	NATO Interoperability and RE-Use Study
ONR	Office of Naval Research
RAO	Response Amplitude Operator
RN	Royal Navy (UK)
RNAS	Royal Navy Air Station (UK)
SAIF	Ship/Air Interface Framework
SHOL	Ship-Helicopter Operational Limit
SMP	Ship Motion Program
SMS	Ship Motion Simulation composed of routines identified as NAV
TCS	Tactical Control Station
UAV	Unmanned Aerial Vehicle
VTOL	Vertical Takeoff and Landing
VTUAV	Vertical takeoff and Landing Unmanned Air Vehicle
WOD	Wind-over-deck

8. REFERENCES

1. White, S., Reading, R. (2001). NATO/PfP HLA Federation of VTOL Operations Supporting Simulation Based Acquisition. 01E-SIW-032. Brussels

2. Applebee, TR., Baitis, AE., Meyers, W.G. (1981) "User's Manual for the Standard Ship Motion Program, SMP" David W. Taylor Naval Ship R&D Centre Report DTNSRDC/SPD-0936-01.

3. Ferrier, Dr. B et al (2016). "Helicopter-Ship Motion Interface Prediction using Simulation Technology Developments and Trial Results from a Dedicated Royal Navy Trial". Proceedings of the Warships Conference. RINA. Bath.

4. Ferrier, Dr. B., Ernst, R., Sehgal, A., Nelson, J. (2014). "MQ-8C (Fire Scout) Launch and Recovery Program Improvements". Proceedings of the American Society of Naval Engineers Launch & Recovery Symposium. Alexandria

5. Ferrier, Dr. B (2014). MQ-8B/C x LCS-4 Deck Characterization using NAVAIR LPD. QR140530. Arlington.

6. Chen, Z, et al (2017). "S-Band Doppler Wave Radar System ". Remote Sensing Journal. MDPI. Vol. 12, No.1 Basel.

7. Belmont, Prof M. et al. (2019). "Deterministic Sea Waves Prediction Using Mixed Space-Time Wave Radar Data". Journal of Atmospheric Oceanic Technology. Vol. 36, No. 5. Boston.

8. Applebee, TR., Baitis, AE., Meyers, W.G. (1981) "User's Manual for the Standard Ship Motion Program, SMP" David W. Taylor Naval Ship R&D Centre Report DTNSRDC/SPD-0936-01.Launch & Recovery Symposium. Alexandria

9. Ferrier, Dr. B. et al (1998). "Simulation and Testing of the Landing Period Designator (LPD) Helicopter Recovery Aid". Naval Engineering Journal. Vol. 110, No 2. Alexandria

10. Ferrier, Dr. B., Ernst, R. (2019). „Development of Enhanced Dynamic Interface Methodology". Proceedings of the Australian Association of Unmanned Systems. Sydney.

11. Applebee, T., Baitis, A.E., Ferrier, Dr. B. (2000). Active Operator Guidance – Landing Period Designator System (AOG_LPD). NCRADA-NSWCCD-99-042. Carderock.

9. AUTHORS' BIOGRAPHIES

Michael Belmont, BSc, PhD, hold a chair at the University of Exeter UK, where he heads the Marine Dynamics group. He pioneered the new discipline of Deterministic Sea Wave Prediction which allows the short-term estimation of the details of wave driven vessel motion. This makes possible the prediction of quiescent intervals in the motion of naval platforms allowing the safe extension of the range of sea states for the launch and recovery of air, sea surface and sub-surface craft. Recently the same principals have been applied to the prediction of air-wakes over vessels. The goal of his research is the ability to allow the real time deterministic prediction of all the major environmental factors affecting maritime launch and recovery operations.

Jacqueline Christmas is an Associate Professor of Machine Learning in the Computer Science Department at the University of Exeter, UK. After many years working in industry, she obtained a PhD in the field of statistical pattern recognition in 2011. The focus of her work is the development of machine learning models for remote sensing, with an emphasis on explain ability and measures of uncertainty, especially for potentially infinite sets of real-time observations, such as are collected by long-term, continuously operating sensors. Her particular interest is in the use of inexpensive, off-the-shelf sensors, such as video, radar and still cameras, to remotely sense difficult dynamic environments in real time. Since 2012 this work has been directed towards the measurement and prediction of sea waves, based on physical oceanographic theory.

Larry Fedrick FRINA MIET completed a fully indentured apprenticeship in MV/HV switchgear design in 1979 and embarked on a career in the construction industry specialising in power generation and distribution, Renewables, system testing and asset management before joining the UK MOD assigned as Royal Fleet Auxiliary heavy electrical Engineer in 2007, conducting numerous overseas assignments, then was appointed Fleet Electrical Equipment Authority for rotating electrical machinery in 2012 before being selected as Team Leader/Eng. Mgr. for Naval Design Partnering and Simulation Based Acquisition projects in 2016 and 2018 respectively. He leads the surface Warship platform pre-concept and concept design and Fleet in-service support design teams, simulation, and modelling projects in conjunction with UK and overseas stakeholders and industry partners for DE&S Ships Acquisition. Military experience includes 20 years RAF reserve forces as a trained fixed wing pilot/instructor.

Bernard Ferrier, PhD, FRINA, FRAeS is a Systems Engineering and Technology Assistant (SETA) in NAVAIR PMA-266 with the primary task of integrating and deployment of Fire Scout MQ-8C and its technologies and defining the operational limitations of the deck environment. Ferrier holds an honourary Professor position at the University of Exeter's College of Engineering. He is the Engineering Head of Hoffman Engineering's Aircraft-Ship Dynamic Interface Lab in Fairfax, VA. Ferrier led the CL227 UAV dynamic interface program at Bombardier in Montréal. Ferrier was a rotor dynamist and project leader of the dynamic interface program of the marinised AH-64 (Apache) and LHX programs at the McDonnell Douglas Helicopter Company. He received his last doctorate in helicopter/ship interface engineering at the École Polytechnique de Montréal (Québec) Canada.

Commander Robert Taylor is the RN Chief Naval Test Advisor and is responsible for the delivery of all Fleet Air Arm Test and Evaluation activities. He trained as a Lynx observer and has completed numerous front line, training, and trials assignments over a 30-year career. His early career centred around flying and included deployments to the Adriatic, the Baltic, and the Falklands, this was interspersed with training assignments on the Lynx Mk3 helicopter and its simulators. It was at this stage he began a career in test flying completing the Aerosystems Course and tours at the Lynx Operational Evaluation Unt and at Boscombe Down. Command tours include as a Lynx Flight Commander, Command of 700W Sqn, the first RN Wildcat Sqn, and of 744NAS, the Mission Systems and Weapon Integration Sqn based at Boscombe Down.Staff roles have centred around NCHQ, predominantly centred round assurance and ship-air integration. He has also completed an operational tour based in US Central Command, working closely with the US Joint Operations Centre and the Coalition support to the Middle East. He has completed an MBA in his spare time and an in-service MSc in Aerospace Systems. He is married to Joanna, a doctor, and they have 2 sons. He is a keen sportsman, keeping fit and enjoying golf and alpine skiing, and is an amateur woodworker and beekeeper.

VIRTUAL REALITY BEYOND DESIGN REVIEWS IN SHIPBUILDING: THE NEED FOR INDUSTRY-TAILORED IMMERSIVE DATA INTERACTION

N Fourrier[1], **M Benaouicha**, SEGULA Technologies, France
G Moreau, IMT Atlantique, France
JM Normand, Nantes Université[1], France

SUMMARY

In the context of shipbuilding, the potential of virtual reality (VR) as a visualization tool has been demonstrated [1, 2]. However, many conditions must be met for VR to be widely adopted in the shipbuilding industry. It primarily must be seamlessly integrated into current workflows. The shipbuilding industry will likely first need to tame its data with efficient models accessible throughout the ship's lifecycle. However, some productivity issues with VR have not yet been identified clearly. While working to develop a VR design tool for ship outfitting, it became clear to us that there is a need for industry-driven development of efficient data interaction and manipulation tools in VR. As an example of such new manipulation tools, we are currently developing a new way to input text in VR that is efficient and easy-to-learn through handwriting. The evaluation of this system highlights the need for more controlled and systematic testing of VR tools in the shipbuilding industry to justify investments and adoption.

1. INTRODUCTION

The popularity of virtual reality (VR) in industry is at an all-time high. Even in shipbuilding, industrials and academics investigate the usefulness of VR, pushed by growing demands from many shipyards around the world. At ICCAS, the topic of mixed reality (MR) has been tackled regularly for at least a dozen years [3-5]. From marketing to visualization, the range of VR potential applications is as vast as the range of operations related to shipbuilding.

In the consumer market, immersion and social interaction are the main reasons for VR to be used. In an industrial setting, immersion can provide better visualization of 3D models and VR's social aspects may be used to design efficient collaborative tools. As such, there is no unique way to use VR and its potential is demonstrated in many fields: sales, design reviews, training and more. However, integrating new and disruptive tools into existing workflows is technologically complex, time-consuming and expensive. The first step, which has already received major contributions, is to define properly all the conditions for a successful integration of VR into current industrial processes.

Design and conception are two core industrial processes and they both require manipulating data and not only visualizing it. Paradoxically, the industry only uses VR for visualization purposes. The usage of VR as a data manipulation tool remains mostly unexplored today. Indeed, this use case requires massive changes in the way industries handle their data. This kind of transformation is already happening though, hidden behind terms like digitalization and Industry 4.0. Using VR as a design tool can then become a new incentive to accelerate the transition towards standardized and easily consumable data. This would make it possible to develop more efficient design tools that can leverage all the advantages VR has to offer.

2. CONDITIONS FOR LARGE-SCALE INDUSTRY ADOPTION

There are many barriers to large-scale adoption of virtual and augmented reality (AR) systems in an industrial setting, and especially in shipbuilding. Defining them properly is necessary as a first step towards better implementations of VR tools.

2.1 COMMUNICATION BETWEEN VR AND CAD APPS

Integrating new tools into established processes requires bidirectional communication and data transfer with existing tools. This is especially difficult when the tools to link have different data structures. This becomes increasingly apparent with innovative tools like VR. Indeed, they are often developed with game engines that represent 3D models in a discrete manner using triangles, called tessellation. This is highly efficient for the real-time rendering that is necessary for video games and VR. However, this model is incompatible with Computer-Aided Design (CAD) data that mostly uses continuous representations like NURBS. The conversion from CAD to VR, from a geometrical standpoint, is possible thanks to many algorithms that aim to discretize and optimize 3D models for real-time rendering [6, 7]. However, when it comes to metadata such as part hierarchy or textures, the transformation often requires some amount of work that cannot be streamlined and depends on both the CAD and VR file formats [8]. There is a push for CAD data structures to represent tessellated geometry, for example with the new versions of STEP AP242 (ISO 10303-242) files but most design software cannot yet interpret this new norm. The CAx-Interoperability Forum regroups the companies actively working on this topic with the aim of accelerating the development of STEP AP 242 translators [9]. Those companies evolve in different industries such as automotive, defence, aerospace and energy. CAD software developers such as Autodesk, Dassault Systèmes, Datakit and others, accompany them in this endeavour. This shows there is a growing demand from every industry for better interoperability of CAD file formats especially when it comes to representing tessellated geometry.

Unity, one of the most commonly used game engines to develop VR applications, acquired Pixyz in 2021, which is a plugin that aims to let users import CAD files directly into the game engine. The import is automatically optimized for 3D rendering and supports files coming from most CAD software. There also exist file conversion software such as CAD Exchanger, that lets users transform CAD file formats into others, including ones compatible with game engines. Also, there are open libraries that aim to ease the development process of conversion software such as Assimp [10]. This type of transformation, no matter the software used, often comes with a partial loss of geometrical data and metadata that prevents two-way data transfer.

Indeed, VR to CAD is even more complex as this conversion cannot be done strictly with only the deteriorated data – because it has been discretized and tessellated – that emanates from the VR system. The process of converting back geometric data from a tessellated format into one that is readable by CAD software mainly relies on interpolation [11, 12] or primitive shape extraction techniques [13]. In some cases, it might be possible to reconstruct fully – including metadata – the CAD data but this would rely on accessing 3D model databases or creating a third-party software that has access to both domains. Some software use this kind of technology to create conversion-less import of CAD models in VR scenes and even allow some amount of bidirectional real-time modifications :

- Exxar CAD integrates with standard CAD applications such as Aveva Marine, CADMatic, Catia... to stream the 3D data directly in a VR review application. Users working on the same model both in a CAD software and in VR can annotate the models. Those annotations propagate in real-time.
- Techviz interfaces with CAD applications like Catia and NX for real-time visualization of 3D models. Modifications made in the CAD software can be visualized while a VR session is ongoing. Some interactions with the models are possible in VR such as measuring or moving parts, but it is not possible to modify the CAD model in VR.
- Mindesk also integrates with many CAD applications like Solidworks, Revit or Rhino. It allows streaming CAD data directly into VR for visualization purposes. Leveraging Rhino's data format, Mindesk allows VR modifications to be reflected in real-time on the CAD model through Unreal Engine, which is another game engine often used for developing VR applications.

These observations highlight the need for industries, including shipbuilding, to tame their data. Data must become easily consumable which means having a unique model used throughout the lifecycle of the project. Having access to up-to-date models is necessary to widen the range of possible VR applications [2, 14], as they are currently limited to non-modifiable visualization when it comes to CAD. In the context of Industry 4.0, many powerful models are emerging:

- Building Information Modelling (BIM): a technology that aims to simulate construction projects in virtual environments. A unique model is constructed for "design, procurement, fabrication, construction [...], operations and maintenance" [15]. The model is accessible by all teams for the whole lifecycle of the project. This model serves as the basis for digital transformation in the architecture and construction industry, which are highly similar to the shipbuilding industry. A recent open-source Unity plugin aims to allow bidirectional communication between BIM authoring platforms and the game engine [16].
- Product Lifecycle Management (PLM): a technology that aims to manage the complete lifecycle of a project from ideation to disposal. This model contains all the necessary information from the specifications, technical solutions, to fabrication and maintenance details. Data is streamlined as to create a "coherent data flow, avoiding redundancies and gaps" [17]. In BIM, the digital mock-up is at the center of the model while PLM also formalizes the processes of modification and validation of the model.
- Digital Twins: the virtual representation of real-world assets with the goal of representing the real time state of said assets. This virtual representation should span the whole lifecycle of the asset. The virtual model can be used for monitoring, diagnostics, design iterations, simulation... to improve productivity. The link between digital twins and VR is a popular research topic [18].
- Industrial Internet of Things (IIoT): application of Internet of Things (turning every object into a connected device) to industry. It enables ubiquitous monitoring and communication between machines for better automation [19]. This kind of technology requires data standardization and is often paired with augmented reality to monitor interactively the state of a machine [20].
- Knowledge-Based Engineering (KBE): aims to increase productivity and reduce costs by automating design tasks through the formalization of expert knowledge [21]. The technology often uses an inference engine to determine if a set of facts validates a set of rules [22]. The rules should be representative of the expert knowledge.

The objective of such technologies and processes is to create a unique data model that is easily consumable during the whole lifecycle of the project: from early designs all the way to the supply chain. It can enable multidisciplinary teams to interact simultaneously with the same data. They can also reduce iteration time and costs if used as decision-helping tools. This makes these systems ideal candidates for application in the ship building industry.

2.2 EASE OF USE AND DEVELOPMENT

The communication between VR and CAD tools is important but is not the only condition preventing wider adoption of VR. Martin [4] stated that VR tools must be easy-to-use even for non-CAD experts. Indeed, the expertise needed to run and be proficient with VR tools nowadays far outweighs the benefits. First, as stated before, the amount of work required to prepare data for VR is non-negligible and non-trivial. While there are solutions to convert the data for visualization as demonstrated in Section 2.1, they often require human intervention or expertise. As of today, teams that want to use VR regularly and efficiently for design reviews purposes would need a dedicated VR expert.

Secondly, the development of VR prototypes is difficult and often lacks collaboration with field experts. This means systems end up missing features or are clunky to use. It might not be a priority to create a perfect user experience because we are still experimenting and trying to prove the potential of VR, but it does not help with convincing experts of VR capabilities. VR development is not easy also because there are many systems that are not all compatible with each other. Indeed, different levels of VR can be experienced depending on the output (head-mounted displays (HMDs), CAVEs, 3D screens…) and input (controllers, navigation, body and gesture tracking) devices [23]. This means some amount of work is required to standardize the experience across platforms.

Finally, when it comes to actual usage, it may be difficult for the less tech-savvy to get the hang of VR. New users must first realize what it means to be immersed in a virtual environment, and they must then learn how to interact with said environment. New users may also fall into the trap of expecting too much from VR. Misconceptions about the capabilities of VR can lead to disappointment, which may harm one's ability to adapt to the technology. We can however expect VR acquaintance to grow as time passes. Nick Clegg, President for Global Affairs at Meta Platforms, recently published a manifesto about the vision of Meta – ex-Facebook – regarding the future of VR [24]. This vision revolves around standardisation and a new social way to experience the internet called metaverses.

With more people acquainted with VR, it will be easier for industry to adopt the technology. Moreover, Meta's short-term objectives are all about releasing new HMDs designed with productivity and industrial use in mind. It is currently pushing for more collaboration with lawmakers, academics and industrials. This is a major opportunity to define properly what is necessary for VR in industry. For the first time, workshops [25, 26] were dedicated to these topics during the IEEE Virtual Reality 2022 conference (the premier international conference on VR) and the push for VR in industry is one of the main factors that will influence the transition towards metaverse-like VR implementations. Such standardization of the VR ecosystem would help greatly the development of industrial applications.

VR applications are often developed using game engines and SDKs that are suited for video games. As such, visualization, interaction and manipulation techniques in virtual environments are vastly different from CAD software. VR apps are not traditionally created with productivity and industrial use in mind, and this shows even when industrials are involved in the making. As such, making VR easy-to-use is probably not sufficient, and new innovative ways to interact with data in VR, that differ radically from the ones used today, must be studied with industry as a priority.

2.3 THE NEED FOR INDUSTRY-TAILORED DATA INTERACTION TECHNIQUES

The main hindrance to adopting VR in industry remains data compatibility. As stated before, major VR and CAD actors globally are actively working on this issue. We can realistically believe it will be resolved in a not-so-far future. We must then anticipate the next problem in adopting VR, which is how to make it productive enough to rival current processes. Indeed, current data interaction and manipulation techniques in VR are not compatible with productivity and industrial use. For example, in a traditional CAD software, a keyboard and mouse setup lets the user select and modify data. This setup is efficient in both interacting with and manipulating data and does not really require any training, as everyone has some experience with the setup. In VR though, while there exist many performant and intuitive ways to select data [27], manipulating and changing it is challenging. Indeed, it is difficult to embark a physical keyboard in a VR setup, even if doing so results in unmatched data entry speeds – measured in words per minutes (WPM) – varying from 40 to 80 WPM [28]. This kind of system however requires specific setups that are difficult, from a logistical and technological standpoint, to integrate with standard HMDs. In general, "plug and play" VR apps have only tried to emulate and virtualize classic – keyboard and mouse setups – interaction methods. This makes sense in a first approximation because the familiarity helps when it comes to understanding VR. However, ultimately, since being immersed in data is wildly different from interacting with it through a flat screen, the virtualization of classic techniques (such as virtual keyboards) proves not to be efficient enough. As such, it is understandable that we cannot expect VR to succeed in an industrial setting – where productivity is paramount – if even basic tasks such as typing text are daunting.

There are a lot of research and commercial works exploring new ways to handle locomotion, interaction and collaboration in virtual reality. However, entertainment industries are the ones driving this exploration. The needs in those fields is vastly different from industrial needs. Industry must be the driver behind the creation of efficient VR data interaction and manipulation techniques. Three factors are primordial when designing VR applications for industrial use: efficiency, ergonomics, and ease-of learning and mastering. Combining these factors and bidirectional communication between VR and CAD will allow, in the end, to use VR as a design tool beyond its traditional usage of visualization and design reviews.

3. VR AS A DESIGN TOOL

There are three main reasons VR would be great for design purposes. First, VR is immersive, which makes it an ideal tool for visualizing difficult situations at 1:1 scale. Indeed, in standard CAD tools, it may happen that some design tasks, and especially in the highly constrained environment of shipbuilding, are rendered difficult because of the less-than-optimal visualization strategies [29]. Secondly, VR is becoming more and more of a social platform that lets its user interact and share experiences with each other. Some amount of standardization in the domain of identity and avatars in VR will happen thanks to the rise of metaverses, which will make it easier to integrate collaborative components into industrial VR tools. Collaborative tools are great for better communication between designers and between teams, which will in turn improve design efficiency and reduce iteration costs. Finally, users in VR are able to interact with the environment in a more intuitive and closer to the real-world manner. As such, VR tools can help the decision-making process. For example, one could verify if doors can be opened by trying them, could walk through maintenance spaces to check if they are accessible, or even simulate the process of evacuation after pushing in an incendiary alarm button.

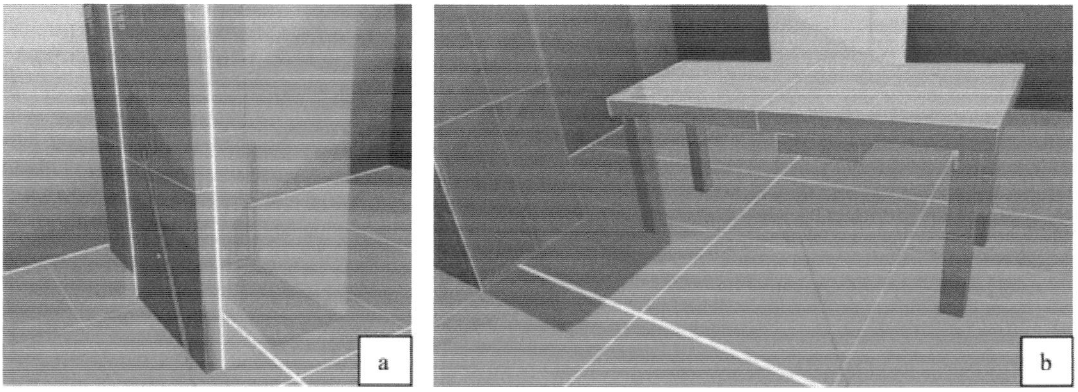

Figure 1 – a) The green zone represents an accessibility constraint. It is associated with an electric box object that has an open door highlighted in yellow. The green zone must remain free to let the door open. b) Whenever another object enters the area, the constraint is detected as being violated and the volume turns red to indicate the problem to the user.

We are leveraging the advantages of VR through the development of an immersive, interactive, and collaborative VR design tool for ship outfitting. Outfitting is a highly multi-disciplinary process that involves many trades such as piping, electrical, plumbing, HVAC (heating, ventilation, and air-conditioning) and even furnishing. This process is often sequential and iterative. Its goal is to ensure that all the different constraints and norms from each trade are respected in contrived spaces. Combining the visualization power of VR with a decision-helping component will make the outfitting process more efficient. The goal of the tool is to let outfitters immerse themselves at 1:1 scale in the technical locals for better visualization and be able to place new objects. Interacting with those objects will let them verify certain properties such as accessibility. Additionally, we are developing a decision-helping tool which goal is to let outfitters know when constraints are not respected.

By working hand-in-hand with outfitting experts, we aim to formalize knowledge [30] and exploit it in real-time in the VR tool. An example is given in Figure 1: placing an electric box in the environment comes with an accessibility constraint. Indeed, it must remain possible to open the door at all times, even after subsequent changes in the environment. The tool under development must be able to verify in real-time that all constraints are respected, and lets the outfitter know if there are any problems. As such, by formalizing knowledge, the amount of work necessary to verify everything is in order should be reduced.

In the other hand, while developing this VR design tool, it became abundantly clear that text entry in VR, which is a common task in design, was not efficient enough for industrial use.

4. A NOVEL TEXT ENTRY TECHNIQUE

When it comes to conception and design, one of the main fundamental tasks is data entry. Indeed, creating and manipulating textual and numerical data is necessary in all design processes. In industry, these processes rely heavily on the keyboard as a mean to input data. Keyboards are overall highly efficient: they reach high words per minute (WPM), are not physically and mentally tiring, and most importantly they require little training since they are so common. In the context of using VR tools for design, creating and interacting with this type of data is not trivial. Three types of virtual keyboards are often used for tools that want to only rely on a standard HMD setup [31]:

- Head/eye pointing: the user uses the movement of its eyes or head to move a pointer on a virtual keyboard. Pressing a button on a controller will select the pointed-at key [32].
- Controller pointing: the user uses a laser-like apparatus associated with a controller to point directly at a virtual keyboard. Pressing a button on a controller will select the pointed-at key [33].
- Drum-like: the user uses virtual drumsticks to hit on a virtual keyboard that reacts to the collisions between the drumsticks and keys. No other input than the collision is required to type a character [34].

Compared to a classic keyboard that has a 75 WPM average typing speed for commonly skilled users [35], these virtual keyboards all have drawbacks that make them not practical for industrial purposes. Head/eye pointing keyboards have low typing speeds (16 WPM) and are physically demanding. Controller pointing keyboards suffer from the same problems in a lesser extent. Finally, drum-like keyboards report the best performance (21WPM), but it is still far from the performance of traditional keyboards, and it remains physically tiring [36]. A lot of work has been done trying to create innovative data input techniques for VR using T9-like keyboards [37] or gestures-based methods [38]. While they can be better when it comes to physical and mental load, they are slower than virtual keyboards for beginners and their mastery requires a lot of training. It is then important to create a VR text input system that is suitable for industry: efficient, ergonomic and easy-to-learn.

4.1 HANDWRITING IN VR

In our work to create an interactive and collaborative VR tool for ship outfitting, we faced the problem described previously that it is hard to do design without performant data input, which does not really exist in the context of virtual reality. The scenarios where we need to input text or numbers are, non-exhaustively, the following:

- Filter databases to find 3D models to import in the virtual environment.
- Find items in large lists.
- Create or modify textual values like object names and annotations.
- Create or modify numerical values like object positions, scales…

All these scenarios are concise but reoccurring typing tasks. As such, we need an input technique that is efficient and reactive mainly for single – or few – word scenarios. In a real-life setting, when asked to provide short textual information without the use of a keyboard, one would grab a piece of paper and a pen. We believe mimicking this type of behaviour in VR could be more efficient than virtualizing keyboards. As such, we developed a system that enables handwriting in VR on a board with a pen. An off-the-shelf handwriting recognition software (HWR), that we integrated into the Unity game engine, transforms the handwritten text in real-time in an exploitable data format as shown in Figure 2. Such a system would indeed answer all the criteria for an industry-suitable input system in VR: handwriting speeds surpass the performance of virtual keyboards for short writing sessions as they average ~38 WPM [39, 40] and requires no training since virtually everyone knows handwriting. Moreover, writing in VR has been gamified many times (The Lab, Half-Life Alyx, DryErase…) and can become an entertaining task, which would reduce drastically the mental load associated with typing in VR.

Nevertheless, writing on a virtual surface that is not present physically comes with two challenges:

- It is necessary to create appropriate feedback for the user to know when the pen is indeed in contact with the virtual whiteboard.
- It can be tiring as it is not possible to rest against the writing surface.

To make the experience more intuitive, we added visual and haptic feedbacks. Visual feedbacks consist of enabling collisions between the virtual pen and the virtual whiteboard, shadows that help estimate the distances to the board, and an adaptive line thickness that varies following the amount of virtual pressure applied on the board. Haptic feedback consists of a varying controller vibration based on the pressure applied to the board. We are also experimenting with visual haptics by representing the contact of the pen with the board by adding a rippling texture on the board, substituting a haptic cue by a visual one. These feedbacks allow for writing that is more natural and lifelike, even if the cues do not perfectly match reality [41]. Adding guides, such as horizontal lines to follow when writing, will maximize the effectiveness of the handwriting recognition software because its performance depends greatly on each character's size and verticality.

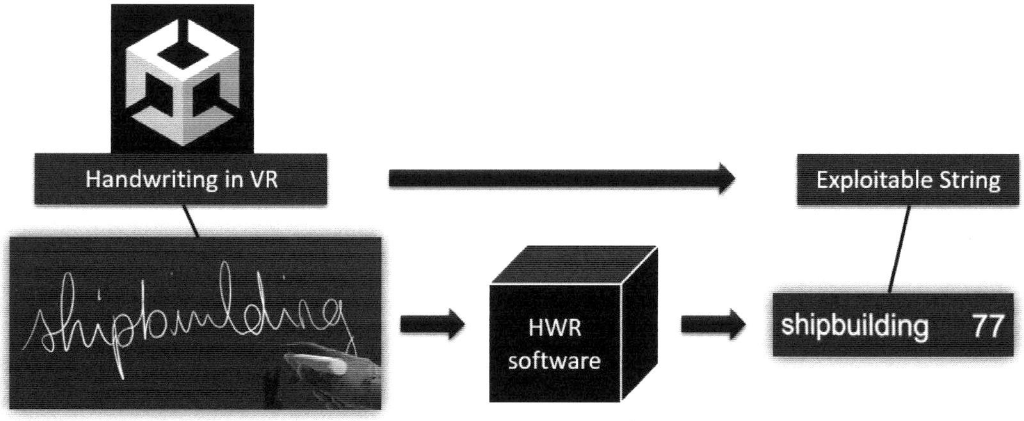

Figure 2 – Handwritten text on a virtual blackboard in a Unity VR scene is transformed into exploitable string data thanks to the HWR software. The number provided with the data represents the recognition confidence of the algorithm on a 0-100 scale.

We are also studying the influence of posture on the physical load of the virtual handwriting task. Indeed, for long writing sessions in a classroom whiteboard setup, not being able to rest on the board makes the task tiring. As such, it is possible to imagine new writing positions where the arm can be at rest either by providing a physical support or by finding an ergonomic position for the board. However, this problem only arises for long writing sessions, which are not the use case we envision for this system. Preliminary testing has demonstrated the potential of the system when it comes to writing speed, ease-of-use and general appreciation compared to virtual keyboards in a MacKenzie type test [42]. We are planning a two-part user study that will first determine implementation details of the system (feedback types, posture…) and then prove the effectiveness of the system in a more industrial task that is related to design in shipbuilding.

4.2 DESIGNING A USER STUDY TO ASSESS INDUSTRIAL USABILITY

One common flaw found in many of the VR tools developed for shipbuilding [3, 5, 29] is that there is no quantification of their usability. It is important to prove the potential of VR in both quantitative and qualitative manners. In the rare cases where user or case studies were conducted [1, 43], either there were not enough participants to be relevant, the study was not structured in a way a quantitative measure could be performed or statistically analysed, or the population of the study was made of students who could not give an appropriate industrial evaluation of the system. If we do not make more effort to provide metrics for the usefulness of VR in industry, it will remain nearly impossible to justify investing massively into this kind of system, especially if no comparisons with current systems are available. The "lack of perceived benefits" described by Morais et al. [14] is in part due to this lack of metrics.

Quantitative measures will always depend on the task; it is not possible to provide a general framework to assess quantitatively the performance of an industrial VR system. However, measuring qualitatively the usability of a system is a well-known exercise [44]. The SUS (System Usability Scale) questionnaire is the most often used: it evaluates 10 items related to usability in a task-abstracted fashion. This evaluation is often conducted in pair with the NASA-TLX questionnaire that aims to quantify the mental and physical loads of a task [45]. Combining these questionnaires should provide an objective metric of a VR tool's potential as well as a way to compare them to the currently used tools. It is important for the questionnaires that the tested population in a user study is representative of the target population. In academics settings, it is often the case user studies are conducted on students. It is debatable whether such results can justify industrial use. In industrial settings however, while technicians and engineers are often involved in the development process of VR tools, proper case studies with quantifiable metrics are rarely performed. As such, more collaboration would benefit greatly the implementation of VR in industry. Indeed, such collaboration can lead to better-controlled environments for testing systems with relevant populations. We will follow these principles when carrying out the studies for the handwriting input system presented in Section 4.1 and the ship outfitting design tool presented in Section 3.

5. CONCLUSIONS

The difficult communication between current tools and processes and VR is the main limiting factor to large-scale VR adoption. Indeed, in industry, processes require smooth and streamlined transitions between the different cycles of a project. Additionally, this makes VR cumbersome to use because of the amount of work necessary to prepare data even for basic visualization. We showed these are popular research topics that industrials and academics are actively trying to resolve.

We argue there are additional limiting factors that have not been yet properly identified when it comes to expanding the scope of VR in industry to tasks like conception and design. Indeed, there is a dire need for more performant, more ergonomic, and easier-to-use data interaction techniques in VR that are suited for industrial use. Moreover, it is also important to start evaluating industrial VR applications in a more rigorous manner in controlled studies that can provide metrics of VR usefulness and usability. Morais & al. stated VR has to "offer a solution that is ten times better than the current way of doing things" [14]. It is important to be able to measure this statement, and compare VR to current systems, to justify massive investments towards industrial adoption. We believe that close collaboration between industry and academia would better advance this topic.

6. REFERENCES

1. J.J. GARCIA AGIS; P. BRETT; A. EBRAHIMI, 'The potential of virtual reality (VR) tools and its application in conceptual ship design', *Conference on Computer Applications and Information Technology in the Maritime Industries*, 2020.
2. J. MARTIN; A. CONNELL, 'Accessible immersive visualisation for shipbuilding', *ICCAS*, 2015.
3. H. LÖDDING; A.FRIEDEWALD; M.HEINIG; S. SCHLEUSENER, 'Virtual Reality Supported Assembly Planning in the Shipbuilding Industry', *Journal of Ship Production and Design*, 2011.
4. J. MARTIN, 'Virtual reality: tool or toy for shipbuilding?', *ICCAS*, 2019.
5. C.-J. CASSAR; R. SIMPSON; N.BRADBEER; G.THOMAS, 'Integrating virtual reality software into the early stages of ship design', *ICCAS*, 2019.
6. J. GUO; F. DING; X. JIA; D.-M. YAN, 'Automatic and high-quality surface mesh generation for CAD models', *Computer-Aided Design*, 2019.
7. A. MAGLO; G.LAVOUÉ; F. DUPONT; C. HUDELOT, '3D Mesh Compression: Survey, Comparisons, and Emerging Trends', *ACM Computing Surveys*, 2015.
8. K. MCHENRY; P. BAJCSY, 'An Overview of 3D Data Content, File Formats and Viewers', *Technical Report for Image Spatial Data Analysis Group*, 2008.
9. 'CAx Interoperability Forum', https://www.cax-if.org/cax/cax_introduction.php.
10. 'Open Asset Import Library (assimp)', https://github.com/assimp/assimp.
11. P. PAL, 'A reconstruction method using geometric subdivision and NURBS interpolation', *International Journal of Advanced Manufacturing Technology*, 2008.
12. P.PAL; R. BALLAV, 'Object shape reconstruction through NURBS surface interpolation', *International Journal of Production Research*, 2007.
13. R. BÉNIÈRE; G.SUBSOL; G. GESQUIÈRE; F. LE BRETON; W. PUECH, 'A comprehensive process of reverse engineering from 3D meshes to CAD models', *Computer-Aided Design*, 2013.
14. D.MORAIS; M. WALDIE; D. LARKINS, 'The Evolution of Virtual Reality in Shipbuilding', *COMPIT*, 2017.
15. S. AZHAR, 'Building Information Modeling (BIM): Trends, Benefits, Risks, and Challenges for the AEC Industry', *Leadership and Management in Engineering*, 2011.
16. K. W. MAY; J. WALSH; R. T. SMITH; N. GU; B. H. THOMAS, 'UNITYREV – Bridging the gap between BIM authoring platforms and game engines by creating a real-time bi-directional exchange of BIM data', *CAADRIA*, 2022.
17. S. TERZI; A. BOURAS; D. DUTTA; M. GARETTI; D. KIRITSIS, 'Product lifecycle management – From its history to its new role', *International Journal of Product Lifecycle Management*, 2010.
18. J. PIRKER; E. LORIA; S. SAFIKHANI; A. KUNZ; S. ROSMANN, 'Immersive Virtual Reality for Virtual and Digital Twins: A Literature Review to Identify State Of The Art and Perspectives', *IEEE Conference on Virtual Reality and 3D User Interfaces Abstracts and Workshops (VRW)*, 2022.
19. E. SISSINI; A. SAIFULLAH; S. HAN; U. JENNEHAG; M. GIDLUND, 'Industrial Internet of Things: Challenges, Opportunities, and Directions', *IEEE Transactions on Industrial Informatics*, 2018.

20. M. HUSINSKY; A. SCHLAGER; A. JALAEEFAR; S. KLIMPFINGER; M.SCHUMACH, 'Situated Visualization of IIoT Data on the Hololens 2', *IEEE Conference on Virtual Reality and 3D User Interfaces Abstracts and Workshops (VRW)*, 2022.
21. W. J. C. VERHAGEN; P. BERMELL-GARCIA; R. E. C. VAN DIJK; R. CURRAN, 'A critical review of Knowledge-Based Engineering: An identification of research challenges', *Advanced Engineering Informatics*, 2012.
22. M. T. SHAHZAD; J. HOFFMANS, 'KBE/AI for Ships Construction – A Feasibility Study', *COMPIT*, 2021.
23. C. ANTHES; R. J. GARCIA-HERNANDEZ; M. WIEDEMANN; D. KRANZLMÜLLER, 'State of the art of virtual reality technology', *IEEE Aerospace Conference*, 2016.
24. N. CLEGG, 'Making the metaverse: What it is, how it will be built, and why it matters', *Medium*, 2022.
25. 'Metaverse as a promise of a bright future? – social interactions in a world of isolation', *IEEE VR Workshops*, https://workshop.htt.events/, 2022.
26. 'Metabuild – Building the Foundations for the Metaverse', *IEEE VR Workshops*, https://metabuild-workshop.herokuapp.com/, 2022.
27. F. ARGELAGUET SANZ; C. ANDUJAR, 'A Survey of 3D Object Selection Techniques for Virtual Environments', *Computer and Graphics*, 2013.
28. D.-M. PHAM; W. STUERZLINGER, 'HawKEY: Efficient and Versatile Text Entry for Virtual Reality', *25th ACM Symposium on Virtual Reality Software and Technology*, 2019.
29. V. ALONSO; R. PÉREZ; L. SANCHEZ; R. TRONSTAD, 'Advantages of using a virtual reality tool in shipbuilding', *COMPIT*, 2012.
30. Z. ZHU; M. VAN TOOREN; G. LA ROCCA, 'A KBE Application for Automatic Aircraft Wire Harness Routing', *53rd AIAA/ASME/ASCE/AHS/ASC Structures, Structural Dynamics and Materials Conference*, 2012.
31. T. J. DUBE; A. S. ARIF, 'Text Entry in Virtual Reality: A Comprehensive Review of the Literature', *Human Computer Interaction. Recognition and Interaction Technologies*, 2019.
32. C. YU; Y. GU; Z. YANG; X. YI; H. LUO; Y. SHI, 'Tap, Dwell or Gesture? Exploring Head-Based Text Entry Techniques for HMDs', *Proceedings of the CHI Conference on Human Factors in Computing Systems*, 2017.
33. M. SPEICHER; A. M. FREIT; P. ZIEGLER; A. KRÜGER, 'Selection-based Text Entry in Virtual Reality', *Proceedings of the CHI Conference on Human Factors in Computing Systems*, 2018.
34. C. BOLETSIS; S. KONGSVIK, 'Text Input in Virtual Reality: A Preliminary Evaluation of the Drum-Like VR Keyboard', *Technologies*, 2019.
35. A. S. ARIF; W. STUERZLINGER, 'Analysis of text entry performance metrics', *IEEE Toronto International Conference Science and Technology for Humanity (TIC-STH)*, 2009.
36. C. BOLETSIS; S. KONGSVIK, 'Controller-based Text-input Techniques for Virtual Reality: An Empirical Comparison', *International Journal of Virtual Reality*, 2019.
37. D. YU; K. FAN; H. ZHANG; D. MONTEIRO; W. XU; H.-N. LIANG, 'PizzaText : Text Entry for Virtual Reality Systems Using Dual Thumbsticks', *IEEE Transactions on Visualization and Computer Graphics*, 2018.
38. S. CHEN; J. WANG; S. GUERRA; N. MITTAL; S. PRAKKAMAKUL, 'Exploring Word-gesture Text Entry Techniques in Virtual Reality', *CHI Conference on Human Factors in Computing Systems*, 2019.
39. N. VAN DREMPT; A. MCCLUSKEY; N. A. LANNIN, 'A review of factors that influence adult handwriting performance', *Australian Occupational Therapy Journal*, 2011.
40. J. SUMMERS; F. CATARO, 'Assessment of handwriting speed and factors influencing written output of university students in examination', *Australian Occupational Therapy Journal*, 2003.
41. N. COOPER; F. MILELLA; C. PINTO; I. CANT; M. WHITE: G. MEYER, 'The effects of substitute multisensory feedback on task performance and the sense of presence in a virtual reality environment', *PLOS ONE*, 2018.
42. I. S. MACKENZIE; R. W. SOUKOREFF, 'Phrase Sets for Evaluating Text Entry Techniques', *CHI : NEW HORIZONS*, 2003.
43. K. NORDBY; S. BØRRESEN; E. GERNEZ, 'Efficient Use of Virtual and Mixed Reality in Conceptual Design of Maritime Work Places', *COMPIT*, 2016.
44. C. KUN; N. ASHAARI, 'Utilitarian or Experiential? An Analysis of Usability Questionnaires', *International Journal of Computer Theory and Engineering*, 2015.
45. S. G. HART; L. E. STAVELAND, 'Development of NASA-TLX (Task Load Index): Results of Empirical and Theoretical Research', *Advances in Psychology*, 1988.

7. AUTHORS BIOGRAPHY

Nicolas FOURRIER holds the current position of PhD student at Segula Technologies in collaboration with AAU (Ambiances, Architectures, Urbanités) laboratory. He is working on a thesis titled "Interactive and collaborative ship outfitting in virtual reality".

Mustapha BENAOUICHA obtained his PhD in Numerical Modelling from La Rochelle University, France in 2007. He is currently Head of the Naval and Energy Engineering Research and Innovation Unit at SEGULA Technologies. He is responsible for the project concerning the use of VR technology in decision support for ship outfitting.

Jean-Marie NORMAND obtained his PhD in Computer Science from Nantes University, France in 2008. He is currently Associate Professor at École Centrale de Nantes, France. His research interests are VR, especially Virtual Embodiment, as well as AR, especially perception and interaction.

Guillaume MOREAU obtained his PhD in Computer Science in 1998 from Rennes 1 University. He is currently full Professor in Computer Science at IMT Atlantique, France. His research interests are Virtual and Augmented reality with a focus on how these environments are perceived.

DESIGN OF A MOBILE APPLICATION TO ASSESS THE STABILITY OF SMALL FISHING BOATS

A Grech La Rosa, C Ryan, G Thomas, University College London, UK
L Huang, Cranfield University, UK
W Hetharia, University of Pattimura, Indonesia
S Riyadi, D Setyawan, I Utama, Institut Teknologi Sepuluh Nopember, Indonesia

SUMMARY

Indonesia is one of the top three countries in the world for fisher fatalities due to a lack of stability monitoring, poor vessel design, and the fishers' poor safety culture. Improved safety through technology, in particular stability awareness, has been identified as a key aspect to prevent capsizing and sinking. Despite the low technology uptake by Indonesian fishers, they are known to make extensive use of smart mobile phones as a primary form of communication. These phones are equipped with various sensors, including motion sensors that can be used by different mobile applications.

This paper proposes the use of a mobile application to aid Indonesian fishers in assessing the stability of their vessel by using results from a roll decay test or realtime roll motion data monitored when operating at sea. For the pre-departure test, the mobile application uses the onboard gyroscope to record the roll angle time series which can then be used to quantify the roll period. An offline calculation is then carried out to approximate the vessel's metacentric height. This is then visually presented using a traffic light system to provide the fishermen with a simple indication of whether the vessel is stable or not. Should the boat be unstable, additional guidance is provided to improve the vessel's stability. The realtime monitoring system alerts the fishers when the thresholds are exceeded, to ensure realtime awareness of increased risk to the stability of the vessel.

NOMENCLATURE

B	Vessel Beam (m)
GM_T	Metacentric Height (m)
k_{xx}	Radius of Gyration (m)
DAQ	Data Acquisition System
FFT	Fast Fourier Transform
GDP	Gross Domestic Product
GUI	Graphical User interface
IOS	iPhone Operating System
$_0$	Roll Natural frequency (rad/s)
PPE	Personal Protective Equipment
SMS	Short Message Service

1. FISHING IN INDONESIA

The livelihoods of many Indonesians rely on fishing [1], [2]. In 2019, 1.83 million fishers were living on the Indonesian Coast [3] as part of a fishing industry that contributed 469.6 trillion Indonesian Rupiah to Indonesia's GDP (Gross Domestic Product) making up a total of 2.8% of the country's economy [3]. In 2021, the fishing industry was the second largest contributor to Indonesia's (GDP) out of the major industries of agriculture, forestry, and fishing.

Fishing struggles with a high fatality rate for fishers at sea. Due to poor reporting, the total number of fatalities relating to fishing is unknown, however between December 2020 and June 2021, due to accidents at sea, 83 fishers were reported missing, alongside 14 confirmed dead and 42 survivors [4]. Other incidents have also been reported such as when 10 fishers went missing after their boat capsized [5], or 2 fishers dying following another capsizing incident [6] or an additional 15 fishers missing following a capsize incident [7].

According to the Lloyd's Register Foundation [1], poor safety culture and a lack of data monitoring have been identified to be some of the main causes for the poor safety record of the Indonesian fishing fleet. Although many fishers have dedicated their careers to fishing, many have done so without recognised training [8], this has resulted in many fishers having a poor understanding of how to mitigate risks through vessel design and proper safety equipment usage and alongside this, high pressure to provide economically for their families, has resulted in improper recognition and management of risks at sea.

In work carried out by Wibawa et al. [9] and Wibawa and Birmingham [10], it was found that fishers showed reluctance to alter their vessel design and operational practices to improve health and safety on board. Aspects relating to occupational health, sanitary, fire and escape were disregarded by the fishers taking part in the study as being minor problems. Vessels were also found to be ill equipped and missing vital equipment such as navigation lights, radar reflectors and lifesaving equipment. The reason for the unwillingness to change was not discussed in detail making it hard to use as a basis on which to recommend operational improvements. The study also highlighted that the enforcement of local regulations relating to the safety onboard their vessels was not being conducted.

The majority of fishing in Indonesia is conducted using small vessels of up to 3GT or less than 12m in length [10]. Traditional vessels such as Purse Seiner, Skipjack Pole and Gillnetter are all examples of widely used vessels for different small and large fish types in Indonesia. Each vessel type has particular features which make fishing easier for the fishers but may not necessarily be in line with safety standards.

Fishing is inherently a dangerous way of earning a living [11]. A global effort has been made to improve safety practices when fishing as well as designing vessels to minimise the fisher's exposure to dangerous practices. Guidelines for undecked vessels of less than 12m in length have been drafted to aid boat builders design and build vessels that are safer to operate however compulsory regulations do not exist as yet [10],[12].

2. MOBILE APPLICATIONS

A study conducted by Saville et al. [13]shows that mobile phones are widely used by fishers in North Jakarta. This is one of the main areas of the capital that has a large population of small-scale fishermen permanently and temporarily based there. 90% of the survey respondents had a mobile phone however no distinction of whether it has smart functionality was recorded. A smartphone is a mobile device having superior processing power, user input and connectivity to a standard mobile device. Dedicated mobile applications can be downloaded and installed to tailor the phone to have additional functionality as required by the user.

The study demonstrated how mobile phones are used by fishers to collaborate between themselves and make others aware of good fishing grounds and contaminated areas. This communication was carried out by the fishers via Short Message Service (SMS) or phone calls. This information used to be shared by word of mouth before the advent of mobile phones. More importantly, this study provided evidence that mobile phones are one of the main causes for the higher income of fishers in recent years [13].

Figure 1 shows the coverage of 2G, 3G, 4G and 5G mobile networks for four service providers available in Indonesia[14]. The maps clearly show how fewer points have been recorded at sea. These results conform with one of the outcomes recorded from Wibawa and Birmingham [10] study that notes the poor mobile coverage away from the shoreline. Without having regular access to a mobile network, fishers cannot rely on cloud-based services while out at sea.

(a) 3Tri

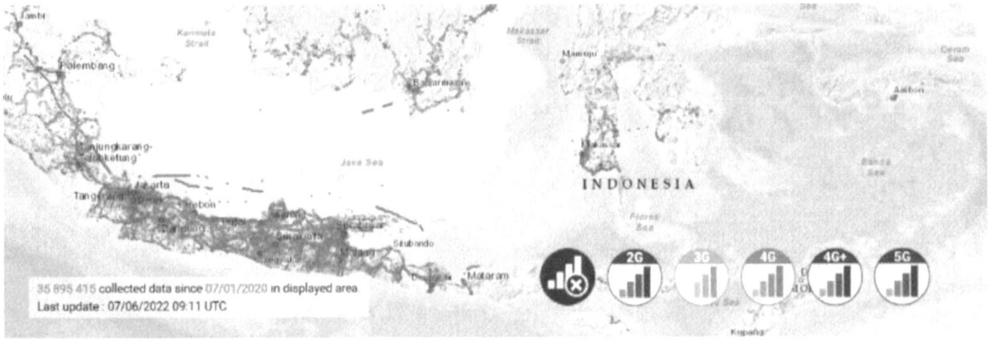

(b) Indosat

(c) XL Axiata

(d) Telkomsel

Figure 1: Mobile network coverage in Indonesia

Smartphones are equipped with 3 types of sensors that can be utilised to measure or provide input applications. These relate to motion, environment and position. These include a gyroscope, accelerometer, light sensor and magnetometer. These sensors are used to provide a seamless and intuitive interaction with the user for a wide range of functions. They are implemented in many applications having a variety of uses including gameplay, fitness tracking and photography.

The degree of accuracy and precision of each sensor found within a mobile phone depends on the particular mobile phone. Zhizhong Ma et al. [15] demonstrate how mobile phone sensors can be used for experimental purposes. Different tests were conducted to determine how well the accelerometer, gyroscope, magnetometer, and GPS within the Google Nexus 4 performed. The results demonstrated that depending on the degree of accuracy required by an application, it would be possible to use those in-built sensors for experimental purposes. While the study considered the implications on battery life when sensors are being used for extensive periods of time, it did not consider the sampling frequency in great detail. Any frequency dependent analysis would require a stable sampling rate.

Figure 2 shows how the number of smartphone users is expected to increase over the next few years in Indonesia [16]. This potential makes it more feasible to develop tailored mobile applications for Indonesian fishers to try and make them more aware of the limitations of their vessels. This concept has already been applied by others where Amrita and

Karthickumar [17] highlight the need to integrate mobile phones into the everyday roles of Indonesian Fishers. Fujita et al. [18] have listed a number of mobile applications that can be used by fishers to monitor their catch, keep records of quotas up to date and to identify fish species.

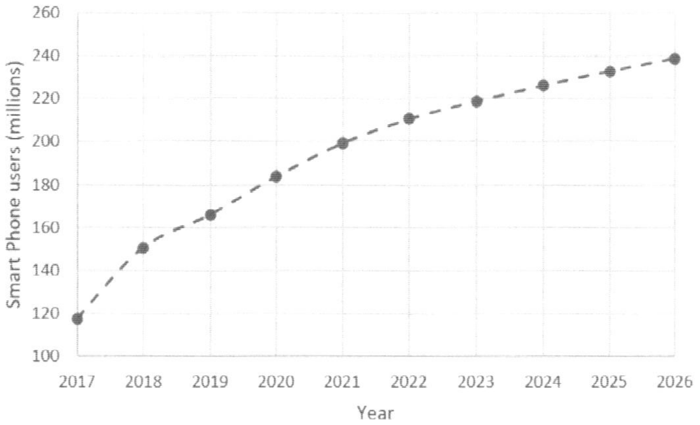

Figure 2: A chart showing the actual and predicted number of smartphone users

3. PRACTICAL STABILITY ASSESSMENT

A vessel's stability is dependent on the design of the vessel's hull form and its loading condition. A key measure of a vessel's intact transverse stability is its metacentric height which is the distance between the vessel's centre of gravity and the metacentre. In the design stage, it is common practice to assess a vessel's stability in an intact and damaged condition to determine whether it meets the stability criteria stipulated by local governments, classification societies and international regulations.

In the case of Indonesian fishing vessels, it is common for fishers not to have documentation relating to their vessel's stability. The design of their vessels as well as loading arrangements and tank layouts are set by intuition and past experience. While it is evident that these boats form part of the native culture, the need to fish further out from the shoreline due to a depletion in fish numbers means that vessels need to be better equipped and designed to safeguard their crew [9].

3.1 STATIC - INTACT

Once a vessel is built, practical ways of assessing its intact transverse stability exist. Two options to assess stability are the inclining experiment and the roll decay test.

3.1 (a) Inclining experiment

The industry standard method of determining a vessel's stability is by conducting an inclining experiment. This involves shifting known weights across the vessel's beam and recording the resultant heel angle. If done meticulously, an accurate reading for the metacentric height can be calculated.

Although this method provides an accurate value for the metacentric height, the procedure itself requires a lot of coordination and resources. Another common challenge relates to the lack of hydrostatic data for Indonesian vessels which would be needed to determine the vessel's displacement. This procedure should be conducted for new vessels or vessels having had major refits. This approach does not lend itself as a practical way for fishers to estimate their metacentric height prior to leaving port.

3.1 (b) Roll decay test

A roll decay test involves measuring the vessel's roll period by producing a large enough excitation force to cause an angle of heel. By releasing the force as quickly as possible, the vessel is free to roll until it comes to rest. The roll frequency $_0$ of the oscillations together with an estimate of the vessel's radius of gyration k_{XX} can be used to estimate the metacentric height by applying the Weiss formulae found in Equation 1.

$$GM_T = \frac{k_{xx}^2 \times \omega_0^2}{g} \quad \text{...Eq 1}$$

This method is not as accurate as an inclining test but if done correctly, it can generate a good estimate to get a clear indication of whether the vessel is stable or not. This procedure was recommended for fishing vessels by Hudspeth et al. [19] where a detailed account of the procedure that should be followed to get reliable results was drawn up. Although the procedure still requires a certain amount of coordination, the resources and analysis needed to get a result are simpler. Authorities such as the Maritime and Coastguard Agency [20] also suggest applying this procedure for fishing vessels under 15m in length.

This process could be simplified further by automating the process in a mobile application. By setting up a smartphone in a fixed position on the vessel, the vessel's oscillation during the roll decay test could be measured via the smartphone's gyroscope. The application would then conduct all calculations necessary to present the user with an estimate of whether the vessel is stable or not. In the event that the vessel is found to be marginally stable or unstable, the user will be prompted with useful recommendations to try and improve the vessel's stability.

3.2 DYNAMIC - INTACT

As a vessel is underway, its stability may be compromised by moving weights, sloshing fluids in partially filled tanks and encountering unfavourable wave systems. Large ships are commonly fitted with operational guidance systems that monitor the vessel's stability and make the master aware of changes that could put the vessel at risk.

Such systems are not usually applicable for smaller scale vessels. The research conducted by Míguez González et al. [21] proposes a methodology to provide crew with realtime stability data that they can easily understand. By measuring the roll amplitude using a motion tracking sensor, the test vessel was assessed in realtime by means of the guidance system and GM_T was calculated [22].

This method assumes that the peak frequency of the vessel roll spectrum coincides with its natural roll frequency. To quantify the roll frequency, a sample of roll data would be analysed using a Fast Fourier Transform (FFT). By assuming that the peak frequency is the natural roll frequency, the Weiss formulae can be used to estimate GM_T.

The results recorded in the trial were found to over-predict GM_T. The reason for the discrepancy was not identified, however the validity of the Weiss formulae was highlighted. The discrepancy in the estimation of GM_T may be linked to the radius of gyration estimation. This was extensively discussed by Míguez González et al. [22] that concluded that the benefit of not needing crew interaction to estimate the vessel's moment of inertia outweighed the numerical error it introduced.

Further work to analyse the vessel in bad weather would be beneficial to understand the influence of external excitation forces on the estimated roll spectra [22].

4. DESIGN OF A STABILITY APPLICATION

The research conducted by Golding and Donaldson [23] demonstrates that the uptake of a mobile application is dependent on various aspects including how useful the application is perceived to be, the target demographic, and how streamlined the process of using the application is. An application's value of utility is often assessed by how well it works and whether it is an improvement to the status quo. In order to achieve these goals, behavioural and design science can be applied.

Kim [21] considers the psychological perspective of designing a mobile application to increase user retention. This study applied a branch of psychology known as environmental psychology that delves into the relationship between a person and the environment where cognitive maps were found to be one way for the person to process information [25]. An effective environment provides a better space for processing information where aspects relating to the mobile application's structure, aesthetics, content, and consideration of interacting with a smaller screen are considered. Similar to Golding and Donaldson [23], this study puts perceived usefulness and perceived likability as key attributes that can impact user retention.

Jayatilleke et al. [26] on the other hand applied a design research model proposed by Ma and Harmon [27] for its iterative nature. The study shows the development of an educational mobile application to work in tandem with printed course material. An important aspect of this process is the close collaboration with key stakeholders related to the development of the mobile application. The five stakeholders identified at the early stage of the design were involved until the final implementation.

Another interesting concept that could be applied to increase engagement, uptake and retention is gamification. This is when non-gaming scenarios are designed to have gaming elements. Basten [28] explains how such approaches could increase usability, trust and motivation. The scope of gamifying an application is wide and varied. Star ratings and virtual badges are two examples of how users could become more involved with the application. More recently Paliokas et al. [29] proposes the application of gamification with augmented reality for digital heritage. The mobile application enhances the on-the-spot museum visitor's experience by means of an augmented reality quiz game. This concept was created to increase the amount of time a visitor interacts with an artefact and was deemed a success. As Basten [28] highlights, it is important to strike a balance when implementing gamification elements in order to not distract the user from the task's main purpose.

4.1 THE APPLICATION

The aim of the mobile application is to provide a simplified platform for a fisher to determine whether their fishing vessel is stable or not. By monitoring the vessel's GM_T before departure and while underway, the fisher would be able to complement their experience with a mobile aid to understand the limitations of their vessel better. By implementing this functionality through the mobile phone, the fisher would not need to buy any additional equipment making this a cost-effective option.

The requirements for the application are:

- To require minimal effort to carry out analysis;
- To deliver a clear and simple message of whether the vessel is stable or not;
- To provide guidance if the vessel has been deemed unstable;
- To require no additional hardware other than a mobile phone.

Djebli et al. [30] proposed the use of a smartphone to conduct inclining tests and roll decay tests. This work shows how the mobile phone's gyroscope was able to successfully measure the vessel's roll angle. The post processing required to estimate the vessel's metacentric height for either method was not conducted within the application and relied on the user to estimate GM_T using other means. This added effort would require additional knowledge to carry out the post processing which may not be compatible with most fishers who do not have a technical background.

Similar to Jayatilleke et al. [26], the design process of the application will involve different stakeholders iteratively designing an application that fishers would want to use. These stakeholders are local researchers, subject matter experts, developers and fishers. Each person's contribution is expected to help meet the requirements for this mobile application.

In light of this, the application will have two main functions. The first function is a pre-departure test to assess the fishing boat's stability while still close to shore. To limit the complexity and required resources in conducting a test, a roll decay test will be programmed in this module instead of an inclining test. The second function relates to continuously monitoring GM_T while the vessel is underway. Although it would be recommended for the user to first conduct a pre-departure test, the continuous monitoring function can be utilised without having conducted the pre-departure test.

Figure 3 shows a mock-up of the home page of the application. It consists of three buttons to navigate to the main functions of the application and the variable inputs page. An additional button to control settings associated with the main variables is needed for the analysis. The minimalist artwork is aimed at keeping the application decluttered and straightforward to operate.

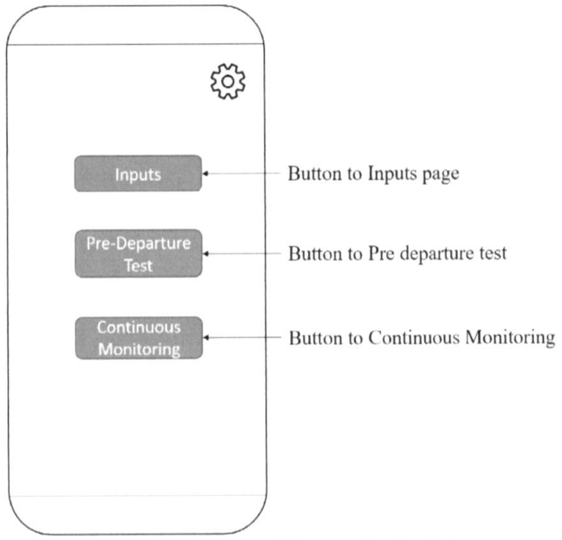

Figure 3: Mock up screen of the Application's homepage

For both functions, the mobile phone's gyroscope will be the main sensor used to conduct the analysis. The coordinate axis of the mobile phone may be seen in Figure 4 [31]. By monitoring the angular motion about the y - axis when the mobile is positioned such that its y - axis is parallel to the vessel's centreline, the vessel's roll motion can be recorded. To ensure that the mobile phone remains in alignment with the vessel's centre line and to prevent the phone from being moved while conducting analysis, the phone should be docked in a housing.

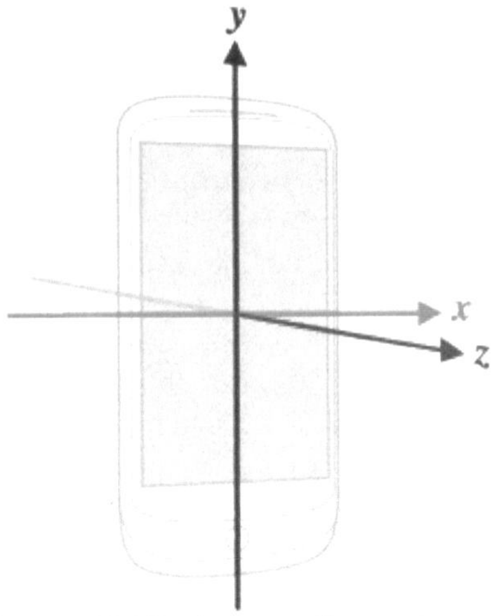

Figure 4: Mobile phone coordinate axis system

Provided that the mobile phone is positioned well on the vessel, the y - axis gyroscope data will correspond to the vessel's roll angle. This means that the roll period can then be calculated by conducting a Fast Fourier Transform (FFT) Analysis on the recorded spectrum. The dominant frequency from the analysis corresponds to the vessel's roll natural frequency.

This application will require the mobile phone to act as a data acquisition system (DAQ) and processing tool. Depending on the number of applications running in the background and the sampling frequency chosen, the computational effort on the phone varies. This has a number of implications including battery consumption, sampling reliability and processing temperature. Since the mobile phones likely to be used by the Indonesian fishers may be classified to be entry level phones, a key driver for the application is to minimise the computational effort. A study to determine the optimal range of sampling frequencies will be conducted however the study conducted by Terada et al. [32] recommended a sampling frequency of 20 Hz.

To use either function, the user would need to input the vessel's beam. This will be used to estimate the roll radius of gyration of the vessel by means of Equation 2. Although the radius of gyration is dependent on the moment of inertia of the vessel's waterplane and loading condition, this approximation makes it possible for fishers to get an approximation without needing to input a comprehensive account of the vessel's characteristics. The study conducted by Grin et al. [33] explains the importance of calculating the radius of gyration for a vessel. While this research focused on merchant vessels, a similar study to get a more refined estimate of the radius of gyration for Indonesian vessels could be carried out in future work. This will make it possible to get a more refined approximation that would improve the accuracy of the mobile application.

$$K_{xx} = 0.4 \times B \qquad \text{...Eq 2}$$

The results of either function will be displayed to the user by means of a visual aid. A traffic light system summarising the outcome of either the pre-departure test or continuous monitoring will be used. Green would signify that the vessel is stable, red that it is unstable and orange that it is unstable or that the analysis is undetermined. Following the recommendations set out by Jayatilleke et al. [26], this system provides an output that most people would be familiar with and is internationally recognised.

The main pages will also be fitted with heel angle indicators. This feature will also change colour depending on the phone's angle of roll. The colour change will be scaled according to a heel angle that has been confirmed to be safe for the vessel. This heel angle is dependent on various design features of the vessel including the freeboard and position of down flooding points. This will help the user be more aware of the vessel's limitations while underway.

4.2 PRE-DEPARTURE TEST

Figure 5 shows a mock-up of the pre-departure test screen. A linear layout was chosen to help make an emphasis on the phone's orientation in relation to the vessel. The preceding screen will display clear instructions of how this function in the application should be used.

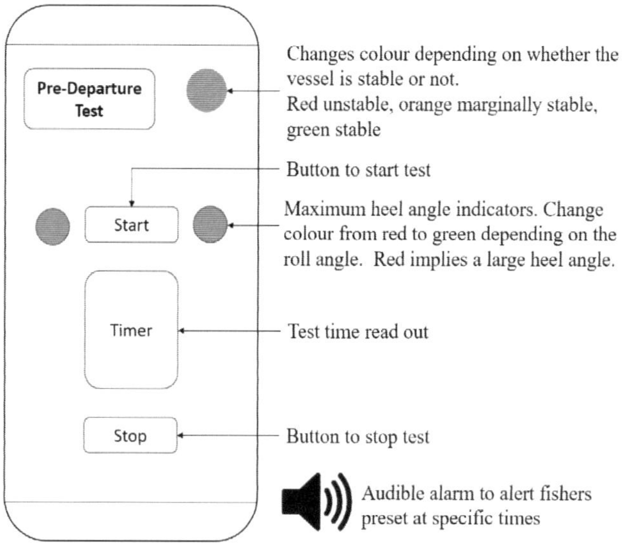

Figure 5: Pre-departure test mock up screen

The features on the screen include the transverse stability indicator, heel angle indicators and the control buttons needed for the test. The heel angle indicator will assist the user in determining whether a sufficient heel angle has been achieved to conduct the test. This should avoid the fishers releasing the vessel prematurely and registering a bad test.

Figure 6 shows the process that a user would need to follow to conduct a pre-departure test. Different approaches could be followed to incline the vessel enough to generate at least three to four roll oscillations. The duration of the test would be set by the user by specifying a test time in the setting. Users will be encouraged to repeat the test to confirm the result recorded.

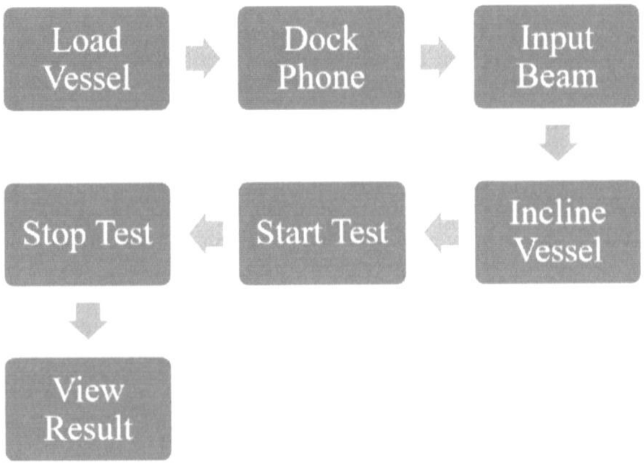

Figure 6: Roll decay user process

A flowchart showing the processes that need to be carried out by the mobile application can be seen in Figure 7. Data first needs to be manipulated to produce a clean signal for the FFT. Following the FFT analysis, the application will have a means of determining whether the test was conducted correctly. Data recorded for good tests would continue to estimate the metacentric height. This value is then translated into a colour to demonstrate the outcome of the test to the user. Those tests that were identified to be improperly conducted will result in an error message appearing requesting the test to be conducted again.

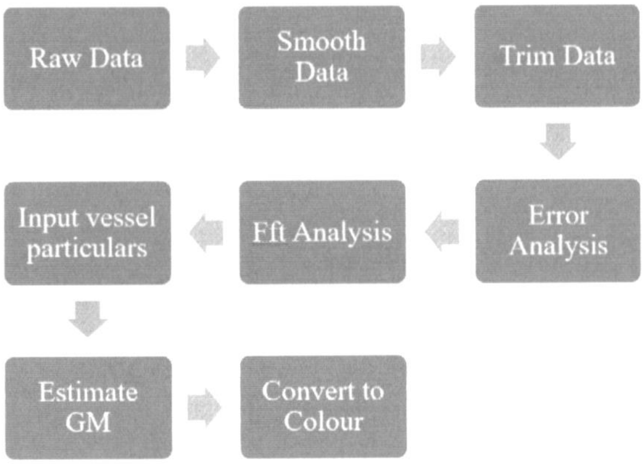

Figure 7: Roll decay code process

4.3 CONTINUOUS MONITORING

Figure 8 shows a mock-up of the Continuous Monitoring screen. Similar to the pre-departure test screen, a linear layout was chosen. The preceding screen will also display clear instructions on how this function should be used.

In addition to the features found in the pre-departure test screen, a message board, live sensor read out and velocity widget will be available. The message board will be used to state whether the vessel is stable or not as well as communicate any error messages to the user. The live sensor read out will show the user how the phone is responding. This will give the user confidence that the phone is operating well and is positioned correctly relative to the vessel's centreline. Since the instrumentation available on small fishing boats is typically minimal, the speed over ground may be shown by using the mobile phone's accelerometer. This will help the fisher understand how fast they are travelling.

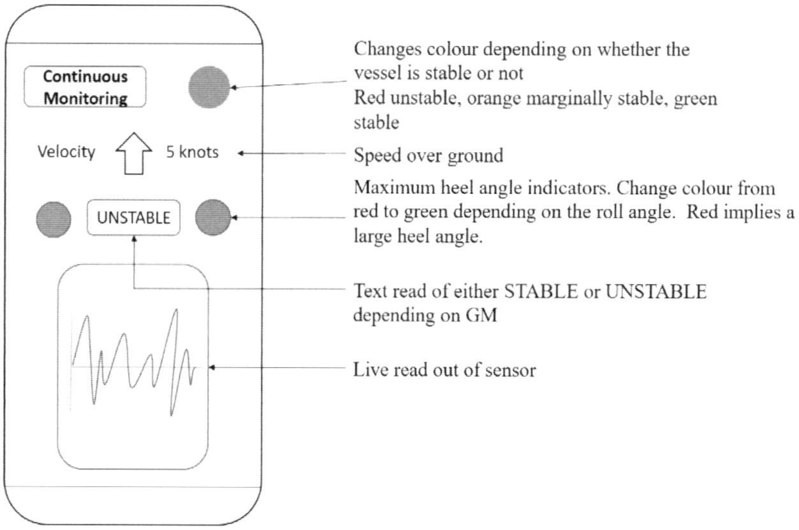

Figure 8: Continuous Monitoring Screen Mock up

Figure 9 shows the process that a user would need to follow to continuously monitor the transverse metacentric height while underway. This process has been designed to require minimal effort from the fisher to get results.

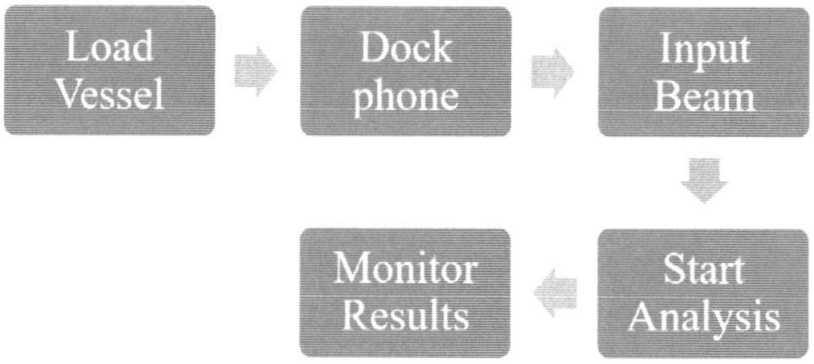

Figure 9: Continuous Monitoring user process

The programming needed for the continuous monitoring function is similar to that carried out in the pre-departure test. A simplification of the procedure recommended by Míguez González et al. [21]; Míguez González et al. [22] will be implemented in the application. A flowchart of the key steps can be seen in Figure 10.

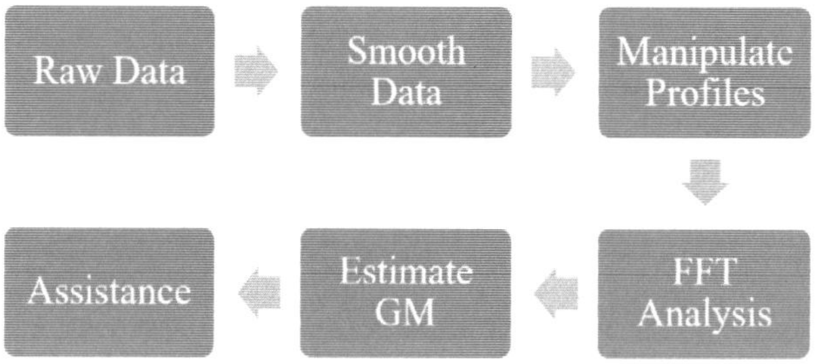

Figure 10: Continuous Monitoring Code process

5. CASE STUDY

To determine how well the concept of utilising a mobile phone application will work, an initial case study using a full-scale small boat was conducted. The boat had a full tank of fuel to have no free surface effects and all unnecessary equipment was removed. A readily available application to record sensor data was downloaded from the Google Play Store. This application did not conduct any post-processing.

The mobile phone used to conduct this preliminary test was a One Plus 6. This mobile phone is equipped with a Qualcomm Snapdragon 845 processor having four 2.8GHz processors and an additional four 1.7GHz processors. The phone has 6Gb of Ram and an array of sensors. The inbuilt gyroscope is a Bosch BMI160 designed to be used for realtime applications. A calibration sequence was carried out on a level service at the test site. During the tests, the mobile phone was placed on a flat wooden bench, parallel to the boat's centreline.

A Python code representing the core functions and processes outlined in Section 4 was created to post process the recorded data. This made it possible to get an understanding of the workflow required to conduct the analysis prior to migrating to Android Studio. The code generated was built in sub-modules to make it possible to use the same elements for both of the application's functions. All test data was post processed after the tests were conducted.

5.1 PRE-DEPARTURE TEST

A pre-departure test was first conducted by the quayside in a sheltered bay. The weather was calm and little wind was present. The test was carried out when no sea traffic was passing.

The boat was loosely moored by the bow and stern such that they did not provide any restriction in roll motion. An excitation force was applied by a person on the quay while another person was onboard the boat controlling the mobile phone. The test required coordination between the people involved to try and achieve good results. The test was repeated three times to record repeated readings.

Figure 11 shows the raw and post processed results for the pre-departure test where the first subplot shows the raw data recorded, the second shows a smoothed plot that's been trimmed to solely consider the roll test and the final plot is the FFT spectral plot.

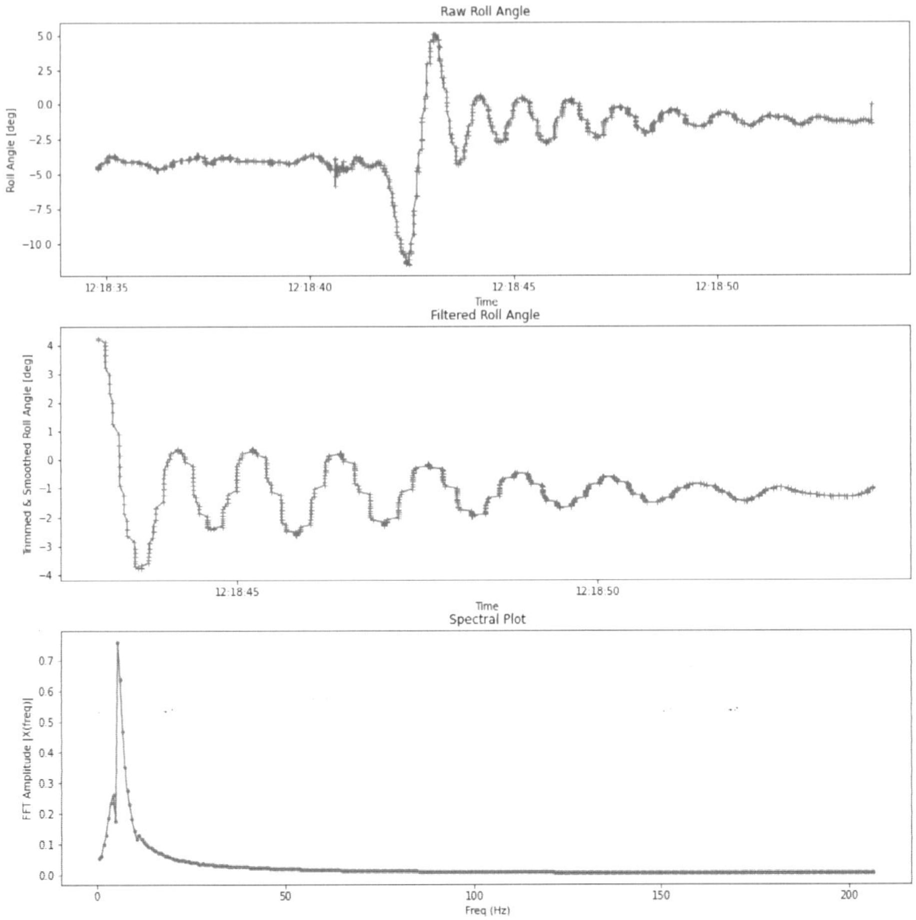

Figure 11: Initial results for small boat

The raw data shows that when the mobile started to record data, it was not registering 0° angle of roll. The mean angle of roll was found to be -4°. This could either mean that the level surface used to calibrate the gyroscope was not perfectly level or the boat had an angle of list. The same portion of data confirms the stillness of the water.

The results show that the roll decay test was conducted correctly where at least four roll oscillations were generated. This made it possible to conduct the rest of the analysis to estimate GM_T.

Another observation is that the final angle of heel did not match the initial angle of list recorded in the initial portion of data. The difference in the initial and final angle of heel is 2°. The reason for this is unknown and will need to be investigated further. A possible cause for this is the shifting of weight of the person controlling the mobile phone. This highlights the importance of ensuring that no weight is moved during the test.

5.2 CONTINUOUS MONITORING

Following the pre-departure test, the moorings were released, and the boat was driven out while continuously monitoring data. No additional weights were added to the boat however the position of the person controlling the mobile phone shifted further aft. The boat was driven in different directions relative to the prevailing wind at different speeds ranging from 5 - 10 knots. The duration of the test lasted approximately one hour.

The recorded data was later post processed using the continuous monitoring code. Unfortunately, the data recorded in this experiment was found to be unusable since the mobile was recording data at irregular time intervals. The cause for this is likely to be related to two main issues. The first relates to the number of processes that the mobile phone was conducting during the test since multiple applications were running in the background. The second relates to the sampling frequency used to record data. Since the application used to record data did not have the ability to set a sampling frequency, the mobile phone's maximum sampling frequency was used which was too high for this application.

6. DISCUSSION

The testing carried out confirms that a mobile phone can be a useful tool in determining the transverse stability of a fishing vessel. The successful use of the application will be dependent on how well the user sets up the mobile phone and carries out testing.

The roll decay test was a success. The mobile phone was able to record the gyroscope data to a file that was later used to approximate the boat's metacentric height. The procedure followed to achieve this is recommended by different authorities including the Maritime and Coastguard Agency [20]. This being said, the practical challenges associated with achieving this aren't necessarily listed. Based on the case study, the main challenges experienced where:

- Mooring the vessel correctly - The loose mooring lines resulted in the vessel drifting, even though the weather was very calm. This made it difficult to apply an excitation force at the quayside. It was also difficult to be certain that no tension in the mooring lines impacted results. The mobile application will need to recommend a mooring strategy to help fishers get the best results;
- Applying an appropriate excitation force - The test was conducted without any specialist tools or equipment to simulate the most basic ways of conducting a test. The mobile application will need to have some way of determining what the excitation force or safe angle of heel is to ensure that the person or team at the quayside can generate it safely. Attention should be taken not to collide with the quayside as it is rolling;
- Position and magnitude of excitation force - Care should be given to try and apply the excitation force in a position that would solely generate roll motion. This would reduce the forces needed to reach the desired angle of heel. The tool may also suggest the expected magnitude of the excitation force depending on which method of application is chosen to effect it;
- Coordination between the people involved - A person coordinating the test must be designated as the master to ensure that the team is working together. The mobile application can also be set to aid the master in conducting the test by means of an audio alarm;
- Waiting for calm water - Knowing when the water is calm enough to conduct the test was also another challenge. The mobile application could be equipped with heel angle thresholds that notify the user when the water is calm enough to start applying the excitation force;
- Self-Correcting right arm - If the boat has a large number of crew onboard, their natural tendency to correct for the change in heel angle as the vessel is rolling may impact results. Smaller boats will be more susceptible to this.

Although the test carried out for the continuous monitoring function did not yield good data, the lessons learnt from the test were important. Considering that the mobile phone used was a fairly high specification example of an android phone, the importance of minimising the computational effort within the code has been noted. To try and improve results further, the mobile application should notify the user to close any unnecessary applications running in the background.

Another important aspect is the need to select the appropriate sampling frequency. Although Terada et al. [32] recommended a sampling frequency of 20Hz, a sampling frequency sensitivity study should be conducted to determine the most appropriate sampling frequency for this application while quantifying the impact on the computational effort for the mobile phone.

The importance of docking the mobile phone was noted in both experimental studies. While it is not imperative to have a docking station, clear guidelines that encourage the use of a docking should be made available. Although the docking station can be as simple as a wooden placeholder to secure the phone correctly, some fishers may opt to consider more elaborate options that include external battery packs and solar panels.

7. FUTURE WORK

The work carried out in this study demonstrates the potential of using mobile phones as sensor suites. The application of using a mobile phone to assess a vessel's stability could help fishers understand the safe operational envelope of their vessel. Apart from developing the application to process the raw data in realtime with a user-friendly Graphical User Interface (GUI), further work could be carried out to investigate other opportunities for this technology. These include the following:

7.1 VALIDATION

Further full-scale tests need to be carried out on an Indonesian vessel having a known metacentric height. The *Yora 02* has been identified to be a potential candidate for the test [34]. By first conducting an inclining test, weight survey and a comprehensive linesplan assessment; a detailed comparison between the different methods could be carried out to quantify the errors associated with using a mobile phone to estimate the metacentric height.

7.2 CAUSE AND EFFECT OF GM_T ESTIMATION

Free surface effects, hoisting nets and lifting operations are some examples of activities that impact the metacentric height. So far, the code produced would be able to quantify a change in GM_T but would not be able to determine the cause of the shift. Should the application be able to determine the most likely cause of the change, the application would be able to provide more detailed and applicable help if the vessel is found to be unstable or marginally stable.

An additional aspect that should be looked into further is the identification of a listing or lolling vessel. This would alert the fisher of a potentially dangerous loading condition and give them time ahead of bad weather to address the situation.

Although the analysis conducted has been focused on intact transverse stability, further work could investigate the possibility of using the mobile application for damaged stability. The numerical analysis carried out by Gao and Vassalos [35] demonstrates how the motions of a damaged vessel are notably different to those of an intact ship. Being able to identify the change of ship motion behaviour immediately may help the fisher take the necessary action to save the vessel and its crew.

7.3 UNSUCCESSFUL PRE-DEPARTURE TESTS

To help the fisher coordinating the pre-departure test be sure that the test was conducted properly, the mobile application should be able to detect when roll decay tests were improperly carried out. This includes instances where:

- an insufficient excitation force was applied;
- the sea was too rough;
- the mooring lines restricted movement;
- the mobile phone was improperly positioned.

Additional tests purposely conducted improperly should be carried out to generate a library of failed roll decay tests. This would make it possible to non-dimensionalise the failed roll decay test and compare them to the results of new roll decay tests. This would make it possible for the application to not only inform the user that the test was improperly conducted but to also give guidance on why it failed to compute.

7.4 RADIUS OF GYRATION

As this study has highlighted, the estimation of the metacentric height in both the pre-departure test and continuous monitoring is dependent on various aspects. This includes the estimate of the radius of gyration. An additional study to get a better approximation of the radius of gyration for Indonesian vessels should be carried out.

7.5 COUPLED MOTION

When a vessel is not operating in beam seas, roll motion is affected by sway and yaw motions [36], [37]. The work conducted by Ali et al. [38] suggests that drift motion increases roll damping. Further work with the application should seek to cater for this coupling to get a better representation of vessel's motion by monitoring the other axis of the accelerometer and gyroscope.

7.6 ADDED CAPABILITY

So far, this application will be mainly making use of the mobile phone's gyroscope. As stated in Section 2, mobile phones are equipped with multiple sensors which could also be incorporated into the application in creative ways to further increase the application's capability for the benefit of the fisher. Additional wireless sensors could be fitted to the vessel that would measure different parameters to further aid the fisher in operating the vessel and ultimately to fish more efficiently.

7.7 DATA ANALYTICAL TECHNIQUES AND MACHINE LEARNING

A more advanced mobile application could be used to predict whether a vessel is entering into dangerous territory by using the data uploaded by other fishers to help predict its expectant performance. This would be carried out with data analysis techniques and machine learning. Such a concept would create a network of fishers helping other fishers.

The same concept could also be used to help fishers fish by making it possible to share information about fish activities in an area. This would build on the suggestions put forward by Saville et al.[13]. Such an approach may also help achieve other national goals associated with sustainable fishing [2].

8. ACKNOWLEDGEMENTS

This work is part of a project that has received funding from the British Council under the Newton Institutional Links Grants - Ensuring the safety of Indonesian seafarers and fishers in the time of COVID-19 and beyond (agreement No. 623457938), in conjunction with the Indonesian Governmental Funding from the Ministry of Education, Culture, and Higher Education (agreement No. 1369/PKS/ITS/2022).

9. REFERENCES

1. Lloyd's Register Foundation, "Insight report on safety in the fishing industry," 2018.3, 2018.
2. C. Warren and D. J. Steenbergen, "Fisheries decline, local livelihoods and conflicted governance: An Indonesian case," Ocean and Coastal Management, vol. 202, p. 105498, 2021, doi: 10.1016/j.ocecoaman.2020.105498.
3. Statista, "Fisheries in Indonesia," 2022. [Online]. Available: https://www.statista.com/study/88836/fisheries-in-indonesia/
4. Voi, "Sad News From Indonesian Waters," VOI, Jakarta, Jun. 21, 2021. [Online]. Available: https://voi.id/en/economy/60586/sad-news-from-indonesian-waters-83-fishermen-missing-in-the-last-6-months
5. The Jakarta Post, "10 missing after fishing boat capsizes off Indonesia," pp. 4–11, Jun. 20, 2020.
6. K. Epa, "Two Indonesian migrants die as overloaded boat capsizes," pp. 1–15, Mar. 21, 2022.
7. D. Widi and R. Articles, "15 Indonesian fishermen missing after boat capsizes," pp. 15–16, 2022.
8. R. Pramoda, I. Muliawan, T. Apriliani, C. M. Witomo, and M. Yulisti, "Competency of fishing boat crew from Indonesia in the framework of standards of training, certification, and watchkeeping for fishing vessel personnel 1995 (STCW-F 1995)," in IOP Conference Series: Earth and Environmental Science, 2021, vol. 860, no. 1. doi: 10.1088/1755-1315/860/1/012088.
9. I. Wibawa, R. W. Birmingham, and M. D. Woodward, "Design of Sustainable Fishing Vessels, Future Challenges for the Indonesian Fisheries," in 12th International Marine Design Conference, 2015, vol. 3, no. May, pp. 357–367.

10. Wi a a and i ingha , o ing a e y Wo ing n i on en on ndone ian Fi hing F ee Ca e Study on Local Fishing Communities in East Java," in Mastic2018 Maritime Safety International Confrence, 2018, no. July.
11. National Research Council, Fishing Vessel Safety: Blueprint for a National Program. Washington, DC: The National Academies Press, 1991. doi: 10.17226/1622.
12. B. Deakin, "Developing Simple Safety Guidance for Fishermen," 9th International Conference on Stability of Ships and Ocean Vehicles, 2015.
13. R. Saville, E. Riani, and K. Hatanaka, "The role of mobile phone among small scale fishermen for life improvement and community support tool in Indonesian coastal area," International Journal of the Bioflux Society, vol. 8, no. 6, pp. 846–854, 2015.
14. NPerf, "3G / 4G / 5G coverage map, Indonesia," nPerf, 2022. https://www.nperf.com/en/map/ID/-/-/signal/?ll=-2.567894216434251 3&lg=118.01999999999998&zoom=5 (accessed Jul. 25, 2022).
15. Zhizhong Ma, Yuansong Qiao, B. Lee, and E. Fallon, "Experimental evaluation of mobile phone sensors," in Irish Signals and Systems Conference, 2013, no. June, pp. 49–49. doi: 10.1049/ic.2013.0047.
16. Statista Digital Market Outlook, "Number of smartphone users in Indonesia from 2017 to 2020 with forecasts until 2026," Statista, 2022. https://www.statista.com/statistics/266729/smartphone-users-in-indonesia/ (accessed Jul. 25, 2022).
17. C. M. Amrita and P. Karthickumar, "Need for mobile application in fishing," International Journal of Science, Environment, vol. 5, no. 5, pp. 2818–2822, 2016, [Online]. Available: www.ijset.net
18. R. Fujita et al., "Technologies for Improving Fisheries Monitoring," San Francisco, 2018. [Online]. Available: https://m.bjbchjg.com/sites/default/files/oceans/Technologies_for_Improving_Fisheries_Monitoring.pdf%0Ahttps://www.edf.org/sites/default/files/oceans/Technologies_for_Improving_Fisheries_Monitoring.pdf%0Ahttps://m.bjbchjg.com/sites/default/files/oceans/Te
19. R. T. Hudspeth, F. Mazenot, W. G. Carter, and T. H. Gentle, "Using the Rolling-Period Test to Estimate Stability in Small Fishing Vessels," 779, 1986.
20. Maritime and Coastguard Agency, "Procedure for Carrying out a Roll or Heel Test to Assess Stability for Fishing Vessel Owners and Skippers," 2009. [Online]. Available: https://assets.publishing.service.gov.uk/government/uploads/system/uploads/attachment_data/file/561857/MGN_503.pdf
21. M. Míguez González, G. Bulian, L. Santiago Caamaño, and V. Díaz Casás, "Towards real-time identification of initial stability from ship roll motion analysis," in Proceedings of the 16th International Ship Stability Workshop (ISSW 2017), 2017, no. June, pp. 221–229. [Online]. Available: https://bit.ly/3C1eUfV
22. M. (University of A. C. Miguez Gonzalez, L. (University of A. C. Santiago Caamaño, and V. (University of A. C. Diaz Casa, "On the applicability of real time stability monitoring for increasing the safety of fishing vessels," in Proceedings of the 13th International Conference on the Stability of Ships and Ocean Vehicles, 2018, vol. 159, no. September, pp. 76–87. doi: 10.1016/j.oceaneng.2018.04.002.
23. P. Golding and O. Donaldson, "A design science approach for creating mobile applications," 2009.
24. J. Kim, "The effect of design characteristics of mobile applications on user retention: An environmental psychology perspective," in 18th Americas Conference on Information Systems 2012, AMCIS 2012, 2012, vol. 6, pp. 4366–4374.
25. a an and a an, Cogni ion and en i on en un ioning in an un e ain o d e Yo aege , 1982.
26. B. G. Jayatilleke, G. R. Ranawaka, C. Wijesekera, and M. C. B. Kumarasinha, "Development of mobile application through design-based research," Asian Association of Open Universities Journal, vol. 13, no. 2, pp. 145–168, 2018, doi: 10.1108/AAOUJ-02-2018-0013.
27. Y. Ma and S. W. Harmon, "A Case Study of Design-Based Research for Creating a Vision Prototype of a Technology-Based Innovative Learning Environment," Journal of Interactive Learning Research, vol. 20, no. 1, pp. 75–93, Jan. 2009, [Online]. Available: https://www.learntechlib.org/p/25226
28. D. Basten, "Gamification," IEEE Software, vol. 34, no. 5, 2017.
29. I. Paliokas et al., "A gamified augmented reality application for digital heritage and tourism," Applied Sciences (Switzerland), vol. 10, no. 21, pp. 1–18, 2020, doi: 10.3390/app10217868.
30. M. A. Djebli, B. Hamoudi, O. Imine, and L. Adjlout, "The application of a smartphone in ship stability experiment," Journal of Marine Science and Application, vol. 14, no. 4, pp. 406–412, 2015, doi: 10.1007/s11804-015-1331-9.
31. "Sensor Coordinate System." https://developer.android.com/guide/topics/sensors/sensors_overview (accessed Jul. 27, 2022).
32. D. Terada, M. Tamashima, I. Nakao, and A. Matsuda, "Estimation of metacentric height using onboard monitoring roll data based on time series analysis," 2016. doi: 10.1007/s00773-018-0552-4.

33. R. Grin, S. F. Ruano, N. Bradbeer, and H. Koelman, "On the prediction of weight distribution and its effect on seakeeping," PRADS 2016 - Proceedings of the 13th International Symposium on PRActical Design of Ships and Other Floating Structures, p. 2016, 2016.
34. L. Huang et al., "Computational study on the potential transmission of COVID-19 virus on an Indonesian fishing vessel," 2022.
35. Q. Gao and D. Vassalos, "Numerical Study of the Roll Decay of Intact and Damaged Ships," in 12th International Ship Stability Workshop, 2011, no. 4, p. pp.277-282.
36. A. R. J. M. Lloyd, Seakeeping: Ship Behaviour in Rough Weather. A.R.J.M. Lloyd, 1998. [Online]. Available: https://books.google.co.uk/books?id=NHQqGwAACAAJ
37. S. N. Das and S. K. Das, "Determination of coupled sway, roll, and yaw motions of a floating body in regular waves," International Journal of Mathematics and Mathematical Sciences, vol. 2004, no. 41, pp. 2181–2197, 2004, doi: 10.1155/S0161171204305363.
38. B. Ali, T. Katayama, and Y. Ikeda, "Roll Damping Characteristics of Fishing Boats With and Without Drift Motion," in International Shipbuilding Progress, 2004, vol. 51, no. 2–3, pp. 237–250.

10. AUTHORS BIOGRAPHY

Andrea Grech La Rosa BEng (Hons)(Melit), MSc, MRINA is a Research Associate at University College London. His research interests include CFD, experimental testing and data analysis.

Mr Christopher Ryan BEng, MSc, is a PhD student of naval architecture at University College London. He was a research assistant in the SEDNA project (https://www.sedna-project.eu/), working on ship performance modelling, strategic research planning, accident analysis, and identifying future trends in Arctic shipping.

Dr Luofeng Huang BEng, MSc, PhD, AFHEA is a Lecturer in Mechanical Engineering at Cranfield University. Previously he completed MSc, PhD and Postdoc studies in the Marine Research Group of University College London.

Professor I Ketut Aria Pria Utama BEng, MSc, PhD, CEng, FRINA, FAIPI, is a Professor of Ship Hydrodynamics in the Department of Naval Architecture, Sepuluh Nopember Institute of Technology. He is fellow and the Vice President of the Asian Region of the Royal Institution of Naval Architects (RINA). He is also a fellow of the Indonesian Academy of Sciences (AIPI). His research interests include ship resistance and propulsion, experimental and computational fluid dynamics, marine renewable energy, and maritime education.

Soegeng Riyadi, ST, MT, MRINA, is Ph.D. student at Department of Naval Architecture Institut Teknologi Sepuluh Nopember (ITS). He has experience designing ships in several shipyards. Previous experience includes reducing ship resistance with foil at the back of the ship and investigating the operational mode of the fleet to achieve optimum fuel consumption.

Dony Setyawan, ST, M.Eng, is a PhD student of naval architecture at Sepuluh Nopember Institute of Technology. He is also a lecturer in the Department of Naval Architecture, Sepuluh Nopember Institute of Technology. His research interests include ship structure and strength, Advanced Ship Material and Ship stability

Dr Wolter Roberth Hetharia, FRINA membership No 00209111, currently holds the position of lecturer at Faculty of Engineering, Department of Naval Architecture, Pattimura University. He is responsible for the courses of ship design and resistance. Several papers have been published concerning ship resistance and fishing boats. He also presented some papers at RINA conference and other international conferences. He and his team executed several research projects funded by Department of Higher Education, Research and Technology.

Professor Giles Thomas is currently the BMT Chair of Maritime Engineering at University College London (UCL). He leads the UCL Marine Research Group and is responsible for their experimental ship and offshore research facilities. He has over 30 years of research and development experience in the fields of hydrodynamics and fluid-structure interaction, specialising in experimental work: both at model scale and full scale. Through his research work he is a three-time recipient of the Medal of Distinction from the Royal Institution of Naval Architects and was awarded the Vice Admiral E. L. Cochrane Award in 2019 by the Society of Naval Architects and Marine Engineers.

VEHICLE HOT SPOT DETECTOR AND DANGEROUS GOODS DETECTOR TO FIRE IGNITION PREVENTION IN RO-RO SHIPS

M Haas, H Brumm, L Wallimannand and **T Oe**, SICK AG, Germany
A Marrero, CENIT Group for Innovation in Multimodal Transport at CIMNE (UPC), Spain

SUMMARY

Nowadays, all cargo (except dangerous goods) is loaded without consideration to the hazards they pose, opening the door to improvements from the fire prevention point of view by means of a custom and objective management of the units based on the potential hazards they involve. The technical approach presented here takes the systems currently used in parking lots, tunnels, tolls or logistics as starting point but applied to the maritime environment. Specifically, both Vehicle Hot Spot Detector (VHD) and Vehicle Dangerous Goods Detector (VDG) have been modified and extended with developments that now consider usual loading units in Ro-Pax, Ro-Ro and Vehicle Carrier ships. Also, a refactoring has been performed aiming at adapting the system to the environmental conditions in which it must work. The dates are used in cloud based new online stowage tools (Stowage Planning Tool) for loading optimization. This enables more safety on board.

NOMENCLATURE

2D	Two-dimensional
3D	Three-dimensional
APU	Application-Processing Unit
DDGG	Dangerous Goods
Hazmat	Hazardous materials
IR	Infra-Red
IRFPA	Infra-Red Focal Plane Array
LADAR	Laser Detection and Ranging
LASH FIRE	Legislative Assessment for Safety Hazards of Fire and Innovations in Ro-ro Ship Environment
LiDAR	Light Detection and Ranging
RADAR	Radio Detection and Ranging
SPT	Stowage Planning Tool
TEMS	Traffic Enhanced Measurement System
TOF	Time-of-flight
VDG	Vehicle Dangerous Goods Detector
VHD	Vehicle Hot Spot Detector
WP	Work Package

1. INTRODUCTION

LASH FIRE project an H2020 funded project (Grant Agreement #814975) in which this publication is framed aims to develop maritime fire safety solutions with innovative technologies, operations and applications. LASH FIRE will provide a basis for the revision of international maritime regulations and gives European industry knowledge to build safer and more competitive ships for sustainable transport [1]. One of the work packages (WP) of this project is Ignition Prevention (WP08). The purpose of the work performed in WP08 is to reduce the ignition of potential fire in cargo due to electrical, mechanical, overheating and other causes. The impact of WP08 is a minimised risk for both ship and crew (and passengers in Ro-Pax ships).

One of the key issues for safety on board of ships is to avoid already overheated vehicles, which may ignite, from entering on-board. In the LASH FIRE project one of the solutions proposed is to identify potential hazards leading to fire ignition situations at terminal and on-board the ship. The outcome of this assessment will be the possibilities for sensor positioning at the terminal and at a ship. To achieve this goal LASH FIRE project, propose to adapt a well-approved and commercially available sensors system, from road systems to the ships boarding procedures and to the maritime environment. The approved detection system from road systems has a wide range to detect objects, heat, clouds of smoke and verify the sensor signals with a library. The system compares the actual measured values and identify possible overheated sections of the detected object as well sources for ignition or spark points.

The goal of this paper is to test the VHD (Vehicle Hot Spot Detector) system installed on a terminal and on a ship in order to prove the concept and sensor-based system. The data collected help to improve the system and the algorithms in order to design a dedicated system for "hot spot detection for vehicles entering a terminal and a ship".
The VHD will be adapted to the maritime needs and with the optimization of the 3D-library for vehicle identification, the segmentation of vehicles entering a ferry and the identification of relevant set points of temperature limits are the main targets. This means if the VHD is a system which allows reducing the danger of ignition sources entering from outside of a ferry, the costs of adaption of the VHD to the maritime needs are paid off and the system can be available for the whole maritime industry under a reasonable investment.

2. BACKGROUND

Systems for the field of road transportation in order to avoid vehicles, which are overheated and bear a potential risk, from entering tunnels has been developed and installed for several years. The principle is to let vehicles drive through a gantry crossing the road and check if some parts of the vehicle are reaching a critical temperature, which is a sign for possible future ignition. If not, the vehicle will continue its journey and enter the tunnel, if yes, it will be stopped before entering the tunnel.
The system called Vehicle Hot Spot Detection is using a combination of two-dimensional scanning and thermal technologies. LIDARs (or scanners) and thermal IR cameras are the core sensors of the system and are delivering the raw data, which are computed locally (data fusion) in order to build a three-dimensional thermo model of a vehicle. The local computer (APU) analyses the thermo model and identifies the hot spots. The road operator is getting the information in the control room and can apply his safety strategy. The VHD system in combination with other sensors such as photo cameras can also deliver further safety related information, for example hazard placard recognition and reading.
The state of the art is to configure a gantry with the right combination of scanners and thermal cameras. The layout of the gantry is depending on the design of the road, ramp to a ship or on the ramp in the ship and the type of the vehicles.
Lots of experience has been collected on road transportation. A vehicle boarding on a ship is a very similar situation as a vehicle entering in a tunnel. In both cases, it is possible to perform a control upfront and apply a safety procedure in order to avoid the boarding in case of high risk.
In the LASH FIRE project, the specific boarding conditions on ships has to become transparent to get an understanding in which directions a VHD system has to be adapted to the ships boarding procedures and to the maritime environment. Another technical challenge is to create real 3D imagines of specific vehicles entering ferries, which are port or ferry specific. New 3D generated images have to be added to the VHD – database to enlarge the database. Based on new generated images of port or ship specific new vehicles a clear identification and localization of hot spots on the vehicles or on parts of the vehicles is one of the clear objectives of the project. [2]
The defined target of the VHD system in the LASH FIRE project is to demonstrate that a three-dimensional thermo model of a vehicle ensures to give relevant information during the boarding procedure.

2.1 DESCRIPTION OF THE TECHNOLOGIES

Under this section the sensor technologies and how the data fusion and the reporting of the computed information to the operator is realized. 2D LIDARs and thermal IR cameras are delivering the raw data, which are computed locally (data fusion) in order to build a 3D thermo model of a vehicle. The local computer (APU) analyses the 3D thermo model and identifies the so-called hot spots. The temperature of the hot spots in combination with a risk assessment strategy (max temperatures limits) allows an operator to take decisions (loading/not loading).

2.1 (a) 2D LiDAR Sensors

LiDAR (Light Detection and Ranging) sensors have already been in global industrial use for decades. The sensors protect people and enable process flows to be automated. They work both indoors and outdoors. Port automation, traffic management systems, and object protection systems are just some examples of the potential applications for this technology [3].
LiDARs are sensors that use lasers to perform non-contact distance measurements and have become an absolute integral part of automation engineering. Its development first began in the form of TOF (time-of-flight) measuring technology. TOF has since been largely replaced by the more accurate terms LADAR (Laser Detection and Ranging) or, most commonly, LiDAR. The terms LADAR and LiDAR are based on the popular term RADAR, which stands for Radio Detection and Ranging [4]

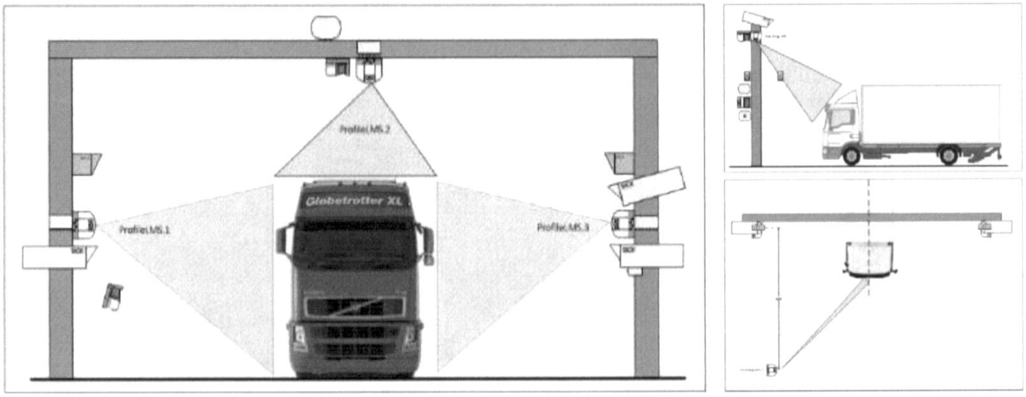

Figure 1: Assembly of the 2D LIDARs. Source SICK AG

2.1 (b) Cargo Identification Technology – 3D Profiles

To get a clear identification of each vehicle passing the terminal where the scanners are located, an addition scanner track the speed of the vehicle and identify when the complete vehicle passed the scanner terminal. This data´s are send to the APU where the raw data´s are calculated by algorithms to a 3D-Model. This 3D-Model is than compared with the 3D-models stored in the library in the APU computer. The result is a clear identification of the specific vehicle. As well, the exact length, height, width and volume, of the vehicle is an output of the APU.

2.1 (c) IR Cameras

Infrared (IR) thermal imaging, also often called thermography for short, is a very rapidly evolving field in science as well as industry owing to the enormous progress made in the last three decades in microsystem technologies of IR detector design, electronics, and computer science. Thermography nowadays is applied in research and development as well as in a variety of different fields in industry, such as non-destructive testing, condition monitoring, and predictive maintenance, reducing energy costs of processes and buildings, detection of gaseous species, and many more areas [5].

2.1 (d) 2D Thermal Images

The Application Processing Unit (APU) computes locally the IR measurements and is generating a 2D thermal image of the vehicle surface temperature. The temperature range of the camera is from 0°C (32 °F) up to 500°C (1112 °F). An ambient temperature sensor installed near the IR camera defines the temperature reference. The IR-camera measures the surface temperature while the vehicle passing the terminal where the scanner and the IR-camera is positioned [6].

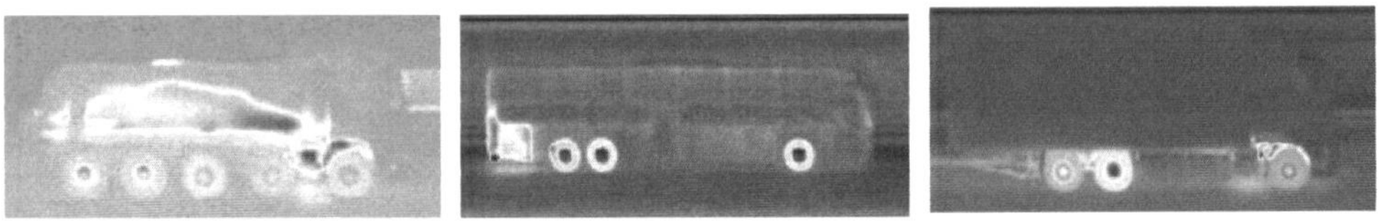

Figure 2: Example of 2D thermal images. Source SICK AG.

2.1 (e) 3D Thermo Model

After computing the 3D profile, which is generated from the data from the scanners and the 2D thermal image form the IR-camera, the APU generates a 3D thermo model (Figure 3). The generated 3D thermo model shows exact, which surface temperature, is on each part of the vehicle and shows the temperature ranges.

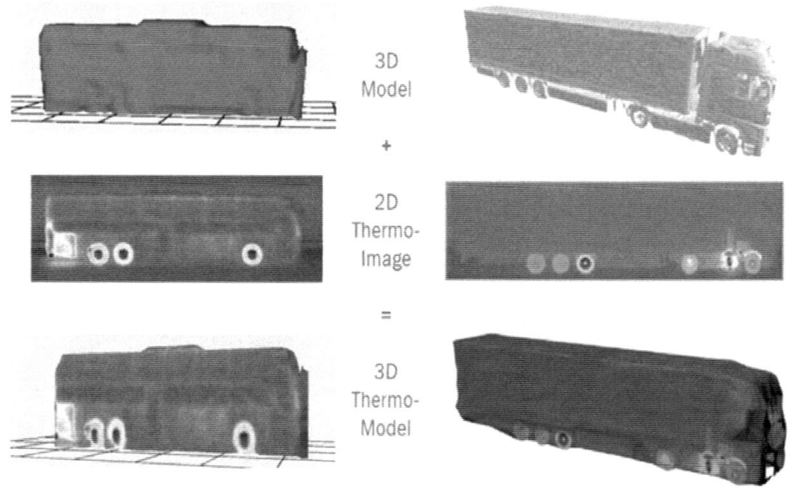

Figure 3: Mixing 3D profile and 2D thermal image to create the 3D thermo mode. Source SICK AG

2.1 (f) Hot Spot Detection

When the 3D thermo model is established, the APU is first classifying the vehicle between the 28 different vehicle classes available in the library. In the APU a grouping of vehicles is predefined. There are pre-defined groups characterize clear specific thermal classes such as truck, bus, van, car, and bikes or miscellaneous which are relevant for the function of the VHD.

For each thermal class, different segments are applied which have their own temperature limits (threshold). The typical segments are the wheels (green), the engine – exhaust pipe (red), the passenger zone (yellow), the loading zone (purple) and the contour (grey). The segmentation allows a quick and fully automated hot spot detection and alarm in case the measured temperature exceeds a defined limit value. It maximizes the sensitivity and minimizes the risk of false alarm.

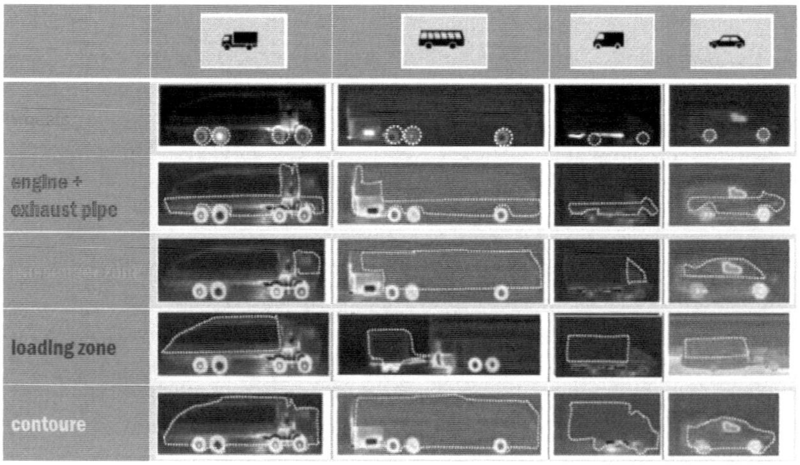

Figure 4: Segmentation of the different vehicle thermal classes - example. Source SICK AG

2.1 (g) Hazardous Goods Plate Recognition (VDG)

Hazardous goods plate is normalized signs visible on a vehicle when transporting hazardous materials. The hazardous goods plate contains two information. On the upper part, the hazard identification number (also called Kemler code) which identifies the risk related to the material and the lower part shows the UN code of the material.

When a truck transporting hazardous materials is passing the detection gantry, a camera detects and recognizes the hazardous goods plate based on its shape and his location on the truck. The system reads its content and deliver the two information: the hazard identification number and the UN material code.

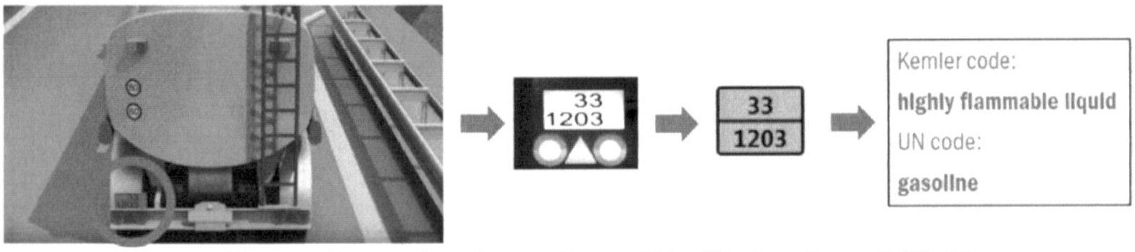

Figure 5: Hazmat placard recognition and identification. Source SICK AG.

2.2 VEHICLE HOR SPOT DETECTION SYSTEM FOR ROAD APPLICATIONS

The VHD system for road applications delivers information that the road operator uses to avoid a vehicle entering a tunnel when the system has detected hot spots, which may ignite a fire.

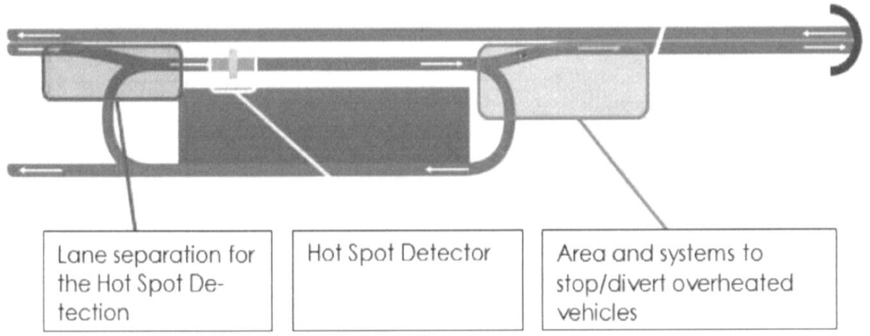

Figure 6: Take-out scheme with VHD system. Source SICK AG.

The road VHD system consists in a gantry crossing the road under which traffic is flowing. The gantry is equipped with 2D LIDARs and thermal cameras. The Application Processing Unit (APU) is located nearby and is in charge to compute all information in order to deliver a 3D thermo model, the classification of the vehicle the segmentation to assign a potential hot spot.
The cabinet on the road-/ terminal side includes the power and signal distribution and the Application Processing Unit (APU) for data processing. From the APU the signals are distributed to a server room, dashboard or local control room.

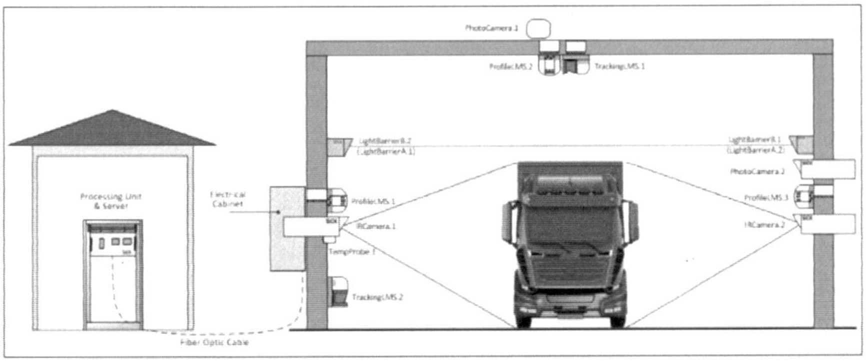

Figure 7: VHD system description. Source SICK AG.

The electrical cabinet mounted near the gantry is powering the sensors and collecting all data in the APU, which is in charge of all calculations and visualizations. A fibre optic connects the electrical cabinet with the customer's system.

Digital outputs are also available to connect to traffic lights or automatic barriers. The browser-based software and the configuration of the system can have adjusted remotely. Based on the access the generated data can visualized on the operator client PC or dashboard to follow during the loading procedure. With access to the whole operator software, the VHD system can be configured and set points changed.

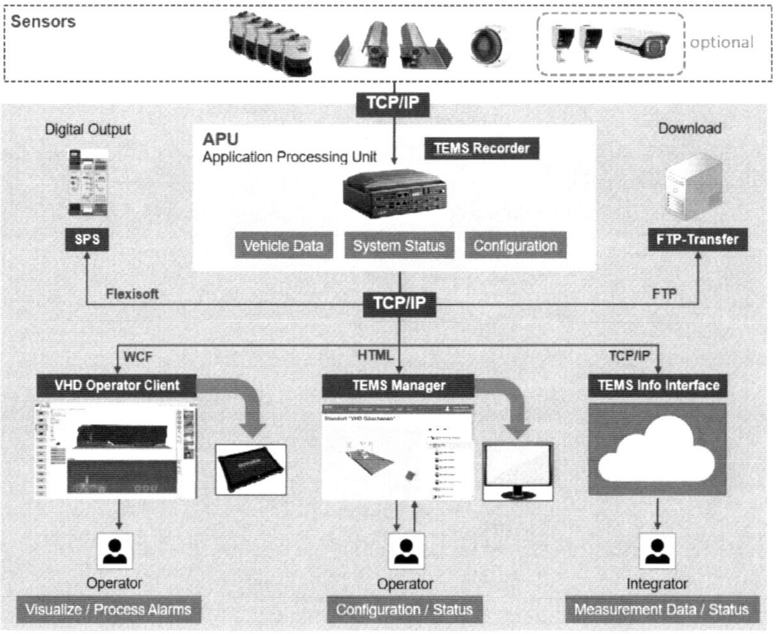

Figure 8: Architecture / Interface layout. Source SICK AG.

The 3D-thermo images give a clear picture of the loading zones and cargo. Hot spots on the surface of the loading zone of tailors, on tanks, on containers or tarpaulin of the vehicle, etc. indicated that mouldering- or open- fires or chemical reactions, but not limited to that, are present inside. The heat of the fire inside of the loading is transferred to the surface. The hot spot can be detected while measuring the surface temperature through the VHD system. Open fires on vehicles or loadings can be detected.

3. VEHICLE HOT SPOT DETECTION AND HAZARDOUS GOODS PLATE RECOGNITION FOR SHIP BOARDING

One way to minimize the risk of fire on-board is to avoid already overheated objects entering the ship as the object itself or surrounding objects might ignite because of too high temperature. The procedure for boarding vehicles on a ship is very similar to monitoring and controlling vehicles entering a tunnel. In both cases, it is possible to add a control step before the vehicle enters the enclosed area and apply a safety procedure in order to avoid or modify boarding in the case of high risk.

For the LASH FIRE the control step was to install a measurement for hot spot detection, conduct a 3D thermal profile and Hazard placard and number plate recognition. With a prove the concept of early hot spot detection, and thermal profiling of vehicles boarding the ship the VHD was tested under real loading procedures on a STENA terminal in Gothenburg [2]. All collected information from the sensor data from the VHD and VDG are going to an PC with the evaluation and operator client software. From the PC the actual data from the online monitored vehicles can be transfer the Stowage Planning Tool (Stowage Planning Tool featuring a fire hazard management that supports the loading process by means of guidelines and recommendations or even suggesting appropriate placement of the cargo based on historical data from the database and additional constraints) [1]. The information's from the VHD and VDG can sent also to a cloud based online ship-loading tool to have relevant information's on one place and centralized. The data´s can than used from the ship operator, ship owner, insurance companies and fire fighters or for immediate use or storage of each identified vehicle.

3.1 HOT SPOT DETECTION

Depending on the port configuration, it is possible to install the sensors on the already existing VHD gantry on board of the ship or on the quay, directly at the entrance point.
Starting with the installation on the terminal side and establish the functionality, data transfer to the (cloud based) boarding tool and an overall get feedback of the functionality, the VHD can paced/mounted, in a further step of the project, direct to the ramp or loading arm of a ferry. This ensures that the VHD also works on board during loading the ferries and under operational conditions when a ship is moored in port.
With an on-board installation, the ship is not dependent upon the port infrastructure and can therefore ensure the highest safety level, when sailing to destinations without high level of infrastructure.

3.2 3D THERMAL PROFILING SYSTEM

Ro-Pax, Ro-Ro and Vehicle Carriers are carrying a numerous type of different vehicles or tugs bring trailer on board, which for some of them cannot travel on open roads. The extension of the VHD vehicle library is an outcome of the project. This means that for ship boarding, the library of VHD was updated to include port specific vehicles as well vehicles, which are not allowed to operate on roads. New 3D thermo profiles were created during the tests. The models became part of the vehicle segmentation and individual set points in term of critical surface temperatures was set in the project to be able to release alarm before port specific vehicles (e.g. tugs) or trailer entering the ferry.
The VHD software is connected between the server and the information's can send in real time to a tablet, e.g. stowage planer, for visualization of the measured vehicles in real time and confirm or process alarms. The clear image and a picture of the detected vehicle helps to identify the vehicle, generating an alarm and illustrate the reason for an alarm. The information´s coming from the VHD give also the volume, size, number of axes to the loading PLC or to the interactive stowage plan.
For dangerous goods, when passing the VHD gantry before entering the ship, the system will recognize (by detecting the hazardous goods plate) and identify the dangerous goods and the vehicle (by reading the vehicle number plate). This information will be made available in the VHD server as well and added to the stowage plan.

3.3 TEST INSTALLATION

A demonstration version of the system was being installed at the Majnabbe terminal (Stena terminal in Gothenburg, Sweden) for test purposes. The VHD gantry was be installed on the quay in order to generate as many measurements as possible. One of the major challenges is to get clear scans and thermal images. If a road offers a linear flow/displacement, the boarding on ships can be much more complex.

The main limitation of the VHD on a terminal or on board is that the library of the vehicle identification is limited to vehicles running on roads. The starting point for the VHD is that the system has a vehicle classification of 28 different classes of vehicles. Currently the VHD has five groups of thermos classes (trucks, bus, van, car, motor bikes). Harbour and ferry specific tugs, trailer with or without cooling-/heating systems, which enter the ferries for loading and unloading purposes, have to be identified with the VHD. The 3D scans will be created with the software in the VHD system and the library with the new models of harbour/ferry specific vehicles has to be extended. In addition, the segmentation in different vehicle zones such as wheels, engine-/exhaust pipe, passenger zone, loading zone, contour and heating/cooling units must be adapted. The VHD must also adapted on hybrid- and electric cars, because this is a new type of a potential source of ignition. Individual temperature limits according to vehicle class allows fully automated alarming and leads to maximized sensitivity. Minimization of false alarm ratio is another goal to get.

Continuous measuring the exact length of the vehicle is crucial for the VHD. Otherwise, the thermal images are squeezed or stretched and for example wheels will not be detected as wheels as they are oval instead of round. The new software algorithm adapted to ship environmental works like a pincer and measures the vehicle position from the front and the back during the measurement which improves the performance of the system.

During the commissioning phase the total length measurement of the vehicle in a correct manner was not possible. The light beam of the scanner, mounted on exit part of the terminal cannot monitor the total length of the vehicle, because the light beams are blocked by the roof of the terminal. For this reason, the terminal was modifying in a way that the measuring of the total length of the vehicle is possible. For this Openings in the roof of the terminal were done and a new alignment of the scanner was possible and carried out. With this modification the measurement with an accuracy less than +/- 400mm of the length of the vehicle will be possible.

So, for LASH FIRE VHD the TEMS (Traffic Enhanced Measurement System) Version was upgrade, the new algorithm will further improve the accuracy of the system and it will decrease the number of faulty alarms due to its new approach to the data evaluation. This made possible to:

- integrate the nameplate and dangerous goods cameras (provide number plate and dangerous goods plate output)
- integrate the 3. IR camera (measure the temperatures from the top) Thermal pictures are collected to evaluate the best temperature threshold and further optimize the detection of dangerous hot spots on cooling units.
- measure the truck length more accurate (with the new developed software feature in the algorithm) The length measurement was improved with a new measurement algorithm. The improvement in accuracy has been confirmed by Stena line. The optimization of the length measuring was therefore completed successfully.

Figure 9: Opening in the roof of the terminal. Source LASH FIRE Project.

The VHD system is designed for harsh outdoor conditions especially for tunnels in the Alps. In winter, salt powder is distributed on the roads and salt spray reaches the components of the VHD system. Temperature gradients in the Alps over the day and night and over the whole year stresses the equipment as well. Beside the harsh climate conditions vibrations through the traffic also hits the VHD system. SICK was started the test with the standard mechanical outdoor protection system, like painting, weather huts, stainless steel brackets and an electronic cabinet, designed for rough outdoor conditions. In the project the protection class (IP rate, type of painting, chloride resistant steel for the brackets) has been analysed.

As a result of the first tests, the first images of the VHD operating in a maritime environment have been obtained successfully.

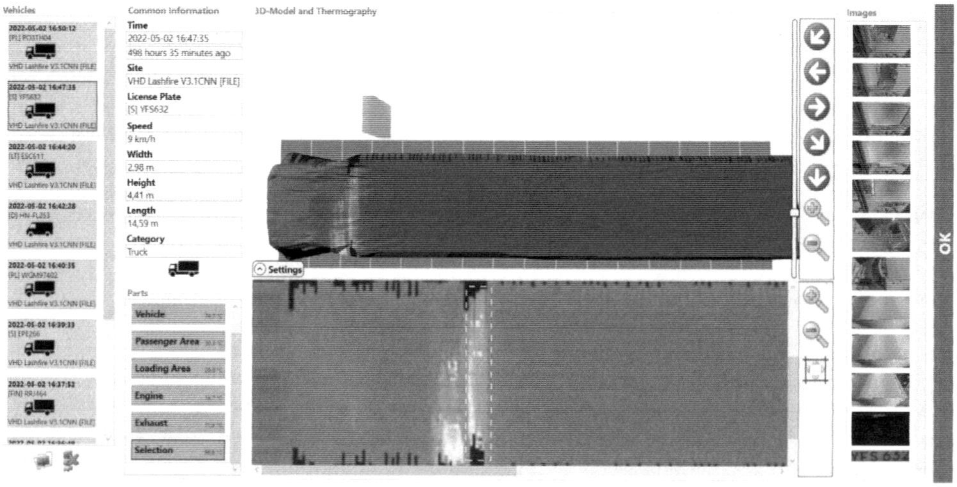

Figure 10: First look at LASH FIRE VHD. Source LASH FIRE Project.

4. CONCLUSIONS

As a result of the extensive experience gathered in the road transport of this type of system, the main objective of this paper, which is to adapt the VHD and VDG system to a maritime environment and ship loading, has been achieved.
The VHD system which is installed in the maritime ambience is a well-engineered sensor system with defined parts and a database for vehicle identification. Operator defined set points of maximum allowed surface temperature for detected hot spots in the software. The basic structure of the VHD can be implemented on nearly each port terminal or on board of a ship. Ship operators with fixed routes or own terminals can install the system in the terminals or on board and integrate the information in their digital logistic management tool.

The information generated from the tests with the VHD close a gap that undetected overheated parts on a vehicle, which are a real source of fire ignition or real fires, can enter a ship. If hot spots are detected automatically before a vehicle enters the ship, the access to the ship can be prohibited while closing a boom. As well, the information which is generated in the VHD can be used online in the boarding stowage plan for optimization of the loading procedure.
One objective of early hot spot detection is to prevent a fire, which is brought from outside of the ship to the ship itself. During scanning of the vehicle also the surface of the load is scanned and a three - dimensional thermo model of the surface of the load (e.g. container, trailer, heater-/ or cooling units) is generated. Such kind of hot spot detection system can also be applied to container terminals to scan the surface of containers during loading a container carrier to avoid fire ignition on board. Container carrier often ship thermo containers where the aggregate can also be damaged, and overheating accrue, and a fire can ignite. Due to goods inside the containers, which release heat, e.g. while damaged or chemical reactions happen or batteries/cotton ignite, the surface of the container can become overheated, and the hot spots are detected when a VHD is in place.

The outcome of the terminal tests with the VHD has provided the basis for online data on size, dimensions and thermographic images for the digital logistics management tool with decision support and risk-based cargo planning. The results of testing this system in a maritime environment will generate technical specifications for possible recommendations for hot spot detection on any type of vehicle entering in Ro-Pax, Ro-Ro and Vehicle Carrier.

As a future line of research, it is proposed to analyse the VHD system installed on the ramp of ships or on-board. A VHD-system placed on the ramp or on the ship, would give a last-minute update on the "units" as they roll by the portal. The closer to departure, the better in the perspective of ship safety. It would require a fast procedure to stop the loaded unit to enter the ship due to the higher "cost" involved with backing off a rejected unit from the ship e.g. slowing down or stopping the loading procedure. Furthermore, in the case of the on-board installation of the system classification societies have to approve the system if it is installed on board and could give recommendations to the flag states that safety on board is increased if a hot spot detection system is in place.

5. ACKNOWLEDGEMENTS

The LASH FIRE project has received funding from the European Union's Horizon 2020 research and innovation programme under Grant Agreement No 814975.

The authors would like also to thank the rest of the partners of the project Consortium for all the interesting discussions that have taken place about the different approaches of the solution in the context of LASH FIRE as a demonstrator prototype. Finally, special thanks to the operators that actively participate in LASH FIRE as partners as well as the members of both MOAG and MAAG Advisory Groups for providing valuable information.

6. REFERENCES

1. LASH FIRE. (2022), Home page, available online: https://lashfire.eu/ (accessed on 13 July 2022).
2. LASH FIRE. Deliverable D08.1 - Definition and parametrization of critical fire hazards, classification of cargoes, transport units, engines, fuels and ships and identification methodologies (June 2021) available online: https://lashfire.eu/ (accessed on 05 June 2022)
3. SICK Sensor Intelligence. 2D LiDAR sensors. Available online: www.sick.com/2D_LiDAR_sensors (accessed on 25 July 2022)
4. Weber, H. (2018). LiDAR SENSOR FUNCTIONALITY AND VARIANTS. SICK AG.
5. Vollmer, M., & Möllmann, K.-P. (2018). Infrared Thermal Imaging, Second Edition. WILEY-VCH Verlag GmbH & Co.KGaA.
6. Rai, M., Maity, T., & Yadav, R. (2017, July Volume 6, Issue 2). Thermal imaging system and its real time applications: a survey. Journal of Engineering Technology (ISSN: 0747-9964), pp. 290-303.

7. AUTHORS BIOGRAPHY

Markus Wolfgang Haas holds the current position of Technical Industry Manager Power & Combustion / Maritime in SICK AG. He is Dipl. Eng. (FH) Process/Environmental Engineering for University of Applied Since Offenburg. He has more than 13 years' experience as project manager in the field of EPC, plant operation and services. More than 11 years of experience in analytical chemistry in SICK AG, in the field of emission monitoring of exhaust gases from large combustion engines for maritime applications. After some years working as technical industry manager for Power and Cement worldwide, now working since more than 8 years as a specialist in emissions from ships and maritime applications with LiDAR sensors.

Hinrich Brumm is completing his master's degree in environmental protection technology in Hamburg, Germany, and started his career in 1997 at the classification society Germanischer Lloyd. In the following 15 years, he was responsible for the maritime approval of engines and products and worked also on various committees, including at the IMO in the development of regulations. From 2010 to 2014 he was Managing Director / Senior Consultant at MARENCON, Germany. Hinrich Brumm held the position of Strategic Industry Manager Maritime & Combustion Engines at SICK AG since 2014.

Lukas Wallimann holds the current position of Project Manager CPM. in SICK AG. He is Certified technician systems engineering (automation), Advanced Federal Diploma of Higher Education. He has more than 8 years of experience in the automation industry and since more than 4 years responsible for product development projects as a project- and product manager. Currently responsible for the project and product management for hot spot detection systems within SICK AG.

Takatsugu Oe holds the current position of Head of Process Automation Sales Department in SICK K.K. He has more than 12 years' experience of sales and service business development in the field of Oil & Gas, Maritime, Chemical, Petrochemical, etc. and sales marketing as a manager. 6 years' experience in SICK K.K. since a launch of Process Automation Sales Department as a responsible head of the business and more than 4 years of sales activity in emission gas monitoring system and automatic identification system for Maritime application.

África Marrero holds, since 2019, the position of Research & Development Project Manager at CENIT, the Group for Innovation in Multimodal Transport of the Transport RTD Area at CIMNE, the International Centre for Numerical Methods in Engineering, a research organization created as a partnership between the Government of Catalonia and the Technical University of Catalonia. Bachelor of Science + Master of Science in Naval and Oceanic Engineering from the Polytechnic University of Cartagena. She is finishing a Master in International Business from Universidad del Atlántico Medio. She began her professional career as a planning and control engineer in submarine construction projects at the Navantia shipyard in Cartagena. Later, she worked as an Associate Ship Manager in Astilleros Canarios (ASTICAN Shipyard), mainly carrying out work control functions in oil rigs.

IoS-OP: INITIATIVES FOR SHIP OPERATION DATA COLLECTION, DISTRIBUTION, AND UTILIZATION

Y Ikeda and **V Sharma**, Ship Data Center, Japan

SUMMARY

This paper introduces the legal arrangements and activities for data sharing within the IoS-OP implemented by ShipDC for the promotion of ship IoT data collection, distribution and utilization.

1. INTRODUCTION

IoS-OP (Internet of Ships Open Platform) is a framework that is composed of a data platform which has mechanisms for data distribution according to common rules that are agreed to within the industry and for onshore storage and distribution of that data, to realise the smooth, fair and equitable distribution of ship operation data. Not only will utilizing this framework make it easy to use data but it will also enable data sharing with a high level of transparency without incurring any disadvantages.

Ship Data Center (hereinafter, "ShipDC") was established in 2015 as a 100% wholly owned subsidiary of Nippon Kaiji Kyokai (NK) to promote the digitization of the industry and provide a space for co-creation. ShipDC provides the data platform as a space for fair and equitable data sharing and as an ecosystem by operating this framework together with private companies. In 2018, the IoS-OP Consortium was launched, and the data distribution business was started through IoS-OP.

This paper describes the legal issues and the activities undertaken through private initiatives for IoS-OP data sharing.

2. IoS-OP

As shown in Figure 1, IoS-OP designates the common rules established for data handling (software side) and the data center (hardware side), which is the onshore storage vault for the data sent from the ships, as the cooperative domain, and provides an environment in which the members can focus on the creation of new forms of innovation which is the resulting competitive domain. In addition, IoS-OP is a framework that increases opportunities for data utilization not only in the maritime industry but also outside of the industry and can contribute to the safe operation and efficiency of ships as well as ESG and SDGs. These activities are being implemented through the IoS-OP Consortium which is the ShipDC members' organization.

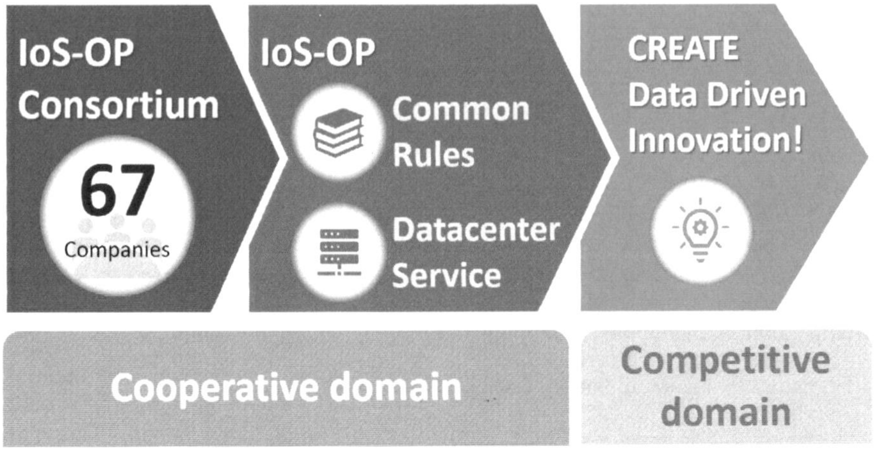

Figure 1. Cooperative and competitive domains

2.1 IoS-OP CONSORTIUM

The IoS-OP Consortium was launched in April 2018 and has mainly worked on arranging the legal rights regarding data utilization and application, examining the utilization and application needs, and promotion, etc. Recently, the Consortium has shifted to solving issues that have become apparent in the actual business usage and activities to promote data utilization and application.

At the time it was launched, the IoS-OP Consortium consisted of 44 member companies, and as of the end of August 2022 that number has expanded to 67 companies. The activities of the IoS-OP Consortium are primarily carried out within the Japanese maritime cluster, but because there are no established restrictions on nationality or business type as requirements for participation, companies outside of Japan and IT-related companies outside of the industry are also participating and expanding their activities worldwide.

The IoS-OP Consortium is structured as follows with the promotion of data distribution, legal reviews and other activities being continuously carried out through private initiatives.

- General meeting which revises terms and conditions and approves business plans and budget plans, etc.
- Steering Committee which carries out the operation principles of the IoS-OP Consortium, matters concerning the operation of the Sub-committee, establishment and operation of the Working Groups (WG), election of leaders, and the amendments or abolition of the IoS-OP Terms of Use, etc.
- Sub-committee which discusses and proposes the operation principles and makes adjustments between each WG, etc.
- Rule Development & Data Governance WG、Solution WG、Business Promotion WG、Human Resources Development WG

2.2 WORKING GROUPS

The IoS-OP Consortium not only consults and discusses so that the basic rules within the actual data distribution are in accordance with the operation but also conducts a diverse range of activities such as planning, hosting seminars and human resource development to promote data utilization and application. These activities are divided among the following four WGs.

- Rule Development & Data Governance WG
  Debates decisions regarding new issues as needed such as the discussion and formulation of rules regarding data utilization and application within IoS-OP and legal arrangements concerning data transactions, etc.
- Solution WG
  Extracts and discusses issues within solutions utilizing IoS-OP, formulates security guidelines to ensure data confidentiality and holds discussions to solve issues from the data users' perspective. Recently, this WG has held review meetings concerning data quality.
- Business Promotion WG
  Plans and implements various seminars for the promotion of data utilization and application as well as the continuous implementation of IoS-OP public relations and dissemination activities. Seminars are held not only within Japan but also Singapore and Taiwan, etc.
- Human Resources Development WG
  Plans and implements human resource development within the industry such as ideathons and operational performance analysis study meetings.

3. COMMON RULES

The common rules regarding data handling (IoS-OP Terms of Use) were repeatedly discussed and formulated within the IoS-OP promotion conference, which is made up of 55 companies from the maritime industry, with a focus on achieving secure data distribution without the stakeholders being placed at a disadvantage. In accordance with the IoS-OP Terms of Use, which are the common rules, the IoS-OP Consortium aims for a data-driven world and provides a secure, safe, fair and equitable space for data sharing.

In formulating the common rules, data ownership, the roles of stakeholders and the disadvantages of data sharing were discussed and organized from both the data providers' and users' perspectives so that neither would be placed at a disadvantage.

3.1 DATA OWNERSHIP

Defining data ownership is the key to controlling data access. In the IoS-OP, data ownership is defined as the authority to utilize data on an agreement basis, and the platform user (PU) is the entity who has the authority to utilize data on the IoS-OP.

For the PU to have the authority to utilize data and provide the data to the IoS-OP, it must establish who, between the stakeholders involved in the data generation and collection process (e.g., the data generator and data collector), has the authority for the utilization of data.

It is often the case that the data collector become PU, however, any entity such as a ship manager, shipyard etc. can become PU.

In general, the data generator is considered to be the shipowner, therefore, if stakeholders other than the shipowner (in many cases, the data collector) wants to become the PU, then consultation and arrangement with the shipowner is required for the establishment of the authority to utilize data, which would be determined based on the contribution to the generation and collection of the data as below:

- Who will own shipboard data collecting device
- Who will bear cost for installation of shipboard data collecting device and/or satellite data communication
- Who will bear cost for maintenance work of data collecting and /or device
- Who is contractor for the commission of PP
- Who has the responsibility of data management
- Who has the responsibility of security management of data

3.2 DEFINING THE ROLES OF STAKEHOLDERS

After clarifying the rights of stakeholders from a legal perspective, it was necessary to clarify the role of each stakeholder according to the data flow from data generation to usage. Therefore, as shown in Figure 2, using the data flow as the basis, the data usage manager (PU: Platform User), shipboard data server seller (PP: Platform Provider), onshore solution provider (SP: Solution Provider), data user for the purpose of contributing to ship operation (SU: Solution User), and the data user for the purpose of improving their products (DB: Data Buyer) were defined.

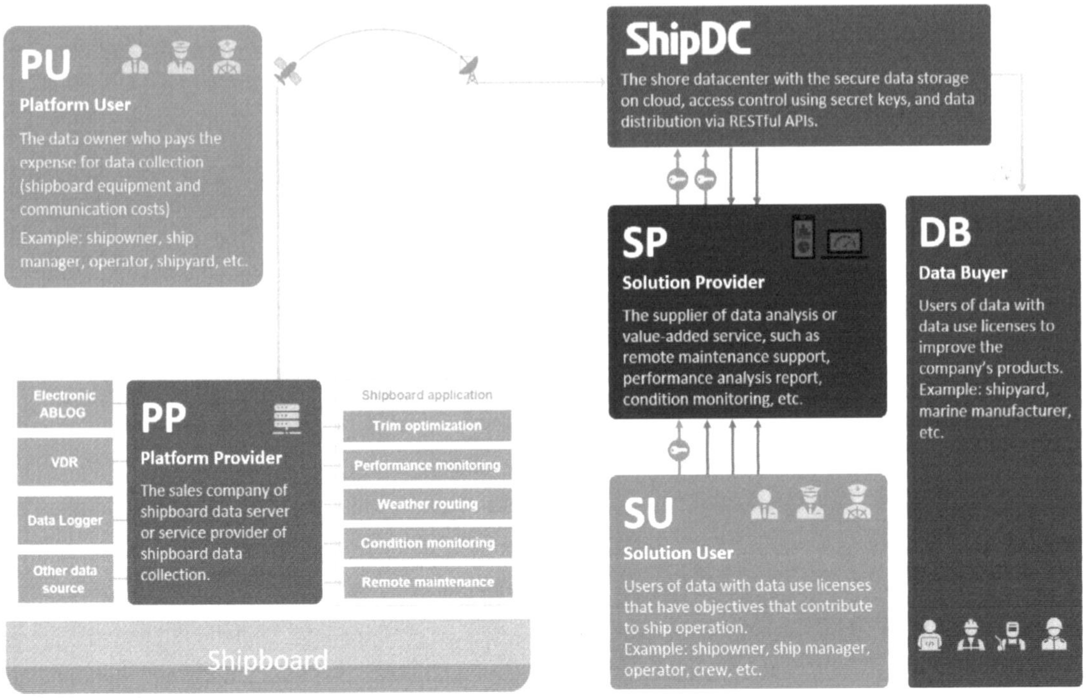

Figure 2. Roles of stakeholders

3.3 MANAGING DATA SHARING DISADVANTAGES

To ensure that the parties who store the data, parties that use the data or businesses that are involved in data collection and distribution can securely share the data, the IoS-OP Terms of Use (common rules) prohibit the sharing of data with

competing companies as shown in Table 1. This means that data cannot be freely shared even by a data manager and it is based on the belief that data should be shared based on public policy.
Furthermore, in the event that new issues regarding data sharing arise, they will be debated in the IoS-OP Consortium as needed.

Business operator attributes		Data Buyer	Solution User
Companies pertaining to the ship	Shipowner and ship manager	May purchase	May be permitted
	Charterer	May purchase	May be permitted
	Ship building yard	May purchase	Not applicable
	Manufacturers of the machinery, equipment, and other onboard devices	May purchase	Not applicable
	Insurance company	May purchase	Not applicable
Competitors of the companies above	Shipowner and ship manager	Not available	Not available
	Charterer	Not available	Not available
	Shipyard	Not available	Not available
	Manufacturers of the machinery, equipment, and other onboard device	Not available	Not available
	Insurance company	Not available	Not available
Third parties	Classification society, research institute, analysis provider, etc.	May purchase	Not available
Where not appliable to the above AND where ShipDC could find that it will not harm the profits for stakeholders based on the advice and reports from Rule Development & Data Governance WG.		May purchase (Conditional)	Not available

Table 1. Data access matrix

3.4 REDUCING THE BURDEN OF CONTRACT WORK

By defining the stakeholders, each of their relations of right and covenants were clarified, and as shown in Figure 3, their required terms were defined. These terms are one of the unique features of IoS-OP. As the IoS-OP users agree to a common set of terms, it eliminates the complexity of individual contracts signed by each user while also providing a space for transparent data provision and usage.

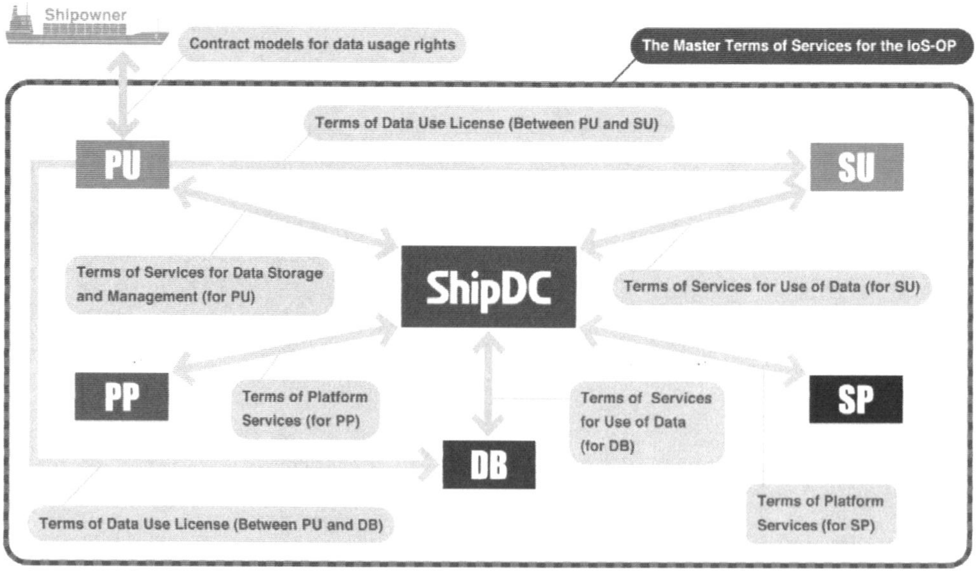

Figure 3. IoS-OP Rules

4. ShipDC AS SHORE DATA PLATFORMER

ShipDC provides not only onshore data storage but also functions for data access management, etc. so as to be able to realise data distribution according to the IoS-OP terms of use. These functions also allow specification of the access range down to the levels of ship, data item and data granularity. Moreover, when it comes to timeseries data or data with timestamps, ShipDC can receive and store any kind of data and provides methods for accessing and downloading from the web system and automatically retrieving data through a Web-API.
Furthermore, ShipDC is compliant with ISO 19848, which means that data stored onshore can be used by either the data channel name used on the ship or a standard name that is compliant with ISO 19848. This reduces the complexity of mapping due to differences in the data names for each ship during data usage.

4.1 ShipDC FUNCTION

As shown in Figure 4, the ShipDC data platform is equipped not only with data storage but also data access controls, data monitoring functions and API-based data delivery functions. In addition, data can be sent via mail attachment and API Post.
In the data access controls, the data usage manager (PU: Platform User) can edit the access settings (vessel, data items, granularity, usage period) of the data user (SU: Solution User), and there is a mechanism for controlling the range of data the SU can retrieve through a data key that is issued at that time. With the data monitoring functions, you can monitor lack of data, exceeded thresholds, and data non-delivery, etc.

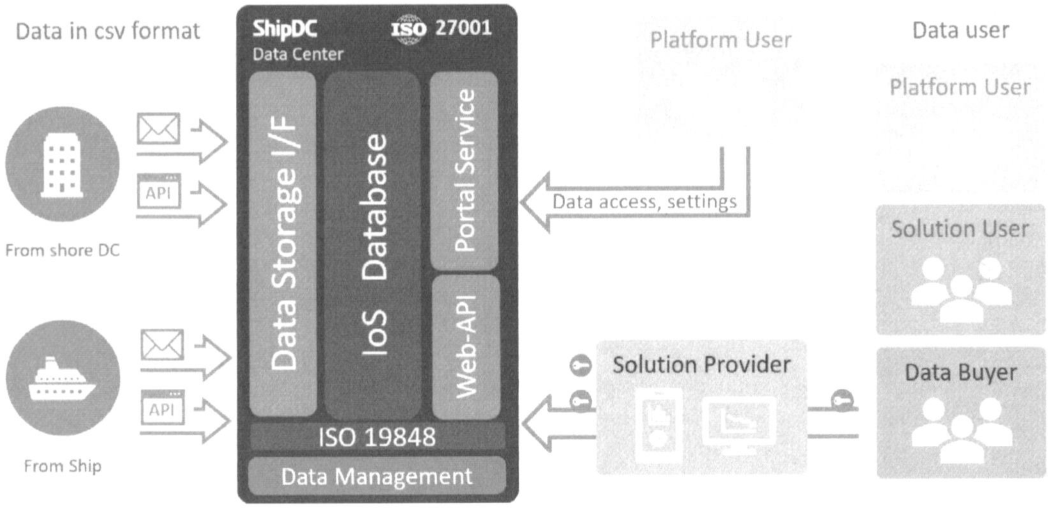

Figure 4. Data flow

4.2 ISO19847 AND ISO19848

The Smart Ship Application Platform Project (SSAP) under the Japan Ship Machinery and Equipment Association (JSMEA) formulated international standards for related technologies to make it easy to develop and introduce onboard and ship-to-shore applications.
Because the ShipDC data platform is compliant with ISO 19848 and the shipboard data servers (PP) are in some cases compliant with ISO 19847, the following provides a simple explanation of each standard.

- ISO19847: Shipboard data servers to share field data at sea
  Regulates the functional and performance requirements for the shipboard data server that collects data from the onboard devices and systems and is used to share the data safely and efficiently. It is primarily composed of the data input/output functional requirements and regulates the data storage functions, input/output interface functions, condition monitoring functions, data backup and restore functions, security requirements, and condition reporting functions.
- ISO19848: Standard data for shipboard machinery and equipment
  It was formulated to improve the connectivity convenience between devices and systems by standardizing the various types of data exchanged between onboard devices on the ships or between systems. The SSAP has created a

list of standard names, dictionary, and data catalogue based on this standard and made them available on the SSAP web site.

5. CONCLUSIONS

The data ownership approach is intended to achieve an orderly data distribution. We believe that making this approach common throughout the world will accelerate the shift to a data-driven world without harming the interests of all stakeholders involved in data distribution. We hope that through these activities, we can help pursue safe ship operation and contribute to the environment while supporting everyone's business activities.

6. AUTHORS BIOGRAPHY

Yasuhiro Ikeda, President and Head of Technical Department, Ship Data Center Co., Ltd.
Vikrant Sharma, Digital Officer, Ship Data Center Co., Ltd.

A ROUTE PLANNING METHOD FOR COASTAL NAVIGATION OF SMALL SHIPS

D G Jeong, M I Roh, Seoul National University, Republic of Korea
K S Kim, University of Ulsan, Republic of Korea
J S Lee, D H Kim, and **W S Jang,** Avikus Co.,Ltd, Republic of Korea

SUMMARY

This study proposes route planning considering the characteristics of small ships sailing on coasts. An electronic chart based on the S-57 standard object and attributes catalogue was used. A Quadtree data structure was used for modelling an environment for route planning. The main idea of Hierarchical Pathfinding A* (HPA*) was adapted to our route planning algorithm. The route planning method proposed in this study can find route for small ships within a short time that the users can understand. In addition, it consumes little computer resources so that there is no difficulty in operating it on the computer of the small ships. In addition, the program was verified by conducting route planning tests in various areas along the coastal areas of Korea.

1. INTRODUCTION

1.1 RESEARCH BACKGROUND

In general, route planning finds the shortest route, which avoids obstacles and can additionally consider multiple purposes. The purpose of route planning in the field of ship navigation varies depending on the type of the ship. For example, large merchant ships mostly require routes of the minimum Fuel Oil Consumption (FOC) for economic reasons. On the other hands, the main characteristic of a small ship is that it navigates a complex coastal area with lots of small obstacles. It is necessary to recognize them with high resolution to avoid small obstacles. In order to express high resolution, the size of the chart increases. Furthermore, the route planning time and computing resources also increase. The route planning time is also an important point, and the specifications of the computer mounted on the ship are also limited. Therefore, we proposed a route planning method that solves the above problems.

1.2 RELATED WORKS

Most of the research on ship route planning was targeted at large ships. Since the route of a large ship greatly affects the economic feasibility of the ship's operation, many researchers aim to plan a route that minimizes FOC. In order to achieve this purpose, there have been attempts to solve the problem in various ways. Tsou et al. [1] used an ant colony algorithm on uniform grid charts considering sea weather. Lee et al. [2] proposed a route planning that minimizes the shortest distance, sea meteorological data, fuel consumption, and vessel sailing time based on a uniform grid chart using a Genetic Algorithm (GA). There are studies that model charts for efficient route planning with various data structures. Wu et al. [3] made the chart with the visibility graph and planned the routes using the Dijkstra algorithm. Lee et al. [4] converted the electronic chart into a quadtree chart and then made a visibility graph based on the quadtree chart and planned the route with the Dijkstra algorithm on the visibility graph. In this study, we propose an efficient data structure for chart and route planning algorithm that mainly considers route planning time and memory usage.

2. QUADTREE CHART FOR ROUTE PLANNING

The electronic chart created under the S-57 standard object and attributes catalogue represents maritime information as an object, and each object has attributes that describe further information about the object. For example, the anchorage area (CODE: 4) is expressed in its region as a polygon, and the attribute contains the type of the anchorage area.
Since the objects in the electronic chart are represented as vector data, it is necessary to convert the electronic chart into a suitable form of chart for route planning. The size of the chart should be small, and the converted chart should accurately represent the complex topography and various objects of the coast. Describing coastal topography in detail with the uniform grid chart used in classical route planning is not appropriate because the size of the chart becomes excessively large in high resolution. To satisfy both requirements, a chart was modelled using a quadtree data structure. It is possible to express geographic information by recursively dividing the region into a quadtree [5]. Regions that have information are described in detail, and regions that do not have information are described in less detail, so complex coasts can be expressed efficiently compared to a uniform grid. The method of converting an electronic chart into a quadtree chart is as follows. The Point Region (PR) method is a method that only considers points. When the quadtree node contains more than c points, the quadtree node is divided and expresses points. PM-Quadtree considers both points and lines. More sophisticated charts can be created in the order of PM-3, PM-2, and PM-1. In this study, charts were created using the PM-

l method, which is a method of dividing if a point or line segment is within a quadtree node until the number of divisions reaches the maximum number of divisions [6].
In contrast to the objects, which are defined as obstacles in this study, there are objects that require the additional process. There are objects that require a higher resolution than a general obstacle. For example, if a pontoon, where ships are actually anchored, is converted to a quadtree chart with a normal minimum node, it will be expressed too roughly so that the ship cannot approach the anchorage area. Making all quadtree charts with a smaller minimum node increases the size of the chart significantly. Therefore, objects which need detailed description should be added to quadtree charts with normal minimum nodes separately. Since the coast has many areas with low water depth, it must be accurately expressed for the safe navigation of ships, so water depth is also an object that requires an additional process. Sounding (CODE:129) is the object which contains water depth and its specific coordinate in the S-57 standard object and attributes catalogue. The average of the sounding within a quadtree node was determined as the water depth of that node. If the water depth of the node is lower than the reference value determined by the draft of the vessel, the node is set as an unnavigable region.

3. ROUTE PLANNING ALGORITHM

HPA* is selected as a route planning algorithm. HPA* is an algorithm developed based on A*. It searches routes on the upper charts, and then it searches routes on the lower charts. [7]. The route planning on the upper chart is deciding which charts to navigate. The route planning on the lower charts determines the detailed route inside a chart. For the route planning applied in this study based on HPA*, it is necessary to create a hierarchy of charts by dividing the quadtree charts. The easiest method of creating a hierarchy of charts is to divide the charts into equally spaced latitude and longitude. Even if a chart of the same area is expressed, charts that express the more complex area become heavier due to the characteristic of the quadtree. When planning a route with equally spaced charts, there is a big difference in route planning time and memory usage. Therefore, we divide the chart based on the number of nodes in the quadtree chart for a consistently low route planning time and memory usage. In order to plan the route between the divided charts, the upper chart is necessary. The upper chart is a graph having the designated entrance as a node in the divided chart. An entrance is the set of quadtree nodes that composes the edge of the divided chart. The entrance is connected with other passable entrances. The lower chart is the divided chart. Route planning is carried out with the upper and lower charts. When the user inputs the starting point and goal point, starting point and goal point are connected with passable entrances at the upper chart. Then, A* determines which entrances on the upper chart will pass. Routes between entrances are also decided by A* on a lower chart. The route planning is completed by integrating each route generated in the lower charts. However, because the center of the quadtree node composes the route, there are awkward bends at the route. Therefore, smoothing is required. Smoothing is performed by checking obstacles between the nodes of the route. If there is no obstacle between the nodes of the route, the corresponding route node is removed and if there is an obstacle between the nodes of the route, leave it as it is.

4. APPLICATION

The route planning proposed in this study was applied in various regions along the coast of Korea, and the results are shown in Figure 1. The route planning time according to the distance is shown in Table 1. In general, the longer the distance, the longer the route creation time. However, it can be seen that the route generation time increases as the route passes through a complex coastal area. It has a memory usage of around 100MB.

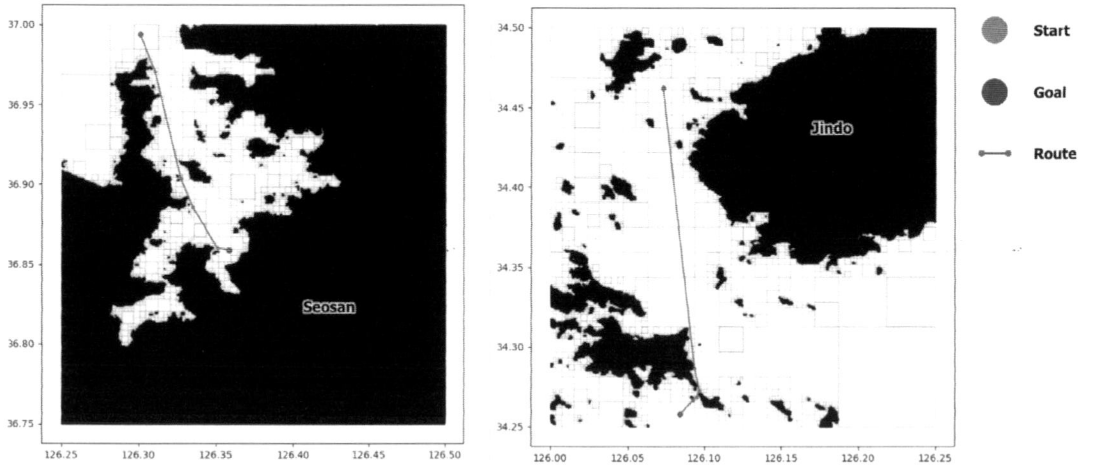

Figure 1: Route planning results

Route planning time(s)	Distance (km)
0.28	0.81
0.85	35.69
3.16	309.87
2.72	512.77
12.86	1622.25

Table 1: Route planning timetable

5. CONCLUSIONS AND FUTURE WORKS

In this study, complex coastal areas and various objects in the sea were made into quadtree charts with a high resolution and small chart size. By applying HPA* as the route planning algorithm, two stages of route planning were carried out: the upper chart and the lower chart. As a result, even long-distance areas were able to generate routes in a short time. Smoothing was performed for abnormal bends of the generated route. The proposed route planning is verified on the coasts of the Republic of Korea.

Further research is needed to be mounted on a small vessel. All objects specified in S-57 do not only object to avoid but also objects to provide further information such as recommended routes or navigation rules. Additional treatments are needed to reflect these in the route plan. Furthermore, it is important to reflect dynamic information such as sea meteorological information on a quadtree chart in real-time.

6. ACKNOWLEDGEMENTS

This work was partially supported by
(a) Avikus Co.,Ltd.(Hyundai Heavy Industries Group), Republic of Korea,
(b) BK21 Plus, Education & Research Center for Offshore Plant Engineers (COPE) of Seoul National University, Republic of Korea, and
(c) Research Institute of Marine Systems Engineering of Seoul National University, Republic of Korea.

7. REFERENCES

1. M. C. Tsou, and H. C. Cheng, "An Ant Colony Algorithm for efficient ship routing." Polish Maritime Research (2013).
2. S. M. Lee, M. I. Roh, K. I. Kim, H. R. Jung, J. J. Park, "Method for a simultaneous determination of the path and the speed for ship route planning problems." Ocean Engineering 157 (2018): 301-312.
3. G. Wu, I. Atilla, T. Tashsin, M. Terziew, and L.C. Wang, "Long-voyage route planning method based on multi-scale visibility graph for autonomous ships." Ocean Engineering 219 (2021): 108242.
4. W.H. Lee, G. H. Choi, and T. W. Kim, "Visibility graph-based path-planning algorithm with quadtree representation." Applied Ocean Research 117 (2021): 102887.
5. H. Samet, "The quadtree and related hierarchical data structures." ACM Computing Surveys (CSUR) 16.2 (1984): 187-260.
6. A. Touir, "ML-Quadtree: The design of an efficient access method for spatial database systems." Journal of King Saud University-Computer and Information Sciences 17 (2004): 45-64.
7. B. Adi, M. Müller, and J. Schaeffer. "Near optimal hierarchical path-finding." J. Game Dev. 1.1 (2004): 1-30.

8. AUTHORS BIOGRAPHY

Dong-Geun Jeong is a graduate student of naval architecture and ocean engineering at Seoul National University, Korea. His research field is route planning of small ships.

Myung-Il Roh is a professor of naval architecture and ocean engineering at Seoul National University, Korea. He holds a B.S., M.S., and Ph.D. in naval architecture and ocean engineering from Seoul National University. His main area of teaching and research interests includes ship and offshore plant design, simulation-based design and production, optimization, CAD, CAM, CAE, CAPP, and so on.

Ki-Su Kim is a professor of naval architecture and ocean engineering at the University of Ulsan, Korea. He holds a M.S., and Ph.D. in naval architecture and ocean engineering from Seoul National University. His main area of teaching and research interests includes optimization of arrangement design, flooding analysis, economical & smart ship technology, and so on.

Jun-Sik Lee is a senior researcher of recreational boat solution team at Avikus Co., Ltd. (Hyundai Heavy Industries Group). He is responsible for research and development of autonomous leisure boat solution.

Dae-Hyuk Kim is a senior researcher of recreational boat solution team at Avikus Co., Ltd. (Hyundai Heavy Industries Group). He is responsible for research and development of autonomous leisure boat solution.

Wang-Seok Jang is a senior researcher of recreational boat solution team at Avikus Co., Ltd. (Hyundai Heavy Industries Group). He is responsible for research and development of autonomous leisure boat solution.

INTERFACE DEVELOPMENT BETWEEN THE 3D CAD SOFTWARE AND THE STRUCTURAL STRENGTH ASSESSMENT SOFTWARE FOR EFFICIENT CLASSIFICATION APPROVAL

J O Kim, J H Lee and **H G Park,** Korean Register, Republic of Korea
T G Jang, NAPA Korea, Republic of Korea

SUMMARY

As the ship design has been increasing complexity with the need to drive optimization and performance at the same time, the use of various computer-aided software during the ship design process is no longer an option for these days. However, most software basically supports their own proprietary file format which can only be used by their own software, and also requires to reiterate modeling the ship design which had already been used in other software. Often these repetitive works become a major hinder to streamline the design process due to the huge cost of time and effort and may have potential risk for human error. In this regard, interfaces to increase productivity between software for data reusability were developed, but the available data was limited in 2D CAD-based. However, as 3D CAD is adopted in the shipyards, it became possible to access and utilize the 3D modeling data containing various information of the ship design. We have developed the interface function that can carry out hull structural strength evaluation for classification approval very quickly and efficiently based on the 3D CAD model. In this paper, Korean Register (KR) present the interface between 3D CAD software, NAPA Designer, and classification's structural strength assessment software, SeaTrust-HullScan [1], that can dramatically improve design productivity.

1. INTRODUCTION

The hull structure design process consists of various processes, and software suitable for each process is used. Due to data of ship in each software is managed in its own file format, the user must input it according to the software manually, even though it is the same hull data. It has been 10 years since shipyards introduced 3D CAD system, and compared to the previous 2D CAD system, realistic expressions and more detailed information can be efficiently managed. The modeling data 3D CAD hull model created by the shipyard must comply with structural strength requirements of the relevant classification, and KR provides SeaTrust-HullScan for structural strength assessment. User must input 3D CAD model into classification software to calculate the rule requirements, and the input of modeling-related data requires a lot of time. Also, a modeling input error by user will lead to occur wrong results of assessment. The changed design information to satisfy the rule requirements is to be reflected back to 3D CAD system. Figure 1 shows structural design workflow between shipyard and classification and the design iteration is very time cusuming work. In this regard, the development of interfaces and common formats for hull modeling data that can improve design efficiency is being actively developed [2] [3] [4] [5] recently.

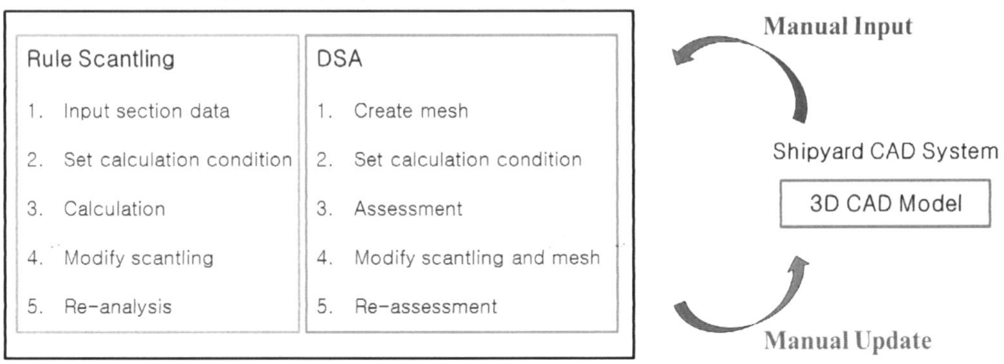

Figure 1: Structural design workflow between shipyard and classification

KR and NAPA have jointly developed an interface between the classification approval software, SeaTrust-HullScan and the shipyard 3D CAD system, NAPA Designer in order to improve productivity of work and streamline user input of ship modeling. The implemented interface works in real time, and ship design information including various modeling information in 3D CAD system can be transferred into SeaTrust-HullScan. Therefore, user can quickly perform calculation of classification requirements through the minimum input. In this study, KR introduces an interface that can greatly improve the accuracy of hull structure design and work productivity.

2. DEVELOPMENT OF INTERFACE

2.1 JOINT DEVELOPMENT PROJECT BETWEEN KR AND NAPA

In 2D CAD system, data recycling through the interface has a limited range due to the characteristics of 2D-based data so that it had a low level of data utilization. However, data in 3D CAD System can be expressed and modelled alike a actual ship and the range of data that can be utilized is greatly expanded. KR had started the research on interface with 3D CAD system that can improve design productivity and decided to develop the interface between NAPA Designer which the most used 3D CAD system in Korea shipyards, and KR's SeaTrust-HullScan. In order to develop the interface function, KR and NAPA have been started to carry out the joint development project from May 2020 and still going on.

2.2 GOAL OF 1ST PHASE

3D CAD system has a large amount of data on ships including modeling data, and it is necessary to determine the data required for the interface among these data. The first goal of the interface is to perform prescriptive assessment of the Common Structural Rule (CSR) [6]. In order to carry out prescriptive assessment, not only modeling data but also additional information needed by the classification regulations are required. Therefore, prescriptive assessment cannot be fully automated through the interface between SeaTrust-HullScan and NAPA Designer, but prescriptive assessment can be evaluated with minimal user input. In this study, the interface that can quickly perform Prescriptive Assessment was developed using various data of NAPA Designer.

3. APPLICABLE INTERFACE DATA

Prescriptive assessment of CSR requires various information including cross-sectional information on a vessel at an evaluation location, and Figure 2 shows the types of data supported in the interface by based on the main functions of Seaturst-HullScan. To exchange these data, first read the data of NAPA Designer and then it must be converted according to the data structure format of SeaTrust-HullScan. It is important to maintain geometric characteristics such as length and curvature information of the own original data in the interface process, especially geometric data such as curve.

Common Data (Common Data)
- General Particular
- Frame Information

Section Creation (Section Creation)
- Geometry Curve
  (Member Type)
- Plate
  (Material, Thickness, Compartment)
- Stiffener
  (Material, Index, Type & Size, Span & Spacing, Compartment)
- EPP
  (Automatically defined by plates & stiffeners)

Compartment (Compartment)
- Basic Geometric Data

Figure 2: Applicable data in interface between SeaTrust-HullScan and NAPA Designer

The purpose and meaning of modeling data may depend on the software used. It is important that the modeling data in CAD system has the same shape and material property information to build a vessel, but classification software has the idealized structural model. For example, in the case of plates and stiffeners, scantlings, shapes, and materials required for ship building are key information in CAD system, but in classification software, they are idealized as beam-column models and orthotropic plate models for structural strength evaluation. In this regard, the modeling data in 3D CAD system is converted into the idealized model required by the classification software when interface.

For Prescriptive Assessment, various data must be organically connected. For example, the compartment information includes geometric and load information. The corrosion allowance and load of the structural member such as plate and stiffener are determined according to its adjacent compartment. When interface between SeaTrust-HullScan and NAPA Designer, the relationship of various data is automatically configured to enable the calculation of classification requirements.

4. WORKFLOW

As mentioned in Section 2.2, the interface for Prescriptive Assessment of NAPA Designer data from SeaTrust-HullScan is developed, and the data flow of the interface is currently a one-way direction from NAPA Designer to SeaTrust-HullScan. User only needs to click the desired data in SeaTrust-HullScan to launch the interface and do not require any additional files or operations. When the interface is executed in SeaTrust-HullScan, NAPA Designer is internally executed in the background to construct the data required for prescriptive assessment in SeaTrust-HullScan.

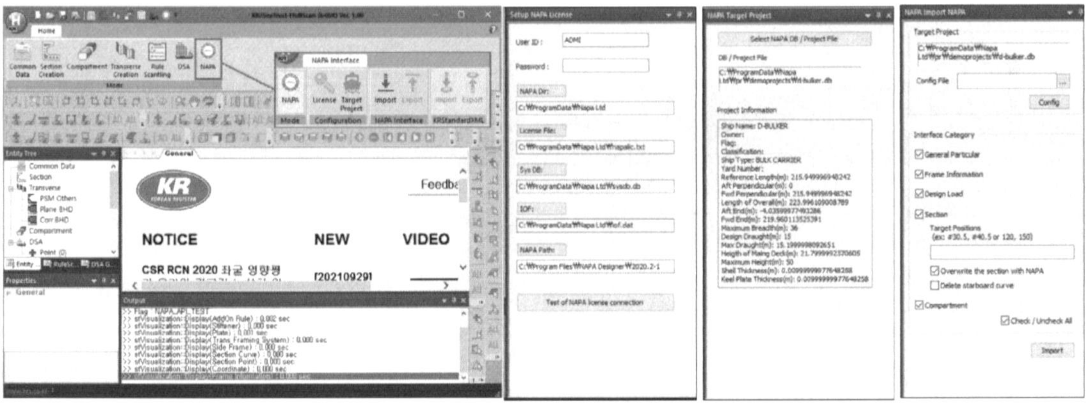

Figure 3: Configuration of interface menu

In SeaTrust-HullScan, NAPA menu is newly configured for the interface and the related functions, and the menu includes license setup, selection of NAPA Project DB, and Import as shown in Figure 3. User can quickly and easily execute the interface functions with just a few clicks within SeaTrust-HullScan.

5. RESULTS OF INTERFACE

5.1 SECTION MODELING DATA

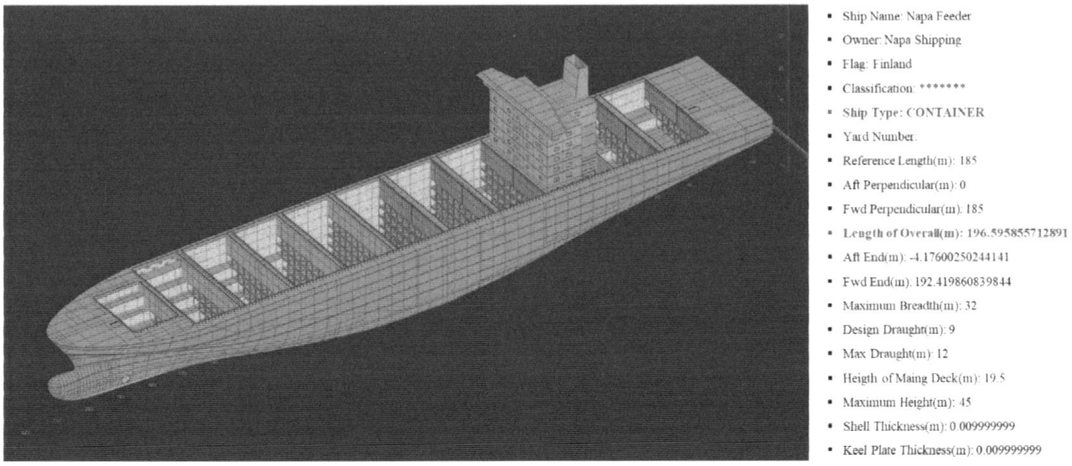

Figure 4: Information of example container ship in NAPA Designer

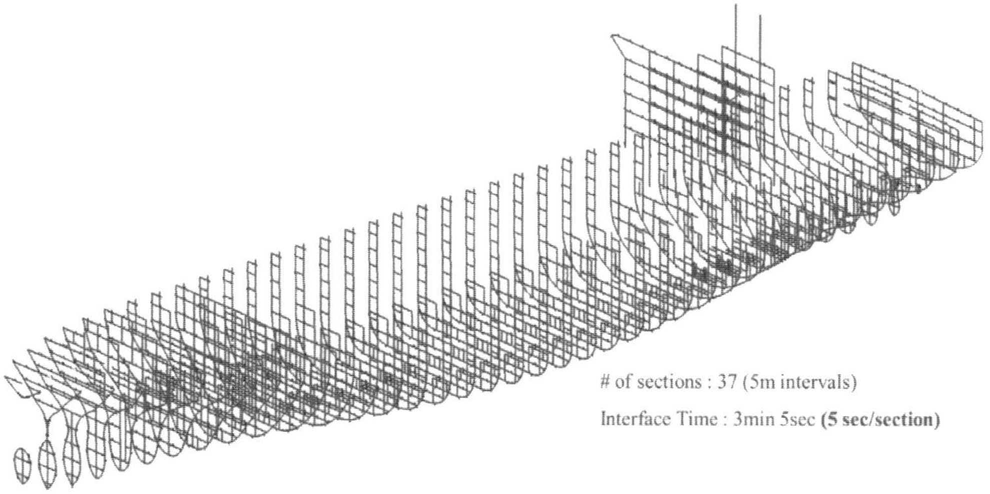

Figure 6: Interface results of container ship in SeaTrust-HullScan

The data handled in the interface can be roughly divided into text (or number) and geometry. In the case of text data, there is no issue in the interface because the corresponding information is moved as is. However, modeling data must be converted according to the software because it is managed differently for each software even if the displayed shape is looked like the same and maintain own geometric characteristic such as shape, length, area, and curvature.

To test the developed interface function, the container example file provided by NAPA Designer was used. Since the cross-section modeling data has various shapes along the longitudinal direction, a total of 37 cross-sections were interfaced at uniform intervals of 5m. Figure 5 and Figure 6 show the detailed information of the example ship in NAPA Designer and the interface result of sectional data in SeaTrust-HullScan, respectively. The interface time took a total of 3min 5sec, and this is about 5sec per a single section. This interface function is expected to dramatically improve productivity compared to the time required for users to manually input modeling data as a conventional method.

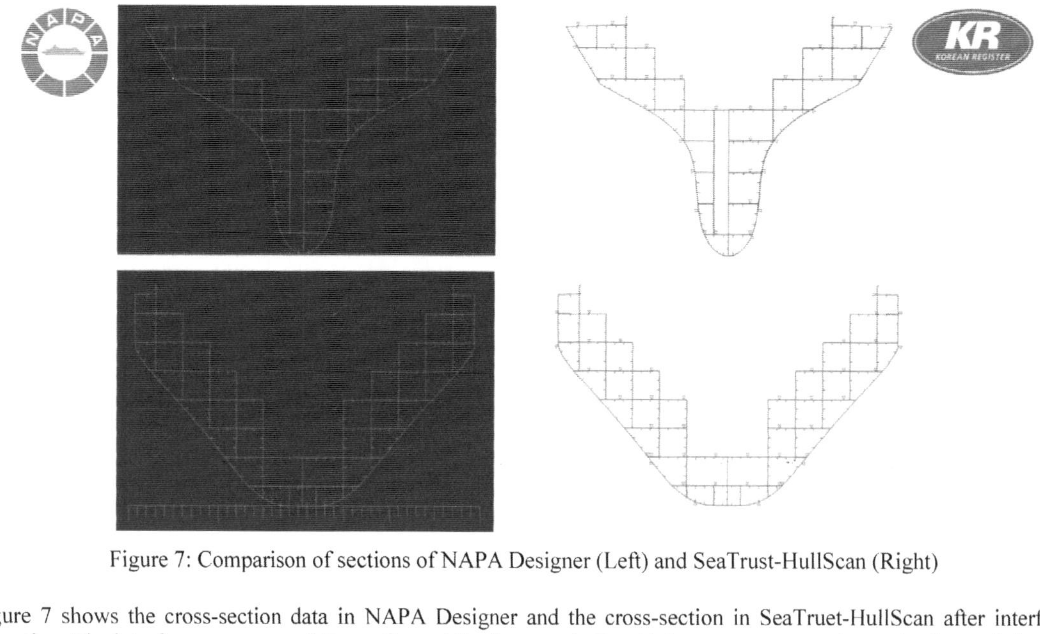

Figure 7: Comparison of sections of NAPA Designer (Left) and SeaTrust-HullScan (Right)

Figure 7 shows the cross-section data in NAPA Designer and the cross-section in SeaTruet-HullScan after interface execution. The interface was successfully performed for the outer hulls with large curvature and complex internal shapes. Figure 8 shows the properly transferred property of the vessel and Elementary Plate Panel for hull structural strength evaluation is automatically defined in SeaTurst-HullScan.

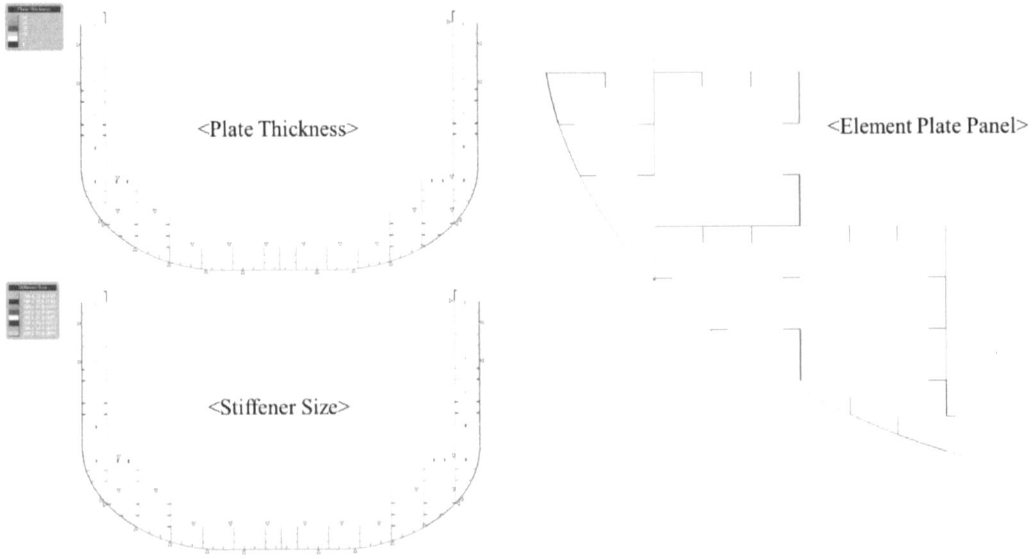

Figure 8: Sectional member's scantling(Left) and the defined 'Elementary Plate Panel'(Right)

5.2 DATA CONSISTANCY

The most important thing in the interface is to keep data consistency. If data information is changed after interface, it loses its value as data. In particular, in the case of ship cross-section modeling, if the shape and length of the curve are changed, the results of the hull structural strength evaluation cannot be trusted anymore. The section modulus of the hull section is determined by the section modeling data, which directly affects the hull girder stress required for structural strength assessment. Recently, shipyards have a high level of design capability, and they meet the design requirements almost closely. Therefore, since small modeling deviation can affect design scantling, data consistency should be constantly maintained during the interface.

In this study, the cross-sectional modeling data modeled by SeaTrust-HullScan and the cross-sectional modeling data through the interface were compared to confirm the data consistency. By comparing the geometric values such as the position of plate and stiffener and the length of the curve, the position of the neutral axis for the hull cross section, and the section modulus at the deck and bottom, it was confirmed that the values were consistent.

6. CONCLUTIONS

As 3D CAD system was introduced to shipyards in earnest, same expression as the actual ship became functional, and the range of data that could be utilized was greatly increased compared to previous 2D CAD system. KR studied the interface that can improve the productivity of the hull design process by using data of 3D CAD system. To go with doing this, KR and NAPA have jointly developed the interface between SeaTrust-HullScan and NAPA Designer and set the first goal of the interface at a level capable of performing prescriptive assessment of CSR.

The developed interface between SeaTrust-HullScan and NAPA Designer can be performed by simple user operations within SeaTrust-HullScan without additional files or programs. Various information required for prescriptive assessment including modeling information was successfully interfaced with their property, and idealization and data required for hull structural strength were automatically configured. Designers can easily import a number of complex cross-section modeling data into SeaTrust-HullScan. It was confirmed that the data consistency was maintained by comparing values such as shape comparison and section modulus for the interfaced data.

KR and NAPA continue to jointly develop the interface, and plan to expand the data range with the classification software. In particular, it is currently a one-way interface from NAPA Designer to SeaTrust-HullScan, but a two-way interface that allows user to check the Prescriptive Assessment results of SesTrust-HullScan in NAPA Designer is under development as shown in Figure 9. KR hopes the developed interface makes user streamline the hull design process and enhance work productivity.

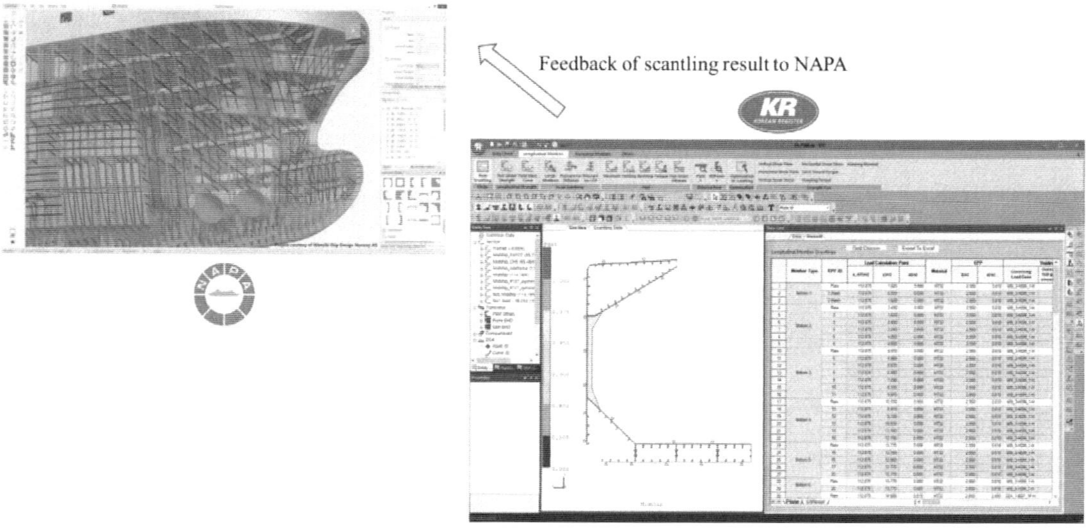

Figure 9: Example of interface from SeaTrust-HullScan to NAPA Designer

7. REFERENCES

1. Park, H.G., "Introduction and development of SeaTrust-HullScan." *Proceedings of KR Software User Conference 2018, Busan, Korea*, 2018.
2. Kim, J.O., Lee J.H., Byun S.H., and Park, H.G., "Development of KR Standard Format(KRStandardXML) for integration of a structural design process." *The society of Naval Architects of Korea*, Annual Autumn Conference, 2021
3. Son, M.J. Lee, J.Y. Park, H.G. Kim, J.O. Woo, J.J. & Lee, J.H., "Development of 3D CAD/CAE Interface in Initial Structural Design Phase of Shipbuilding." *Transactions of the Society of CAD/CAM Engineers*, Vol. 21, No. 2, pp. 186-195, 2016
4. Hamada, K. Hiraki, T. Sasaki, Y. & Ozaki, M., "The development of speeds (smart platform of enhanced engineering data for shipping and shipbuilding) and innovative use of ship 3D data across borders using CADs." *International Conference on Computer Applications in Shipbuilding 2019*, Rotterdam, Netherlands, 24-26 September 2019.
5. u , C ae, u , yani , i e ing, a , o ini, i aya, G. Yu, K. & Juttu, L., "MODEL-BASED APPROVAL - THE OPEN CLASS 3D EXCHANGE (OCX) STANDARD." *International Conference on Computer Applications in Shipbuilding 2019*, Rotterdam, Netherlands, 24-26 September 2019.
6. IACS, "Common Structural Rules for Bulk Carrier and Oil Tankers 01 JAN 2022," Issue: International Association of Classification Societies, 2022

8. AUTHORS BIOGRAPHY

JongOh Kim holds the current position of Principal Surveyor at Korean Register. He is Ph.D. student in naval architecture and ocean engineering department of Pusan National University and obtained M.S. Degree in automotive engineering of Hanyang University. He is responsible for strength assessment, CAD interface, and system kernels in SeaTrust-HullScan.

JoungHyun Lee holds the current position of Senior Surveyor at Korean Register. obtained B.S. Degree in chemistry of Soonchunhyang University. He is responsible for Direct Strength Assessment, Pre-process, and post-process in SeaTrust-HullScan.

HoGyun Park holds the current position of the manager of ICT Solution Team at Korean Register. He obtained his M.S. Degree in naval architecture and ocean engineering of University of Ulsan. His research interests include 3D model-based approval and shipbuilding design software.

TeukGyu Jang holds B.S. Degree in naval architecture and ocean engineering of Pusan National University and holds of lead technical consultant of NAPA's Design Solutions.

A METHOD OF VARIABLE RECOGNITION AND CONNECTION FOR REVIEWING SHIP REGULATIONS

M C Kong, M I Roh, K S Kim, Seoul National University, Republic of Korea
J Kim, H Park, and **J Lee,** Korea Register, Republic of Korea

SUMMARY

CSR, one of the most referenced documents in the classification, is a rule established for the safe construction of bulk carriers and oil tankers. This document contains not only extensive content but also contains complex equations, figures, and tables. In particular, some variables in the equations represent the requirements of the ships, and they are often defined with other variables. Therefore, even an expert in the review task has the inconvenience of having to search through the CSR several times to find the meaning of the desired variable. Accordingly, we devised an algorithm that automatically recognizes variables in a document and identifies the relationship between each variable. Consequently, we developed a program to visualize this information for the user effectively. This study achieved variable recognition accuracy of F1-score=0.96 for CSR, and its effectiveness was confirmed by being applied to related fields.

1. INTRODUCTION

1.1 RESEARCH BACKGROUND

The classification reviews whether the shipbuilder has designed a ship following the rules to approve the construction. These rules are established by the international association of classification societies (IACS) and are revised yearly. Each rule sets out the requirements (mainly related to safety and environmental issues) to be observed when building a ship. Therefore, the reviewers of the classification should check several values to determine whether each part of the ship satisfies the requirements. Since these values are calculated automatically, it is difficult to know how they were derived. In addition, reviewers should know all the definitions of equations and variables for detailed analysis in addition to calculated values. The documents containing the rules for large ships are so extensive that even experts need to refer to them frequently. In addition, the definitions of each variable are located on various pages in the document and are interrelated, so there are more pages to find.

In this study, we proposed a series of methods for automatically recognizing and visualizing the relationship between complex variables described above. Generally, rulebooks are converted into PDF format and uploaded for easy distribution. However, in this process, valuable information (variable names, formulas, and table structures, etc.) from the document disappears, and the data that can be obtained becomes limited. Therefore, we proposed a method for recognizing and organizing variables by utilizing the minimum amount of data obtained from PDF. To extract data from PDF, we utilized a public library called PDF Miner [1].

The process for recognizing variables and defining a connection relationship in this study is shown in Figure 1. This study is divided into three main parts. First, since it is necessary to accurately extract the desired component to recognize the variable in the PDF document, we found an appropriate method to extract the component and pre-process it. Next, the documents were indexed according to the table of contents by using the table of contents extraction result. Meanwhile, variables were extracted from the document using the text extraction result. Combining these two results completes the indexing of variables within the document. Finally, using the indexed results in the document, we developed a program to identify the connection between variables and visualize them.

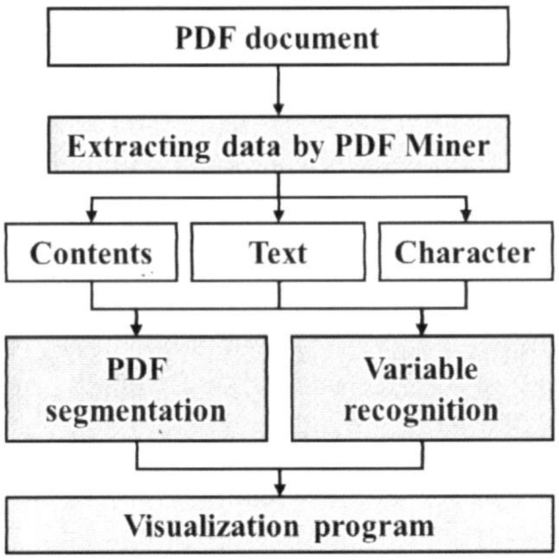

Figure 1: Overall process of variable recognition and connection

1.2 RELATED WORKS

Research on extracting data from PDF using PDF Miner is steadily progressing in various fields. For example, there are various ways to use the PDF, such as reconstructing a PDF through the extracted data and turning it into a database [2] or recognizing the structure of specific components such as tables in a document and automatically saving them [3]. In particular, in deep learning, which is gaining popularity recently, there is a possibility that misrecognition or non-recognition may occur if the recognition model is not properly trained because the recognition model mostly depends on the training data or the training process. Therefore, when recognizing tables or formulas, maximizing utilization of the raw data in a document has the potential to produce better performance than using a deep learning model.
Consequently, this study proposed a method of recognizing and indexing equation structures and variables using information extracted from PDFs through PDF Miner.

2. THEORETICAL BACKGROUND

2.1 RECOGNITION TARGET

CSR, the subject of this study, is an abbreviation for common structural rules and describes the rules to be followed when building bulk carriers and oil tankers. The document has various components such as figures, tables, table of contents, and equations. The format is not standardized, and each chapter/section has a characteristic. Table 1 shows the components included in CSR 2021. As shown in the table, since CSR consists of many characters and formulas, it is necessary to designate the target to be extracted for variable recognition accurately.

Table 1: Number of components in CSR 2021

Components	Number of components
Pages	822
Characters	1,141,414
Figure/Table	36,207
Table of contents	1,063
Equations	1,929

2.2 EXTRACTION TARGET

PDF Miner separates the components within the PDF and manages them like layers. In this study, each layer is called an element. We can extract the element's in-page location (bounding box) and a string if it contains a string. In general, an element called LTFigure refers to a figure within a PDF, and LTTextbox refers to a paragraph. However, PDF Miner separates these layers by itself, so the referenced object may be different depending on the document. Therefore, we must figure out first what the elements parsed by PDF Miner on the CSR document point out. Table 2 shows the elements extracted by PDF Miner and the corresponding objects and information that we can extract from CSR. Consequently, we segmented PDFs and recognized variables using the extracted elements and data.

Table 2: Extracted data of CSR using PDF Miner

Elements	Components in CSR	Extracted data
LTTextbox	Paragraph	page, bounding box, string
LTTextline	Text line	page, bounding box, string
LTChar	Character	page, bounding box, character
LTRect	Figure, Table	page, bounding box

3. PDF SEGMENTATION

In addition to the function of extracting elements, PDF Miner can extract the table of contents existing in PDF. To know where a variable's location belongs in the table of contents, we must first segment the document. We extracted the title of the table of contents and the page the table of contents indicates by extracting the part of the table of contents that you can usually see in a PDF document viewer. The title also exists on the page where the table of contents indicates. However, it may be expressed slightly differently with a symbol such as an index number or colon. Segmenting a PDF requires a perfect match between the title of the table of contents and the title within the page. Therefore, we applied an algorithm called Levenshtein-distance [4] and used it to match two slightly different titles.
Levenshtein-distance is an algorithm that tells the difference between two strings, indicating how much correction is needed to make them equal. For example, the difference between "other" and "there" is to delete "o" from "other" and add "e", so the difference between the two becomes 2. The calculation method of the Levenshtein-distance algorithm is shown in Equation 1 below.

$$lev(a,b)=\begin{cases}|a| & \text{if } |b|=0,\\ |b| & \text{if } |a|=0,\\ lev(tail(a),\ tail(b)) & \text{if } a[0]=b[0],\\ 1+min\begin{cases}lev(tail(a),\ b)\\ lev(a,\ tail(b))\\ lev(tail(a),\ tail(b))\end{cases} & otherwise.\end{cases} \quad (1)$$

$|x|$: Length of the string x
$x[0]$: The first letter of the string x
$tail(x)$: The string except for the first letter of the string x

Using this algorithm, we could know that the part with the smallest difference between the table of contents and the text lines on the page is the beginning of the table of contents. Therefore, we segmented PDF according to the table of contents.

4. VARIABLE RECOGNITION

We used the positional relationship between characters to recognize variables from a text line. Foremost, a character element called LTChar, the smallest extraction unit, was extracted to utilize the positional relationship. An example of the result can be seen in Figure **2** below. As we can see, each character has a different bounding box size depending on the font size. However, we found unnecessary special characters mixed with the characters. Therefore, we identified the Unicode range of unnecessary elements and removed them all.

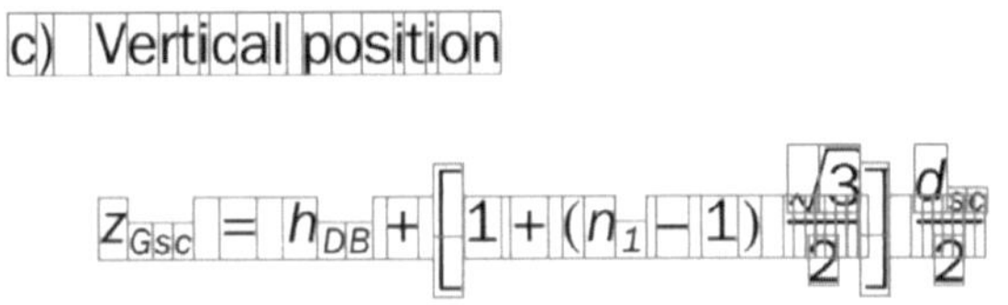

Figure 2: Extracted result of LTChar element

Next, since each text line has a different height, it is necessary to separate the layers according to the height of the character. In addition, since the formula also has a fractional form, separating the layers has the effect of automatically separating the fractional area.

Within a CSR, a variable usually consists of one character followed by n subscripts. Therefore, separating the superscripts/subscripts is necessary to recognize a variable. First, we find the maximum and minimum height of the bounding box in each layer and separate the superscripts/subscripts based on the ratio of this height to the character's height. Next, if the separated result was closer to the top, it was classified as a superscript, and if it was closer to the bottom, it was classified as a subscript. Figure 3 shows the result of subscript separation in one layer.

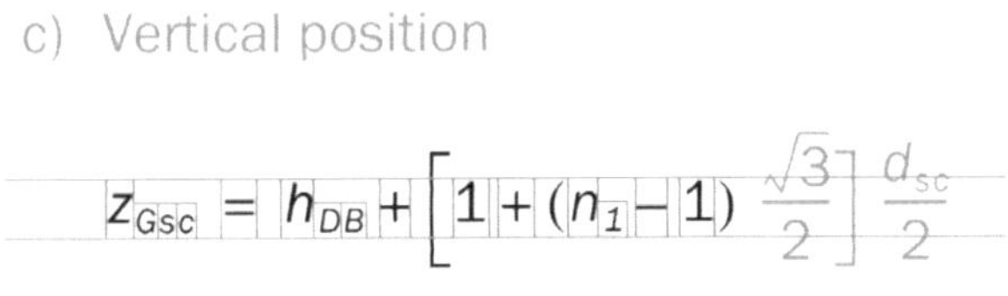

Figure 3: Result of superscript/subscript separation

Repeating the above process for each layer can separate subscripts from all layers. This study considered one normal character with n subscripts a variable. Generally, variables composed of single characters are difficult to distinguish from normal text, and most of them are constants with fixed meanings such as g or m, so they are excluded from recognition. Finally, variable recognition is completed, as shown in Figure 4.

c) Vertical position

$$z_{Gsc} = h_{DB} + \left[1 + (n_1 - 1)\frac{\sqrt{3}}{2}\right]\frac{d_{sc}}{2}$$

Figure 4: Final result of variable recognition

To judge the accuracy of this algorithm, we labeled the variables in the CSR. Labeling was performed at intervals of 10 pages, and an index that could obtain recognition accuracy for 1,200 variables was created. The accuracy evaluated through this index is shown in the table.

Accuracy	Value
Precision	0.95 (1,137/1,193)
Recall	0.96 (1,137/1,181)
F1-score	0.96 (2 * Precision * Recall / (Precision + Recall))

5. VARIABLE CONNECTION

Since variables are often defined by the relationship between other variables, it is necessary to know the definition of the variable in order to understand the connection relationship. In CSR, a variable is defined in text or equation form. Generally, if it is defined by text, there is a colon symbol on the right side of the variable, and if an equation defines it, there is an equal symbol. Therefore, we could know the definition position of each variable through this. Especially, the variables to the right of the equal sign in the equation usually constitute the variable to the left of the equal sign. Consequently, the connection relationship between variables could be defined.
We took advantage of this relationship to create a program that effectively shows the user the definition of a variable. If a variable is composed of other variables, it is displayed in a tree structure so that the definition can be easily understood. At the same time, it is possible to know where each definition is included in the table of contents by combining the position of each definition and variable with the segmentation result of Section 3. The result of the connection relationship visualized through this program can be seen in Figure 5. Variables may or may not be defined as text or equation, so if there is no definition, they are displayed as blank.

Variable	Eq_Page	Eq_Location	Txt_Page	Txt_Location
˅ z_gsc	240	Part1/Chapter4/Section6/4/1	242	Part1/Chapter4/Section6/4/4
d_sc	-	-	221	Part1/Chapter4/Section6
˅ h_db	236	Part1/Chapter4/Section6/3/1	341	Part1/Chapter5/Section1/3/5
K_cf	222	Part1/Chapter4/Section6	222	Part1/Chapter4/Section6
P_bfs	236	Part1/Chapter4/Section6/3/1	237	Part1/Chapter4/Section6/3/1
˅ z_c	223	Part1/Chapter4/Section6	250	Part1/Chapter4/Section6/7/1
› h_c	228	Part1/Chapter4/Section6/2/3	221	Part1/Chapter4/Section6
h_db	236	Part1/Chapter4/Section6/3/1	341	Part1/Chapter5/Section1/3/5
› z_f	235	Part1/Chapter4/Section6/3/1	750	Part2/Chapter1/Section3/4/1
› p_c	231	Part1/Chapter4/Section6/2/3	749	Part2/Chapter1/Section3/4/1
n_1	-	-	426	Part1/Chapter7/Section2/4/4

Figure 5: Capture of variable tree

6. CONCLUSIONS

In this study, we made a program that recognizes variables in CSR and visualizes the connection relationship by utilizing the PDF Miner library that can extract data from PDF-format documents. To divide the document according to the table of contents, we applied the Levenshtein-distance algorithm to match the table of contents and page titles. To recognize the variables in the document, we utilized the size of the characters and the positional relationship between other characters. Through the variable recognition method proposed in this study, a high recognition rate of F1-score = 0.93 was achieved for 1,200 variables in CSR. To obtain a connection relationship between variables, we created a tree structure between related variables by recognizing the definitions of variables. Consequently, we designed a program that effectively visualizes variables' location and connection relationships to the user.
Further research is needed to increase the recognition rate of variables and definitions. In this study, variables composed of single characters were excluded because of the low importance and difficulty of recognition. Therefore, in the future, we plan to add an algorithm that distinguishes it from a regular string to make it possible to recognize all variables in the document. In addition, higher accuracy can be obtained by optimizing the various tolerances used to recognize variables in this study. Lastly, if the definition of a variable is in the form of a table, it is impossible to recognize it, so we plan to devise a new algorithm.

7. ACKNOWLEDGEMENTS

This work is an expansion of our previous studies [5] and was partially supported by
(a) Korean Register, Republic of Korea,
(b) BK21 Plus, Education & Research Center for Offshore Plant Engineers (COPE) of Seoul National University, Republic of Korea,
(c) Research Institute of Marine Systems Engineering of Seoul National University, Republic of Korea, and
(d) Korea Institute for Advancement of Technology (KIAT) grant funded by the Korea Government (MOTIE) (No. P0001968, The Competency Development Program for Industry Specialist).

8. REFERENCES

1. S. Yusuke, "Pdfminer.six," Github, 2019. https://github.com/pdfminer/pdfminer.six.
2. H. Chao and J. Fan, "Layout and content extraction for PDF documents," Lect. Notes Comput. Sci. (including Subser. Lect. Notes Artif. Intell. Lect. Notes Bioinformatics), vol. 3163, 2004, doi: 10.1007/978-3-540-28640-0_20.
3. M. Ohta, R. Yamada, T. Kanazawa, and A. Takasu, "A cell-detection-based table-structure recognition method," 2019, doi: 10.1145/3342558.3345412.
4. V. Levenshtein, "Binary codes capable of correcting deletions, insertions, and reversals," Sov. Phys. Dokl., vol. 10, no. 8, 1966.
5. M. C. Kong, M. I. Roh, K. S. Kim, H. G. Park, and J. O. Kim, "Variable Recognition and Indexing Method for PDF Documents Using Components Extraction," in Proceedings of the Society of Naval Architects of Korea, SNAK, 2021, p. 659.

9. AUTHORS BIOGRAPHY

Min-Chul Kong is a graduate student of naval architecture and ocean engineering at Seoul National University, Korea. His research field is data mining, AR/VR for ships and offshore structures, and deep learning for object detection

Myung-Il Roh is a professor of naval architecture and ocean engineering at Seoul National University, Korea. He holds a M.S., and Ph.D. in naval architecture and ocean engineering from Seoul National University. His main area of teaching and research interests includes ship and offshore plant design, simulation-based design and production, optimization, CAD, CAM, CAE, CAPP, and so on.

Ki-Su Kim is a professor of naval architecture and ocean engineering at the University of Ulsan, Korea. He holds a M.S., and Ph.D. in naval architecture and ocean engineering from Seoul National University. His main area of teaching and research interests includes optimization of arrangement design, flooding analysis, economical & smart ship technology, and so on.

Jongoh Kim is a principal surveyor of ICT solution team at Korean Register. He is responsible for research and development of smart ship technologies including machineries, data science, VR/AR, etc.

Hogyun Park is a general manager of ICT solution team at Korean Register. He is responsible for research and development of smart ship technologies including machineries, data science, VR/AR, etc.

Jeongyoul Lee is a senior vice president/head of digital technology center at Korean Register. He is responsible for research and development of smart ship technologies including machineries, data science, VR/AR, etc.

ACCELERATING JAPANESE SHIPBUILDING DX ADOPTION

D Larkins, S Crook, and **G Goulanian**, SSI, Canada

SUMMARY

Japanese shipbuilding has long been recognized for the production and on-time delivery of high-quality, innovative ships. Japanese shipyards will continue to remain competitive in an ideal future because of these factors. However, shipbuilding in Japan faces several internal and external challenges that put this future at risk: a generation of highly skilled experts retiring from the workforce, a reliance on institutional knowledge, more competition from abroad, and a demand for new innovative ship types. In addition, Japanese shipbuilders have not evolved their businesses to take advantage of the process enhancements digital technologies support, further adding to the risk of falling behind.

This paper defines DX in the context of shipbuilding and examines how Japanese shipbuilders can accelerate DX within their organizations to stay globally competitive for the long term and overcome industry-wide risks. This paper will also offer a starting point for a DX framework that will help the Japanese shipbuilding industry avoid the digital cliff.

NOMENCLATURE

CAD	Computer Aided Design
CAM	Computer Aided Manufacturing
DX	Digital Transformation
ERP	Enterprise Resource Management
MRP	Material Requirements Planning
PLM	Product Lifecycle Management

1. INTRODUCTION

Japanese shipbuilders are recognized worldwide for delivering high-quality ships as part of a long history of shipbuilding excellence. However, those shipbuilders are now facing internal and external challenges that are allowing other shipbuilding nations to catch up. These challenges include a generation of highly skilled experts retiring from the workforce, a reliance on institutional knowledge, more competition from abroad, and a demand for new innovative ship types. Using technology as part of a digital transformation (DX) will allow Japanese shipbuilders to build upon their reputation and maintain a competitive advantage.

Shipbuilding is not the only industry in Japan to face these challenges. Throughout the covid-19 pandemic, the Japanese government has introduced a series of DX initiatives that aim to accelerate the pace of transformation in Japanese society [1]. Examining the approaches taken by the Japanese public sector due to the pandemic can inform a starting point for shipbuilders within Japan, particularly because, in some cases, the government is addressing the same or similar challenges to those currently being experienced by shipyards.

2. CURRENT STATE OF JAPANESE SHIPBUILDING

2.1 STRUCTURE AND FEATURES OF JAPANESE SHIPBUILDING

Shipbuilding has a centuries-long tradition in Japan. More recently, in the second half of the 20th century, Japan rose to be a leader in shipbuilding from both a volume (number and size of ships produced) and quality. However, as the shipbuilding industry matured in both South Korea and China, Japan has lost this leadership position from an order book perspective [2].

The importance and depth of the shipbuilding industry in Japan can be illustrated through the impact of booms and slumps in shipbuilding on the broader Japanese economy. According to the Maritime Bureau within Japan's Ministry of Land, Infrastructure, Transport, and Tourism (MLIT), a change in output within the shipbuilding industry is reflected three-fold in the total economy, ranking it sixth out of 49 industries [3]. The MLIT also defines the industry as a model of a sustainable and fair maritime industry [3].

Major shipyards, like Imabari Shipbuilding, Tsuneishi Holdings, and Oshima Shipbuilding Company to name a few, mainly construct bulk carriers, container ships, and tankers [2]. However, most shipyards in Japan (around 700) concentrate on the construction and repair of smaller ships.

Japanese shipbuilders do continue to invest in technology and research and development (R&D). This is primarily driven through a need to transition further up in the shipbuilding value chain, to continue meeting international standards, and take advantage of efficiency enhancements made available through new innovations. For example, between 2006 and 2011, Japanese shipbuilders increased R&D spending two-fold to 12.6 billion JPY (157 million USD) [4]. Taking Mitsubishi Heavy Industries (MHI) as an example, the organization's 2012 R&D spending was focused on green shipbuilding development and improving the energy efficiency of their designs [4].

Apart from shipbuilders, several organizations are focused on specifically R&D for the shipbuilding industry. These organizations provide an innovation foundation that the shipbuilding industry can pull from. Apart from traditional sources like universities, the National Maritime Research Institute (NMRI) and Japan Ship Technology Research Association (JSTRA) are two organizations focused on shipbuilding and maritime R&D in Japan [4]. In section 4, the authors cover some of the research work currently being undertaken by the teams there.

From a production technology perspective, most shipbuilders in Japan continue to perform engineering tasks in 2D. Shipyards are focused on hull and structure work, while outfitting is outsourced to other Japanese companies. Often shipyards will have developed proprietary or in-house production design software which is only loosely connected to detailed design. Typically, these packages are also completed in 2D.

As it stands right now, most design and engineering for mid-size shipyards in Japan is done in 2D. This work is either done in specific products like MATES (if done in house at the shipyard) or in AutoCAD. There is no intelligence in these models, no part level data, the drawings do not supply easily accessible information for feeding a digital shipyard. The information in the provided 2D drawings is extensive and incredibly rich, but it is also more error-prone as it involves extensive human input. Both the quality and the accessibility of information are vital in a digital shipyard.

Shipyards in Japan have also traditionally dedicated significant resources, both monetary and time, to customizing and building out proprietary solutions for their organizations. This leaves the shipyard responsible for maintenance, upgrades, and ensuring future compatibility.

Part of the challenge that makes information difficult to use lies in the extensive use of outsourcing within the Japanese industry. Most mid-size shipyards will not do their own design and engineering work, particularly for piping and HVAC systems, instead choosing to outsource it to a variety of design and engineering companies. This is typically done in 2D, completely separate from the Hull Structure modeling. Processes and organizations that are this disconnected can (of course) benefit from overall management efficiencies when going digital. However, without changing the information that gets delivered and increasing the amount of intelligent information gathered, implementing any level of change management or controls will be very difficult.

Japanese shipyards typically rely on an unbelievable amount of information provided in the drawings that get issued to the production workers. This is part of the reason why Japanese vessels are regarded worldwide as having such high-quality. This information, useful only to very experienced production staff, is used to actually build the ships. The average age of the workers is very high - with past estimates that around 50% of shipbuilding employees were over the age of 50 – and without that expertise, their current process would be very difficult [4]. This is perhaps not a reason to need a new CAD system. Still, it does support the need for a digital shipbuilding initiative and a need to think differently/holistically about the whole approach.

With all these factors in mind, there is a need for Japanese shipbuilders to look at what they do, not just do the exact same thing with better tools. So much of the eco-system of Japanese shipbuilders, combined with the changing landscape and emerging disrupters doesn't fit these new realities.

The types of information already being created in design and engineering, who consumed it, and how it is used in the shipyard were generally assumed to be good enough as-is. On the surface, this appeared to make sense. After all, Japanese shipbuilders were (and still are) recognized as amongst the best in the world. The design and engineering departments provide an extraordinary level of detail to production that allows for the rapid and accurate fabrication and assembly of each ship. These shipyards were rightly focused, or so it seemed, on managing and improving the processes around how they performed the business of shipbuilding *today*.

Focusing on how you do business today can be effective, particularly in the context of the pressures shipbuilders face, provided the world doesn't change around you. Shipbuilding, as a conservative industry dealing with extraordinarily complex challenges, hasn't historically experienced the same level of disruption as other industries. In the past, shipbuilders could focus on today, knowing tomorrow would look very similar. Trends that were so disruptive in other areas were kept at bay: globalization, workforce migration, digitization, automation, material science, advanced production technologies, environmental awareness.

Explore how some of the trends affecting the Japanese shipbuilding market (and others) create a need to take a holistic view of the Japanese digital shipyard. This view must consider design and engineering, process, change and resource management, and the means of production.

Without this more holistic view towards the future, Japanese shipyards are at risk of getting left behind by competition from abroad – despite the high-quality vessels they produce today.

2.2 COMPARISON TO ROW

High quality and relatively quick delivery times have throughout history been regarded as key characteristics of the Japanese shipbuilding industry. However, in recent years, that advantage has been eroded by shipbuilders from other nations, particularly South Korea and China, and reduced the market share of Japanese shipbuilders. Identifying the key aspects that have resulted in this market shift and allowed others to catch up may offer clues for Japanese shipbuilding.

In general, compared to shipyards in China and South Korea, Japanese shipyards are constrained by smaller capacity limits and higher labor and production costs [5]. In terms of the types of ships produced, Japanese shipyards favour higher complexity vessels like passenger ships and liquified natural gas (LNG) carriers, compared to bulk carriers or container ships that dominate Chinese and South Korean shipbuilding [2].

From a technological perspective, yards outside of Japan have embraced 3D and other new technologies for far longer than those within Japan. These shipyards are able to use these technologies to increase quality and shorten delivery times – eroding the advantage of key Japanese hallmarks. They are achieving this through improved visibility of engineering errors, more efficient change management processes, and improved information management as part of their larger digital transformation approaches. It is unlikely that foreign competitors will be able to completely close the quality gap to Japanese shipyards; however, the price differential will continue to pull ship owners towards lower cost offerings, despite a quality difference.

2.3 INTERNAL CHALLENGES

In many respects Japan is an extremely technologically advanced nation, but even within its own borders the notion that Japan is falling behind from a digital transformation perspective is widespread. In a 2020 report, Fujitsu Research Institute outlined and spoke of the risk that Japan is facing as it approached a "digital cliff." The report illustrated the digital cliff using the following example:

"By 2025, the shortage of human resources in IT will double to about 430,000. More than 60% of core systems will have been in operation for more than 21 years. The mainframes will also be deteriorating. The challenges Japanese companies face now are expected to become more serious at once." [6]

This shortage of IT resources and aging infrastructure is acutely felt within the Japanese shipbuilding industry. A 2018 report by the Japanese Ministry of Economy, Trade, and Industry (METI) further expanded on the aging infrastructure problem facing Japanese companies, stating that in many cases, the existing digital systems Japanese organizations are using are quickly becoming legacy systems [7]. Working with unsupported systems results in increased amounts of troubleshooting, compatibility issues, and limits the pool of workers skilled in these platforms. Further, it limits what other systems can be implemented at an organization without extensive and proprietary customization.

2.4 EXTERNAL CHALLENGES

Japanese shipyards continue to become more global. They are hiring an increasing amount of contractors and employees from abroad. Working with new third-parties can introduce risks to consistency and quality. It is important for Japanese shipbuilders to easily communicate and share engineering and project information with any sub-contractor to minimize this risk. The ability to easily share work packages and shipyard-specific standards can further mitigate this risk.

3. DX IN THE CONTEXT OF SHIPBUILDING

Shipbuilding executives need to identify the goal they are trying to achieve, break initiatives into manageable chunks, and take control over their information. Fundamentally, what that means is ensuring that the tools and processes within the shipyard can integrate and give your teams a source of truth they can trust throughout the lifecycle of a vessel.

So, while many shipyards today can claim to have an architecture that is 'digital,' it is often severely fragmented and inefficient. Indeed, the industry has found that even after significant investments in an efficient 3D modeling tool and a world-class ERP/MRP system, many process inefficiencies are maintained in the shipyard, impeding the organization from moving forward in its digital transformation journey.

Unlocking information within the shipyard is vital for process improvement and competitiveness. The adoption of a new digital platform that can manage diverse data sources and sets outside of the CAD/CAM/MRP/ERP domain has become a critical requirement. This is the key to transformation in shipbuilding.

Shipbuilders already know that there is no one-size-fits-all system out there that will fix those issues and have embraced a combination of 3D CAD and ERP to satisfy their needs. However, the missing piece is a lack of connective tissue between these platforms, also known as the platform of platforms approach. Introducing an information platform that bridges the gap between these systems and provides configuration control and visibility to the entire organization has emerged as the way forward.

The digital shipyard is often treated as a marketing buzzword, but more narrowly, it can be defined as the implementation of digital and innovative Industry 4.0 technologies in a shipbuilding environment. There is no single technology or tool that enables the digital shipyard. The complexity of modern shipbuilding ensures that this will remain the case in the future.

Similarly, because every shipyard is different, there is no single metric to measure the degree of digital shipbuilding. There have been attempts to calculate digital maturity through the use of heuristics; however, for most implementations, a rigorous appraisal of maturity is unlikely to prove useful [8]. What is more useful is assessing the return on investment from the implementation of technologies and innovations.

Shipyards do not seek a digital shipyard for the sake of it. They are looking for a better way to approach their core business. Identifying the specific areas where digital technologies can influence shipbuilding gives us a framework to approach which enablers will be most effective.

4. CURRENT DX APPROACHES OF JAPANESE SHIPBUILDERS

The Japanese shipbuilding industry has experimented and considered digital technologies, with research organizations and a select number of shipyards leading innovation and R&D efforts. For example, the National Maritime Research Institute (NMRI) and the Japan Ship Technology Research Association (JSTRA).

NMRI is, among many activities, demonstrating leadership in digital transformation for the Japanese shipbuilding industry. The Digital Transformation project team at NMRI is currently researching "digital twin technology for hull structure, main engine and navigation digital twin technology, and digital information base technology for hull form design and ship construction" [9]. The Japan Ship Technology Research Association is working on visualization system in shipyards, IoT, additive manufacturing and their applicability within Japanese shipyards [10].

The Japanese National Diet's passing of the *Act on Strengthening Maritime Industries* in 2021 is a further attempt to take a policy approach to promoting DX and increase the Japanese maritime industries' competitiveness among its peers [11]. Looking at shipbuilding in particular, the Act aims at addressing the following challenges:

> 1. Responding to large-lot orders through joint design, sales, and construction by multiple companies; 2. Integrating production systems and operating multiple locations in a unified manner; 3. Responding to shipowners' needs in detail by installing cutting-edge technologies, improving on-site productivity, and developing and marketing zero-emission ships and autonomous vessels [11].

By providing investment to Japanese shipbuilders, the Ministry of Land, Infrastructure, Transport and Tourism hopes that shipyards will be better resourced to address those challenges. In particular, additional public investment may make Japan's shipbuilding industry more competitive with heavily state-supported Chinese and Korean shipyards [11].

In an effort to understand the investments Japanese shipyards were making in technology, SSI sponsored a market study in 2019 to better understand the current DX approach and focus of mid-size Japanese shipbuilders when looking at the next stage of their digital transformation. We started with a significant list of Japanese shipbuilders, and in the end, received enough information to include 17 shipyards in our conclusions. This was a blind study, and the results are composed of feedback from individuals at different levels of seniority and positions from within each shipyard. While the market study is not a rigorous analysis of any individual organization, it does illustrate larger trends present in the market as a whole. In addition, this study is only a reflection of the pre-pandemic situation.

In only 35% of the 17 shipyards were the respondents aware of a planned upgrade of some sort to their enterprise information systems including PLM, PDM, ERP or MRP systems. Of those who were aware of planned upgrades, only one in three reported that the shipyard had an approved budget in place. Note that respondents were not asked about plans to undergo a digital transformation, but simply about plans to upgrade the systems they use to manage information in the shipyard.

While 35% may seem low, especially while concerns about the economy and other factors related to the post-pandemic recovery are top of mind, only 24% of the shipyards surveyed had any plans in place to upgrade CAD/CAM systems and of those four that did only two had an approved budget for the plans. Respondents were not asked if they were planning on changing CAD/CAM systems as part of a larger digital transformation, only if they had planned any upgrades at all.

It can be argued that Japanese shipbuilders are concentrating on optimizing the business and systems they have at hand today; however, a reluctance to plan for system upgrades is a further demonstration of the legacy system

trend observed by METI. The longer the gap between digital infrastructure investments, the more likely shipbuilders will – willingly or not – be using unsupported or outdated software solutions.
Focusing on how you do business today can be effective, particularly in the context of the pressures shipbuilders face, provided the world doesn't change around you. Shipbuilding, as a conservative industry dealing with extraordinarily complex challenges, hasn't historically experienced the same level of disruption as other industries. In the past, shipbuilders could focus on today, knowing tomorrow would look very similar. Trends that were so disruptive in other areas were kept at bay: globalization, workforce migration, digitization, automation, material science, advanced production technologies, environmental awareness.
Only a few short years later and every leading shipbuilder in the world is talking about most, if not all, of these disrupters. There is a realization that the business of shipbuilding is changing and that shipbuilders must look at what they do, not just how they do it.

5. EXAMPLES FROM JAPANESE PUBLIC SECTOR DURING COVID-19

No segment has more quickly realized that it is time to change what they do and how they do it than the Japanese public sector throughout the covid-19 pandemic. The shock of the pandemic has spurred various parts of Japan's government to re-invigorate or start DX initiatives. Even as early as April 2020, the Japanese Minister of Justice spoke on the importance of adopting AI and ICT technologies in response to the pandemic [1]. Two examples are the move to electronic contracts and the creation and promotion of a Digital Governance Code.
Japanese business transactions typically rely on *Hankos*, or physical carved seals, to be used to certify contracts. The physical nature of the seal requires contracts to be printed, stamped, and mailed when needed. During the first wave of the covid-19 pandemic, lockdowns forced Japanese society to isolate and reduce social contacts – as in most of the world.
The need to remain physically distanced forced Japanese society to consider contracts without a carved seal and look towards electronic contracts. Recognizing this, in June 2020, the Japanese government revised their existing guidelines on carved seals, making them optional for private companies and contracts involving public-private transactions. Further, they expanded the acceptance of contracts certified by email, digital signature, and other digital methods [1].
Another example of how the Japanese public sector has embraced DX is through the Digital Governance Code. The project was initiated in 2019 by METI through a review of the development of such a code. In that review, METI laid out five key steps for successful digital governance: establishing and sharing a vision for growth, developing a digital strategy to realize the vision, establishing IT systems and collaborating with stakeholders, appropriately allocating digital management resources, and implanting and evaluating digital strategies [12]. In November 2020, the certification system for the code was announced. This system is a Japanese developed framework that companies from all sectors, including shipbuilding, can use as a starting point for their own DX initiatives [13].

6. DEFINING AND APPLYING A DX FRAMEWORK FOR JAPANESE SHIPBUILDERS

These DX initiatives and concepts led by the Japanese public sector are a great starting point for a Japanese approach to transformation. However, when assessing the need for a holistic approach to implementing a digital shipyard strategy in Japan, we first must assess the current situation and look at what has made the industry successful to date. For example, the production of high quality 2D class approval and engineering drawings that fit the Japanese way of building a ship.
Moving to a 3D and digital-first environment, or integrated engineering model, cannot happen at the risk of losing the current advantages and expertise that exist within Japanese shipyards. Any digital initiatives will need to add the advantages of 3D technology, work with the existing processes and systems that are in place, and continue to work with the Japanese approach to engineering information.
A DX framework for Japanese shipbuilders must also support an increased amount of design flexibility, while maintaining the internal efficiencies within engineering departments. The relatively higher complexity of ships typically constructed in Japan further lend themselves to a desire from ship owners for increased customization.
One strategy to implementing the DX approaches is to implement a unified engineering environment. In particular, one that is 3D first, addresses some Japanese specific shipbuilding challenges, and is focused on addressing the goals of the shipyard in question.

6.1 IDENTIFYING DX GOALS

Shipbuilders looking at DX are well aware of the benefits that digital systems can bring to their organizations. Streamlined nesting workflows, the ability to visualize a design through a virtual reality headset, or automatically extract class drawings from a model are just a few examples of the individual activities that a more digital shipyard can perform. However, in many cases, the implementation of these activities is not driven by the business needs of the shipyard.

Identifying the specific goals of the shipyard, that can be supported through DX, is a more holistic approach. Shipbuilders who work towards establishing a clear understanding of their goals for digital innovations, analyze their current reality and the roadblocks ahead of them, and take incremental steps to achieve their goals will be the best positioned for future success. However, some specifics of the Japanese shipbuilding industry will demand unique goals.

For example, while Japanese shipyards are extremely capable of producing complex, one-off or small run vessels, the majority of ships produced (particularly by gross tonnage) in Japan are cargo ships, bulk and ore carriers, and tankers. This focus has meant that, at an industry-wide scale, more standardization is present than is typically seen within shipyards [4]. The result is that there is less of a business need to better manage and change from downstream of the design process. Despite this, ship owners are increasingly demanding vessels that more closely meet their operational needs. It stands then, that one business goal for Japanese shipbuilders to consider will be increased design flexibility.

6.2 STREAMLINING THE FLOW OF INFORMATION

With siloed systems, it is extremely difficult for the organization to have a 360-degree look at all the available information. Ensuring that information can move across the design office, shipyard, and any subcontractors freely provides a foundation for process improvement, competitiveness, and to achieve the shipyard's goals. The adoption of a digital platform that can manage diverse data sources and sets outside of the CAD/CAM/MRP/ERP domain is a key requirement to achieving that flow of information.

The unique nature, complexity, and scale of shipbuilding ensures that there will never be a single system that can effectively manage the entire process. The unique demands of shipbuilding have also resulted in shipyards custom developing tools to serve their own specific needs. Because of this, shipyards live within a world where platforms need to talk to other platforms – commonly referred to as platform of platforms [14]. To support this approach, a source of truth, built with an open architecture that is capable of easy integration to all of the other platforms in place is crucial. These connections make the flow of information possible, and provide the ability for any stakeholder to understand the status of a part, block, or entire project.

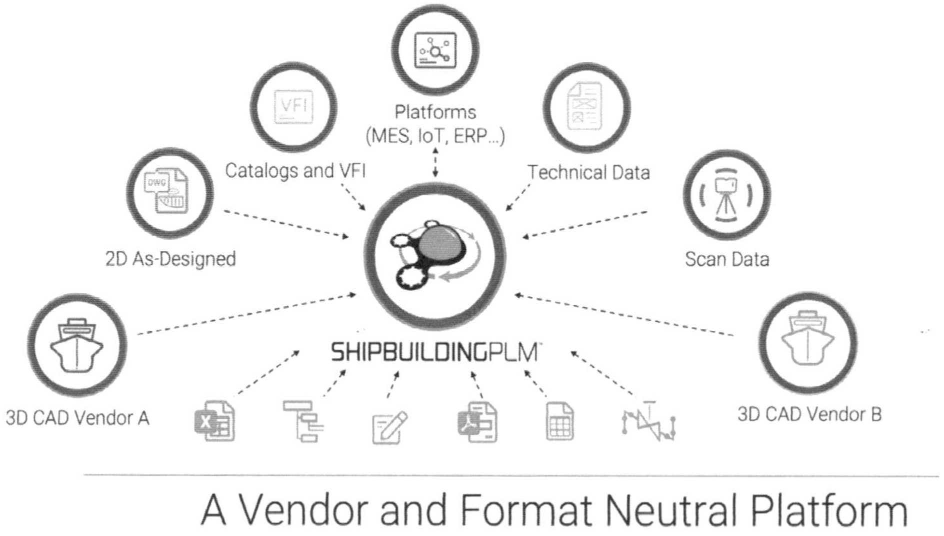

Figure 1. An information platform for shipbuilding must bring together a wide variety of data sources.

For any system to be worth adopting by a shipyard today, it must demonstrate value from day one and be implementable within months. ShipbuildingPLM is SSI's platform for managing and understanding diverse shipbuilding data sets that satisfies these criteria and is a vertical solution built on top of Aras' Innovator platform. Using ShipbuildingPLM as an example, we can isolate how an information platform interacts with the rest of the shipyard's systems to deliver an integrated view across disciplines and understand product data. The contrast is particularly striking when compared to a traditional architecture that only supports data retrieval.

Each configuration will be different depending on the individual needs of the shipyard, but in general, information like the product hierarchy, BOMs (bill of materials), 3D viewables, and MI (manufacturing instructions) will be pulled in from the engineering tool. ShipbuildingPLM will make that information available to the organization for engineering and planning review and share it with the ERP or MRP system. The ERP or MRP will then feed back data like the order status for parts to the PLM in support of manufacturing and change management. In all cases, the two-way nature of synchronization ensures that every discipline has confidence that it is working with the most up-to-date information available.

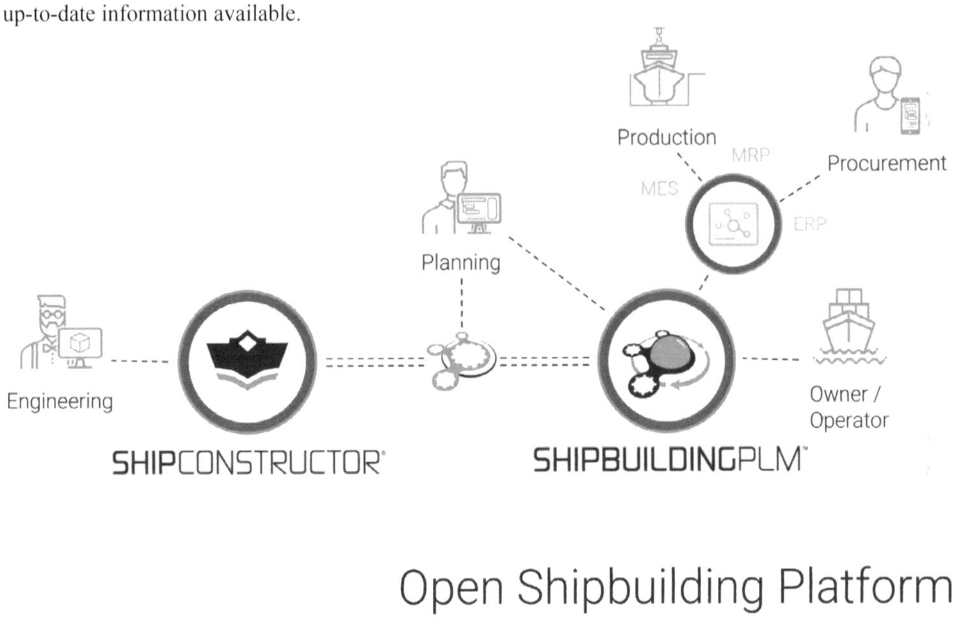

Figure 2. The architecture of an open shipbuilding platform.

6.3 ENABLING DIGITAL TWIN CAPABILITIES

Possessing a single source of truth enables a shipyard to take advantage of digital twin capabilities to support their business goals. Broadly defined, for shipbuilding in particular, a digital twin is a digital representation of a specific configuration of a ship, or multiple related configurations, where the information captured is indistinguishable from the physical ship at a particular point in its lifecycle, and where the information captured is sufficient to serve the purpose for which the digital twin exists.

While 2D drawings may possess the same level of detail as a 3D model, the inability to easily relate, track revisions, and look at the metadata of elements within the drawing make them unsuitable as a foundation for a digital twin. Taking a digital-first approach, we can briefly examine how a true digital twin can be applied in the case of a refit on an existing vessel.

As a starting point, it is necessary to understand the existing on-board reality of a vessel. If as-built drawings or an as-built model exists, that can be used as a starting point; however, often an accurate view will need to be captured even when an as-built is present.

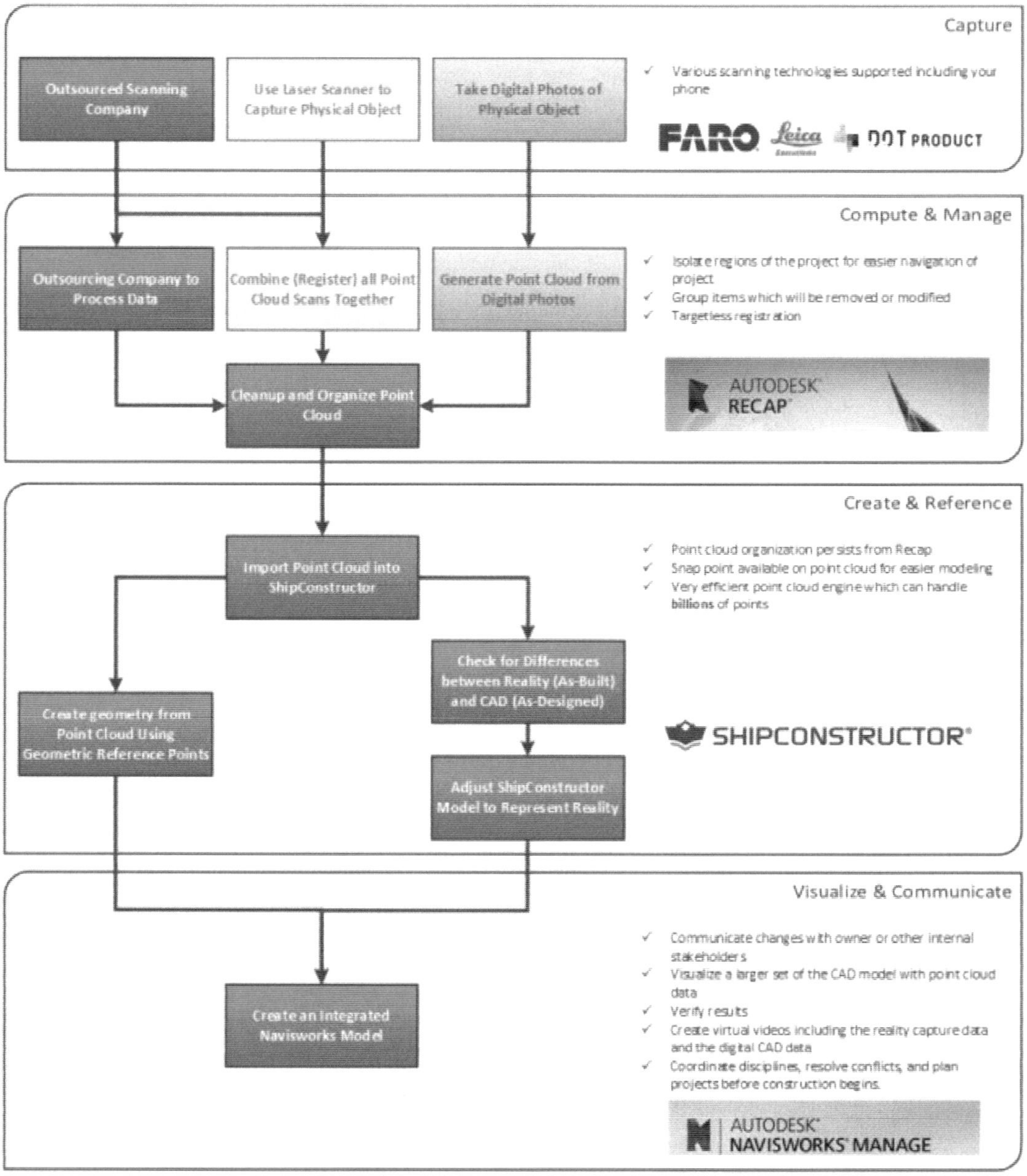

Figure 3. A basic workflow for reality capture using ShipConstructor.

Reality capture, using either point clouds or photogrammetry, are both solutions that can be used to digitally capture 3D space. Both point clouds and photogrammetry are usually created using a scanning device mounted on a tripod that scans the compartment. The tripod is moved throughout the vessel and the information stitched together to create the digital representation of the ship. There is often a misconception that this technology creates a digital twin. However, scans only capture what they can see, leaving unknowns behind equipment, linings, and insulations. There are also no properties associated with it. This technology is best used with the other elements that make up the digital twin to validate the design, identify missing information and to identify clashes when designing refit solution.

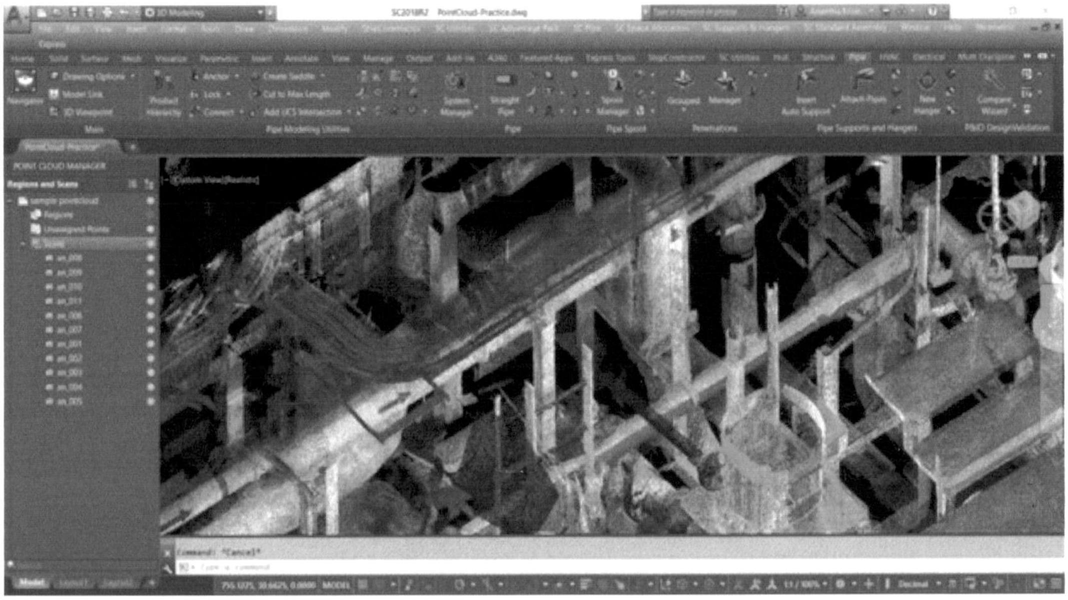

Figure 4. Point clouds overlayed with the product model in ShipConstructor

As shown in figure 4, the point clouds are easily overlayed with the product model in the CAD system and modelling has begun on a pipe system flagged for an upgrade and shown in green. Within the CAD system, pipe centre lines can be identified and automatically generated in the scan to help rapidly model anything required. Integrating the point cloud information with the CAD model and associating it with the rest of the contextual information that exists provides a through-line for the lifecycle of the vessel. All together, this information can be considered as the digital twin.

6.4 FUTURE-PROOFING INFORMATION ACCESSIBILITY

The source of truth for the shipyard must also serve as a foundation for the future, both to satisfy the shipyards' internal requirements and those of their clients. For the shipyard, a foundational platform ensures that implementing the latest digital innovations is based on the value they bring to the organization, rather than based on technical or skill limitations like proprietary software or the retirement of a skilled worker.

7. CONCLUSION

The existence of a way through is becoming apparent to those in the industry, identifying first-steps and the most important areas to focus on is the upcoming challenge. In the next part of this series, we'll look at other trends and technological disrupters that will drive Japanese shipyards towards a needed digital transformation. 3D, materials technology, AI, and innovative production techniques will all work to reaffirm Japan as a leader in the industry and address future challenges before they become industry-wide problems.

8. REFERENCES

1. J. Iida, "Digital Transformation vs. COVID-19: The Case of Japan," *Digital Law Journal,* vol. 1, no. 2, pp. 8-16, 2020.

2. M. Kamola-Cie i , Change in he o a hi ui ding ndu y on he a e o e e ed a e Worldwide in the 21st Century," *European Research Studies Journal,* vol. XXIV, no. 2B, pp. 91-112, 2021.

3. S. Otsubo, "Maritime Cluster in Japan - Shipbuilding and WP6," OECD, Paris, 2016.

4. OECD, "Peer Review of the Japanese Shipbuilding Industry," OECD, Paris, 2012.
5. S. K. Tan, "Race in the Shipbuilding Industry: Cases of South Korea, Japan and China," *International Journal of East Asian Studies,* vol. 6, no. 1, pp. 65-81, 2017.
6. Fujitsu Research Institute, "The Latest Trends in DX Policy and Future Outlooks," 2020. [Online]. Available: https://www.fujitsu.com/jp/group/fri/knowledge/opinion/consul/2020/2020-1-3.html.

7. Ministry of Economy, Trade and Industry, "DX Report-Overcoming the IT System "Cliff in 2025" and Full-scale Deployment of DX," 7 September 2018. [Online]. Available: https://www.meti.go.jp/shingikai/mono_info_service/digital_transformation/20180907_report.html.

8. T. Pang, J. Pelaez Restrepo, C.-T. Cheng, A. Yasin, H. Lim and M. Miletic, "Developing a Digital Twin and Digital Thread Framework for an 'Industry 4.0' Shipyard," *Applied Sciences,* vol. 11, no. 1097, 2021.

9. National Maritime Research Institute, "Digital Twin Technology for the Digital Transformation of Maritime Technology," [Online]. Available: https://www.nmri.go.jp/en/study/research_organization/digital/digital_en.pdf. [Accessed 18 July 2022].

10. Japan Ship Technology Research Association, "R&D (Research and Development)," [Online]. Available: https://www.jstra.jp/en/R&D/. [Accessed 18 July 2022].

11. S. Otsubo, "Enactment of the Act on Strengthening Maritime Industries and Its Significance," 20 November 2021. [Online]. Available: https://www.spf.org/en/opri/newsletter/511_1.html?full=511_1. [Accessed 18 July 2022].

12. Expert Committee on Digital Governance, "Consideration for Developing Digital Governance Code," September 2019. [Online]. Available: https://www.meti.go.jp/shingikai/mono_info_service/digital_governance/pdf/report_001.pdf.

13. Ministry of Economy, Trade and Industry, "Digital Governance Code," 9 November 2020. [Online]. Available: https://www.meti.go.jp/shingikai/mono_info_service/dgs5/pdf/20201109_01.pdf.

14. D. Morais, M. Waldie and N. Danese, "Open Architecture Applications: The Key to Best-of-Breed Solutions," in *COMPIT 2016*, Lecce, 2016.

9. AUTHORS' BIOGRAPHIES

Darren Larkins is a co-CEO at SSI. He combines over 22 years of experience in the development, marketing, sales and implementation of marine systems with knowledge gained from shipbuilders and offshore experts around the world.
Simon Crook IEng MRINA is a Solution Specialist at SSI. He has 20 years of experience in naval architecture and marine engineering with a strong technical skill-set and a proven record in the management and development of professional, multidisciplinary design packages.
Greg Goulanian is a Shipbuilding Solution Evangelist at SSI. He is responsible for showcasing SSI's innovative software solutions, global success stories, and the future of shipbuilding through papers, presentations, and various forms of media.

FROM POINT CLOUD TO CAD-MODEL BASED ON AI

J Luetzenberger, M Grau, J Stjepandic and **C Zerbst**, PROSTEP AG, Germany

SUMMARY

Most vessels undergo modernization and overhaul during their lifecycle. A model representing the as-is status is a pre-requite to plan these activities. Especially when it comes to overhauling a piping system, it is crucial to know their exact spatial layout. Otherwise, the existing parts and the newly delivered parts may not fit together.

Capturing the surfaces of the parts using a 3D scanner and generating a point cloud representing the as-is situation is state of the art. The point cloud is then loaded into a CAD system to represent the existing parts before designing the new parts and ensure they fit together seamlessly. This requires remodelling at least the connection points of the existing parts manually. Remodelling is both time-consuming as well as error-prone, so we'd like to avoid it.

In the DigiTwin research project [1], we developed a solution for this. It features a fully automatic generation of a CAD model based on recognizing piping components from the point cloud using artificial intelligence. The main task is running object recognition from the point cloud. It is crucial to find the correct components and detect their position. Based on that, we generate native CAD models for typical shipbuilding tools. This enables using the automatically generated CAD models for the as-is status as reference geometry for the updated design. In our paper, we'd like to present the steps needed to do so and also the result we achieved.

NOMENCLATURE

AI	Artificial Intelligence	IP	Intellectual Property
CAD	Computer Aided Design	ML	Machine Learning
FPSO	Floating Production Storage and Offloading Unit	P&ID	Piping & Identification Diagram

1. INTRODUCTION

Vessels such as cruise liners, FPSOs or comparable are products with high complexity. It´s operation-phase is often lasting for decades. The development process of such complex products is multi-disciplinary, and a synchronization or alignment of different processes or disciplines is needed. A prominent example for the alignment of different disciplines is the routing process of pipelines. A vessel such as a cruise liner is comparable to a small town. It includes a piping system for fresh and grey water, sewage and further. The according pipes need to be routed almost throughout the entire ship.

In a first step, the piping system needs to be defined from a process perspective. Which medium needs to be transferred? From which point to which point? What pressure or flow rate is needed? Is a pump needed to achieve the needed pressure? Is heating or cooling necessary? Etc. This way the process is defined, and the corresponding requirements are translated into technical specifications.

The piping and instrumentation diagram (P&ID) is the technical schematic representation of a pipe system. However, spatial information is not covered. Within authoring systems, the pipeline is routed throughout the entire ship as part of the digital 3D ship model. Thereby the pipes are crossing decks and bulkheads. At the penetration points, nothing else (e.g. power cabinet) can be placed. The current planning status is always updated and synchronized between the routing and the electrical department for example. That makes clear, that during the routing process a close alignment or synchronization between the involved disciplines is mandatory to ensure consistency. The resulting ship model covers the final planning stage of the piping system and spatial information respectively.

During the operation along the long lifecycle of such products maintenance, repair and overhaul become necessary for various reasons. The usual documentation of the piping system is the P&ID, which is delivered with the ship to the operator. So far documentation is still paper-based and changes are often only poorly documented. During maintenance activities, technical personnel are forced to find one specific piece of equipment from the P&ID within the ship or vice versa. On the piping diagram significant equipment, instruments or fittings are connected by lines representing the piping branches between them. However, the schematic representation does not provide information about the routing in space, their arrangement and structure. Technicians must follow the pipes to find the component of interest in order to maintain it. Hereby as stated above the pipes are routed through walls across the entire ship. This way maintenance activities are time-consuming and cost intensive.

Especially, when it comes to overhauling, a model representing the actual state of the vessel covering spatial information is a pre-requisite for planning the activities. The CAD models from the ship development provide spatial information. However, they are usually not delivered with the ship to the operator. And even if the CAD models were delivered, they would not be updated according to the changes carried out during maintenance activities and therefore they would soon render obsolete. To get a reliable basis for planning these activities, the current state needs to be captured and digitalized. Nowadays this is realized by a laser scan and manual modelling according to scan afterwards. The laser scans the surfaces of a piping system and its surroundings and stores it as a point cloud. This point cloud can be imported into authoring systems and a designer remodels the parts of interest by placing the parts within the point cloud. This way a 3D model of the as-is state can be achieved. Nevertheless, this approach is tedious and slow due to the huge amount of bulk point data.

Modernization activities with regards to piping systems, it is getting even more complex. Usually, not the whole piping system is going to be replaced. Parts are kept and others are newly designed. The designers must ensure that both old and new ones fit together. Therefore, designers remodel at least the connection point between the legacy portion to be kept and the new portion to be planned. The planning of the new portion is carried out based on these remodelled models and ensures, that the new portions can be seamlessly connected to the remaining portions.

Both presented scenarios of maintenance and especially overhauling are examples for beneficial use of maintained and actual digital representation of a ship. The concept of digital twin is a digital representation of a physical product and further provides the potential to reflect usage information into the development process and improve future product generations. However, the digital twin needs to be maintained regarding geometrical and spatial information in an efficient and economical manner [1]. To support this process PROSTEP has developed a solution to derive CAD-models from point cloud through an object recognition process [2].

This paper outlines the generation of a digital twin model of an already existing ship. The process of object recognition is presented, and a sample is presented. Following this introduction, the background is given in chapter 2. While chapter 3 describes the general process, its application to a real-world example is shown in chapter 4.

2. BACKGROUND

The foundation for AI-based object recognition from point cloud was laid in the German national research project DigiTwin [1]. The objective was to generate a digital twin of an existing manufacturing hall. Therefore, the as-is state has been captured by laser scans and stored as a point cloud. This is the input for an automated object recognition process to detect manufacturing machinery (such as drilling or milling machines), the floor, walls, etc. After the object recognition the identified machines are compared with a database including various machines and their CAD-models. Once a specific machine has been found, this can be placed within an authoring system. To find the exact position in space a bounding box approach was implemented. The origin of a bounding box was derived from the point cloud and the extensions (length, width, depth) have been derived as well. Based on that information the CAD models from the database are positioned correctly in the design space. Adding all identified machines in place within a common model produces a digital twin model of the manufacturing hall. Downstream processes such as material flow analysis or layout optimization become possible this way.

Within the research project, PROSTEP developed a solution for the object recognition process. Suitable machine learning (ML) algorithms have been elaborated and developed, data has been prepared for the training of neural networks and the training executed. Further, a database with available machines has been built and algorithms for the analysis of the geometrical information as well as the position in space have been implemented. The detection of the machine within the point cloud and the position of the correct machines in space was enabled.

Based on that results PROSTEP improved the solution for the object recognition of piping systems. While the manufacturing assets are single solitary objects, a piping system is largely a combination of seamlessly connected parts. To distinguish between these parts and recognize every single part even though they are seamlessly connected, is a higher complexity compared to the original application. In addition, the form factor with regards to the objects to be recognized changed significantly. Manufacturing assets usually have a cubic form factor while many components of piping systems have a rather extruded shape, such as pipes or elbows. For those reasons, an adaptation and further development of the algorithms were necessary and appropriate training of the neural networks beforehand is a pre-requisite [3].

Especially with regards to the training of the neural networks a reliable set of training data is mandatory. Although all components such as straight pipes, elbows, reducers, tees etc. are standardized products, the possible combinations of them are nearly infinite. That increases the complexity significantly. Point clouds for the training must be prepared accurately.

These prepared point clouds are the input for the training of the neural networks. The patterns of equipment, pipes, elbows, fittings, tees etc. need to be trained in various conditions to ensure a reliable recognition process of the piping components.

To trace back CAD models from point cloud spatial as well as geometrical information is needed. The most important information about a piping system is the centreline. It enables the correct position and orientation in space of each component. Additionally, the extension in terms of dimensions is mandatory for the generation of a CAD model. Once the component has been successfully isolated from the point cloud data it is not too difficult to derive key parameters like diameter, length, height and width from point cloud. This way the basis for the generation of CAD models representing the as-is state can be achieved.

3. PROCESS - OBJECT RECOGNITION BASED ON ARTIFICIAL INTELLIGENCE

As mentioned above the objective is to trace back CAD models of existing piping systems from point clouds. To achieve this goal a point cloud which represents the current state of piping systems is needed. A 360° high performance laser scanner captures the surfaces of the piping system and its surrounds and generates high density point clouds. To capture all areas and to avoid scan shadows etc. usually several scans from different positions are carried out. Afterwards, these clouds are joined together and aligned to each other through a highly automated process. This is called registration and provides a comprehensive point cloud consisting of the merged scans. This point cloud is the input for the object recognition process of PROSTEP's 3DigitalTwin solution [4, 5].

3.1 POINT CLOUD REDUCTION

High density point clouds comprise a huge amount of data. This can range from several gigabytes to several terabytes depending on the dimensions of the scanned area and density of points. Loading and processing such a large amount of data e.g. in a CAD system brings it to its limits in terms of managing the point clouds. Likewise, the duration of the object recognition process increases significantly with the amount of data. Therefore, in the first step the density is reduced to a reasonable size to achieve reliable object recognition on the one hand and make the point cloud usable in an efficient manner on the other hand. The reduced point cloud can be stored in formats such as e57 and loaded into CAD systems, in which they can be managed more efficiently.

3.2 SEGMENTATION 1 – ISOLATION OF THE PIPING SYSTEM

The reduced point cloud is the input for the initial object recognition process, called segmentation. It is processed in two steps. During the first segmentation step the cloud is analysed and the points are assigned to different object classes. In this way, the equipment, the piping and the structural elements can be separated from each other.

Figure 1: Segmented point cloud in equipment, piping and construction

Figure 1 shows the result of segmentation 1. The AI recognized the classes equipment (cyan), piping (blue) and structure elements (yellow). Each of the mentioned classes is stored and accordingly available as individual point clouds. These can be imported into CAD systems together. That enables the user to have the full point cloud to a manageable extent since it is already reduced or to make them visible or invisible and apply some filtering mechanism this way. Through that everything can be seen, when the context is needed or disturbing parts can be made invisible, e.g. structure elements and equipment can be made invisible when the focus is on the piping.

3.3 SEGMENTATION 2 – RECOGNITION OF PIPE COMPONENTS

The second segmentation step analyses the piping. Input is the piping point cloud from the previous process step, segmentation 1. The AI algorithms recognize the different components the piping system is composed of.

Figure 2: Segmentation 2 - Recognition of pipe components

Figure 2 shows the piping point cloud as result of the first process step, segmentation 1 on the left side. After processing the second segmentation, the components are assigned or dedicated to six different types:

1. Straight pipe (green)
2. Elbow (blue)
3. Reducer (cyan)
4. Tee (yellow)
5. Flange (magenta)
6. Valve (lilac)

As mentioned above the different types are stored in individual point clouds as well. Respectively it is possible to import all point clouds together in one CAD system and to apply a filter by making them visible or invisible. Of course, it is possible to combine the point clouds of segmentation 1 with the equipment and the structure point cloud. A more detailed filtering can be achieved this way.

3.4 CLUSTERING - SPATIAL AND GEOMETRICAL INFORMATION

The correct recognition of the different components is the first and most relevant step. However, to ensure that CAD models can be generated from it, spatial and geometrical information is crucial. The object recognition during the first steps is a process on a single point level. To derive the needed spatial and geometrical information the single points need to be brought in context to each other. During the clustering, all points, which describe one component are isolated. Based on that the dimension such as width, length, height or diameter are derived. Further, the centreline can be calculated. It is the red line throughout the entire piping system and allows to specify the exact position of each component in space.

3.5 GENERATION OF CAD MODELS

The spatial and geometrical information is the input for the generation of CAD models. Most CAD systems which allow the routing or planning of piping systems also provide component catalogues. These catalogues provide various components according to specific specifications that can be applied. Besides geometrical information also pipe classes, material, and other non-geometric information is available from here.

Based on the diameter and the component type a suitable component can be selected automatically from the catalogues. Only the piping specification to be applied needs to be chosen by the user. The centreline provides information about the position in space. It enables to place the parts correctly.

4. FROM POINT CLOUD TO CAD-MODEL

So far the general process from point cloud to the generation of a CAD model has been presented. For the proof of concept a piping system belonging to biogas plant was used. In the course of this example CADMATIC has been chosen as one of the CAD systems widely used in shipbuilding industry.

4.1 REDUCTION OF DATA VOLUME

To capture the as is state of the biogas plant a high-performance scanner was used. The scan was conducted from various positions and the different point clouds have been registered to one complete point cloud of the plant. As mentioned above, due to the high density of the scans the data size is huge as well. The reduction in case of the present piping system including surroundings has been reduced significantly from approx. 5,5GB to 18MB or of more than 99%.

Figure 3: Reduced point cloud of a piping system incl. surroundings and equipment

Figure 3 shows the reduced scan of the plant. This comes with two benefits. The reduced point cloud increases efficiency of the object recognition process on the one hand and the manageability of the point cloud within CAD systems on the other hand. Especially for cloud-based applications a reduction of the data size is a plus.

4.2 SEGMENTATION 1 - RECOGNITION OF THE PIPING SYSTEM

Input for the AI-based object recognition process is the reduced point cloud. During the first process step – Segmentation 1 – the points are dedicated to three main classes:

- Equipment: Pump,
- Pipeline: Straight pipes, elbows, tees and valves
- Structure: All points which cannot be dedicated to one of the classes above

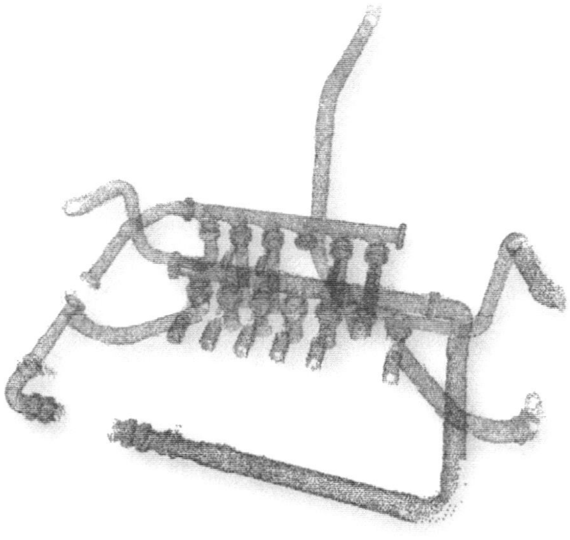

Figure 4: Result first segmentation – piping system

Each class is stored in a single point cloud. That enables the user to load them individually into an CAD system. However, with regards to the process the focus relies on the piping itself and its extraction from the point cloud. Figure 4 shows the segmented piping system without any equipment or structure. Within the piping empty space can be recognized. At that position are equipment placed which are stored in the equipment point cloud.

4.3 SEGMENTATION 2 - RECOGNITION OF COMPONENTS

The segmented piping point cloud (Figure 4) is the basis for the second segmentation step. Here another ML algorithm is applied, which enables a more detailed recognition. The neuronal network is trained to distinguish between the component classes pipe (green), elbow (blue), tee (yellow), reducer (turquoise), flange (magenta) and valve (lilac):

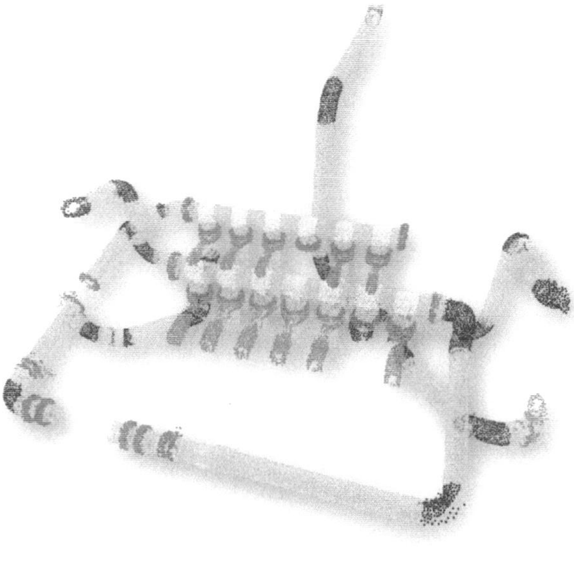

Figure 5: Result segmentation 2 - component classes

Like in segmentation 1 also here the different classes can be stored in individual point clouds, which enables a filtering by loading all point clouds into the CAD system and making them visible or invisible as described above.

4.4 CLUSTERING – COMPONENTS, DIMENSIONS, POSITION IN SPACE

To enable a generation of CAD models from the point cloud, the geometrical dimensions and the position in space are crucial information. To derive them, first points that are describing the surface of one component need to be clustered. Based on this, geometrical dimensions such as diameter, length, width or height can be derived.

Figure 6: Segmented point cloud incl. centre lines

The most important information with regards to a correct positioning in space of the components is the centre line. Figure 6 illustrates the derived centrelines within the segmented point cloud. These vectors enable not only the correct position, but also the orientation of each component. All mentioned parameters derived from point cloud are derived in a neutral csv-file.

4.5 GENERATION OF CAD-MODEL IN CADMATIC

CADMATIC is a typical representative of authoring systems for shipbuilding industry. Also, it provides the module to plan and route piping systems. The piping parts are loaded from provided catalogues within CADMATIC. According to the applied specification the parts can be chosen from the catalogues and placed within the design space.

For the 3DigitalTwin solution, a reverse engineering approach is needed. Through the object recognition process the components of the piping system were identified and the geometrical information, as well as the position in space, were defined. This way all parameters to select the correct part and load it from the provided catalogue are given.

In the first step, the parameters are checked and optimized with regard to their consistency. Therefore, the specification to be applied needs to be selected beforehand. The part can be mapped to the available pipe material or standard component according to the chosen specification. However, it needs to be considered, that the laser scanner is capturing the surface of the piping system and respectively the outer diameter of pipes for example. The pipe provided by the catalogue is defined by their nominal diameter. Here additional mapping is mandatory.

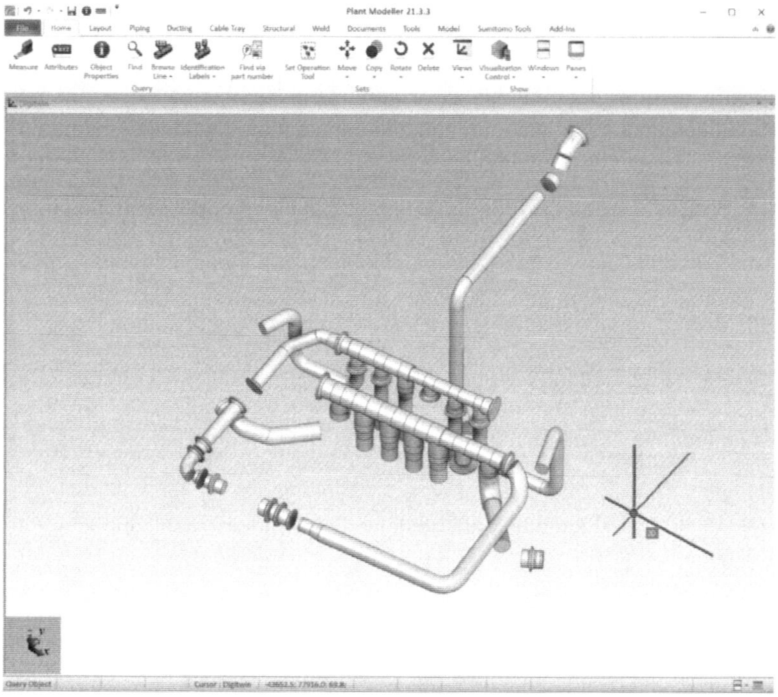

Figure 7: Generated CAD-model in CADMATIC

After the recognized components and catalogue parts have been successfully mapped the CAD model can be generated automatically as shown in Figure 7. Components dedicated to the class of equipment, such as pumps, heat exchangers etc. are not considered in the model and need to be placed manually. Such models usually are provided by the manufacturer or supplier of these equipment.

In the example shown in Figure 7 the part catalogue according to CADMATIC´s standard specification has been applied. This deviates to the original specification which has been used for the planning. Therefore, some components are not loaded correctly and deviations to the original CAD model appear. However, if the correct specification and the according catalogue respectively, this specification can be applied, and more precise model is generated in an automated manner.

5. CONCLUSIONS

Manual remodelling of pipe systems based on point clouds is time-consuming and therefore expensive work. However, it is necessary, especially for the planning of overhauling. A maintained digital ship model is appreciated not only for overhauling but serves benefits for many tasks during the lifecycle of a ship. The digital twin, which is a digital representation of a physical product, is a promising approach here, but so far a technological approach for efficient and economical creation and maintenance of this model with regards to spatial and geometrical information were missing.

3DigitalTwin is a solution developed by PROSTEP to overcome this gap. An object recognition process based on artificial intelligence enables the segmentation of a point cloud or more precise the analysis of a piping system. Depending on the needs of a user various benefits are provided. In a first step the data volume of the point cloud is reduced, this makes the also huge scans manageable within the CAD system. In addition, the point cloud is segmented into component classes such as straight pipe, elbows, reducer, etc. These type classes can be stored in single point clouds respectively. This enables users to achieve a comprehensive point cloud by loading all single type point clouds together on the one hand. By making the different point clouds visible or invisible a filtering mechanism is enabled. Disturbing parts can be made invisible and manual modelling can be executed more efficiently.

Based on the segmented point clouds the position in space and its geometrical extension can be derived. This information enables a mapping with part catalogues of CAD systems. Various specifications can be applied this way and a CAD model in a native format is provided. The authors proved that tracing back a CAD model from the point cloud has been achieved.

Enabling this technology also for additional CAD systems such as AVEVA Marine or AutoCAD is under development. The presented solution is currently available through PROSTEP´s OpenDESC CAD conversion service.

6. REFERENCES

1. e andi , o e , e e ogni ion e hod in a ui n i on en , in e andi e a ed DigiTwin: An Approach for Production Process Optimization in a Built Environment, Springer Nature (2021).
2. au, W o o , en e ge , e andi , u o a ed gene a ion o a digi a in o a o e an y using 3D scan and artificial intelligence, Advances in Transdisciplinary Engineering, 2021, Vol. 16, pp. 93-102.3. E. Agapaki, I. Brilakis, CLOI-NET: Class segmentation of industrial facilities' point cloud datasets, Advanced Engineering Informatics, 2020, Vol. 45, 101121.
4. M. Sommer, K. Seiffert, Scan Methods and Tools for Reconstruction of Built Environments as Basis for Digital Twins, in J. e andi e a ed igi in n oa h o odu ion o e i i a ion in a ui Environment, Springer Nature (2021).
5. P. Kremer, J. Lützenberger, F. Müller, e andi , An Approach for the Incremental Update of a Digital Twin of a Process Plant, Advances in Transdisciplinary Engineering, 2022, Vol. 22, in press.

7. AUTHORS BIOGRAPHY

Johannes Lützenberger, Dipl.-Ing. holds the current position of Technical Account Manager Plant at PROSTEP. As a mechanical engineer, he has been involved in several PLM-projects in various industries. In his role as Product Manager 3DigitalTwin he is responsible for the development of the service for an AI-based object recognition in point clouds.

Dr.-Ing. Matthias Grau is currently Account Manager Shipbuilding Industry. As a Naval Architect, he has been involved in various shipyards as well as in CAD/PLM system development on the vendor side. In his role as a PLM Strategy Consultant, he also gained experience in other industries such as aerospace and automotive.

Dr. Josip Stjepandic holds the current position of Senior Manager at PROSTEP AG. He is responsible for new solutions and markets. His previous experience includes development of solutions for data exchange and knowledge-based engineering.

Dr.-Ing. Carsten Zerbst PMP holds the current position of Manager System Integration at PROSTEP. He is responsible to define and implement solutions for maritime customers. He obtained a doctorate from the Technical University Hamburg-Harburg in the area of onboard data integration.

DIGITAL TRANSFORMATION THROUGH DATA-DRIVEN AND DATA-CENTRIC APPROACHES WITH ARTIFICIAL INTELLIGENCE

A Maimun, C L Siow, and **J Khairuddin**, Universiti Teknologi Malaysia, Malaysia
K Hiekata, University of Tokyo, Japan

SUMMARY

Data-driven approach emphasises on extracting information and insights from data into deducing the next course of action. Whilst data-centric approach focuses on exploiting domain knowledge based on existing data and their relationships into producing near accurate output. They serve as the key to the shipping industry's transformation which defines innovations and the use of digital technologies to deliver processes, product quality, and organisational improvements. Digital transformation plays an important role in ensuring the industry remains competitive and efficient. This work reviews the potentials, and challenges of adopting the approaches. Case studies are explored at the UTM Marine Technology centre on the design, simulation and analysis tools, intelligent towing tank, AI and IoT technologies applications. They present significant impact on the existing processes and productivities with focus on the data collection, modelling, validation, and optimization. Therefore, the findings open a new venture into expanding the use cases, technologies, and processes improvements.

1. INTRODUCTION

Digital transformation plays a significant role in the information systems strategy especially in investigating the impact of information technologies and their deployment. It is defined as a value-added process and performed to improve the organisation and its business activities by applying significant changes to the operating processes and leveraging on the digital technologies and applications [1, 2].

In general, digital transformation is explored as the means for an organisation to remain competitive and efficient [3]. This is important especially for large, complex, low-volume and high value industries such as the shipping industry. Key technologies include automation, Internet of Things (IoT), big data, simulation, edge computing, extended reality (XR) and artificial intelligence (AI) [4, 5, 6, 7]. Furthermore, integrating the digital transformation with AI or specifically machine learning (ML) amplifies the overall implementation and impact [8]. This is particular to the processes which involve data-driven decision-making and data-centric inferences that require minimal human intervention.

These technologies are seen to gain wide interests in the shipping industry and diverse areas. The drivers are: 1) organisational - cost reduction, streamlining operations, minimising lead and turnaround time goals, 2) technological - new emerging technological application and processing big data, and 3) external - customer-driven, competitive environment, regulatory requirements, improving collaboration and data transparency [6].

Recently, Sánchez-Sotano et al., [4] reported major industrial adoption in transforming the organisational, processes, people and implementing innovative solutions. It focuses on exploiting digital technologies in delivering product and service innovation, client-centric approach, and value. Specifically, Guangrong [5] described the big data and AI application in finding optimal ship systems design and operations through a multi-level collaborative design, operation, and management platform.

Another digital transformation deployment is reported by Cil et al., [9] on the shipbuilding processes improvement and monitoring through the smart manufacturing and IoT technologies. Emphasising on the business model changes, Tkalich et al., [10] studies on how to effectively transform a traditional business model through organisational changes and concurrent engineering.

It is observed that the advantages from digital transformation benefitted across the organisational processes, value creation and chain. The works show wide use cases, applicability and scalability. However, such transformations are very challenging due to its disruptive nature, the organisational, technological, and external barriers [6, 9, 11, 10]. It is reflected by the lack of progress in the industry especially in applying the data-driven and data-centric design and processes.

This work is devised to discuss the impact of the approaches in exploring the digital transformation concept. It is structured as the following: Section 2 describes the data-driven and data-centric approaches in the shipping industry. Section 3 discusses the knowledge representation and methodologies. Section 4 describes the use case explored in this study. Section 5 discusses the impact of the work. Finally, section 6 concludes and summarises the research work.

2. DATA-DRIVEN AND DATA-CENTRIC SHIP DESIGN AND BUILDING

Data-driven approach refers to the activities that are performed based on the characteristics and the outcomes of the extracted information gained from the data [12]. Gaspar [13] and Montáns [14] described the data-driven approach as the means to extract information and insights from a design and related processes. It is proposed to facilitate data collection, processing, analysis, and manipulation throughout the product lifecycle into producing effective processes and quality design.

While Pan et al., [15] and Reijers et al., [16] explained the data-centric approach as a method that focuses on exploiting domain specific knowledge using existing or newly generated data and processes. The approach is emphasised on the validated data from the working setup, experiment, or simulation along with the data-driven approach applied to enhance the process-oriented output. Therefore, while the data-driven approach concept makes use of the data to gain further insight, the data-centric approach is applied on the established data and domain knowledge to deterministically produce the outcomes.

2.1 DATA-DRIVEN APPROACH

Gaspar [13] proposed a collaborative data-driven ship design approach to facilitate the studies on the ship design tasks. The design information and knowledge are generated, analysed, kept, shared and reused within a single platform. It serves to handle key design processes into producing conceptual and preliminary designs based on the existing ship design module, data, and empirical formulas.

Fardelas and Kim [17] explored the data-driven approach in handling the ship design complexities, producing, and selecting best design parameters. Based on the design data, they applied an AI method to probabilistically infer best design parameters that are influenced by set conditions and criteria. Therefore, the design functional requirements, design parameters and their relationships can be identified, assessed, and used in supporting the decision-making process.

While in shipbuilding, Lim et al., [18] highlight the challenges in exploring new technologies and aligning the requirements between the regulatory bodies, business, and transformational needs. On top of the already conventional nature of the industry, such challenges have imposed further constraints. Additionally, Bao et al., [19] present data-driven shipbuilding process planning and management using real-time data. It is implemented by consolidating the shipbuilding and equipment data to build and deploy ML models. It is used to assess the event-driven activities and equipment status in predicting the action plans.

The data-driven approach observed enables for quick, lightweight and near accurate simulation and verification. It is proven significant especially when dealing with new problems and domains. The work such as by Mittendorf et al., [20] has shown good performances in comparison to the standard methods. However, the practical use for the data-driven approach is challenging, factored by the reliability of the data collection process, the amount and quality of the data, appropriate knowledge representation and models' performance.

2.1. DATA-CENTRIC APPROACH

Zub and Nesin [21], and Dushko et al., [22] works on the Electronic Information Model for a Ship (EIMS) system explored the data collection, synchronisation, analysis, sharing and retrieval within a heterogeneous environment. It presents the concurrent ship design and development processes similar to Papanikolaou et al., [23]. The concept serves as a cross-platform integration tool that implements data-centric and systematic design development. It provides the means for interfacing and integrating the data-driven, simulation, physics, and engineering processes.

Pan et al., [15] discussed the current data-centric approach integrating simulation, ML, and statistical methods. These include the digital twin, hybrid ML and dynamical systems models' application, probabilistic numeric and programming for simulation, generative modelling and simulations, and real-time simulation-based methods. Applying the concept, Guangrong [5] proposed a data-centric-based platform for design, operation, and maintenance of ship systems built on top of knowledge and data-based systems.

Emphasising on the knowledge management, collaborative and data-centric approaches are applied using AI-based solutions to enable effective data and information flows, decision-making as well as the integrated and concurrent engineering activities [5, 24]. As the key component in the digital transformation, knowledge management focuses on systematically converting tacit expert knowledge into explicit knowledge representation. It is aimed at integrating and reusing the established knowledge for data flow, automatic processing, and AI.

The data-centric approach makes use of the knowledge representation model, data flow and processing that represents the design problems and rules as depicted in Figure 1. The approach requires integrating big data from multiple sources such as the IoT devices, simulations, and existing databases within the ship design and building processes. While this is feasible through computer technology, the issues remain on how to ensure generalised, robust, and reusable processes can be established.

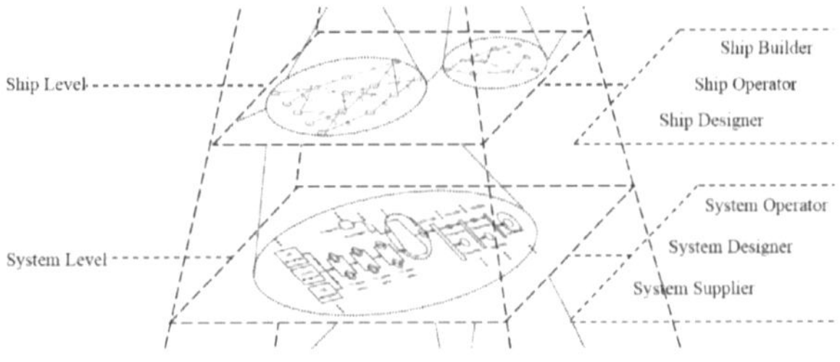

Figure 1. Ship and systems levels collaboration environment and knowledge representation [5]

3. KNOWLEDGE REPRESENTATION AND AI

It is understood that the knowledge representation role is to express the experts' domain and knowledge into a representable form that can be reused using the information systems and technologies. Often the process requires a systematic approach that is consistent and adoptable to wide design problems and use cases.

There are numerous methods that can be used in abstracting knowledge. From a canonical perspective, knowledge can be abstracted through; 1) declarative or descriptive, 2) procedural, 3) heuristic and 4) structural knowledge models [25]. This work focuses on the practical examples demonstrated on the ship design and building processes structured knowledge representation using graph theory [26, 27, 28, 29]. It refers to the knowledge graph that represents and organises the knowledge entities or objects, the relationships between them and their properties respectively.

Based on the graph theory, entities are represented as vertices (V) with their interrelationships defined as edge (E). Additionally, the E property can be numerically defined as weight (ω) describing the strength of the relationship. Mathematically, a directed graph (G) can be defined as $G = (V, E, \omega)$. From the knowledge graph development perspective, the process is initiated by knowledge extraction, followed by the knowledge construction (representation) and management [30, 31]. Particularly, this work focuses on the top-down-based knowledge graph that is built based on a well-defined knowledge domain. Examples of the other applications are discussed in the survey by Zhou [32].

3.1. QFD-AD METHOD FOR KNOWLEDGE GRAPH

Based on the ship design domain and current scope, this work proposes the integrated QFD-AD method [33] to systematically; 1) identify related data, 2) construct ontology and extract knowledge entities, relations, and properties, 3) map the ontology to the extracted knowledge and then 4) construct, store, and visualise knowledge graph.

Applying the QFD-AD method, the ship design knowledge graph development is performed based on the workflow as in Figure 2. Step 1 is executed by establishing the ship design requirements and performing the ship design data analysis. It serves to establish the domain of interest, data, and analysis method for knowledge extraction. In this step, the Axiomatic Design (AD) is introduced within the Quality Function Deployment (QFD) framework to systematically establish the design goals and top-level customer needs (CN).

Step 2 continues with the domain-centric process modelling and extended functional analysis. It is executed using the AD method to determine the top-level design functional requirements (FR) and design parameters (DP) in reference to the established CN. The example for the ship design domain is presented as in Figure 3 describing the top-level system architecture and knowledge entities. Past models developed by the domain experts and principal ship design parameters can also be used as the starting point for the knowledge graph development.

Based on the initial information, step 3 proceeds with the multivariate data analysis to extract the entities relations, usually using advanced statistical and ML methods. Based on the set domain as in Figure 3, the process continues with AD functional decompositions and data features engineering to further extract additional entities, relation, and properties. It further derives the detail, lower-level FR and DP across the AD domains as depicted in Figure 4. This is performed iteratively and continuously as the knowledge graph and data grow. This initial knowledge extraction functions to initiate the knowledge graph development and to be maintained as the knowledge changes and evolves.

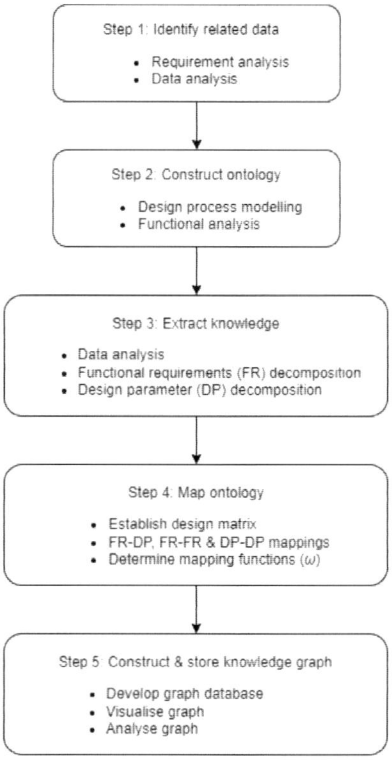

Figure 2. Proposed knowledge graph development using QFD-AD method

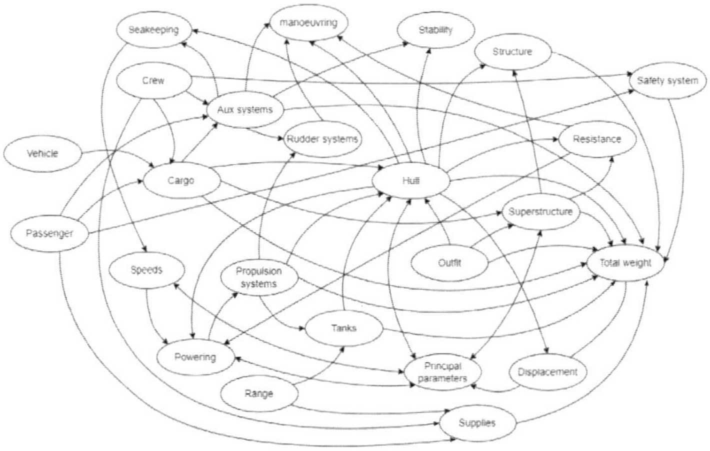

Figure 3. Proposed passenger ship design attributes

Figure 4 describes the process for identifying the data and requirement analysis that serve to establish the domain of interest. It proceeded with the AD functional analysis to construct the ontology [34]. The FR and DP decomposition continue to develop the ontology while concurrently extracting and mapping the design knowledge thus the foundation for step 4. Jiang et al., [35] described the ontology layer, in this case the FR as the knowledge representation model layer while the extracted knowledge layer or instance layer, in this case the DP as the data layer that holds the entities.

Once the preliminary ontology and extracted knowledge has been mapped, they are then organised using the QFD house of quality (HoQ) design matrix as presented in Figure 5. It functions to facilitate the knowledge organisation, analysing the entities, relations, and properties as well as to enable multi-criteria decision-making (MCDM) methods in determining the numeric entities relationship, ω.

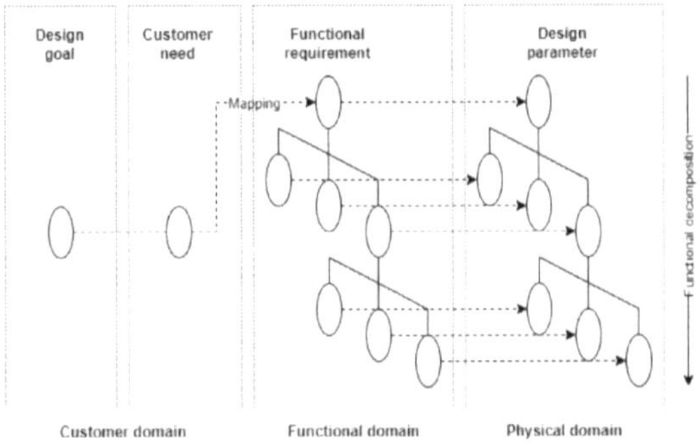

Figure 4. Proposed QFD-AD framework, functional decompositions, and mappings

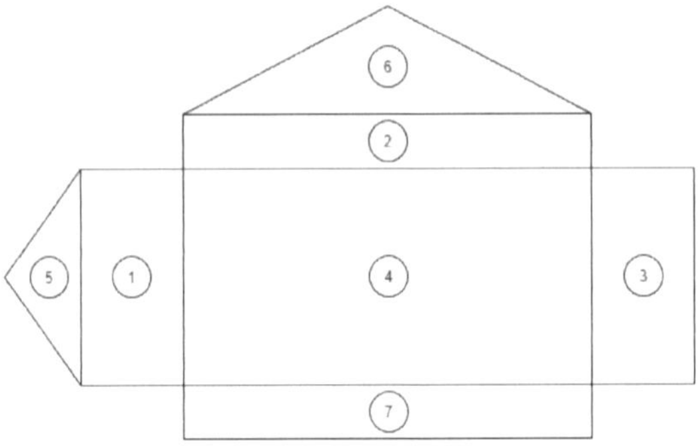

Figure 5. Generic QFD HoQ chart

Based on Figure 5, ① is for organising the FR, ② is for organising the DP, ③ is for the FR ranking, ④ is for the FR – DP mapping functions, ⑤ is for the FR – FR relationship, ⑥ is for DP – DP relationships. In practice, the FR – FR and FR – DP and DP – DP relationships ω are defined based on preferences or through the MCDM. Using the data-driven approach, the FR – DP and DP – DP relationships ω can also be determined through the empirical, statistical or ML method. Finally, the FR and DP ranks are determined in ③ and ⑦ respectively based on the relative ω. Other than the relations, this process also further extracts information on the FR and DP conflicts and significance based on their comparisons.

Essentially, the mappings from Figures 4 and 5 produce the knowledge graph with the entities, relations and properties that can be stored in the graph database system. At the same time, the mappings and knowledge graph can be represented as modules showing the process flow that can be used for the development of future applications.

3.2. GRAPH DATA SCIENCE

Analysing a knowledge graph with large connected, complex, and high dimensionality data requires an efficient approach in terms of complexity, computational power, and processing speeds. They are aimed to facilitate: 1) graph statistics for basic measurements and interpretation, 2) graph analytics along with queries and algorithms for analyses and 3) graph ML for decision support application.

Graph statistics focuses on analysing the connections between entities. Often, it is used together with the graph algorithms to explain the overall graph's structure and metrics. The algorithms are designed to investigate the entities connections and their characteristics such as the node distributions, centrality, and adjacency at the node, edge, and overall graph level [36]. Additionally, it provides the means for graph analytics such as the node degree measures, path analysis, clusters, partitions, and other predictive analysis [36, 37, 38].

While the ML on graph application emphasises on learning the node representations capability to make predictions and uncovering new insights [39, 40]. At the same time, it is also practically aimed at improving ML predictions and workflow using graph algorithms as well as developing the algorithm in supporting the decision-making processes. It is observed that there are growing interests in the knowledge graph application especially toward the digital transformation [41].

4. USE CASE

This work demonstrates a locally developed native graph database using neo4j Desktop version 1.4.15 with graph DBMS version 4.4.5 for the passenger ship design knowledge graph development, and visualisation. The plugins used in this work are the APOC 4.4.0.6 and Graph Data Science library 2.0.3. The Neo4j graph database is selected due to its index-free adjacency feature and query performance compared to the RDMS.

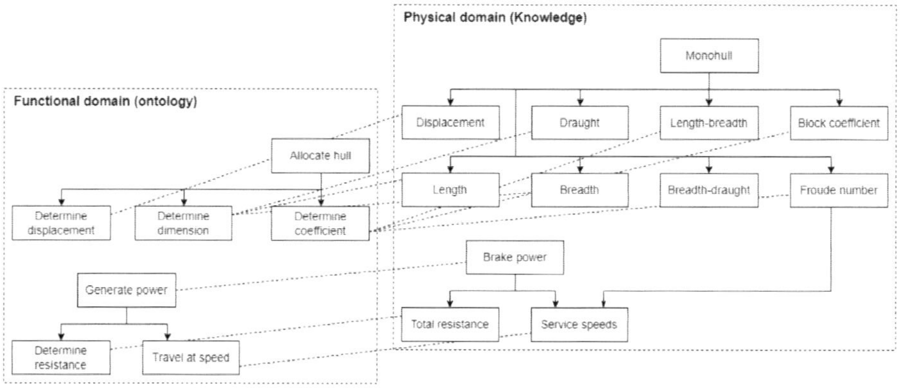

Figure 6. Partial QFD-AD based Knowledge graph

The design knowledge is modelled using the proposed QFD-AD method and developed as a labelled property graph (LPG). The data used for this study is collected from the RINA significant ships, significant small ships and the UTM towing tank tests. The data structure is predefined based on the outcome from the QFD-AD and knowledge modelling processes. Then, the graph database is developed based on the ontology and extracted knowledge as in Figure 6.

The ontology is used to create domain-specific use cases while extracted knowledge is used to create the data model for the graph. The instance model with sample data is then developed and executed to test the use cases and the data model's applicability. Ideally, the data model would satisfactorily answer the questions arising from the use cases. From the QFD-AD context, these would satisfy the FR – DP and thus automatically satisfy the established CN, predetermined set design goals and objectives.

Practically, the graph can be improved and extended through refactoring, identifying, and creating more entities, relations, and properties. For example, the superstructure, cargo, and others as described in Figure 3. For the initial graph, the data model is developed based on Figure 6 and presented in Figure 7. It shows 14 nodes with specific labels as described and 22 relationships. Based on the figures, the entities are represented by the design parameters "nouns" and the relation is described by the "verbs" based on the ontology. The graph also shows two prominent entities "dimension", "displacement", "coefficient" and "power" that defines the ship hull.

The data model is then tested using the use cases and the instance models for verification based on the sample data. Conveniently, the graph is also referred to as the basis for computer-based software development. In this project, the knowledge graph and the data model are used for the ML model and data-centric power prediction application as reported in [24, 42]. It is developed as a tool to support the preliminary ship design process and decision-making.

The sample outcome from the data model testing is presented in Figure 8. It verifies the data quality, ontology, data model and overall knowledge graph development. In addition, this shows the tool's suitability for effective knowledge representation and management in supporting the digital transformation. Natively, additional queries and processing can be readily performed using the built-in tool for the data-driven and data-centric processes [36]

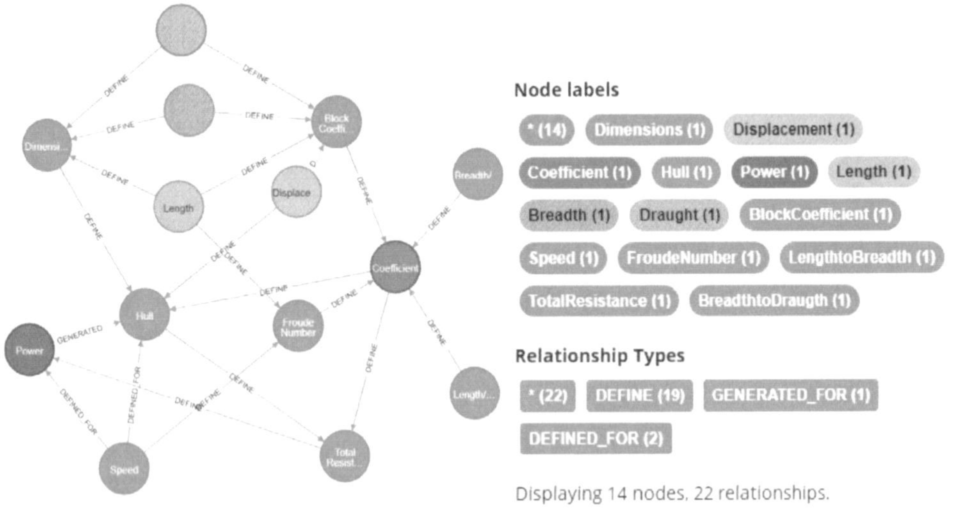

Figure 7. Initial ship design knowledge graph data model using Neo4j

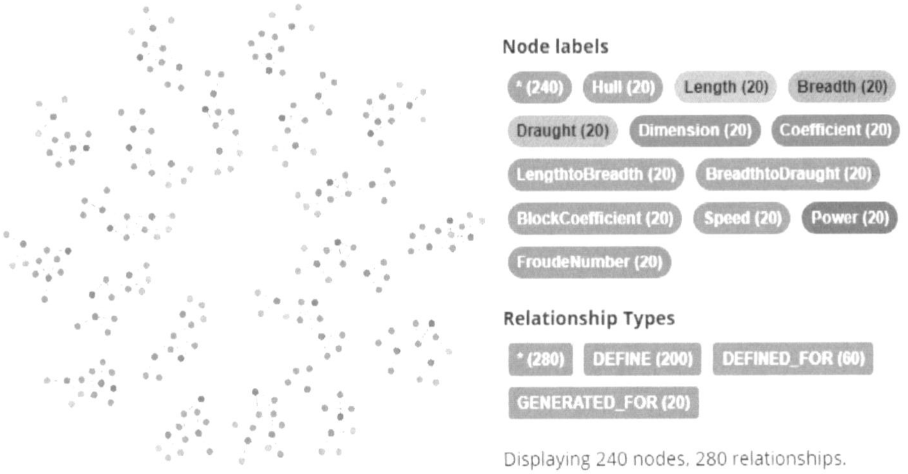

Figure 8. Instance models for passenger ship design knowledge graph

5. IMPACT

The initiative taken in exploring new approaches in the ship design domain presents considerable challenges especially in adopting and integrating new technologies due to the existing practices and processes constraints. The deployment of these approaches and technologies are disruptive and can be risky. However, the trade-off is viewed as very significant based on the successful implementation from other industries.

Data has now become more accessible. This makes the data-driven approach to be the best starting point to transform the current paradigm and practices into advocating the use of the computer and information technologies. Applying the approach along with ML, knowledge can be evoked and systematically represented. It enables models and simulation to be developed, validated, and implemented effectively to the real industrial cases at a fraction of the time, cost, and risk. These present significant advantages compared to the conventional approach.

Consequently, the data-centric approach is viewed as the succeeding step to the data-driven approach. Based on the new insights, data can be used to replace the existing rules and thus to support the design development and decision-making processes. With AI and computer-based tools, the represented and new knowledge can be reused to support routine tasks and automation. This crucially reduces the design development cycle by quick response to change, providing recommendations and reliable design based on pre-validated models.

While the knowledge representation and AI are not totally new to the shipping industry, systematic and formal alternative methodologies in supporting their implementation are still lacking and require further work. It is hoped that the increasing interest in transforming the industry will accelerate the advancement of these technologies in the industry.

6. CONCLUSION

This work discusses the conceptual review of the data-driven and data-centric approaches to the digital transformation with AI in the shipping industry. It emphasised on modelling and representing knowledge for analysis and reuse. This work explored integrated QFD-AD as a formal methodology to model and represent the ship design knowledge that can be used to develop design and decision-making support tools.

Additionally, this work demonstrates the ship design knowledge representation and analysis using the LPG, Neo4j graph database and the passenger ship design data. It is observed that the QFD-AD is capable of minimising the number iteration for developing the ontology and extracting knowledge. Yet, the outcome still requires refactoring and guidance from standard practices though minimal. Moreover, the method applicability in supporting the resource description framework (RDF) is not investigated in this study thus the needed future work.

It is found that implementing computer-based tools into the complex processes such as in the shipping industry carry huge potential in terms of exploring and deploying assistive technologies, tools, and proof of concept. Another challenge observed is due to the level of expertise needed for their adoption. However, this can be overcome by the support from the organisation. Moreover, the technology advancement shows the trend for less to no-code tools to support minimal disruptions during the deployment.

Importantly, this work recognises and discusses the applicable methods, techniques, and tools for the data-driven and data-centric approach with AI for the shipping industry digital transformation. It also acknowledges the potential technology and alternative approach for effective ship design and building processes as well as their further improvement.

7. ACKNOWLEDGEMENT

We would like to thank the Universiti Teknologi Malaysia (UTM) for supporting this work through the Research University Grant (RUG), project number: Q. J130000.3851.19J88.

8. REFERENCE

1. G. Vial, "Understanding digital transformation: A review and a research," *Journal of Strategic Information Systems,* vol. 28, no. 2, pp. 118 - 144, 2019.

2. S. Kraus, S. Durst, J. J. Ferreira, P. Veiga, N. Kailer and A. Weinmann, "Digital transformation in business and management research: An overview of the current status quo," *International Journal of Information Management,* vol. 63, p. 102466, 2022.

3. G. George and S. J. D. Schillebeeckx, "Digital transformation, sustainability, and purpose in the multinational enterprise," *Journal of World Business,* vol. 57, p. 101326, 2022.

4. A. Sánchez-Sotano, A. Cerezo-Narváez, F. Abad-Fraga, A. Pastor-Fernández and J. Salguero-Gómez, "Trends of Digital Transformation in Shipbuilding Sector," in *New Trends in the Use of Artificial Intelligence for the Industry 4.0,* London, IntechOpen, 2020, pp. 3 - 25.

5. Z. Guangrong, "Intelligent Design and Operation of Ship Energy Systems combining Big Data and AI," in *17th Conference on Computer Applications and Information Technology in the Maritime Industries*, Pavone, Italy, 2018.

6. i an, o i , en i e i and u iha , igi a an o a ion in he a i i e an port sector," *Technological Forecasting and Social Change,* vol. 170, p. 120879, 2021.

7. B. P. Sullivan, S. Desai, J. Sole, M. Rossi, L. Ramundo and S. Terzi, "Maritime 4.0–opportunities in digitalization and advanced manufacturing for vessel development," *Procedia manufacturing ,* vol. 42, pp. 246 - 253, 2020.

8. J. Holmström, "From AI to digital transformation: The AI readiness framework," *Business Horizons,* vol. 65, pp. 329 - 339, 2022.

9. I. Cil, F. Arisoy, H. Kilinc, E. Özgürbüz, A. Y. Cil and E. Uysal, "Challenges and trends in shipbuilding industry: digitization of SEDEF shipyard in Turkey," in *2021 5th International Symposium on Multidisciplinary Studies and Innovative Technologies (ISMSIT)*, Ankara, Turkey, 2021.

10. A. Tkalich, M. Mikalsen, N. B. Moe and T. Sporsem, "Digital transformation of a traditional business model: A case study from the maritime industry," *ECIS 2021 RESEARCH-IN-PROGRESS PAPERS,* vol. 42, 2021.

11. V. Babica, D. Sceulovs and E. Rustenova, "Digitalization in Maritime Industry: Prospects and Pitfalls," in *ICTE in Transportation and Logistics 2019*, Virtual, Latvia, 2019.

12. A. Bertoni, "Data-driven design in concept development: systematic review and missed opportunities," in *Proceedings of the 16th Design Society: DESIGN Conference*, Cavtat. Crotia, 2020.

13. H. M. Gaspar, "Data-driven ship design," in *17th Conference on Computer Applications and Information Technology in the Maritime Industries (COMPIT 2018)*, Pavone. Italy, 2018.

14. F. J. Montáns, F. Chinesta, R. Gómez-Bombarelli and J. N. Kutz, "Data-driven modeling and learning in science and engineering," *Comptes Rendus Mécanique,* vol. 347, no. 11, pp. 845 - 855, 2019.

15. I. Pan, L. R. Mason and O. K. Matar, "Data-centric Engineering: integrating simulation, machine learning and statistics. Challenges and opportunities," *Chemical Engineering Science,* 2022.

16. H. A. Reijers, I. Venderfeesten, M. G. A. Plomp, P. Van Gorp, D. Fahland, W. L. M. van der Crommert and H. D. D. Garcia, "Evaluating data-centric process approaches: Does the human factor factor in?," *Software & Systems Modeling,* vol. 16, pp. 649 - 662, 2017.

17. G. Fardelas and S. G. Kim, "Data-driven sustainable ship design using Axiomatic Design and Bayesian Network Model," *IOP Conference Series: Materials Science and Engineering,* 2021.

18. J. H. Lim, J. H. Kim and J. H. Huh, "Recent trends and proposed response strategies of international standards related to shipbuilding equipment big data integration platform," *Quality & Quantity,* pp. 1 - 22, 2022.

19. J. Bao, X. Zheng, J. Zhang, X. Ji and J. Zhang, "Data-driven process planning for shipbuilding," *AI EDAM,* vol. 32, no. 1, pp. 122 - 130, 2018.

20. M. Mittendorf, U. D. Nielson and H. B. Bingham, "Data-driven prediction of added-wave resistance on ships in oblique waves—A comparison between tree-based ensemble methods and artificial neural networks," *Applied Ocean Research,* vol. 118, p. 102964, 2022.

21. O. V. Zub and D. U. Nesin, "Hierarchical structure of the electronic information model of a ship," *Journal of Physics: Conference Series,* vol. 1333, p. 042032, 2019.

22. V. R. Dushko, D. U. Nesin and A. V. Tihonchuk, "The basic principles of creating an information model in PDM-systems as a means of vessel lifecycle management," *Journal of Physics: Conference Series,* vol. 1661, p. 012170, 2020.

23. A. Papanikolaou, M. Flikkema, S. Harries, J. Marzi, R. Le Néna, S. Torben and A. Yrjänäinen, "A Holistic Approach to Ship Design: Tools and Applications," in *SNAME Maritime Convention*, Tacoma, Washington. USA, 2019.

24. A. Maimun, K. Hiekata, J. Khairuddin, C. L. Siow and A. Ali, "Estimation of Ship Powering in Preliminary Ship Design Using Graph Theory and Machine Learning Method," in *International Conference on Design and Concurrent Engineering 2021 & Manufacturing Systems Conference 2021*, Japan, 2021.

25. Y. Yang, Y. Zhuang and Y. Pan, "Multiple knowledge representation for big data artificial intelligence: framework, applications, and case studies," *Frontiers of Information Technology & Electronic Engineering,* vol. 22, no. 12, pp. 1551 - 1558, 2021.

26. R. Arendt and E. Van Uden, "A decision-making module for aiding ship system automation design: A knowledge-based approach," *Expert Systems with Applications,* vol. 38, no. 1, pp. 410 - 416, 2011.

27. Y. H. Wu and H. J. Shaw, "Document based knowledge base engineering method for ship basic design," *Ocean Engineering,* vol. 38, pp. 1508 - 1521, 2011.

28. J. Dong, X. Jing, J. Liu, H. Li, X. Cao, C. Du, J. Li and L. Li, "Process knowledge graph modeling techniques and application methods for ship heterogeneous models," *Scientific Reports,* pp. 1 - 14, 2022.

29. E. A. E. Dechateau, P. de Vos and S. van Leeuwen, "Early stage routing of distributed ship service systems for vulnerability reduction," in *The 13th International Marine Design Conference (IMDC 2018)*, Espoo, Finland, 2018.

30. Z. Zhao, S. K. Han and I. M. So, "Architecture of knowledge graph construction techniques," *International Journal of Pure and Applied Mathematics,* vol. 118, no. 19, pp. 1869 - 1883, 2018.

31. a a au ai and o h, "Defining a Knowledge Graph Development Process Through a Systematic Review," *ACM Transactions on Software Engineering and Methodology,* 2022.

32. X. Zhou, "A Survey on Application of Knowledge Graph," *Journal of Physics: Conference Series,* vol. 1487, p. 012016, 2020.

33. J. Khairuddin, A. Maimun and C. L. Siow, "Ship systems synthesis and analysis using holistic design approach: The QFD-AD method," *IOP Conference Series: Materials Science and Engineering,* 2020.

34. J. T. Foley and D. S. Cochran, "Manufacturing System Design Decomposition: An Ontology for Data Analytics and System Design Evaluation," *Procedia CIRP,* vol. 60, pp. 175 - 180, 2017.

35. W. Jiang, Y. Wang, J. Hu, L. Guan and Z. Zhu, "Construction of Substation Engineering Design Knowledge Graph Based on “Ontology Seven-step Method”," in *4th International Conference on Energy, Electrical and Power Engineering (CEEPE)*, Chongqing, China, 2021.

36. Neo4j, "The Neo4j Graph Data Science Library Manual v2.1," Neo4j, 2022. [Online]. Available: https://neo4j.com/docs/graph-data-science/current/algorithms/. [Accessed 8 August 2022].

37. J. Powell and M. Hopkins, A librarian's guide to graphs, data and the semantic web, Oxford, UK: Elsevier, 2015.

38. I. Robinson, J. Webber and E. Eifrem, Graph databases: new opportunities for connected data, Sebastopol, CA: O'Reilly, 2015.

39. M. Needham and A. E. Hodler, Graph algorithms: practical examples in Apache Spark and Neo4j, Sebastapol, CA: O'Reilly, 2019.

40. C. Stamile, A. Marzullo and E. Deusebio, Graph Machine Learning: Take graph data to the next level by applying machine learning techniques and algorithms, Birmingham, UK: Packt, 2021.

41. S. R. Bader, I. Grangel-Gonzalez, P. Nanjappa, M. E. Vidal and M. Maleshkova, "A knowledge graph for industry 4.0," in *European Semantic Web Conference*, Online, 2020.

42. J. Khairuddin, A. Maimun, K. Hiekata, C. L. Siow and A. Ali, "Web application with data centric approach to ship powering prediction using deep learning," *Software Impacts,* 2022.

PERFORMANCE OPTIMISATION OF SOLID OXIDE FUEL CELLS THROUGH WASTE HEAT RECOVERY SYSTEMS FOR MARINE APPLICATIONS

P Manias, C McKinlay, D A Hudson and **S R Turnock,** University of Southampton, UK

SUMMARY

With the ever-increasing demand for emission reductions within the maritime industry, a global effort is being made to decrease the carbon footprint of vessels. This study investigates the introduction of 10.5 MW SOFC, as an integral component of a 6000-passenger cruise vessel's powertrain, to provide and cover the total energy demand on board whilst in port and the 'hotel load' during voyages. Furthermore, effective ways of utilizing the otherwise wasted heat from the exhaust are investigated, to increase the powertrain system's overall efficiency and subsequently decrease its carbon footprint, through the utilization of a Waste Heat Recovery System. Results showed 7,078 tonnes reduction of LNG fuel consumption and 31,919 tCO_2e emission savings annually when running the SOFCs on LNG, with the recoverable waste heat from the SOFCs being less than from the current LNG generator sets, requiring additional boiler power to meet a 4 GWh of annual heat deficit.

NOMENCLATURE

υ	Kinematic viscosity (N s m^{-2})
ρ	Density of water (kg m^{-3})
P	Pressure (N m^{-2})
Cp	Specific Heat Capacity at constant pressure ($KJ.kg^{-1}.K^{-1}$)
R	Gas Constant ($KJ.kg^{-1}.K^{-1}$)
MW	Megawatts
WHRS	Waste Heat Recovery System
GT	Gross Tonnage
HFO	Heavy Fuel Oil
MDO	Marine Diesel Oil
CH_4	Methane
LNG	Liquified Natural Gas
SOFC	Solid Oxide Fuel Cells
PEMFC	Proton Exchange Membrane Fuel Cells
ICE	Internal Combustion Engines
h	Latent Heat of Vaporisation
GWh	GigaWatt Hours
°C	Degrees Celsius
VBDC	Cylinder Volume at Bottom Dead Centre
VTDC	Cylinder Volume at Top Dead Centre
SFC	Specific Fuel Consumption
SOL	Start Of Life

1. INTRODUCTION

The environmental impact of emissions from burning fossil fuels is a major global concern and the maritime sector is not exempt from this. Some ship operators have made progress towards reducing emissions in recent years by switching from traditional fuels, such as heavy fuel oil (HFO) and marine diesel oil (MDO), to liquefied natural gas (LNG). LNG is primarily comprised of methane (CH_4) and given that it has a higher hydrogen-to-carbon ratio the resulting carbon dioxide emissions are lower than for other fossil fuels [1].

Although the burning of LNG can be considered an improvement, it does not provide alone a clear pathway towards the aim of powering a large ship without any greenhouse gas emissions. There are several targets for the shipping industry to reduce emissions, for example the UK government's Clean Maritime Plan requires all new ships to be zero emission ready by 2025[2]. Several studies have suggested that true zero emission shipping will be achieved through the use of hydrogen (or other e-fuels) and fuel cells. [1], [3-6]. Fuel cells are electrochemical devices that convert the chemical potential energy of a fuel into electrical power and tend to have significantly higher efficiencies than internal combustion engines [7]. Unfortunately, hydrogen is not currently commercially available in either the quantities required to power a large cruise ship or the necessary port locations, hence, this is not currently an option for large ships in the very near term.

The technology considered in this study is that of solid oxide fuel cells (SOFC). These do not require as high level purity fuel as other leading fuel cell technologies, such as proton exchange membrane fuel cells (PEMFC) [4]. Subsequently, when considering the inclusion of a reforming process to convert the feedstock into hydrogen, this allows for the utilisation of other fuels in the case of SOFCs. Moreover, an advantage of SOFCs is that they can reach efficiency levels of over 65% which is significantly higher than current combustion methods (around 20 to 40% efficiency [5]). However, due to its electrolyte and architectural structure, this fuel cell type usually operates at high temperatures, in excess of 500°C[1,8].

Given the high operating temperature of SOFCs, this study investigated the utilisation of this heat, that would otherwise be wasted through the exhaust stacks, for several tasks on-board the vessel, such as hot water, heat pumps and mainly steam production. Consequently, the main objective for employing a waste heat recovery system (WHRS), is to reduce fuel consumption and subsequent emissions.

To aid this investigation, a typical large cruise ship was used as a case study. Data has been provided for both electrical and heat demand. A preliminary study found that 10.5 MW of SOFCs combined with 5 MWh of batteries would be sufficient to meet the objective of delivering all of the in-port demand. The concept of operation for this setup included using the fuel cells to provide a constant baseload whilst at sea. An operating profile was produced for the fuel cells, batteries and generators from this preliminary study at three-minutely intervals which was used as the basis for the heat output calculations in this study.

2. AIM & OBJECTIVES

The aim of this work is to evaluate the extent at which the constant heat rejection of the SOFCs and Internal Combustion Engine (ICE) generators could be utilised for several activities on board. To achieve this, a relation has been established between the exhaust's heat rejection rate and the predicted heating demand, in terms of available energy to be used on board, for various activities such as steam production.

The purpose of this modelling is to provide a better understanding of a future powertrain proposition that involves the use of fuel cells, internal combustion engine powered generators and batteries, operating simultaneously to cover the power demand on board a vessel, in the interest of providing a better emission and efficiency profile than the sole use of fossil fuel powered generators. This involves the investigation for certain weak points of the system, due to its complexity, since it is still in prototype stages, as well as areas of improvement that would make this hybrid system more suitable for vessel applications.

Consequently, the objectives of this project included developing models for the heat output of SOFCs and generators. To evaluate the total amount of recoverable waste heat from two concepts: the currently used all-combustion system and the use of a SOFC-battery hybrid to meet the baseload. Finally, a further objective was then to calculate the overall impact that a WHRS would have on annual power output, fuel consumption, and emissions.

3. BACKGROUND OF MARINE WHRS & FUEL CELL APPLICATIONS

Since the industrial revolution, many have made an effort to reduce their energy requirements to minimise their operational expenditure. In recent years, an emphasis is being given to reducing greenhouse gas emissions, to preserve the environment, as reflected in the Clean Maritime Plan and the International Energy Agency's aim to reach net zero carbon by 2050. [2,9]

During the early 18th century, when industrial machinery primarily ran on steam that was formed by burning coal to heat up water, an emphasis was given on how to better utilise the excess heat that would be otherwise wasted from the flue gases. The first application of such a device that could indeed process this exhaust heat output, can be traced back to 150 years ago with the patented name "economiser" [10]. This device is, in basic principles, a heat exchanger which transfers heat from the high temperature flue gas to a lower temperature fluid within a closed circuit.

Consequently, it was a matter of time until economisers were utilised onboard vessels to salvage heat that would otherwise be wasted through the exhaust stacks. These provided high temperature water for on board activities instead of a separate fossil fuel powered boiler, since 51% of the fuel energy consumed by the generators is converted directly to heat [11]. As a result, economisers can provide a better overall efficiency and reduce the energy requirements on board, in turn providing significant cost savings for the vessel's operators.

An economiser is the biggest and most important aspect of a system that is designed to recover waste heat from the exhaust gases, referred to as a waste heat recovery system (WHRS) [12,13]. Studies have been conducted on how to optimise the application of a WHRS on board a vessel with current figures showing upwards of 10% efficiency increase overall [14]. This is achieved through building high temperature and pressure steam by recovering heat from the main ICE generators' hot exhaust gases, to spin a turbine connected to a generator which, in turn, provides electricity [12].

A typical powertrain on a ship will use internal combustion engine (ICE) to provide either mechanical propulsion or to generate electricity (referred to as generators or gensets). An alternative to these, proposed here is to use SOFCs to produce electricity on board [7]. McKinlay et al [6], argue that fuel cells could be used in shipping, as a way of reducing harmful emissions when compared to ICE gensets. Other studies show that fuel cells could yield higher efficiency for producing electricity over the ICE gensets [7,15], yet, none of these studies are specific to SOFCs onboard vessels and they do not consider the implementation of WHRS.

Wang et al [16], conducted a study with regards to employing WHRS for fuel cells on board vessels, in order to convert heat energy back to electricity, emphasizing efficiency increase and emission reduction rather than the utilisation of heat for on board activities. For the same reason, the purpose of this paper is to study how heat from these types of fuel cells could be utilized to provide heating for steam formation on a large cruise ship and whether this method of steam formation would be viable to replace auxiliary gas boilers.

4. MODELLING OF WASTE HEAT RECOVERY SYSTEM

The first step towards modelling the WHRS, would be to attain data from a vessel during operation which came in the form of thousands of data taken from a vessel throughout a years' worth of operating time in three-minute time increments. This dataset contained certain values such as generator power output and speed over ground, which would later be used to model a power and emission profile for that specific vessel. The vessel upon which the calculations are based is a 6,600-passenger capacity cruise ship, with 183,000 GT and 62 MW of installed power, fuelled by LNG [17].

The output of this modelling effort is to predict values such as power output based on current demand, consumption, and exhaust emissivity, for different types of green-house gasses. This, in turn, allows the investigation of how a hybrid dual fuel generator and fuel cell powertrain system accompanied by batteries, would be designed, and optimised to cope with rapid peaks and drops of onboard power demand, while keeping up with any additional requirement when underway.

By so doing, it would then be decided whether to have a fixed power output from the fuel cells or a fluctuating one, to cover the baseload of the vessel whilst in port. In the end, the later method of covering power demand was decided to be the optimal one, as it was discovered that average port baseload would vary throughout the year, from 8 to 10 MW. Not only that, but as the concept of the fuel cell powertrain was to minimise port emissions, the use of SOFCs would be preferred over generator sets, which meant that either their power output would have to fluctuate based on the demand or large batteries would be required which would take up a lot of space on board, to cover additional demand fluctuations.

To continue, having the power output requirements based on the demand, meant that the next stage of the Waste Heat Recovery Modelling, would be to evaluate the efficiency of each system and further looking into utilising this heat rejection, so as to boost the system's efficiency overall. To achieve this, fuel cell specifications were provided by the manufacturer of the fuel cell powertrain and gave an indication of exhaust gas flow and temperature for given power output value. This was also the case with the generator sets, where the manufacturer supplied the subsequent emissions, exhaust flow and temperature after the turbocharger, as well as the cooling system's heat rejection.

Independent of the manufacturers' provided specifications, a simulation of operation was developed in both cases to evaluate the performance of both systems and develop performance curves, resulting from the load each system is subjected to. This was required, as the energy demand on board is not consistent and the performance of the powertrain is, therefore, fluctuating.

Finally, having all parameters described above, the heat output in terms of steam production, was estimated throughout one season of the vessel's operation. This was done to decide whether there was a need to install a separate gas fired boiler to cover the demand and to see if there was a benefit overall, emission-wise.

4.1 Data analysis

The first part of calculations involved the estimation of battery size and fuel cell power output. To do so, datasets containing recordings for a full year of operation were combined and converted, enabling the extraction of values such as total power output per given time. That said, the datasets contained numerous time gaps, which can be attributed to malfunctioning sensors, the recording system being offline or due to other maintenance carried out, disabling data provision.

Using these values, a power baseload would be established, with the aim of having the fuel cell powertrain target that level of power, always. As it turned out, the power baseload when in port, is fluctuating depending on the season and place the vessel would be located at, as each season and location had different climates and the heating, ventilation and air conditioning requirements would match that in terms of demand.

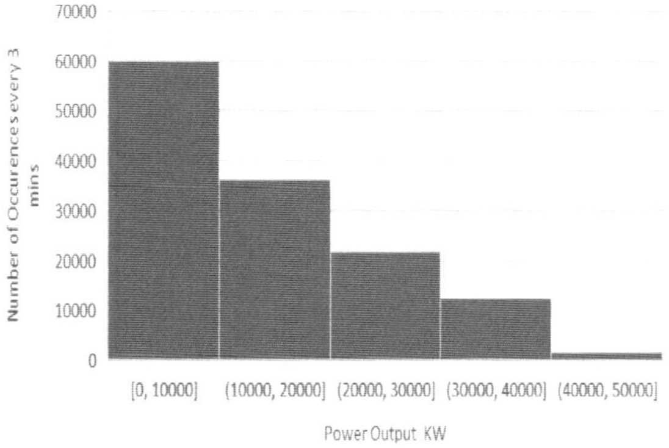

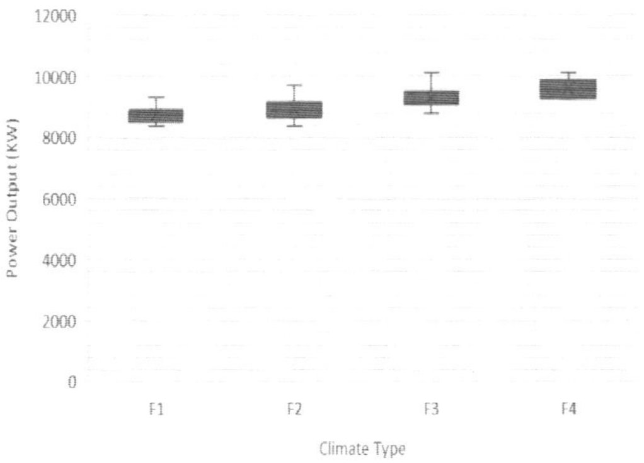

Figure 1: Power Output Variation throughout the year 2019 & Average Port Load variation depending on climate

An observation was made that hotter climates correlated to a higher power requirement, and this can be attributed to higher air-conditioning, cold-water requirements and increased service and hotel demand on board, as observed in Figure 1 and Table 1, where the increasing number of climate type is associated with increasing temperature and humidity.

Table 1: Average port load for each climate type

Climate type	Average Port Load (kW)
F1	8,697
F2	8,820
F3	9,411
F4	10,048

To handle this changeable baseload, it was decided that the fuel cells would operate at a variable power output. This can be seen in Figures 1 & 2, where the demand during each port station and climate, could vary significantly over time, making the installation of batteries necessary to cope with rapid changes of power demand.

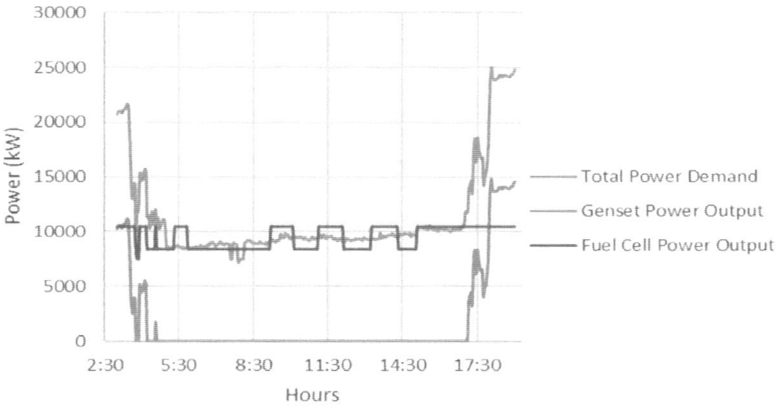

Figure 2: Typical power demand variation during a port visit and exit in climate type F4, including each system's power output variation

An aim of this project was to meet all in-port power demand using a hybrid combination system of fuel cell and battery. The model was setup to run the SOFCs at a constant output where possible, however they would be ramped up or down to avoid overcharging or undercharging the batteries. This is shown in Figure 2, as the fuel cells and generator-sets work in parallel when underway, and the FCs are operating solely when in port. The linearity with regards to the ramping rate response of the fuel cells can also be observed and compared to the response rate of the generators, which is significantly quicker. That said, this preliminary study shows a 7,078 tonnes reduction of LNG fuel consumption and 31,919 tCO_2e emission savings annually when running the SOFCs on LNG.As such, knowing the power output of each system, meant that the exhaust flow and temperature would be estimated for each power producer respectively and cross referenced with the manufacturers' specifications.

4.2 Generator set modelling

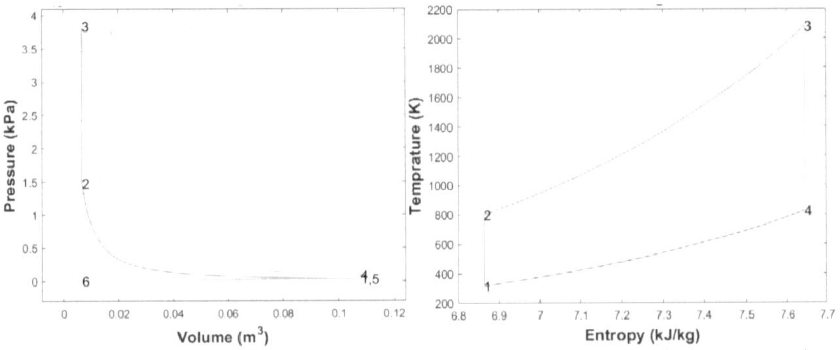

Figure 3: P-V diagram (left) and T-S diagram (right) of the VM46 DF during Otto cycle.

The generators employed in this case, are two sets of MaK VM 46 dual fuel internal combustion engines, totalling an overall power output of 61.6 MW. As per the manufacturer's specifications, this type of engine is meant to be run on Otto Cycle whilst using LNG as fuel [18], with a minor burst of diesel pilot fuel to initiate combustion. Since this vessel is meant to be run mainly on LNG, the simulation of operation for this engine was solely based on the thermodynamic principles for an engine running on this type of cycle.

An Otto cycle, 4-stroke engine, is comprised of the following processes and order; isentropic compression, constant-volume heat addition, isentropic expansion, and finally constant-volume heat rejection [19,20]. The relevant equations used for each process and input to the coding script, can be found in Table 2:

Table 2: Thermodynamic equations used in Matlab

Inlet Air Mass	$M_m = (P_1 \cdot V_1)/(R \cdot T_1)$
Compression Ratio (r_c)	16.5
Specific Heat Ratio (K)	1.35
Gas Constant (R)	0.287 (kJ/kg.K)
Specific Heat for Constant Volume of Air (C_V)	0.821 (kJ/kg.K)
Specific Heat for Constant Pressure of Air (C_P)	$C_P = R + C_v$
Isentropic Compression Process(1-2)	
Temperature at Compression Start (T_1)	$T_1 = (T_0) + 273$
Pressure at Compression Start (P_1)	$P_1 = P_0\ i\ P_{Turbo}$
Volume at Compression Start (V_1)	$V_1 = V_{BDC}$
Temperature at Compression End (T_2)	$T_2 = (T_1)(r_C{}^{(K-1)})$
Pressure at Compression End (P_2)	$P_2 = (P_1)(r_C{}^{K})$
Volume at Compression End (V_2)	$V_2 = V_{TDC}$
Combustion Process (2-3)	
Temperature at Combustion End (T_3)	$T_3 = ((FHV_{NG} \cdot eta_c)/(AF_{NG} \cdot C_v)) + T_2$
Pressure at Combustion End (P_3)	$P_3 = (P_2/T_2) \cdot T_3$
Volume at Combustion End (V_3)	$V_3 = V_2$
Expansion Process (3-4)	
Temperature at Expansion End (T_4)	$T_4 = T_3((1/r_C)^{(K-1)})$
Pressure at Expansion End (P_4)	$P_4 = P_3((1/r_C)^{K})$
Volume at Expansion End (V_4)	$V_4 = (M_m \cdot R \cdot T_4)/P_4$
Exhaust Process (4-5)	
Temperature at Exhaust End (T_5)	$T_5 = T_4$
Pressure at Exhaust End (P_5)	$P_5 = P_{Exhaust}$
Volume at Exhaust End (V_5)	$V_5 = V_{TDC}$
Power, Torque & Turbo Performance Formulas	
Work Output (Compression)	$W_{12} = M_m \cdot C_v(T_1 - T_2)$
Heat Input (Combustion)	$Q_{23} = M_m \cdot FHV_{NG} \cdot eta_c$
Work Output (Expansion)	$W_{34} = M_m \cdot C_v(T_3 - T_4)$

Work Output (Exhaust)	$Q_{45} = M_m \cdot C_v(T_5 - T_4)$
Heat Input (Qin)	$Q_{in} = M_f \cdot FHV_{NG}$
Net Work Output (Wi)	$W_i = W_{12} + W_{23} + W_{34}$
Thermal Efficiency (eta)	$eta_t = W_i/Q_{in}$
Indicated Mean Effective Pressure	$imep = W_i/V_d$
Brake Mean Effective Pressure	$bmep = imep \cdot eta_m$
Brake Work Output per Cylinder	$W_b = eta_m \cdot W_i$
Brake Power Output at Specific RPM	$P_b = W_b \cdot N \cdot (n/(60^2))$
Brake Torque Output at Specific RPM	$Taw = (P_b/(2\pi(n/60))) \cdot 1000$
Exhaust Flow	$M_{exFlow} = M_{ex} \cdot N(n/(60^2))$
Turbo Outlet Pressure	$T_{Turbot\_out} = T_4 - (T_{Turboc\_out} - T_0)(C_{pa}/C_{pex})$

Following the development of the initial model for the engine system, it was tested against the manufacturer's data [18]. Results, shown in Table 3, indicate that the calculated results were within accepted percentage of error limits for most fields, except for the exhaust temperature, which was lower than expected. That said, minor adjustments were made on this simulation to closely reflect the temperature output of the turbocharger to those specified in the manufacturer's manual. As such, the next step is to verify the solid oxide fuel cell output values provided.

Table 3: Comparison of engine model to data provided for this specific engine

	MaK VM 46 DF data	Calculated
Mean Effective Pressure	22.2 bar	22.3 bar
Power Output	15,440 kW	15,530 kW
Overall efficiency	50-55 %	56.8 %
Specific fuel Consumption	173 g/kWh	177.5 g/kWh
Exhaust Flow	28 kg/s	27.4 kg/s
Exhaust Temperature	C	C
CO_2 Emissions	480 g/kWh	488 g/kWh
Percentage Error	Green <15%	Orange >15%
RPM	514	
P turbo Ambient	3 bar	
T_0	C	

4.3 Solid Oxide Fuel Cell Modelling

As discussed in the introduction, the purpose of the fuel cell installation, would be to provide the baseload power for a 6,000+ passenger cruise ship, including meeting all energy demand whilst in port. A MatLab simulation was developed, to assess the performance and given power output as well as consumption of the Solid Oxide Fuel Cells (SOFC) and compare that to the manufacturer's data. This simulation is based upon the main working principles of a fuel cell system seen in Figure 4.

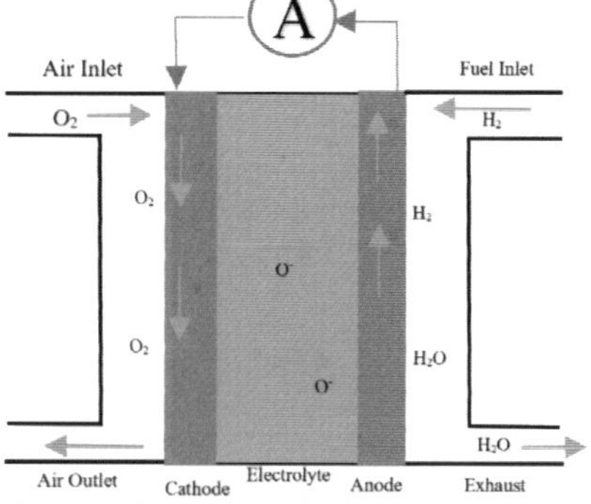

Figure 4: Illustration of a fuel cell's operation

A fuel cell operates by providing a fuel, in this case LNG reformed into hydrogen, which reacts with oxygen ions to form water molecules through an ion-conducting electrolyte between the anode and cathode. For SOFCs, the electrolyte is a non-porous solid metal oxide. Due to the oxygen ion formation, an electrochemical potential difference is induced within the system, which causes ions to travel through the electrolyte, as they are attracted to their ionic counterparts and form the aforementioned water molecules [8]. In this study, a methane reformer is employed, utilising the heat coming out of the fuel cells to reform methane in hydrogen fuel for the cells.

This whole chemical process provides the electrical power of the fuel cell. Many parameters had to be considered during modelling, such as: internal cell resistance; fuel utilisation factor; fuel cell plate thickness and cross-sectional area; the required voltage and amperage output to be compatible with current machinery; the total power required; the heat developed from the reactions occurring within the fuel cell system. These factors also dictate the size of the fuel cells and their daily fuel consumption.

The main part of the code included the following equations [8], [21]:

$$W_{\text{out}} = \Delta G = RTln(Ko) - RTln(Kp) \qquad \textbf{(1)}$$

where, W_{out} is the work extracted by the cell for a given for a given product of partial pressures (Kp), based on the Gibb's energy principle, R is the gas constant, T_{ref} is the reference temperature of the cell, and the equilibrium constant K_o is $exp(DG/(R\cdot T_{\text{ref}}))$;

$$Voc = \frac{W_{\text{out}}}{nelF} \qquad (2)$$

where, V_{oc} is the open cell voltage, ne i he nu e o e e on e hange i hin he ea ion H H and F is the Faraday Constant;

$$P_{\text{stack}} = V_{\text{cell}} \cdot I_{\text{s}} \qquad (3)$$

where P_{stack} is the power output of each cell that depends on the fixed voltage of the cell (V_{cell}) and the cell's current supply I_s which are found through the following equations;

$$I_{\text{s}} = nel\, F\, \dot{m}_{\text{H2}} \qquad (4)$$

where I_s the combined amperage output, directly related to the fuel consumption;

$$V_{\text{cell}} = V_{\text{oc}} - R_{\text{i}} - V_{\text{Act}} - V_{\text{Con}} \qquad (5)$$

where i is the cell's current R is the ohmic resistance, $VCon$ is the concentration loss and $VAct$ is the activation loss; and

$$V_{\text{Act}} = bln(i), \qquad V_{\text{Con}} = me^{ni} \qquad (6 \,\&\, 7)$$

where, b is the electrodes activation potential constant, m and n represent the concentration loss constant and index respectively.

These formulae constitute the main skeleton of the code and are responsible for the results generated in Figures 5 & 6, which compare the predicted results with the manufacturer ones' at the FCs SOL. The same code was used for all the solid oxide fuel cell system loadings, with the main parameters being altered. For example, the operating temperature was a variable which affects the Gibb's free energy change constant, as well as the enthalpy change of formation for the water molecule formation and finally, the fuel utilisation factor. [21]

Table 4: Solid Oxide Fuel Sell Simulation Predictions at SOL

Target Power (kW)	Efficiency %	SFC LNG (g/kWh)
5250	61.9	116.4
7875	65.5	115.2
8925	64	114.3
10500	62.5	118.2

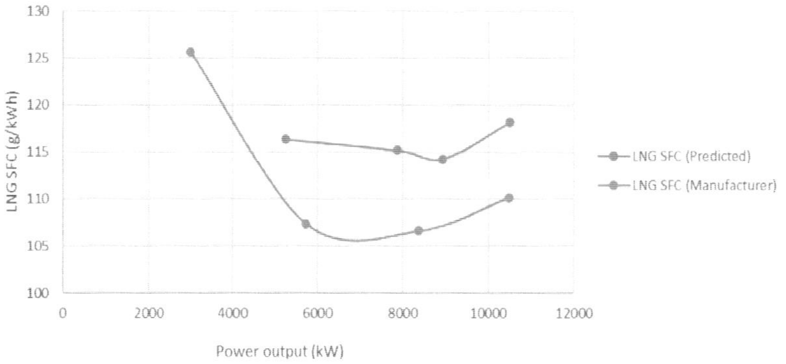

Figure 5: LNG Specific Fuel Consumption vs Power Output

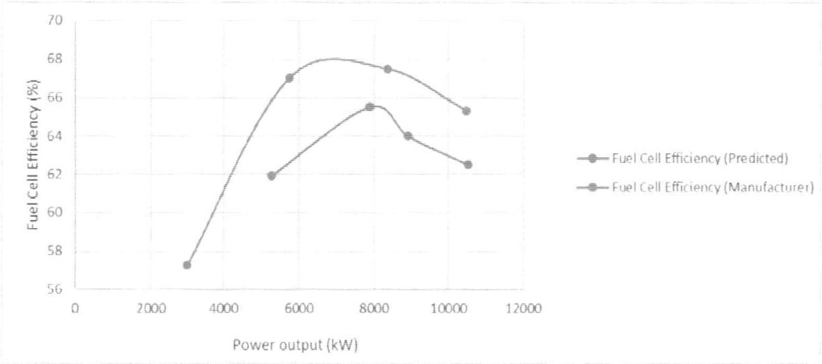

Figure 6: Efficiency vs Power Output

4.4 Heat Rejection Modelling

Each system has different ways of rejecting heat. In both systems, the main point of heat rejection would be the exhaust gasses, however, the Generator sets can also provide heat through their coolant lines, as cooling water is required to maintain the generator sets at a safe operating temperature. Figure 7 displays the architecture of the generator and SOFC waste heat recovery system.

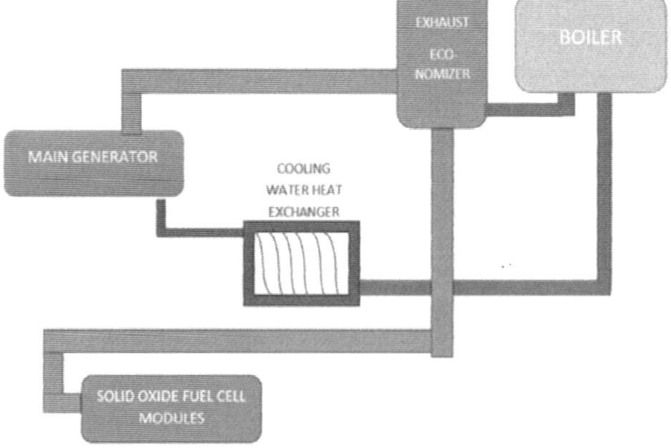

Figure 7: Waste Heat Recovery System Architecture

As observed, the generator sets provide two individual sources of heat for the boiler to utilise while the fuel cells only provide one. To calculate the rate of heat rejection, in terms of power, the exhaust flow of both systems and its temperature is required. An additional parameter for the generators would be the coolant's temperature and rate of circulation through the heat exchangers. Considering these, the following equation was used:

$$Heat\ Power\ Loss = M_g \cdot C_p\,(T_{\text{in}} - T_{\text{out}}) \qquad (8)$$

where, M_g is the mass flow of exhaust gas / coolant, C_p is the specific heat capacity of exhaust gas and coolant respectively, T_{in} is the inlet temperature of exhaust gas or coolant and finally T_{out} is the outlet temperature of exhaust gas or coolant.

The exhaust mass flow rate and temperature for both systems, where supplied by their respective manufacturer. From these data, a relation between load, power output and these parameters was established in Figure 8:

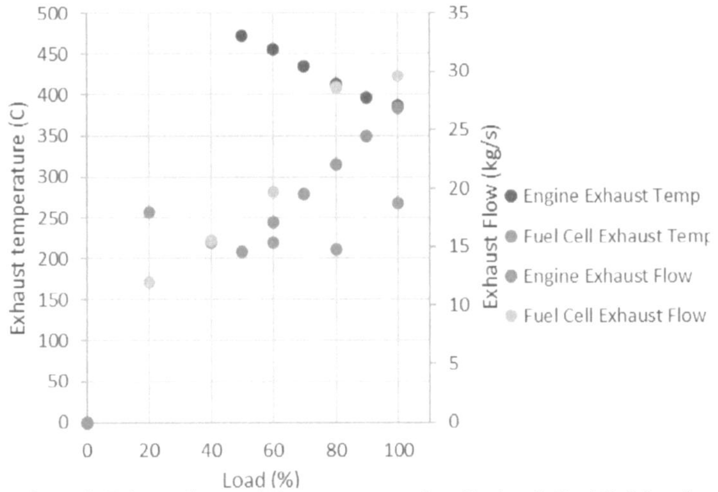

Figure 8: Exhaust flow and Temperature against Engine & Fuel Cell Load

Using Figure 8, the exhaust temperature and mass flow rate could be predicted at any given time increment based on the data provided. Subsequently, the total rate of heat rejection could be estimated with the use of thermodynamic equations. In the model changes are assumed to occur relatively slowly so static values can be used.

4.5 Heat Calculations

To estimate the rate of heat production within the system, the following equations were used:

$$\dot{Q}_{enginexh} = \dot{m}_{g\_ex\_engines} \cdot Cp_{ex} \cdot (\Delta T)_{out} \qquad (9)$$

where, $\dot{m}_{g\_ex\_engines}$ is mass flow of exhaust gas ($kg.s^{-1}$) and Cp_{ex} = 1.082 ($kJ.kg^{-1}.K^{-1}$) is the specific heat capacity of exhaust gas.

$$\dot{Q}_{engine\ coolant\ out} = \dot{m}_{coolant} \cdot Cp_{cow} \cdot (\Delta T) \qquad (10)$$

here, $\dot{m}_{coolant}$ is the mass flow of exhaust coolant ($kg.s^{-1}$), $\dot{m}_{sw}$ is the Mass flow of Steam/Water ($kg.s^{-1}$) and the specific heat capacity of coolant and water is Cp_{cow} = 4.18 ($KJ.kg^{-1}.K^{-1}$);

$$\dot{Q}_{engine\ out} = \dot{Q}_{engine\ exh\ out} + \dot{Q}_{engine\ coolant\ out} \qquad (11),$$

$$\dot{Q}_{fuel\ cell\ exh\ out} = \dot{m}_{ex\_fuel\ cells} \cdot Cp_{\_ex}(\Delta T) \qquad (12),$$

$$\dot{Q}_{cooling\ water} = \dot{m}_{\_sw} \cdot Cp_{co_w}(\Delta T) \qquad (13),$$

$$\dot{Q}_{enthalpy\ of\ vaporization} = \dot{m}_{sw} \cdot \Delta h \qquad (14),$$

where, the Latent heat of Vaporization (Δh) at 7 Bar is 2064 ($kJ.kg^{-1}$) and the Specific heat capacity of Steam (Cp_s) at 7 bar and 167°C is 2.4 ($kJ.kg^{-1}.K^{-1}$)

$$\dot{Q}_{steam\ heat-up} = \dot{m}_{\_sw} \cdot Cp_S \cdot (\Delta T) \qquad (15),$$

$$\dot{Q}_{steam} = \dot{Q}_{engine\ out} + \dot{Q}_{fuel\ cell\ exh\ out} - \dot{Q}_{cooling\ water} - \dot{Q}_{enthalpy\ of\ vaporization} - \dot{Q}_{steam\ heat-up} \qquad (16).$$

The mass flow of steam is the result of the predicted steam demand displayed in Figure 9 where the steam output in terms of Total power output is observed for this cruise ship class.

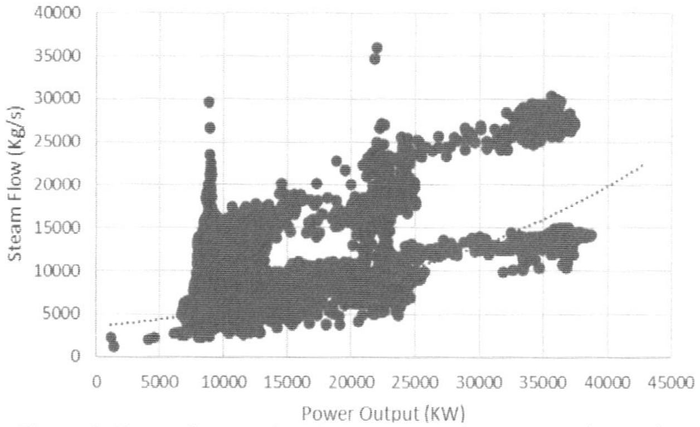

Figure 9: Steam flow against Power output for one year in service

Using the data displayed in Figure 9, an equation was developed in order to estimate the average subsequent water feed through the steam production system that would be turned into steam, for any given power output value of the powertrain, assuming that all water fed is turned into steam, as part of the conservation of mass law.

The final parameter to be considered, is the temperature outlet of the economisers for each system respectively. This is a critical aspect of the design for the entire WHR system, as it limits the amount of heat that can be scavenged by the exhaust gases to avoid the formation of water vapor condensation and excessive backpressure within the system. Considering this, it was decided that the temperature of the outlet for the economizers would be set at 140 °C and 220 °C for the fuel cell and generator exhaust streams respectively, as per the manufacturer's recommendations of the current economiser system for the vessel used in this study.

5. RESULTS & DISCUSSION

Using the equations displayed in sections 4.4 and 4.5, the rate of heat scavenging from all potential sources of the powertrain system can be estimated and converted into steam power equivalent. A single trip has been used as a case study, including a representation of the cruise ship going into port & staying overnight. This can be used to directly compare the steam production capabilities for different powertrain combinations.

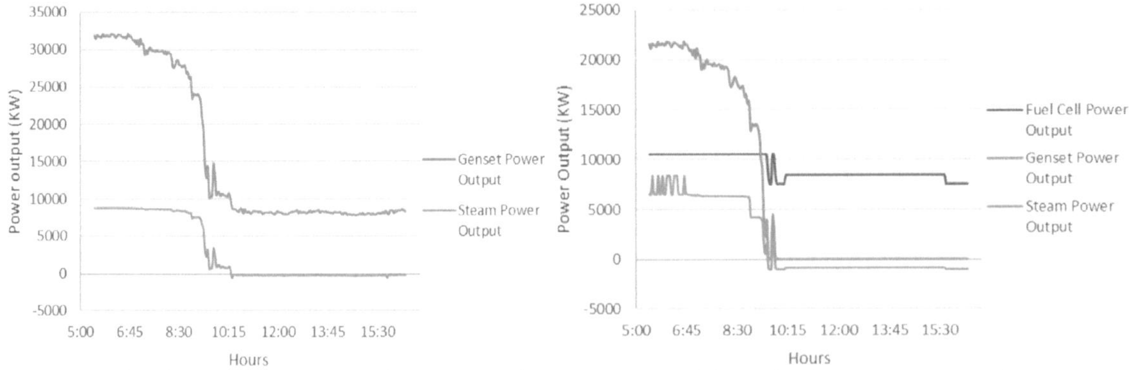

Figure 10: Graphs comparing steam production capabilities between the original ICE-Generator powertrain (**left**) and the new hybrid powertrain with Generators and Fuel cells (**right**)

In Figure 10, one can observe that for this trip, the vessel can produce enough steam for the biggest part from the positive steam power output. At specific time increments, there appears to be a steam power 'deficit', a result of subtracting the predicted steam demand from the heat output available from the economisers. This requires additional power to produce steam, just to meet that demand, set by the feedwater requirements against the total power output of the vessel, discussed in section 4.5.

This is attributed to the way the generators are set to run, where subjecting a single generator to higher loads rather than having two sharing this power demand at lower loads is favoured by the vessel's operators, as this way engines run more efficiently [18], utilise less fuel and have lower exhaust temperatures, also shown in Figure 8.

Putting this in terms of annual operation, having a powertrain solely comprised by ICE gensets means that an additional 22 tonnes of LNG would have to be burnt, in order for the average predicted steam demand to be covered. When comparing this to the estimated annual consumption of 26,472 tonnes of LNG to power that vessel, the additional 22 tonnes are insignificant.

In the case of substituting one of the ICE gensets for a 10.5 MW solid oxide fuel cell stack, the same trip is used as an example of how the rate of heat scavenging and subsequent steam production can be affected. It is observable from Figure 10, that employing batteries to compensate for the lack of immediate response from the fuel cells, translates to operating at an almost constant load, unlike the generators, which also allows them to run in an efficient matter. That said, it is shown in the same figure that there are sustained periods where the steam energy output is less than the heat demand of the cruise ship. This steam power "deficit" is evident by the negative values shown in the graph when the generators are turned off, specifically whilst in port when the fuel cells are expected to cover most of the energy load.

As discussed in section 4.3, this SOFC system can run on LNG which is internally reformed into hydrogen. As such, this system balance of plant manages the temperatures internally and is configured to efficiently generate electrical power from the fuel, with an average exhaust temperature of 230°C, whereas the gensets are closer to 380°C [18]. For this proposed SOFC system configuration to meet the expected average steam demand, additional heating power would have to be provided by the natural gas fired boilers on board, which is estimated to be an addition of 330 tonnes to the 9,288 tonnes of annual LNG consumption required to operate the 10.5 MW SOFCs. Further investigations are required to determine what might be a more optimum exhaust temperature that might remove the need for an additional boiler.

6. CONCLUSION

In this study the concept of using solid oxide fuel cells, powered by LNG, to provide a baseload for a cruise ship's power demand and to meet all demand whilst in port was investigated. A preliminary study found that this required 10.5 MW capacity of SOFCs, together with (5 MWh) of batteries to handle peaks in demand in port. The fuel cells assessed in this study provide a significantly better efficiency overall compared to the current LNG gensets. An amount of heat from the fuel cell exhaust is fed back into the steam reforming processes for the LNG fuel source to convert it into hydrogen. Consequently, this process restricts the amount of heat that can be scavenged by the economisers. The proposed SOFC based system is more efficient and produces fewer emissions than the gensets. The proposed concept reduces the total fuel consumption of the cruise ship by up to 27% and the carbon dioxide equivalent emissions by up to 35%.

Simulations were run over 12 months based on data from a cruise ship. Results showed that with a peak overall efficiency configuration, this SOFC system is not capable of providing enough heat to produce steam all of the time, specifically when in port, relative to the predicted demands for this vessel type. The SOFC system could in the future be configured to elevate exhaust temperatures with a small sacrifice in efficiency.

Without this slight adjustment in the configuration, the resulting heat deficit would require other steam producing methods to fulfil the expected average steam demand on board the vessel, amounting to 330 tonnes of additional LNG fuel annual consumption by employing a typical gas boiler compared to the 9,288 tonnes required to operate the 10.5 MW solid oxide fuel cells.

7. ACKNOWLEDGEMENTS

This paper is the result of a project completed under the Innovate UK research grant, with title "Zero-Carbon Base Load Power For Large Ships" and project number 10011959, as part of the Clean Maritime Demonstration Competition.

8. REFERENCES

[1] E. Lindstad, G. S. Eskeland, A. Rialland and A. Valland, "Decarbonizing maritime transport: The importance of engine technology and regulations for LNG to serve as a transition fuel," in *Sustainability*, vol. 12, 2020, p. 8793.

[2] "Clean Maritime Plan," London, UK, 2019.

[3] M. Cheliotis, E. Boulougouris, N. L. Trivyza, G. Theotokatos, G. Livanos, G. Mantalos, et al., "Review on the Safe Use of Ammonia Fuel Cells in the Maritime Industry," in *Energies*, vol. 14, 2021, p. 3023.

[4] McKinlay Charles, Hudson A. Dominic, Turnock J. Stephen, "A Comparison of Hydrogen and Ammonia for Future Long Distance Shipping Fuels," in *LNG/LPG and Alternative Fuels*, The Royal Institue of Naval Architects, 2020.

[5] McKinlay Charles, Turnock Stephen R, Hudson Dominic A, "Route to zero emission shipping: hydrogen, ammonia or methanol?," in *International Journal of Hydrogen Energy*, vol. 14, 2021, pp. 28282 - 28297.

[6] C.J. McKinlay, S.R. Turnock, D.A Hudson, "Fuel Cells for Shipping: To meet on board demand and reduce emissions," in *Energy Reports*, vol. 7, Elsevier, 2021, pp. 63-70.

[7] Lukas Kistner, Fritjof L.Schubert, Christine Minke, Astrid Bensmann, Richard Hanke-Rauschenbach, "Techno-economic and Environmental Comparison of Internal Combustion Engines and Solid Oxide Fuel Cells for Ship Applications," Journal of Power Sources, vol. 508, 2021.

[8] EG&G Technical Services, Inc., "SOLID OXIDE FUEL CELLS," in Fuel Cell Handbook, Morgantown, West Virginia, U.S. Department of Energy, 1992, pp. 197-245.

[9] I. E. A. (IEA), "iea.org," 2022. [Online]. Available: https://www.iea.org/reports/net-zero-by-2050. [Accessed 19 May 2022].

[10] C. French, "Economizers," in Plant Engineer's Reference Book, Butterworth Heinemann, 2002, pp. 31.1-31.7.

[11] Khaled Senarya, Adel Tawfikb, Elsayed Hegazyb, Amr Alia, "Development of a waste heat recovery system onboard LNG carrier to meet IMO regulations," in Alexandria Engineering Journal, Elsevier, 2016, pp. 1951-1960.

[12] Hussam Jouhara, Navid Khordehgah, Sulaiman Almahmoud, Bertrand Delpech, Amisha Chauhan, Savvas A.Tassou, "Waste Heat Recovery Systems," in Thermal Science and Engineering Progress, Elsevier, 2018, pp. 268-289.

[13] M. Marine, "Waste Heat Recovery Systems," MAN Diesel & Turbo, Copenhagen, 2014.

[14] Francesco Baldi, Cecilia Gabrielii, "A feasibility analysis of waste heat recovery systems for marine applications," Energy, vol. 80, pp. 654-665, 2015.

[15] J. J. de-Troya, C. Alvarez, C. Fernández-Garrido, and L. Carral, "Analysing the possibilities of using fuel cells in ships," in International Journal of Hydrogen Energy, 2016, pp. 2853-2866.

[16] Zhe Wangab, Huaiqiang Chena, Rui Xiaa, Fenghui Hanab, Yulong Jia, Wenjian Caic, "Energy, exergy and economy (3E) investigation of a SOFC-GT-ORC waste heat recovery system for green power ships," Thermal Science and Engineering Progress, vol. 32, 2022.

[17] L. Register, "classdirect.lr.org," 2022. [Online]. Available: https://classdirect.lr.org/assets/IHS9781865/details . [Accessed 20th June 2022].

[18] C. Marine, " Caterpillar Motoren, MaK Dual Fuel VM46 Engine," 2016. [Online]. Available: https://www.makmed.fr/en/emission-solutions/dual-fuel/. [Accessed 15 10 2021].

[19] E. Richardson, "Thermodynamics: Combustion Processes & Internal Combustion Engines," in SESM 2017 Thermodynamics, University of Southampton, 2019, pp. Chapters 5-6.

[20] C. L. Proctor, " Internal Combustion Engines," in Encyclopedia of Physical Science and Technology, California, Academic Press, 2003, pp. 33-45.

[21] U. G.Bossel, "Final report on SOFC Data- Facts & Figures," Swiss Federal Office of Energy,, Berne, 1992.

9. AUTHORS BIOGRAPHY

Panagiotis Manias is a research technician at the University of Southampton from which he graduated with master's in mechanical engineering. He initially took part in a concept design project of a ballast free liquid hydrogen carrier where he developed the powertrain for that particular vessel and was later involved in the CMDC projects, as part of the energy and emissions modelling. Earlier to this, some industry experience in the technical department of a company was gained, including drydock maintenance, as part of summer jobs.

Charles McKinlay is a PhD researcher at the University of Southampton, as part of the CDT in Energy Storage and its Applications programme. His research is focused on using dynamic energy system modelling to demonstrate a viable pathway to zero emission shipping. He recently collaborated on 3 Clean Maritime Demonstration Competition projects. His background also includes Mathematics and Energy Engineering.

Professor Stephen Turnock, FRINA is Head of Civil, Maritime and Environmental Engineering at the University of Southampton. He has long standing interests in decarbonisation of shipping and in particular analysis and modelling of alternative powertrains including the use of hybrid battery systems.

Professor Dominic Hudson is the Shell Professor of Ship Safety and Efficiency at the University of Southampton and leads the Centre for Maritime Futures. He has research interests in all aspects of ship hydrodynamics, particularly ship design and operation for energy efficiency, and maritime decarbonisation. He has published over 150 papers in journals and proceedings of refereed conferences. He is a Chartered Engineer and member of RINA. He is presently a member of the RINA IMO Committee.

International Conference on
Computer Applications in Shipbuilding
13th-15th of September 2022, Yokohama, Japan

DEVELOPMENT OF PLM SYSTEM FOR PRODUCTION PLANNING AND PRODUCTION CONTROL IN SHIPBUILDING

K Matsuo, M Morishita and T Taniguchi, National Maritime Research Institute, Japan

SUMMARY

Data integration has become an important issue to develop complex products in shipbuilding. This paper introduces a development of a prototype system for PLM systems in shipbuilding. The PLM system is developed for the purpose of precise production planning and the production management in the shop floor. Especially, the PLM system is considered as a use case to provide input data which is necessary for production simulators that reproduce precise work procedures at the shop floor. The data structure in the PLM system is reviewed to provide the necessary data to the production simulator in the beginning of this paper. The data is represented as BOM (Bill of Materials) and BOP (Bill of Process), which are associated with product information, process information, and facility information, and is implemented in the PLM system. The data flow from the PLM system to the production simulator is introduced, then a series of demonstrations from the PLM system to the execution of the production simulator are presented, followed by a discussion of the effects and the future issues of the proposed data structure and the PLM system.

1. INTRODUCTION

Data integration is a more important issue for the shipbuilding industry in order to respond to shorter delivery times, flexible specification changes, and complicated product development. NMRI has been conducting an R&D project on data integration between design and manufacturing in the shipbuilding industry. The data integration between design and manufacturing will contribute to build a front-loading design system in design stage and a high-precision production simulation system in manufacturing stage.

As a concrete way of data integration between design and manufacturing, NMRI are proceeding with research and development on a PLM system for shipbuilding that uses BOM (Bill of Materials) and BOP (Bill of Process). The PLM system systematically manages not only product information but also process information. For this purpose, NMRI are conducting research and development on data flows that organize the data generated in the shipbuilding process and flow them into the BOM and BOP in the PLM system.

2. System Configuration and Operation Flow around the PLM System

The system configuration and system operation around the PLM system are explained with respect to the data integration between design and manufacturing (Figure 1). Information on products and processes is sequentially generated in a series of processes in shipbuilding. This information is generated and managed by CAD systems, dedicated systems such as process planning systems, and various types of paper drawings. Shipbuilding has been carried out by compiling the necessary information for each operation and each department in charge of the work and creating the information as drawings. It is necessary to manage information on products and process in an integrated manner, when accurately representing manufacturing within a system. Therefore, it is necessary to design the necessary data structure in advance and link the information dispersed throughout the shipyard. The PLM system in this study determines the data items necessary for production planning and production control in advance and manages the data in an integrated manner by connecting to systems that generate such data. The PLM system is connected to a production simulator, various production planning systems, and production control systems.

When data to be managed in the PLM system are generated in the design process, these data are collected in a PLM system (Figure 1 (1)). The data structure is expressed in the form of BOM (Bill of Materials) and BOP (Bill of Process), and product information and process information are managed in an integrated manner (Figure 1 (2)). Production planning and production control are performed using the database constructed in the PLM system. In this study, assuming the utilization of these data to a production simulation, the data structure is designed for granular production planning and production control, so that production simulation can be operated by referring to the database in the PLM system (Figure 1 (3)). Production plans and schedule plans can be made by using the simulation results, and detailed work procedures can be created for a shop floor. In addition, the database in the PLM system is directly used for production planning, schedule planning, production control and material procurement management at a shop floor. When block division or production line are changed as a result of production planning, schedule planning, the data in the PLM system is modified and updated by changing product and process information in the design system.

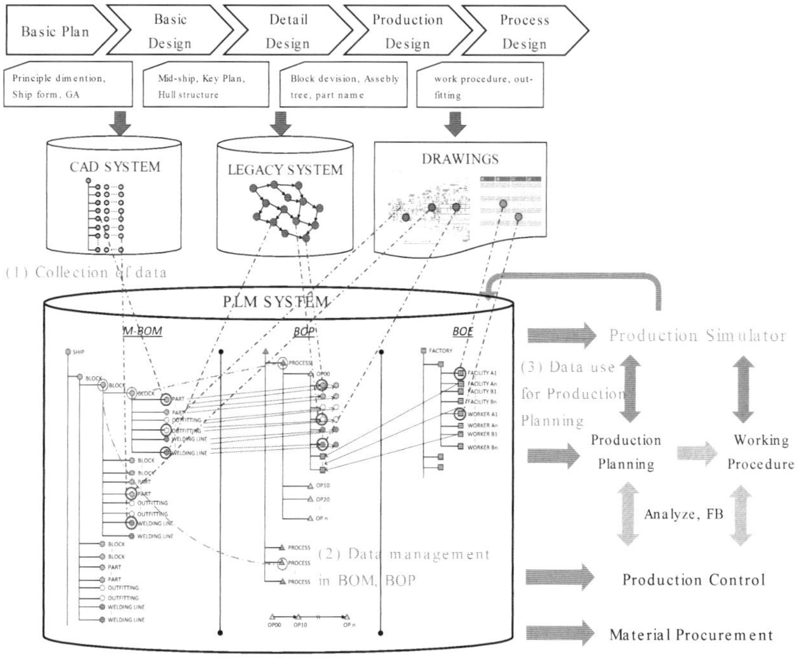

Figure 1: Systems and information flow around the PLM system.

3. Data Structure and Data Standardization in PLM system

In this study, data structures based on BOM, and BOP are investigated for the integrated data management of product, process, and facility in shipbuilding. Product information is represented by BOM, and process information is represented by BOP. This chapter explains the proposed BOM and BOP data structures. Figure 2 shows an overview of the proposed BOM and BOP data structures.

3.1 DATA STRUCTURE FOR PRODUCT INFORMATION

The BOM represents the parts of a product in a manufacturing hierarchy. In the case of a ship's hull, the precedence of the assembly is represented by a hierarchical structure. The final product (ship) is placed at the top of the hierarchical structure, and the hierarchical structure is expanded until it becomes a single part. In shipbuilding, a block assembling tree is assumed as a typical hierarchical structure. The registration of parts on the hierarchical structure is not only for hull parts, but also for outfitting items or working pieces to be installed at the time of the block. The weld lines that occur during the assembling of the block are registered as parts. By appropriately expressing the parts required for manufacturing in the BOM, the work related to the relevant parts can be explicitly expressed in the BOP.

3.2 DATA STRUCTURE FOR PROCESS INFORMATION

The BOP represents process information, in other words, information on how to make a product. The BOP proposed in this study defines a process and operations that comprise each process. Each operation defines the parts required for that operation and the available facilities. The necessary parts are referenced from the BOM defined in the previous section, and the available facility is referenced from the separately defined BOE (Bill of Equipment). The BOP does not handle manufacturing instruction information such as "when". This information is handled in a MES (Manufacturing Execution System), or an ERP (Enterprise Resources Planning) system connected to the PLM system.

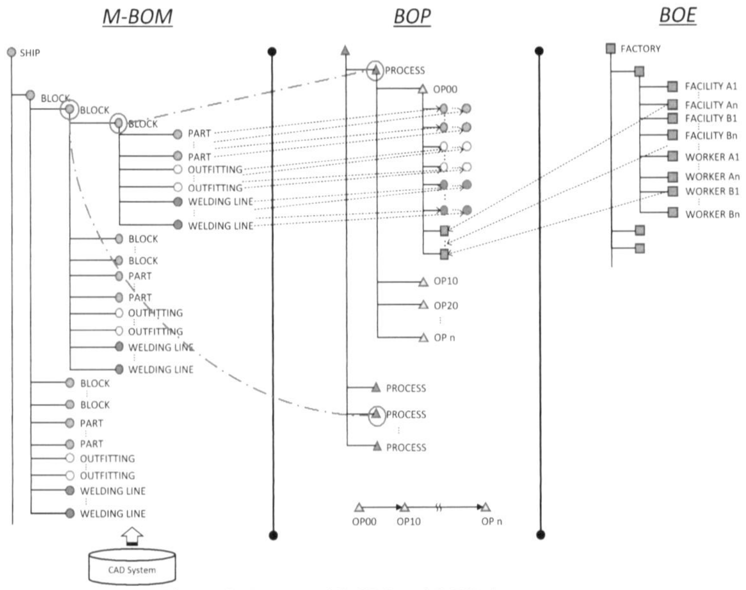

Figure 2: Proposed BOM and BOP data structures.

4. Implementation of BOM, BOP in PLM system

Based on the data structure proposed in Chapter 3, BOM and BOP are implemented in a PLM system. Assuming an actual ship, the model is constructed using a commercially available shipbuilding CAD system and PLM system. Siemens NX Ship was used as the shipbuilding CAD system and Siemens Teamcenter as the PLM system in this modelling.

4.1 MODELLING OF BOM

The BOM is created as product information, assuming an actual ship. In this verification, the entire hull is modelled using a shipbuilding CAD system (Figure 3). The BOM is constructed in the PLM system. Blocks, hull parts, and weld lines are defined according to the block assembling tree as parts generated by the CAD system. For hull parts, shape information (managed in JT file format) and attribute information (part name, connected parts etc.) of the parts modelled by the CAD system are registered in the PLM system.

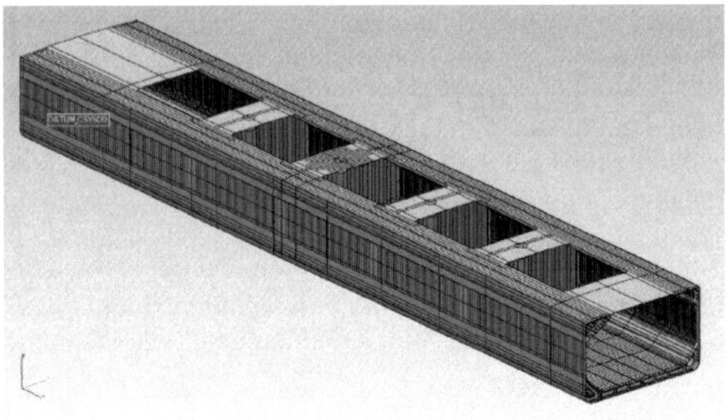

Figure 3: Proposed BOM and BOP data structures.

4.2 MODELLING OF BOP

The BOP is created as process information for the modelled hull blocks. In this verification, the BOP is created for the subject block, assuming large assembly, medium assembly, sub assembly of the block as processes. The operations in each process are defined as material distribution, part fitting, welding, and so on. When processes and operations in a factory are systematized, the processes and operations can be made into templates. In this study, the BOP are created by analysing actual factory operations and creating templates for operations in process. Specifically, the operations for the sub-assembly process are templated as follows: OP10 (material distribution) => OP20 (fitting) => OP30 (welding) => OP40 (material distribution) => OP50 (gas heating) => OP60 (fitting on the reverse side) => OP70 (welding on the reverse side) => OP80 (material distribution), and the products and facilities corresponding to each operation are registered. Various operations are represented using product information on weld lines. In other words, not only welding work but also gas heating work for distortion removal is represented by weld line. The available facilities are also defined for each operation. The facility information is defined in BOE, including information on available workers. In this verification, in order to correctly represent interference in the use of facility in the production simulation, major equipment such as cranes and welding machines etc. are defined in the BOE. The BOP created based on the above is shown in Figure 4.

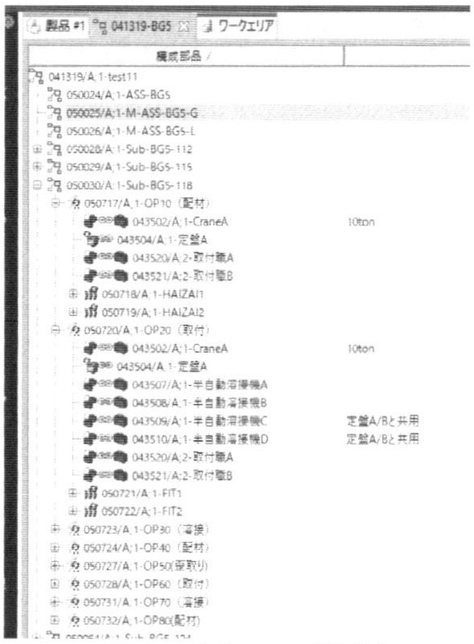

Figure 4: Image of BOP.

4.3 DATA CONNECTION WITH PRODUCTION SIMULATOR

The BOM and BOP data modelled in the PLM system are connected to the production simulator developed by NMRI, and the validity of the data structure are verified.

The production simulator uses a multi-agent system to reproduce the shipbuilding work in detail, including the incidental work of the worker. Based on the process and operation information defined in the PLM system, verification is conducted whether the detailed work of the worker can be reproduced in the production simulator. Regarding the data connection with the production simulator for the sub-assembly process, it is confirmed that the work of the sub-assembly process can be reproduced in the production simulator while using the information in the PLM system as input data (Figure 5, 6). As a result, a series of data flow from the data in the PLM system to the production simulator is organized.

Figure 5: Image of Production Simulator

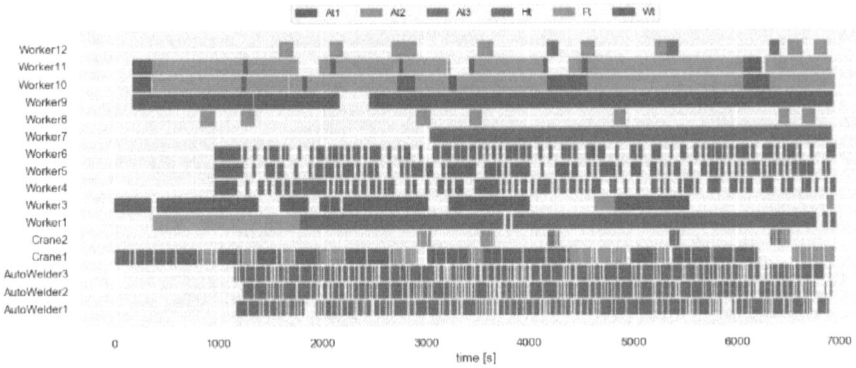

Figure 6: Gantt chart of each worker by production simulation

5. CONCLUSIONS

This paper introduces the development and utilization of a shipbuilding data structure for the purpose of production planning and production control in shop floor with fine granularity, as an effective utilization of design data for manufacturing. The data structure is expressed as BOM and BOP, and it is confirmed that product information, process information, and facility information can be managed centrally. A series of data flows from the BOM and the BOP data to the production simulator developed by NMRI for a detailed reproduction of worker's work in a shop floor is also verified and demonstrated in this paper.

In the future, we will continue to organize data structures for various shipbuilding processes and work on standardization of data structures for data integration between design and manufacturing in shipbuilding. In addition, we will continue to improve the environment around the PLM system so that the production simulator can be operated as soon as the data is prepared in the PLM system.

6. ACKNOWLEDGMENT

This research was conducted in collaboration with JMU Systems Corporation. We would like to express our deepest gratitude to all parties involved.

7. REFERENCES

1. T. Taniguchi, M. Takezawa. "Development of High-Precision Production Simulation for Shipbuilding Based on Multi-Agent System " *Conference proceedings, the Japan Society of Naval Architects and Ocean Engineers* (33), 327-333, 2021-11.

2. K. Matsuo, M. Morishita, K. Hiyoku, M. Sueoka, T. Taniguchi, M. Takezawa. "A Fundamental Study on PLM System for Shipbuilding" *Conference proceedings, the Japan Society of Naval Architects and Ocean Engineers* (34), 2022-5.

8. AUTHORS BIOGRAPHY

Dr. Kohei Matsuo holds the current position of Head of Industrial System Research Group at National Maritime Research Institute of National Institute of Maritime, Port and Aviation Technology. He is responsible for manufacturing engineering of shipbuilding. His experience includes various approaches on ICT development for shipbuilding process.

Dr. Mizuki Morishita holds the current position of Researcher at National Maritime Research Institute of National Institute of Maritime, Port and Aviation Technology. He is responsible for research and implementation of data integration for shipbuilding.

Dr. Tomoyuki Taniguchi holds the current position of Chief Researcher at National Maritime Research Institute of National Institute of Maritime, Port and Aviation Technology. He is responsible for research and implementation of the utilization of computational science and technology in production and design on shipbuilding.

DYNAMIC MODELLING OF AMMONIA CRACKERS AND HYDROGEN PEM FUEL CELLS FOR SHIPPING APPLICATIONS

CJ McKinlay, P Manias, SR Turnock and **DA Hudson,** University of Southampton, UK

SUMMARY

The pathway towards sustainable maritime propulsion is unclear. One concept is using ammonia as a fuel. This study investigated using an ammonia cracker (a device to convert ammonia to hydrogen) and a Proton Exchange Membrane fuel cell. This requires a hydrogen buffer tank and purification, each process has been dynamically modelled. Simulations show the power demand for different vessel types (cargo and survey) for several routes. This enabled analysis of ammonia demand and system requirements during: in port; manoeuvring; cruising; surveying. The same simulations ran for other potential setups to facilitate a fair comparison of the technical challenges, storage requirements, and support system requirements (e.g. batteries). Results showed that the total system plus fuel for this concept for a case study cargo voyage would weigh 150 tonnes and require 586 m^3 of volume. For comparison, a liquid hydrogen system for the same voyage weighed 35 tonnes and required 222 m^3.

GRAPHICAL ABSTRACT

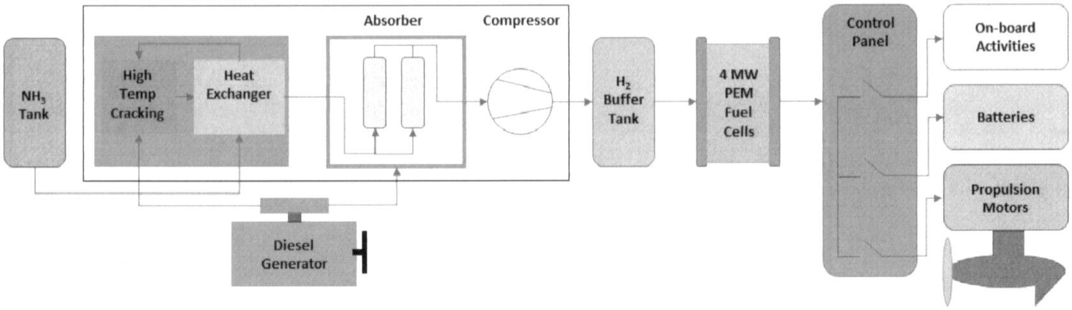

NOMENCLATURE

78m	Reference vessel used in this study of length 78 metres
CO_2	Carbon Dioxide
DP	Dynamic Positioning
h	Hours
H_2	Hydrogen
Kts	Knots (1 knot = 1.94 ms-1)
kW	Kilowatts
LARS	Launch and Recovery System
MDO	Marine Diesel Oil
NH_3	Ammonia
NOx	Nitrogen oxides
Nm	Nautical miles (1 nm = 1.852 km)
PEM	Proton Exchange Membrane (type of fuel cell)
ROV	Remotely Operated Vessel
SoC	State of charge (batteries)

1. INTRODUCTION

Current emissions levels from the maritime sector are unsustainable. This is reflected in several targets set by various organisations, for example the International Maritime Organisation (IMO) targets a 50% reduction in shipping emissions by 2050 [1] and the UK government's Clean Maritime Plan states that all new built vessels should be zero-emission ready by 2025 [2]. This cannot be achieved using current fossil fuels (unless there are major advances in carbon capture technology) and many believe that a transition to alternative fuels is necessary [3-5].

One substance that has been gaining a lot of attention from the maritime industry as a potential alternative fuel type is ammonia. Ammonia (NH_3) is primarily used as a fertiliser in today's market and is unproven as a fuel source, but it has no carbon content and therefore has the potential to deliver ship propulsion with zero tank-to-wake carbon emissions. Although other pollutants, namely nitrogen oxides (NOx), may be a concern [6].

Fuel cells are an electro-chemical device used to convert the chemical potential energy of a molecule (typically hydrogen) into electricity. These are known to be more efficient than combustion engines or generators which are currently deployed onboard most vessels. Many believe that this efficiency gain could be key to counteracting the lower energy densities of alternative fuels to fossil fuels [6]. The proton-exchange membrane (PEM) fuel cell is one of the leading technology types for converting hydrogen to electricity, advantages include low operating temperatures and relatively high overall efficiency, although a drawback is that very high hydrogen purity (>99.99%) is required to avoid performance degradation [7]. Ammonia crackers are a device that can extract the hydrogen content of ammonia these achieve high efficiency but do require high operating temperatures [8].

This study has used a 78-metre-long vessel, currently under construction, as a case study. This vessel has been designed for two possible functions, to move cargo and to conduct scientific surveys. It is likely that the energy demand and major loads of these two operating patterns will be quite different. As this is a new vessel, no historical power consumption data was available, hence part of this study has been to develop a model for approximating demands for voyage types.

1.1 AIMS AND OBJECTIVES

The aim of this study has been to investigate the technical feasibility of using ammonia as a fuel in conjunction with an ammonia cracker and a PEM fuel cell to meet all the energy demand of the 78m vessel.

To achieve this, an objective was to develop energy demand profiles to represent typical voyages for either cargo or survey modes. The second objective was to develop dynamic models for PEM fuel cells, ammonia crackers and other required devices, based on ramp rates and efficiency curves. Simulations could then be run to determine the system size requirements to meet all demand. A further objective was to calculate fuel consumption, such that results could be compared to other alternative fuel concepts such as liquid hydrogen.

2. BACKGROUND

Ammonia is being considered as a potential future maritime fuel by many in the industry [9]. It has zero carbon content and can be used as a fuel source via internal combustion engines. However, the use of a pilot fuel is required to achieve combustion due to the low flammability of ammonia, hence still relying on fossil fuels to some extent [10]. Reiter and Kong investigated ammonia-diesel dual fuel engines, they found that when ammonia comprised of 40% of the fuel there was no significant impact in NOx and hydrocarbon emissions [11]. Although, at 95% ammonia, a significant increase in specific fuel consumption is observed, as well as ammonia slip and unburnt hydrocarbon emissions [12]. Ammonia use in internal combustion engines can significantly impact emissions due to the formation of the N_2O gas [13], which is considered to be 300 times worse than CO_2 in terms of greenhouse effects [14].

Many in the maritime industry perceive ammonia as an effective carrier of hydrogen molecules [10] as liquid storage can be achieved at -34°C [15] compared to -240°C for pure hydrogen [16]. Though, ammonia is also a toxic substance and its safe handling of as a maritime fuel is unproven. Ammonia can be detected in very small concentrations (5 ppm) in the air due to its strong pungent odour [17], it is colourless and can be lethal when ingested. When the effects of a potential ammonia spillage in the marine environment are considered, its toxicity can severely impact marine wildlife by having fatal consequences [18].

When using PEM fuel cells, as in this study, ammonia must be cracked into hydrogen and then purified to be used as fuel. Even with low levels of fuel contamination, ammonia poisons the electrodes and degrades the electrolyte within the fuel cell reducing performance [19-20]. Solid oxide fuel cells (SOFC) by contrast, can operate without requiring external cracking or extensive purification of fuel [21]. Several studies have been conducted with regards to ammonia cracking applications, many of them being aided by some form of metallic catalyst enrichment [22] and some considered the utilisation of portable PEM cells [23].

During the cracking process, ammonia is dissociated into hydrogen and nitrogen gas. Nitrogen is a noble gas and consists roughly 78% of the atmospheric air [24], this implies that no harmful emissions are developed during this process. However, ammonia itself, has a significant number of emissions related to its production, as it is currently heavily relied upon fossil fuels. This is partially attributed to the common practice of producing hydrogen via steam methane reforming, as opposed to using electrolysis with renewably sourced energy, resulting in carbon emissions [25]. Furthermore, the conversion of hydrogen to ammonia requires the Haber-Bosch process, which is known to be very energy intensive [6]. Ammonia production accounts for 1.8% of total global CO_2 emissions for its current market [26]. When considering that the maritime contributes 2.89% of total global CO_2 emission production [27] for the entire industry, it is possible that the transition to ammonia would actually increase emissions, unless a more sustainable production method is developed.

Some studies have been conducted on the application of ammonia cracking mechanism to feed PEM fuel cells [28], and others have considered the use of hydrogen PEM fuel cells for marine applications [29, 30], but this paper is the first to specifically test the concept of ammonia cracking to PEM fuel cells for maritime applications. Furthermore, a novel model has been developed for the energy requirements of a vessel's powertrain system, including the energy profile of fuel cells, ammonia crackers and batteries.

3. METHOD: POWER PROFILE BUILDING

The concept of this study was to power the 78m vessel using of proton exchange membrane fuel cells, using an ammonia cracker and an internal combustion engine-generator set (genset). In this case, the purpose of the genset was solely to power the ammonia cracker and would not meet any of the vessel's demand directly.

As this is a new vessel, there was no available power demand data. Therefore, it was first necessary to develop a model to generate power demand profiles specific to this 78m vessel accounting for factors such as speed and load variation. Separate models are required for the two operating profiles considered in this study, cargo and survey, as their main energy loads are quite different.

3.1 DATA OVERVIEW

To build a realistic power profile building model, two datasets were gathered from separate sources each representing a different vessel type.

The first source of data was generated by towing tank tests on a scale model of the 78 m vessel through various speeds and wave heights, this dataset will be referred to as DS1. Based on these data, equations were derived to represent a propulsion power input for variable sea states and speeds during sailing. This was a useful initial starting point, although this model only represented the sailing mode of operation, for the purpose of this exercise it was necessary to also consider modes such as manoeuvring and entering port.

The second source of data recorded from an LNG carrier, referred to as DS2. This vessel has 8 MW of installed power, which is considerable larger than the 78m vessel, although data could be scaled down to resemble the target output.

This dataset is recorded from a real vessel under operation. Although, it is challenging to establish a relationship between speed and power, as the wave states change constantly and the generators, behave more like switches to cope with demand onboard, rather than fluctuating their power output (similar to direct drive applications). This vessel has a different hull design and operates in different sea states than the 78m vessel, however this dataset was valuable as it includes recordings for all modes of operation.

3.2 MODES OF OPERATION

During operation, a vessel undergoes many activities that have an impact on power demand, either direct propulsion loads or onboard auxiliary demand. Hence, separate power profiles are required to replicate the different operating modes: port idle, port stay, manoeuvring, sailing, and entering port. A separate mode was developed for surveying, this includes different loads such as the deployment of submersible remotely operated vessels (ROVs).

3.2 (a) Port Stay mode

During the "port stay" mode of operation, the vessel is moored in port and any propulsion devices are disabled, with no major activities carried out onboard. As such, it is expected to have a minimum hotel demand as well as some power requirements for winches, cranes and pumps. Consequently, this mode of operation has the lowest power demand of all. The minimum rated hotel demand has been modelled as 100 kW, winches and other equipment on board would add an extra 50 to 100 kW, and finally heat and freshwater pumps would be around 50 kW with a maximum of 100 kW.

3.2 (b) Port Idle mode

The "port idle" is the succession of "port stay", here the vessel is ready to depart from port and major activities are being carried out on board. This requires auxiliary devices to be on standby for the main power supply to start up. Moreover, major mooring equipment is expected to operate constantly as the vessel is slowly being unmoored, hotel demand is rising as the crew is moving within the vessel constantly, and all navigation and control equipment are switched on. Consequently, the powering requirements during idling would be considerably higher than when staying in port overnight and peak power demand has been modelled as 700 kW.

3.2 (c) Manoeuvring mode

Manoeuvring is the act of moving in and around ports at slow speeds. This is one of the most difficult modes of operation to predict and model accurately [31] and there are several factors to consider such as space limitations and weather conditions (wind and wave amplitude). The power demand is significantly increased from the previous states, as it is constantly required by the vessel to "break" momentum with short bursts of propulsive power, varying considerably. Consequently, the power profile would include many peaks and valleys over a short period of time, as the vessel is expected to undergo manoeuvring actions until it leaves the port.

To model this operational state, data from DS2 were used, where the vessel would undergo manoeuvring actions for a certain period, which were scaled down to the modelled installed power for the 78m vessel.

3.2 (d) Sailing (Underway) mode

The "sailing mode" is relevant to the vessel's expected power demand whilst sailing. This mode is expected to have the lowest variation, as the vessel's main propulsion devices are set to operate at a specific power input. This varies slightly due to flow characteristics around the hull, depending on wake effects. That said, the vessel is expected to have increased propulsive demand when the wave amplitude and wind speed is higher than at calm sea state, if it is required to keep the same speed. Finally, any activities on deck are halted, as the vessel is in motion, leaving the hotel demand high as the crew is located mostly within the accommodation.

The model of propulsion for the sailing mode was developed based on the DS2 data fluctuations but scaled down using the DS1 profile. Furthermore, the power profile can be scaled for different speeds based on the cubic relationship between power and speed that was calculated specifically for this 78m vessel.

3.2 (e) Entering port mode

The "Entering port" mode is relevant to when the vessel starts entering a port or when approaching an area regulated by port authorities, to reduce wake, pollution, and noise. As such the vessel is required to slow down, lowering the overall power demand, as it approaches its final position to dock. The trends for this mode are developed from DS2.

3.2 (f) Surveying mode

The vessel's operator advised that a 30 day long surveying mission would be representative of typical operating conditions and should include the following operations:

- 3 days transiting to the desired location at 10 Knots
- 9 days spent in total whilst in transit to the desired research site, travelling at 6-8 Knots
- 12 days surveying using dynamic positioning
- 6 days in total spent while on stand-by

The purpose of a surveying mission is usually to record the sea floor of a specific area of interest. To achieve this, an ROV is deployed by a launch and recovery system (LARS) that uses winches. This submersible vessel then scans a section of the sea floor and upon completion, it is recovered back to the research vessel. The ROV is not deployed for the entirety of the surveying mission, instead it is deployed for a specific amount of time, around 40% of the total mission. When the ROV is deployed, it constantly obtains electrical power from the research vessel via a cord, while the research vessel itself maintains its position on the surface dynamically using its own propellers and thrusters, known as "dynamic positioning".

Several modelling assumptions have been made, for example the 12 days of surveying research would include the equivalent of 5 days ROV deployment. Further assumptions include: LARS is used for ROV deployment and recovery only; hotel demand is between 120 and 150 kW; a single ROV is used throughout the mission. Furthermore, the power demand from dynamic positioning (DP) feature is assumed to fluctuate considerably, similar to the manoeuvring mode. Basically, short bursts of power are required to maintain a set position due to wakes, wind directions changes, and below surface currents. The DP power demand fluctuates from 350 kW to almost 1 MW [32]. Using these assumptions and power tables provided by the operator, the power profile and breakdown shown in Figure 1 is developed.

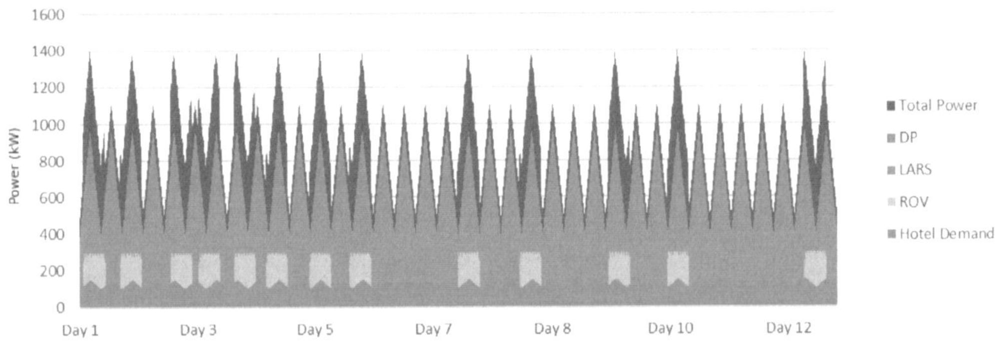

Figure 1. Power breakdown of all equipment whilst carrying out surveying.

3.3 ROUTE DEVELOPING

Using the operating modes from section 3.2, a model has been developed to generate a power profile for a cargo voyage based on a series of conditions. The variables for this model are: distance (nm); target speed(knots); port stop duration(h); port idle duration (h); hotel load (kW). This was used to generate the power profiles used in sections 5.1 and 5.2.

4. METHOD: ENERGY SYSTEM MODELLING

The concept for propulsion in this study comprises of PEM fuel cells to provide onboard electrical power, with lithium-ion batteries to cope with rapid changes in power demand. The main fuel source for this concept is liquid ammonia. Ammonia is converted to hydrogen using an ammonia cracker, although this process is energy intensive, requiring an additional power source (diesel generator).

4.1 AMMONIA CRACKER

An ammonia cracker is a device that splits up ammonia molecules into nitrogen and hydrogen, this requires a heat energy input [33]. There are two main types of hydrogen crackers, a high-temperature cracker and a low-temperature "catalyst-aided" cracker, with the use of a catalyst reducing the temperature requirement for the cracking process. For this study, a catalyst-aided cracker has been modelled based on liaison with a technology developer. A visualisation of the ammonia cracking process is shown in Figure 2.

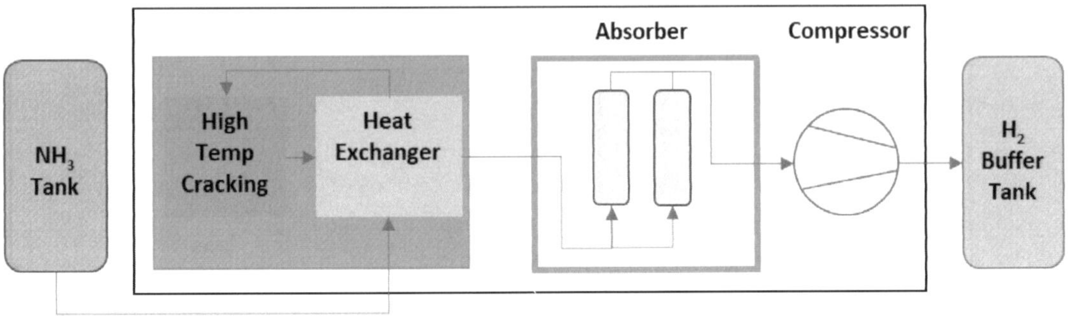

Figure 2. Ammonia cracking plant illustration for a high temperature ammonia cracker.

The process of high-temperature cracking includes:

- Evaporation of ammonia
- Preheat of ammonia
- Heating of ammonia at 850 °C
- Hydrogen Formation
- Precooling of Hydrogen at 500 °C
- Cooling of hydrogen at 100 °C
- Compression of Hydrogen at 6 bar
- Cooling at -50 °C
- Storage for later use

The catalyst-aided cracking process is similar but requires one less hydrogen cooling stage:

- Evaporation of ammonia
- Vaporisation and preheat of ammonia
- Heating of ammonia at 450 °C
- Hydrogen Formation
- Precooling of Hydrogen at 150 °C
- Compression of Hydrogen at 6 bar
- Cooling at -50 °C
- Storage for later use

To estimate the total amount of energy required to completely crack ammonia into hydrogen, each process is calculated separately, and the final energy demand is the result of their sum. The largest energy loads are the ammonia heater, the cooling of hydrogen, and hydrogen compression. The mass of the fluid is calculated based on the hydrogen demand of the Proton Exchange Membrane (PEM) fuel cells and the hydrogen loss of 20% during its purification process [34]. The energy required to heat ammonia is calculated with the equation shown below:

$$Heat\ input = M_{\mathrm{NH3}} \cdot C_{\mathrm{p}}(T_{\mathrm{in}} - T_{\mathrm{out}})$$

Where:

- M_{NH3} = Mass flow of ammonia fuel input
- C_{p} = Specific heat capacity of ammonia (2.175 kJ/kg.K)
- T_{in} = Inlet temperature of ammonia after heat exchanger
- T_{out} = Outlet temperature of ammonia at cracking temp

For hydrogen cooling and compression, enthalpy tables at different temperature and pressures are used [35]. Considering the pressure and temperature change of each stage, the final energy input can be calculated using the following equation:

$$Energy\ input = M_{\mathrm{H2}}(h_{\mathrm{in}} - h_{\mathrm{out}})$$

Where:

- M_{H2} = Mass flow of hydrogen input, subsequent of ammonia
- h_{in} = Specific enthalpy of hydrogen prior to compression and cooling at specific temperature and pressure
- h_{out} = Specific enthalpy of hydrogen after compression and cooling at specific temperature and pressure

Using the equations above, Table 1 was developed showing the power demand for each ammonia cracking method and fuel cell power output value respectively.

Table 1: Estimation of Ammonia cracking power requirements based on fuel cell power output.

Fuel cell power output [kW]	Consumption of hydrogen [kg/s]	Ammonia required [kg/s]	Power required without a catalyst [kW]	Power required using catalysts [kW]
1000	0.017	0.10	460	245
1500	0.025	0.15	676	360
2000	0.036	0.21	965	514
2500	0.047	0.28	1325	681

It is observable from Table 1 that the power required is significant in comparison to the fuel cell output. It would be possible to design an energy system that was self-sustaining after start up, where some of the fuel cell output was also used to power the ammonia cracker. However, results show that this would require 48.2 to 53.5% of the power output for a high-temperature system and 25 to 27.8% for a catalyst assisted cracker. For this reason, this study has included a separate diesel generator to crack the ammonia, sized at 25% of the fuel cells stack's rated power. Although, as the system now uses some fossil fuel combustion there will be some resulting emissions.

4.2 DYNAMIC SYSTEM MODELLING

It has been established that the energy system tested in this study will be comprised of several components, including fuel cells, batteries, ammonia storage, and a diesel generator. One of the objectives of this study is to determine the size requirements of these components for each tested voyage. To facilitate this, a model of each component type has been developed. These models are "dynamic" meaning that they simulate power output and fuel consumption in the time domain, for this study outputs have been simulated at 3-minute intervals. Components, such as fuel cells, are modelled based on ramp rates and efficiency curves based on typical equipment.

As discussed in section 3.3, a "route builder" model has been set up such that the user can select a list of parameters and variables. Here, values for the fuel cell stack's rated power (in kW) and the battery capacity (in kWh) can also be inputted. Thereafter, simulations can be run to test whether this proposed setup can meet all the power demand for this given voyage and to calculate the exact fuel consumption, both in terms of hydrogen to the fuel cells and the feed fuels: ammonia, and diesel. Batteries do not require fuel, although there would be some losses in the charging and discharging processes. This has been accounted for, with a round trip efficiency of 90% assumed for Lithium-ion batteries. Furthermore, the state of charge (SoC) has also been modelled at 3-minute intervals.

Simulations are written such that the fuel cells would deliver a constant output for as long as possible. Although, the fuel cells have been ramped up to full power when the SoC of the batteries dropped below 30%, similarly the fuel cell was dropped down to 25% of the rated power when the SoC rose above 85%. This is to mitigate the chances of overcharging or undercharging the batteries, which can increase their degradation rate considerably.

5. RESULTS

To demonstrate the capability of this energy system model, three separate case study voyages have been tested. Two of these are for a typical domestic cargo voyage (under different operating conditions) and one of these is for the survey mode. All results are for the 78m vessel, and the ammonia-fed system described.

5.1 SCENARIO 1: SOUTHAMPTON TO LIVERPOOL AT 14 KNOTS

The first case study was a route between two major UK ports, Southampton and Liverpool. By considering typical seafaring routes in the English Channel and Irish Sea, the total distance was approximated as 540 nm (1000 km). The initial target speed for the vessel was set at 14 knots (7.2 m/s), with assumed times of 2 hours for "port stay" and 30 minutes for "port idle". The route builder model was used (see sections 3.2 and 3.3) and the resulting power profile is shown in Figure 3.

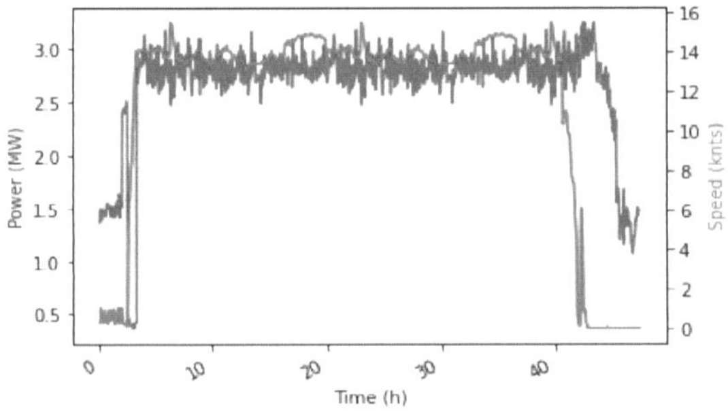

Figure 3: Power (blue) and speed (red) plot by time for Southampton to Liverpool route with a target speed of 14 knots.

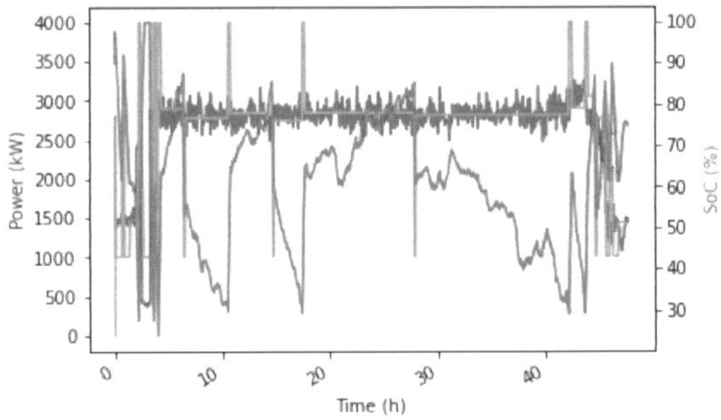

Figure 4: Simulated 4 MW PEM fuel cell output (orange) and battery state of charge (red) compared to energy demand for Southampton to Liverpool route with a target speed of 14 knots.

The power profile in Figure 3 is comprised of both propulsion demand and hotel load. Several different fuel cell and battery system setups were tested, and the smallest system found to capable of meeting all of the energy demand was comprised of: a 4 MW PEM fuel cell stack; ammonia cracker; 1 MW diesel generator (to power cracker); 800 kWh batteries. Results of the simulated fuel cell output are shown in Figure 4.

Where the fuel cell output exceeded the power demand, this energy was used to charge the batteries. Thereafter, when the power demand was higher than the fuel cell output, the batteries were used to meet these peaks. As the batteries are constantly being charged or discharged, it was necessary to model the SoC to ensure there was enough capacity remaining to handle these chances in load, shown in Figure 4. Key results are shown in Table 2.

5.2 SCENARIO 2: SOUTHAMPTON TO LIVERPOOL AT 12 KNOTS

The dynamic energy system model developed in this study is a tool that analyse the implications of several different changes in the energy management, this not only includes changing the technology setup, but also the operating conditions the system is subjected to. To best understand the impact of an individual change to the concept of operation, it is necessary to only change one variable and keep all other conditions the same. In "Scenario 2" all variables, such as the route and the in-port durations, have remained constant except from the target sailing speed that has been reduced from 14 knots to 12 knots. The resulting power profile is shown in Figure 5.

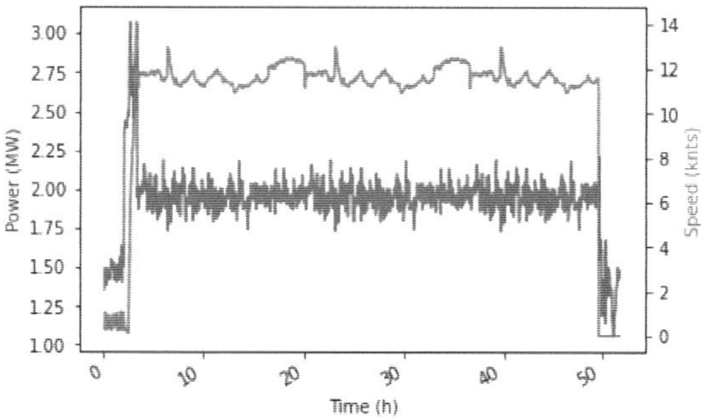

Figure 5: Power (blue) and speed (red) plot by time for Southampton to Liverpool route with a target speed of 12 knots.

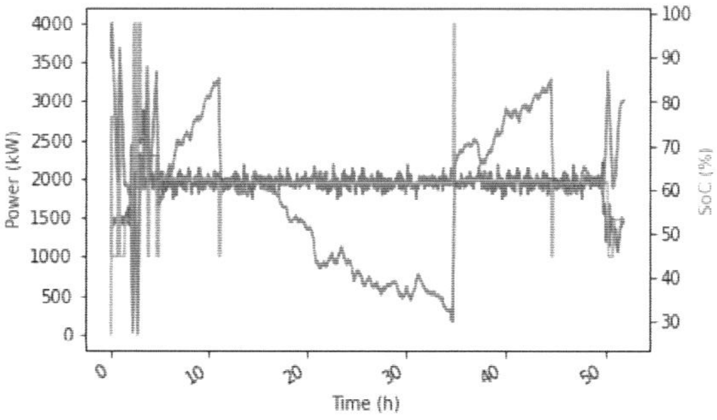

Figure 6: Simulated 4 MW PEM fuel cell output (orange) and battery state of charge (red) compared to energy demand for Southampton to Liverpool route with a target speed of 12 knots.

Comparing Figure 5 to Figure 3 shows that a relatively small change (2 knots) has caused a significant reduction to power demand whilst sailing, reducing the total energy demand by 20.6% (see Table 2). A simulation was run to meet this power demand with the same setup as in Scenario 1 (4 MW fuel cells with 800 kWh batteries), results are shown in Figure 6.

It is observable from Figure 6 that the fuel cells operate at a constant output for longer periods that in Figure 4. Also, the fuel cell produces approximately 50% of its maximum power output for most of the duration, this is within the power range that a PEM fuel cell is most efficient, hence reducing overall fuel consumption. A full comparison of the key results from Scenarios 1 and 2 simulations is shown in Table 2.

Table 2: Key details for the cargo voyages from Southampton to Liverpool.

Key details	Scenario 1	Scenario 2
Total energy [MWh]	126	100
Distance [nm]	548	548
Time [h]	47.5	51.8
Target speed [knots]	14	12
Actual average speed [knots]	11.5	10.6
Hydrogen consumption [kg]	7,371	5,423
Ammonia consumption [kg]	52,436	38,581
Diesel consumption [kg]	6,309	4,994

The total voyage energy demand has reduced by 20.6% but, due to the more favourable fuel cell operating conditions resulting from the power profile changes, the fuel consumption has reduced by a larger percentage of 26.4% and 20.8% for ammonia and diesel respectively. Furthermore, this efficiency gain has been achieved with only a 4.3 hour (9%) increase to transit duration.

Alternative fuels, such as ammonia or hydrogen, are likely to require more storage capacity and, at least in the short term, be more expensive to run that fossil fuels. Therefore, it will become more valuable to be able to accurately predict fuel consumption and to find methods of improving efficiency, such as in shown here.

5.3 SCENARIO 3: SURVEY

The final scenario tested in this study was designed to simulate the typical operation of a survey mission. This was modelled as the compilation of the following modes:

- 2 hours in port idle state
- 3 days travelling at 10 kts
- 9 days travelling at max 8 kts (close to the survey site)
- 12 days undergoing surveying mission with 5 days' worth of ROV deployment (as described in section 3.2(f))

The resultant power profile is shown in Figure 7. The total duration for the survey mission was 521 hours (21.7 days), during this period the total transit distance was 2,046 nm and the energy demand totalled 522 MWh.

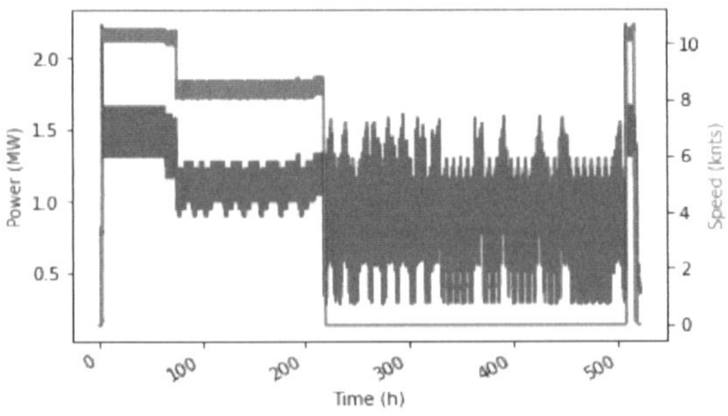

Figure 7: Power (blue) and speed (red) by time plot for modelled "typical" survey.

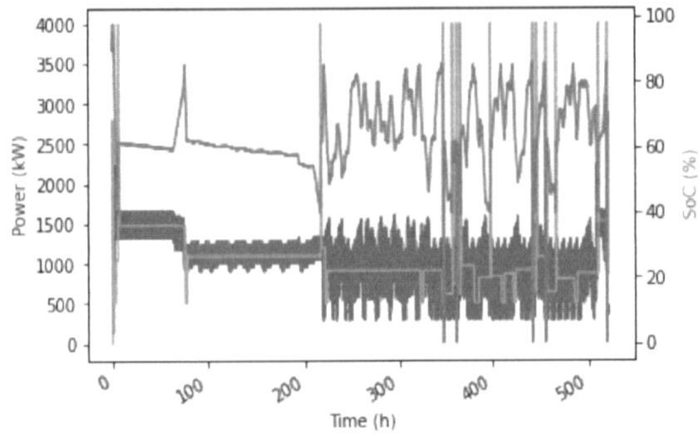

Figure 8: Simulated fuel cell output (orange) and battery state of charge (red) compared to energy demand for modelled "typical" survey for a setup of 4 MW PEM fuel cells and 5.5 MWh Li-ion batteries.

This survey mission has a considerably different operating profile to the cargo voyages tested in scenarios 1 and 2, however it was found that the same fuel cell capacity (4 MW PEM) would be sufficient to meet all of the demand, including propulsion and all major surveying loads. The main challenge was sizing of the battery system, due to the large fluctuations in demand during the surveying mode. Simulations have been repeated by increasing the battery capacity until all energy demand could be met with zero excess energy (where fuel cell output exceeds the demand, but the batteries are fully charged). Due to the large fluctuations in demand, the minimum battery capacity capable of delivering this was 5.5 MWh, results of this simulation are shown in Figure 8. Fuel consumption results are shown in Table 3.

For maritime applications, batteries are known to be relatively large, heavy, and expensive compared to other propulsion sources [3]. Therefore, it would be preferable to minimise the battery contribution to a fuel cell and battery hybrid. The model developed in this project can be used to test changes to operational patterns and energy management systems to minimise the battery requirements and further optimise the system.

5.4 FUEL STORAGE METHODS COMPARISON

To assess the suitability of this study's propulsion concept (ammonia cracked to feed PEM fuel cells) an additional concept has been developed and tested for these three scenarios. For this new concept, pure hydrogen (stored as a liquid) would be the primary fuel source and, therefore, could feed the PEM fuel cells without the need for additional stages, such as cracking and purification. As the PEM fuel cells still provide the main source of electricity, the battery requirements would be the same.

Table 3: Consumption values of each fuel type for the separate case studies. Densities assumed: 71 kg/m^3 for liquid hydrogen; 603 kg/m^3 for ammonia; 900 kg/m^3 for diesel (MDO).

	Ammonia concept						Liquid hydrogen concept	
	Fuel mass			Fuel volume			Fuel mass	Fuel volume
Scenario	NH_3 [kg]	MDO [kg]	Total [kg]	NH_3 [m^3]	MDO [m^3]	Total [m^3]	H_2 [kg]	Liquid H_2 [m^3]
1	52,436	6,309	58,745	87	7	94	7,371	104
2	38,581	4,994	43,575	64	6	70	5,423	76
3	198,375	27,379	225,754	329	30	359	27,886	393

Table 3 shows that the liquid hydrogen concept causes an 87 to 88% reduction in the required fuel mass and an increase of 8 to 9% to fuel volume requirements. However, it is important to, not only to consider the requirements of the fuel itself but, also the system infrastructure requirement.

Table 4: Mass and volume requirements for the total energy storage system (before fuel cells) for the two concepts considered. Calculations are based on typical equipment specifications.

	Ammonia concept								Liquid hydrogen concept	
	System mass				System volume				System mass	System vol
	NH_3 fuel + tank	Cracker system	MDO fuel + tank + genset	Total	NH_3 tank	Cracker system	MDO tank + genset	Total	H_2 fuel + tank	Liquid H_2 tank
Scenario	[t]	[t]	[t]	[t]	[m^3]	[m^3]	[m^3]	[m^3]	[t]	[m^3]
1	54.3	82	28.8	165	96	470	47	613	50	316
2	40.7	82	27.2	150	71	470	45	586	35	222
3	206.7	82	55.7	344	359	470	93	922	172	994

The results in Table 4 show that selection of the liquid hydrogen system would have significant weight and space savings for the cargo scenarios. For scenario 1, the pre-fuel-cell system's mass is reduced by 70% and volume by 48%, for scenario 2 these values are 77% and 62% respectively. For the survey vessel, there is still a significant mass reduction (50%), although the volume requirements for the liquid hydrogen system in this case were 8% larger. This shows that for smaller energy applications, the liquid hydrogen is clearly more feasible in terms of design requirements. As the energy demand begins to increase, the ammonia system may have a marginal volumetric advantage, although the additional weight requirements may negate these benefits. Additionally, as the liquid hydrogen concept requires no cracking or purification, there is no need for a diesel generator, hence this concept and be considered zero emission from tank to wake.

7. CONCLUSIONS

In this study power profiles have been developed for three scenarios. For the example cargo voyage of Southampton to Liverpool, results showed that reducing the average sailing speed by two knots caused fuel consumption savings of 26.4% yet only extended the duration by 4 hours 21 minutes (9.2%). This demonstrates the value of the dynamic modelling system to accurately calculate the trade-offs between the two approaches.

For the surveying mission, large fluctuations in load are expected, especially during dynamic positioning. Results showed that 5.5 MWh of battery storage would be required to meet peaks in demand. By comparison, only 0.8 MWh of battery capacity is required for the first two case studies for the same FC size of 4 MW. The volume and mass of the additional energy requirements is significant. As such, saving weight and space when it comes to the fuelling system, would be valuable to achieve better overall efficiency.

One of the main objectives for this study was to assess the feasibility of a specific powertrain which used ammonia as the primary fuel source and used PEM fuel cells to generate electricity. This also required a hydrogen cracker and purifier (powered by a diesel generator). For comparison, a second concept was introduced, which would use the same FC-battery hybrid, but used cryogenically cooled liquid hydrogen as the only fuel. For the scenarios tested (2 cargo and 1 survey), the volume of fuel required for ammonia and diesel combined, is 8 to 9% less than that required for liquid hydrogen. When considering the volume of the extra equipment for the ammonia fuel case (including an ammonia cracker, two storage tanks rather than one, a diesel generator, and extra piping) it was shown that the volume required for the liquid hydrogen system was actually 48% and 62% lower for the respective cargo vessel scenarios. The ammonia setup fared marginally better (8%) in terms of volume required for the survey scenario. However, there are additional advantages of the liquid hydrogen, for example this simpler system requires no diesel generator and hence produces zero emissions from tank-to-wake. Furthermore, hydrogen is non-toxic (ammonia is a highly toxic) and the system is between 50% and 77% lighter, making it a more suitable fuel option for all scenarios tested.

8. ACKNOWLEDGEMENTS

This research was conducted as part of the Clean Maritime Demonstration Competition (competition code 2103_DFT_CMD_S1), funded by the Department for Transport and delivered in partnership with Innovate UK.

The authors gratefully acknowledge financial support from the UK Engineering and Physical Sciences Research Council (EPSRC) through the Centre for Doctoral Training in Energy Storage and its Applications (Grant No. EP/L016818/1).

9. REFERENCES

1. Marine Environmental Protection Committee, "Fourth IMO GHG Study," International Maritime Organisation, 2020.

2. Department for Transport, "Clean Maritime Plan," 2019.
3. C. J. McKinlay, S. R. Turnock, and D. A. Hudson, "A Comparison of Hydrogen and Ammonia for Future Long Distance Shipping Fuels," *LNG/LPG and Alternative Fuels by the Royal Institute of Naval Architects,* 30th January 2020 2020.

4. J. E. Buckingham, "Future Fuels for Commercial Shipping," presented at the LNG/LPG and alternative fuels, London, 2020.

5. M.-C. Chiong *et al.*, "Challenges and opportunities of marine propulsion with alternative fuels," *Renewable and Sustainable Energy Reviews,* vol. 149, p. 111397, 2021.

6. C. J. McKinlay, S. R. Turnock, and D. A. Hudson, "Route to zero emission shipping: hydrogen, ammonia or methanol?," *International Journal of Hydrogen Energy,* vol. 46, pp. 28282 - 28297, 2021, doi: https://doi.org/10.1016/j.ijhydene.2021.06.066.

7. F. A. Uribe, S. Gottesfeld, and T. A. Zawodzinski, "Effect of ammonia as potential fuel impurity on proton exchange membrane fuel cell performance," *Journal of the Electrochemical Society,* vol. 149, no. 3, p. A293, 2002.

8. S. Giddey, S. Badwal, C. Munnings, and M. Dolan, "Ammonia as a renewable energy transportation media," *ACS Sustainable Chemistry & Engineering,* vol. 5, no. 11, pp. 10231-10239, 2017.

9. "Ammonia as a marine fuel," in *DNV GL, Group Technology & Research*: White Paper 2020, 2020.

10. T. S. Valentin Scharl, "Ignition and combustion characteristics of diesel piloted ammonia injections," in *Fuel Communications*, vol. 11: Elsevier, 2022.

11. K. S. Reiter AJ, "Diesel engine operation using ammonia as a Carbon free fuel source," in *ASME 2010 Internal Combustion Engine Division Fall Technical Conference*, Texas, USA, 2011.

12. K. S. Reiter AJ, "Combustion and emissions characteristics of compression-ignition engine using dual ammonia-diesel fuel," in *Fuel*: Elsevier, 2011, pp. 87-97.

13. Y. Niki, Y. Nitta, H. Sekiguchi, K. Hirata, "Emission and combustion characteristics of diesel engine fumigated with ammonia," in *Proceedings of the ASME internal combustion engine fall technical conference*. New York: The American Society of Mechanical Engineers, 2019.

14. "U.S. E.P.A. : Overview of Greenhouse Gases," ed, 2022.

15. N. de Vries, "Safe and effective application of ammonia as a marine fuel," Delft University of Technology C-Job, 2019.

16. M. Hirscher, "Handbook of hydrogen storage," *Topics in applied physics,* vol. 12, 2010.

17. S. C. E. Tsang, y a , an i a denho en, he o i ion o a onia in de a oni ing a i i e indu y an overview and perspective: part II, Costs, safety and environmental performance and the future prospects for ammonia in shipping," in *Johnson Matthey Technology Review*: Johnson Matthey, 2021, pp. 291-300.

18. B. Z. Omer Berkehan Inal, Cengiz Deniz, "Investigation on the decarbonization of shipping: An approach to hydrogen and ammonia," in *International Journal of Hydrogen Energy*: Elsevier, 2022, pp. 19888-19900.

19. "Poisoning of proton exchange membrane fuel cells by contaminants and impurities: review of mechanisms, effects, and mitigation strategies," in *J Power Sources*, 2019, pp. 21-48.

20. "Direct ammonia solid-oxide fuel cells: A review of progress and prospects," in *International Journal of Hydrogen Energy*, 2021, pp. 35365-35384.

21. H. M. Adam Wojcik, Ioannis Damopoulos, JanVan herle, "Ammonia as a fuel in solid oxide fuel cells," in *Journal of Power Sources*: Elsevier, 2003, pp. 342-348.

22. "Miniature NH3 cracker based on microfibrous entrapped Ni-CeO2/Al2O3 catalyst monolith for portable fuel cell power supplies," in *International Journal of Hydrogen Energy* Elsevier, 2009, pp. 1710-1716.

23. "Novel microfibrous composite bed reactor: high efficiency H2 production from NH3 with potential for portable fuel cell power supplies," *Lab on a Chip,* no. 1, pp. 133-140, 2007.

24. "NASA Climate," ed: NASA, 2016.

25. S. Y. B. Muhammad Usman, "Thermodynamic assessment of chemical looping combustion and solar thermal methane cracking-based integrated system for green ammonia production," in *Thermal Science and Engineering Progress*. Doha, Qatar: Elsevier, 2020.

26. "Ammonia: zero carbon fertiliser, fuel and energy store," London, UK, 2020.

27. S. H. Jasper Faber, Shuang Zhang, Paula Pereda, Bryan Comer, Elena Hauerhof, Wendela Schim van der Loeff , Tristan Smith, Yan Zhang, Hiroyuko Kosaka, Masaki Adachi, Jean-Marc Bonello, Connor Galbraith, Ziheng Gong, Koichi Hirata, David Hummel, "Fourth IMO GHG Study 2020: Executive Summary," IMO, London, UK, 2021.

28. G. Jeerh, M. Zhang, and S. Tao, "Recent progress in ammonia fuel cells and their potential applications," *Journal of Materials Chemistry A,* vol. 9, no. 2, pp. 727-752, 2021.

29. C. McKinlay, S. Turnock, and D. Hudson, "Fuel cells for shipping: To meet on-board auxiliary demand and reduce emissions," *Energy Reports,* vol. 7, pp. 63-70, 2021.

30. T. Leo, J. Durango, and E. Navarro, "Exergy analysis of PEM fuel cells for marine applications," *Energy,* vol. 35, no. 2, pp. 1164-1171, 2010.

31. S. Sutulo and C. Guedes Soares, "Mathematical models for simulation of manoeuvring performance of ships," *Marine technology and engineering,* pp. 661-698, 2011.

32. "A78 Power System Summary," Ocean Infinity, Southampton, 2022.

33. T. Brown, " Ammonia Energy Association." [Online]. Available: https://www.ammoniaenergy.org/articles/ammonia-cracking-to-high-purity-hydrogen-for-pem-fuel-cells-in-d

34. O. G. I. (OXGRIN), "Low temperature Ammonia Cracker," ed. Oxford, Uk, Oxford, 2022.

35. H. A. R. CENTER, "Hydrogen Tools: Hydrogen Enthalpy at different temperatures and pressures." [Online]. Available: https://h2tools.org/hyarc/hydrogen-data/hydrogen-enthalpy-different-temperatures-and-pressures

10. AUTHORS BIOGRAPHY

Charles McKinlay is a PhD researcher at the University of Southampton, as part of the CDT in Energy Storage and its Applications programme. His research is focused on using dynamic energy system modelling to demonstrate a viable pathway to zero emission shipping. He recently collaborated on 3 Clean Maritime Demonstration Competition projects. His background also includes Mathematics and Energy Engineering.

Panagiotis Manias is a research technician at the University of Southampton from which he graduated with master's in mechanical engineering. He initially took part in a concept design project of a ballast free liquid hydrogen carrier where he developed the powertrain for that particular vessel and was later involved in the CMDC projects, as part of the energy and emissions modelling. Earlier to this, some industry experience in the technical department of a company was gained, including drydock maintenance, as part of summer jobs.

Professor Stephen Turnock, FRINA is Head of Civil, Maritime and Environmental Engineering at the University of Southampton. He has long standing interests in decarbonisation of shipping and in particular analysis and modelling of alternative power trains including the use of hybrid battery systems.

Professor Dominic Hudson is the Shell Professor of Ship Safety and Efficiency at the University of Southampton. He has research interests in all aspects of ship hydrodynamics, particularly ship design and operation for energy efficiency. He has published over 140 papers in journals and proceedings of refereed conferences. He is a Chartered Engineer and member of RINA. He is presently a member of the 29th ITTC Specialist Committee on Ships in Operation at Sea and the RINA IMO Committee.

DEVELOPMENT OF A PIPE FABRICATION PROCESS DETERMINATION SYSTEM USING GRAPH DATABASE AND PROCESS SIMULATION

Y Mihara and **T Mitsuyuki**, Yokohama National University, Japan

SUMMARY

As the importance of piping in shipbuilding increases, streamlining pipe fabrication processes has become one of the most critical issues. However, since pipe fabrication processes are complicated, it has not been adequately digitized and modeled. This paper proposed a system to automatically determine the pipe fabrication process considering the design of the target pipe and various information in a factory. In the developed system, tasks are generated using the information on the parts, and these relationships are registered in a graph database. This system also generates all conceivable fabrication processes by allocating workplaces to each extracted task. The optimal fabrication process is selected by executing the production simulation using each fabrication process which is generated from the information on the graph database and target factory. We will also show the result that the developed system is applied to three virtual pipe fabrication factories.

1. INTRODUCTION

As the importance of piping outfitting in shipbuilding increases, the rationalization of pipe fabrication has become one of the most critical issues. With the recent introduction of digital twin systems in shipyards, the digitization of the fabrication process, which is the input for the system, is required to apply such systems to pipe fabrication factories. However, the pipe fabrication process has not been adequately digitized until now. It is challenging to create data for the pipe fabrication of each product, in which components are processed and assembled under complex flows in a factory, because of the excessive labor required for manual operations and the complexity of the procedure of the process determination to automate it.

Along with the increasing adoption of digital technology in shipyards, modeling and simulation of the shipbuilding process have been developed [1][2], and automatic production planning systems using simulation are evaluated for their practicality. Okubo et al. applied the developed system to constructing a hull block in a virtual shipyard to evaluate its practicality [3]. Lee et al. applied their system to a real shipyard production plan and quantitatively evaluated the resulting improvement compared to a conventional plan [4]. However, data management technology in shipyards is still under research, and one of the issues is the collection and preparation of input data for these systems. On the other hand, in fields aside from shipbuilding, methods for generating and determining production processes have been proposed for processing or assembling. Koguchi et al. used a connectivity graph for sheet metal parts and a process generation tree to automate process planning for sheet metal bending with a handling robot [5][6]. Sae-Tang et al. used a precedence matrix and a corresponding directed graph to represent precedence constraints among operations in determining the operation sequence to minimize the number of tool changes [7]. Ando et al. proposed a manufacturing feature recognition method for multiple process plan generation using the Extended Super Relation Graph (Extended SRG) method and the FSR method [8]. Kasahara et al. generated a network of semi-products and solved it as the shortest path problem to obtain the machining sequence that gives the shortest machining time [9]. Ichikizaki et al. proposed an exhaustive derivation method for feasible intermediate products, including intermediate product constraint, consecution constraint, and precedence constraint [10]. Zhu et al. described a CAD model using an Attributed Adjacency Graph (AAG) and defined machining features using the AAG in an automatic process planning system for a multitasking machine [11]. However, no system can be applied directly to determining complex tree-type processing and assembling processes, such as pipe fabrication processes in which individual fabrication procedures and factory designs are considered.

Therefore, we developed a system to automatically determine pipe fabrication processes considering the target factory's situation. The system first generates all candidate fabrication processes from information on parts and their connection relationships and registers them in a graph database. Then, the developed system selects the optimal fabrication process by conducting a production process simulation using the information on each process and the target factory.

This paper presents an overview of the developed system and reports the case study of applying the developed system to the virtual pipe fabrication process.

2. DEVELOPED SYSTEM

The purpose of the developed system is to automatically determine the optimal Bill of Process (BOP) from product information through simulation-based evaluation for pipe fabrication. BOP is fabrication process information that represents process information combined with resource information such as facility. In this paper, BOP consists of a task network and some information about each task. The five components of BOP in this system are shown below.

- Targeted parts of each task
- Content of each task
- A workplace where each task is performed
- Working time of each task
- Candidate relationships among tasks

"Content of each task" is what to do in the task, for example, "cutting tube" or "tube-flange welding", or it with more detailed information. "Candidate relationships among tasks" means which tasks will be predecessor or successor of the target task.

2.1 SYSTEM OVERVIEW

Figure 1 shows an overview of the developed system. The input information to the system consists of product information like drawings, factory information, and workers' information. First, the developed system extracts candidate draft BOPs comprehensively from the product information on the target product and the factory information on the target factory. Next, create the data model from each draft BOP. Finally, the developed system conducts the production process simulation using each data model. We can determine the optimal BOP by evaluating the results of the production process simulations. Note that a graph database is used to extract the draft BOPs and create the data models.

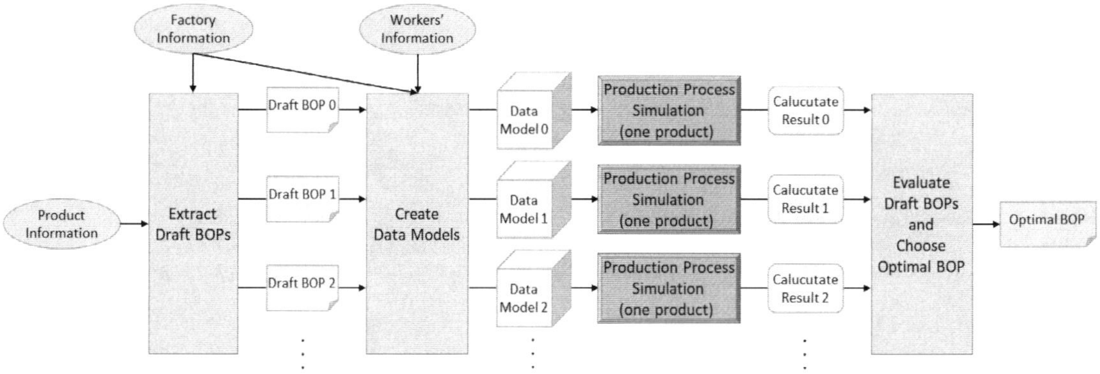

Figure 1. Overview of the developed system

2.2 INPUT INFORMATION

There are three types of input information to the system: product information, factory information, and workers' information.

2.2.1 Product Information

Product information corresponds to the information described in the product design drawings. The contents are shown below.

- Product name
- Part information
- Connection relationship information between parts
- Positional relationship information between parts

"Part information" includes information on the part type (TUBE, FLANGE, ELBOW, etc.), diameter, thickness, material, cut length (tube only), bending times (tube only), and part ID. "Connection relationship information between parts" includes the part ID of the part to which each part is connected, whether the connection is welded or only tack-welded, and information on the connection angle. "Connection relationship information between parts" shows a part to which the part connects. In the case of branch connections, it also includes information on the main tube to its branch tube relationship. "Positional relationship information between parts" is information on the length between points of the product. It consists of the maximum distance from each weld line to each part (positive in the flow direction) and the rotation radius of each part for each weld line. Figure 2 shows "positional relationship information between parts" about part B and weld line (a) as an example.

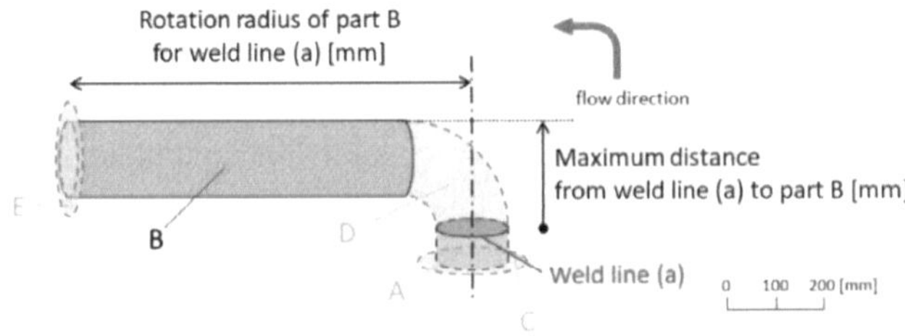

Figure 2. Positional relationship information between parts

2.2.2 Factory Information

Factory information includes information about the target factory's workplace layout, placed facility, and tasks to be performed in the target factory. The contents are as follows.

- Condition of task occurrence
- Name of workplace
- Relationship between workplaces
- Content of tasks that can be performed at each workplace
- Working time at each workplace
- Placed facility
- Constrain by component size limit of each facility

"Condition of task occurrence" is the conditions under which tasks are generated by the factory in response to product information. "One tube cutting task is generated for one tube part" and "one tube-flange tack-welding task and one tube-flange welding task are generated for one tube-flange welding connection" are examples of "condition of task occurrence". "Relationship between workplaces" indicates the upper or downer stream of components flow in the workplaces. "Content of tasks that can be performed at each workplace" is, for example, "tube-flange tack-welding (diameter: 200A, thickness: 8.2 mm)", "bracket welding (weld-length: 50 mm)" or "bender bending (1 point, diameter: 150A)". This paper assumes that "working time at each workplace" can be different for the same content tasks at different workplaces because the placed facilities differ from workplace to workplace. "Constrain by component size limit of each facility" is information on the maximum length and the maximum rotation radius of the assembly which can be mounted on the facility assuming rotating facilities such as positioners.

2.2.3 Workers' Information

Workers' information consists of two components: "name of worker" and "name of facilities each worker can handle".

2.3 EXTRACTION OF DRAFT BOPs

2.3.1 Procedure for Extracting Draft BOPs

The procedure for extracting draft BOPs is shown below.

1. Register "part information" and "connection relationship information between parts" in the graph database.
2. Search the graph database and generate tasks.
3. Combine workplace assignments for the tasks.
4. Eliminate the combinations based on "constrain by component size limit of each facility".
5. Assign working time information to the tasks.

As a simple example, the following is a description of the procedure for extracting draft BOPs for a product, which consists of three parts (tube, flange, and elbow) as shown in Figure 3 (a), in a pipe factory shown in Figure 4.

First, from the product information, the "part information" and the "connection relationship information between parts" are formatted into a graph as shown in Figure 3 (b) and registered in the graph database. Each node of the graph is a part of the product, and the properties of the nodes correspond to the "part information" of the product information. The relations of the graph correspond to the "connection relationship information between parts" of the product information.

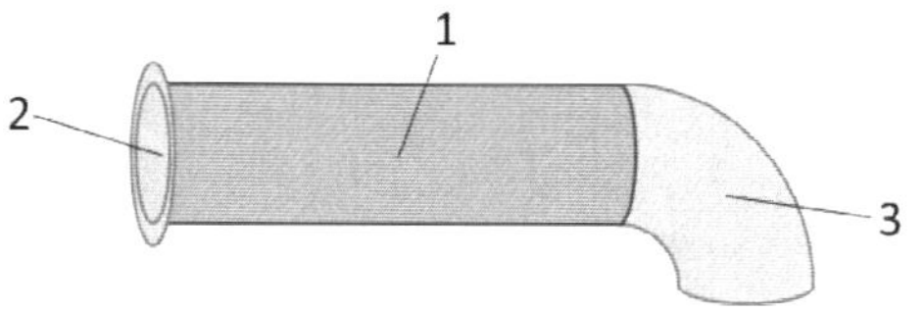

Part Information

ID	Type	Diameter	Thickness	Material	Length	Bending Times
1	TUBE	300A	10.3	STPG370	1100	0
2	FLANGE	300A	-	-	-	-
3	ELBOW	300A	10.3	STPG370	-	-

(a) Drawing and part information

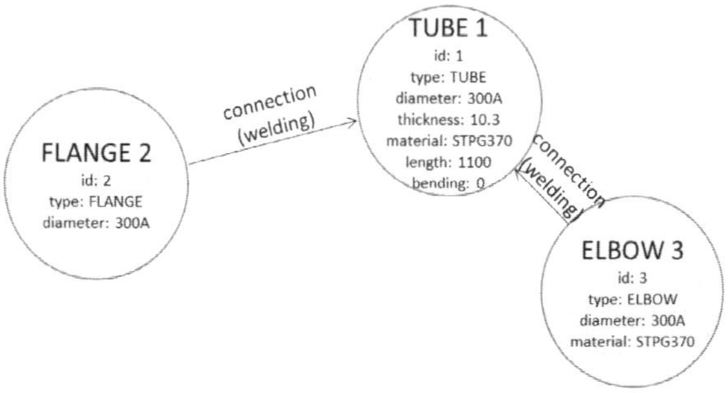

(b) Graph data

Figure 3. Example product

Work Content	Dia-meter	Thick-ness	Material	Work-ing Time
Tube-Flange Tack-welding	200A	8.2	STPG370	3
Tube-Flange Tack-welding	300A	10.3	STPG370	4
Tube-Elbow Tack-welding	200A	8.2	STPG370	12
Tube-Elbow Tack-welding	300A	10.3	STPG370	15

Work Content	Dia-meter	Thick-ness	Material	Work-ing Time
Tube-Flange Tack-welding	300A	10.3	STPG370	10
Tube-Elbow Tack-welding	200A	8.2	STPG370	18
Tube-Elbow Tack-welding	300A	10.3	STPG370	20

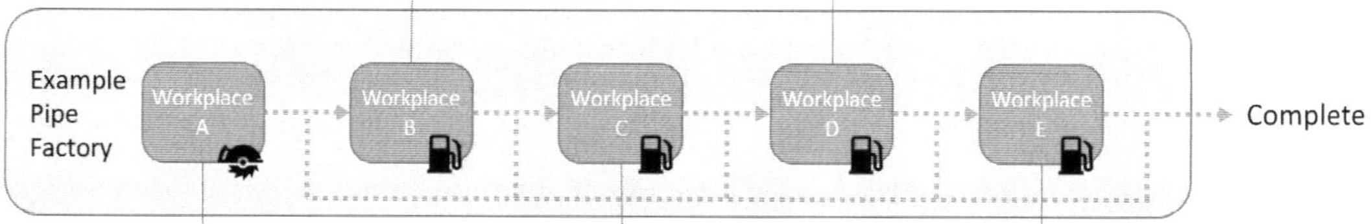

Work Content	Dia-meter	Thick-ness	Material	Work-ing Time
Tube Cutting	200A	8.2	STPG370	5
Tube Cutting	300A	10.3	STPG370	6

Work Content	Dia-meter	Thick-ness	Material	Work-ing Time
Tube-Flange Welding	200A	8.2	STPG370	20
Tube-Flange Welding	300A	10.3	STPG370	30

Work Content	Dia-meter	Thick-ness	Material	Work-ing Time
Tube-Flange Welding	300A	10.3	STPG370	25
Tube-Elbow Welding	200A	8.2	STPG370	30
Tube-Elbow Welding	300A	10.3	STPG370	40

Figure 4. Example pipe factory

Next, based on the "condition of task occurrence" of the factory information, the created graph data is searched to detect one tube part, one tube-flange welding connection, and one tube-elbow welding connection. A tube cutting task is generated for a tube part, and a tack-welding task and a welding task are generated for each welding connection. The generated tasks are assigned the targeted parts' ID, work content, diameter, thickness, and material information from the properties of the part nodes.

Then, the workplaces are combined with the generated tasks using the "content of tasks that can be performed at each workplace" of the factory information. The tube cutting task can be performed only at Workplace A. Therefore, Workplace A is allocated for the tube cutting task for part 1. In addition, the tube-flange tack-welding task with a diameter of 300A, a thickness of 10.3 mm, and material of STPG 370 can be performed at Workplace B or D, and such tube-flange welding task can be performed at Workplace C or E. Accordingly, considering that the welding task is performed after the tack-welding task, the combinations of workplaces for performing the tack-welding and welding tasks for the 1-2 connection are three patterns: Workplace B and C, B and E, and D and E. Similarly, the combinations of workplaces for performing the tube-elbow tack-welding and welding tasks for the 1-3 connection are two patterns: Workplace B-E and D-E. Consequently, there are six feasible combinations of workplaces in this example.

Since there is no "constraint by component size limit of each facility" in the example pipe factory, step 4 is skipped.

Finally, for each workplace combination, working time information is assigned to each task using the "working time at each workplace" of the factory information.

From the above, six draft BOPs are extracted in this example. Table1 shows those (i) ~ (vi) draft BOPs. The number in each cell in Table 1 means the part IDs of parts targeted at the workplace, and the values given in parentheses are the working time [minute] of the tasks.

Table 1. Extracted draft BOPs

	Draft BOP					
Workplace	i	ii	iii	iv	v	vi
A (cutting)	1 (6)	1 (6)	1 (6)	1 (6)	1 (6)	1 (6)
B (tack-welding)	1-2 (4) 1-3 (15)	1-2 (4)	1-2 (4) 1-3 (15)	1-2 (4)	1-3 (15)	-
C (welding)	1-2 (30)	1-2 (30)	-	--		-
D (tack-welding)	-	1-3 (20)	-	1-3 (20)	1-2 (10)	1-2 (10) 1-3 (20)
E (welding)	1-3 (40)	1-3 (40)	1-2 (20) 1-3 (40)	1-2 (20) 1-3 (40)	1-2 (20) 1-3 (40)	1-2 (20) 1-3 (40)

2.3.2 Eliminate Combinations based on "Constrain by Component Size Limit of Each Facility"

When a rotating facility such as a positioner and an automatic welding machine at a workplace has a "constrain by component size limit of each facility", it is necessary to judge whether the assembly that will be allocated at the workplace satisfies the constrain.

Whether "constrain by component size limit of each facility" is satisfied is judged by deducing the parts which are the members of the target assembly and calculating the length and rotating radius of the assembly for each weld line welded on the facility. Deducing the parts consisting of the assembly is easy because the relationship between the workplaces is unidirectional. The length of the assembly is calculated by the difference between the maximum and minimum distances from the target weld line to the parts consisting of the assembly using the "positional relationship information between parts" of the product information. The rotating radius of the assembly is the maximum rotation radius of the parts consisting of the assembly for the weld line.

As an example, judge the weldability of the A-D weld line (a) of the assembly consisting of flange C, tube A, elbow D, and tube B of the product shown in Figure 2, using the facility at Workplace Z, which the length limit is 1000 mm, and the rotating radius limit is 300 mm. Since the maximum distance from the weld line (a) to elbow D (or tube B) is +215 mm and the maximum distance from the weld line (a) to flange C is -66 mm, the length of the assembly is 281 mm. Thus, the assembly satisfies the length limit of the facility at Workplace Z. The rotating radius of the assembly is 835 mm since the rotating radius to tube B for weld line (a) is 835 mm. Therefore, the assembly does not satisfy the rotation radius limit of the facility at Workplace Z. So, it is judged that the assembly cannot be mounted on the facility in the direction to weld the weld line (a) at Workplace Z.

2.4 EVALUATION OF DRAFT BOPs AND BOP DETERMINATION USING SIMULATION

The developed system creates the data model from each draft BOP extracted in the previous section and executes the production process simulation targeting only one product to evaluate the draft BOPs. For data modeling and production process simulation using the draft BOP and factory and workers' information, we use the same method of Okubo et al. [3]. The developed system evaluates the draft BOPs by multi-objective optimization using the following indicators.

- The sum of the working time of the tasks is small.
- The production time obtained from the simulation result is small.
- The use of workplaces that tend to become bottlenecks is small.

3. CASE STUDY

The developed system was applied to virtual pipe factories and a pipe product. The virtual pipe factories are one baseline factory (Pipe Factory I) and two factories (Pipe Factory II and III) that differ in some settings from the baseline factory. Figure 5 shows the layout of the workplaces, facilities, and workers in the pipe factories. Table 2 shows the working time [minute] for each content of the task at the workplace in each pipe factory. Pipe Factory II has the same arrangement of workplaces and workers and the same number of facilities placed at each workplace as in Pipe Factory I, but the performance of the facilities differs from that of Pipe Factory I. In Pipe Factory II, the cutting machines' performance is lower, and each welding machine's performance is higher. In other words, the working time in Pipe Factory II differs from in Pipe Factory I. The working time for the pipe cutting tasks at Workplace A is 50 % longer than in Pipe Factory I. The working time for the assembling and welding tasks at Workplace C to F is about 27% (19 ~ 44%, depending on the task) shorter than in Pipe Factory I. The reason why the shortening rate of the working time depends on the tasks is that the time required to transport the assembly and prepare for welding, which is included in the task, is not shortened by the welding machine's performance, so the shortening rate becomes higher as the baseline working time is longer. In Pipe Factory III, the arrangement of workplaces and workers and the performance of the facilities placed at each workplace are the same as in Pipe Factory I, and the working time for each task is also the same. But the number of welding machines placed at Workplace C or D is one each, compared to two each in Pipe Factory I.

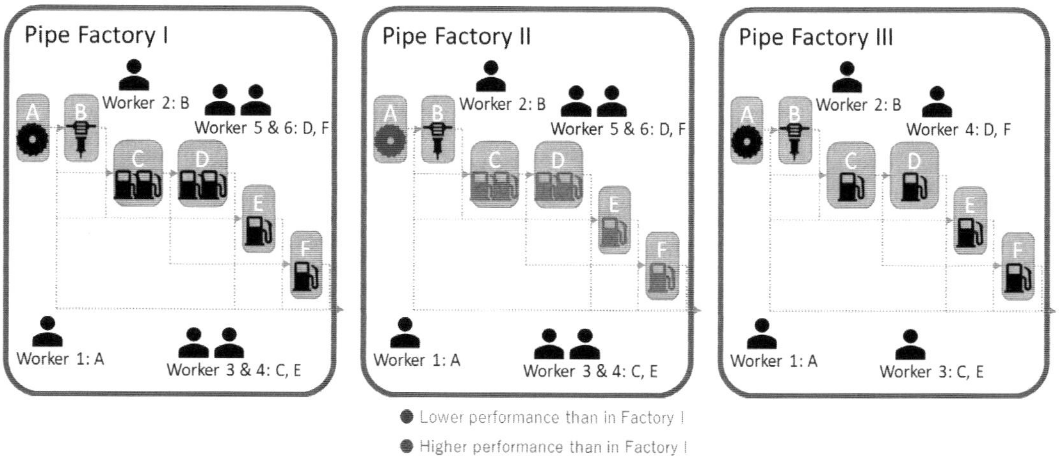

(a) Pipe Factory I (b) Pipe Factory II (c) Pipe Factory III

Figure 5. Pipe Factories

Table 2. Working time in Pipe Factory I and II

Workplace	Content of task		Pipe Factory		
	Work content	Diameter	I	II	III
A	Tube cutting	300A	10	15	10
A	Tube cutting	150A	10	15	10
B	Branch drilling	300A	22	22	22
B	Branch cutting	150A	17	17	17
C	Tube-flange tack-welding	300A	12	9	12
C	Tube-flange tack-welding	150A	12	9	12
C	Tube-elbow tack-welding	300A	15	10	15
C	Branch tack-welding	300A-150A	30	17	30
D	Tube-flange welding	300A	25	19	25
D	Tube-flange welding	150A	21	17	21
D	Tube-elbow welding	300A	25	19	25
D	Branch welding	300A-150A	40	30	40
E	Tube-flange tack-welding	300A	12	9	12
E	Tube-flange tack-welding	150A	12	9	12
E	Tube-elbow tack-welding	300A	15	10	15
E	Branch tack-welding	300A-150A	30	17	30
F	Tube-flange welding	300A	25	19	25
F	Tube-flange welding	150A	21	17	21
F	Tube-elbow welding	300A	25	19	25
F	Branch welding	300A-150A	40	30	40

In this chapter, we determine the optimal BOP for Product X of each pipe factory and compare the differences in the optimal BOPs. Figure 6 (a) shows Product X, and Figure 6 (b) shows its graph data. Note that the BOP evaluation indicator is only the production time obtained from the simulation result.

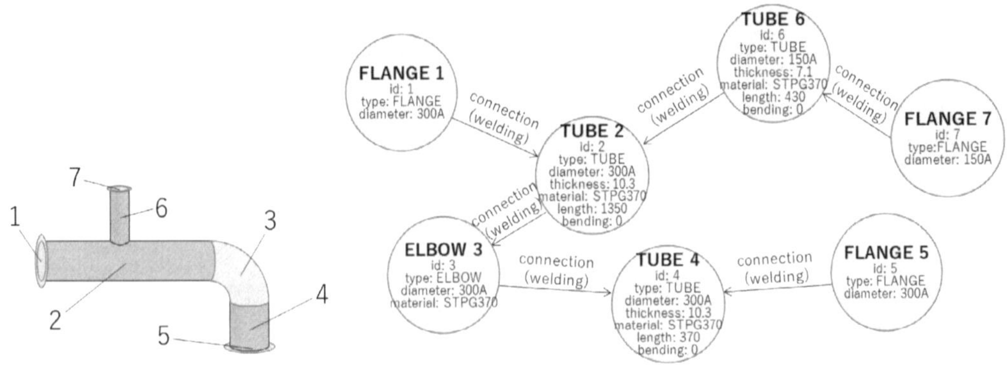

(a) Drawing (b) Graph data

Figure 6. Product X

For each pipe factory, there were 729 draft BOPs extracted. Table 3 shows the determined BOP of each pipe factory. The number in each cell in Table 3 means the part ID of parts targeted at the workplace, and the values given in parentheses are the working time [minute] of the tasks targeting the part or assembly. In Pipe Factory I, the optimal BOP shows that three assemblies are assembled at Workplace E: an assembly with parts 1 and 2 welded, an assembly with parts 3, 4, and 5 welded, and an assembly with parts 6 and 7 welded. In Pipe Factory II, the optimal BOP shows to assemble three assemblies at Workplace E: an assembly with parts 1, 2, and 3 welded, an assembly with parts 4 and 5 welded, and an assembly with parts 6 and 7 welded. In Pipe Factory III, the optimal BOP shows to assemble two assemblies at Workplace E: an assembly with parts 1, 2, 6, and 7 welded and an assembly with parts 3, 4, and 5 welded. In other words, the difference between the BOP of Pipe Factory I and that of Pipe Factory II is which part elbow 3 connects to in the sub-assembling. The difference between the BOP of Pipe Factory I and that of Pipe Factory III is whether the branch assembling of tube 6 to tube 2 is performed in the sub-assembling.

We confirmed that the system determines the BOP reflecting the factory information from this case study.

Table 3. Determined BOP of each pipe factory

Workplace		Pipe Factory I			Pipe Factory II			Pipe Factory III		
A	(Cutting)	2 (10)	6 (10)	4 (10)	2 (15)	6 (15)	4 (15)	2 (10)	6 (10)	4 (10)
B	(Processing)	2 (22)	6 (17)	-	2 (22)	6 (17)	-	2 (22)	6 (17)	-
C	(Sub-Assembling)	1-2 (12)	6-7 (12)	3-4-5 (27)	1-2-3 (19)	6-7 (9)	4-5 (9)	1-2-6-7 (54)		3-4-5 (27)
D	(Sub-Welding)	1-2 (25)	6-7 (21)	3-4-5 (50)	1-2-3 (38)	6-7 (17)	4-5 (19)	1-2-6-7 (86)		3-4-5 (50)
E	(Assembling)	All (45)			All (27)			All (15)		
F	(Welding)	All (65)			All (49)			All (25)		

4. CONCLUSION

In this study, we developed a system that automatically determines the fabrication process for pipe products considering the design of the factory. Specifically, with product information, factory information, and workers' information as inputs, the system determined the optimal Bill of Process (BOP), which consists of a task network and some information about each task such as the target components and workplace. The optimal BOP was automatically determined by exhaustively extracting feasible draft BOPs for products using a graph database and evaluating the extracted draft BOPs using simulation. The developed system was applied to a virtual pipe fabrication process considering the different virtual factories, and it was confirmed that the system determines BOP reflecting factory information. Results show that the developed system can automatically determine the optimal BOP from the information on the target pipe, factory, and workers. We can execute effective pipe production planning by using the information on optimal BOP.

6. REFERENCES

1. AOYAMA, K., NOMOTO, T. & WATANABE, K., 'Development of Shipyard Simulation System based on Petri Net', *Journal of the Society of Naval Architects of Japan 1997, 795–807,* 1997.

2. MITSUYUKI, T., HIEKATA, K., MATSUBARA, H., Yamato, H. & Moser, B., 'Study on Installation of Production Facilities Using Shipbuilding Process Simulation', *Journal of the Japan Society of Naval Architects and Ocean Engineers 24, 291–298,* 2016.

3. OKUBO, T. & MITSUYUKI, T., 'Ship Production Planning Using Shipbuilding System Modeling and Discrete Time Process Simulation', *Journal of Marine Science and Engineering 10, 176,* 2022.

4. LEE, Y.G., JU, S. & WOO, J.H., 'Simulation-based planning system for shipbuilding', *International Journal of Computer Integrated Manufacturing 33, 626–641,* 2020.

5. KOGUCHI, A., AOMURA, S. & IGOSHI, M., 'The Automated Process Planning for Sheet Metal Bending by Handling Robot', *Transactions of the Japan Society of Mechanical Engineers Series C 66, 2060–2067,* 2000.

6. KOGUCHI, A. & AOMURA, S., 'Automated Process Planning for Sheet Metal Bending by Handling Robot', *Journal of the Japan Society for Precision Engineering 68, 602–607,* 2002.

7. SAE-TANG, A., IGOSHI, M., AOMURA, S. & ARAI, E., 'Determination of Operation Sequences with Minimum Tool-changes in Mechanical Machining', *JSME International Journal Series C Mechanical Systems, Machine Elements and Manufacturing 44, 254–260,* 2001.

8. ANDO, K., MULJADI, H. & OGAWA, M., 'Manufacturing Feature Recognition Method for the Generation of Multiple Process Plans', *JSME International Journal Series C Mechanical Systems, Machine Elements and Manufacturing 48, 269–277,* 2005.

9. KASAHARA, K., ISHIMORI, S. & IKEGAMI, A., 'Study on Supporting System for Process Planning Using an Algorithm of Shortest Path Problem (1st Report)', *Journal of the Japan Society for Precision Engineering 76, 1188–1193,* 2010.

10. ICHIKIZAKI, O., KIKUCHI, H., TAKAGI, S., IKEGAMI, A. & SHINODA, S., 'A Study on an Effective Assembly Sequences Generation Method Using Constraints', *Journal of Japan Industrial Management Association 70, 71–80,* 2019.

11. ZHU, J., KATO, M., TANAKA, T., YOSHIOKA, H. & SAITO, Y., 'Graph based automatic process planning system for multi-tasking machine', *Journal of Advanced Mechanical Design, Systems, and Manufacturing 9, JAMDSM0034–JAMDSM0034,* 2015.

7. AUTHORS BIOGRAPHY

Yui MIHARA is a master's course student in the Department of Mechanical Engineering, Materials Science, and Ocean Engineering, Graduate School of Engineering Science, Yokohama National University, Japan.

Taiga MITSUYUKI holds the current position of Associate Professor at the Faculty of Engineering, Division of Systems Research, Yokohama National University, Japan.

APPLICATION OF MIXED REALITY TECHNOLOGY TO SHIP MANUFACTURING PROCESS

Y Mimori and **T Hiraki,** Mitsubishi Shipbuilding Co., Ltd., Japan

SUMMARY

In recent years, the evolution of so-called XR technologies and devices has been remarkable, and they have become widespread and familiar especially in the gaming and amusement industries. XR technology is also being used in some manufacturing industry and is expected to be widely used at design and production sites as a means to support the Japanese manufacturing industry, which is expected to face a labour shortage due to a declining population.
Since the 2000s, Mitsubishi Shipbuilding has promoted digitalization of production sites by utilizing the 3D-Viewer developed in-house. The 3D-Viewer has dedicated functions to support shipbuilding design and manufacturing processes. As the next step, we are working to further improve productivity and quality by introducing MR technology to our production sites so that workers can experience a more realistic virtual space.
In this paper, we introduce FitViewer MR, a shipbuilding MR system that works on Microsoft's MR device Hololens2.

1. INTRODUCTION

The evolution of so-called XRs such as VR, AR, and MR is remarkable, and they have become increasingly familiar due to the improvement of device performance and the lowering of cost in recent years. It is mainly used in home game consoles, corporate PR and amusement facilities, and is gradually being used in the manufacturing industry.
XR technology is expected to be widely used at design and production sites as a means to support the Japanese manufacturing industry, which is expected to face a labor shortage due to a declining population.
Because VR is a virtual environment, it works well in design reviews, such as a digital mock-up, regardless of where it's used.
On the other hand, AR/MR is a technology that fuses the virtual world into the real world, so being in the field is important, and the virtual 3D design model can be projected onto the real world, and the necessary information can be displayed to support the user's work.
Since the 1980s, Mitsubishi Shipbuilding has developed MATES (Mitsubishi Advanced Engineering system of Ships), a 3D-CAD system for shipbuilding, and has continuously made improvements toward the realization of CIM for shipbuilding. In order to make more effective use of 3D design information created by MATES in a wider range of fields, we have been developing HullViewer / Fitviewer, a 3D-viewer for shipbuilding since the 2000s [1]. Currently, some manufacturing work can be performed by referencing the 3D viewer instead of the 2D drawing.
The 3D-Viewer can display a large number of hull and outfitting parts simultaneously, and it is also used to review design and workability to reduce design alternation.
As the next step, we are introducing MR technology to our production sites to further improve productivity and quality.
For this purpose, we developed a shipbuilding MR system (FitViewer MR) that runs on Microsoft's MR device Hololens2.
In this paper, we introduce FitViewer MR.

2. FITVIEWER MR OVERVIEW

2.1 FITVIEWER MR OVERVIEW

FitViewer MR is a mixed reality system that uses Microsoft's Hololens 2 to bring Mates and other 3D-CAD design models into the real world and overlay them on a 1: 1 scale. Since FitViewer MR can be used hands-free, users can work while viewing 3D models and design information displayed on the lens, which is expected to improve the efficiency of production activities in the shipyard. Figure 1 shows the image of FitViewer MR.
FitViewer MR provides controls for displaying layers and palettes, displaying properties, searching parts, and measuring dimensions. By using these functions, for example, in the outfitting work on a hull block or on board, it is possible to smoothly carry out the installation work while checking the position and direction by referring to the piping model superimposed on the actual hull block.
In the case of repair work, the completed state can be easily confirmed by overlaying an image after repair work onto the ship, and a highly accurate preliminary survey can be performed.
Since the system operation is performed by selecting a menu displayed in the space, the operation can be performed while wearing the gloves, and the labor of putting on and taking off the gloves can be eliminated.

Figure 1: Image of FitViewer MR

2.2 OVERVIEW OF HOLOLENS 2

HoloLens 2 is Microsoft's mixed-reality device. This is a stand-alone PC with an integrated processor, such as a CPU, and a battery. It uses an optical see-through display that overlays the image of the display on the lens onto the real world that can be seen through the lens.
The Hololens2 ' s cameras and depth sensors constantly sense information about your surroundings, and a self-positioning technology called SLAM (Simultaneous Localization and Mapping) creates a 3D map of your surroundings in real time, combining it with accelerometers and other Inertial Measurement Units (IMUs) to present a virtual model in real space with high accuracy.

2.3 SYSTEM CONFIGURATION

Figure 2 shows the system configuration of FitViewer MR. In order to display virtual information in FitViewer MR, three pieces of information are required: a 3D model representing a shape, attribute information corresponding to the model, and area information for aligning the real and virtual world.
For 3D models and attribute information, import a MATES or other 3D-CAD model into FitViewer, and then select the required range and export.
The area information for alignment is output by specifying the position of a marker (QR code) using a dedicated function of FitViewer.
By aligning with the marker, the piping model can be adapted to the actual hull block without any problem even if the hull block is arranged upside down.
It is easy and fast to prepare ahead of time by providing the dedicated function to create data for FitViewer MR in the FitViewer that workers are familiar with. Ease of preparation is an important part of running a system.

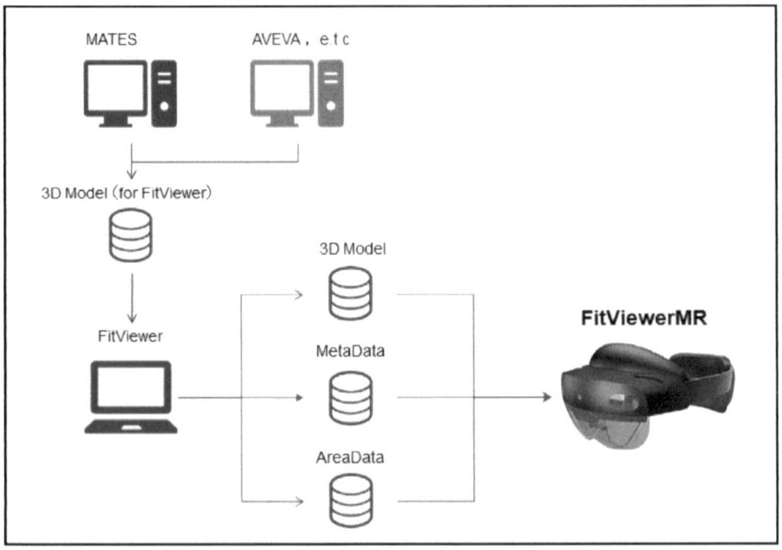

Figure 2: System configuration of FitViewer MR

2.4 OVERVIEW OF MATES

MATES is a 3D-CAD for shipbuilding developed by Mitsubishi Heavy Industries and is used in many shipyards and engineering companies including the Mitsubishi Heavy Industries Group. MATES has been in operation for more than 30 years and is compatible with the structures of various ship types that have been built, such as LNGC and bulk carrier.
It has an easy-to-use interface for shipbuilding designers, various automatic definition functions, and a quality check function incorporating design knowledge. High-precision parts information and 3D-Viewer data can be output from the MATES database.

2.5 OVERVIEW OF HULLVIEWER / FITVIEWER

HullViewer / FitViewer we developed is a 3D-Viewer for shipbuilding we developed with a light three-dimensional display and excellent operability.
The 3D-Viewer is much easier to understand than a 2D drawing. The finished shape or intermediate product state can be easily imagined, and it is useful for design review, prior examination of the construction method, installation work in the production site, etc. It is very easy to operate and can be used by production workers who are not familiar with computers. The 3D-Viewer can import data not only from MATES but also from other shipbuilding 3D-CAD and general-purpose 3D-CAD, and it is used in many other shipyards.

3. FEATURES OF FITVIEWER MR

3.1 SUPERIMPOSED DISPLAY OF 3D MODELS

The most important and fundamental feature of FitViewer MR is the superimposed display of 3D models. Load a FitViewer MR 3D model exported from FitViewer and overlay the virtual model on the real world.
Figure 3 shows an example of superimposing a virtual model (piping and support) on the real world (hull block).
By displaying the parts to be installed in the future on the hull block, workers can experience the completed condition in advance, deepen their understanding of the work, and prevent installation errors.
Alignment between the real world and the virtual model is achieved by markers (QR codes) pasted on the real objects. The virtual model can be displayed at a correct position in the real world by performing coordinate transformation of the virtual model based on coordinate information given to each marker.
FitViewer MR also has an occlusion feature that takes into account the positional relationship between the real world and the virtual model and does not display the virtual model behind the real object. For example, the piping model behind a wall is not displayed. This allows the worker to have a more realistic experience.

Figure 3: Superposing a virtual model on the real world

3.2 DISPLAY CONTROL

Optionally, the worker can select the contents of the virtual model to display. Virtual 3D models are managed on layers of types such as piping, duct, and equipment, and can be turned on and off by layer.
It is also possible to control the display by the assembly tree of the product, so that the worker can smoothly display only the parts which he works.

3.3 PROPERTY DISPLAY

In FitViewer MR, virtual 3D models not only have 3D geometry, but also attribute information about the model (part). The stored attribute information varies depending on the type of model. In the case of a pipe, for example, information such as a pipe number, heat insulation information, and a diameter are stored.
By selecting a model, the property window shown in Figure 4 is displayed, and the attribute information can be checked.

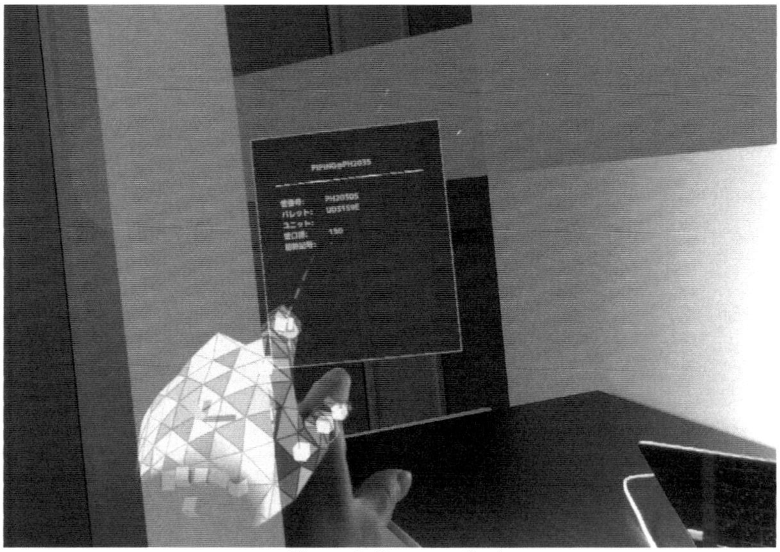

Figure 4: Property display

Part names and design comments are overlaid on the 3D model. Figure 5 shows an example of display of design comments and piping numbers.

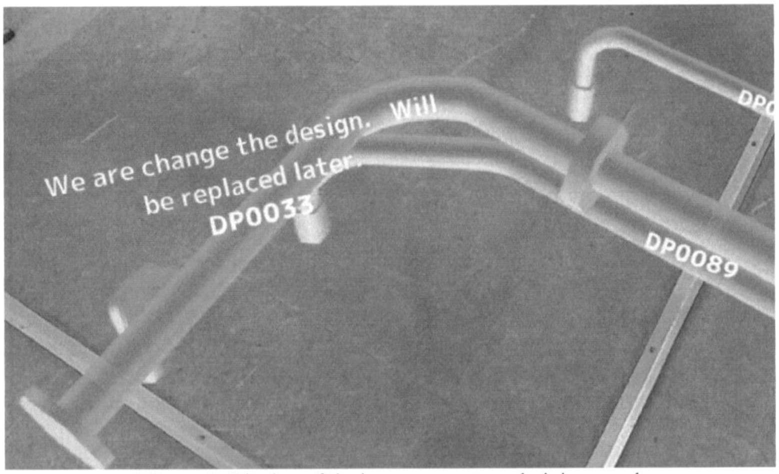

Figure 5: Display of design comments and piping numbers

3.4 PIPING COLOR CHANGE FUNCTION

The piping model displayed in the Fit Viewer MR is usually color-coded for each piping system. It can be changed to color-coded for each heat insulation symbol as needed. Figure 6 shows an example of color-coded piping for each heat insulation symbol.
The worker can prevent work error and improve efficiency by switching the display color code according to the contents of his work. The ability to change the display color of 3D models by attribute type will be enhanced in the future.

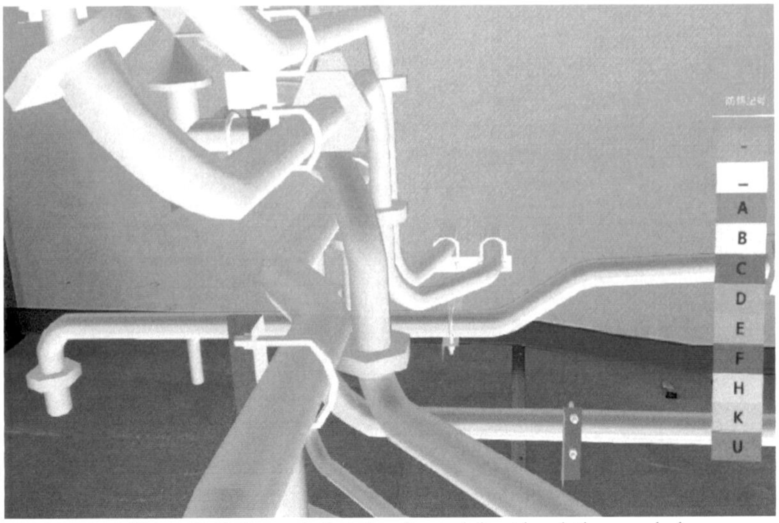

Figure 6: Color-coded piping for each heat insulation symbol

3.5 SEARCH FUNCTION

There's also a feature that support to search the location to install the outfitting part. When the worker specifies the name of the part, an arrow is overlaid to indicate the direction from the current location. As the worker moves in the indicated direction and approaches the target location, the specified part is highlighted in red and a mark indicating its location is superimposed on it. Figure 7 shows an example of the search function.

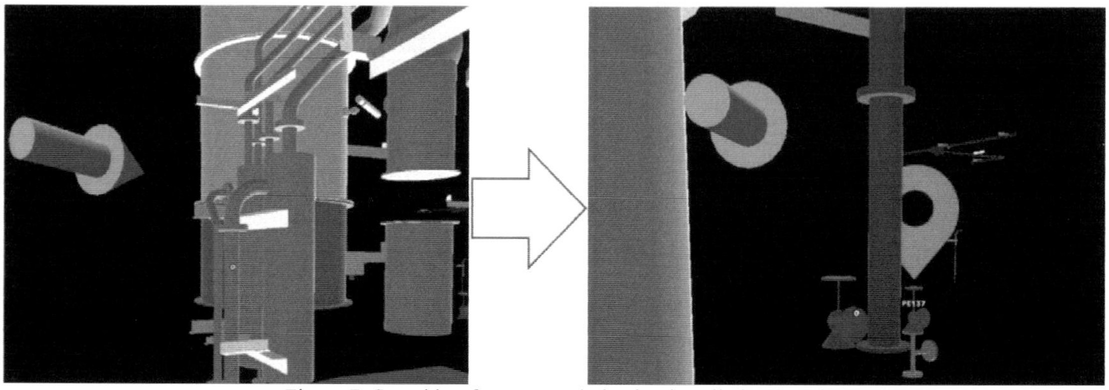

Figure 7: Searching for parts and viewing installation locations

3.6 PHOTO TAKING

Taking a photo that a worker is seeing through the lens of HoloLens 2 (a virtual model overlaid on the real world), it can be used to share information and provide documentation with relevant persons. Figure 8 shows an example of photo taking function.

Figure 8: Photo taking

3.7 DISTANCE MEASUREMENT

By spatial measurement function, it is supported to measure not only distances between real-world objects, but also between real-world objects and virtual models. It can measure the distance between two selected points both in real and virtual world. It can be used in various situations such as confirmation of the clearance between new equipment (virtual model) and hull structure (actual hull) in repair work, as well as confirmation of the position during installation work. Figure 9 shows an example of measuring distance.

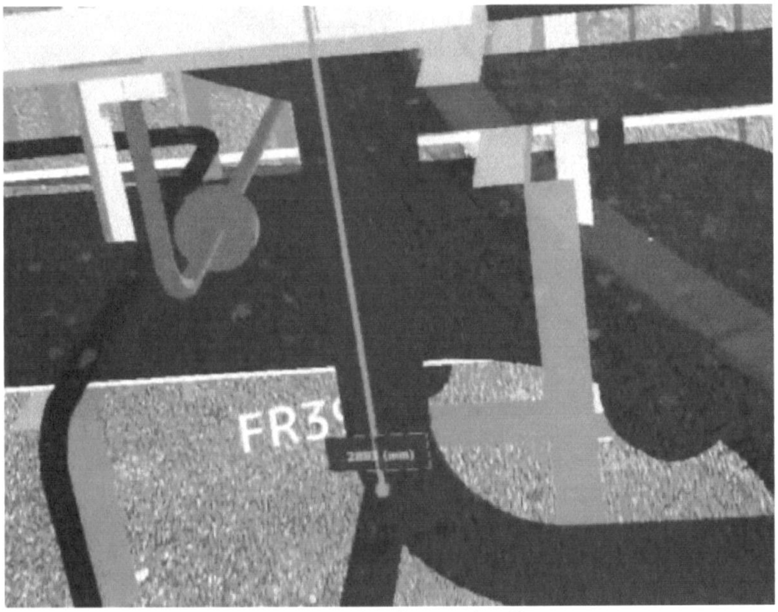

Figure 9: Example of measuring distance

4. SYSTEM EVALUATION

We took FitViewer MR to the production site to evaluate the system suitability. The trial scene is shown in Figure 10.

Figure 10: Trial of FitViewer MR on the production site

The evaluation items are as follows.

1 Usefulness
- Since it can be used hands-free, it is possible to work while wearing it.

- Since the display position may deviate by 0.5%, it is applicable only to tasks that allow the deviation. When the data size is large, it may not be possible to display the model, and it is necessary to reduce data, or use more advanced devices available in the future.
- Measuring the distance of an object in the real world depends on the accuracy of the spatial mesh scanned by Hololens2 and may be less accurate depending on the shape of the object.

2 Visibility
- It's narrower than a human's field of vision, but it is sufficient for work.
- The visibility is not good due to the influence of outdoor ambient light, so it is necessary to take measures such as sticking black film on the lens.
- For models outside the area of spatial tracking, since the position is not corrected, the deviation of the display tends to increase as the distance increases.

3 Safety
- Because the real world through the lens is clearly visible, there are no major safety issues. However, it is preferable to flip up the lens when moving.
- Since it is difficult to use with general helmets, it is better to use a special helmet with built-in hololens2.

4 Workability
- System operation is possible even with gloves on. However, black gloves (including very dirty gloves) cannot be used because Hololens2 does not recognize black.
- The noise around the production site is so loud that the sound of the system is hard to hear. The operation with voice recognition system would also be difficult.

5 Robustness
- No robustness tests were performed. It is a general precision machine, and it seems to be susceptible to physical damage. The dustproof and waterproof performance is as announced by the manufacturer.

6 Continuous operability
- Continuous use limits battery life to about half a day. All-day use requires a quick charge during the lunch break or an optional mobile battery.
- Wearing it for a long time does not cause fatigue or 3D motion sickness. There may be individual differences.

5. CONCLUSIONS

This paper introduces FitViewer MR, a shipbuilding MR system that runs on Microsoft's Hololens2. MR was a dream technology several years ago, but it has been put into practical use due to recent technological advances. FitViewer MR, which utilizes MR technology, will continue to be actively used in various stages of shipbuilding. We will continue to enhance the functions of FitViewer MR based on feedback from actual use cases to improve shipbuilding productivity and quality.

6. REFERENCES

1. T.Hiraki and Y.Mimori " Digital Manufacturing Based on 3D Design Information " *International Conference on Computer Applications in Shipbuilding 2015.*

7. AUTHORS BIOGRAPHY

Yuji Mimori holds the current position of Engineering Manager at Marine Engineering Center, Ship & Ocean Engineering Department, Mitsubishi Shipbuilding Co., Ltd. He is responsible for MATES 3D-Viewer development.

Tokimasa Hiraki holds the current position of Manager at Marine Engineering Center, Ship & Ocean Engineering Department, Mitsubishi Shipbuilding Co., Ltd. He is responsible for MATES system development.

ZERO INFRASTRUCTURE GEOLOCATION OF NEARBY FIRST RESPONDERS ON RO-RO VESSELS

P Mpeis and **D Zeinalipour-Yazti,** University of Cyprus, Cyprus
J Bleye Vicario, Centro Jovellanos, Spain

SUMMARY

The early suppression of fires on ro-ro vessels requires rapid fire identification as a fire of medium growth exponentially reaches 50kW after only 1 minute. Fire patrol members (e.g., able seamen) are asked to act as first responders in such fire incident cases. They do however lack the necessary digital technology for immediate localization, verification and coordination with the bridge and other first responders. Indoor localization requires dense referencing systems (such as Wi-Fi, UWB, Bluetooth antennas), but these technologies require expensive installations and maintenance. Also, Satellite-based indoor localization is obstructed by the bulky steel structures of vessels, so this doesn't work either. Within the LASH FIRE project, an H2020 funded project (Grant Agreement #814975) in which this publication is framed, research has been carried out to develop a ground-breaking localization technology that requires zero infrastructure using computer vision on commodity smartphone devices attached to the gear of first responders. The developed solution comprises of three steps: (i) Training, where vessel owners supply video recordings that are processed on a deep learning data center to produce an accurate computer vision machine learning model; (ii) Logging, where a mobile app allows referencing non-movable objects to the *(x,y,deck)* coordinates of a vessel; and (iii) Localization, where first responders localize on a digital map. Additionally, in case a sparse communication network is available, first responders can share their location, emergency messages and heat scan images with nearby first responders and the bridge. Our proposed technology is shown to be 80% and 90% accurate for localization and tracking scenarios, respectively, in a study we carried out with video traces from a real ro-ro vessel. The overall developed Smart Alert System (SMAS), streamlines the lengthy fire verification, coordination, and reaction process in the early stages of a fire, improving fire safety.
Keywords: localization, first responders, fire safety, early suppression

1. INTRODUCTION

The early suppression of fires on ro-ro vessels requires rapid fire identification as a fire of medium growth exponentially reaches 50kW after only 1 minute. Fire patrol members (e.g., able seamen) are asked to act as first responders in such fire incident cases. Even though a variety of effective technologies are currently in deployment (e.g., VHF/UHF communication, push alarms, checkpoint RFID readers, heat scanners, fixed telephony, and temperature dashboards with drencher knobs at the bridge), a unified digital tool for immediate local localization but also verification and coordination with the bridge and other first responders is not available.

LASH FIRE is an international EU-funded research project aiming to significantly reduce the risk of fires on board ro-ro ships. A particular focus in the project is the development and validation of smart technical solutions for quick first response and effective fighting of fires in their initial stage. One challenging task was the development of an innovative geo-positioning technology (i.e., longitude, latitude, and deck with area level accuracy). To this end, we develope and demonstrate a ground-breaking localization system that requires zero infrastructure, reducing costs and maintenance. Our solution uses static visual elements of vessel spaces as reference points that can be recognized by commodity smartphone cameras (e.g., deck patterns, bulkhead patterns, hoses, fixed installations, signs, control buttons). The spatial location of vessel objects is collected as a one-off process and can then be utilized by any first responder.

Our localization subsystem does not require any sort of communication infrastructure, as the localization function is executed on the smartphone device with offline data and is, as such, considered a zero-infrastructure solution.

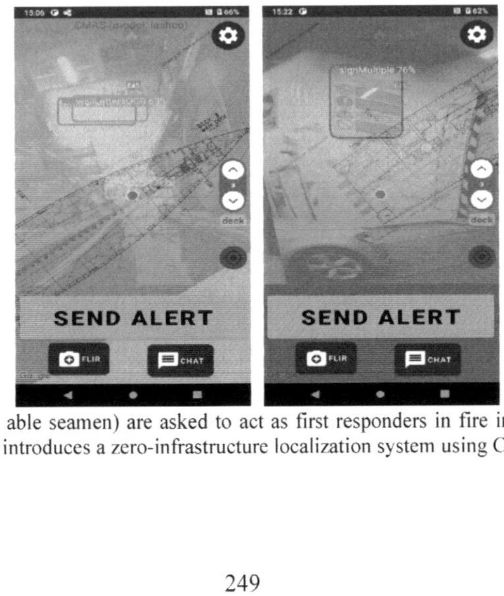

Figure 1: Fire patrol members (e.g., able seamen) are asked to act as first responders in fire incidents. We develop the *Smart Alert System (SMAS)* for smartphones that introduces a zero-infrastructure localization system using Computer Vision (CV).

Our proposed localization system relies on three stages:

- *Training:* vessel owners supply video recordings of vessel interior spaces that are consumed by a deep learning computing server that learns to recognize static vessel objects. This process is carried out once per vessel type (or vessel family, in case of vessels with similar objects);
- *Logging:* technology providers or vessel owners walk around the vessel collecting objects and associating them with locations (i.e., by long-pressing on a map), yielding a *Fingerprint DataBase (FDB)*. Logging is carried out once per unique vessel; and
- *Localization:* The first responders utilize a smartphone application that uses FDB to find their location when necessary.

In the context of this work, we present our zero-infrastructure localization algorithm, named *Surface,* whose aim is to provide area-level localization (i.e., within 10 meters). The name is inspired by the literal meaning of the word, namely our aim is to uncover *"the continuous set of objects that have length and breadth but no thickness"* and organize those objects together with their spatial coordinates *(x,y,deck)* in a database we name *Fingerprint Database (FDB)*. When a user *(u)* aims to localize, *u* queries in a transparent manner through a mobile app the set of objects currently stored in *FDB* retrieving the *(x,y,deck)* of the highest-ranked result.

Our Surface algorithm uses a combination of Global and Local pruning strategies, where the *Surface Global (SG)* algorithm is responsible to find a location from the complete *FDB*, while the *Surface Local (SL)* algorithm is responsible to find the next location from *FDB* based on the prior *(x,y,deck)*. The latter is used for tracking scenarios, where a continuous location over time is necessary. Realizing the SG and SL algorithms in a scalable and portable manner requires a high performance embedded spatial data management environment that we achieved using SQLite and innovative spatial algorithms.

Besides the core geo-positioning technology, our aim was also to develop a prototype vessel indoor location information system that will provide fire intelligence during patrol operations (for cases a sparse data communication network is available). For the above purpose, we developed a fully functional vessel communication software system, coined *Smart Alert System (SMAS),* which integrates our zero-infrastructure localization technology to a variety of subsystems (e.g., alert, nearest neighbor multimedia chat) allowing first responders to exchange messages and data (e.g., heat scans or images as shown in Figure 2). Although data communication networks might be widely available in ro-ro spaces, *dense deployment* of radio antennas necessary to provide accurate localization will remain an open problem because of installation and maintenance costs. Surface aims to fill this gap as it requires no infrastructure.

Additionally, by equipping first responders with powerful mobile computing devices has many benefits as it will allow them to increase their *cyber-physical senses* (i.e., multiple sensing devices, like heat scanner or measuring apps), *be informed* (e.g., carrying bulky manuals and maps in digital form), *be intelligent* (e.g., carrying deep learning neural networks that can recognize and track objects), *be location-aware* (i.e., localization, navigation and tracking of mobile and static assets) and *be connected* (with the bridge and other personnel, discarding possibly outdated communication gear). All these dimensions, packed in a tiny device, will increase fire safety to a new level by the means of state-of-the-art computer vision technology that has proven itself [7,8,9] and that is for the same reason also unobtrusive, with a low learning curve, adaptable through software and economically-viable for massive deployment.

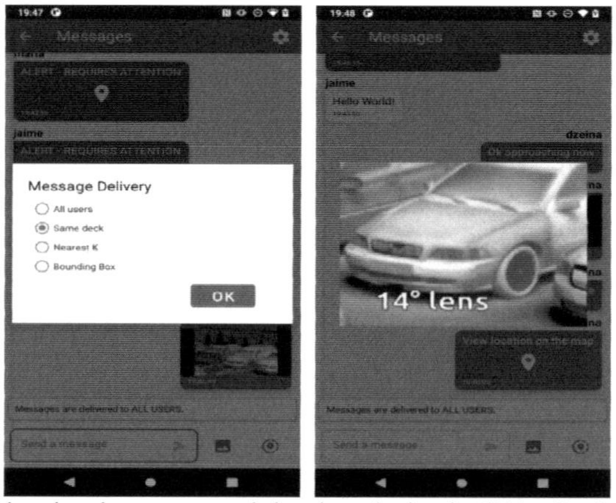

Figure 2: (left) SMAS also provides a location-aware vessel chat channel using the CV localization technology we develop; (right) integration with built-in heat scanners digitizes unnecessary gear and make the images available to nearby first responders.

To assess the correctness and usefulness of our propositions, we carried out an experimental study with extensive video footage from a real ro-ro vessel. We use these for a variety of tests and experiments in the laboratory showing a high accuracy of 80% and 90%, respectively, for localization and tracking. Our study shows that SMAS is an extremely promising technology that promotes situational awareness beyond the current state. There is no other technology that can provide infrastructure-free localization on a vessel nowadays. Our solutions are developed with open and free technology having no barriers-of-entry and a low cost of operation and maintenance.

2. RELATED WORK

The core technology for localization services is the type of hardware enabling the localization process [1,2,3,4,5]. This section surveys the current state.

2.1 NEW-INFRASTRUCTURE LOCALIZATION

These require the deployment of additional dedicated equipment, including proprietary transmitters, beacons, antennas, and cabling, for the provision of location signals. Industrial solutions in this field are termed *Real-Time Locating Systems (RTLS)* and current vendors specialize in specific markets for tracking assets using *Active or Passive RFID*, *Bluetooth Low Energy (BLE Smart) beacons* (e.g., Apple's iBeacon, Estimote.com context stickers), *Wi-Fi beacons* (e.g., Aeroscout.com Active RFID Tags and Kontakt.io cloud/BLE beacon) or *Ultra-Wide-Band (UWB)* chips (e.g., DecaWave.com UWB transceiver offering a 10cm localization accuracy) Assisted-GPS products (e.g., from Qualcomm) or multi-constellation solutions like CSR's SiRFusion chipset that employs information from GPS, Galileo, GLONASS and Compass satellites. These can improve availability in urban canyons and some indoor environments. However, to meet typical horizontal and vertical accuracy requirements indoors, both solutions fuse satellite measurements with other complementary source of information including cellular and/or Wi-Fi and/or multiple inertial sensors [1].

On the other hand, a prominent academic effort is the *Epsilon* system by Microsoft Research, which relies on LEDs that flash in millisecond intervals, so that only a smartphone camera can pick up the pulses (e.g., one commercial effort that uses this idea is Bytelight.com). Subsequently, a smartphone user can be localized on the intersection of circles defined by those distances (i.e., *multi-lateration*). The emerging Li-Fi standard extends the idea of LEDs into communications. Another academic effort is the *ALPS* system by Carnegie Mellon University, which employs ultrasound audio signals captured by the smartphone's integrated microphone.

In theory, all solutions have the potential to achieve sub-meter level accuracy at high deployment densities. However, these solutions raise scalability issues in case of large indoor spaces. For example, Indoo.rs required 300 StickNFind beacons (i.e., estimated at 15,000 USD) to provide guidance to visually blind people at the 60,000 m2 Terminal 2 of the San Francisco Intl. Airport, USA. Additionally, there are costs associated with maintaining the batteries of beacons, tuning their signal levels, transmission frequencies as well as interference issues.

2.2 EXISTING-INFRASTRUCTURE LOCALIZATION

This category includes systems that use location-dependent measurements from existing wireless communication infrastructure, such as Wi-Fi access points and cellular base stations. Field tests revealed that these can attain localization accuracy that is comparable or even better than costly new infrastructure-based systems [2].

In this context, our team has over the years developed the Anyplace Indoor Information System [1,2,3], which is a Wi-Fi fingerprint localization system that won many awards for its utility and accuracy. Anyplace has traditionally only focused on Wi-Fi localization but will also support CV localization after this work given the fact that dense Wi-Fi installations are not frequently available on ro-ro vessel and other scenarios. In Figure 3, we show how the indoor model of a real ro-ro vessel used in this study was mapped using our Anyplace Architect tool (https://anyplace.cs.ucy.ac.cy).

In this category we can also classify a special type of RTLS that emerge by enterprise wireless LAN vendors. Companies such as Cisco, Ericsson, Ekahau.com, Arubanetworks.com and Aerohive.com allow enterprises to manage their enterprise WLAN networks but also offer location tracking extensions to their services. Also, Wi-Fi routers with proprietary operating systems (e.g., OpenWrt.org) can be configured for these scenarios to acquire the signal intensity of users and localize them.

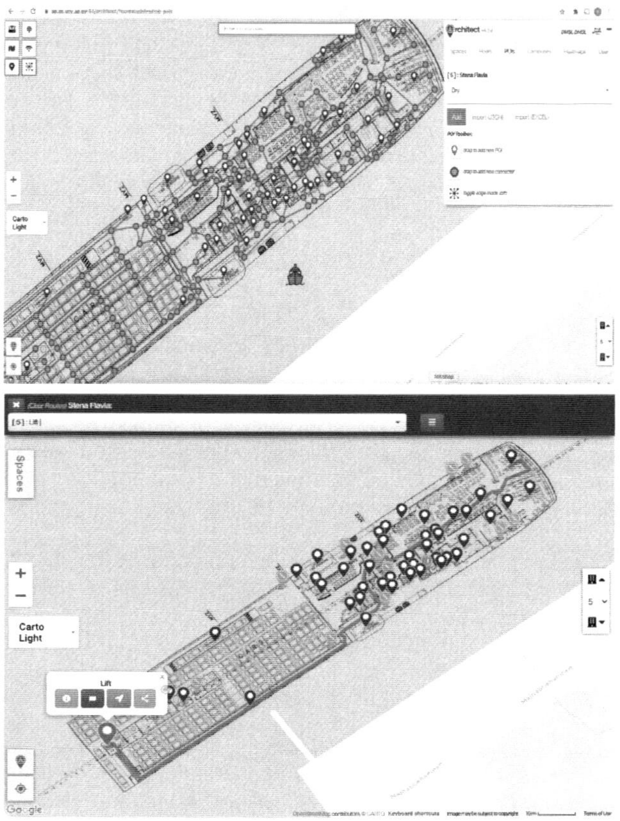

Figure 3: (top) Mapping the ro-ro vessel in our evaluation using our Anyplace Architect tool; (bottom) Obtaining navigation instructions and information search in Anyplace Viewer.

2.3 ZERO-INFRASTRUCTURE LOCALIZATION

These are used in environments where there is no localization infrastructure available. In this category we find Magnetic localization systems (e.g., IndoorAtlas), which take advantage of the magnetic field anomalies typical of indoor settings by using them as distinctive place recognition signatures. This sort of technology offers accuracy without any hardware requirements and a relatively low total cost of ownership. On the other hand, these sorts of signatures work only for stationary steel structures in specific coordinates but not for vessels that are always moving to different locations. As such, magnetic localization is not applicable when it comes to mobile indoor spaces like vessels.

Infrastructure-free localization often also refers to solely IMU-based approaches that have been extensively studied in the context of Pedestrian Dead Reckoning (PDR) systems. Particularly, sensory data reported by *Inertial Measurement Units (IMU)*, including accelerometers, gyroscopes, and digital compasses (e.g., CyweeKIOS or WiFiSLAM) [1]. Such sensor modules are either integrated into modern consumer electronics, or attached externally on the human body, i.e., head, back, waist or foot mounted while processing occurs by low-power co-processor (e.g., motion coprocessors on iPhones). IMU solutions can be used to provide relative location of a mobile but is known to suffer from drifting (i.e., even the slightest localization or orientation error builds up over short future windows yielding high location errors). As such, IMU solutions are known to require an infrastructure to correct the location signal over time. In Anyplace we traditionally utilize Wi-Fi to correct the IMU signal. Likewise in this work, we will consider in the future a hybrid tracking algorithm by the means of fusing the CV signals to the IMU signals, to cope with the drifting problem of IMU.

The SMAS zero-infrastructure localization system developed in this work, proposes the Surface algorithm, which exploits novel data management ranking algorithms with deep learning models trained on ro-ro vessel footage to carry out accurate zero-infrastructure indoor localization with smartphones.

3. SURFACE ALGORITHM

3.1 OVERVIEW

In this section we describe our Surface algorithm, explaining its rational through various examples. Our Surface algorithm uses a combination of Global and Local pruning strategies, where the *Surface Global (SG)* algorithm is responsible to find a location from the complete *FDB*; while the *Surface Local (SL)* algorithm is responsible to find the next location from *FDB* based on the prior *(x,y,deck)*. These algorithms deploy sophisticated object ranking functions founded on the concepts enumerated below and that are implemented with the expressive power of the *Standard Query Language (SQL):*

- *Multiset Subtraction (SG & SL algorithms),* where we scan through the *Fingerprint Database (FDB)* and identify the (*x,y,dec*k) locations that have their set of objects more closely to the query object-set.
- *Global Partitioned Frequency Counting (SG & SL algorithms)*, which captures the frequency a given object has after the FDB creation stage (i.e., after logging). This way, the Surface algorithm knows which objects are of high importance and takes those objects into the ranking process of candidates. The objective is to return the (*x,y,deck*) triples whose object-set resembles more closely the object-set of the query. Clustering of the frequencies is carried out with spatial hashing.
- *Spatial Partitioning of Fingerprints (SG & SL algorithms)*, which clusters close-by fingerprints based on a system-derived clustering parameter (in our setting 10 meters) that allows more accurate ranking of location similarity results.
- *Bounding Box Filtering of Fingerprints (SL algorithm)*, which applies to the case of the SL algorithm and tracking scenarios where location estimates are aimed to be at most 100 meters and +/- 1 deck apart. If the bounding box filter is too aggressive yielding no results, the global localization algorithm alleviates the problem and finds temporarily the best result until a better estimate can be made.

3.2 DATA PROCESSING CONCEPTS

In this section we explain the underlying concepts of the Surface algorithm in further detail.

3.2 (a) Multi-set Subtraction

A *set* in Mathematics is generally defined as an unordered collection of distinct objects. In the case of objects captured during CV logging, a given object might appear multiple times (e.g., an area contains multiple drenchers). As such, we relax the discussion and adopt the notion of a *multiset,* which allows the repetition of objects in the collection. A multiset difference is generally defined in the set builder notation as $A - B = \{x \mid x \in A \text{ and } x \notin B\}$, considering that the set is permitted to contain duplicates.

The below example shows the basic object subtraction operation that takes place in Surface.

Example Multiset Subtraction (Query – FDB Record):
- {drencher, charger} – {drencher, charger} = ∅
  Dissimilarity = 0
- {drencher, charger} – {drencher, charger, door} = ∅
  Dissimilarity = 0
- {drencher, charger} – {drencher} = {charger}
  Dissimilarity = 1
- {door, door, door} – {door} = {door, door} Dissimilarity = 2
- {door} – {door, door, door} = ∅
  Dissimilarity = 0

3.2 (b) Global Partitioned Frequency Counting

The Fingerprint view in the SMAS data layer contains all the (*x,y,deck*) locations that have been collected during the SMAS Logging stage. For an object, whether unique or frequent, we might have a dozen of fingerprints depending on the coverage effort during logging and the physical space characteristics. Effectively, the number of fingerprints is clearly not a good indication of whether an object is unique and whether this uniqueness can be exploited to yield a higher accuracy localization.

To alleviate this problem, we decided to carry out a spatial hashing of objects to locations and create the so-called *OBJECT_COUNT* view. This allows identifying how unique an object is in relation to its location. For example, if a *signGather* object appears in three locations then this will have a frequency counter of 3, otherwise if these three locations map to the same location the frequency counter will be 1. This also simplifies logging, as we can collect multiple fingerprints and retain stable frequency counters.

In order to normalize the importance of OBJECT_COUNT objects we average over the total number of objects yielding the OBJECT_FREQUENCY table. This was implemented as a materialized table, as opposed to view, to avoid computing the view for each query (i.e., for performance reasons).

3.2 (c) Spatial Partitioning of Fingerprints

Even though the original OBJECT_COUNT idea clusters fingerprints on (*oid,x,y,deck*), where *oid* is the unique object identifier, one problem is that the fingerprints will rarely be precisely on the same (*x,y,deck*) because the location we use in the logging process has very high precision (e.g., we use Google Maps that uses WGS84 with 1 meter accuracy.)

The below shows one example of an (*x,y*) point we represent in the SMAS application as WGS84 location (longitude, latitude) pair: (57.695137769363, 11.911948062479). With 4 decimal places of precision, we get a clustering of ~11 meters. The above clustering will be exploited in both the SG and the SL algorithms. In Table 2 we present the OBJECT_FREQUENCY table for the top-5 (most rare) and lower-5 (most common) objects.

3.2 (d) Bounding Rectangle Filtering of Fingerprints

One problem with the discussion so far is that we consider object ranking to be a global task, namely an object contributes the same way to a ranking whether it appears a few meters apart from the prior location or whether it is 3 decks above. To this end, we endanger resolving object-sets to irrelevant locations on the vessel in tracking scenarios (i.e., where the localization task takes place every few seconds or after the inertial sensor triggers a re-computation.)

To alleviate this problem, we aim to rank object similarities based on the prior location of a user (*prevX, prevY, prevDeck*). To do so, we create a bounding rectangle around the prior location and carry out a counting and ranking of objects in that area only. This allows finding the closest most relevant object set to the query, which effectively will not be very far from the prior location. In our case, we set the bounding rectangle to be 100 meters and not more than +/- 1 decks, but these parameters are configurable. In case this threshold does not yield any relevant fingerprint (e.g., a user entering an elevator and coming out a few decks away), we automatically run the global ranking SG algorithm that will find the closest match through the complete vessel.

In order to visualize the Bounding Rectangle, consider the example in Figure 4. Here we show the FDB clustered using the Spatial Partitioning we introduced earlier. The sample also shows how the bounding rectangle drawn around the prior location reduces the prospective database fingerprints that will be examined for relevance to the query fingerprint. We observe that the prospective objects and the query object are now in spatial locality, as opposed to having the complete fingerprint database as the search object space. This means that the next location will be derived from the bounding rectangle and not from the complete spatial database, which makes the localization more accurate and usually also faster.

3.3 LOCALIZATION ALGORITHMS

3.3 (a) The Surface Global (SG) Algorithm

The SG algorithm is an algorithm that can be executed without any prior localization state. SG will search through the fingerprint database FDB to find the closest match that can be in any of the vessel decks. SG deploys multiset subtraction, global partitioned frequency counting and spatial partitioning of fingerprints to rank and return the closest fingerprint to the localization query. Given that SG is a less restrictive case of the Surface Local (SL) Algorithm, we present next, we will defer its detailed presentation for Section 3.3 (c) where we present both algorithms in a more elaborate fashion.

3.3 (b) The Surface Local (SL) Algorithm

This algorithm requires the prior location of a user (*prevX, prevY and prevDeck*) to find the most relevant fingerprints from FDB using the concept of Bounding Box Filtering of Fingerprints. SL searches for the closest dissimilar objects in a bounding box that is system-defined (e.g., 100m and not more than +/- 1 decks). Figure 4 presents an example: We observe that the bounding rectangle prunes the object space to a small subset of results, yielding a sub-ranking of results within this box.

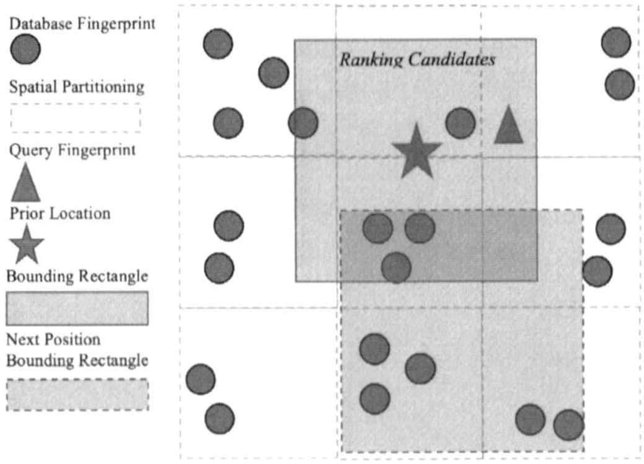

Figure 4: Example with Bounding Rectangle Filtering of Fingerprints in the Surface Local (SL) Algorithm.

The SL algorithm is particularly suited for tracking scenarios where fingerprints need to be dynamically re-ranked based on where a user currently is. Of course, such a sub-ranking might not always yield a result. As such, the SL algorithm will invoke the SG algorithm (global ranking) as fallback case in case it yields an Ø set.

3.3 (c) Surface Implementation in SQLite

Figure 5 presents the implementation of the Surface algorithm in SQLite, which is a C-language library that implements a small, fast, self-contained, high-reliability, full-featured, SQL database engine. SQLite is the most deployed database in the world as it is embedded in smartphone operating systems SDKs (e.g., iOS and Android), web browsers (Chrome, Safari), Web application frameworks (e.g., drupal, Django) and others. For brevity we will explain the common components of SL and SG together and finally focus on the additional restrictive aspect of the SL algorithm.

```
1  SELECT F.fid, F.X, F.Y, F.deck,
2         ABS(x - '$prevX') as xDiff, ABS(y - '$prevY') as yDiff, ABS(deck - '$prevDeck') as deckDiff,
3         (SELECT IFNULL(COUNT(*),0) FROM
4    ( -- Ranking Criterion A: Multi set Subtraction as
5    SELECT ROW_NUMBER() OVER (PARTITION BY FLT.oid) AS RowNum, FLT.oid
6    FROM FINGERPRINT_LOCALIZE_TEMP FLT WHERE FLT.uid='$uid'
7    EXCEPT
8    SELECT ROW_NUMBER() OVER (PARTITION BY FO.oid) AS RowNum, FO.oid
9    FROM FINGERPRINT_OBJECT FO WHERE FO.fid=F.fid
10   )
11 ) AS dissimilarity, (SELECT IFNULL(AVG(weight),1) FROM
12   ( -- Ranking Criterion B: Multi set Subtraction on Global Partitioned Frequency Counting with Spatial Partitioning of Fingerprints
13   SELECT ROW_NUMBER() OVER (PARTITION BY FLT.oid) AS RowNum, FLT.oid, OF.weight as weight
14   FROM FINGERPRINT_LOCALIZE_TEMP FLT, OBJECT_FREQUENCY OF
15   WHERE FLT.oid = OF.oid and FLT.uid='$uid'
16   EXCEPT
17   SELECT ROW_NUMBER() OVER (PARTITION BY FO.oid) AS RowNum, FO.oid, OF.weight as weight
18   FROM FINGERPRINT_OBJECT FO, OBJECT_FREQUENCY OF
19   WHERE FO.oid = OF.oid and FO.fid=F.fid
20   )
21 ) AS weight
22 FROM FINGERPRINT F
23 -- Bounding Rectangle Filtering of Fingerprints
24 WHERE F.modelid='$modelid' and F.buid='$buid' and
25       (x between '$prevX' - '$smas_db_location_bound_meters' and '$prevX' + '$smas_db_location_bound_meters') and
26       (y between '$prevY' - '$smas_db_location_bound_meters' and '$prevY' + '$smas_db_location_bound_meters')
27       and (deck between '$prevDeck' - 1 and '$prevDeck' + 1)
28 -- keep only x,y,deck rounded by 10m bounding box
29 GROUP BY ROUND(F.X,'$smas_db_location_bound_rounding'), ROUND(F.Y,'$smas_db_location_bound_rounding'), F.deck
30 -- include results that have at least 1 common object
31 HAVING dissimilarity < (SELECT COUNT(*) from FINGERPRINT_LOCALIZE_TEMP FLT WHERE FLT.uid='$uid')
32 -- rank by dissimilarity, then OBJECT_FREQUENCY weight than by distance from prior location
33 ORDER BY dissimilarity, weight, ABS(x - '$prevX') + ABS(y - '$prevY') + ABS(deck - '$prevDeck') ASC
34 -- return highest ranked result only
35 LIMIT 1;
```

Figure 5: Implementation of the Surface Local (SL) Algorithm in SQLite as part of the SMAS backend.

The common parts of both algorithms are the following: (i) the computation of a dissimilarity score using the notation of multiset subtraction. This is achieved as part of the pre-last predicate in the outermost SELECT block (lines 4-10); (ii) the computation of the frequency ranking given higher ranking precedence to rare objects. This is achieved as part of the last predicate in the outermost SELECT block (i.e., lines 12-20). The clause IFNULL(AVG(weight),1) aims to penalize objects in the ranking process for objects that are part of the Query but not the Database. For all cases, given that ranking is a global process, which might yield a large result set, we filter the results set on two predicates:

- **Partitioning:** We exploit the spatial partitioning of fingerprints that we discussed in object frequency discussion (line 29).
- **Overlap:** We require at least one object overlap between the Query and the FDB (line 31).
- **Sorting:** We sort the results by importance (line 33).
- **Top-1:** We return the highest-ranked result (line 35).

Contrary to the SG algorithm, the SL algorithm computes the following two additional concepts:

- **Bounding Rectangle Filtering of Fingerprints**: that filters results using the bounding box we described earlier (line 24).
- **Query Deviation**: that provides the deviation of the query from the returned result (line 2).

3.3 (d) Discussion

Structured Query Language (SQL) is a standardized programming language that is used to manage relational databases and perform various operations on the data in them. We use this powerful programming language as a novel paradigm to implement a state-of-the-art algorithm, where localization algorithms are traditionally implemented in imperative programming languages (featuring control loops and other programming constructs). Expressing the localization algorithm in SQL has the following benefits:

- *Declarative:* It exploits a declarative query language that has expressive power without getting into the implementation details – we describe what we want not how we want it. This allows quickly prototyping complex ideas that would have otherwise taken numerous man-hours to be developed in an imperative host language like Kotlin (i.e., the SMAS front-end layer).
- *Relational:* It allows expressing the localization task as a sequence of set-theoretic operators on a relationally-complete language (i.e., SQL), which is founded on the mathematical pillars of relational algebra upon which the complete relational data management field is founded and succesful.
- *Structured:* Perceiving the data in relations with relationships allows easier development of ideas as data is organized in tabular form with constraints and foreign keys.
- *Performance:* It allows testing important aspects (like query response time) early in the algorithm development stage and applying respective remedies if needed (e.g., indices, view materialization). I/O performance tuning now becomes a first-class citizen.
- *Portability:* The code runs directly both in the cloud on an extremely powerful server and on a low-end device. More importantly, it runs on the SMAS android app written in Android that is natively supported by SQLite and SQL.
- *Data-driven Algorithm Design:* We are able to design the algorithm by looking at the data through various SQL predicates. We found this extremely powerful in designing and progressing the logic of our algorithm.

In the future we aim to deploy more complex spatial operators by deploying a designated extension (e.g., SpatiaLite). This will provide more expressive power in capturing the importance of further optimization criterions, e.g.., closeness of objects based on rectangle size, orientation, and map-matching for grounding localization requests to the underlying indoor graph topology we maintain but also more study with capturing in low-light conditions (e.g., first responder with torch for which our preliminary findings suggest that we can retain good localization accuracy).

4 EXPERIMENTAL EVALUATION

To assess the correctness and performance of the SMAS architecture and implementation, we carried out a remote study, where we collect CV logs for the vessel based on video footage that the operator provided us. We use these logs to carry out a variety of tests and experiments in the laboratory (i.e., in vitro).

4.1 METHODOLOGY

Dataset: We use 14 videos captured by a commercial ro-ro operator as part of the LASH FIRE project with a total size of 11GB (i.e., approximately 800MB per file). The footage is from one specific ro-ro vessel that we name *VESSEL* for the remainder of this work. The dataset had a total length of 1 hour and thirty minutes (i.e., each video was 6.55 minutes with standard deviation 2.83 minutes.) The videos capture a variety of areas like, Open ro-ro, Closed ro-ro, Weather deck, PAX public areas, Cabin areas but not any Crew areas. The focus of the research has been mainly the Closed ro-ro, Open ro-ro and Weather deck, which are the decks 3 and 4 of the given vessel, as such, the bulk of footage comes from these areas. Additionally in the above areas there is no privacy concern for the usage of the camera localization system. Results from deploying our system on the real vessel might be presented in the future as we plan to carry out an on-board study of SMAS very soon.

Table 1: VESSEL Video traces used for training and localization.

Video ID	Size (MB)	Length (min)	Deck	Area	Annotation
V01	637.88	8.51	3,4	Closed ro-ro	ro-ro, elevator, stairs
V02	640.88	8.51	4	Closed ro-ro, Open ro-ro	Empty, Loading
V03	1698.15	3.56	4	Closed ro-ro	Loading
V04	638.80	8.51	4	Closed ro-ro	Loading
V05	643.77	8.51	4	Closed ro-ro	Loading
V06	644.05	8.51	4	Closed ro-ro	Loading
V07	385.17	5.12	4	Closed ro-ro	Empty
V08	50.79	0.42	5	PAX public	Open Deck, Reception, ASSY A
V09	546.31	7.3	5	PAX public	ASSY A, Restaurant, Open Deck, Reception, Passenger Deck
V10	129.56	1.44	5	PAX public	ASSY B, Open Deck
V11	476.28	6.29	6,5	PAX public, Cabin	cabin, stairs, restaurant, ASSY A loop
V12	649.69	8.52	5,6	PAX public	restaurant, reception, open-deck, stairs
V13	3521.50	8.09	3	Closed ro-ro	cars, trailers, containers
V14	639.02	8.51	4	Closed ro-ro, Open ro-ro	containers, cars

CV Model Training: Computer Vision Annotation Tool (CVAT) is a free, open source, web-based image and video annotation tool which is used for labeling data for computer vision algorithms. Originally developed by Intel, CVAT is designed for use by a professional data annotation team, with a user interface optimized for computer vision annotation tasks. To execute CVAT in our environment we use a dedicated deep learning server, namely an HP DL380 Gen10 with 80 logical processors and a powerful NVIDIA V100 card. This card reduces training time down to a few hours from several days or weeks we required initially on Google's free Colab environment (https://colab.research.google.com/).

Additionally, by processing our data locally allows us to improve I/O performance as handling large video traces over a slow network can become the bottleneck. The videos were annotated by a team of 5 persons over a period of 4 weeks with several iterations, to refine the quality of the constructed LASHCO neural network model we built for VESSEL (24MB with approximately 100 classes). Our builds were initially encoded in YOLO [6]. One problem is that the YOLO models are not sufficiently efficient on smartphones, as such, we export the models in Tensorflow Lite [8], which is a mobile library for deploying models on mobile, microcontrollers and other edge devices. To reduce the size of the trained Neural Networks (NN), we chose to apply the quantization method during training. This has to do with the number of bits used to represent the floating point numeric in the NN models. Under normal circumstances, all FPN (Floating Point Number) computations are carried out on 32-bit byte sequences. With quantization however to 16-bit, we managed to dramatically reduce the NN model size by over 50%. This size reduction had an impact on the performance of the model with no great loss in accuracy.

Localization Hardware: We use a standard led monitor that shows the video files playing with the VLC tool. For the logging and localization tasks, we use the Caterpillar S62 smartphone, which was selected as the experimental apparatus for its rugged properties, battery lifetime and embedded heat camera. Particularly, the device features the following specs: Android 10, Qualcomm Snapdragon 660 (Qualcomm Kryo 260 CPU, Octa-core CPU, 64-bit, 1.95 GHz to 2.2 GHz), 4GB RAM, 128GB ROM, 4000 mAh non-removable Lithium-Ion battery, Image Signal Processor Qualcomm Spectra 160 image signal processor, 14-bit, 2x Image Signal Processor (ISP), Single Camera, MFNR, ZSL, 30fps: Up to 25 MP, Hybrid Autofocus, Optical Zoom, Qualcomm Clear Sight camera features, Zero Shutter Lag. Even though the description of the camera sounds very specific, the system works well with any smartphone having a capable camera.

Tuning Remote Logging: The video player was configured in the Logging experiments at "Medium" speed to provide our camera system enough time to capture the various objects. Generally, due to varying lighting conditions as well as inherent pixelation and discretization of a computer monitor output, we observe that our logger and localization engines did not recognize as many objects as they would recognize if the input came directly from the camera. We observed this in the laboratory with both UCYCO, which was trained by us, and COCO [9] that comes pre-trained. As such, the logging and localization accuracy provides a worst-case bound on the accuracy we will obtain in a live environment. We aim to validate this claim in the on-board study where logging and localization will happen directly through the camera-system of our smartphone.

4.2 LOGGING TIME EVALUATION

The purpose of this evaluation was to observe the effort necessary to collect object-sets through the SMAS Logger, which subsequently will enable zero infrastructure localization on a vessel. For this series, our evaluation metric is expressed mainly in logging time, but we do also measure the number of objects that were collected per deck. Figure 6 present some example fingerprints in closed ro-ro, PAX public and shows how these are marked on a map (that subsequently registers the reading in FDB). We include fingerprints with varying light intensities as well as camera visibility obstructions (e.g., corridors between trucks).

Figure 6: CV fingerprints collected from VESSEL and entered to the Fingerprint Database (FDB) through a map interface and OCR example shown last.

Results: In Figure 7 (top), we show the logging time required to model the VESSEL. Deck 5 required the most time by far, with a logging time of a bit more than 2 hours. This was the most diverse deck of the vessel in terms of room types, consisting of 4 restaurant areas, a bar, a lounge space, a reception hall, 2 outdoor spaces, amongst others. Deck 4 required about 10 mins, while Deck 3 and Deck 7 required less than 30 minutes. Deck 6 did not have a lengthy capturing and required just 10 minutes. For each spot it required roughly 3 mins. Given that we cluster close-by fingerprints in the OBJECT_FREQUENCY view, collecting many fingerprints per location is good.

In Figure 7 (bottom), we visualize the fingerprint objects that were captured during the modeling of VESSEL. In total about 460 objects were stored in 5 different decks. Those were clustered in 81 geographic locations. More than half of the objects were assigned in Deck 5, which was the most diverse deck in terms of space types. Deck 4 had a bit more than 100 objects, while the remaining 3 decks had 25 fingerprinted objects on average.

In Table 2 we present the top-5 and lower-5 objects identified in our logging task as part of the OBJECT_FREQUENCY view. The given view ranks objects by rarity using by using our WGS84 11-meter clustering method. These weights determine the ranking of localization results after the dissimilarity criterion. Note that all objects with the OCR are passed through Google's ML kit on-device *Optical Character Recognition (OCR)* library (available in multiple languages), which can refine an object recognition. For example, for Figure 6 (last) it reads *"STB PILOT DOOR BUNKER TTION"*, missing only the "station". OCR will be used further in the future.

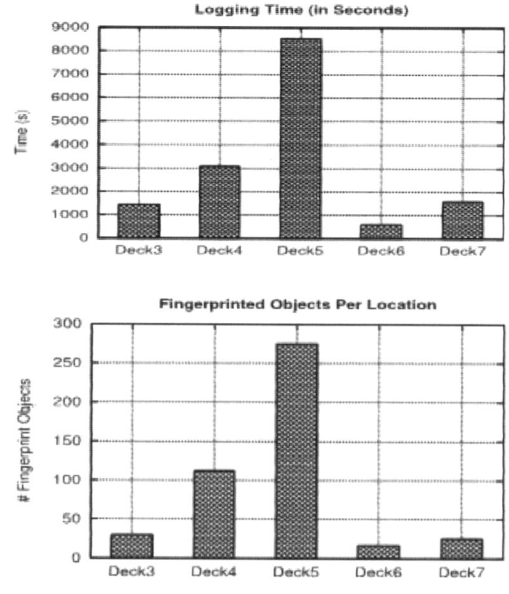

Figure 7: (top) Time necessary to perform the logging task per deck on VESSEL using footage from 14 videos; (right) Fingerprint Objects collected per deck.

Table 2: OBJECT_FREQUENCY View after Logging

Object ID (oid)	Object Description	Weight (rare to frequent)
98	signBathroomDisabled	0.0052631
95	signFireSafetyLever	0.0052631
93	wallSingleDigitOCR	0.0052631
92	specificTankDeck	0.0052631
91	signNumberOCR	0.0052631
...	...	...
49	wallPatternRedWhite	0.0315789
4	signExitPersonLetters	0.0315789
69	signLetterOCR	0.0368421
43	doorBlue	0.0368421
52	wallFireExtinguisherEmbedded	0.0473684

4.3 LOCALIZATION EVALUATION

The purpose of this evaluation was to remotely assess the CV localization system when localizing without any prior knowledge of the user's location (e.g., for the first time). Particularly, we test the effectiveness of the SG algorithm. Our aim is to observe whether Surface can distinguish in each of the 15 scenes the correct location within 10 meters (i.e., room or area-level accuracy).

Metric: We use *Localization Accuracy (A)* as the metric for our evaluation. Generally, accuracy measures the closeness of measurements to the true value of the quantity being measured. In our case, an expert user judges the closeness of the result. Specifically, the expert user judges that the localization query (i.e., $Result_i$) is successful (i.e., $Result_i = Correct_i$), when the user localizes within 10 meters of the ground truth (i.e., $Correct_i$). The above is repeated *n* times (in our experiments n=10). Below we provide the definition of our accuracy using a set-theoretic notation, where Result indicates the returned location and Correct location the ground truth.

$$A = (\sum_{i=1}^{n} \frac{|Result_i \cap Correct_i|}{|Correct_i|})/n * 100$$

Description: For this experiment we have used footage from 6 different videos. We have used 15 different scenes in total, with two thirds concerning cargo/ro-ro areas and one third concerning passenger areas (PAX). We made this decision as our project focused on the former areas. Table 3 lists the 7 different dimensions that each scene had, as well a short description. These include whether the scene was used in training or not, the deck number, the video source, the lighting conditions of the scene, the objects that were contained by the scene (e.g., cars, trucks, passengers, or mixed), the density of those objects in the scene, and finally whether the camera that captured the footage was moving or not.

Table 3: Scenes used in SG Localization Experiment

#	Video	Short Description	Deck	ML Trained	Type	Lighting	Density	Mobility
S01	V13	Lift	3	NO	mixed	lights	full	static
S02	V13	Cargo Office	3	NO	mixed	lights	semi-full	static
S03	V13	Funnel (front)	3	NO	mixed	lights	semi-full	moving
S04	V13	Between-Trucks-180	3	NO	mixed	low	full	moving
S05	V14	outdoor equipment	4	YES	mixed	high	full	static
S06	V14	cross walk	4	YES	mixed	lights	full	moving
S07	V14	wall walk front	4	YES	mixed	low	full	moving
S08	V02	cross walk	4	YES	mixed	lights	full	moving
S09	V14	wall left	4	NO	truck	low	full	moving
S10	V14	free walk (middle)	4	yes	empty	day-light	empty	moving
S11	V14	Reception	4	yes	passenger	high	semi-full	static
S12	V09	Restaurant	5	yes	passenger	lights	empty	moving
S13	V09	Open Deck	5	yes	passenger	high	empty	static
S14	V12	Stairs	5	yes	passenger	lights	empty	moving
S15	V11	Cabins	6	yes	passenger	low	semi-full	moving

The first scene, (i.e., S01) the user was in front of a lift in Deck 3, with many objects recognized. In S02, the user was standing in front of the Cargo Office in Deck 3, but only a few objects were recognized. In S03, the user was in front of walls that are surrounding a funnel, recognizing a few labels on that wall. In S04, the user was walking between trucks, recognizing some labels on the front of the vessel. In S05, the user was recognizing some outdoor equipment. In S06 and in S08, the user was crossing between the lanes that are used for parking trucks or cars, recognizing some labels and doors on the outer wall of the vessel. In S07, the user was walking in open ro-ro space recognizing some labels and doors on the wall. In S09, the user was walking between trucks and the left-outer wall of the vessel, recognizing some equipment. In S10, the user was freely walking in Deck 4, recognizing some patterns, bins, labels, and other objects in the wall. In S11, the user was walking in front of the Information Desk in the vessel's Reception, recognizing it. In S14, the user was walking in the restaurant/lounge bar area, recognizing specific table and chair types. In S13, the user was outdoors, in the Open-Deck area, recognizing life-saving equipment. In S14, the user was passing in front of in the main stairs close to the restaurant in Deck 5, recognizing them. Finally, in S15, the user was in the cabin rooms of Deck 6, recognizing some labels and some lights that are present only in cabin corridors. Video V13, contained around half of those scenes, while the remaining scenes were distributed to 5 other videos. Quite importantly, V13 was not used for building our ML model, as such, was a good test of our constructed model.

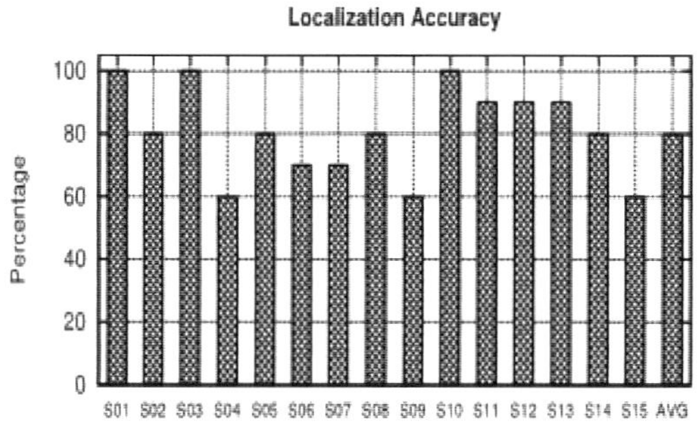

Figure 8: Localization Accuracy for the 15 scenes of Table 3 as well as the average (last column) showing that Surface exposes an 80% accuracy on average.

Results: Figure 8 shows the localization success percentage for each of the 15 Scenes (labeled S01-S15), as well the average. In more than two thirds of the scenes we were able to achieve between 80%-100% accuracy, with an average of 80%, in a total of 150 localization attempts. For about two thirds of the scenes, we were able to achieve a remarkable 80% success, while for 3 scenes we achieved a perfect score. On average, we achieved an 80% localization success. Scenes S04, S09, and S15 were the ones with the lowest percentage, that was still 60%. In S04, the user was walking in a lane that had high-height trucks on both sides and was not able to recognize a necessary number of objects before attempting to localize (for such scenarios the SL algorithm will be beneficial). There is a similar scenario in S09, with the only difference that the user was walking in a lane that had trucks on the one side and the vessel's wall on the other. Lastly, in S15, the lighting and stabilization conditions for the footage were not ideal, hence the CV system was not able to readily identify the objects.

The most contributing factors in these 3 cases were the lighting conditions, monitor reflection (due to the remote execution of this experiment), as well the naturally low number of objects in those areas. We have scheduled an onboard study on a ro-ro vessel to verify these claims and will report new results in the future.

4.4 TRACKING EVALUATION

The purpose of this evaluation is to remotely assess the Computer Vision localization system when localizing on a particular trajectory. Particularly, we test the effectiveness of the SL algorithm in a tracking scenario (without IMU and without a particle filter fusion). We have used two different trajectories. We used footage from two different videos, namely V13 (closed ro-ro) and V09 (Public PAX), following two different trajectories playing at normal speed. Each trajectory had 5 points, and for each point we performed the localization 5 times. Figure 9 shows visually Trajectory 1 for V13 (closed ro-ro).

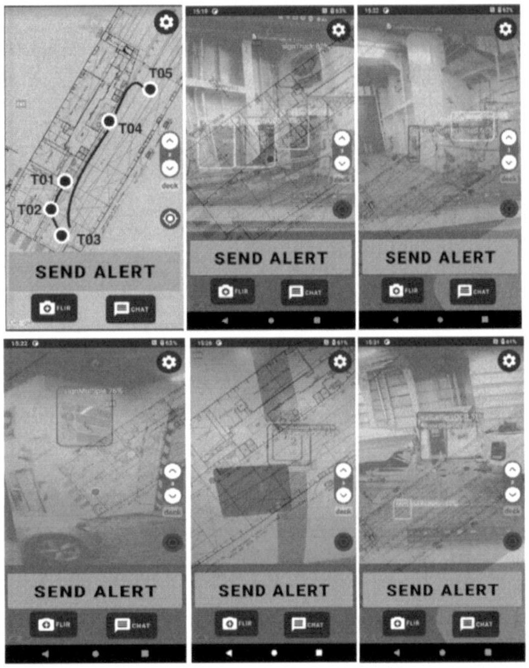

Figure 9: Trajectory 1 (with V13) used in Tracking Experiment

Figure 10 shows the Trajectory Accuracy results for the two trajectories. The average accuracy percentage was 88% and 92% for Trajectory 1 and 2, respectively.

We have achieved a perfect score for 3 out of the 5 points of Trajectory 1, and 4 out of the 5 points for Trajectory 2. The lowest point in Trajectory 1 was T05, at 60%, where the user was not able to recognize some floor objects due to fast video movement.

The lowest point in Trajectory 2 was T01, and similarly at 60%, where the user was capturing only one out of the two objects. As a result, the localization algorithm pointed the user to a nearby location, instead of the Drivers Restaurant. Capturing in a live scenario would have presumably corrected this issue and we will verify this claim in the planned onboard study.

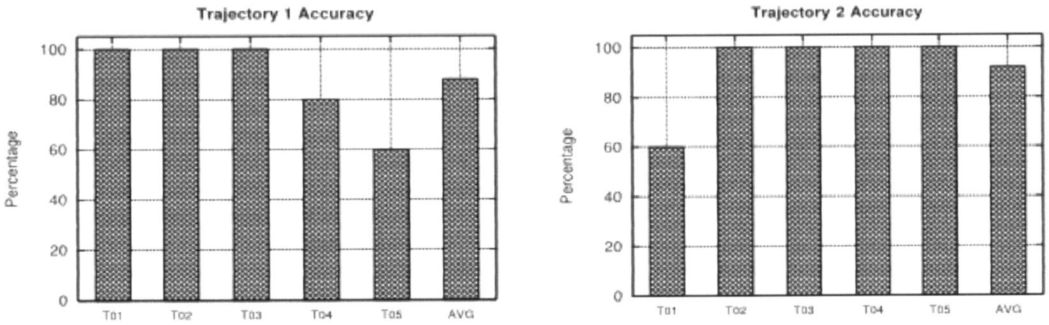

Figure 10: (top) Trajectory 1 with V13 and (bottom) Trajectory 2 (with V09); in the Tracking Experiment. The average accuracy in both trajectories was 90%.

4.5 SCALABILITY EVALUATION

The purpose of this evaluation was to assess the scalability of the localization algorithms. In particular, our aim was to observe the localization response time as the number of fingerprints increases. The remote study had around 0.5K object fingerprints mapped to 81 individual locations. In this experiment we uniformly and randomly generate 7 different object fingerprint datasets at much larger scales and observe the response time of the localization algorithm. We chose the SL algorithm with a random prior point and randomly chosen objects, as this exposes the more complex case for our algorithm (i.e., in case SL does not yield any result, the SG algorithm is executed.)

In Figure 11 we visualize the query response time in milliseconds (ms) for the 7 datasets. As shown, Surface scales linearly with the dataset size. For the real dataset (1k points), we required less than 10ms. For 10k points we required about 50ms, which is sub-linear. Between 10k and 50k the increase is linear, as we observe ~250ms. This is within the latency of the CV object detection system, so we've verified that the localization algorithm will not be on the critical path of the evaluation even for very large Fingerprint Database scenarios capturing complete vessel spaces. For reference we mention that the CV object detection system requires on the S61 around 400-500ms and 200-300ms on the S62. For more powerful smartphones this can be less than 100ms [9], but localization is not anticipated to run so frequently.

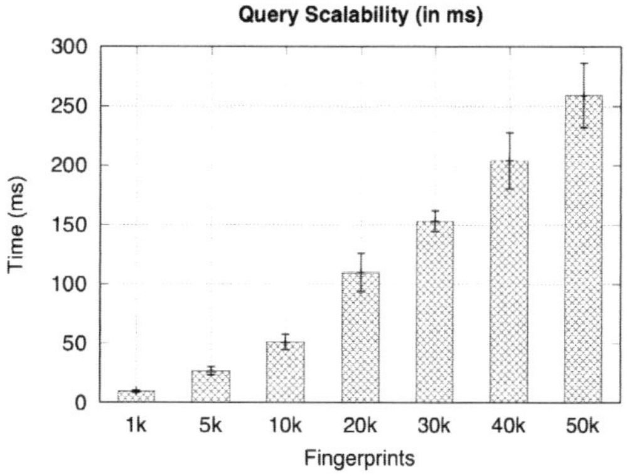

Figure 11: The Surface localization algorithm scales linearly with the increase in the size of the Fingerprint Database (FDB).

6. CONCLUSIONS

In this paper we present a ground-breaking localization technology that requires zero infrastructure using computer vision on commodity smartphone devices attached to the gear of first responders. Our solution deploys a data-driven localization algorithm, coined Surface, which operates with global and local optimizations. Surface has been found to be 80% and 90% accurate for localization and tracking scenarios, respectively, in a study we carried out with data from a real ro-ro vessel. The developed Smart Alert System (SMAS) streamlines the lengthy fire verification, coordination, and reaction process in the early stages of a fire, improving fire safety. Our developments are expected to have a long-lasting impact on the problem of zero-infrastructure localization on ro-ro vessels, which effectively will have an impact in the complete identification-tracking-positioning spectrum, namely live fire detection and localization, live monitoring, and tactical support, monitoring of cargo, quality control and optimization of cargo load and distribution.

7. ACKNOWLEDGEMENTS

The LASH FIRE project has received funding from the European Union's Horizon 2020 research and innovation programme under grant agreement No 814975.

8. REFERENCES

1. Demetrios Zeinalipour-Yazti, Christos Laoudias, Kyriakos Georgiou and Georgios Chatzimiloudis, "Internet-based Indoor Navigation Services", IEEE Internet Computing (IC '17), IEEE Computer Society, Vol. 21, Iss. 4, pp. 54-63, Los Alamitos, CA, USA, 2017.

2. Paschalis Mpeis, Thierry Roussel, Manish Kumar, Constantinos Costa, Christos Laoudias, Denis Capot-Ray Demetrios Zeinalipour-Yazti, "The Anyplace 4.0 IoT Localization Architecture", Proceedings of the 21st IEEE International Conference on Mobile Data Management (MDM '20), IEEE Computer Society, pp. 218-225, doi: 10.1109/MDM48529.2020.00045, June 30 – July 3, 2020, Versailles, France, 2020.

3. Christos Laoudias, Artyom Nikitin, Panagiotis Karras, Moustafa Youssef, Demetrios Zeinalipour-Yazti, "Indoor Quality-of-Position Visual Assessment using Crowdsourced Fingerprint Maps", ACM Transactions on Spatial Algorithms and Systems (TSAS '21), Association for Computing Machinery, Vol. 7, Iss. 2, New York, NY, USA, 2021.

4. Rathin C. Shit, Suraj Sharma, Deepak Puthal, and Albert Y. Zomaya, "Location of things (lot): A review and taxonomy of sensors localization in iot infrastructure", IEEE Communications Surveys & Tutorials, vol. 20, no. 3, pp. 2028–2061, 2018.

5. Andreas Konstantinidis, Panagiotis Irakleous, Zacharias Georgiou, Demetrios Zeinalipour-Yazti, and Panos K. Chrysanthis, "Iot data prefetching in indoor navigation SOAs", ACM Transactions on Internet Technology, vol. 19, no. 1, pp. 1–21, 2018.

6. Alexey Bochkovskiy, Chien-Yao Wang, Hong-Yuan Mark Liao, "YOLOv4: Optimal Speed and Accuracy of Object Detection", 2021 IEEE/CVF Conference on Computer Vision and Pattern Recognition (CVPR), pp. 13024-13033, doi: 10.1109/CVPR46437.2021.01283, 2021.

7. Tesla Autopilot. Future of Driving https://www.tesla.com/autopilot

8. Tensorflow Lite. Deploy machine learning models on mobile and edge devices. https://www.tensorflow.org/lite

9. COCO: Common Objects in Context https://cocodataset.org

9. AUTHORS BIOGRAPHY

Paschalis Mpeis is a Postdoc Researcher at the Department of Computer Science at the University of Cyprus and holds a PhD from the University of Edinburgh, UK. His research includes mobile systems. He is the lead developer of the open source SMAS app.

Jaime Bleye Vicario is a Seaman and Nautical Engineer. Is the Head of the firefighting area at Jovellanos Training Centre, Salvamento Maritimo (SAS). He has specialized in marine firefighting operations, publishing a "Manual about interventions of land firefighters on board vessels" edited by the Port Authority in Spain. He has working in previous EU Projects and now is leading WP06 "Manual interventions" in the LASH FIRE project.

Demetrios Zeinalipour-Yazti is an Associate Professor of Computer Science at the University of Cyprus and holds a PhD from the University of California – Riverside, USA. His research includes data management in systems and networks, particularly mobile data management. He is an ACM Distinguished Speaker, a Senior Member of ACM, a Senior Member of IEEE, and a Member of USENIX, a Humboldt Fellow (Germany) and a Marie-Curie Fellow (EU). He is the technical leader of the SMAS system and the lead develop of the SMAS backend.

SEMI-AUTOMATIC DISTRIBUTED SHIP SERVICE SYSTEMS ROUTING FRAMEWORK FOR SUBMARINE EARLY-STAGE DESIGN

M H Mukti, R J Pawling, and **D J Andrews,** University College London, UK

SUMMARY

Part of the complexity of submarine design is that such needs are typified by extensive and densely engineered distributed ship service system(s) (DS3). Even though submarine early-stage design is the phase to address major DS3 choices in the design process, in initial sizing and submarine synthesis reliance is often made on "past practice" and simple vessel displacement-based weight algorithms. Such an approach not only limits the ability of the concept designer to consider the impact of different DS3 options but also inhibits the opportunity (or necessity) to undertake Requirements Elucidation, specifically for DS3. One of the greatest challenges of DS3 synthesis is the development of the routing model of cables, ducting, and piping which can be time-consuming undertaken in Early-Stage Ship Design (ESSD). Thus, this paper proposes a novel method for early-stage routing of DS3 for a submarine. The approach uses network descriptions of the DS3 topology and computer-aided ship design (CASD) to produce DS3 routing rapidly in ESSD. The usefulness of the approach as an early-stage design tool is demonstrated by using it to design the routing for various DS3 of a conventionally powered submarine.

1. INTRODUCTION

Part of submarine design complexity is that they are typified by extensive and densely engineered distributed systems (Andrews, 2017) as shown in Figure 1. Submarine systems, distributed systems, or service systems are termed in this paper as Distributed Ship Service System(s) (DS3), namely "a collection of connected components that provide a service from one or multiple sources to multiple users, via connections throughout the ship, directed towards defined functions, supporting specific operations of the vessel" (Mukti et al., 2021). Therefore, each DS3 is composed of two main parts: components and connections.

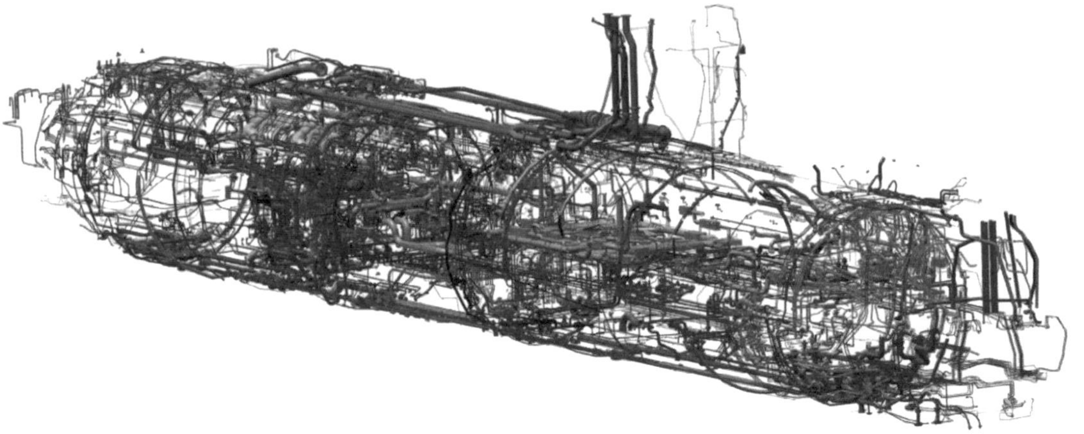

Figure 1: Example of the complexity of selected distributed systems runs on the project A26 submarine (SAAB, 2021)

The paper begins with a brief discussion on early-stage submarine design and DS3 issues, followed by the investigation of DS3 modelling using a Computer Aided Ship Design (CASD) tool. After that, the description of a novel approach to tackling the DS3 issues is outlined. The proposed method is then applied to a case study of a conventional submarine. This is followed by a discussion of that case study's results. Finally, conclusions are drawn, and ongoing work presented.

2. APPROACHES TO EARLY-STAGE SUBMARINE DESIGN

Early-Stage Ship Design (ESSD), encompassing three overlapping stages defined by Andrews (1994), aims to achieve the Requirement Elucidation process (Andrews, 2018): Concept Exploration; Concept Studies; and Concept Design. Concept Exploration is a stage where a wide-range divergent of initial ideas on solutions to meet the initial outline requirements should be explored. Such an exploration could be based on the design solution space, which Andrews (2018) suggests has three main axes: capability, related to the Speed, Strength, Seakeeping, Stability ("S^4") performance (Andrews, 2021);

packaging, the operational capacity of the vessel; and technology, the overall technology incorporated in the design. Then some selected designs are developed in sufficient level of detail through Concept Studies investigating issues that are likely to be significant size or cost drivers in the design. Before the Feasibility Phase, the Concept Design is conducted, working up the selected baseline design or possible two distinct options by performing trade-off studies of cost-capability refining the set of matching requirements and highlighting design risks. The following three subsections discuss the main approaches to submarine ESD.

2.1 DECISION MAKING PROCESS OF COMPLEX VESSELS

Many important submarine design decisions are made either consciously or unconsciously in the Concept Phase, however such major design decisions (see Figure 2) are often not questioned nor acknowledged by design stakeholders, furthermore these decisions are subject to change in a properly conducted Requirement Elucidation process (Andrews, 2018). Andrews (2018), in turn, suggests the whole ship design process, which is nicely illustrated in Figure 3 should not only capture the "wickedness" of design, and the complexity of the task but also encapsulates major decisions the designer must undertake to design a complex vessel (Andrews and Andrews, 2022).

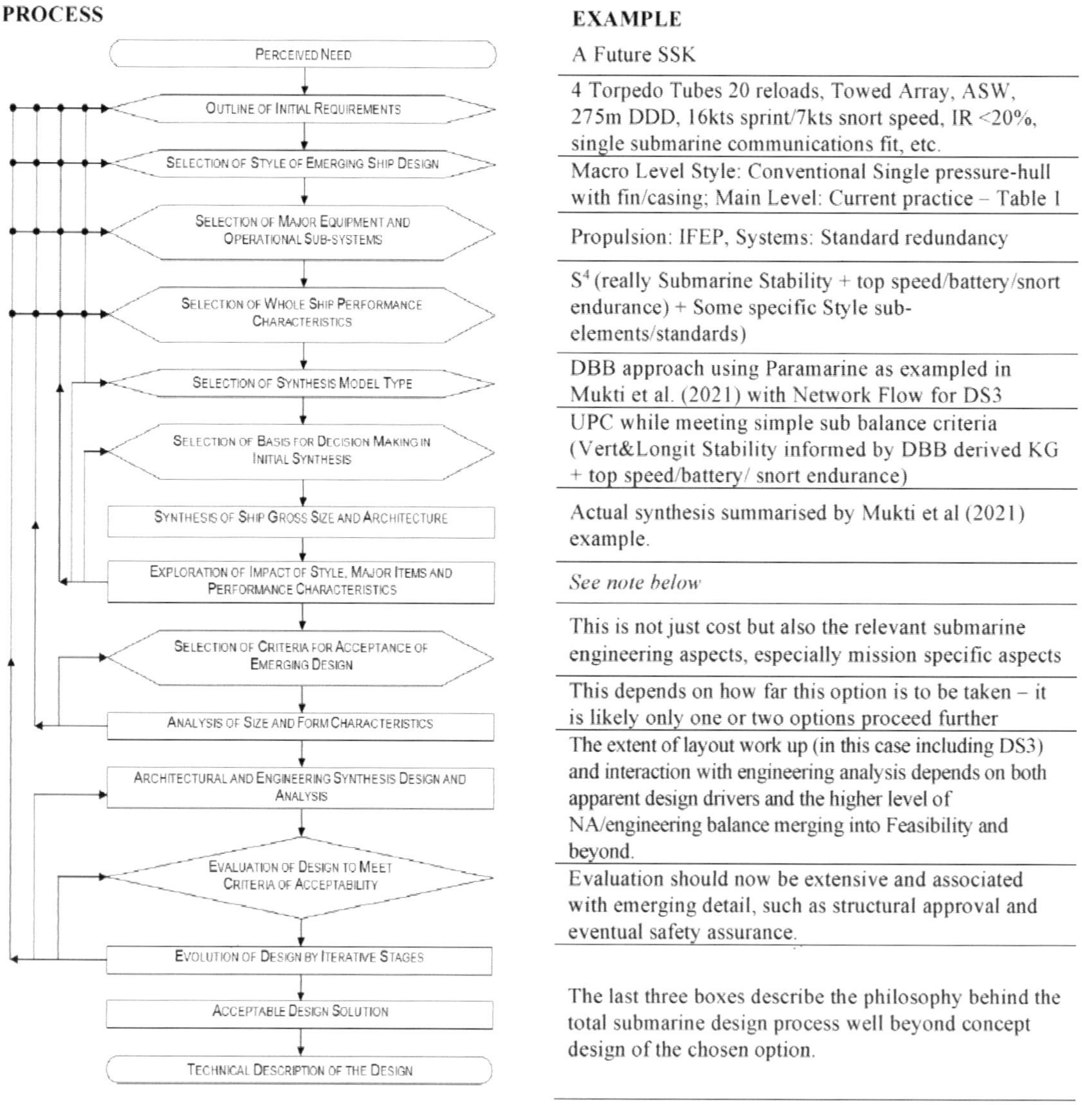

Figure 2: A partial representation of ESSD for a SSK example when in support of Requirement Elucidation with example steps (Andrews, 2021)

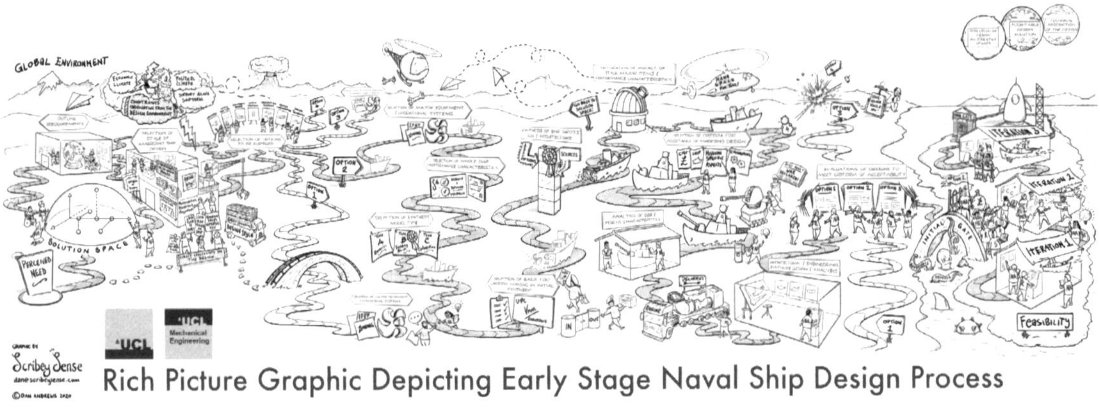

Figure 3: The illustration of decision making sequence for complex vessels (Andrews and Andrews, 2022) © Dan Andrews. Outlined in Figure 4 and Appendix A (Andrews, 2018) and for a submarine example in Figure 4 of Andrews (2021)

The first nine steps in the 'decision making' approach illustrated in Figures 2 and 3 were considered sufficient to meet the ESD scope of the paper. The next subsection discusses design approaches that could potentially accommodate both the synthesis of the whole submarine as well as that for the DS3. This leads to a discussion on the selection of the synthesis model type, part of the 'decision making' approach (Andrews, 2018).

2.2 TRADITIONAL NUMERICAL SYNTHESIS

The approach to initial submarine synthesis used at UCL for the annual post-MSc submarine design and acquisition course (SDAC) was adopted using the sequential design procedure given by Burcher & Rydill (1994). The procedure utlises an initial submarine size can then be used to parametrically estimate the numerical size of component design features based on mathematical relationships with coefficients suggested by Burcher and Rydill (1994) or scaled based on the UCL submarine course data (UCL-NAME, 2014). The term 'numerical' here refers to 'crude' (or 'gross') initial estimation of weight and space of the submarine design, which does not reflect more (architectural) detail. Therefore, this first numerically balanced design implies the highly sensitive longitudinal moments or vertical balance indication have yet to be considered.

The outline of a sequential submarine design procedure by Burcher and Rydill (1994) also provides some possible design algorithms with coefficients, which may be developed based upon 'rules of thumb' drawn from their hands-on submarine design experience using UK Royal Navy practice, which suggests such rules of thumb are likely to be different from navy to navy (e.g., US Navy (Arentzen and Mandel, 1960) and MIT professional summer programme (Jackson, 1992)). This, in turn, implies the 'traditional' characteristic in any such procedure. Likewise, if such algorithms are largely used in the design procedure, it reflects a narrow evolutionary design approach or even a "type ship" approach (Andrews, 2018). Such a crude numerical based synthesis sizing therefore cannot reflect how the spaces onboard are configured beyond duplicating the 'type ship' design style.

Therefore, it was considered that the adoption of just this procedure was insufficient and could not deal with specific physical models of DS3. So submarine design needs to incorporate the 'architectural' or the configurational aspect of the submarine to enable the physical models of DS3 to be considered early in the design process, as is discussed in the next subsection.

2.3 THE UCL DESIGN BUILDING BLOCK APPROACH

The limitation of the traditional numerical synthesis to facilitate requirements elucidation was first raised in the 1980s by Andrews (1981), who then demonstrated an architecturally driven ship synthesis (Andrews, 1984), which was fully integrated in the early 90s to the submarine case with the internal architecture organised functionally, (e.g., Fight, Move, Float, and Infrastructure - FMFI) as opposed to the traditional weight breakdown structure (Andrews et al., 1996). This approach, known as the UCL Design Building Block (DBB) approach (Andrews and Dicks, 1997), is now a proven design method and was implemented as the SURface CONcept (SURFCON) module (for both surface ships and submarines) in the sophisticated fully three-dimensional (3D), commercial naval architectural Computer-Aided Design (CAD) software Paramarine™ (Qinetiq, 2019), coded by the Graphic Research Corporation (GRC) (Andrews and Pawling, 2003). Other commercial 3D CASD software might be in existence but were not considered to readily facilitate the UCL DBB design philosophy.

The implementation of the DBB approach in Paramarine™ (Qinetiq, 2019) provides an object-oriented and top-down approach that allows discrete objects to be modelled and manipulated to different levels of granularity. These objects can have information in a form of string and numerical data, such as weight and geometry and even can be assigned different sizing algorithms (Pawling, 2007). One can start to develop a small number of coarse models as indicated by 'Space (Geometry) Definition'. These models can be based on equipment databases, including new equipment that is under development, reflecting the technology and configurational innovations implicit in commencing a 'ship design' process that positively fosters 'Radical Ideas'.

As the design progresses, the coarse 'ship' model of a few Super Building Blocks (SBB), which do not fully populate the enclosed volume is expanded, broken down into more detailed definitions as necessary. This reflects the building block design phases for surface ship design (e.g., topside and major feature design phase and super building block-based design phase) (Andrews and Pawling, 2008). From these assembled blocks the ship design can be manipulated and assessed under an accumulative block, called a Master Building Block (MBB) (Andrews and Pawling, 2003), until the design is balance i.e., reach an acceptable performance that can go beyond S^4 concerns (Andrews, 2018).

However, there were several drawbacks in implementing such a sophisticated (fully 3-D), high-fidelity, high-capability Computer-Aided Design (CAD) modelling tool in Early-Stage Submarine Design, especially the difficulties in modelling or creating each of the numerous features and individually placing them. The latter task can be considered laborious and demanding, especially if detailed modelling must be carried out after each design change and iteration (Andrews et al., 2009). Such modelling effort can be referred as to the Gulfs of Execution and Evaluation (see Figure 4), which highlights the level of effort required to make a system correctly perform the desired task (Pawling, 2007; Norman, 2013).

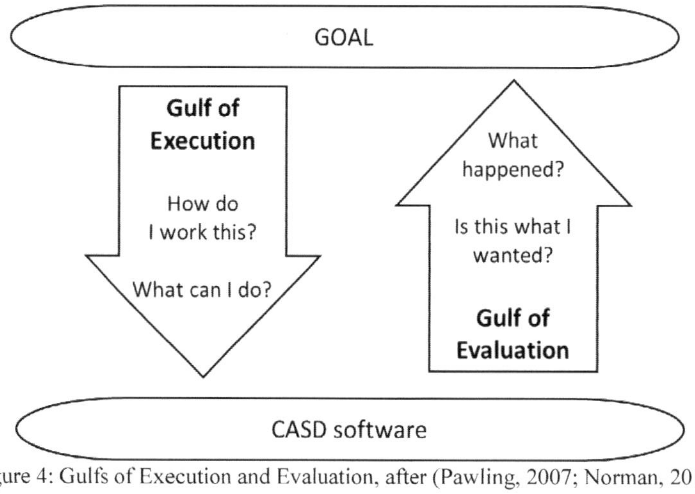

Figure 4: Gulfs of Execution and Evaluation, after (Pawling, 2007; Norman, 2013)

Therefore, the 3D based synthesis was then reduced to what might be called '2.5D' to allow a simpler architecturally oriented design tool than SURFCON Paramarine to be developed in-house for specifically surface ship research and education. This has been referred as to the UCL JavaScript layout exploration tools (Pawling et al., 2015; Kouriampalis et al., 2021). In reporting the current paper, an alternative solution needed to be developed without creating a further separate or standalone design tool such as the UCL Javascript tool. Such a tool needed to sacrifice many advantages of using 3D based synthesis and be fully able to provide a 3D informed dialogue, which Paramarine facilitated. Unfortunately, it was found necessary to have the 3D capability to explore submarine DS3 in ESSD.

Thus, the sophisticated 3D based synthesis UCL DBB approach in SURFCON Paramarine were investigated. This is outlined in the next section.

2. INVESTIGATING SUBMARINE SYSTEMS ROUTING MODELLING IN PARAMARINE

Paramarine can be used to model distributed ship systems routing or connections that were not exploited in the previous UCL DBB research at a PhD level (Pawling, 2007; Purton, 2016). In Paramarine, DS3 can be modelled using a design building block called "**system**" object. As shown in Figure 5, a "**system**" object contains a "**system_diagram**" and two "**placeholder**" objects: components and connections. Some DS3 components could be inserted under components placeholder whereas DS3 connections under connections placeholder. These connections can then be visualised as a simple line diagram using the "**system_diagram**" object.

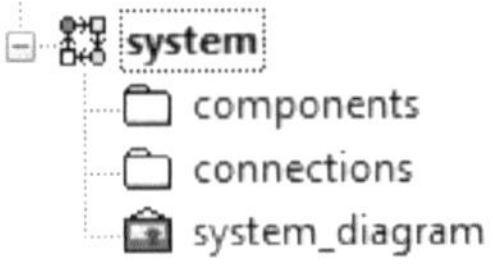

Figure 5: System object in Paramarine (Qinetiq, 2019)

But before a connection can be modelled there was a significant Gulf of Execution (Figure 4) need to be done. One of them was to define the connection or line specification using a "**service_line_specification**" object, some detailed connection properties needed to be known or assumed. These were the definition of the cabling shape (e.g., circular or rectangular) and allowable bending radius as shown in Figure 6.

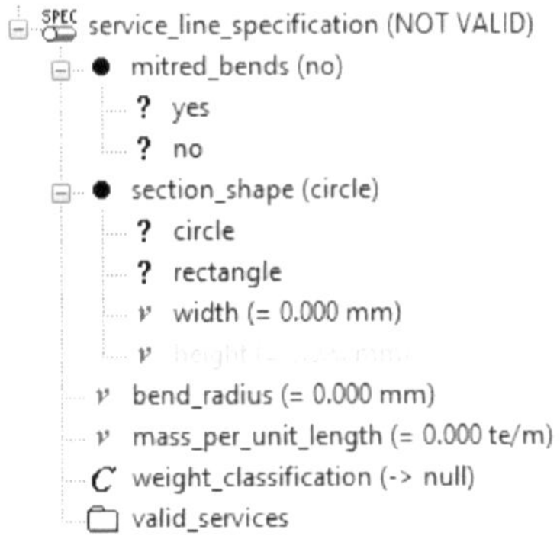

Figure 6: Service line specification input in Paramarine (Qinetiq, 2019)

When it comes to DS3 specification, the Paramarine™ V8.x and the 2019 update 2 only provide four different specifications, namely electrical AC and DC, chilled water, and ventilation. Hence, these are the only available options to model cabling, piping, and trunking as shown in Figure 7.

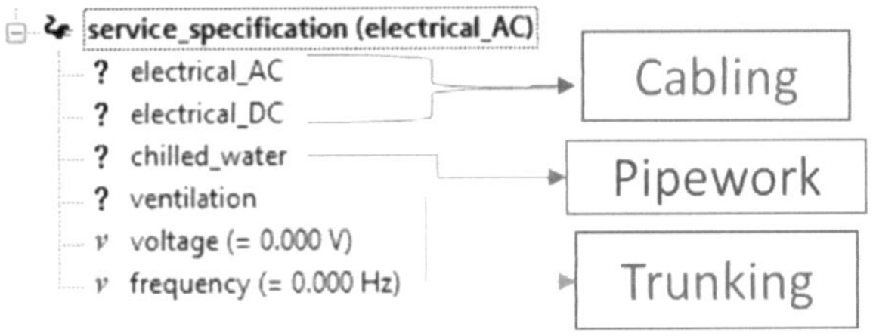

Figure 7: Service specification input in Paramarine

Paramarine has the capability to perform automatic routing to some extent. This is demonstrated using an arbitrary example as follows. Let say there are two building blocks at component level, A and B, in purple on different decks in green with unique X, Y, and Z location as shown in Figure 8 (left). Paramarine then can seek the shortest path between these two building blocks and thus, as shown in Figure 8 (right), a connection in green can be generated.

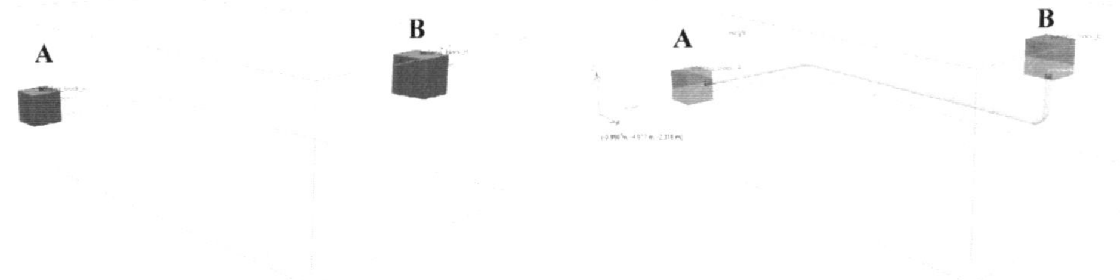

Figure 8: DS3 models using Paramarine showing two components A and B in purple (left) are connected by a connection in green (right), at this point the number of clicks required were 40 clicks

At this point, the position of both building block component A or B can still be manipulated using Paramarine's graphic interface "drag and drop" approach while the routing is adjusted automatically. This is illustrated in Figure 9.

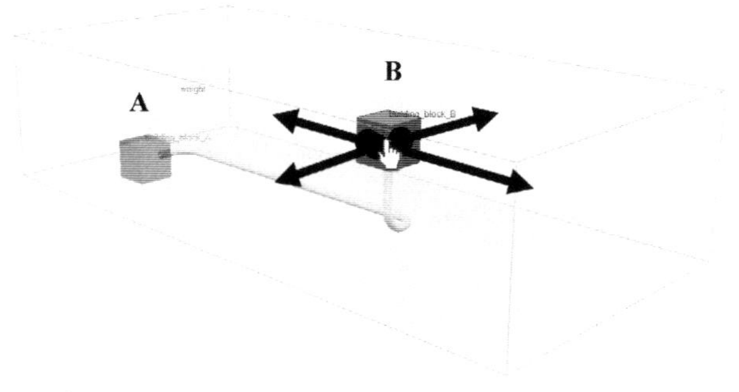

Figure 9: Flexibility in Paramarine DS3 modelling

However, when another building block C is added into the model, the automated routing does not detect the presence of that building block and thus a clash occurs as shown in Figure 10.

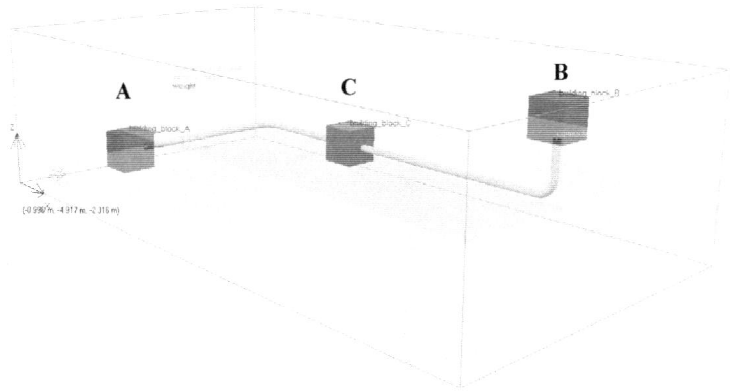

Figure 10: Automatic routing does not detect clash with Building Block C object

Automatic routing tools exist (Gottschalk et al., 1996; Asmara, 2013; Gongora, 2019), but are for detailed design. Thus, several methods were investigated to avoid clashes at early-stage design. The first method is using more inputs to define the routing using "**service_line**" object, which allows segment by segment definition as shown in Figure 11.

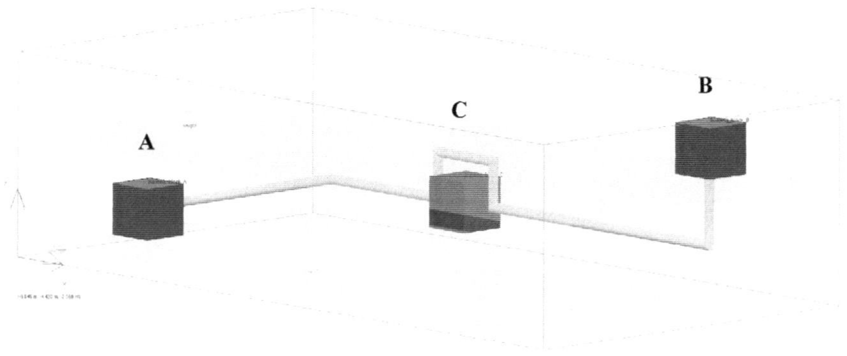

Figure 11: Segment by segment routing in Paramarine

Another method makes use of highway ("**service_highway**" object in Paramarine). A highway can be allocated to avoid building block C, and then the connection is 'pointed' to that highway resulting a 'forced' route as shown in Figure 12.

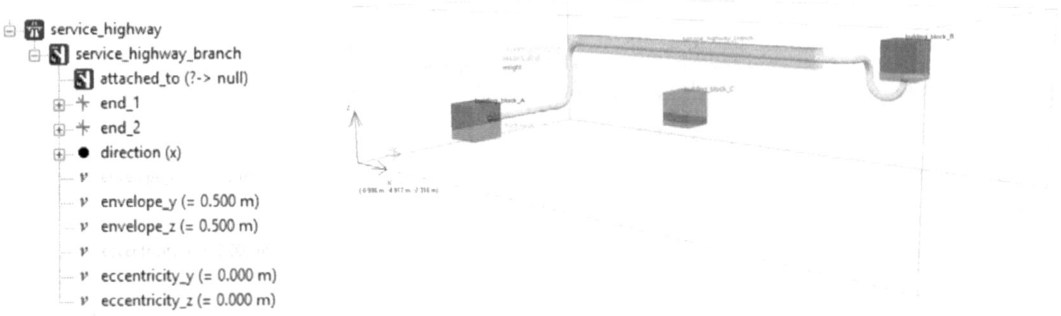

Figure 12: The use of highway routing (orange) in Paramarine

Various sets of highway configurations were investigated for submarine DS3 problem as shown in Figure 13. Figure 13 (top) provides a more detailed model forcing the automatic routing to follow highway attached to the bulkheads. Figure 13 (bottom) is less constrained but is sufficient to minimise the number of routing clashes in ESSD.

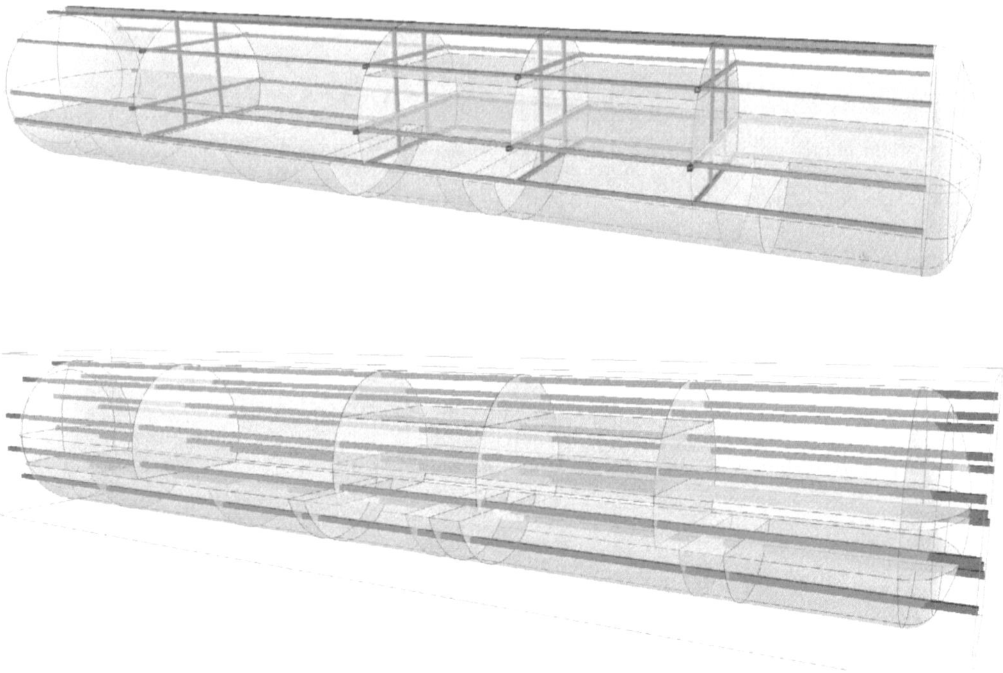

Figure 13: Multiple DS3 highway network in purple modelled using Paramarine, showing a longitudinal highway configuration (bottom) and a complex highway configuration (top)

It is also possible to model a structure duct (Wilgenhof and Reijmers, 2010) in Paramarine, which is a connection that follows the curve of the pressure hull as shown in Figure 14.

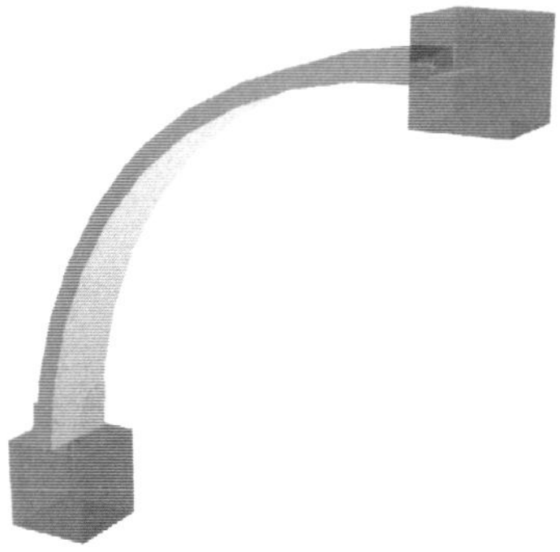

Figure 14: Structure duct modelling in Paramarine

Having investigated the capability of Paramarine to model DS3 connection using an arbitrary example, this capability was implemented in a submarine case study presented in the recent IJME paper (Mukti et al., 2021). As shown in Figure 15, the model was reached with 277 DS3 building blocks with numerical geometric data.

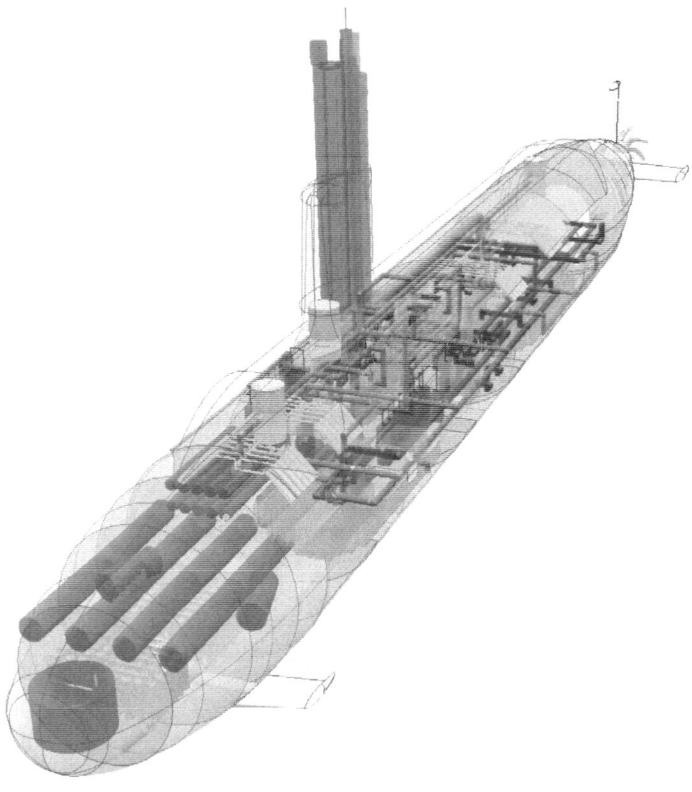

Figure 15: Refined physical model of a submarine case study (Mukti et al., 2019, 2021) in SURFCON Paramarine

In summary, the complexity of the Gulfs of Execution and Evaluation for submarine DS3 synthesis is captured in Figure 16. This exercise furthermore reveals that there is a trade-off between the number of inputs and the routing output. If high quality routings are desired (e.g., no clash, accurate bending radius), the number of inputs required are increased, which can be time consuming. Although such inputs can be pre-defined or be hard coded into the tool (Chalfant, 2022) this makes the tool black box, i.e., it is difficult to determine the causal link between the inputs or design rules and the resulting routing generated. This paper presents an automated approach but it still a human-centred and style-driven approach.

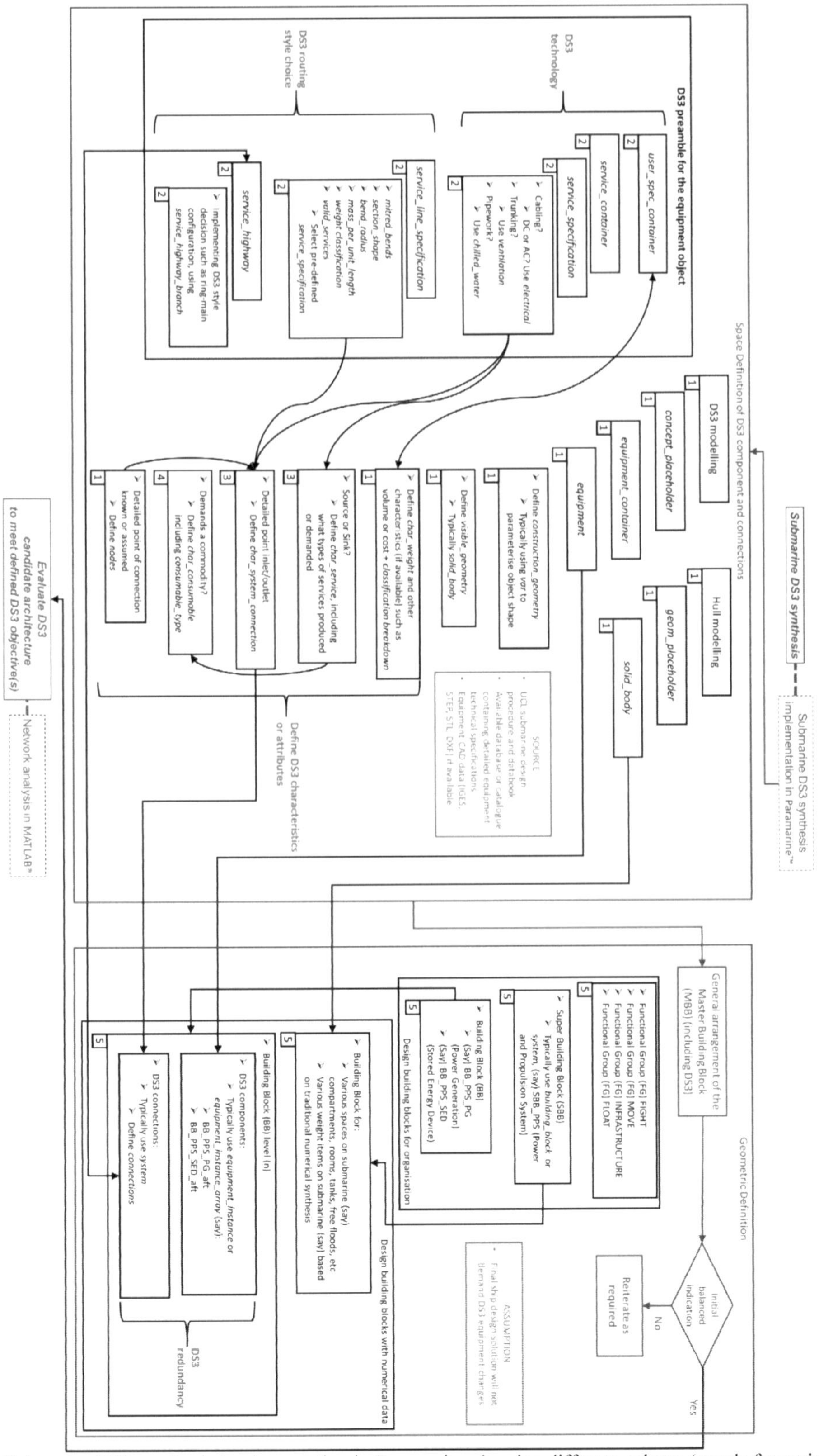

Figure 16: Submarine DS3 synthesis implementation in Paramarine showing different colours (purple for main tasks, red for DS3 modelling, green for assumptions, and blue for the network analysis task)

This consists of five steps (see the left-bottom of each box): Step 1 is to create objects for modelling the hull and DS3 components in Paramarine, shown in italic, such as "**geom_placeholder**" and "**equipment_container**" (Qinetiq, 2019). DS3 equipment physical size including creating objects to host the possible connection point (x, y, z) of equipment; Step 2 to set up the DS3 connection(s) of a DS3 component adding more objects in Paramarine, such as "**user_spec_container**", and "**service_container**" (Qinetiq, 2019); Step 3 is to provide input for DS3 component as a source or sink and define what relevant DS3 connections could be connected to the DS3 component in terms of DS3 technology; Step 4 is to define the numerical commodity demand of the equipment (if applicable); Step 5 is to perform DS3 modelling using SURFCON module in Paramarine, which consists of creating design building block objects that are descriptive (or for 'organisation' (Pawling, 2007)) and building block with numerical data, not only for defining space and weight on the submarine but also for the DS3 components and connections.

3. THE NETWORK BLOCK APPROACH

The implementation of the UCL Design Building Block approach in Paramarine with its SURFCON (DBB) module is highly flexible and so useful for DS3 synthesis (see Figure 15). The algorithms and assumptions were part of the input and thus the implementation of the UCL Design Building Block approach in Paramarine was not a black-box process. However, such a glass box approach is demanding every time a new design is developed. Furthermore, integrating the network-based sizing approach with the submarine design process using SURFCON Paramarine added significant effort which was found to inhibit exploring DS3 options in ESSD.

Therefore, an approach termed the Network Block Approach (NBA) was proposed. The NBA consisted of frameworks, methods, and design tools that employed a strategy to 'intercept' data flow before being inputted to Paramarine to a set of spreadsheet Excel inputs. The main objective of the development of the NBA was to create an integrated design procedure that incorporates a network and simulation based DS3 sizing approach called SUBFLOW (see Mukti (2022)) and the UCL Design Building Block approach. Unlike the previous SUBFLOW in the recent IJME paper (Mukti et al., 2021), the current SUBFLOW, part of the Network Block Approach (NBA) (Mukti, 2022) was not used to optimise DS3 design nor the overall submarine design. Thus, compared to other network-based approaches (Shields et al., 2017; Duchateau et al., 2018; Parsons, 2021), the optimisation technique in SUBFLOW was only used to solve the energy balance through linear programming. This was consistent with requiring either a simple measure of merit (MoM) or even subjective combinations of MoMs required in multicriteria approaches (Andrews, 2018).

On one hand, SUBFLOW requires design data at a specific level of design granularities (Mukti et al., 2021). On the other hand, the UCL Design Building Block approach facilitates the development of a new design ab initio at the level of design granularity required for DS3 synthesis. Therefore, the architecture of the NBA must consider the two distinct design philosophies. To merge the advantages of both approaches, the NBA was developed based on DevOps software practice that is a blend of two different activities, 'Development' and 'Operations' (Hüttermann, 2012). In this case, the 'Development' represents the implementation of the UCL Design Building Block approach in Paramarine and the 'Operations' represents the DS3 synthesis using SUBFLOW in MATLAB. This leads to an infinity loop diagram presented in Figure 17, which shows the iterative nature of the approach and can be terminated once the design is considered sufficiently naval architecturally balanced.

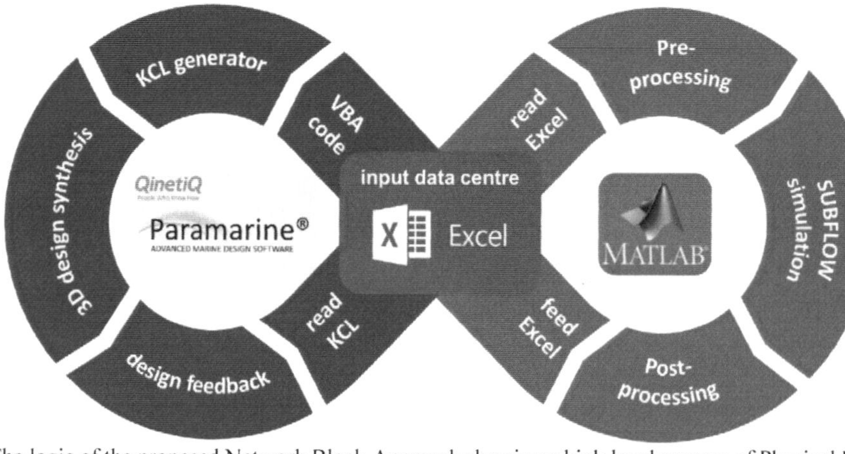

Figure 17: The logic of the proposed Network Block Approach showing a high level process of Physical Loop method in purple and Logical Loop method in blue

The input data centre in Figure 17 consisted of several programs, summarised in Figure 18 and Table 1. Each program was developed as a worksheet with its distinct cells layout, which could be read both by Paramarine for rapid modelling of objects, and by MATLAB, for automatically generating codes to perform SUBFLOW simulations. With this approach, the designer could focus and readily manipulate the architecture of the vessel and perform SUBFLOW simulations without needing to address the Gulf of Execution (Figure 4) in Paramarine and MATLAB.

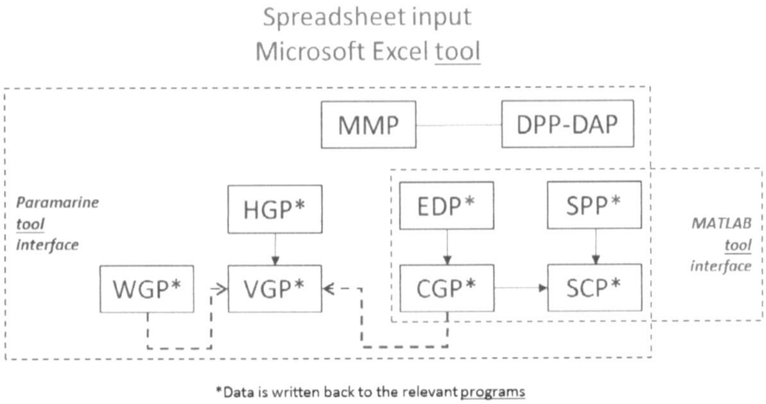

Figure 18: The detailed breakdown of the input data centre, showing multiple programs in green, Paramarine interface in purple and MATLAB interface in blue

Table 1: Summary of programs in the Network Block Approach (see Figure 18)

Program	Description	Function
MMP	Main Menu Program	Execution menu to compile all programs
DPP	Design Preamble Program	Hardcoded design setup
DAP	Design Analysis Program	Hardcoded analysis setup
HGP	Hull Geometry Program	Input for hull size
VGP	Volume Granularity Program	Input for spaces
WGP	Weight Granularity Program	Input for weight
EDP	Equipment Database Program	Input for equipment data
CGP	Component Granularity Program	Input for DS3 components for arrangement and SUBFLOW
SPP	System Preamble Program	Input for DS3 connections
SCP	System Connection Program	Input for DS3 connection and SUBFLOW

The NBA is not just an Excel tool (Mukti et al., 2022b), it comes with extensive frameworks (Mukti et al., 2022c). The main framework specific for DS3 routing is outlined in the following subsection.

3.1 DISTRIBUTED SYSTEMS ROUTING FRAMEWORK

The issue with the DS3 modelling, more specifically, DS3 routing in Paramarine has been shown in Section 2, which led to two different systems routing frameworks. The first framework employed automatic routing in Paramarine that was termed in the current research as the 'point to point' framework. However, if, for instance, the designer wanted to model a more detailed routing could be used, namely 'highway' framework, combining automatic routing and highway consisting multiple longitudinal system runs (see Figure 19).

The input procedure for the point-to-point framework can commence by developing the design to the third level of granularity (i.e., component granularity). Hence, two design building blocks (at the component level) could be defined as a connection. This was found to provide a quick routing approach commensurate with early-stage design but would have made the routings of the system overlap each other as the DS3 component building block had yet to be arranged in unique x, y, and z coordinates. This is demonstrated in the next section.

The procedure for the highway framework was akin to the point-to-point framework. However, this procedure requires the selection of a system highway, which hosts multiple system runs, as an additional input and thus the automatic 'point to point' DS3 routing can be guided by the selected highway. The highway framework consisted of some pre-defined longitudinal lines from forward to aft of the vessel and could be modelled as a highway object in Paramarine (see Figure 12). In this approach, the designer needed to first define how many highways of a given distributed system there were on the vessel. The system connection could then be assigned to one of the highways available, as shown in Figure 19. To define highways, allowances needed to be assumed for the cross-sectional space taken up by pressure hull ring frame stiffeners. This framework allowed the designer to investigate how much space needed to be reserved for systems routing early in the design process and before the layout become too cramped further downstream in the design. The advantages this now provides are discussed in greater detail in Section 4.

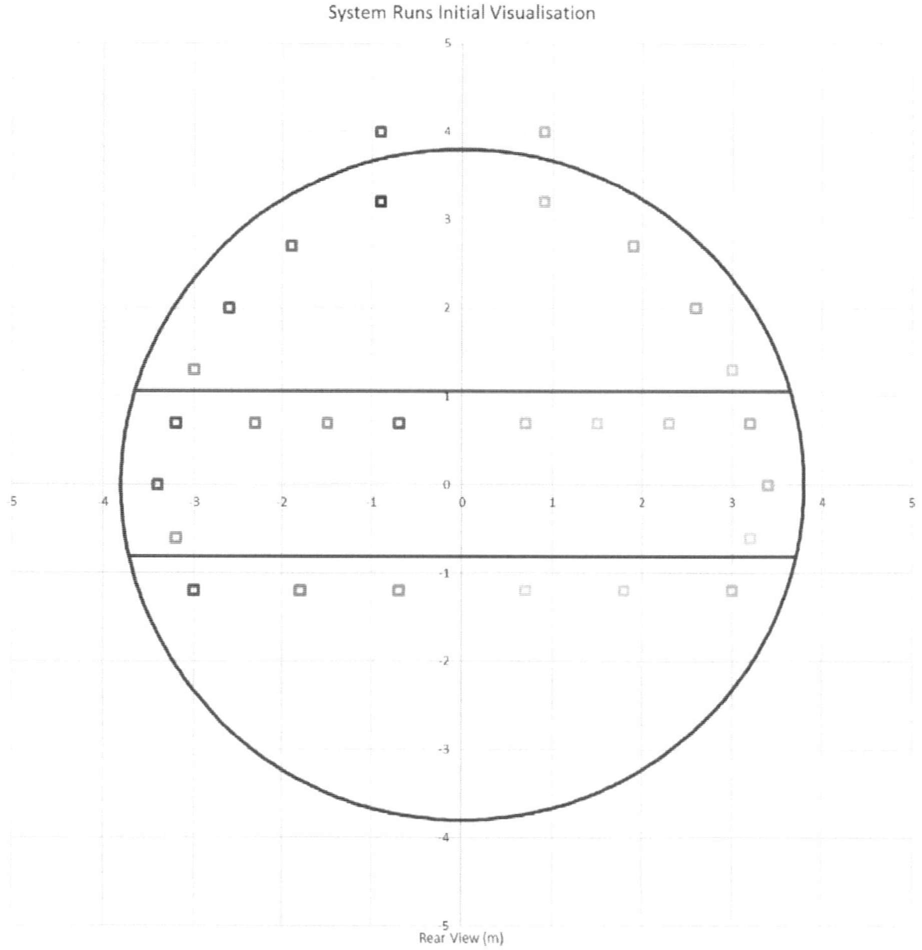

Figure 19: Initial highways visualisation and setup in System Preamble Program (SCP) for a typical submarine pressure hull

4. SUBMARINE CASE STUDY

The main objective of this section is to show that the new approach not only could provide more accurate or better-budgeted allowances for the weight and space of DS3 but also could inform Requirement Elucidation regarding choosing the style of various DS3 in the design.

4.1 CASE STUDY SETUP

To test whether the new approach could capture the style choices of DS3 at component granularity level and could be validated with available data, a case study was developed with the payload and style choices akin to the ocean-going 2500 tonne generic submarine (SSK) (Mukti and Randall, 2017) extracted from the database used in the annual UCL submarine design exercise (UCL-NAME, 2014) and is summarised in Table 2.

Table 2: The realisation for submarine case study following the decision making sequence for complex vessels outlined in detail in Figure 4 and Appendix of Andrews (2018) in a similar manner to the submarine example in Figure 4 of Andrews (2021) - detailed equipment spelt out in Mukti (2022)

Process Step	Selection Decision / Realisation for Submarine Case Study	
Perceived need	Demonstrate the application of the Network Block Approach (NBA)	
Outline of initial requirements	Initial SSK key parameters desired (capabilities)	
	A selected set of payload equipment (see (Mukti, 2022) for acronyms)	
	Payload Equipment	**Unit/Set**
	600 mm tubes (OD)	6
	Spearfish-UK	18
	Air turbines	2
	Bandfish - UK	2
	SPA4 RFOM-81	1
	R3 RFOM-100	1
	EW4 RFOM-72	1
	C5 RFOM-53	1
	CMS7 RFOM-85	1
	Initial Performance	**Value**
	Complement	46 personnel
	Operational environment density range	1-1.025 te/m^3
	Design diving depth	250 m
	Patrol length	49 days
Selection of style of emerging ship design	**Style Level**	**Choice**
	Macro Level	Non-nuclear (SSK)
	Main Level	Diesel-electric power plant
		Medium size ocean-going submarine
		Single hull with casing
		Three watertight bulkheads
	Micro Level	Detailed DS3 styles
Selection of major equipment and operational sub-systems	DS3 configurations (see (Mukti, 2022))	
Selection of whole ship performance characteristics	**General Performance**	**Value**
	Sprint speed	20 knots (2 hrs)
	Submerged speed	5 knots (17 hrs)
	Snort speed	6.5 knots (5 hrs)
	Indiscretion ratio	22 %
	Sub Level	**Value**
	For DS3	Subsections
Selection of synthesis model type	The Network Block Approach (NBA) ((Mukti et al., 2022c)	

4.2 INPUTS FOR DISTRIBUTED SYSTEMS ROUTING

As the development of the DS3 network was not intended to 'bottom out the preferred design' i.e., it required to be open to revision, simplifying assumptions were made for each DS3 technology. This kept the network simple for ESSD application while avoiding the loss of sufficient DS3 complexity (see Figure 20). Therefore, although each DS3 network was developed based on the relevant system line diagram, some nodes were modelled without yet having geometric data. They were produced to complete the network structure and to aid the DS3 routing decisions. Furthermore, the detailed components, which were not explicitly modelled in this network, were accommodated using typical design margins. This input could be manually adjusted using the System Connection Program (SCP) and Component Granularity Program (CGP) (see Table 1).

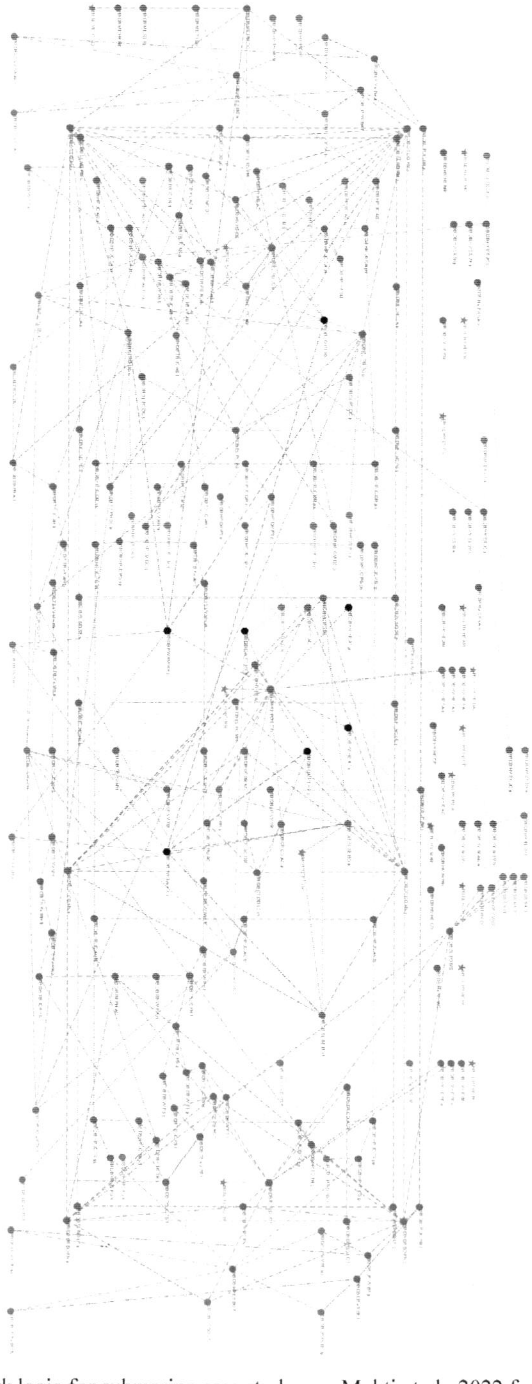

Figure 20: Overall DS3 network logic for submarine case study, see Mukti et al., 2022 for the description of the colour coding used for different types of nodes and arcs for various DS3 technologies (see (Mukti, 2022) for the shape and acronyms of the nodes)

4.3 INITIAL ROUTING RESULTS

Before the selected DS3 can be properly sized, the indication of the overall DS3 network logic in Figure 20 could also be translated into physical definition rapidly by the new programs (Table 1). This was done by first assigning the DS3 components to the various spaces on the vessel. For initial routing results, some simplifications in the inputs for Component Granularity Program (CGP) and System Connection Program (SCP) were made just for the visualisation, enabling the designer to simultaneously develop the DS3 design physically and logically. Firstly, the program was set to employ point to point, which meant the 468 connections overlapped each other. Secondly, the size of all connections was adjusted so they can be reasonably visible in the 3D model. Thirdly, the 231 DS3 components were assigned at the centroid of each object's volume, which are modelled as simple spheres. These simplifications resulting in a visualisation of all 14 DS3 routings are given in Figure 21.

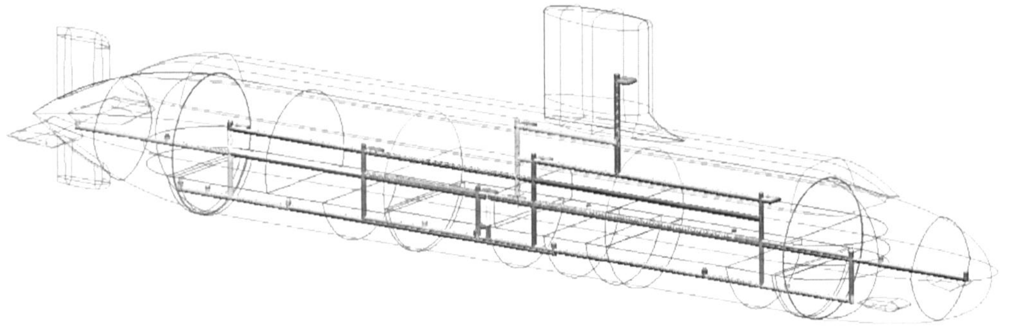

Figure 21: Initial routing showing DS3 components situated at the centre of each volume space, which had been defined using CGP, SPP, and SCP (see Table 2)

4.4 DISTRIBUTED SYSTEMS FRAMEWORK APPLICATION

A more refined Paramarine model was produced after DS3 sizing using SUBFLOW (Mukti et al., 2021) to demonstrate the applicability of the novel DS3 routing framework outlined in Subsection 3.1. As shown in Figure 22, this was done using the graphical interface in Paramarine by arranging each DS3 component, compartment by compartment. Thus, 230 unique x, y, z locations for the DS3 components could be identified and stored in the Component Granularity Program (CGP). Furthermore, the appropriate system highways (Subsection 3.1) could be added as additional inputs to the existing 698 items of DS3 data in the System Connection Program (SCP), as illustrated in Figure 23.

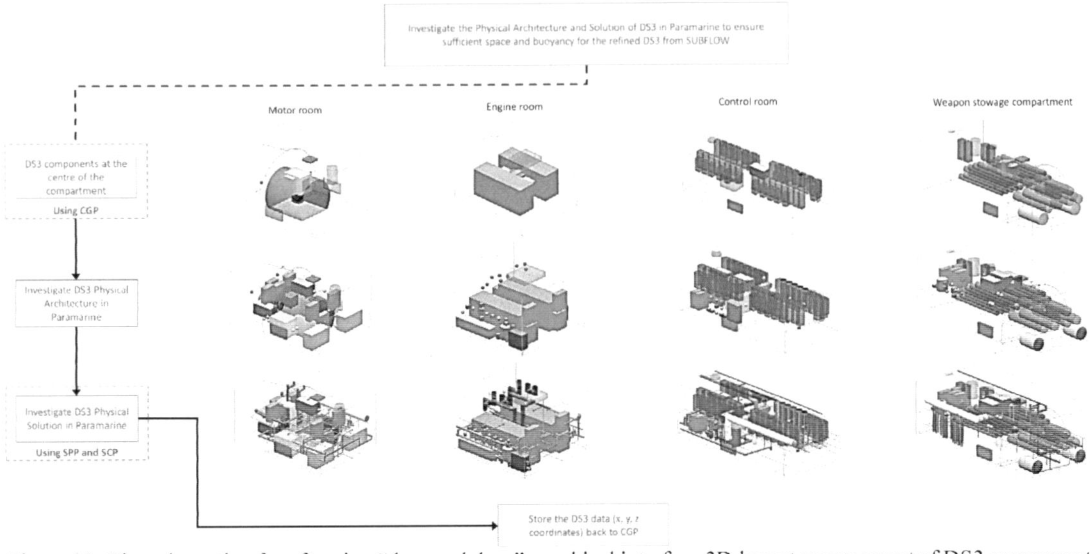

Figure 22: The schematic of performing "drag and drop" graphical interface 3D layout arrangement of DS3 components at ESSD fine stage in Paramarine, showing gradual step by step arrangement of procedure compartment by compartment

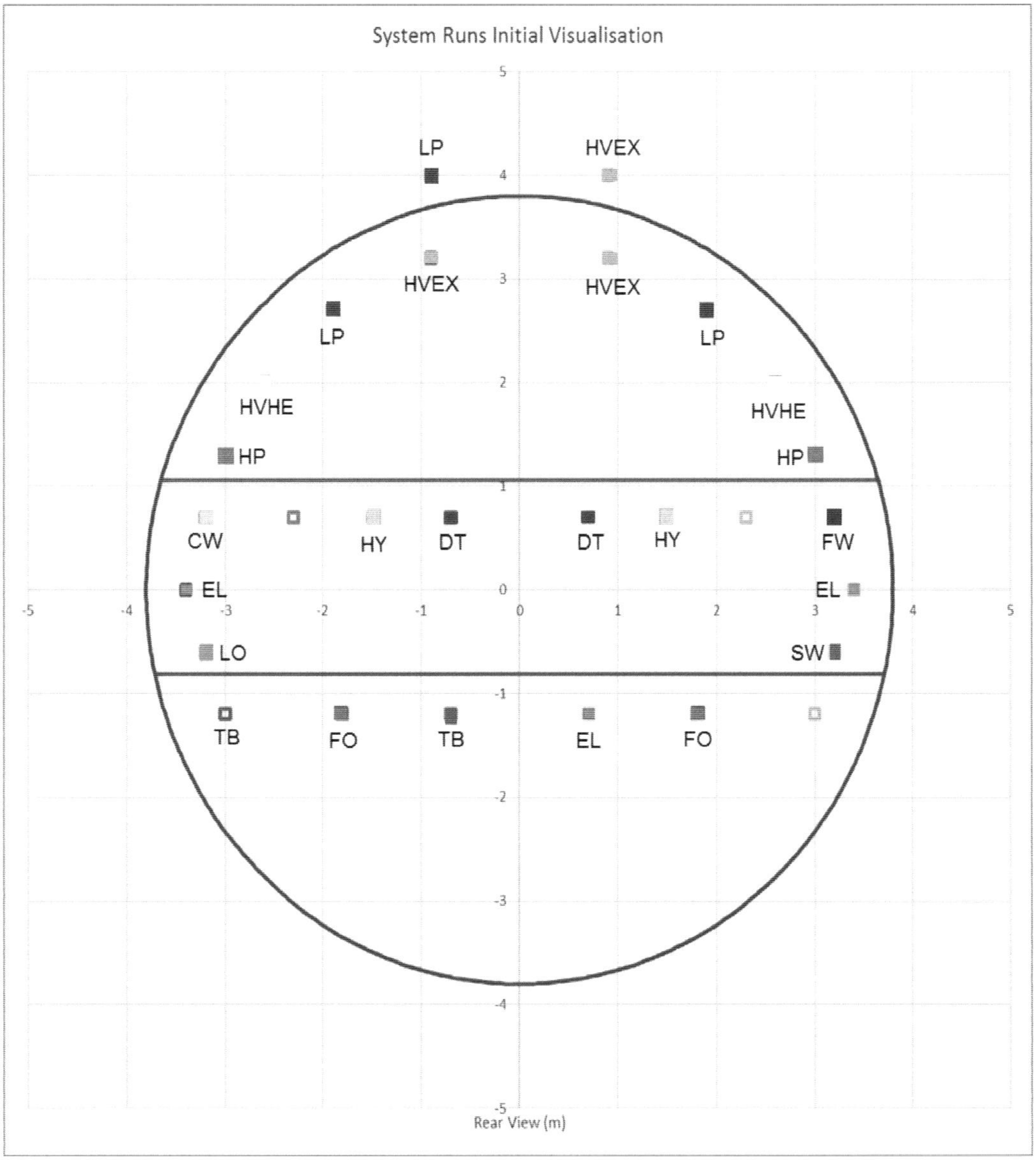

Figure 23: Initial highways of the submarine case study defined using System Preamble Program (SPP) part of the Network Block Approach (NBA) (see Mukti et al., 2022c for DS3 identifiers)

The refined overall submarine model is given in Figure 24, showing gradual steps in manipulating DS3 components in the 3D layout. Starting from major components based on the FMFI breakdown as in Figure 24 (top), followed by the electrical 'EL', mechanical 'ME', and data 'DT' systems in Figure 24 (middle), and finally, the rest of the systems listed in Mukti et al. (2022c) were arranged, as shown in Figure 24 (bottom).

5. DISCUSSION

If compartment density were used and the design progress was stopped at a level of detail such as that given in Figure 21, this then raised the question as to whether a more automated approach would be suitable for this level of detailed arrangement of DS3 components. This level of detail is likely to be performed by the design organisation downstream in the subsequent design phases. This would be in preference to using an automated arrangement approach for arranging such DS3 components in ESSD, such as Chalfant (2022) employed. As a complex vessel, many of the stakeholders in submarine design, including specialist engineers, are likely to have conflicting visions that must be resolved in the eventual design solution and thus detailed arrangement should not be made by default or determined by a black box toolset.

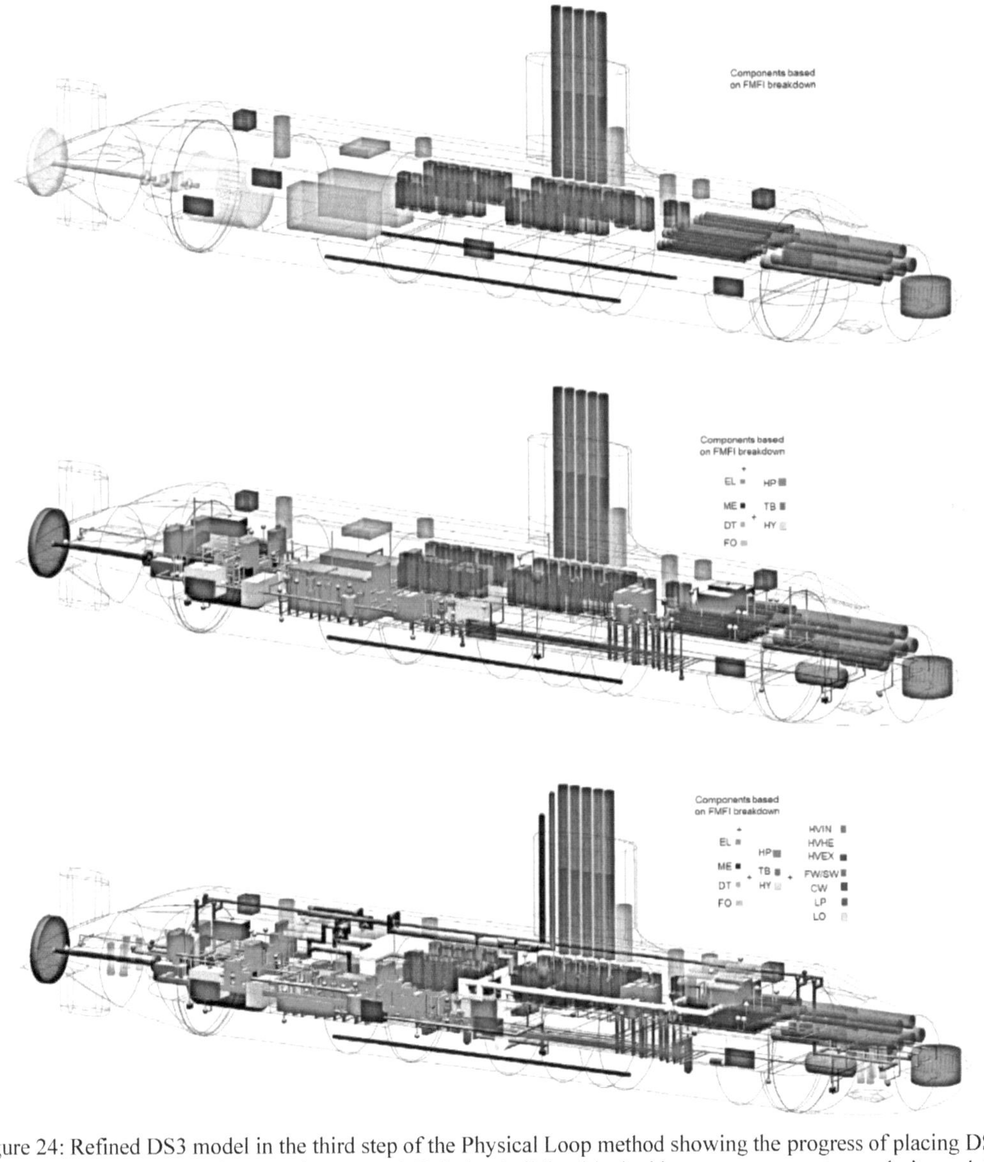

Figure 24: Refined DS3 model in the third step of the Physical Loop method showing the progress of placing DS3 system by system in the 3D layout. There were 365 design building block objects at component granularity and 472 design building block objects at connection granularity level. The latter shows 14 DS3 top level routings

The proposed approach demonstrated that the initial DS3 routing would be amenable to automation or (preferably) a semi-automatic approach. In the latter case the designer can still have the authority to define where the highways are located (Figure 23). This gave an estimate of at least 1.5 km of cabling, 280 m of trunking, and 2.8 km of piping required to build such a submarine DS3. However, the case study revealed a realistic routing (as shown in Figure 1) remained difficult to be made at the concept. This was due to requiring input data (beyond the overall DS3 network in Figure 20) and design work (such as required to model complex shapes, such as structure duct routing, and many small cables or piping), unlikely to be available or seen necessary in ESSD.

This also suggests that such modelling refine meant is then likely to require margins to be addressable in the relevant programs within the Network Block Approach (NBA). However, the quality of 3D based semi-automatic DS3 routing produced using the NBA in the Paramarine-SURFCON could be said more plausible than the current 2.5D routing approaches on the surface ship application (Figure 25). Furthermore, another potential advantage of an initial definition of the DS3 element shown by the case study presented in Figure 25 (bottom), was to reveal whether there was adequate space for various DS3 connections and the lessening of potential routing clashes when it came to allocating detailed DS3 components within relevant compartments. This would clearly benefit design development later in the design.

Physical solution (Goodrum et al., 2017)

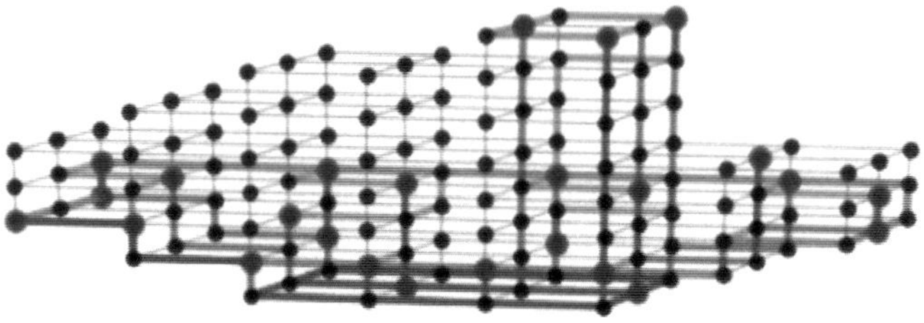

Early routing (Duchateau et al., 2018)

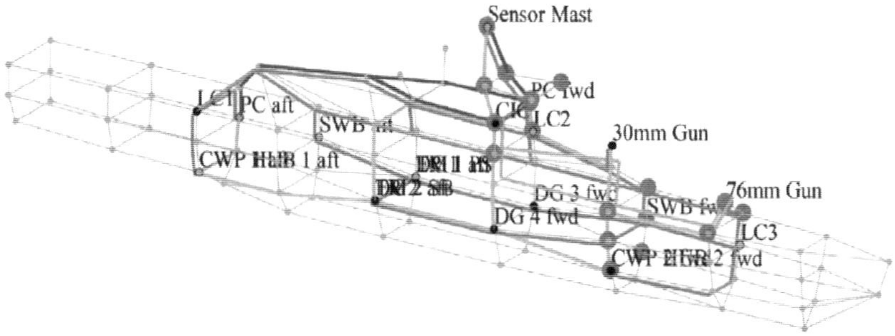

The Architecture Flow Optimisation (AFO) approach (Brown, 2020)

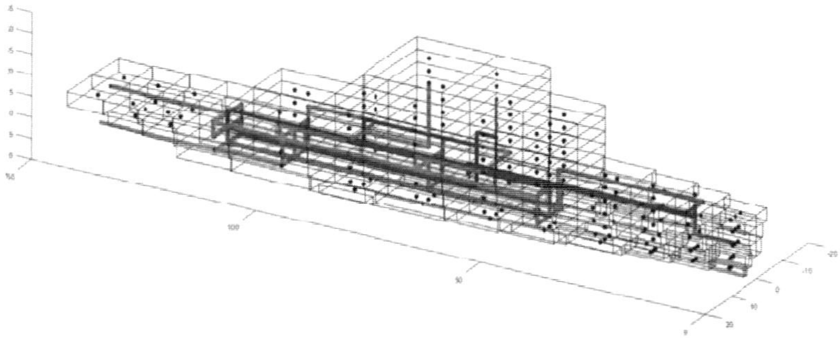

Smart Ship Systems Design (Chalfant, 2022)

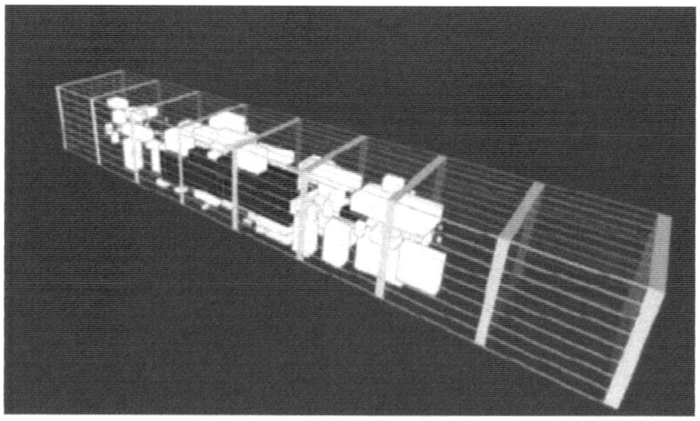

The Network Block approach (Mukti, 2022)

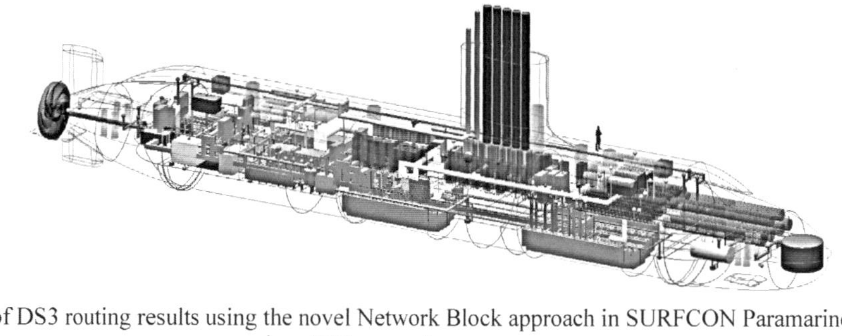

Figure 25: Comparison of DS3 routing results using the novel Network Block approach in SURFCON Paramarine with previous DS3 research

6. CONCLUSIONS AND RECOMMENDATIONS

Despite the benefits identified above, realistic routing remains difficult to be made in ESSD, as it required more design input effort and tends to bottom out the preferred submarine design, which assumes the final submarine design solution will not demand many DS3 components and connections changes from the initial Concept Design — itself primarily driven by the exploratory nature of Requirement Elucidation. Furthermore, although some alternative approaches have been proposed (such as the use of margins and numerical nodes), still, to accurately size various DS3, with specific style choices, would require the use of an extensive equipment database.

The case study in Section 4 is considered to reveal at ESD the acknowledged high dependency of submarines on their distributed ship systems. The DS3 network in Section 4 could still be developed to a greater level to model more of the total distributed ship systems of a submarine, such as degaussing system or salvage blow system, which would introduce many more pumps and valves than in the case study modelled. This raises the issue as to whether the greater detail, beyond that presented in Section 4, might not be seen to overwhelm the designer with DS3 information in ESSD. However, some DS3, such as plumbing and water system, were not seen to present issues as they do not need to be modelled to further levels of design granularity, and so the use of a simple scaling approach was sufficient.

Validating the accuracy of DS3 modelling would require further comparison of the routing results with several real submarine designs. However, given the lack of such detailed data in the public domain due to military or commercially sensitivity, the level of effort to perform such validation would have been significant, if at all possible, in an academic environment. However, in industry, such a validation ought to be undertaken using weight and space breakdown data from previous real submarine designs available to government or submarine designers and builders.

The DS3 components were arranged manually not only so that they did not clash with each other but also in consideration of DS3 aspects, such as Sustainability (Mukti et al., 2022a). Still, many clashes would require more detailed arrangement and routing modelling, which are unlikely to be resolved in ESSD. Alternatively, this could be mitigated if a more robust DS3 automatic routing was readily available in SURFCON Paramarine. The model shown in Figure 25 (bottom) was considered sufficient or a step closer to show the extent of the complexity of the physical architecture of DS3 which can be modelled rapidly with flexibility in ESSD, using the novel tools within the Network Block approach.

8. ACKNOWLEDGEMENTS

The travel grant funded by Larry Fredrick from the UK Ministry of Defence is gratefully acknowledged.

9. REFERENCES

1. Andrews, D. and Andrews, D. (2022) "A New Way to Visualise the Human Element in A Complex Engineering Design Environment." The Naval Architect January 2021.

2. Andrews, D.J. (1981) Creative Ship Design. Trans. RINA, 123. doi:10.3940/rina.trans.1981

3. Andrews, D.J. (1984) Synthesis in Ship Design. Ph.D. thesis, University College London.

4. Andrews, D.J. (1994) Preliminary Warship Design. Trans. RINA, 136. doi:10.3940/rina.sbt.1994

5. Andrews, D.J. (2018) The Sophistication of Early-Stage Design for Complex Vessels. RINA IJME Special Edition, 472 Part A. doi:10.3940/rina.ijme.2018.SE.472.

6. Andrews, D.J. (2021) Who Says There Are No Real Choices in Submarine Design? RINA Submarine Conference, June 2021. RINA.

7. Andrews, D.J., Casarosa, L. and Pawling, R. (2009) "Integrating Simulation and Computer Aided Ship Design Software and Processes." RINA ICCAS in September 2009. September 2009. RINA. doi:10.3940/rina.iccas.2009.53.

8. Andrews, D.J., Cudmore, A.C., Humble, P., et al. (1996) "SUBCON - A New Approach to Submarine Concept Design." RINA Warships Proc. Symp. Naval Submarines 5: The Total Weapons System. London, UK, June 1996. RINA. doi:10.3940/rina.warship.1996.13.

9. Andrews, D.J. and Dicks, C.A. (1997) "The Building Block Design Methodology Applied to Advanced Naval Ship Design." Proc. IMDC. Newcastle, June 1997. International Marine Design Conference (IMDC).

10. Andrews, D.J. and Pawling, R.J. (2003) "SURFCON A 21st Century Ship Design Tool." The 8th International Marine Design Conference. Athens, Greece, May 2003. IMDC.

11. Andrews, D.J. and Pawling, R.J. (2008) A Case Study in Preliminary Ship Design. RINA IJME. doi:10.3940/rina.ijme.2008.a3.139.

12. Arentzen, E.S. and Mandel, P. (1960) Naval Architectural Aspects of Submarine Design. Trans. SNAME, 68.

13. Asmara, A. (2013) Pipe routing framework for detailed ship design. Ph.D. thesis, Delft University of Technology.

14. Brown, A.J. (2020) Design of Marine Engineering Systems in Ship Concept Design. The Marine Engineering Series. SNAME. Available at: https://www.sname.org/marine-engineering.

15. Burcher, R. and Rydill, L. (1994) Concepts in Submarine Design. Cambridge Ocean Technology Series. Cambridge [England]: Cambridge University Press.

16. Chalfant, J.S. (2015) Early-Stage Design for Electric Ship. Proceedings of the IEEE, 103 (12): 2252–2266. doi:10.1109/JPROC.2015.2459672.

17. Chalfant, J.S. (2022) "Advancing Automation in Early-Stage Navy Ship System Design." International Marine Design Conference (IMDC). Vancouver, Canada, June 2022.

18. Duchateau, E.A.E., de Vos, P. and van Leeuwen, S. (2018) "Early-stage routing of distrubuted ship service systems for vulnerability reduction." Marine design XIII: proceedings of the 13th International Marine Design Conference (IMDC 2018). Helsinki, Finland, June 2018.

19. Gongora, R. de (2019) "Practical Use Cases of Artificial Intelligence in the Ship Design Stage." ICCAS Rotterdam, Netherlands, September 2019. RINA

20. Goodrum, C.J., Shields, C.P.F. and Singer, D.J. (2017) "Investigating the Impact of Distributed System Routing Densities on Vessel Operability." ICCAS 2017: International Conference on Computer Applications in Shipbuilding. Singapore, September 2017.

21. Gottschalk, S., Lin, M.C. and Manocha, D. (1996) "OBBTree: a hierarchical structure for rapid interference detection." Proceedings of the 23rd annual conference on Computer graphics and interactive techniques - SIGGRAPH '96. Not Known, 1996. ACM Press. pp. 171–180. doi:10.1145/237170.237244.

22. Hüttermann, M. (2012) DevOps for developers. The expert's voice in web development. Apress. New York, NY.

23. Jackson, C.H.A. (1992) Fundamentals of Submarine Concept Design. SNAME Transactions, 100: 419–448.

24. Kouriampalis, N., Pawling, R.J. and Andrews, D.J. (2021) Modelling the operational effects of deploying and retrieving a fleet of uninhabited vehicles on the design of dedicated naval surface ships. Ocean Engineering, 219: 108274. doi:10.1016/j.oceaneng.2020.108274.

25. Mukti, M.H. (2022) A Network-Based Design Synthesis of Distributed Ship Services Systems for a Non-Nuclear Powered Submarine in Early Stage Design. Ph.D. thesis, University College London.

26. Mukti, M.H., Pawling, R.J. and Andrews, D.J. (2021) Distributed Ship Service Systems Architecture in The Early Stages of Designing Physically Large and Complex Vessels: The Submarine Case. IJME, 163. doi:https://doi.org/10.5750/ijme.v163iA2.755.

27. Mukti, M.H., Pawling, R.J. and Andrews, D.J. (2022a) "Design for Support in the Initial Design Using the Network Block Approach: The Submarine Case." RINA Warship. Bristol, UK, June 2022. RINA.

28. Mukti, M.H., Pawling, R.J. and Andrews, D.J. (2022b) Development of an Early-Stage Design Tool for Rapid of Distributed Ship Service Systems Modelling in Paramarine – a Submarine Case Study. COMPIT Pontignano, Italy, June 2022.

29. Mukti, M.H., Pawling, R.J. and Andrews, D.J. (2022c) "The Network Block Approach Applied to the Initial Design of Submarine Distributed Ship Service Systems." The 14th Annual International Marine Design Conference. Vancouver, Canada, 2022.

30. Mukti, M.H., Pawling, R.J., Savage, C., et al. (2019) "Distributed Ship Service Systems Architecture in the Early Stages of Physically Large and Complex Products." ICCAS Rotterdam, Netherlands, September 2019. RINA.

31. Mukti, M.H. and Randall, R.E. (2017) "Graphic Method for Improved Indonesian Navy Submarine Design Acquisition; the Investigation of Small (Midget) Naval Submarine Development." RINA Warship: Naval Submarines & UUVs. Bath, UK, June 2017. doi:10.3940/rina.ws.2017.07.

32. Norman, D. (2013) The design of everyday things. Revised and expanded edition. Cambridge, Massachusetts: MIT Press.

33. Parsons, M.A. (2021) Network-Based Naval Ship Distributed System Design and Mission Effectiveness using Dynamic Architecture Flow Optimization. Ph.D. thesis, Virginia Tech.

34. Pawling, R.J. (2007) The Application of the Design Building Block Approach to Innovative Ship Design. Ph.D. thesis, University College London.

35. Pawling, R.J., Piperakis, A. and Andrews, D. (2015) "Developing Architecturally Oriented Concept Ship Design Tools for Research and Education." 12th International Marine Design Conference 2015. Tokyo, Japan, May 2015.

36. Purton, I.M. (2016) Concept Exploration for a Novel Submarine Concept Using Innovative Computer-Based Research Approaches and Tools. Ph.D. thesis, University College London.

37. Qinetiq (2019) Paramarine. Available at: https://paramarine.qinetiq.com/products/paramarine/index.aspx (Accessed: 8 July 2019).

38. SAAB (2021) A submarine in space | Stories | Saab. Available at: https://www.saab.com/newsroom/stories/2020/july/a-submarine-in-space (Accessed: 13 August 2021).

39. Shields, C.P.F., Rigterink, D.T. and Singer, D.J. (2017) Investigating physical solutions in the architectural design of distributed ship service systems. Ocean Engineering, 135: 236–245. doi:10.1016/j.oceaneng.2017.02.037.

40. UCL-NAME (2014) Submarine Data Book. London: UCL Department of Mechanical Engineering, Naval Architecture and Marine Engineering.

41. Wilgenhof, J. and Reijmers, J. (2010) "Ventilation system design and application in a new submarine" UDT 2010: u o e nde ea e en e e hno ogy Ha u g, e any, une 1

9. AUTHORS BIOGRAPHY

Dr Muhammad Hary Mukti is a Post-Doctoral Researcher at UCL UK investigating computer-based tools to synthesise distributed systems in the early stages of complex products. Prior to this he obtained an MEng in Ocean Engineering (in the subsea project and underwater vehicle design) from Texas A&M University and accomplished a 2N Professional Summer Course at Massachusetts Institute of Technology USA. In 2022 he was awarded the W H C Nicholas Prize for the best paper published in the RINA Transactions for describing his research in the field of engineering distributed systems design.

Dr Rachel Pawling is a Lecturer in Ship Design at UCL, teaching the subject of ship design to undergraduate and postgraduate students. She obtained her PhD in computer aided ship design at UCL in 2007 as part of the UCL Design Research Centre and has subsequently continued her research in projects funded by the EC, UK and US governments. She also awarded the W H C Nicholas Prize in 2010 and Samuel Baxter Prize in 2009.

Prof David Andrews, FREng, PhD, FRINA was given a new Chair in Engineering Design at UCL in 2000, following his early retirement as a Director the UK MoD. His early career in the RCNC included postings in in-service and new construction submarine design sections. He was subsequently Head of Preliminary Design responsible for the initial studies, including on the Future Attack Submarine. In 1993-98 he was appointed MoD Professor of Naval Architecture at UCL where he directed the MSc in Naval Architecture and the post-MSc Submarine Design Course. In 2001 he set up a new Design Research Centre in the Department of Mechanical Engineering, which is focused on CASD and the design of complex systems. He is a Fellow of RINA and was a Vice President and was awarded the Froude Medal in 2020. In 2000 he was elected a Fellow of the Royal Academy of Engineering and in 2021 awarded the David Taylor Medal by SNAME.

*** Formatting Issue - Best Available Paper/Graphic ***

AN EXPANDED APPLICATION OF BASIC SHIP-PLANNING SUPPORT SYSTEM USING BIG DATA IN MARITIME LOGISTICS FOR PANAMAX AND CAPESIZE DRY BULK CARRIERS

D A F Muzhoffar, K Hamada, T Saito, Hiroshima University, Japan
Y Wada, National Maritime Research Institute, Japan
H Harada, Tsuneishi Shipbuilding Co., Ltd., Japan

SUMMARY

Fluctuations in global industrial cargo require continuous ship advancement. Meanwhile, digitized shipping generates big data, which enables diverse analyses. Thus, we developed a basic ship planning support system that utilizes big data in maritime logistics. Prior studies enhanced the allocation algorithm to include shipping cost (COST) and greenhouse gas (GHG) features, although it was only applied to Capesize dry bulk carriers operating in the time-charter contract. This study presents an expansion of the applicability to include Panamax dry bulk carriers. A cargo movement model was developed to accommodate monthly cargo movement. Moreover, the allocation algorithm was broadened to consider voyage-charter contracts. This development enables operation-level optimization, assuming that ships operate in time- and voyage-charter contracts. Finally, the ships in demand can be analyzed through the defined scenarios, namely ship replacement while preserving existing ship allocation, optimization of ship allocation using only existing ships, and the presence of new ships.

NOMENCLATURE

a_i	Daily average fuel consumption of auxiliary engine of Ship i (t/d)
b_i	Daily average fuel consumption of the boiler of Ship i (t/d)
CF	Heavy fuel oil (HFO) carbon factors (assumed constants: 3.114 t-CO2/t-fuel)
cf	Marine diesel oil (MDO) carbon factors (assumed constants: 3.206 t-CO2/t-fuel)
CM_{ij}	Cargo movement carried by Ship i while serving route j (t)
D_i	Maximum draught of Ship i (m)
d_{ij}	Actual draught of Ship i while serving route j (m)
dc_i	Daily depreciation cost of Ship i (USD/d)
FP	HFO prices (assumed constants: 300 USD/t-fuel)
fp	MDO prices (assumed constants: 600 USD/t-fuel)
I	Set of available ships
i	An individual ship
J	Set of available routes
j	An individual route
M	Set of twelve months in a year
m	An individual month
m'_{ij}	Merit for each ship i serving route j
n_{ij}	Possible annual trip number
ηw_{ij}	Correction factors of weather while serving route j (assumed constant:0.909 [1,2])
ηf_{ij}	Correction factors of fouling while serving route j (assumed constant:0.917 [1,2])
oc_i	Daily operation cost of Ship i (USD/d)
P_i	Maximum main engine power of Ship i (kW)
sc_l	Stockpile cost unit (iron ore:0.012; coal:0.042; grain:0.186; others:0.186 USD/t/d) [3]
sfc_i	Specific fuel oil consumption of the main engine of Ship i (g/kWh)
st_{ij}	Round-trip sailing time of Ship i (d)
V_i	Design speed of Ship i (kn)
v_{ij}	Average sailing speed while serving route j of Ship i (kn)
$w_{i'j}$	Key performance index (KPI) for each ship i' serving route j
yd	Assumed days in a year: 365 days
y_{ij}, w_{ij}	KPI for each ship i serving route j
y_{ijm}	Monthly KPI for each ship i serving route j
$\bar{y}_j$	Average KPI for route j

*** Formatting Issue - Best Available Paper/Graphic ***

1. INTRODUCTION

Worldwide economic growth promotes huge demand for industrial cargo transport, including ship movement and replacement in the maritime industry [4]. This emphasizes the importance of continuously developing adequate ship specifications. Meanwhile, the digitized business model generates a vast data stream [5-7]. The utilization of these data is significant for various analyses in diverse industries. In the maritime industry, digitized shipping through an automatic identification system (AIS) enables geospatial monitoring of ships, together with formerly analog-stored data, such as ship, port, and route data [8-11]. This accessible big data has allowed various studies to be conducted [12-15]. Specifically, data-driven studies have conducted demand forecasting analyses for both regional and global scales, such as the cargo throughput, shipping, and shipbuilding market [16-28]; however, few studies have focused on examining the demand for new ship specifications [29,30].

A number of studies have constructed long-term models for predicting cargo demand using a combination of ship and port data. Yang and Chang [16] studied forecasting the demand for container throughput in five ports in Taiwan using a mixed- e i ion neu a ne o Fu he o e, ugo i e a [17] discussed the aspects that influence maritime shipping markets, focusing on defining the three key economic concepts to analyze supply and demand in the freight market: supply, demand, and equilibrium prices. Akar and Esmer [18] investigated container volume forecasting of container ports in Turkey using a proposed multi-regression analysis. i i a y, u e a [19] examined the accuracy of four discrete forecasting models to forecast annual container throughput in the ports of Istanbul, Izmir, and Mersin. Moreover, Jia et al. [20] estimated the payload of bulk carriers based on AIS-reported draught data by adopting a multifactor regression model. Similarly, Zhou and Hu [21] examined AIS-based iron ore trade volume estimation using a constructed back-propagation neural network. Kanamoto et al. [22] discussed the applicability of the logit model to forecast future dry bulk cargo movement using AIS data for dry bulk carriers.

In the case of shipping and newbuilding markets, Prochazka et al. [23] analyzed the effectiveness of AIS data to comprehend the characteristics of the crude oil tanker spot-market charterer using a defined spatial pattern. Using statistical data, Sharma and Sha [24] predicted a newbuilding index using a formulated forecasting model and case studies. Similarly, Wada et al. [25-27] suggested a system dynamics model to forecast shipbuilding market demand and long-term greenhouse gas (GHG) emission reduction measures by applying the proposed demand-forecasting model. Finally, Lee and Jung [28] proposed an autoregressive model to forecast the order quantity of container ships and bulk carriers. From the aforementioned studies, it can be observed that no study has been conducted on operation-level demand forecasting to understand the benefits of new ship specifications in the future.

Therefore, the authors [29,30] proposed a basic ship planning support system that integrates big data into maritime logistics. However, the system was built to consider only iron ore bulk carriers operating on limited routes. Moreover, we previously broadened the scope of simulations to accommodate iron ore and coal routes worldwide, with the extended shipping cost (COST) and GHG attributes as allocation algorithm considerations [31]. Hence, the feasibility of simulating operation-level ship allocation in a time-charter contract using the proposed ship allocation algorithm was explored.

Furthermore, the objective of this study was to expand the application of previously built systems to accommodate Panamax dry bulk carriers through the following improvements:

- Wider target ships, including Panamax and Capesize dry bulk carriers.
- The global network model is expanded to include the routes served by these target ships, worldwide routes of iron ore, coal, grain, and others.
- The cargo movement model is extended to include the monthly port-to-port cargo movement of iron ore, coal, grain, and others.
- A newly developed algorithm for reconstructing ship allocation assumes that ships operate in time- and voyage-charter contracts.

*** Formatting Issue - Best Available Paper/Graphic ***

2. BASIC CONCEPT

2.1 DATA USED IN THIS STUDY

This study extracted and integrated the following data from our basic ship planning support system.

- Ship movement data: This dataset presents dynamic data of the ships' port calling and its attributes, including ports of arrival and departure and draught information. Typical ship movement data comprises an entry in a port-to-port manner. We used AXSDry ship movement data from January 2013 to December 2020 [32].
- Ship data: This dataset presents static data of ship attributes, such as the International Marine Organization (IMO) number, principal particulars, design speed, main engine power, and year of construction. We adopted Sea-web Ships data covering 11,744 entries [33].
- Port data: This dataset presents static data on port properties, such as port name, ship size limitation, latitude–longitude coordinates, and handled cargo type. We used Sea-web Ports data covering 2098 ports [34,35].
- Route data: This dataset presents static data on sailing routes and the distances between ports. We obtained these from the IHS Fairplay [34,35]

2.2 BASIC SHIP-PLANNING SUPPORT SYSTEM IN PREVIOUS STUDY

The authors [29,30] proposed a maritime logistics database (MLDB) and ship allocation model in a previous study. These were assembled in a previously built basic ship-planning support system. In this context, MLDB integrates the available big data in maritime logistics, such as ship movement, ship, port, and route data. The MLDB was defined as a relational database; for example, the operating conditions can be clarified by integrating ship movement data and ship data. Error cleaning was performed to ensure the reliability of the data. Figure 1 shows the assembled discrete models and ship allocation algorithm in our previous basic ship-planning support system [31]. We define the following models using accessible big data in maritime logistics:

- The global network model characterizes each port and route served by the target ships.
- The cargo movement model describes the worldwide port-to-port cargo movement demand (CARGO) served by target ships.
- The ship model represents individual target ships' data and predicts its operation conditions.
- The ship allocation algorithm calculates GHG emissions [1,2] and COST [2,4,36,37] using the results of the above models, and further reconstructs the ship allocation, which is the core of our basic ship-planning support system. The total GHG emissions consist of the GHG emitted from the ship's main engine, auxiliary engine, and boiler, whereas the total COST includes the costs of ship fuel, operation, depreciation, and stockpile.

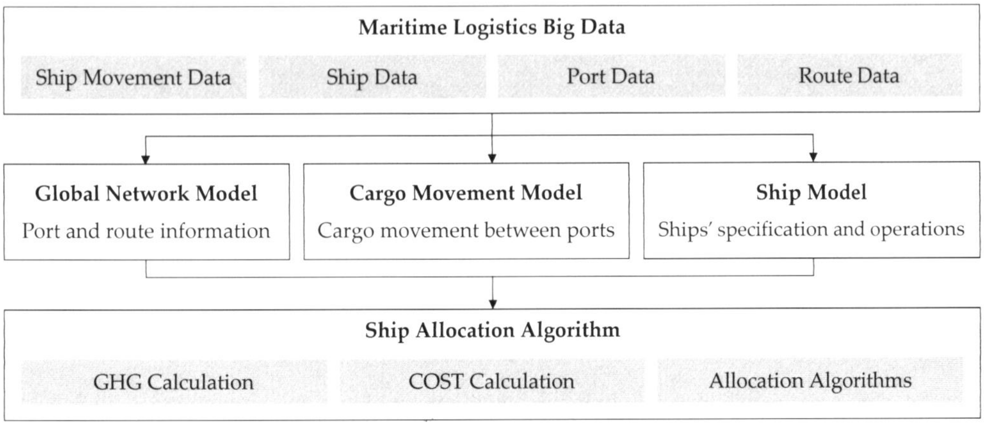

Figure 1. System configuration [31].

The changes in ship allocation can be determined by inputting the CARGO, existing ship and route lists, and future scenarios into the allocation algorithm. A previous study introduced new ships in the input of the simulation; hence, new ships that replace existing ships were also identified, making it possible to assess competitive ships in demand.

*** Formatting Issue - Best Available Paper/Graphic ***

In this study, the following improvements were developed to attain an extended application to include the Panamax dry bulk carrier:

- An expansion of the global network model: This model is expanded to include worldwide ports served by the target ships.
- An extension of the cargo movement model: This model is extended to include the monthly port-level CARGO served by the target ships, comprising the cargo types of iron ore, coal, grain, and others.
- Ship allocation scheme: A new allocation algorithm was developed to simulate the ship allocation, assuming the ships were operated in a voyage-time charter scheme, which was previously limited to a time-charter scheme.

3. MODEL DEVELOPMENT

3.1 GLOBAL NETWORK MODEL

This model was developed in a manner similar to that in our prior study [31]. This model defines each port and route served by Panamax and Capesize dry bulk carriers. We extracted 1234 ports visited by such carriers from the 2018 worldwide ship movement data. These ports were classified into main ports and port clusters. The main ports represent an individual port with significant port calling numbers, whereas port clusters include all other ports. The defined port and route network includes 147 main ports and the remaining 1087 ports as 85 port clusters, as shown in Figure 2.

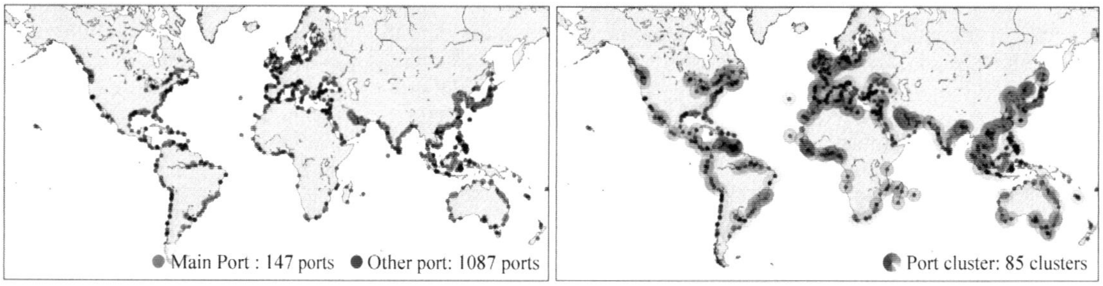

Figure 2. Worldwide ports of the global network model of Panamax and Capesize dry bulk carriers.

This model also comprises the dimensions of each port and the sailing distances between the ports. Port limitations are defined based on port data and the dimensions of the largest served ship. The sailing distances were extracted from the IHS Fairplay Ports and Terminals Guide [35]. Finally, this model comprised 3338 sailing routes of Panamax and Capesize dry bulk carriers.

3.2 CARGO MOVEMENT MODEL

This model defines the worldwide port-to-port CARGO served by the Panamax–MiniCape dry bulk carriers (deadweight tonnage (DWT) of 65,000–140,000). In contrast to our previous study [31], this study defined the CARGO between ports directly from the reported operational-level voyage intake (cargo movement, cm_{ijm}) in the AXSDry ship movement data [32].

Moreover, cargo type was categorized based on the arrival ports of the data. These data record discrete cargo types at the operation level, as iron ore, coal, grain, and others, making it feasible to group certain cargo types into specific time ranges. Thus, origin-destination (OD) tables for each cargo type were collected monthly, as shown in Table 1 [32]. The data enclosed in these OD tables were used as CARGO in further case studies.

*** Formatting Issue - Best Available Paper/Graphic ***

Table 1. A sample of origin-destination (OD) table.

Origin Port	Destination Port	Cargo Type	Cargo Movement Demand (MT) [1]											
			Jan	Feb	Mar	Apr	May	Jun	Jul	Aug	Sep	Oct	Nov	Dec
Newcastle	Japan-B	Coal	1.16	0.87	1.52	1.09	0.72	0.86	1.22	1.40	1.09	0.81	1.10	1.05
Cape Preston	Ningbo	Iron Ore	0.74	0.52	0.65	0.64	0.42	0.85	0.64	0.84	0.74	0.64	0.94	0.44
Newcastle	Taichung	Coal	0.57	0.24	0.50	0.50	0.42	0.83	0.63	0.47	0.32	0.72	0.31	0.22
Fujairah	Mesaieed	Others	0.23	0.36	0.42	0.54	0.52	0.39	0.69	0.28	0.44	0.65	0.51	0.62
Santos	BI. Khomeini	Grain	0.77	0.33	0.49	0.34	***na***	0.35	***na***	0.28	0.58	0.91	0.29	0.21
Newcastle	Kaohsiung	Coal	0.24	0.36	0.32	0.56	0.26	0.33	0.60	0.40	0.40	0.18	0.59	0.32
Gove	Longkou	Others	0.43	0.42	0.23	0.39	0.32	0.52	0.52	0.58	***na***	0.26	0.33	***na***

[1] *Jan:* January; *Feb:* February; *Mar:* March; *Apr:* April; *May*: May, *Jun*: June; *Jul*: July; *Aug*: August; *Sep*: September; *Oct*: October; *Nov*: November; *Dec*: December; ***na***: not available–no cargo movement.

3.3 SHIP MODEL

This model contains individual ship data and operating conditions. We extracted a list of Panamax–MiniCape dry bulk carriers, which consisted of 2479 ships that operated in 2018 worldwide ship movement data [32]. This model encloses each ship's specifications, namely principal particulars (DWT, length, breadth, and draught) and performance (main engine power and design speed).

Moreover, these ships were tied to their operating conditions: draught rate, average sailing speed, and port stay time. In the previous system [31], we adopted a deep learning (DL) model to predict the operating conditions for each ship. However, the developed model is limited to a single model for worldwide operation. Therefore, this study proposes an estimation by adopting a route-based average (rAVG) for laden and ballast voyages, allowing flexible operating condition on a port-to-port basis.

The average operating conditions of the worldwide routes (AVG) were used as the baselines to confirm the benefits of the proposed method. Table 2 suggests that rAVG has a lower mean absolute error (MAE) than AVG. In the case of draught rate, the rAVG yielded an advancement of 55 % and 35 % in laden and ballast voyages, respectively, compared to the AVG. These values prove to be an ambitious accuracy compared to the DL model, which gave 59 % and 17 % higher precision than the AVG for both voyage modes. Furthermore, we used the rAVG operating conditions to calculate the COST and GHG emissions variables for each target ship.

Table 2. Mean absolute error of prediction methods of operating conditions.

Operation Condition	Mean Absoulute Error (MAE) in							
	This Study				Prior Study [31]			
	Route-Base Average (rAVG)		Worldwide Average (AVG)		Deep Learning Model		Worldwide Average (AVG)	
	Laden Voyage	Ballast Voyage	Laden Voyage	Ballast Voyage	Laden Voyage	Ballast Voyage	Laden Voyage	Ballast Voyage
Draught rate (%)	2.133	1.374	4.714	2.12	1.686	3.923	4.115	4.721
Avg. sailing speed (kn)	1.049	1.037	1.468	1.248	1.181	1.427	1.318	1.434
Port staying time (d)	0.431	0.516	0.885	0.882	1.273	2.439	1.819	3.129

4. SHIP ALLOCATION ALGORITHM

4.1 GHG CALCULATION

Similar to our previous study [31], we calculated the total GHG referred to in prior IMO GHG studies [1,2]. The entire GHG emissions were from the main engine and auxiliary machinery (auxiliary engine and boiler). This calculation uses ship data [33] and port data [34] generated from the global network model and ship model results. We calculated the fuel consumption of the main engine (A_{ij}) and auxiliary machinery (B_{ij}) of ship i operating on route j (t) by referring to IMO GHG studies [1,2,31,38,39], as follows:

$$A_{ij} = \left(\frac{P_i \times \left({}^{d_{ij}}/_{D_i} \right)^{\left(\frac{2}{3}\right)} \times \left({}^{v_{ij}}/_{V_i} \right)^3}{\eta w_{ij} \times \eta f_{ij}} \times sfc_i \times st_{ij} \right), \quad B_{ij} = \left((a_i + b_i) \times st_{ij} \right) \quad \forall i \in I, \quad \forall j \in J, \tag{1}$$

Finally, we defined the total GHG emissions (GH_{ij}) for ship i operating in route j (t/y or t/m) as follows:

$$GH_{ij} = \left((A_{ij} \times CF) + (B_{ij} \times cf) \right) \times n_{ij} \quad \forall i \in I, \quad \forall j \in J, \tag{2}$$

4.2 COST CALCULATION

The method used to calculate the total COST was similar to that used in our previous study [31]. The total COST comprises the following cost attributes:

- Fuel cost: This describes the cost of fuel needed to sail from the origin to destination ports.
- Operation cost: This describes the cost of operating the ship, comprising crew, consumables, repairs, and insurance costs.
- Depreciation cost: This represents the reduction in the ship resale price after a year.
- Stockpile cost: This depicts the cost of stockpiling transported cargo at the destination port until the cargo is consumed.

The fuel cost (FC_{ij}) for ship i serving route j (USD) is calculated from the total GHG emissions equation by replacing the GHG emissions carbon factor (CF, cf) with fuel prices (FP, fp), as follows:

$$FC_{ij} = \left((A_{ij} \times FP) + (B_{ij} \times fp) \right) \quad \forall i \in I, \quad \forall j \in J, \tag{3}$$

Response surfaces were composed to define the daily operation and depreciation costs using the operation constants, newbuilding prices, and second-hand prices [2,4,36,37,40]. Figure 3 shows the response surfaces of the daily operation and depreciation costs for a given ship DWT and ship age. Thus, the operation and depreciation costs (OC_{ij}, DC_{ij}) for ship i operating on route j (USD) are calculated using the following equation:

$$OC_{ij} = (oc_i \times st_{ij}), \quad DC_{ij} = (dc_i \times st_{ij}) \quad \forall i \in I, \quad \forall j \in J, \tag{4}$$

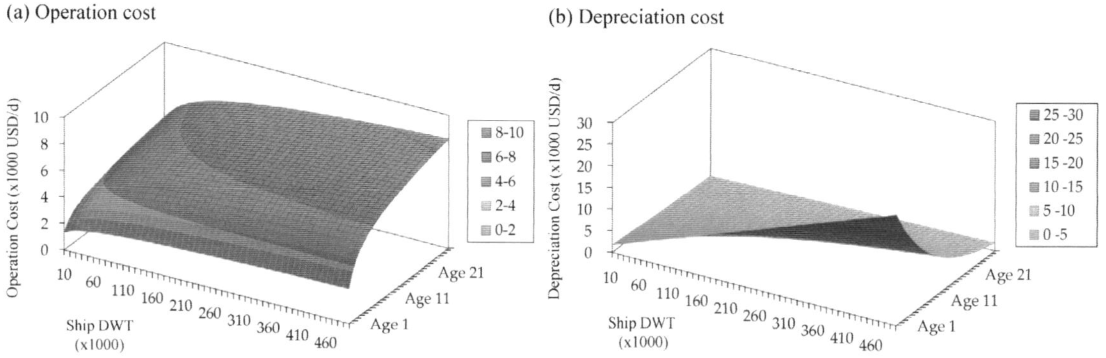

Figure 3. Response surfaces of bulk carrier operation and depreciation costs [31].

Furthermore, this study sets the stockpile cost as in our previous study [31]. The stockpile cost (SC_{ij}) for a ship i serving route j (USD) is determined by referring to the daily destination port's CARGO (CD_l, t/d), as follows:

$$CD_l = \left(\textstyle\sum_k (CA_{kl})\right) \times \left(\frac{1}{yd}\right) \quad \forall (j,l) \in J, \tag{5}$$

$$SC_{ij} = \left(\sum_{x=0}^{\left\lfloor \frac{CM_{ij}}{CD_l} \right\rfloor} CM_{ij} - (CD_l \times x) \right) \times sc_l \quad \forall i \in I, \quad \forall (j,l) \in J, \tag{6}$$

Finally, the monthly total COST (TC_{ijm}) for ship i operating on route j (USD/y or USD/m) was obtained as follows:

$$TC_{ijm} = \left(FC_{ij} + OC_{ij} + SC_{ij} + DC_{ij}\right) \times n_{ijm} \quad \forall i \in I, \quad \forall j \in J, \quad \forall m \in M, \tag{7}$$

4.3 ALLOCATION ALGORITHM

The following are the defined algorithms that were used to simulate ship allocation in a previous study [31]:

- Algorithm 1: Ship replacement while preserving the existing ship allocation.
- Algorithm 2: An optimization for reconstructing ship allocation in time-charter contract.

Additionally, we introduced Algorithm 3, which is identical to Algorithm 2 but used for reconstructing ship allocation in voyage-charter contracts. We propose two optimization schemes: COST- and GHG-optimized. Herein, we provide an example of a COST-optimized case.

4.3 (a) Algorithm 1: Ship Replacement while Preserving the Existing Ship Allocation

Algorithm 1 proposes an optimization by offering a direct clone of the existing ship, a new ship with exact specifications, and operating conditions that serve the same annual allocations. In this scheme, we offer a new ship i' with a higher efficiency (admiralty coefficient (ADM) of the year 2021) for each existing ship i. An existing ship is replaced if the new ship offers lower COST or GHG emissions. We calculated the annual COST ($TC_{ij}, TC_{i'j}$) per ton cargo ($CM_{ij}, CM_{i'j}$) of ship i and offered ship i'. Further, we acquired the global merit (W_{ij}) for the new ship i' replacing an existing ship i as follows:

$$w_{ij} = \left(\frac{TC_{ij}}{CM_{ij}}\right), \quad w_{i'j} = \left(\frac{TC_{i'j}}{CM_{i'j}}\right) \quad \forall (i, i') \in I, \quad \forall j \in J, \tag{8}$$

$$W_{ij} = \begin{cases} w_{ij} \; if \left(w_{ij} - w_{i'j}\right) < 0 \\ w_{i'j} \; if \left(w_{ij} - w_{i'j}\right) > 0 \end{cases} \quad \forall (i, i') \in I, \quad \forall j \in J, \tag{9}$$

4.3 (b) Algorithm 2: An Optimization for Reconstructing Ship Allocation in Time-Charter Contract

In Algorithm 2, we adopted a greedy algorithm to reconstruct ship allocation, assuming that the ships operated in an annual time-charter contract manner. This algorithm sequentially allocates ships with the highest global merit to transport the worldwide CARGO, aiming to achieve a global reduction in total COST or GHG emissions for each ship and route. Additionally, ships were allocated by conforming to the dimensions of the ports. The global merit (M) for each ship i operating on route j is obtained as follows:

$$y_{ij} = \left(\frac{TC_{ij}}{CM_{ij}}\right), \quad \bar{y}_j = \frac{1}{I} \times \textstyle\sum_i \left(y_{ij}\right) \quad \forall i \in I, \quad \forall j \in J, \tag{10}$$

$$m'_{ij} = \left(\bar{y}_j - y_{ij}\right), \quad M = max\left(m'_{ij}\right) \quad \forall i \in I, \quad \forall j \in J, \tag{11}$$

A set of ships and routes with global merit M is considered the most COST effective for carrying CARGO (CM_{ij}). Subsequently, operation-level ship allocation is obtained as follows:

- The annual CARGO of route j (CA_j) was subtracted from the annual cargo carried by that ship (CM_{ij}).
- These processes were repeated until all CARGO were designated to a particular ship.
- Annual CARGO, total COST, and GHG emissions were determined for each operation.

4.3 (c) Algorithm 3: An Optimization for Reconstructing Ship Allocation in Voyage-Charter Contract

Algorithm 3 is based on the greedy algorithm approach, similar to Algorithm 2. This algorithm was created to simulate the ship bidding process, assuming that the ships operated in a monthly voyage-charter contract. This algorithm inputs the defined variables on a monthly basis for both the OD tables and calculated COST and GHG emissions.

The global merit selection (M) in this algorithm is similar to that discussed in the previous section and is obtained by replacing the annual COST (TC_{ij}) with the monthly COST (TC_{ijm}) in the KPI calculation as follows:

$$y_{ijm} = \left(\frac{TC_{ijm}}{cm_{ijm}}\right) \quad \forall i \in I, \quad \forall j \in J, \quad \forall m \in M, \tag{12}$$

5. CASE STUDIES INTENDED FOR PANAMAX DRY BULK CARRIER

5.1 OVERVIEW OF CASE STUDIES

In the previous study, the authors [29–31] presented the case studies considering worldwide ship allocation. These tend to give wide perspective towards the implementation of our system. Given that worldwide applicability have been explored, this study conduct the case study of ships operated by specific ship operator.

We conducted case studies using Algorithms 2 and 3, intended for the Panamax dry bulk carrier of a specific ship operator. The aim was to confirm the different results delivered by the proposed algorithms. The simulation objectives were as follows:

- Target ship: Panamax dry bulk carriers of Ship Operator **A** (DWT 74,000-93,000, 32 ships).
- Cargo types: Iron ore, coal, grain, and others.
- Route: Worldwide (sailing routes served by target ship).
- Operation period: 2018.
- The assumed fuel attributes (specific fuel consumption, g/kWh; emissions factor, t-CO2/t-fuel; fuel prices, USD) are HFO (175–185; 3.114; 300–375) and MDO (daily average fuel consumption [1,2]; 3.206; 600).

Table 3 presents the size categories consisting of ship lists of Ship Operator **A**. Herein, we are able to reconstruct the ship allocation considering COST- or GHG-optimized using the proposed algorithms. Because GHG emissions are linear to fuel consumption, this section discusses the actual ship allocation and several case studies focusing on COST-optimized cases of ship allocation using only existing ships in the time-charter (case 1-1) and voyage-charter (case 1-2) contracts, and ship allocation with new ship presence in voyage-charter contracts (case 2).

Table 3. Size categories consisted in ship list of Ship Operator **A**.

Ship Size Category	Ship Numbers	*DWT* [1]	*L* [1] (m)	*B* [1] (m)	Avg. Ship Age
A	8	75,000	225	32.3	17.1
B	10	82,000	229	32.3	8.5
C	9	85,000	229	35.0	5.9
D	4	93,000	229	38.0	10.3

[1] typical *DWT*: deadweight tonnage; *L*: length; *B*: breadth.

5.2 ACTUAL SHIP ALLOCATION

The cargo types in the actual ship allocation of Ship Operator **A** are shown in Figure 4, indicating that the majority of transported cargo types were grain and coal, where the major CARGO is in the Brazil–China routes.

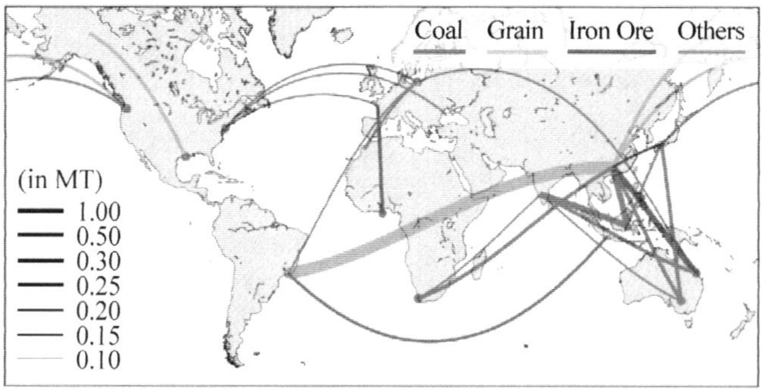

Figure 4. Visualization of actual ship allocation of Panamax dry bulk carrier of Ship Operator **A**.

5.3 CASE STUDY 1: SHIP ALLOCATION OPTIMIZATION USING ONLY EXISTING SHIPS

5.3 (a) Definition of Case Study 1

This simulation is arranged to evaluate the proposed Algorithm 3, which is an optimization for reconstructing ship allocation in voyage-charter contracts. Worldwide, Ship Operator **A** operated 32 ships carrying four types of cargo, mainly in the voyage-charter scheme. Hence, the total COST and GHG emissions resulting from the simulations are observed.

This case study covered ship allocation optimization using only the existing ships of Ship Operator **A**, with the following parameters:

- Method: Allocation Algorithms 2 and 3: optimization for reconstructing ship allocation in time and voyage-charter contracts.
- Offered ships: The existing ships operating in the actual ship allocation.

5.3 (b) Results of Case Study 1

The results obtained for case study 1 are listed in Table 4. Case study 1-1 used Algorithm 2 to allocate a ship by assuming it operated in a time-charter contract; therefore, a ship went round-trip, resulting in similar sailing distances back and forth to the origin port. Thus, this algorithm results in a longer ballast sailing day compared to the actual condition and fails to transport all CARGO. Hence, this scheme is considered unsuitable for simulating the actual contract that occurs in the actual ship allocation.

Table 4. Results of case study 1.

Case Study	Annual CARGO (MT)	Annual Operation Numbers	Annual GHG Emissions (MT)	Annual COST (Mil. USD)	Avg. Annual Laden Sailing Days (d)	Avg. Annual Ballast Sailing Days (d)
Actual Ship Allocation	13.7	191	0.427	139.3	180.4	148.6
Case 1-1: Time-Charter Optimization using only Existing Ships	**11.3**	**156**	0.411	141.3	153.8	**178.7**
Case 1-2: Voyage-Charter Optimization using only Existing Ships	13.7	191	0.398	133.7	169.9	**139.5**

Subsequently, case 1-2, using Algorithm 3, allocated a ship by assuming that it operated in a voyage-charter contract; thus, it made it possible for a ship to serve the next closer sailing route after its trip to a destination port. This algorithm results in a shorter ballast sailing day compared to the results of Algorithm 2. In conclusion, we can observe that all the CARGO is transported, and there is reduction in total COST and GHG emissions.

Specifically, Figure 5 presents the economic days of the sample ship. At the operation level, we confirmed that using Algorithm 3 (case 1-2) resulted in ballast sailing days similar to the actual ship allocation.

*** Formatting Issue - Best Available Paper/Graphic ***

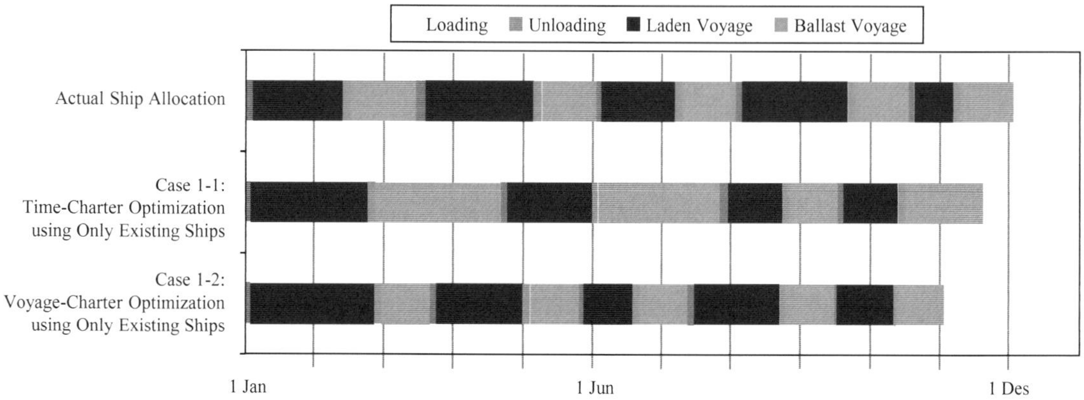

Figure 5. Economic days of a sample ship of Ship Operator **A**.

5.4 CASE STUDY 2: SHIP ALLOCATION OPTIMIZATION WITH AN INSTANCE OF NEW SHIP

5.4 (a) Definition of Case Study 2

This simulation was conducted to examine the demanded new ship specifications by inputting new ships into the simulation. We sampled new ships from each size category that occurred during the actual ship allocation (see Table 3). These ships were proposed to have a higher efficiency for a lower main engine power. This was determined by calculating the ADM [33,41,42], as shown in Figure 6. These ships can be allocated indefinitely, allowing the replacement of the existing ship operations. Hence, ships with high competitiveness against the existing ships can be observed.

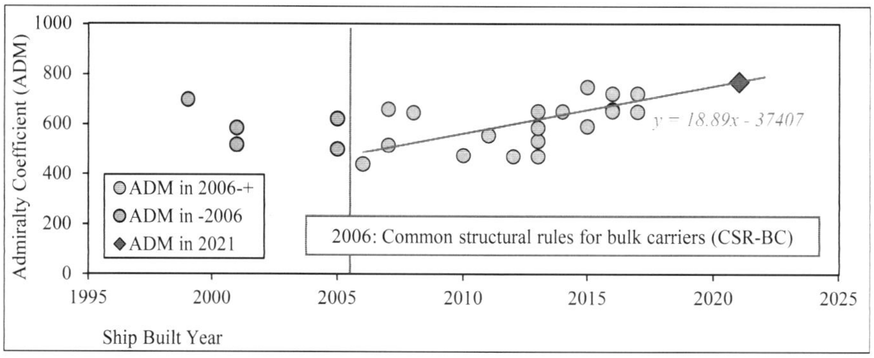

Figure 6. Admiralty coefficient of ships of Ship Operator **A**.

Extended from case study 1, this case study proposes new ships for COST-optimized ship allocation in voyage-charter contracts, following these parameters:

- Method: Allocation algorithm 3: Optimization for reconstructing ship allocation in voyage-charter contracts.
- Offered ships: Existing ships in case study 1 and four new ships sampled from existing ships (Table 5).

Table 5. Proposed new ship specifications.

Defined New Ship	DWT [1]	L [1] (m)	B [1] (m)	d [1] (m)	v [1] (m)	ME [2] (kW)	Age [3]
A	75,000	225	32.3	14.0	14.5	**7746**	1
B	82,000	229	32.3	14.5	14.5	**8023**	
C	85,000	229	35.0	14.5	14.5	**8464**	
D	93,000	229	38.0	14.9	14.5	**9105**	

[1] new ships' DWT: deadweight tonnage; L: length; B: breadth; d: draught; v: design speed; [2] ME: reduced main engine power; [3] Age: assumed ship age.

*** Formatting Issue - Best Available Paper/Graphic ***

5.4 (b) Results of Case Study 2

As new ships can be allocated indefinitely, case study 2 optimized ship allocation by allowing existing ships to be replaced by new ships. Changes in the composition of the ship allocation can be observed, as shown in Table 6.

Table 6. Allocated ship and operation numbers in case study 2.

Ship Size Category	Actual Ship Allocation		Case 2: Voyage-Charter Optimization with New Ships Presence	
	Ship Numbers	Operation Numbers	Ship Numbers	Operation Numbers **(Changes)**
A	8	48	9	71 (**+23**)
B	10	59	11	66 (+7)
C	9	44	6	38 (-**6**)
D	4	30	3	12 (**-18**)
Other	1	10	1	4 (**-6**)
Total	32	191	30	191

In this context, an increase in the number of ships allocated to size categories A and B can be observed. This condition, as a result of significant routes, was served by ships with these size categories, as shown in Figure 7.

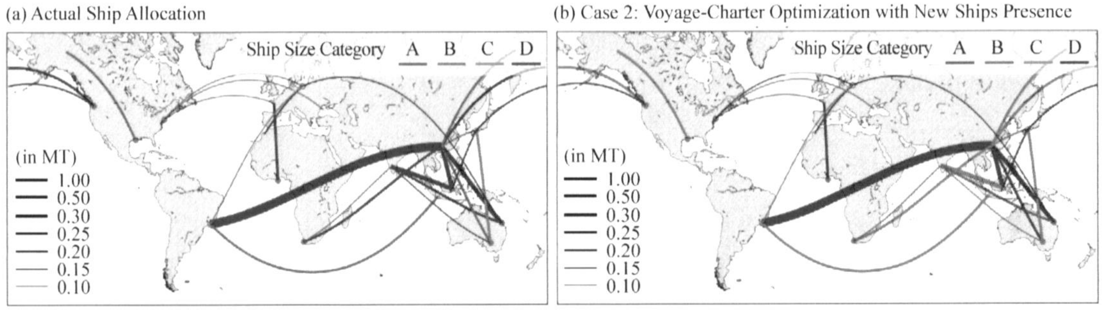

Figure 7. Visualization of actual ship allocation and ship allocation optimization of case study 2.

Moreover, considering the ship ages in actual ship allocation, ships with size category A are fairly older (more than 15 years old); thus, this emphasizes the importance of introducing new ships with higher efficiency, as shown in Figure 8.

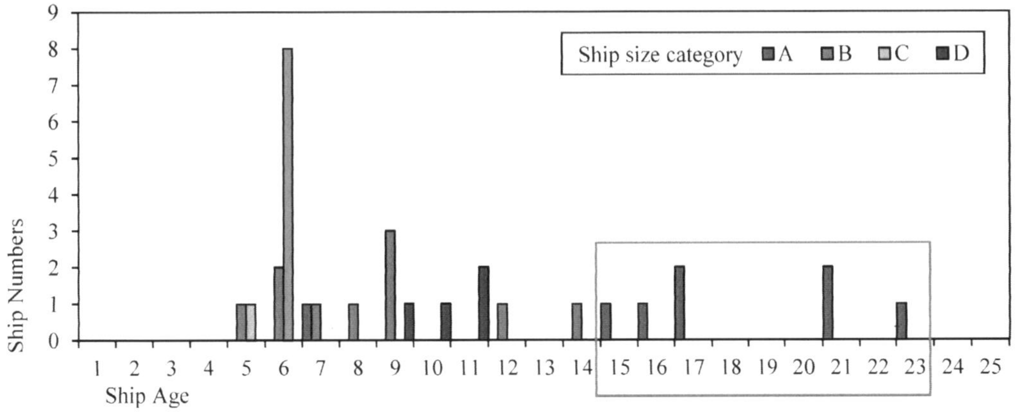

Figure 8. Ship age of ships in the actual ship allocation of Ship Operator **A**.

*** Formatting Issue - Best Available Paper/Graphic ***

6. CONCLUSIONS

This study introduced an extended application of our basic ship planning support system. We expanded the scope of our simulations to other ship size categories, leading to a wider understanding of the applicability of the developed system. Moreover, we proposed a new allocation algorithm to consider the optimization, assuming the ships operated in the time- and voyage-charter schemes. A similar system configuration was adopted from our prior study, which consisted of three discrete models and a defined algorithm to reconstruct the ship allocation: the global network model, cargo movement model, ship model, and ship allocation algorithm. The global network model describes the main ports and the port clusters. The cargo movement model defines the cargo movement and types. The ship model specifies the variables for each ship. The ship allocation algorithm calculates the COST and GHG emissions and reconstructs the ship allocation by adopting a greedy algorithm. This study proposed distinctive algorithms: ship replacement while preserving the existing ship allocation, and optimization for reconstructing the ship allocation in time- and voyage-charter contract schemes. Furthermore, we conducted discrete simulations using the COST- and GHG-optimized schemes. Using the developed system, we recreated operation-level ship allocation considering diverse scenarios. We confirmed that significant reductions in total COST and GHG emissions were not achievable through simulations using only existing ships. Finally, we verified the demand for new ship specifications by presenting a new ship instance. However, this study is only applicable to the presented scope. Further application of the proposed system for the accumulated ship size categories, such as Handymax–Capesize dry bulk carriers, are considered as future research directions. In addition, the importance of weather routing in defining the weather correction factor in fuel consumption calculations can be explored further [43].

7. ACKNOWLEDGMENTS

This research was supported by JSPS KAKENHI (Grant Numbers JP20H02371 and JP21K04501). The authors express their gratitude for this support.

8. REFERENCES

1. Faber, J, S Hanayama, S Zhang, P Pereda, B Comer, E Hauerhof, and H Yuan. "Fourth IMO Greenhouse Gas Study." Retrieved from the International Maritime Organization website: https://docs. imo. org (2020). Accessed 2022.
2. Smith, TWP, JP Jalkanen, BA Anderson, JJ Corbett, J Faber, S Hanayama, E O'keeffe, et al. "Third IMO Greenhouse Gas Study." (2015).
3. "Japan Port Association: Port Logistics Information." https://www.phaj.or.jp/distribution/14port /price.html. Accessed 2022.
4. Stopford, Martin. Maritime Economics 3e. Routledge, 2008.
5. Fan, Jianqing, Fang Han, and Han Liu. "Challenges of Big Data Analysis." *National science review* 1, no. 2 (2014): 293-314.
6. Alharthi, Abdulkhaliq, Vlad Krotov, and Michael Bowman. "Addressing Barriers to Big Data." *Business Horizons* 60, no. 3 (2017): 285-92.
7. Saggi, Mandeep Kaur, and Sushma Jain. "A Survey Towards an Integration of Big Data Analytics to Big Insights for Value-Creation." *Information Processing & Management* 54, no. 5 (2018): 758-90.
8. IMO. "AIS Transponders." https://www.imo.org/en/OurWork/Safety/Pages/AIS.aspx. Accessed 2022.
9. Svanberg, Martin, Vendela Santén, Axel Hörteborn, Henrik Holm, and Christian Finnsgård. "AIS in Maritime Research." *Marine Policy* 106 (2019): 103520.
10. Bauk, SI. "A Review of Navdat and Vdes as Upgrades of Maritime Communication Systems." *Advances in Marine Navigation and Safety of Sea Transportation* (2019): 81-82.
11. Voyage Data Recorders." https://www.imo.org/en/OurWork/Safety/Pages/VDR.aspx. Accessed 2022.
12. Munim, Ziaul Haque, Mariia Dushenko, Veronica Jaramillo Jimenez, Mohammad Hassan Shakil, and Marius Imset. "Big Data and Artificial Intelligence in the Maritime Industry: A Bibliometric Review and Future Research Directions." *Maritime Policy & Management* 47, no. 5 (2020): 577-97.
13. i o i , a i , a io i i e i , and ne ado i ig a a in he a i i e ndu y *NAŠE MORE: znanstveni časopis za more i pomorstvo* 65, no. 1 (2018): 56-62.
14. Zaman, Ibna, Kayvan Pazouki, Rose Norman, Shervin Younessi, and Shirley Coleman. "Challenges and Opportunities of Big Data Analytics for Upcoming Regulations and Future Transformation of the Shipping Industry." *Procedia engineering* 194 (2017): 537-44.
15. Sanchez-Gonzalez, Pedro-Luis, David Díaz-Gutiérrez, Teresa J Leo, and Luis R Núñez-Rivas. "Toward Digitalization of Maritime Transport?". *Sensors* 19, no. 4 (2019): 926.
16. Yang, Cheng-Hong, and Po-Yin Chang. "Forecasting the Demand for Container Throughput Using a Mixed-Precision Neural Architecture Based on Cnn–Lstm." *Mathematics* 8, no. 10 (2020): 1784.

*** Formatting Issue - Best Available Paper/Graphic ***

17. ugo i , en, i a o adina, and na e i Had i Fa o n uen ing he Fo a ion o F eigh a e on Maritime Shipping Markets." *Pomorstvo* 29, no. 1 (2015): 23-29.
18. Akar, Orkut, and Soner Esmer. "Cargo Demand Analysis of Container Terminals in Turkey." *Journal of ETA Maritime Science* 3, no. 2 (2015): 117-22.
19. u , i , eh e inan Y d , and e in u u ydin i a ion o Con aine a i a ea o y Using Several Soft Computing Methods: A Case of Turkish Seaports." *Discrete Dynamics in Nature and Society* 2017 (2017).
20. Jia, Haiying, Vishnu Prakash, and Tristan Smith. "Estimating Vessel Payloads in Bulk Shipping Using AIS Data." *International Journal of Shipping and Transport Logistics* 11, no. 1 (2019): 25-40.
21. Zhou, X, and Q Hu. "Estimation of Shipment Size in Seaborne Iron Ore Trade." *TransNav: International Journal on Marine Navigation and Safety of Sea Transportation* 13 (2019).
22. Kanamoto, Kei, Liwen Murong, Minato Nakashima, and Ryuichi Shibasaki. "Can Maritime Big Data Be Applied to Shipping Industry Analysis? Focussing on Commodities and Vessel Sizes of Dry Bulk Carriers." *Maritime Economics & Logistics* 23, no. 2 (2021): 211-36.
23. Prochazka, Vít, Roar Adland, and François-Charles Wolff. "Contracting Decisions in the Crude Oil Transportation Market: Evidence from Fixtures Matched with AIS Data." *Transportation Research Part A: Policy and Practice* 130 (2019): 37-53.
24. Sharma, R, and OP Sha. "Development of an Integrated Market Forecasting Model for Shipping, and Shipbuilding Parameters." Paper presented at the *Proceedings of the RINA—International Conference on Computer Applications in Shipbuilding*, Portsmouth, UK, 2007.
25. Wada, Yujiro, Kunihiro Hamada, and Noritaka Hirata. "Shipbuilding Capacity Optimization Using Shipbuilding Demand Forecasting Model." *Journal of Marine Science and Technology* (2021): 1-19.
26. Wada, Yujiro, Kunihiro Hamada, Noritaka Hirata, Kazutaka Seki, and Shinji Yamada. "A System Dynamics Model for Shipbuilding Demand Forecasting." *Journal of Marine Science and Technology* 23, no. 2 (2018): 236-52.
27. Wada, Yujiro, Tatsumi Yamamura, Kunihiro Hamada, and Shinnosuke Wanaka. "Evaluation of Ghg Emission Measures Based on Shipping and Shipbuilding Market Forecasting." *Sustainability* 13, no. 5 (2021): 2760.
28. Lee, Sangwon, and Inhwan Jung. "Development of a Platform Using Big Data-Based Artificial Intelligence to Predict New Demand of Shipbuilding." *The Journal of The Institute of Internet, Broadcasting and Communication* 19, no. 1 (2019): 171-78.
29. Arifin, Mohammad Danil, Kunihiro Hamada, Noritaka Hirata, Kai Ihara, and Yuki Koide. "Development of Ship Allocation Models Using Marine Logistics Data and Its Application to Bulk Carrier Demand Forecasting and Basic Planning Support." *Journal of the Japan Society of Naval Architects and Ocean Engineers* 27 (2018): 139-48.
30. Arifin, MD, K Hamada, N Hirata, K Ihara, and Y Koide. "A Study on the Support System of Ship Basic Planning by Using Marine Logistics Big Data." Paper presented at the Proceeding of the *International Conference on Computer Applications in Shipbuilding* (ICCAS), 2017.
31. Muzhoffar, Dimas Angga Fakhri, Kunihiro Hamada, Yujiro Wada, Yusuke Miyake, and Shun Kawamura. "Basic Ship-Planning Support System Using Big Data in Maritime Logistics for Simulating Demand Generation." *Journal of Marine Science and Engineering* 10, no. 2 (2022): 186.
32. "AXSMarine / AXSDry." https://public.axsmarine.com/axsdry/. Accessed 2022.
33. "IHS Markit Sea-Web Ships." https://maritime.ihs.com/entitlementportal/home/information/seaweb_ships. Accessed 2022.
34. "IHS Markit Sea-Web Ports." https://maritime.ihs.com/EntitlementPortal/Home/Information/Seaweb_Ports. Accessed 2022.
35. IHS Fairplay Ports and Terminals Guide 2013 - 2014 (Book and Cd Set). Jane's Information Group, 2012. Accessed 2022.
36. Greiner, Richard. "Ship Operating Costs: Current and Future Trends." In Technical Report. Moore Stephens LLP. Singapore, 2013.
37. Asariotis, Regina, Hassiba Benamara, Hannes Finkenbrink, Jan Hoffmann, Jennifer Lavelle, Maria Misovicova, Vincent Valentine, and Frida Youssef. "Review of Maritime Transport, 2011." 2011.
38. hang, Xiunian, a ine iu ee a , and a a ay i Co d Chain hi ing ode Choi e i h n i on en a and Financial Perspectives." *Transportation Research Part D: Transport and Environment* 87 (2020): 102537.
39. en u ini, iada, a a ay i , Christos A Kontovas, and Allan Larsen. "The Multi-Port Berth Allocation Problem with Speed Optimization and Emission Considerations." *Transportation Research Part D: Transport and Environment* 54 (2017): 142-59.
40. "Clarksons Research Shipping Intelligence Network." https://sin.clarksons.net/. Accessed 2022.
41. Papanikolaou, Apostolos. A Holistic Approach to Ship Design. Springer, 2019.
42. "International Association of Classification Societies (IACS)." https://www.iacs.org.uk/publications/common-structural-rules/. Accessed 2022.
43. Perera, Lokukaluge P, and C Guedes Soares. "Weather Routing and Safe Ship Handling in the Future of Shipping." *Ocean Engineering* 130 (2017): 684-95.

VOYAGE OPTIMIZATION WITH WIND PROPULSION

V Paakkari, Norsepower Ltd, Finland
H Wang and **C Stigler** Napa Ltd, Finland

SUMMARY

Wind propulsion has been proven to produce significant fuel savings on commercial vessels. The typical savings potential of wind propulsion on conventional cargo ships is 5-25%. On the other hand, voyage optimization and weather routing has been shown to produce similarly large fuel savings in the order of 15%. Wind is an intermittent energy source, and it often occurs together with waves, making ships installed with wind propulsion especially interesting from weather routing point of view. This means that a holistic approach is needed when conducting weather routing for ships with wind propulsion.

Using detailed simulations, this paper studies how the fuel savings potential of wind propulsion can be further improved by coupling it with state-of-the-art voyage optimization software. Via case study examples, it is shown that the savings potential of wind propulsion can be more than doubled by using voyage optimization.

1. INTRODUCTION

The shipping industry is facing increasing pressure to search for alternatives for common fossil fuels. There are multiple drivers for this development. First and perhaps most prominently, regulatory frameworks such as EEDI/EEXI and CII issued by IMO require ships to emit less and be more energy efficient. At the same time the oil prices have been steadily increasing after the pandemic. At the time of writing, the Brent quality crude oil is priced above $100 per barrel. Furthermore, there is new demand from the end customers and cargo owners to decarbonize the logistics chain. Finally, many financial institutions have integrated climate considerations into their lending considerations in frameworks such as Poseidon Principles [6].

There are various strategies to accomplish a shift from fossil fuel-based transportation to sustainable logistics chain. Much of the discussion has been revolving around the energy efficiency, which is of high importance. It is however impossible to reach carbon neutrality, or even significant decarbonization with efficiency alone. Some form of energy production is needed. This paper deals with the use of direct wind propulsion as means to supply the ship with significant source of thrust to propel the ship. As wind is an intermittent energy source, it is highly beneficial to couple the wind propulsion technology with modern voyage optimization tools to gain the maximal benefits from wind without added losses due to waves etc.

The purpose of this study is to investigate and demonstrate the additional benefits of voyage optimization on ships using wind assisted propulsion, namely Rotor Sails. The impact is studied on multiple global routes representing some of the busiest trading.

2. METHOD

In this simulation, 384 sailed voyages (using AIS data) are used as comparative baselines. These voyages are selected based on the departure-destination of their sailed routes, which are US – EUR, EUR – US, CN – SA, SA – CN, CN – AUS, AUS – CN. This simulation is conducted using NAPA Voyage Optimization, which provides solutions for improving operational efficiency by optimizing route and speed profiles for any sea passage. To study the impact of the Rotor Sails on the potential fuel saving for ships, a comparative simulation experiment is conducted with 6 types of routing methods listed in Table 1. The experiment is based on historical voyage data, which is indicated in Index 1 row.

Table 1. Routing types of the simulation experiment

Index	With Rotor Sails	Route type	Operation type
1	×	baseline	baseline
2	×	baseline	constant RPM
3	×	optimized	constant RPM
4	✓	baseline	baseline
5	✓	baseline	constant RPM
6	✓	optimized	constant RPM

· baseline: the data is retrieved from real voyages
· optimized: route is optimized by NAPA Voyage Optimization tools

- constant RPM: study ships sail with constant RPM while keeping the same arrival time as the baseline case

The performance model used for wind propulsion is the same that has been discussed earlier in [3]. The model is based on theoretical considerations, and it has been shown to produce results that match with the operational experience and long-term performance measurement.

The weather data considered in this simulation consists of wind, wind waves, swell and ocean currents.

An example of simulated routes on voyage US-EUR are shown in figure 1. It is seen that the original route (route collected from AIS) follows the rhumb line, whereas optimized routes are closer to great circle route, although not strictly. It is also seen that the introduction of Rotor Sails allows the route to be offset further away from the original. All of the routes have the same arrival time.

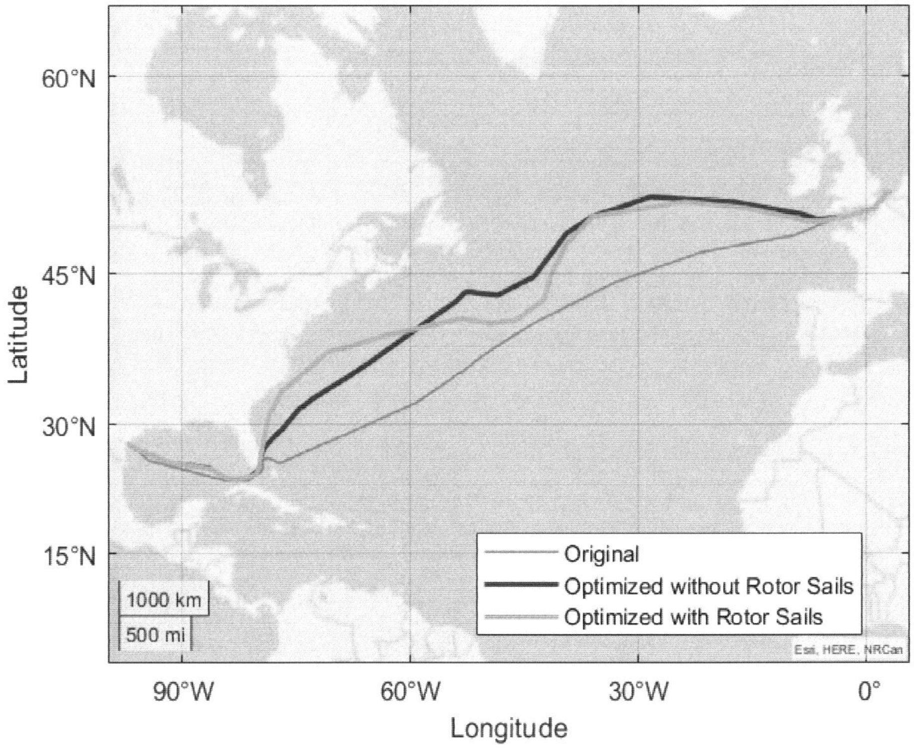

3. RESULTS

The influence of different voyage optimization schemes on the fuel economy of the vessels studied is presented in figure 2. It can be seen, that both voyage optimization and Rotor Sails are able to produce significant savings on their own. The average savings potential of Rotor Sails for the configurations studied is 7% when the ships operation is not adjusted to accommodate the Rotor Sails. With voyage optimization only the savings potential is 17% (when assuming a constant engine RPM). However, when these two technologies are combined, the accumulative savings potential reaches 24%. It is worth noting that the average contribution of Rotor Sails to the total propulsion power is rather modest in the cases studied. It can be anticipated that as the contribution of the wind propulsion increases, also the importance and benefit of the voyage optimization coupled with wind propulsion increases. Typically, the proportion of wind propulsion depends on the number and size of the sails installed. For example, the vessel considered in this study for the Persian Gulf – China route is a VLCC equipped with three 35m tall Rotor Sails with a diameter of 5m. Various concept studies have been made indicating that such vessels are feasible to accommodate at least six Rotor Sail units of such size [1], [2]. Recent analysis has shown that if the contribution of wind propulsion is modest, the savings potential can be estimated to increase linearly with increasing effective sail area. [4] While an exact threshold at which the savings should be considered "modest" is difficult to determine, the level at the current study (7% contribution to propulsion) is comparable to levels for which simple prediction methods have been proven useful [4].

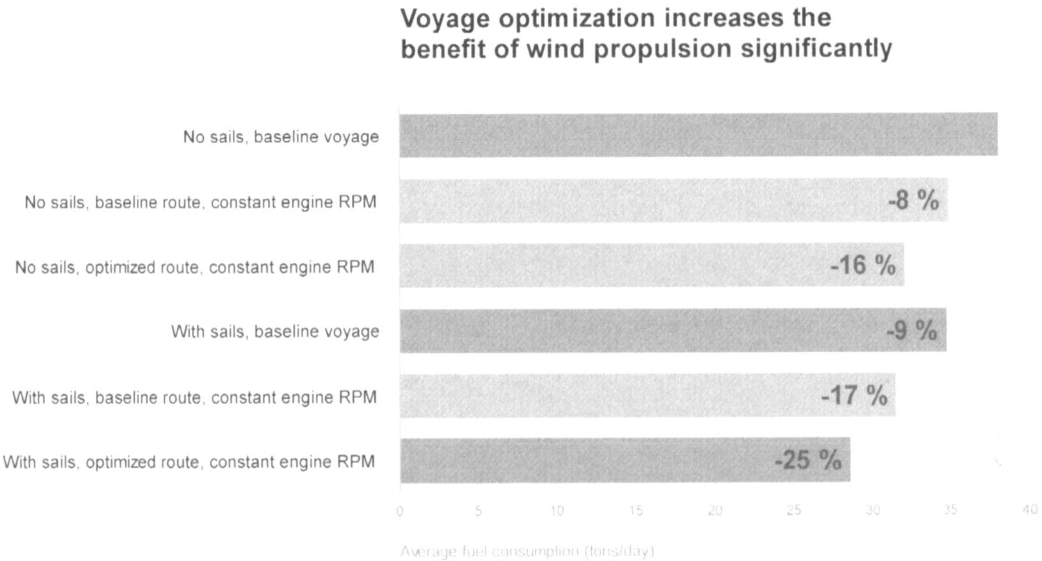

Figure 2 Impact of voyage optimization and 3 units of 35x5 Rotor Sails on the average fuel consumption.

As discussed in the introduction, a typical feature of wind propulsion is its intermittency. This means that the available energy in wind varies over time. Furthermore, the prevailing wind conditions vary over the location. At one place wind speeds are typically higher than in another. Similarly, the prevailing wind directions may be different from place to place. It is then important to consider differences of wind propulsion and voyage optimization potential on different routes.

Figure 3 illustrates the simulated savings potential on different routes. To allow comparison between routes of different length, the results are shown in a form of total fuel consumption divided by the total number of days spent on the voyages. This means any potential differences in arrival times are also taken into account. The savings in figure 3 correspond to cases "With sails, baseline voyage" and "With sails, optimized route, constant engine RPM" of figure 2 compared to case "No sails, baseline voyage".

Figure 1 Savings potential on different routes with Rotor Sails and Rotor Sails coupled with voyage optimization.

It can be seen, that especially without voyage optimization (grey bars), the potential of wind propulsion is highly dependent on the route selected. It is well known that the typical wind speeds near the equator are lower than higher north or south. It is then not surprising that the potential on route Persian Gulf to China has the lowest potential and North Atlantic crossing the highest.

The results indicate that the potential of wind propulsion can be more than doubled when it is coupled with voyage optimization. This large increase in the savings potential is something that is very relevant for the overall ship design and thus it suggests that operational considerations and voyage optimization should be considered as an integral part of ship design cycle.

It is very interesting to note that voyage optimization evens out the differences between "good routes" and "weak routes" significantly. Similar findings were reported by Mason [5]. It then appears that voyage optimization supports especially those ships with wind propulsion that may have the weakest business case without it. This particular notion could indicate importance of voyage optimization in expanding the commercially viable market of wind propulsion. This could mean that voyage optimization can play important role in being an enabler technology for installing wind propulsion and as such contributing to improvement of energy efficiency of global fleet.

4. CONCLUSIONS

In this paper, results from NAPA Voyage Optimization simulations with wind propulsion were presented. It was shown that the voyage optimization can significantly improve the savings potential on wind propulsion and work as an enabler for wind propulsion technology. The results also indicate that savings are increased the most at conditions that would otherwise provide mediocre benefit with wind propulsion.

As the carbon footprint requirements for shipping get more stringent, it is likely that the wind propulsion becomes more integral part of the ships energy system. The higher contribution of wind propulsion highlights even further the benefits of voyage optimization. Furthermore, this opens new opportunities and interesting challenges: when the wind propulsion corresponds to larger parts of ships total propulsion, it is conceivable that the ships' conventional engines can be downscaled. This would lead to the business case of wind propulsion being partly covered by the expense saved on the fuel engine. However, to achieve this, it is necessary to introduce the operational considerations at the design stage of the vessel to ensure the at the ship is capable to fulfil its mission requirements. For instance, that the passage times can be kept as required with high enough confidence level.

5. REFERENCES

1. Norsepower: https://www.norsepower.com/post/norsepower-collaborates-with-sws-for-cost-and-energy-efficient-vlcc-designs Retrieved 01.07.2022

2. SARSILA Severi. Comparison of Achieved Fuel Savings on Ships by Different Wind Propulsion Systems Using a 3-DoF Resistance Model. 2021

3. Paakkari et. Al. ROTOR SAIL GHG REDUCTION POTENTIAL, MODELLING AND SEA

4. TRIAL VALIDATION; *The 5th Int. Conference on Innovation in High Performance Sailing Yachts and Sail-Assisted Ship Propulsion, Gothenburg, Sweden. 2020*

5. IMO MEPC 76/INF.30

6. MASON James C. Quantifying voyage optimisation with wind-assisted ship propulsion: a new climate mitigation strategy for shipping. 2021

7. Poseidon principles, https://www.poseidonprinciples.org/finance/about/, Retrieved 01.07.2022

6. AUTHORS BIOGRAPHY

Ville Paakkari holds the current position of Sales Engineer responsible for performance modelling and analysis of Rotor Sail systems at Norsepower.
Helong Wang is a senior software develop at Shipping Solution Department of Napa.
Claus Stigler is a Product Owner of NAPA Voyage Optimization at NAPA Shipping Solutions

DEVELOPMENT OF CAD BASED STABILITY CALCULATION SOFTWARE FOR PURE-LOSS OF STABILITY AND PARAMETRIC ROLLING FAILURE MODE OF 2ND GENERATION INTACT STABILITY CRITERIA

J Y Park, M S Jang and H G Park, Korean Register, Republic of Korea

SUMMARY

Although the IMO mandates IS Code Part A to build and operate ships, ship safety is not guaranteed due to the problem of dynamic stability. So, IMO prepared the second-generation stability standard with five modes. However, since the software supporting the second-generation stability calculation is insufficient, CAD-based stability calculation SW that supports the second-generation stability calculation was developed. And in this paper, we calculated the change in resilience by waves, and calculated the restoration performance in Pure-Loss of Stability and Parametric Rolling Failure Mode.

NOMENCLATURE

[Symbol]	[Definition] [(Unit)]
L_{BP}	ship length between perpendiculars(m)
B	moulded ship breadth(m)
D	moulded ship depth(m)
d	moulded mean ship draught(m)
LCG	longitudinal centre of gravity from aft perpendicular(m)
C_b	block coefficient
GM	metacentric height with free surface correction(m)
VCG	vertical centre of gravity above baseline(m)
KB	vertical centre of buoyancy above baseline(m)
V_S	ship service speed(m/s)

1. INTRODUCTION

Although the IMO mandates IS Code Part A to build and operate ships, ship safety that satisfy intact stability criteria is not guaranteed due to the problem of dynamic stability. So, IMO prepared the second-generation stability standard with five modes as below.

Table 1: Regulation for 2nd Generation Intact Stability

	Pure loss of stability	Parametric Roll	Broaching	Dead ship Condition	Excessive acceleration
Definition of criteria (L1, L2)	SDC 2/WP.4 Annex 1	SDC 2/WP.4 Annex 2	SDC 2/WP.4 Annex 3	SDC 2/WP.5 Annex 1	SDC 2/WP.5 Annex 2
Explanatory notes	SDC 2/WP.5 Annex 3	SDC 2/WP.5 Annex 4	SDC 2/WP.5 Annex 5	SDC 2/WP.5 Annex 6	SDC 2/WP.5 Annex 7

The pure loss of stability refers to a phenomenon in which the stability of the hull is rapidly decreased when the midship point of the hull is equal to the wave height. The parametric roll refers to a resonance phenomenon that occurs when the period of a wave incident on the hull is half of the resonance period of a general rolling. The dead ship condition refers to a phenomenon caused by a roll that occurs when a wave enters from the side of the hull when the ship's engine fails. The excessive acceleration occurs in the restoring force due to GM, metacentric height, and the phenomenon of cargo dropping out of the ship is called excessive speeding.
And, although the application range of the second-generation stability standard is 24m, calculations in this paper were performed in the pure stability mode and the parametric roll mode, which are the vulnerability modes mainly found in large ships. Pure loss of stability and parametric roll failure modes are consequences of the variation of the restoring moment in longitudinal waves. Pure loss of stability is due to a single-wave effect and Parametric rolling is from repeated encounter wave. So, stability calculation SW is needed to compute the restoring moment in various waves.
In this paper, CAD based software under development with stability calculation capability is introduced, and calculations according to the 2nd generation stability criteria (pure loss of stability mode and parametric roll mode) are performed.

Although goal for software under development is the calculation modes of Level 1 and Level 2 of the second-generation stability criteria, In this study, the calculation for Level 1 was performed.

2. THEORITICAL BACKGROUND

IMO has identified and developed five vulnerability modes caused by waves encountered during operation of ships. In each vulnerability mode, checks are performed in the order of Level1, Level2, DA (Direct Assessment), and OG (Operation Guideline). If Level 1 is not satisfied, Level 2 calculation must be performed. If Level 2 standards are not met, DA (Direct Assessment) must be performed through simulation. If this also is not satisfied, the OG (Operation Guideline) that limits the operating conditions must be followed as below.

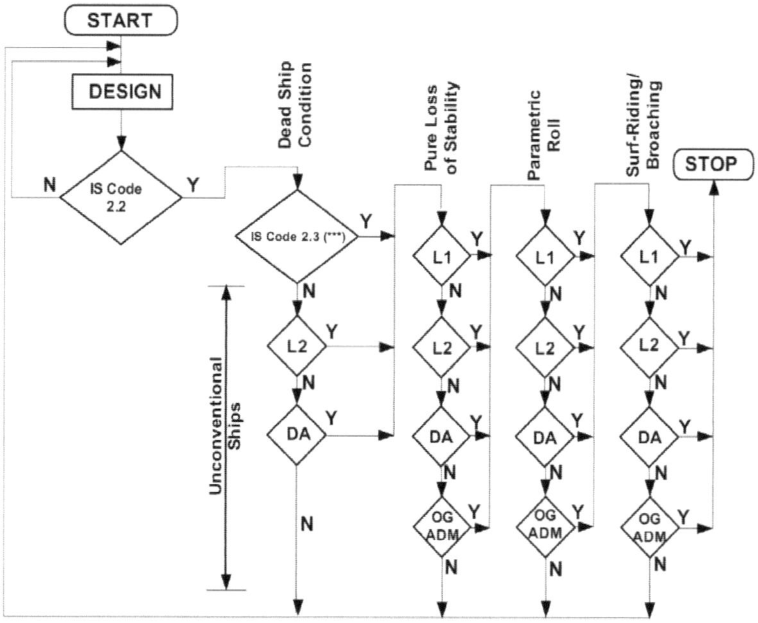

Figure 1: Process for 2nd Generation Intact Stability

2.1 PURE LOSS OF STABILITY

A change for water plane area occurs in the submerged part of the hull due to the wave encountered during the operation of the ship, and the change in the submerged part has effect on the restoring force of the ship. So, It is need to be check. As shown in the figure below, However, the geometry of the waterplane, which is no longer flat, also significantly changes. Hence, its inertia and the associated metacentric heights vary with the wave encounter. Consequently, a risk of insufficient transverse stability exists possibly leading to large heel angle or even to capsizing.

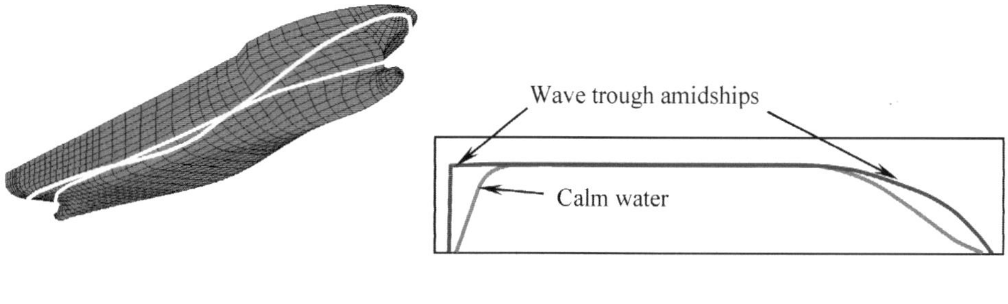

Figure 2: Change of Water Plane Area

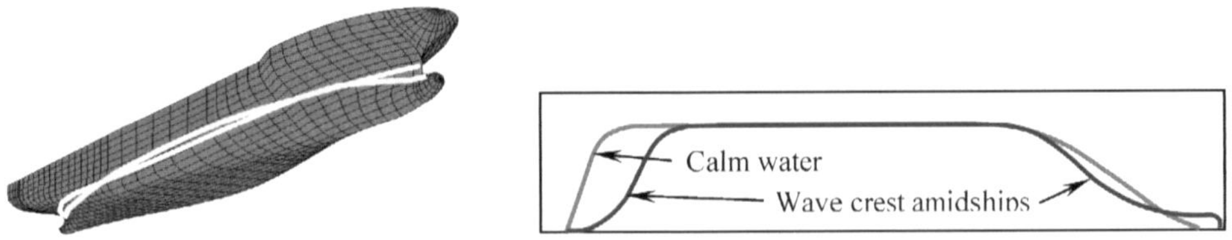

Figure 3: Change of Water Plane Area

2.2 PARAMETRIC ROLL

Changed restoring force, such as pure loss of stability, will have a great influence on the stability of the transverse roll. However pure loss of stability is due to the single wave effect that have same period with ship length. Parametric rolling is from parametric resonance also called parametric pumping. This phenomenon is a sort of well-known characteristic of oscillating systems having variations in their stiffness constant. A ship's parametric roll is a resonance phenomenon that occurs when the period incident on the hull is 1/2 of the general rolling period. And it is distinguished from the general rolling resonance phenomenon. Unlike the rolling that occurs during the transverse wave, parametric roll occurs when the ship encounters a wave corresponding to 1/2 times the transverse resonant period among the longitudinal Waves like as below.

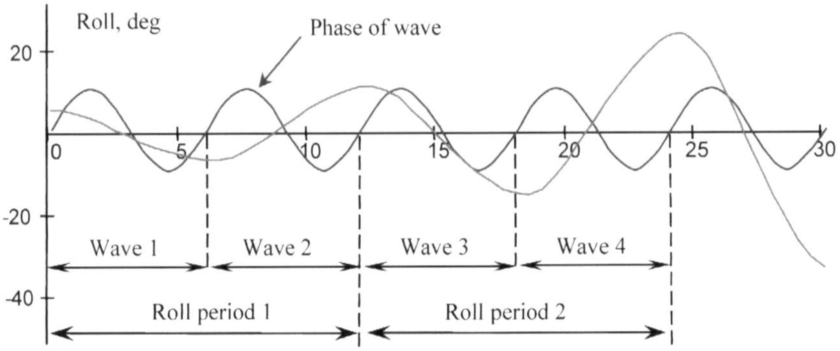

Figure 4: Wave for Roll and Encounter Wave

3. SOFTWARE

For the first level and second level should be implemented by shipyard with adequate stability software. DSA (Direct Stability Assessment) needs highly complex calculation software and know-how. And software for Level 1 and 2 is differentiated for SW for DSA. So, CAD based stability calculation SW for level1 and level2 is developing with main feature as below.

- CAD based Generation Hull Form and Compartment
- 2D Sketch Draw for Hull form and Compartment
- Calculation Hydrostatics for Hull and Compartment
- Calculation (Intact Stability, Longitudinal Strength) Module
- 2nd Generation Stability Calculation Module

For Hull form generation, it supports feature to create and define network curve set, Meshed hull form generation feature using networked curve set is prepared. Network curve set help to design 3D free surface model intuitively, and lines and section curve can be exported to utilize at the design stage. And network curve from other SW can be imported to create

3D model and calculate hydrostatics. 3D model of Hull form utilizes 3D mesh model to secure suitable quality and calculation speed.

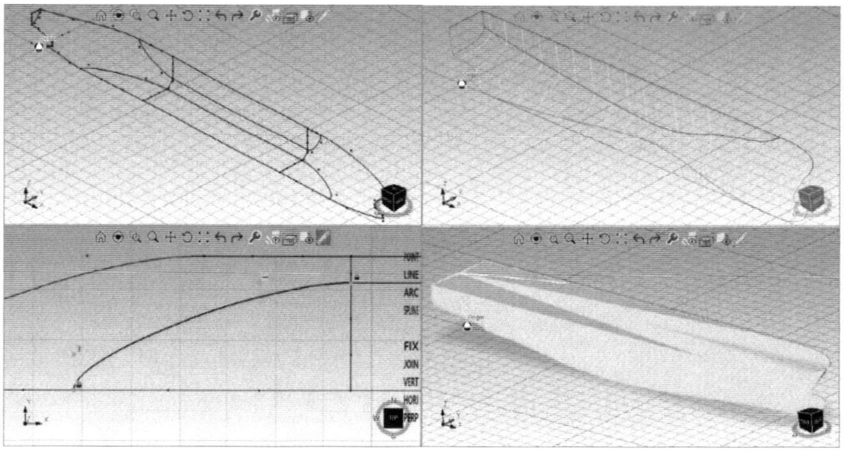

Figure 5: Example for Hull Generation and Sketch Feature

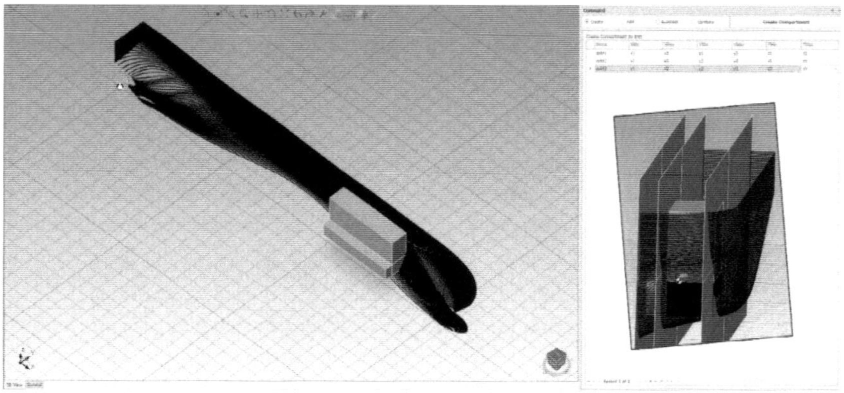

Figure 6: Example for Compartment Generation

In addition, Compartment was defined by using the manufactured meshed hull form and the surface for the vessel division. Surface for ship compartment acts as a parameter for compartment definition, and the defined compartment can be used for basic performance calculation.

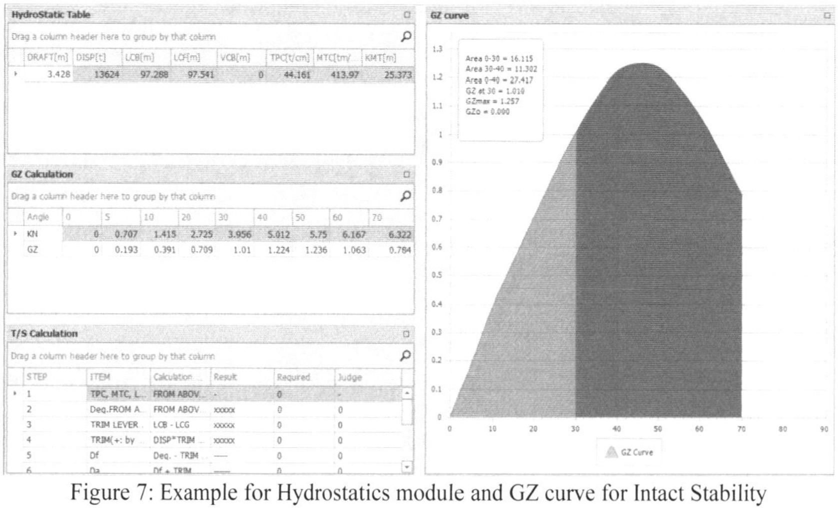

Figure 7: Example for Hydrostatics module and GZ curve for Intact Stability

Hydrostatics were performed as follows by using the defined hull form and Compartment. And the calculation of intact stability and 2nd generation stability using hydrostatics data was performed as follows.

Figure 8: Example for 2nd Generation Intact Stability

4. Result

2.1 PURE LOSS OF STABILITY(LEVEL1)

For Pure loss of Stability(Level1), When ship's Froude number(Fn) is exceed, minimum GM should be exceed 0.05 as below. In here, I_L is the moment of inertia of waterplane at lowest draft. And the lowest draught is come from below formula.

$$GM_{min} = KB + \frac{I_L}{\nabla} - KG \qquad d_L = d - min(d - 0.25d_{full}, \frac{LS_W}{2})$$

The following data(in Explanatory notes by IMO) of a containership at the full load condition are used:

L_{BP}	262.0m	ship length between perpendiculars
L	262.0m	ship length defined in 2008 IS Code
B	40.0 m	moulded ship breadth
D	24.45m	moulded ship depth
d	11.5m	moulded mean ship draught
t	0 m	Initial trim
LCG	125.52m	longitudinal centre of gravity from aft perpendicular
C_b	0.559	block coefficient
GM	1.965 m	metacentric height with free surface correction
VCG	18.4 m	vertical centre of gravity above baseline
KB	6.54 m	vertical centre of buoyancy above baseline
V_S	12.165 m/s	ship service speed

The following results were obtained under the above conditions, and it was confirmed that the values were similar.

Table 2: Result for Pure Loss of Stability

	in Explanatory note	in Software
Result	-1.982	-1.9818

2.2 PARAMETRIC ROLL (LEVEL1)

For Parametric Roll (Level 1), The first level of parametric roll requires that the non-dimensional amplitude of the variation of the metacentric height in waves ($\Delta GM/GM$) is less than a coefficient R_{PR}. This coefficient depends largely on bilge keel area and to a lesser extent on the midship section coefficient (denoted by C_m). The calculation method is described in SDC 2/WP.4. The value of R_{PR} is $0.17 + 0.425(100A_K/(L_{BP}B))$.

$$\Delta GM = \frac{I_H - I_L}{2\nabla}$$

The following results were obtained under the above conditions, and it was confirmed that the values were similar.

The following data(in Explanatory notes by IMO) of a containership at the full load condition are used:

L_{BP}	262.0 m	ship length between perpendiculars
Lf	262.0 m	ship length defined in 2008 IS Code
B	40.0 m	moulded ship breadth
D	24.45 m	moulded ship depth
d	12.34 m	moulded mean ship draught
t	0 m	trim
LCG	124.7 m	longitudinal centre of gravity from aft perpendicular
C_b	0.576	block coefficient
C_m	0.962	midship section coefficient
GM	1.965 m	metacentric height with free surface correction
$T_{\emptyset}$	25.7 s	ship natural roll period
VCG	18.37 m	vertical centre of gravity above baseline
l_{BK}/L_{BP}	0.292	bilge keel length normalized with ship length between perpendiculars
b_{BK}/B	0.0100	bilge keel width normalized with moulded ship breadth
V_S	12.861 m/s	ship service speed

Table 3: Result for Parametric Roll

	in Explanatory note	in Software
Result	0.8844	0.8847

5. CONCLUSION

In this study, stability calculation software under development with various features (hull form generation, compartment generation, 2D Sketch Draw for Hull form and Compartment and hydrostatics calculation module using network curve set) was introduced. And Level 1 which is a part of the second-generation stability standard, was calculated by using the software under development and the result was confirmed. However, In order to utilize this software at the stage of design, Results up to Level 2 are required, and more various test cases are required. Furthermore, it is necessary to check other vulnerability modes other than Pure Loss of Stability and Parametric Roll.

6. REFERENCES

1. International Maritime Organization (IMO). (2009a). International Code of Intact Stability. Development of Second Generation Intact Stability Criteria, London.

2. International Maritime Organization (IMO). (2009b). Information Collected by the Intersessional Correspondence Group on Intact Stability. SLF52/INF.2, Submitted by USA, London, England.

3. International Maritime Organization (IMO). (2012). Proposal of Revision of Updated Draft Vulnerability Criteria of Levels 1 and 2 for the Failure Modes of Pure Loss of Stability and Parametric Roll. SLF ISCG 54/19, Submitted by Japan, London, England.

4. International Maritime Organization (IMO). (2013). Information Collected by the Correspondence Group on Intact Stability Regarding the Second Generation Intact Stability Criteria Development, SDC 1/INF.8, Submitted by Japan, London, England.

5. International Maritime Organization (IMO). (2016a). Finalization of Second Generation Intact Stability Criteria, Amendments to Part B of the 2008 IS Code on Towing, Lifting and Anchor Handling Operations. SDC 3/WP.5, 3rd Session.

6. International Maritime Organization (IMO). (2016b). Observations Regarding the Feasibility of the Current Version of the Level 2 Criterion for Parametric roll in the Second Generation Intact Stability Criteria. Submitted by Sweden, London England.

7. François Grinnaert. Analysis and implementation of second generation criteria in a stability computer code. Fluid mechanics [physics.class-ph]. Université de Bretagne occidentale - Brest, 2017. English.

8. Jaeho Chung 1, DongMin Shin 2, Won-Don Kim 3 and Byung YoungMoon , "Current Status of the 2nd Generation of Intact Stability : Investigation of the Pure Loss of Stability and Parametric Roll Mode", Journal of Ocean Engineering and Technology 34(2), 55-65 April, 2020

7. AUTHORS BIOGRAPHY

Jinyoung Park is presently naval architecture researcher in ICT Solution Team of Korean Register providing domain expertise and knowledge to customers. With previous experience in ship structure strength assessment, Naval architect software development has continued to meet customer's needs

MiSeon Jang holds the current position of Surveyor in ICT Solution Team of Korean Register. She obtained her M.S. Degree in naval architecture and ocean engineering of Korea Maritime & Ocean University. Interest is Design Automation and 3D Reconstruction.

HoGyun Park holds the current position of the manager of ICT Solution Team at Korean Register. He obtained his M.S. Degree in naval architecture and ocean engineering of University of Ulsan. His research interests include 3D model-based approval and shipbuilding design software.

A NEW APPROACH FOR USING CAD AND PLM INTEGRATION

R Perez Fernandez, Siemens Digital Industry Software, Spain

SUMMARY

During the last decades, a tight and efficient integration between the CAD and the PLM systems used in complex shipbuilding products, definition, manufacturing and operation has become a more and more important matter, in order to ensure design efficiency, avoid design errors, prevent wasting time in using wrong data or, simply, to provide a common environment in which different organizations and partners working in the same shipbuilding design programs must share their data and follow common procedures.
When we talk about CAD and PLM integration, we must distinguish between CAD systems that use individual files to store the data (files that must be put under a PLM system control to be properly managed) and CAD applications that use their own database as a data repository.
The first one is a classic mechanical CAD to PLM integration approach, while the second is actually a proper case of systems integration, as both applications make use of their own database and both of them have to be maintained linked and synchronized, in order to guarantee data consistency and, as well, to ensure that the principle of the PLM as the "single point of truth" is maintained.
In this paper, we will be focus on the database CAD systems approach, the one used for a complete integration between FORAN and Teamcenter, where there is an increment in productivity and ease of use go together when using these new capabilities.

NOMENCLATURE

CAD	Computer Aided Design
ECN	Engineering Change Notice
PLM	Product Lifecycle Management

1. INTRODUCTION

During the last decades, a tight and efficient integration between the CAD and the PLM systems used in complex shipbuilding products, definition, manufacturing and operation has become a more and more important matter, in order to ensure design efficiency, avoid design errors, prevent wasting time in using wrong data or, simply, to provide a common environment in which different organizations and partners working in the same shipbuilding design programs must share their data and follow common procedures.
Some software suppliers provide both tools, CAD and PLM, and a high level of integration between both worlds but, in many cases, using a single software supplier is not feasible due to different reasons (*Perez and Penas, 2015*):

- The CAD tool that a company uses is strongly implemented in one of the program partners, while the PLM system to share the data is provided by a different supplier.
- The optimal and more productive CAD tool comes from a supplier that either does not provide a PLM system or it does not fit the expected requirements.

Whatever the reason, transferring the data between CAD and PLM is a strong requirement, and depending on the integration level reached, the efficiency in data handling can be higher or lower.

2. DATABASE VS. FILE BASED CAD SYSTEMS

When we talk about CAD and PLM integration, we must distinguish between CAD systems that use individual files to store the data (files that must be put under a PLM system control to be properly managed) and CAD applications that use their own database as a data repository.
The first one is a classic mechanical CAD to PLM integration approach, while the second is actually a proper case of systems integration, as both applications make use of their own database and both of them have to be maintained linked and synchronized, in order to guarantee data consistency and, as well, to ensure that the principle of the PLM as the "single point of truth" is maintained.
In this paper, we will be centred in the database CAD systems approach.

3. INTEGRATION VS. GATEWAY

3.1 WHAT CAN BE UNDERSTOOD AS AN INTEGRATION BETWEEN TWO APPLICATIONS

Two applications can be communicated in many different ways, and communication level can be tighter or loose.
In any case, we consider the existing communication level between two applications can only be considered an "integration" if both the following conditions are accomplished:

- Updating the data in one of the systems makes that the data will also be updated in the other system without any further user action. Saving data is performed only once and will be updated in both systems, with any additional user task.
- The data update is performed in both systems in real time, without any significant time delay.

According to this, systems in which the user first saves the data in one system but then some further manual action has to be performed to update the data in the other, cannot be considered as an integration, but just a gateway.
Likewise, systems in which the user saves the data only once in one of the systems, but the other one is updated by a delayed in time automatic task (e.g. by some automatic nightly executed trigger that "publishes" all modified data to the other system), is neither an integration.

3.2 CAD AND PLM LEVELS OF INTEGRATION

As a general view, the level of integration that may exist between a CAD and a PLM system can be classified into the following levels:

- Level 1. CAD data is manually exported and imported into PLM system, using generic export/import procedures. Unidirectional CAD PLM. This cannot be considered properly as an integration or a gateway, but just an export/import procedure. This is the approach normally used when no direct communication capability exists between previous systems.
- Level 2. CAD data exported and imported into PLM by some automatic or manual on-demand ad-hoc dedicated procedures invoked from either the CAD or PLM tools. Unidirectiona C PLM. This ensures a better data management quality than level 1 approach, but still cannot be considered yet as an integration, and is normally called a Gateway.
- Level 3. Similar to Level 2 approach, but adding bidirectional capability CAD PLM, by enabling some PLM to CAD data transfer, to apply to CAD system some changes that have been produced in PLM. Like in CAD to PLM data transfer, PLM to CAD communication can be launched either by some automatic tasks (normally invoked as a result of certain events, like approval processes) or by manual on-demand ones.
- Level 4. Having the bidirectional capabilities of Level 3 approach but adding a sort of interoperability capabilities. Enable direct access to PLM system embedded into CAD tool that allow performing actions in PLM on entities selected in CAD, or vice-versa, performing CAD actions on entities that have been identified and selected in PLM and we have loaded in the CAD scene for working on them. This paper will be mostly dedicated to describing this interoperability functionality.
- Level 5. Considering all the capabilities available in a level 4 approach, we add on them the function of saving data in PLM system at the moment we do it in CAD, in real time, without having to dedicate a subsequent either automatic or manual on-demand task to "publish" the CAD data into PLM.

Taking into account the criteria described in the former section about what can be considered as a "full integration" capability between two different applications, it is easy to conclude that only a Level 5 approach can actually be accepted as a real "integration".

4. CAD-PLM INTEROPERABILITY ELEMENTS AND UTILITIES

To better illustrate these concepts, let us take as a reference the integration between FORAN and Teamcenter.

4.1 CAD PRODUCT TREE

A full tree available access to the whole CAD model. In FORAN system, it is called FORAN Product Tree or, FPTree. From this utility, we can:

- Load model data in the 3D scene.
- Access entity properties in PLM manager.
- Display status of entities in PLM (created, published, pending to be published, synchronized, locked ...).
- In the case of a project in a series of ships, the applicability range of the entities (*Perez and Penas, 2016*).
- Launch entity publishing to update data in PLM for it.
- Launch an external web browser to access entity data in PLM.
- Drag & drop the entities to access data in PLM inside an internal browser.
- Drag & drop the entities into the scene.

4.2 EMBEDDED WEB BROWSER IN CAD TOOL WITH ACCESS TO PLM

A web browser with access to PLM server via web is available inside CAD modules, to allow access to the PLM data for the selected CAD entities, as well as to interact with CAD scene by selecting objects navigating into the CAD scene. The browser allows the access to two possible servers only:

- The PLM server connection. It is the main access allowed. It is previously configured, and the user will have access to it. This is to avoid accessing other PLM servers with different data than those mapped than represent the CAD model.

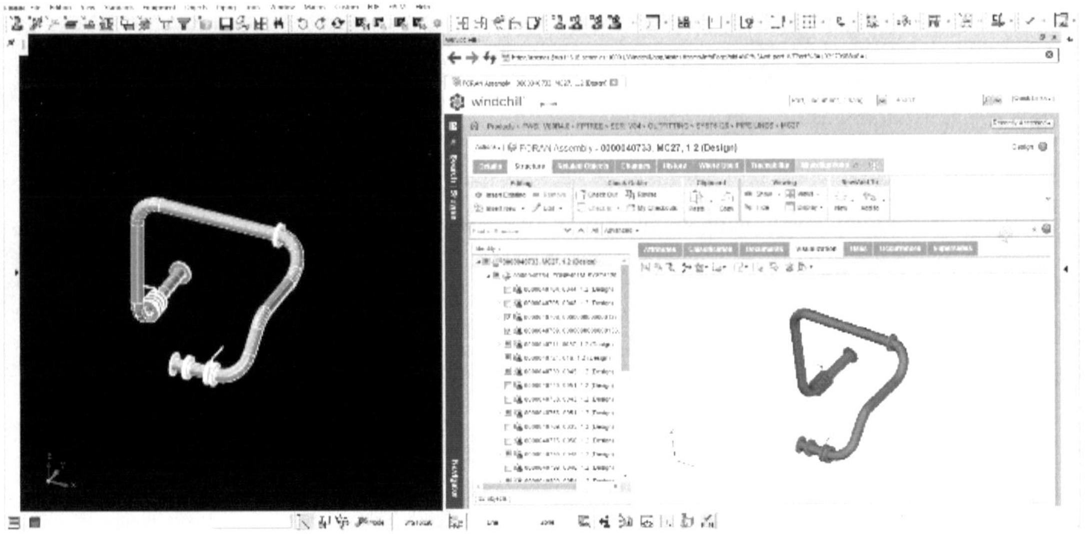

Figure 2: FORAN module including internal access to PLM (Windchill) server

- The FORAN FPLM Web Server. In the case of having synchronization (events connection from PLM to FORAN) enabled, the internal web browser will also be allowed to access that FORAN web server, in order to visualize synchronization results directly from that server.

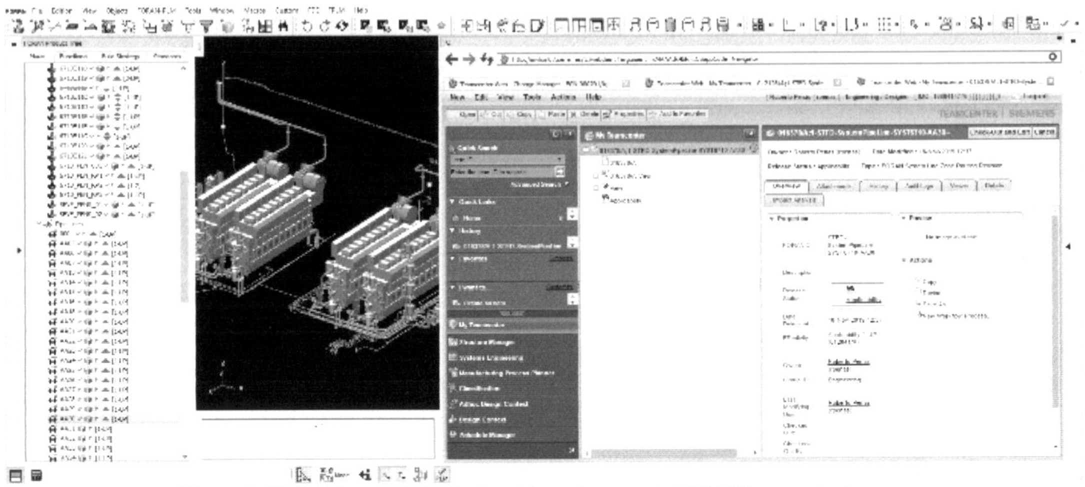

Figure 3: FORAN module including internal access to PLM (Teamcenter) server

4.3 INTEGRATED UTILITIES IN THE CAD SESSION

The different elements appear by default, in the following places:

- The FORAN Product Tree at the left.
- The 3D scene in the middle.

- The internal PLM access embedded browser at the right, although it can be undocked and located anywhere in the screen.

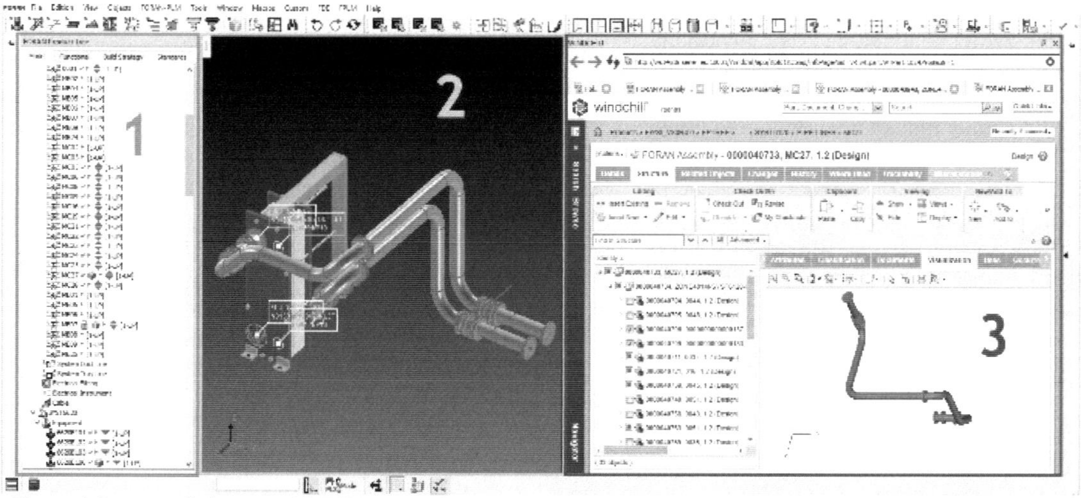

Figure 4: Integrated elements: (1) FORAN Product Tree (2). FORAN 3D scene. (3) Internal web browser with direct access to PLM server (example with Windchill)

5. CAD-PLM INTEROPERABILITY ACTIONS

5.1 DATA FLOW

The following diagram shows the different interoperability actions that could take place.

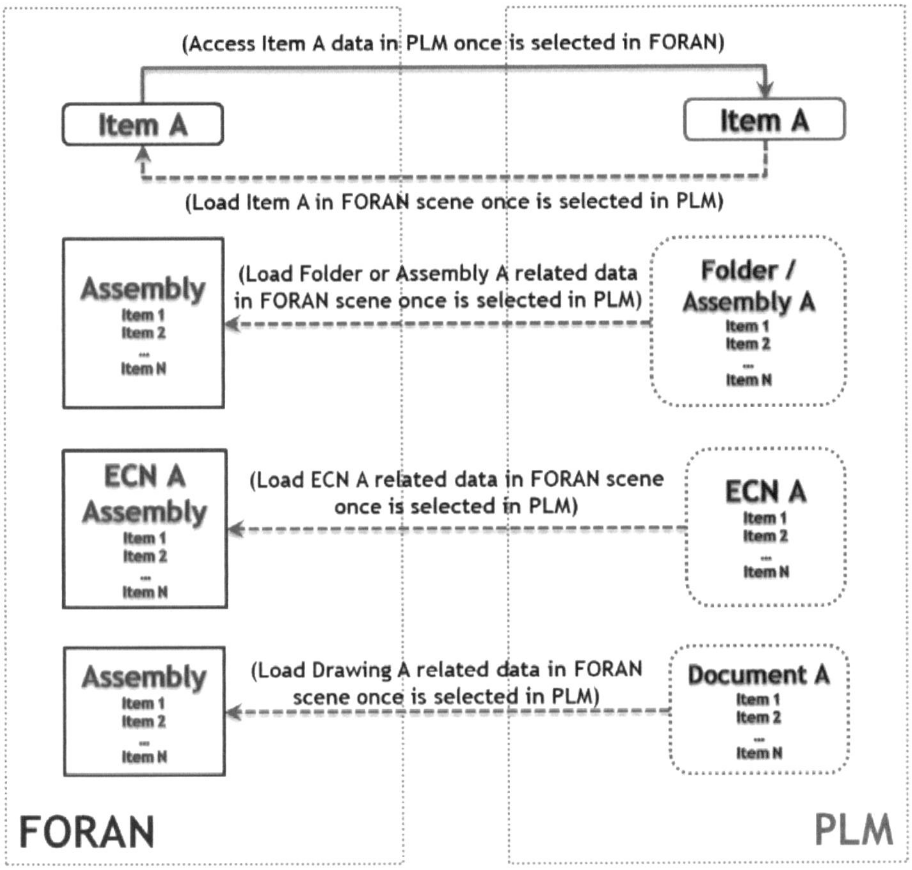

Figure 5: CAD & PLM interoperability flow

5.2 DESCRIPTION OF RELATED ACTIONS

5.2 (a) Direct access to data in PLM for an entity identified in CAD

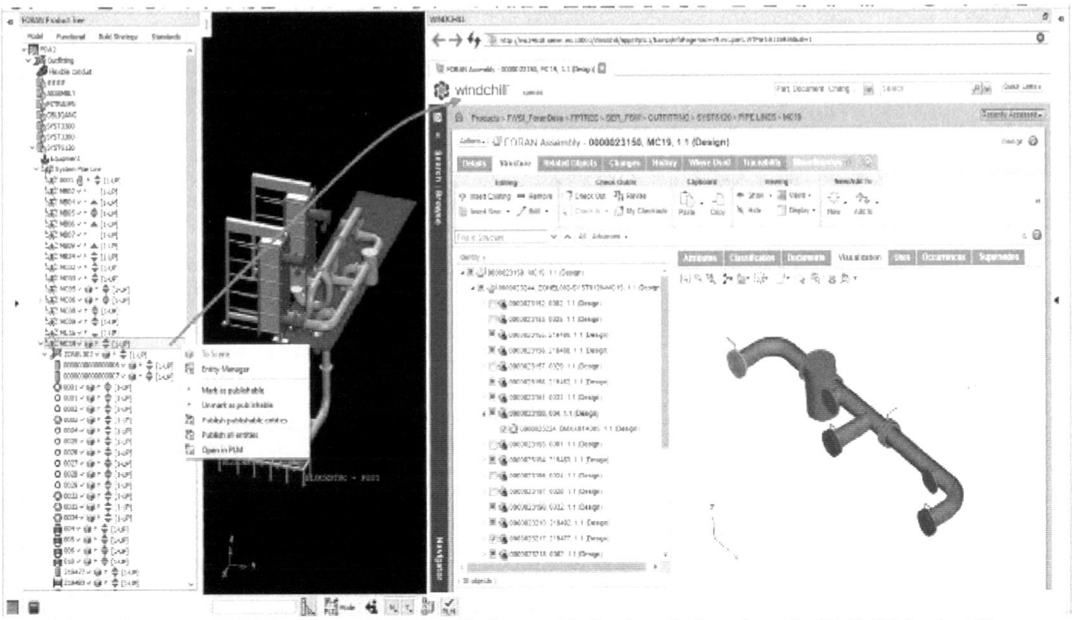

Figure 6: Access PLM data for a CAD identified assembly by drag & drop from the FORAN Product Tree

Identify a CAD entity either displayed in the 3D scene or selected in the CAD Product Tree, a directly open that entity properties page in PLM system. The entity can be either a single item or an assembly. Data is accessed in PLM browser window by performing a drag & drop action from CAD.

5.2 (b) Access both CAD and PLM data from a single click in CAD scene

Select an item in the CAD scene and the system, at the same time, will show the CAD data for it in the CAD properties page, as well as access to PLM properties page for that item.

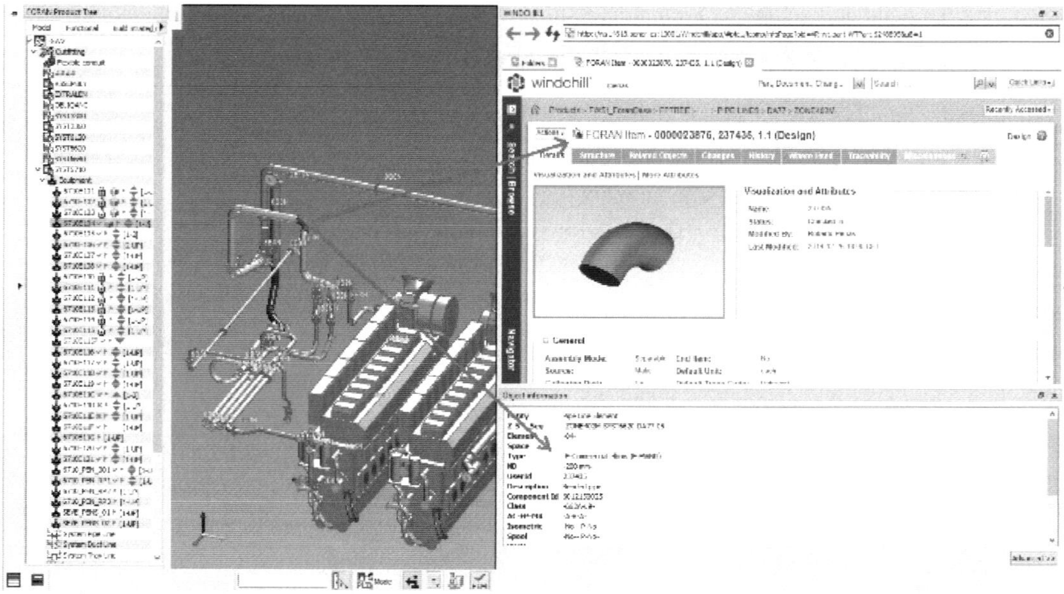

Figure 7: Select an item in CAD scene and directly show simultaneously the item data in CAD and PLM

5.2 (c) Load data in CAD 3D scene by selecting items in PLM

Identify an entity in PLM browser window and load it into CAD scene by drag & drop from the page URL or from any hyperlink.

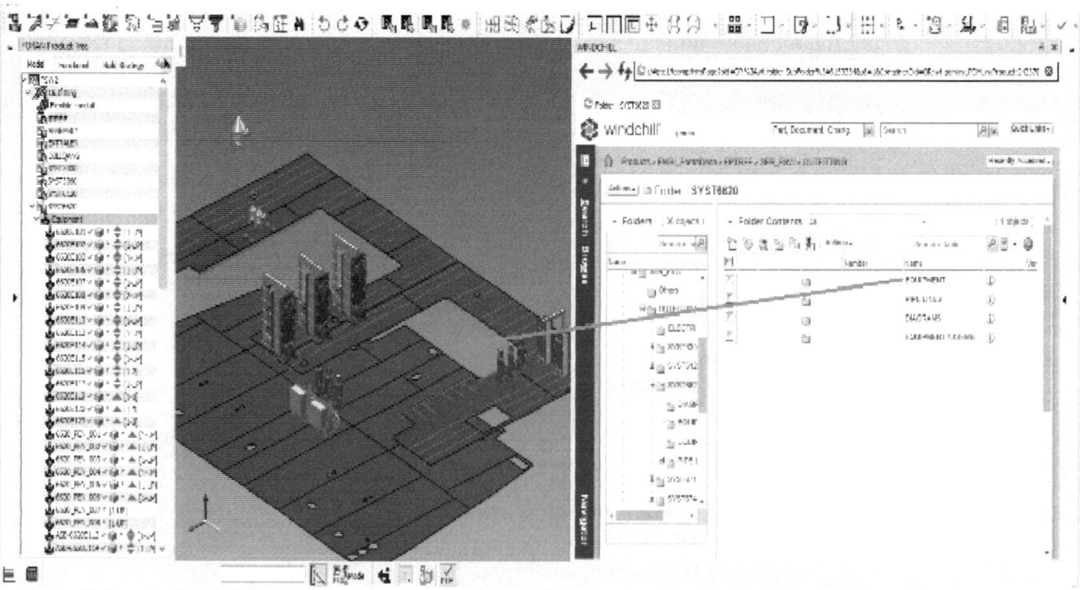

Figure 8: Load a PLM (Windchill) folder contained items into CAD scene, by drag & drop from a folder hyperlink

5.2 (d) Load the related PLM ECN data in CAD 3D scene by selecting the ECN in PLM

Identify an ECN in PLM window and load its related items into CAD scene by drag & drop either from the ECN page URL or from any hyperlink to it. ECN related entities can be affected, impacted or solution ones.

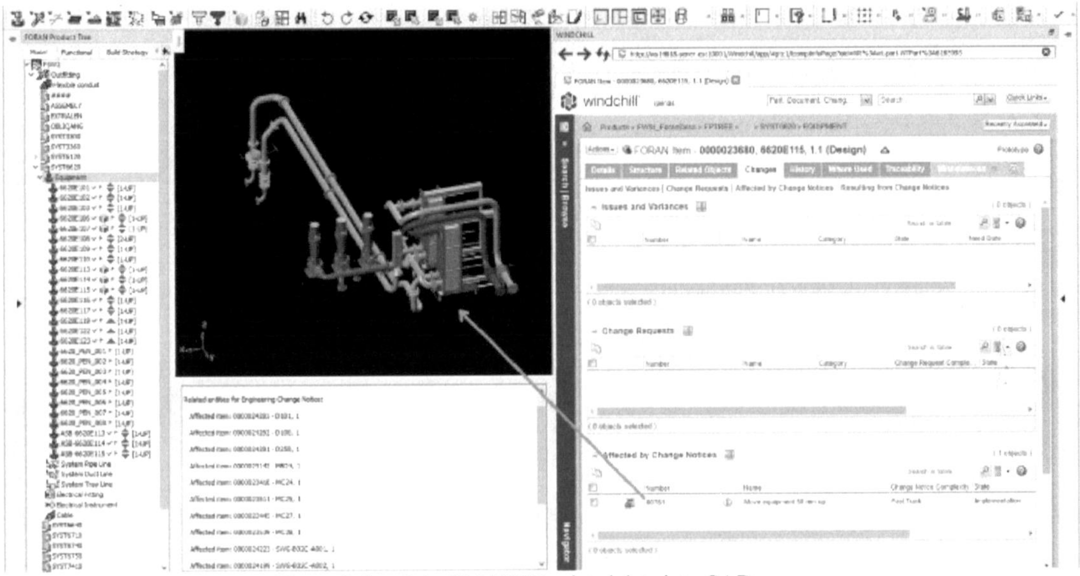

Figure 9: Load the PLM ECN related data into CAD scene

5.2 (e) Load the related PLM document or drawing described data in CAD 3D scene by selecting the document in PLM

Identify a document or a drawing in PLM window and load its described items into CAD scene by drag & drop from the document page URL or a hyperlink to it.

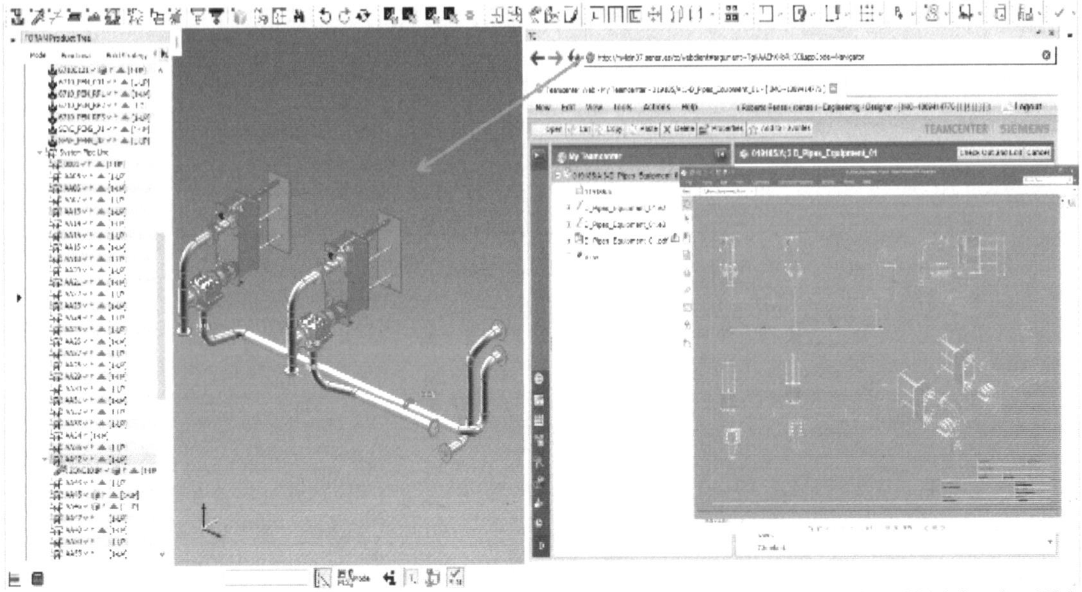

Figure 10: Load a PLM (Teamcenter) drawing related data in CAD 3D scene by drag & drop from PLM drawing URL

6. REACHING INTEGRATION LEVEL 5

The maximum level of integration can be reached as soon as it is possible to launch publishing to PLM when the data is stored in CAD repository DB. The user could store the modification changes performed and, at the same time, the system will launch the publishing of those changes into PLM system.
In order to reduce the returning times of CAD module control, when a user stores the data, publishing to PLM could be launched in a separate thread, so that publish results could then be communicated to CAD database by invoking the appropriate CAD API functions for this purpose. Working that way, the user would recover the session control, while a task is triggered to update the changes to PLM.
In addition, the existing synchronization tasks (those launched from PLM to CAD) can be produced in complete real time, by launching synchronization tasks associated to certain event triggers, like maturity changes performed by a PLM workflow.

7. CONCLUSIONS

The experience in the usage of these new interoperability tools between CAD and PLM shows an extremely positive response on users. Some of the clear advantages and benefits are listed below:

- Increase time saving by avoiding working with the wrong data.
- Increase time saving by allowing selecting the data only once (in the usual single point of truth: the PLM system) and perform the data loading in CAD with the items already identified in the PLM, instead of having to locate the data again in the CAD tool.
- Allow data loading in CAD according to item selections performed in the PLM:
  - Data included in a PLM folder.
  - A set of items included in a product structure created and updated in the PLM, not necessarily linked to any criteria in CAD.
  - Data related to an Engineering Change Notice (ECN).
  - Set of items described by a particular drawing or document whose relationships have been defined in the PLM, not in the CAD.
  - Immediate and direct access to the PLM data just published from the CAD, without having to search the published data in the PLM.
  - Simultaneous access to CAD and PLM data once an item is selected from the CAD scene.
  - Perform any PLM action from the internal browser, without having to set an external session with PLM.

User friendliness has been dramatically increased. The accumulated experiences show that, once users got used to these new tools, they do not want to go back to previous software releases missing them.
In addition, productivity and ease of use go together when using these new capabilities.
As well, these new tools are and could be available with any PLM system including an appropriate web access. They are currently fully available from FORAN system with *Windchill* and *Teamcenter* PLM systems.

8. REFERENCES

1. PENAS, R., SANCHEZ, L., GOMEZ, A. and PASTOR, L. (2009). *A neutral framework for the integration of CAD in product model lifecycle systems*. ICCAS, Shanghai, Sep. 2009.

2. PENAS, R. (2013). *An Advanced Step in Integrating Database-Controlled Shipbuilding CAD with PLM Systems.* ICCAS, Busan, Sep. 2013.

3. PEREZ, R. and PENAS, R. (2015). *Integration between shipbuilding CAD Systems and a generic PLM tool in naval projects*. Computer Science and Applications, Vol. 2, N. 5, pp. 181-191.

4. PEREZ, R. and PENAS, R. (2016). *How to manage sister ships on a shipbuilding CAD tool. The applicability concept*. Brodogradnja, Vol. 67, N. 4, pp. 49-59.

9. AUTHORS BIOGRAPHY

Rodrigo Perez, Ph.D. is Global PFS Director (PES Marine) at *SIEMENS Digital Industry Software.* He is also professor in several subjects at the *Marine Engineering School* of the *Technical University of Madrid.*

THE BENEFITS OF HAVING A BEST-IN-CLASS SHIPBUILDING TOOL

R Perez Fernandez, Siemens Digital Industry Software, Spain

SUMMARY

With the acquisition of FORAN by Siemens, it is expected to enhance the marine design and engineering capabilities soon. This is the perfect marriage, FORAN, the shipbuilding CAD tool with the longest tradition, and Siemens, always at the state of the art of the technology.
This integration will provide a comprehensive portfolio of software and services related to all facets of a ship's lifecycle, including concept design, production and operations, as well as optimized service lifecycle management.
We have become a key partner for the shipbuilding Industry, and although we are a global leader, we need to retain our customers and gain new ones, and in order to do that, we need to work harder providing them the best solutions.
By creating a best-in-class tool for shipbuilding, leveraging the team's industry expertise, shipyards and technical offices will be able to have a better control of their ship design and manufacturing process as they transition to Shipyard 4.0.
There is a lot of work out-there, waiting for us to achieve a greater success, big customers with big needs, and Siemens will provide the best solution as possible.

NOMENCLATURE

AI	*Artificial Intelligence*
AR	*Augmented Reality*
CAD	*Computer Aided Design*
CAE	*Computer Aided Engineering*
CAM	*Computer Aided Manufacturing*
IoT	*Internet of Things*
MR	*Mixed Reality*
RFID	*Radio Frequency IDentification*
VR	*Virtual Reality*

1. INTRODUCTION

The aim of this paper is to explain to the readers some ideas about the benefits of having a best-in-class shipbuilding tool (compound or not by a series of other tools).
Nowadays, shipyards, technical offices, Universities and research centers require cutting edge technology, they are demanding the best of the best, and with a focus on achieving Shipyard 4.0 and sustainability targets, Siemens aims to support marine customers in embracing a digital twin approach for designing and building vessels. By adding our extensive tools to the Xcelerator portfolio and leveraging the team's industry expertise, we will be able to offer commercial and naval shipbuilding customers better control of their ship design and manufacturing process as they transition to *Industry 4.0*.
The *Industry 4.0* transformation implicates a huge collection of interrelated and joined technologies, which can be analysed independently, but must be applied as a complete integrated implementation in each industrial field, moreover, in the *Computer Aided Design, Manufacturing & Engineering* (CAD/CAM/CAE), from now on referred just as *CAD*, development industry.
Each technology exposes a set of boundaries which cannot be differentiate evidently from its neighbouring technology.
Virtual Reality (VR), *Augmented Reality* (AR) and *Mixed Reality* (MR) are closely associated to the *Digital Twin* and interconnected with the *Big Data*, which is produced by the *CAD* tools and all surrounding solutions, which applies some *cloud/edge/fog* computing to this data in a merged technology between finite state machines and *Artificial Intelligence* (AI) cognitive processes (*Benayas-Ayuso and Pérez, 2019*).
To perform in an agile manner all these computing, it requires a network which support different connection ways to add special devices, i.e. *Internet of Things* (IoT), which can access to the data, creating and modifying it, in a different layer which affects to the basic information layer created by the *CAD* System in the shipyard.
This network should be secured, *cybersecurity*, but open to allow distributed work, which must be step controlled in a manner that records any modification of each working step done in an open, transparent, trusted and non-modifiable working method for all actors involved in process, like shipyard, engineering offices, classification society and ship owner, *blockchain*.
Results of the design should be easy integrated with future building ways like *3D* printing, generating printing orders directly from the *CAD* model.

Shipbuilding engineering phases involve design and production, but an integrated *Industry 4.0 CAD* System should also be involved in operation and maintenance.
When a ship comes for a reparation, sometimes the model is not available for this operation of maintenance.
In reparation and maintenance steps, replicating the full engine room or any other local in the *CAD* System could be a nightmare, unless *CAD* tool has an *AI* processing toolkit which, from a cloud of points, can recreate *CAD* equivalent items which can be converted with a minor user intervention, and lesser as the *AI* learns, in a *CAD* integrated and full modifiable design.

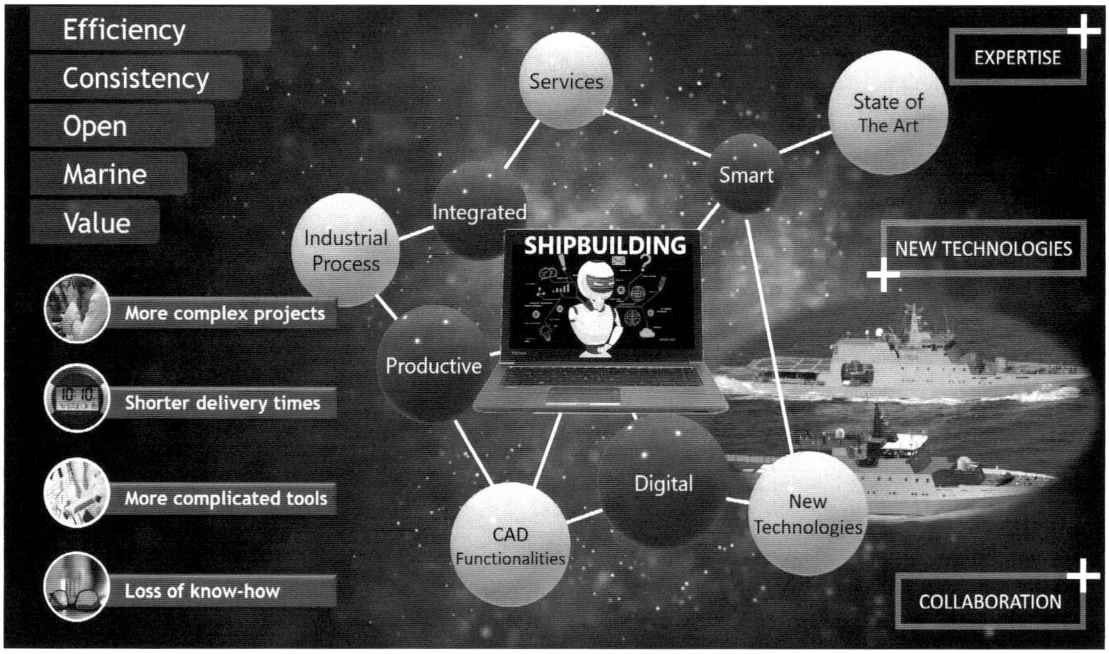

Figure 1: Related Industry 4.0 Technologies in a Shipbuilding CAD environment

At the end, this is just a short summary of the *Industry 4.0* technologies, which can be applied to a *CAD* System (as shown in *Figure 1*), included in, or as an integrated surrounding solution, or as information producers for the evolutive design process.

2. A BRIEF HISTORY OF INDUSTRY 4.0

To be able to understand how *Industry 4.0* became today's buzzword, a look at its predecessors might give us a perspective on how this revolution in particular is different.

- **The First Industrial Revolution**. The industrial revolution in Britain came in to introduce machines into production by the end of the 18th century (1760-1840). This included going from manual production to the use of steam-powered engines and water as a source of power. This helped agriculture greatly and the term factory became a little popular. One of the industries that benefited a lot from such changes is the textile industry and was the first to adopt such methods. It also constituted a huge part of the British economy at the time.
- **The Second Industrial Revolution** dates between 1870 and 1914, although some of its characteristics date back to the 1850 and introduced pre-existing systems such as telegraphs and railroads into industries. Perhaps the defining characteristic of that period was the introduction of mass production as a primary means to production in general. The electrification of factories contributed hugely to production rates. The mass production of steel helped introduce railways into the system, which consequently contributed to mass production. Innovations in chemistry, such as the invention of the synthetic dye, also mark such period as chemistry was in a rather primitive state then. However, such revolutionary approaches to industry were put to an end with the start of *World War I*. Mass production, of course, was not put to an end, but only developments within the same context were made and none of which can be called industrial revolutions.
- **The Third Industrial Revolution** is much more familiar to us than the rest as most people living today are familiar with industries leaning on digital technologies in production. However, the third industrial revolution is dated between 1950 and 1970. It is often referred to as the *Digital Revolution* and came about the change from analogical and

mechanical systems to digital ones. Others call it the *Information Age* too. The third revolution was, and still is, a direct result of the huge development in computers and information and communication technology.

- **The Fourth Industrial Revolution** takes the automation of manufacturing processes to a new level by introducing customized and flexible mass production technologies. This means that machines will operate independently or cooperate with humans in creating a customer-oriented production field that constantly works on maintaining itself. The machine rather becomes an independent entity that is able to collect data, analyse it, and advise upon it. This becomes possible by introducing self-optimization, self-cognition, and self-customization into the industry. The manufacturers will be able to communicate with computers rather than operate them. All of this will be possible through *AI* or *IoT* for example (*Muñoz and Perez, 2017*).

As regards *AI*, it can be said that there is no single origin of the concept known as *AI* and therefore there is no consensus to define that concept, but to be able to understand any of the definitions that apply, it is convenient to know at least briefly the most relevant facts and some of the milestones in its history.

It can be considered that *AI* was born as a philosophical study on human intelligence based on the concern of man to imitate the behaviour of other beings with capabilities beyond the reach of human beings (such as flying or diving), reaching the point of trying to imitate itself. In this sense, it can be said that *AI* is the search to imitate human intelligence. It is clear that it has not yet been completely achieved, but it is also increasingly true that we are closer.

The first man who became aware of his own existence and was able to think, surely wondered how his thought would work and would conclude the idea superior creator, an intelligent being capable of creating another one. In this sense, the idea of a virtual design for intelligence is as old as human thought.

n l , he C e h i e a e a e *Capek, 2017*), published a science fiction stage play called *Rossum's Universal Robots*. The play is about a company that builds artificial organic humans in order to lighten the workload of other people. Although in the play these artificial men are called robots, they have more to do with the modern concept of android or clone. They are creatures that can be passed as humans and have the gift of being able to think.

The English mathematician Turing publishes an article entitled *Computing Machinery and Intelligence* that opens the doors to *AI* (*Turing, 1950*). The article itself began with the simple question: *Can machines think?*

Later Turing proposed a method to evaluate if the machines can think, which got to know itself like the Turing test. The test, or *Imitation game*, as it was called in the document, was presented as a simple test to judge if machines could think.

1956 *Dartmouth conference* convened by McCarthy and where the term *AI* was coined. The conference was attended by researchers from *Carnegie Mellon University* and *IBM*, including Minsky, Newell and Simon. In this conference extremely optimistic forecasts for the next ten years were made that were never fulfilled, which caused the almost total abandonment of the investigations during fifteen years, known today as *AI* winter (*Aggarwal, 2018*).

In the other side, the concept *IoT* was born in the research centre *Auto ID* in *MIT* abound 1999 in research for realizing the object identification using radio frequency, *Radio Frequency IDentification* (RFID). The idea of the research was to identify all the objects and all the people in some way permitting some of the objects could exchange data with others through internet. The object identification was the prerequisite of *IoT*.

3. SHIP DESIGN PHASES

Ship design follows a series of stages, *Figure 2*. These stages are commonly denoted as conceptual design, contractual design, basic design, detail design, manufacture or production and maintenance:

- <u>Initial design, concept design</u>: CAD systems incorporate solutions for arbitrary hull forms and for converting forms generated in third-party software. FORAN offers an innovative approach for the definition of the general arrangement of the ship, allowing the user to set the most important spaces and volumes intuitively. All the naval architecture calculations can be done at this stage. The main output is the General Arrangement plan, and information for the contract phase.
- <u>Basic design, class design</u>: CAD systems work in 2D or 3D environments. FORAN generates ship 3D models at early design stages in short time. Using the same software from the beginning of the project yields significant time and cost savings. FORAN allows defining ship structures with only the primary elements and positioning the main equipment in its final position. This eliminates unnecessary errors and costs and makes material and weight calculations very accurate.
- <u>Detail design</u>: A CAD system is a great tool for defining the detail model of the ship, including information of structure, outfitting and electrical design. Whether we already have a 3D model generated in FORAN or the 3D model starts at this stage, FORAN provides many advanced, automatic and intuitive features to define all the structural elements, equipment, HVAC, supports, auxiliary structures, electrical design, electronics, etc.
- <u>Manufacturing, assembly and operation</u>: CAD systems allow extracting all the information necessary for manufacture and assembly, completely adapted to the production practices of the shipyard. Later in the life cycle, the CAD model is very useful in conversions, repairs, etc.

The use of new technologies in different ship design stages will be a great benefit for ship design offices, shipyards and ship-owners and will improve hugely the production process.

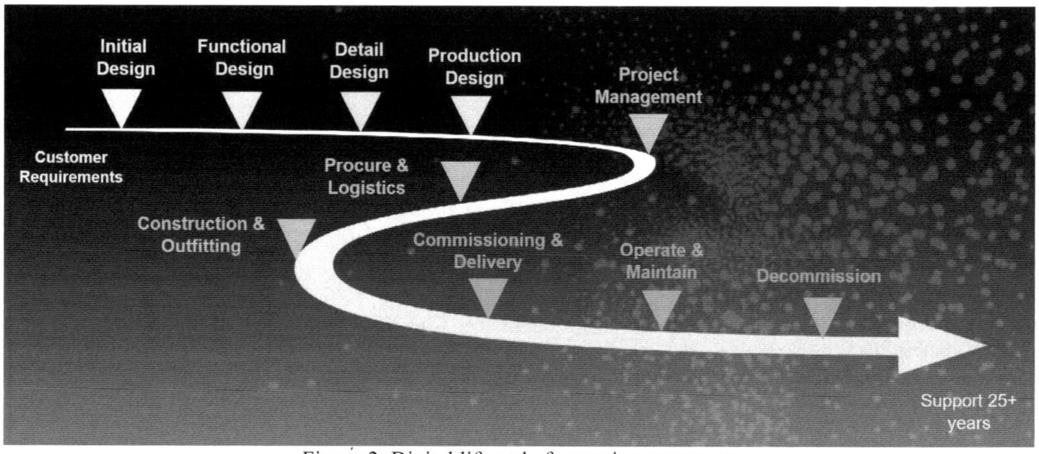

Figure 2: Digital lifecycle for marine customers

4. DIGITAL TWIN

There are industries where this concept is widely applied, but shipbuilding industry is still starting with it. Aero-generators, energy platforms, airplane engines are nowadays the most common cases where we can find *Digital Twin* applied to, with two possible approaches: create a *Digital Twin* of the product and create a *Digital Twin* of a complete factory and control the facility through its *Digital Twin*. An example is *Azure Digital Twin*, a *Microsoft* service announced by the end of 2018. According to a Gartner´s study, *Digital Twin* is used by the *24%* of the organizations with *IoT* technologies in production or with undergoing *IoT* projects and in the next three years, another *42%* is planning to implement it.
Many technologies, most of them described in this paper, are necessary to make possible a real effective *Digital Twin*, but *IoT* could be considered the most important, without forgetting the most basic and important technology necessary to make the *Industry 4.0* works, the communications infrastructure and, as always when we talk about data, security.
Digital Twin means a continuous and bidirectional exchange of Data between the real asset and the model, with the objective to have a real-time synchronization between those worlds (*Perez and Péter, 2019*).
Virtual to *Real Asset* data flow usually comes from a *Control Centre* that manages the data provided from the real world reflected into de virtual design. That Control Centre could interact with the real asset with a direct command executed through a console that will be converted in a digital signal that activates an actuator or by transmitting an order to a human operator.
Real Asset to *Virtual* data flow will be possible with a combination of sensors sending information through the communication infrastructure and human operators sending data by using *Apps* normally designed for devices that allow mobility and wireless connection.
Human or automatic processes, if the virtual asset includes *AI* or some automatic program that simply has an output depending on some input parameters, can manage both communication flows.
At the basis of the *Digital Twin*, the model is a must.
There are different ways to have it:

- **Spherical photographs**: it is a *360º* horizontal and 180º vertical photograph that can be reproduced in an interactive way into a *3D* representation. There are special cameras that can take this kind of photographs, but it is also possible to generate a spherical photograph with standard cameras and specific software that can compose all those pictures taken from a fix position rotating the camera and taking all the necessary pictures to cover the complete surroundings. This is the most economical way to have a digital representation of the infrastructure. This solution is adopted when it is necessary to include the *Digital Twin* for *Industry 4.0* purposes and there is no *3D model* of the facilities.
  This solution is also the fastest one, but the quality of the results and the possibilities are much more reduced than the other solutions.
  One of the most obvious disadvantages is that it is not valid for taking any kind of measurement.
  This technology only provides a conditioned visualization and navigation environment since the possibilities of movement are limited to the movement from one sphere to other following specific directions.
  Nonetheless, it could be enough for cases where it is only necessary a navigable virtual environment where it is possible to attach information provided by sensors, for example.
- ***3D Scanning***: This method is widely explained in the paragraph that talks about *3D* Printing.

- ***3D CAD model***: This is the best option for a new infrastructure, or for a facility that already has the *3D model* that can be easily updated to current situation and be used as *Digital Twin* without excessive effort.

The combination of *3D CAD model* and *3D* Scanning is possibly the most interesting. With both technologies, it is possible to measure, and this is a very important advantage.
Shipyard is not a static scenario, that is why the best option is a *Digital Twin* based in *3D models* generated with a *CAD* System and when necessary, with models generated by using *3D* Scanning technologies by transforming the clouds of points in volumes and surfaces.
This *Digital Twin* made of independent geometries, will be the best virtual environment to simulate and optimize processes, but also the best way to synchronize reality and the virtual world.
There are many applications for the *Digital Twin* and some of them are listed below:

- Remote monitoring and control of assets.
- Simulation of processes in a virtual environment.
- Training purposes.
- An updated *3D model* of the facilities.
- Personnel, supply, assets real time control monitoring.
- Register of all the historical data and the possibility to analyse all that information in order to extract conclusions and decide actions to improve performance, for example.
- By combining historical data and algorithms, it will be possible to generate predictive models that can alert us about a dangerous or anomalous situation. The *Digital Twin* will provide the knowledge of all the dependencies between all the elements of the facility and the information about the monitored parameters. With that global information it will be possible not only the analyse situation of an asset but also predict the repercussions in the complete chain where that asset belongs to.

This concept can be implemented in an easiest way if *all the stakeholders* adopt *Digital Twin*; in fact, this is the real future of this concept, create a Plug and Play *Digital Twin* ecosystem. It means that an asset will be ready to be implemented into a *Digital Twin* facility in a natural way, just install the physical asset into the facility, make the login of the new asset into the data system to acquire the information that comes from sensors and include the *3D model* provided by the vendor into the virtual facility. The *3D model* will include properties, parameters a direct link to the information provided by its sensors and will include operational information and predictive models that will be working during its operational life in order to be possible an early detection of problems.
The *Digital Twin* of the asset will also include (*Perez at al., 2018*).

- Digital manuals to be consulted by using *AR,* infographic animations that provide step-by-step instructions for disassembly and inspection, etc.
- Programmed alerts for periodical inspections and maintenance operations.
- The possibility for manufacturers to link to the data of that asset, analyse its operating data to make real time recommendations and use that information to design new assets with improved characteristics.

An example of this *Plug and Play* ecosystem: if a *Shipyard 4.0* needs storage racks, the provider of that asset will sell us a storage rack with its *Digital Twin*, which should include the *3D model* to be directly implemented into the *Digital Twin* of the facility.

Figure 3: Idealistic Representation of the Digital Twin for a marine customer

To reach this *Digital Twin* ecosystem the key factor is the definition of Standards, a common language to make possible an easy and fast integration of the real and virtual assets into a more complex *Digital Twin*, as in the maritime sector (*Figure 3*).
It is important to assume that a fully integrated *Digital Twin* is not always reachable, and it will be frequent to find partial implementations that consists just in data acquisition from machines in order to analyse and use for different purposes, but once reached this first step, it will be easier to move in the *Digital Twin* direction.

5. CONCLUSIONS

After the review of all this technology, it is possible to have an idea of the magnitude and complexity of the changes that *Industry 4.0* is demanding.
In the middle of this ecosystem, we will find a *3D model*, created mainly with *CAD* Tools and this means that the importance of *CAD* will be even more capital than today. The interface between the *3D model* and the rest of the *Industry 4.0* is the Digital Twin, the link between the reality and the virtual world, both living concurrently during the evolution of the construction and extending this world of possibilities to the entire lifecycle of the product.
CAD tools, as an important part of all this environment, should also evolve to be easily linked to these technologies, but also it is important to be adapted to the new generation of users that demands a different kind of interfaces and workflow and this will require an important effort.
This big world is now open, and Shipbuilding is starting in it, but the potential is clear and to remain competitive is mandatory to study how these technologies can improve the benefits by upgrading processes, resources, workflow, cooperation between stakeholders... the tools are here, now it is necessary to study and analyse our particular case. There is no global solution for all industries, not even a global solution for a particular industry, but digitalization is on hand, and it is mature enough to start implementation and be part of our strategy for the future.
Finally, we would like to transmit an optimistic vision of our future, with Siemens enhancing the marine design and engineering capabilities, providing a complete solution for all the ship design stages from conceptual design to operation and maintenance, paying special consideration to the Industry 4.0, embracing the Digital Twin Concept for designing and building ships.
The only limit to our success of tomorrow will be our doubts of today.

6. REFERENCES

1. Aggarwal, A. (2018). *"The Birth of AI and the first AI Hype Cycle"*.

2. Benayas-Ayuso, A. and Perez Fernandez, R. (2019). *"What should shipbuilding expect from the CAD/CAM systems of the future?"*. The Naval Architect magazine. April. pp. 28-31.

3. Capek, K. (2017). *"R.U.R. (Robots Universales Rossum)"*. Books Mablaz.

4. Muñoz, J.A. and Perez Fernandez, R. (2017). *"CAD tools for designing smart ships in the world of the Internet of Things"* SMART SHIP 2017, 24 and 25 January.

5. Perez Fernandez, R.; Benayas-Ayuso, A. and Perez-Arribas, F. (2018). *"Data management for smart ship or how to reduce machine learning cost in IoS applications"*. SMART SHIP, 23 and 24 January.

6. Perez Fernandez, R. and Péter Cosma, E. (2019). *"The use of CAD systems to manage modularity in multi-role warships"*. Conference Warship 2019.

7. Turing, A.M. (1950). *"Computing Machinery and Intelligence"*. Mind 49, pp.433-460.

7. AUTHORS BIOGRAPHY

Rodrigo Perez, Ph.D. is Global PFD Director (PES Marine) at *SIEMENS Digital Industry Software*. He is also professor in several subjects at the *Marine Engineering School* of the *Technical University of Madrid*.

*** Formatting Issue - Best Available Paper/Graphic ***

INTERCEPTOR EFFECTS ON A 3D RECTANGULAR PLATES IN A CALM WATER BY USING COMPUTATIONAL FLUID DYNAMICS (CFD)

AMF Putra and **H Suzuki**, Osaka University, Japan

SUMMARY

Interceptor is an additional device designed to improve boat efficiency by altering its running attitude. Interceptors are recently used for trim control during the boat's trip which give a significant reduction in water resistance, resulting in improved efficiencies. This paper focused on the hydrodynamics effects on resistance and trim of 3D rectangular plate in addition of interceptors in calm water by using CFD. Investigation has been carried out by using OpenFoam code. The results of this study meet a good agreement with most of previous works related to interceptor's study. This study found that *Angle of Attack* (*AoA*) employment at the interceptor with typical blade depth height shows better lift to drag ratio in comparison with interceptor without *AoA* employment. Moreover, those configurations also produced better trim angle reduction which resulting better running attitude. This study also shows that the efficiency of the interceptor is related to the boundary layer thickness.

1. INTRODUCTION

Interceptor is an additional device designed to improve boat efficiency by altering its running attitude. The interceptor is typically installed at the bottom of the boat's transom. The interceptor blades or plate can create a precise pressure distribution beneath the boat; the blades run vertically and create a high lift effect upwards. Interceptors are recently used for trim control and steering, in order to reduce uneven motions of the boats, especially in the early stages of the trip. Controlling trim during the trip can result in a significant reduction in water resistance, resulting in improved efficiencies.

Some previous works in interceptors shows some phenomena and advantages on boat's performance and efficiencies. A study show that the interceptors demonstrate highly effectiveness in reducing resistance of the model with a percentage of 10% - 18%, accompanied by reduced sinkage and trim [1]. Another experimental study of interceptor's effectiveness on high-speed monohull and catamaran and resulting total resistance reduction up to 19 %. Moreover, combination of interceptors with another device and investigation on various combination of interceptors also show the effect of the in e e o on he oa he e e o he i y e o en a and in e e o o ina ion on e i an e reduction on planning boats has been investigated [3]. This combination shows better resistance reduction up to 21% in comparison with interceptor and/or stern flap only. The study on hydrodynamic effects of trim interceptors on highspeed hulls to analyze the geometrical characteristics with numerical methods declared that the deployment height of the in e e o ade ha a e i e e e on he e ien y It is also found that the interceptor's blade height from the an o edge o a e en age gene a y e een 1 and o he W o he oa o u i i e a ositive pressure gradient at the stern and change the trim [5]. Two unconventional interceptors' system have been introduced [6]. The Double Interceptor System (DIS), and the Split Interceptors (SI) which is consists of the usual straight edge located in the traditional position but fixed at a distance h from the bottom. The DIS have shown the best performances; comparing the results with a bare hull at the same CG, up to 25% of resistance reductions were obtained. Various configurations of trim tab, interceptor, transom wedge, and integrated trim tab and transom wedge investigation on a model test of a deep-V catamaran have been carried out and the interceptor showed the greatest efficiency and showed a fuel reduction of 5% and a trim reduction of 1.2° at design speed [7]. There was an indication that the efficiency of the interceptor related to the boundary layer thickness dependent on the wetted surface and speed [8]. In addition, it is also suggested that interceptor blade depth must be adjusted related to the operation speed. It is also important to give more attention at d/H ratio in designing an interceptor, where d is the interceptor height and h is the boundary layer thickness at the transom [9]. Moreover, some studies proved that pressure on the stern bottom increased in proportion to the interceptor height [10]. The height of the interceptors was more critical at high speeds. Therefore, it is important to determine the appropriate interceptor height in controlling the running attitude of the highspeed planing-hull under various operating conditions. Nevertheless, further investigation in designing interceptor still need to be carried out to obtain some other alternatives on its applications.

Planing-hull geometry mostly used in previous works to investigate interceptor efficiency. This is reasonable, considering the main use of interceptor nowadays is in the planing-hull. The Computational Fluid Dynamics (CFD) method has been u ed o udy he o a o a ound a in e e o ed o a a ounda y, hi h e e en ed he hyd odyna i o around the bottom of a ship and was known as the most idealized model of interceptors ever used. The results show us the relationships between the hydrodynamic forces and the interceptor's main parameters [11]. Further investigation of the hydrodynamic effects of interceptors on a 2-D flat plate based on both CFD and experimental approaches also have been carried out [12]. The results shows that the increase of pressure at the end of the flat plate and the lift force coefficient at high angles of attack in addition of interceptor were proportional to the interceptor height. Specifically, this study focused on the hydrodynamics effects on resistance and trim of 3D rectangular plate in addition of interceptors in calm water by using CFD. The bottom shape of the rectangular plate is flat and tilted in order to approach planing-hull geometry. Rectangular shaped used in this study to obtain and understand the effect of interceptors in more general hull geometries. Various configuration of blade deployment, by depth and angle of attack, were used in the simulations. Investigation has been carried out by using Open source CFD code, OpenFoam. Results of current study will be used by author as reference for further implementation and development of stepped-interceptors design for high-speed planing-hull boats.

2. CFD OPENFOAM

OpenFoam is a multi-physics open-source software package which written in C++ library used in the solution of partial differential equations (PDEs) and ordinary differential equations (ODEs). The application of openFoam code is often used to solve marine hydrodynamics problems. Various capabilities of OpenFOAM in marine hydrodynamics have been presented. Some hydrodynamics parameters such as resistance and seakeeping results shows good agreement with experimental results [13].

OpenFoam (ver. 19) have been used to solve numerical simulations in this study. Finite Volume Method (FVM) used to solve and discretize incompressible Navier Stokes equations in a multi-phase (water and air) scheme. The turbulent is modelled by k - ω SST for the current simulations. To model the domain's multiple phases, a Volume of Fluid (VOF) method is used. The VOF method does not explicitly track the phase interface, but instead assumes a continuum in each cell; that is, in our current setup, each cell consists of a given volume of water, air, or solid, and the aggregate properties o he e a e o u ed a ed on he a ion o ea h ha e 1 he a go i h de ou e e u e and velocity y o ining a adi iona e u e i i i h i ing o e a o and e i-Implicit Method for Pressure- in ed ua ion he e Fo a i i ua ion , he n e Foa o e i u ed, i h an i i i Fi -order Euler scheme for pseudo time integration; for steady-state simulations, this low-order pseudo-time integration is satisfactory. For unsteady cases (pure sway, pure yaw and drift motion), the interDyMFoam solver is used with dynamic deforming mesh [15]. The mesh deforms based on the movement of the geometry, without topological change. The displacement of e h oin i a ua ed y o ing he o o ing a a ee ua ion a ea h i e e

$$\nabla.\left(\gamma \nabla x_g\right) = 0$$

where x_g is movement of mesh point; γ is the diffusivity coefficients based on the square inverse of distance between the geometry and mesh point. First order implicit Euler scheme is used for time integration.

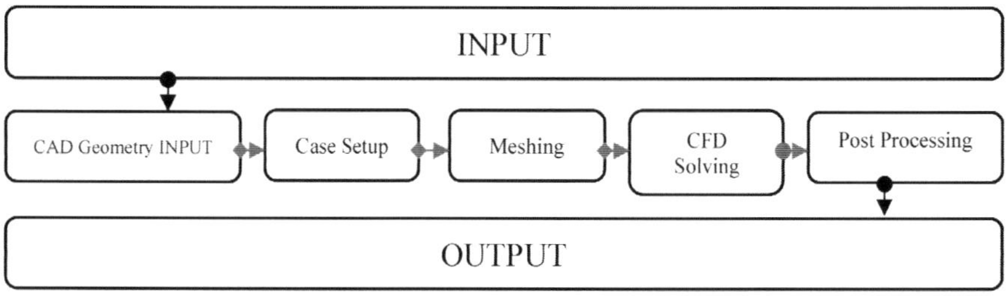

Figure 1 OpenFoam workflow for marine applications.

Figure above shows the openFoam workflow for marine applications [16]. The workflow begins with geometry creation, case setup, meshing, CFD solving and post processing.

3. SIMULATION SETUP

3.1 GEOMETRY

For current studies, the investigation has been carried out by using CFD only. Considering the future works will be experimental and numerical studies, thus the geometry of rectangular plate will be adjusted nearly the capabilities of towing tank of Department of Naval Architecture and Ocean Engineering, Osaka University, Japan. The towing tank carriage can accelerate for up to 3.5 m/s. In order to obtain high-speed state from a 32-meter length of rectangular plate at around 27 knots, with Froude number (Fn i , he ode a a ed o 1 1 he a i u a di en ion o he re angu a a e i a o o

***** Formatting Issue - Best Available Paper/Graphic *****

Table 1 Rectangular plate dimension.

Characteristics	Actual (m)	Model (m)
eng h o Wa e ine W	32	2.03
Breadth (B)	7	
Depth (H)	3.5	0.22
Draught (T)	1.35	0.09
Maximum Speed (V)	13.89 m/s	3.5 m/s

The ode a e i y e n i h ongi udina Cen e o a i y C i o an o he Cen e o a is 0.11 m from the bottom of rectangular plate. The simulations were performed based on the variation of speed including different deployment heights of the interceptors and interceptor angle of attack (AoA). In addition, there was also e o ed a a e o di e en C o a ion o ain ain ode a i i y in addi ion o in e e o

Table 2 Simulation geometry configurations

Configurations	Interceptor blade height	Interceptor AoA (0)	C o AP (m)
Barehull	-	-	0.625
Interceptor Addition 1	W	0	0.625
Interceptor Addition 2	W	0	0.625
Interceptor Addition 2 - C	W	0	0.575
Interceptor Addition 2 AoA 1	W	-	0.625
Interceptor Addition 2 AoA 2	W		0.625

The interceptors were fully mounted at the rectangular plate transom edge. The configurations of interceptor blade height and AoA shown at table 2 above. The AoA were set in largely angle in opposite direction to expect significant different from the resul he in e e o ade heigh a e o W and W hi e ing a a ing he a i u suggestion from previous works to deeply understand the difference between those configurations.

3.2 , Y C

Figure 3 sho he o u a ion do ain a ge do ain ea ed o a oid e e ion e e on he ounda ie he o e i y e i , hu ha ode ed he ounda ie in hi u en i u a ion di ided in o e en a he in e , ou e , bottom, atmosphere, side, midplane and the hull. More information about the boundary setting is presented in the table 3.

Table 3 Boundary condition of the simulation.

	Inlet	Outlet	Atmosphere	Hull
alpha.water	fixedValue	variableHeightFlowRate	inletOutlet	interfaceCompression
k	fixedValue	inletOutlet	inletOutlet	Wa Fun ion
nut	fixedValue	zeroGradient	zeroGradient	nu Wa Fun ion
omega	fixedValue	inletOutlet	inletOutlet	o egaWa Fun ion
pointDisplacement	fixedValue	fixedValue	fixedValue	calculated
p_rgh	fixedFluxPressure	zeroGradient	prghPressure	fixedFluxPressure
U	fixedValue	outletPhaseMeanVelocity	pressureInletOutletVelocity	o ingWa e o i y

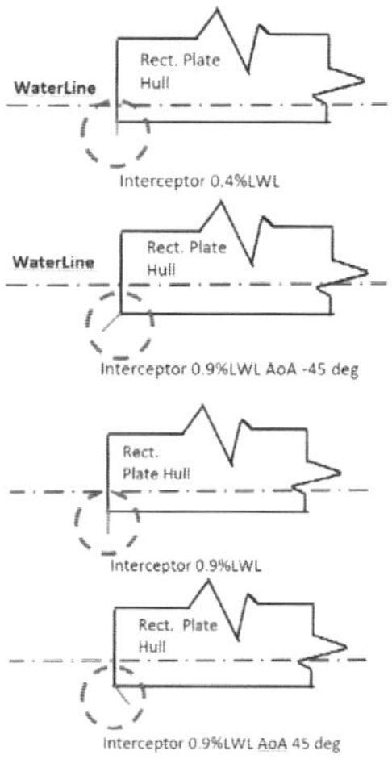

Figure 2 Interceptor configurations.

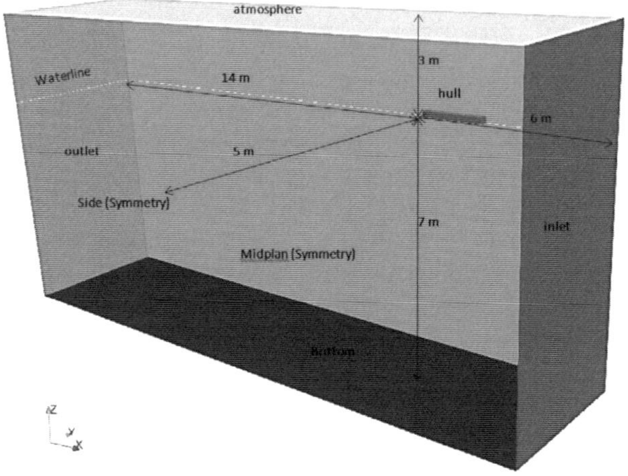

Figure 3 Computation domain and boundaries.

The interFoam o e , a ed on agoi h , a u ed in u en i u a ion i u a ion e e un o l seconds with the time step is 1 x 10- and adjustable during running time, with maximum time step is 2 x 10^{-3}. The Courant number was set to maximum 5. As stated previously, the turbulent is modelled by for the current simulations.

3.3 MESH

The snappyHexMesh is a utility provided by OpenFoam and was used in the current simulation. Dynamic meshing and 6DoF was enabled in this simulation. However, the move was set free for pitch (y rotation) and heave (z translation) only. The other degrees of freedom were locked in order to allows solver correctly to obtain trim and sinkage values. By enabling dynamic meshes, models are allowing the mesh to change during a simulation. Since the cells are changing in time when enabling dynamic meshes, thus time step should properly choose to avoid resulting large volume changes between iterations.

Figure shows the mesh results generated by snappyHexMesh utility in openFoam. The left side figure is the results of rectangular plate without an interceptor. The rectangular plate in addition of interceptor's mesh in each configuration shown below.

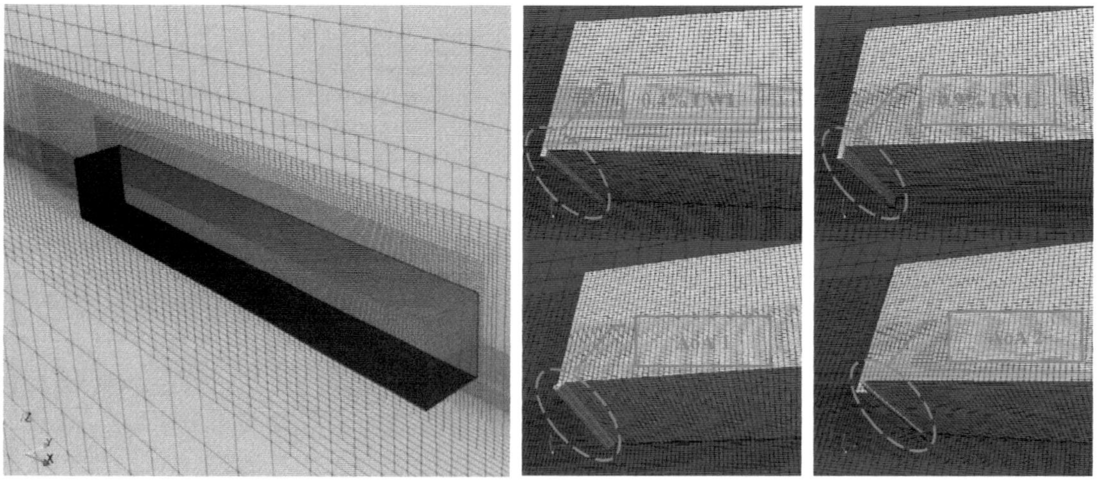

Figure Rectangular plate mesh.

*** Formatting Issue - Best Available Paper/Graphic ***

MESH INDEPENDENCE STUDY

The mesh independence study was performed by three different meshes. A coarse mesh (grid 1) composed of about 0.96 x 10^6 cells, a medium mesh (grid 2) composed of about 1.5 x 10^6 cells and fine mesh (grid 3) composed of about 2.8 x 10^6 cells. Initially, grid 1 was used to obtain the solution. Convergence was reached when the scaled residuals were reduced to 10^{-} for continuity. Furthermore, the values of force coefficient ensured the convergence and stability of the solution throughout the iterative process.

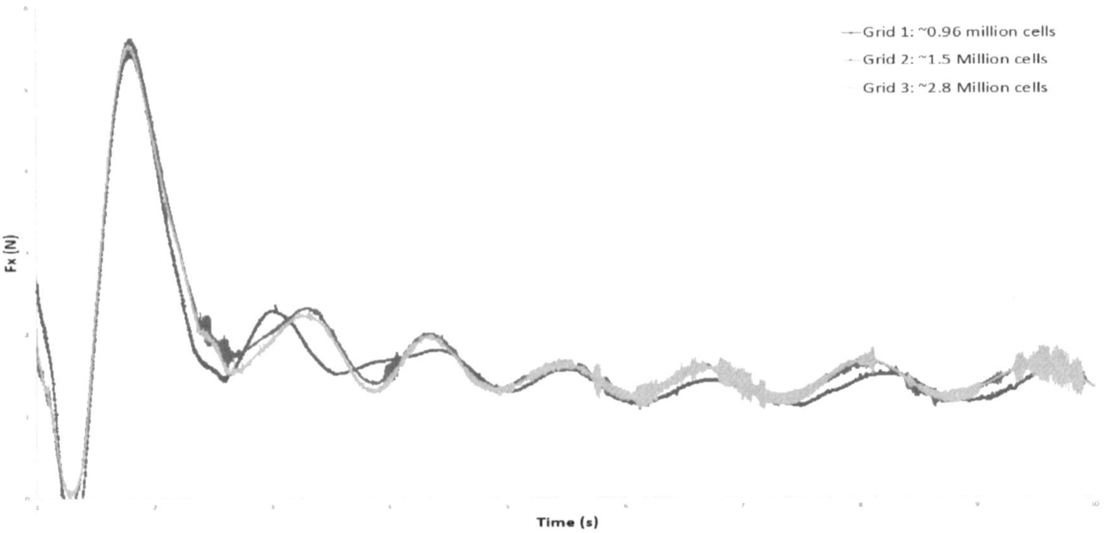

Figure 5 Grid independence study; Force on the rectangular plate bottom for three different grids.

The grid refinement was used to reduce discretization errors and ensure that the solution was only dependent on the physics of the problem and the boundary conditions used. The force values at the bottom of the rectangular plate were used to compare the different grids. The grid which shows relatively lower discretization errors using the lowest number of cells then observed. Figure 5. above shows that grid 3 is having a good agreement with grid 2, however, grid 2 then further employed in this simulation since it is thought to have a ea ona e nu e o e and i no a i ing u h o he solution accuracy as compared with grid 1.

4. RESULTS

.1 C F F C

Figure 6 (a) show the results for rectangular plate resistance in various configurations. In the speed range < 2 m/s, the resistance of rectangular plate without addition of interceptor gives smaller value in comparison with other configurations. Nevertheless, its resistance value increasing in the speed range > 2 m/s and becomes the highest among other configurations. The lowest total resistance value found in configuration of interceptor 2 AoA 1, which gives reduction up to around 30 % in comparison with rectangular plate bare hull. The resistance of rectangular plate in addition of interceptors shows similar pattern, which is increasing in the speed range < 2.5 m/s and decreasing after that. It also found ha di e en C o a ion ay a e e i an e a ue igni i antly even tough composed with typical configurations.

***** Formatting Issue - Best Available Paper/Graphic *****

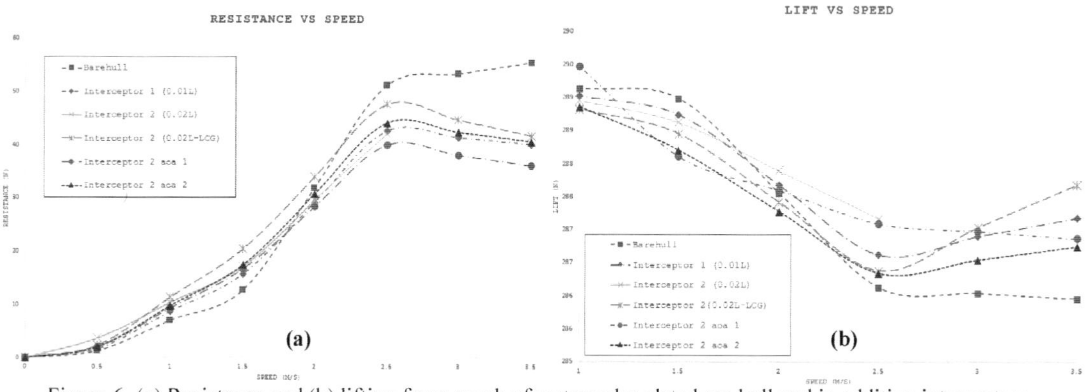

Figure 6 (a) Resistance and (b) lifting force graph of rectangular plate bare hull and in addition interceptors.

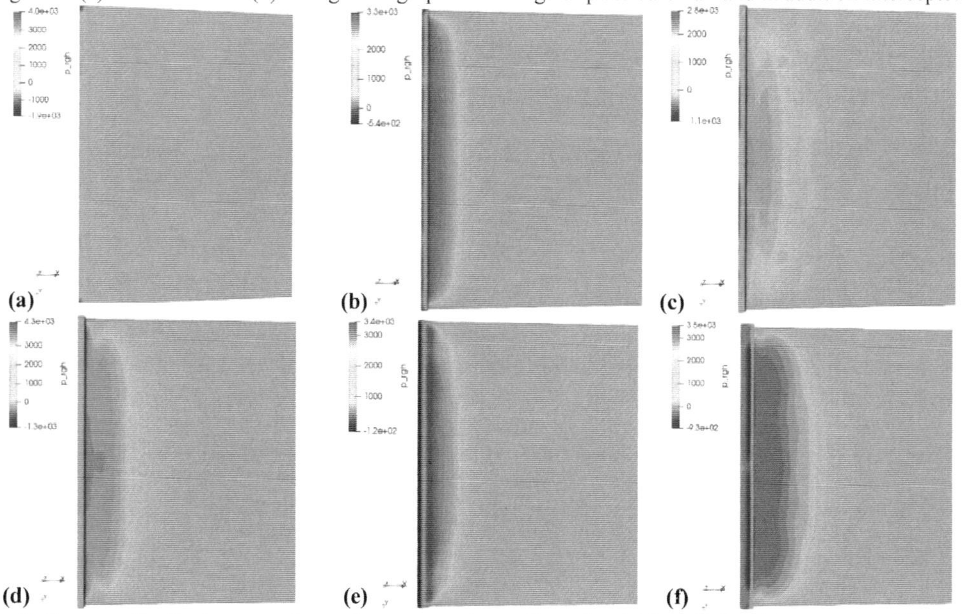

Figure 7 Pressure distribution around stern bottom area in configuration (a) bare hull, (b) interceptor addition 1, (d) interceptor addition 2 – C , e in e e o addi ion o 1, in e e o addi ion o at speed 3.5 m/s and (c) interceptor addition 2 at speed 2.5 m/s

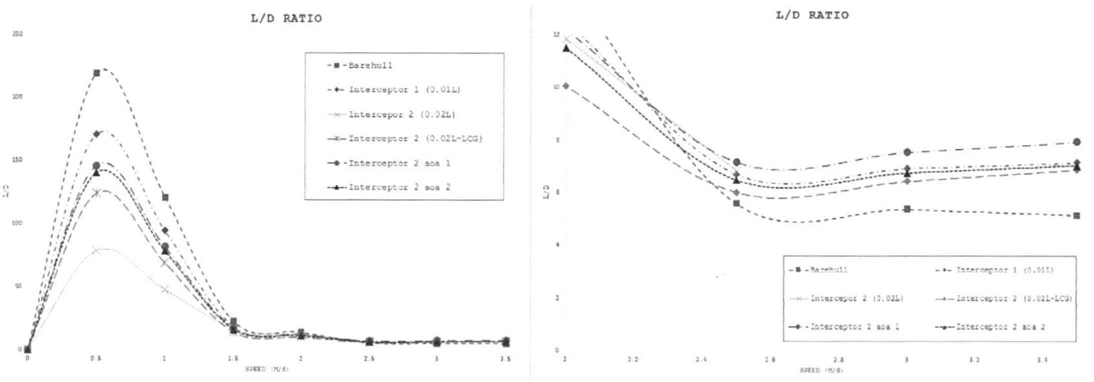

Figure 8 i o ag a io o e angu a a e a e hu and in addi ion in e e o

Initially the interceptor is an appendage, from the graph can be further observed that at the early stages or at low speed, all configurations with addition of interceptor presents higher resistance values in comparison with the configuration without interceptor addition. Vice versa, at the higher speed, the interceptors create a high lift effect upwards which raising stern part of the hull and eventually reducing overall resistance component. As shown in the graph, at the higher speed all configuration with interceptor presenting reduction of resistance value. Even more, that resistance value become lower than the bare hull resistance. However, the effectiveness of interceptor addition in reducing resistance convinced by observing resistance graph.

As mentioned previously, at certain speed addition of interceptor gives an extra lift in each configuration of rectangular plate. The results for rectangular plate lifting force in various configuration shows in figure 6 (b). Neglecting the drag force, the configuration of interceptor 2 – C in e e o ade heigh W i h C o e u the highest lifting force. Different with another configuration, this configuration has different C o i ion, which is closer to the AP in comparison with other configurations. However, the configuration of interceptor 1 obtain better lifting o e o a e i h o he on igu a ion i h a e C o i ion he e angu a a e a e hu i ing o e end o ag and even decreasing at speed > 2.5 m/s.

Figure 7 shows the pressure distribution at stern bottom area of the rectangular plate in each configuration at speed 3.5 m/s, except for configuration of interceptor addition 2, figure 7 (c), which showing the results at speed 2.5 m/s. The bottom pressure of the bare hull, figure 7 (a), around stern area is the lowest in comparison with another configuration. This is in correlation with the lifting force graph which also show the lowest value. Moreover, the pressure occurred at the bottom stern in all configurations with an addition of interceptor strongly convinced that additional lift force, drag force and moment have created.

The ratio of lift force to drag force can be an appropriate relationship for evaluating interceptor's efficiency [12]. Figure 8 shows the drag and lift ratio of rectangular plate. Furthermore, ratio of the drag force and lift force produced by the configuration of interceptor 2 AoA 1 is the highest. Vice versa, rectangular plate without addition of interceptor is less effectives. Among all configuration, we can say that the proportion of AoA and interceptor blade height of the configuration interceptor 2 AoA 1, gives better efficiencies.

MOTION AND TRIM

The sinkage value based on centre of gravity (CoG) height from the rectangular plate and he C o e en in a iou configuration presented in figure 9 (a). Trim condition comparison of the rectangular plate in each configuration depicted in figure 9 (b). The motion of rectangular plate at the configuration interceptor 2 – C ho i i a i y a e n i h he configuration interceptor 2 AoA 2, especially in CoG height and the trim angle. By this we can assume that the combination o e oy en in e e o ade o ha o e a ion i h he C ini ia o i ion a he a e ade heigh o eo e , these two configurations also presenting better reduction in term of rectangular plates trim angle, which resulting better running attitude control in comparison with other configurations.

The effectiveness of an interceptor addition to reducing trim angle in comparison with the bare hull can be clearly seen in the figure 9 (c). The bare hull trim angle keeps increasing, vice versa, all interceptor configurations reduce at speed > 2.5 m/s. Nevertheless, deployment of an interceptor blades (height and AoA) must be wisely considered. As shown by configuration of interceptor 2, this configuration has failed to obtain the advantages of an interceptors. The simulation results shows that this configuration produce too much lift at the stern part of rectangular plate thus high moment occurred which eventually pushing the bow into the water and sinking shortly after reaching speed ~2.5 m/s.

***** Formatting Issue - Best Available Paper/Graphic *****

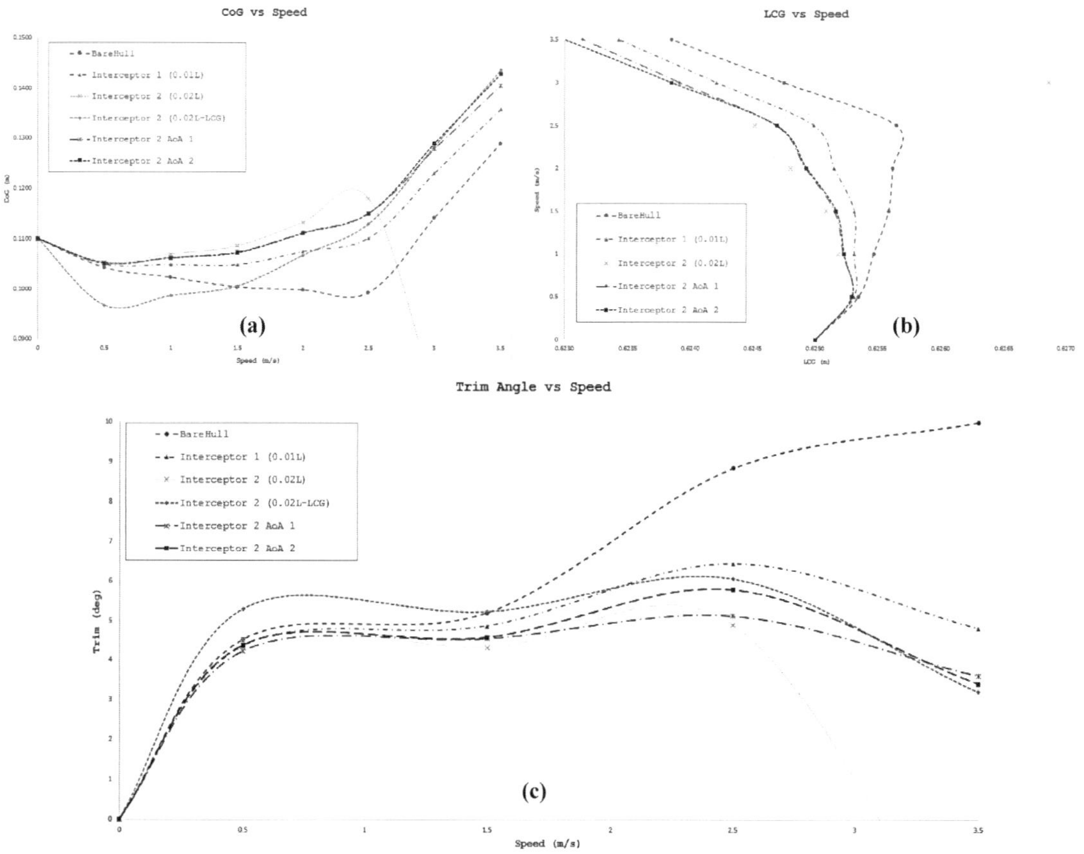

Figure 9. CoG a , C and trim (c) monitoring graph of rectangular plate.

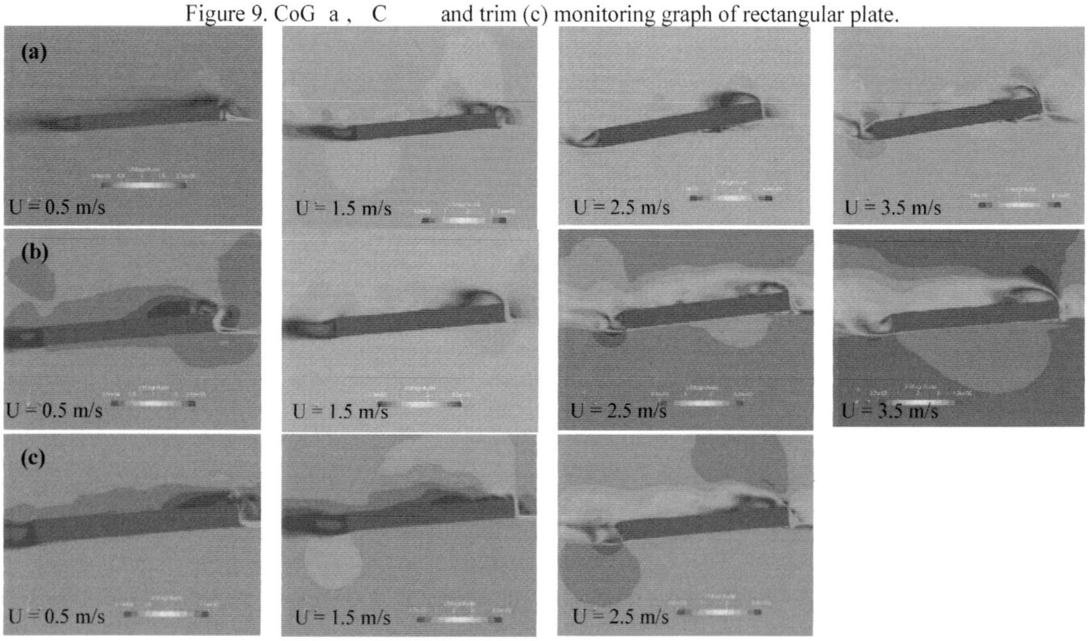

***** Formatting Issue - Best Available Paper/Graphic *****

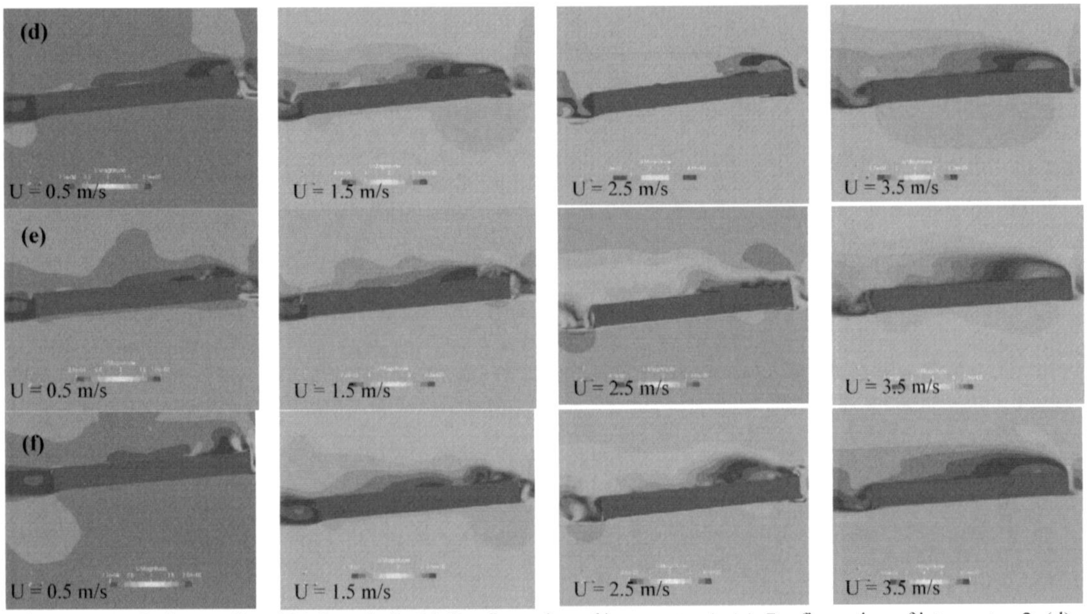

Figure 10. Velocity profile of (a) Bare hull, (b) Configuration of interceptor 1, (c) Configuration of interceptor 2, (d) Configuration of interceptor 2 – C , (e) Configuration of interceptor 2 AoA 1 and (f) Configuration of interceptor 2 AoA 2.

5. INTERCEPTOR 'S EFFICIENCY AND BOUNDARY LAYER THICKNESS

When a ea o a e o e a o id u a e, a hin aye a ed he ounda y aye o he o e o ity changes as it passes through this layer. The fluid's velocity relative to the surface is zero at the surface. The velocity of the flow then rapidly increases from zero to near the velocity of the mainstream. The boundary layer velocity thickness defined as the distance between the solid body and the point where the viscous flow velocity equals 99 % of the freestream velocity. Basically, the boundary layer thickness around rectangular plate can be roughly investigated by observing its velocity profile. Figure 10 (a – f) shows velocity profile in each configuration of a rectangular plate with and without addition of interceptors.

From figure 10, can be observed that the interceptor is located within the boundary layer, changes in flow speed and trim condition cause changes in boundary layer thickness, which directly affects the interceptor's efficiency. The boundary layer thickness decreases from speed 0.5 – 2.5 m/s, which resulting the distance between interceptor height and freestream velocity gradient also decreases. This condition leads the flow velocity near interceptor also increases. Drag force is proportional to velocity squared, in results, during this stage the drag is increases and less lift produced. Vice versa, when the trim angle reduces at speed > 2.5 m/s, boundary layer thickness increases, the flow velocity near interceptor decreases and resulting increment of lifting force.

6. CONCLUSIONS

The effectiveness of interceptor addition on a rectangular plate has been investigated in this study by using CFD simulation. The results of this study generally meet a good agreement with most of previous works related to interceptor's study. It is found that different AoA configuration may affecting interceptor performances. In this study, AoA employment at the interceptor with typical blade depth height shows better lift to drag ratio in comparison with interceptor without AoA employment. Moreover, those configurations also produced better trim angle reduction which resulting better running attitude. This study also shows that the efficiency of the interceptor is related to the boundary layer thickness.

*** Formatting Issue - Best Available Paper/Graphic ***

7. REFERENCES

1. Day, A.H., Cooper, C. 2011 *An experimental study of interceptors for drag reduction on high-performance sailing yachts*. Ocean. Eng. 38 (8 - 9), 983 –

2. Karimi, Saeed Saif, M Abbaspoor, 2013 *An experimental study of interceptor's effectiveness on hydrodynamic performance of high-speed planing crafts*, Polish Maritime Research 2 (78) 2013 Vol 20; pp. 21-29 1 o -2013-0013

3. ai, F , H ang, *Study on the Compound Effects of Interceptor with Stern Flap for Two Fast Monohulls* ean IEEE Techno-o ean , o , 1 - 1028.

Ghassemi, H., Mansouri, M., Zaferanlouei, S. 2011 *Interceptor hydrodynamic analysis for handling trim control problems in the high-speed crafts* o e h , e 1 h doi o g 1 11 11 6650. a C e h nginee ing ien e

5. e u a, F 11 *Experimental study on interceptor's effectiveness*. Diss. Univ. degli studi Napoli "Federico II".

6. e u a and en a 1 *Experimental Investigation On Conventional And Unconventional Interceptors*, Trans RINA, Vol 153, Part B2, Intl J Small Craft Tech, Jul- e o 1 ina i 1 1

7. Seo, K.C., Gopakumar, N., Atlar, M. 2013 *Experimental investigation of dynamic trim control devices in fast speed vessel*. Journal. Korean Navig. Port Res. 37 (2), 137–1

8. Avci and Barlas, 2016 *The Usage Of Interceptors In High Speed Crafts*, Conference paper

9. Mansoori, M., Fernandes, A.C. 2017 *Hydrodynamics of the interceptor analysis via both ultrareduced model test and dynamic computational fluid dynamics simulation*. J. Offshore Mech. Arct. Eng. 139 (2), 021101(1e15)

10. W eo e a *An experimental study on the stern bottom pressure distribution of a high-speed planing vessel with and without interceptors*. International Journal of Naval Architecture and Ocean Engineering 12 691 – 698

11. Brizzolara S. 2003 *Hydrodynamic Analysis Of Interceptors With CFD Methods*, Proceedings of The 7th n e na iona Con e en e on Fa ea an o a ion, Fa , hia, a y, -56

12. Mansoori, M., Fernandes, A.C. 2015 *Hydrodynamics of the interceptor on a 2-D flat plate by CFD and experiments*, ou na o Hyd odyna i , 1 , 1 - 1 1 1 1 1-6058(15)60555-8

13. u e i and a a 1 *OpenFOAM in Marine Hydrodynamics*. 21st sympossium Sorta.

1 C W Hi and i ho 1 1 *Volume of uid (VOF) method for the dynamics of free boundaries*. Journal of Co u a iona hy i , 1 1

15. H. Jasak. *Dynamic mesh handling in OpenFOAM* n h e o a e ien e ee ing n uding he New Ho i on Fo u and e o a e o i ion, age 1,

16. en a , 1 *Marine CFD Applicarion using OpenFOAM*, CINECA

8. AUTHORS BIOGRAPHY

AMF Putra is working for Indonesian National Research Agency (BRIN) from 2009. He is currently pursuing Ph.D degree with Osaka University, Japan. His research interest includes ship engineering and computational fluid dynamics.

Suzuki Hiroyoshi is currently a Professor in the Division of Global Architecture, of Naval Architecture and Ocean Engineering, Graduate School of Engineering, Osaka University, Japan. His research areas are in manufacturing technologies, ship engineering and fluid engineering.

STOWAGE PLANNING TOOL SUPPORTING FIRE RISK MANAGEMENT: A NEW WAY FOR CARGO DISTRIBUTION

F Rodero and **A Marrero**, CENIT Group for Innovation in Multimodal Transport, Spain
E S Petersen, Norwegian University of Science and Technology, Norway

SUMMARY

Feeding from the outcomes obtained after performing a risk analysis based on historical data to prevent cargo ignition on Ro-Ro, Ro-Pax and Car-Carrier ships, this paper describes the foundation for the development of a tool supporting fire risk management during the cargo distribution stage of the stowage process. There, cargo other than dangerous goods are loaded without consideration to the hazards they may involve, leaving room for improvements from the ignition prevention perspective. An approach to both design and specification is presented, aiming to manage all the optimization actions needed to obtain the best placement of cargo units, understanding best as the combination that minimizes the overall risk of fire. This way, a description of inputs being used and some constraints taken into consideration together with an explanation about how the tool can interact with existing software of operators are included.

NOMENCLATURE

DG	Dangerous goods
EMCIP	European Marine Casualty Information Platform by EMSA
EMSA	European Maritime Safety Agency
FRMC	Fire Resource Management Centre (LASH FIRE Project component)
GISIS	Global Integrated Shipping Information System by IMO
HCD	Human Centred Design
HMI	Human-machine Interface
IMDG	International Maritime Dangerous Goods Code
IMO	International Maritime Organization
ITS	Intelligent Transportation System
LSA	Life-saving appliances
MAAG	Maritime Authorities Advisory Group of LASH FIRE project
MOAG	Maritime Operators Advisory Group of LASH FIRE project
RA	Risk Assessment
RTD	Research and Technology Development
SOLAS	International Convention for Safety of Life at Sea
SPT	Stowage Planning Tool
SW	Software
VBA	Visual Basic for Applications
VHD	Vehicle hot-spot detector

1. INTRODUCTION: PROBLEM DEFINITION AND OBJECTIVE

The works described in this publication have been carried out (or are ongoing) in the frame of LASH FIRE [1], an international research project (see Acknowledgements section for funding information), running from September 2019 to August 2023, aiming to significantly reduce the risk of fires on board ro-ro ships by means of the development of technological solutions. That is, to provide a recognized technical basis for the revision of international IMO regulations, which greatly enhances fire prevention and ensures independent management of fires on ro-ro ships in current and future fire safety challenges. To be more specific, this paper builds on and summarizes the public deliverable D08.1 [2] already available and the upcoming D08.4; both documents as a result of the tasks in action A, "Automatic screening and management of cargo fire hazards" of the Work Package 8, "Ignition Prevention".

Today, cargo units other than dangerous goods are loaded without consideration to the hazards they involve, leaving room for improving the stowage process from the ignition prevention perspective. Statistical analysis reveals that since a significant number of the fires start in the carried cargo [3], improvements should focus on a more effective management of the units, making cargo placement in ro-ro spaces more optimal against fire hazards than currently done while minimizing the overall risk before departing.

Therefore, the objective of this paper is to present the foundation of the development of a stowage planning tool including fire management features. The software supporting this feature will include a cargo distribution algorithm powered by the knowledge from the past as a tangible outcome of the lessons learnt; in other words, taking advantage of a risk assessment based on historical data.

2. TECHNICAL APPROACH

As briefly introduced, one of the key elements of the software is the risk assessment of the cargo units being loaded to the ship. While dangerous goods will be managed fulfilling existing IMDG regulations [4] from IMO, other units will be distributed along the decks considering the criticality and severity of reported incidents involving the corresponding type of the cargo.

The stowage planning tool features fire hazard management supporting the loading process by means of guidelines and recommendations or even suggesting appropriate placement of the cargo based on historical data and additional constraints. This tool is actually a set of applications that work together, aiming to identify and detect the potential risks associated to a loading plan, given a certain ship deck configuration. The main components are listed as follows:

- The core component is an algorithm that searches for the combination of cargo placements that avoid high risk and minimizes the risk overall. This will include, for example, not only finding places where certain types of cargo should be placed (i.e., because of the need to be under surveillance) but also where it should not (i.e., preventing proximity to other cargo that may drive to potential hazards).
- End users interact with the previous component using a visual interface that manages external events and user actions as needed.

The distribution algorithm working behind the scenes implements a step-by-step process for every single cargo unit, from its initial risk scoring until its final placement. The score is basically a quantitative representation of the intrinsic risk, firstly, and of the risk because of other units nearby, secondly. The risk assessment, as one of these steps, sets the score based on historical data and provides guidelines and recommendations for the placement. Thus, the global score (as the summation of the score of all units of the ship) is the parameter being minimized through an iterative swapping of units searching the smallest possible value.

The software components are designed considering that they will operate as a plug-in that serves existing SW from the operators. This interaction is depicted in the architecture diagram shown in Figure 1.

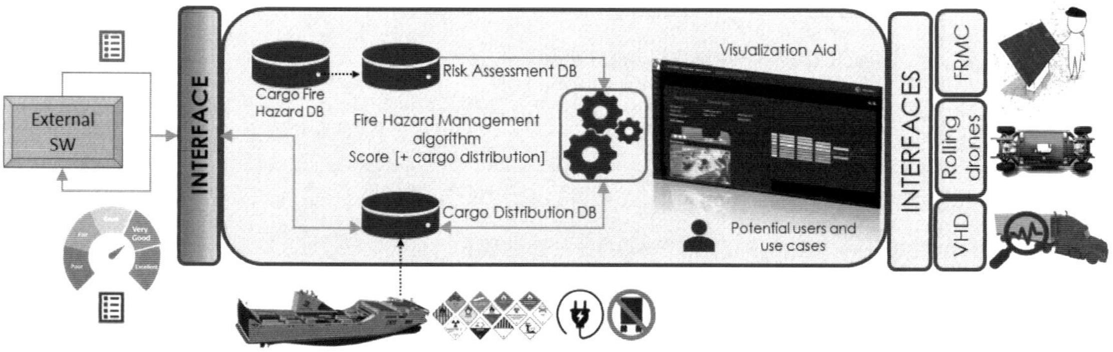

Figure 1: Overview of the software architecture

On the left side, existing applications (i.e., booking system) from the operators feed the software with a suggested stowage plan, which is evaluated as described further in this document. Then, a response is sent back, composed of the score of the suggested stowage plan and, optionally, an alternative distribution of the cargo units which should have a better score in terms of fire risk.

The algorithm for hazard minimization is the core component in charge of performing all necessary steps that implement both scoring and cargo distribution features. It relies in two databases, one storing risk assessment calculations and the other supporting management related to needed information about units, ship layout, constraints and so on. Regarding the risk assessment database, it feeds from statistical analysis applied to the data gathered in the cargo fire hazard database.

The visualization aid acts as the point where considered end users interact with the stowage planning tool providing them with a graphically user-friendly interface which has been developed using a Human-Centred Design (HCD) approach. According to ISO 9241-210 [5], the usage of HCD leads to significant benefits for all the stakeholders involved with a

particular product, the reasoning being that systems with good usability presumably will be more successful in the market. In the specific case of the maritime domain, the development of navigation systems is one such example, where the usage of HCD believably led to a beneficial market position [6].
Finally, the system expects to interact with other LASH FIRE components such as the Vehicle Hot Spot Detector (VHD), the rolling drones inspecting the cargo during the voyage or the Fire Resource Management Centre, where data from many systems is centralized in order to support firefighting. In the case of the VHD, for each single unit that must be loaded into the ship, an alarm may be triggered depending on the status of the cargo, which is used to modify its initial score and to decide if the suggested placement for the unit must change accordingly this new hazard information.

3. BACKGROUND OF THE UNDERLYING RISK ASSESSMENT

The algorithm has a built-in component in charge of evaluating the associated risk of a cargo unit based on historical data. It is a result of a long process starting with a compilation from many maritime sources searching for reports of incidents over the last 25 years and fulfilling two conditions: a) the fire origin must be located in a ro-ro space and b) the fire must be started in the vehicle or the cargo transported by.

Sources from which data has been gathered are composed of both private and public organizations, research institutes and journals or ship operators, including FSI 21/5 [7], GISIS [8], ForeSea [9] or EMCIP [R10], among others. For a more detailed information about this data collection, please refer to [R11] where there is also a description of the multiple correspondence analysis performed in order to search correlations and patterns to find out potential relationships between involved variables in order to define the main causes of cargo fire incidents in ro-ro spaces.

At this stage, as shown in Figure 2, the methodology includes manual (interactive) steps in order to filter, select and standardize the information coming from the different sources and automatic (non-interactive) steps supported by VBA and *Python* scripting to structure and create an *ad-hoc* database storing historical cargo fire hazards.

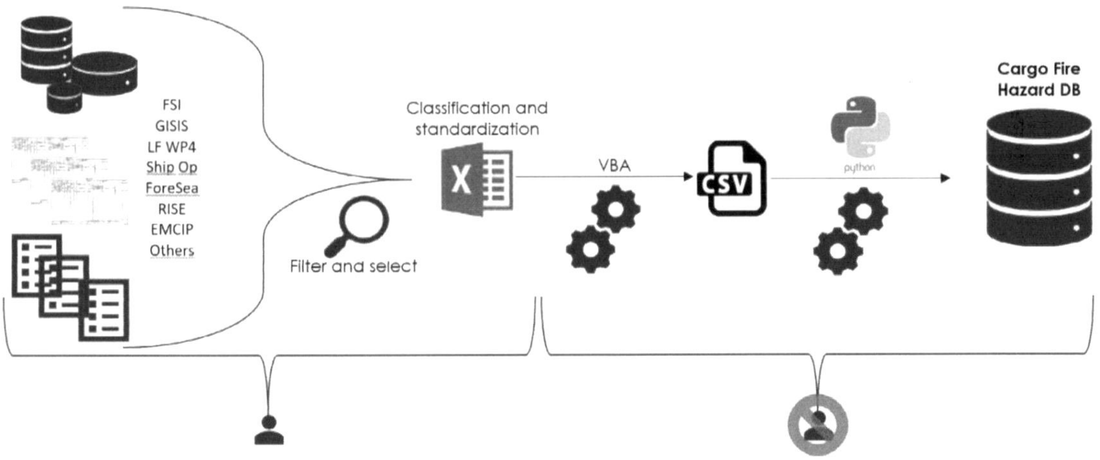

Figure 2: Methodology used for data collection of the Fire Cargo Hazard database

Once the so-called Fire Cargo Hazard database has been created, it becomes the starting point for the risk assessment (RA), which has been developed using criticality analysis techniques aiming at identifying the cargo that poses a potential risk in order to reduce the probability of ignition by avoiding extreme risk values and minimizing the risk overall. Calculations of the risk score for each cargo type are based in frequency and severity as deeply explained in [12] following the steps of Figure 3.

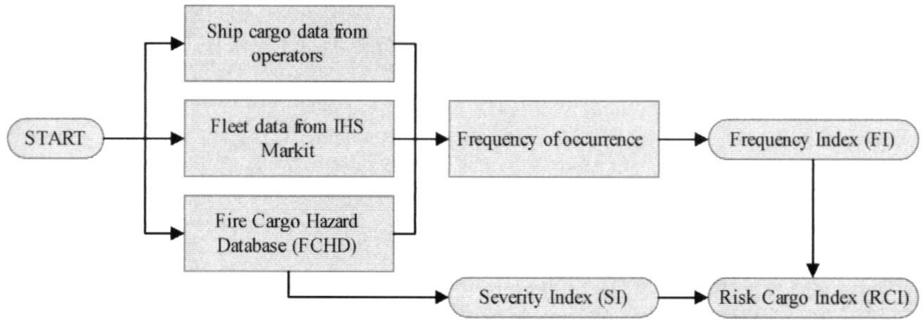

Figure 3: Steps of the calculations made during the Risk Assessment [12]

4. STOWAGE PLANNING TOOL

4.1 INPUTS

Most of the inputs used by the software are structured in terms of logical units of information and relationships representing interactions between them. Such this structure can be described as an entity-relationship diagram which directly provides a helpful tool to design databases, a powerful software component supporting read and write operations to shared data in a fast and reliable way. Entities and relations are shown in Figure 4.

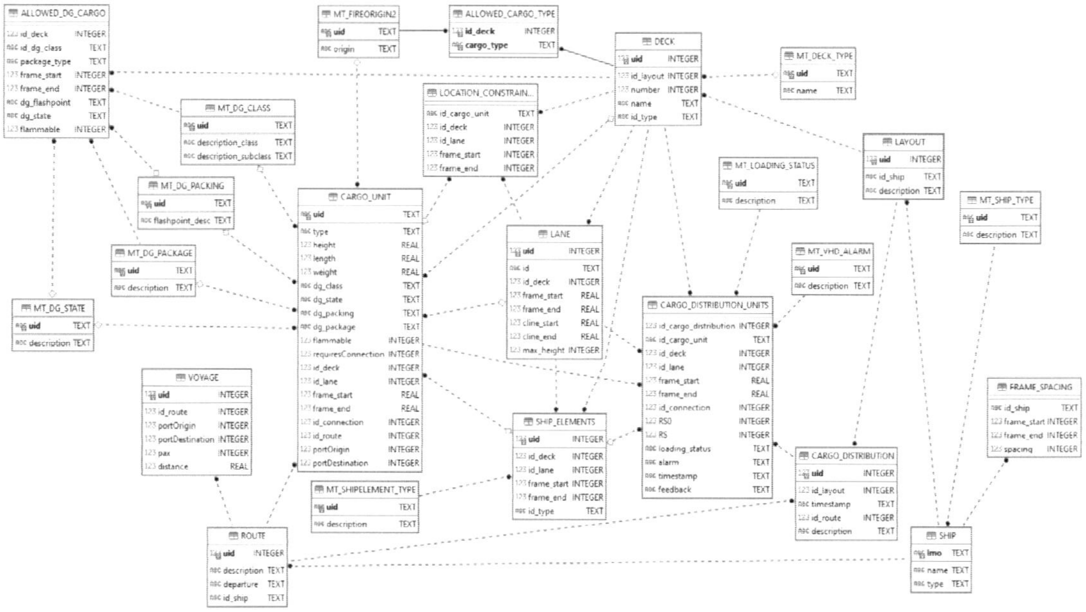

Figure 4: Data model of the Stowage Planning Tool

A set of master tables exist in the data model to store static values of the application such as the DG classification, the types of ships considered, the alarms coming from the VHD or others used to manage the loading process.
In addition to the information stored in the database, the design of the software also includes a configuration file with a set of parameters that give a mechanism to customize the behavior of the algorithm avoiding hard-coded programming, which is not a best practice even in prototypes or demonstrators. These parameters are used to define what is the ship and the layout to be used, what is the exact route in order to consider the corresponding port calls or others concerning the accuracy of some calculations or the customization of safety distance between units during their distribution.

4.1 (a) Cargo characterization and allocation constraints

Information about cargo units retrieved from external software is mapped to the structure defined by the table *CARGO_UNIT* of the data model. Physical characteristics are used to manage in which decks (besides height of the unit, suitability of a deck is also filtered by the cargo type) a cargo may fit and, once selected a deck, what is the remaining space in a given lane after placement (slot is defined in terms of the frames where the unit is located).

All this information is associated to a specific route, which is used to know the distance between the origin and destination ports and if the unit must be placed in a special location because it must be loaded or unloaded in an intermediate port call. Support for port calls is limited to the most usual situations as recommended by the operators of the Consortium and are managed using special values for the *portOrigin* and *portDestination* attributes of the table *VOYAGE* (0 for the initial port, -1 for the final destination and values from 1 to N for intermediate calls, if any) which can be graphically represented with the diagram shown in Figure 5.

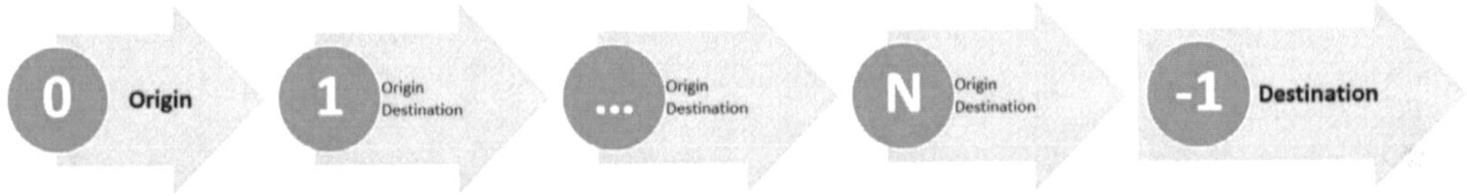

Figure 5: Support of port calls using *portOrigin* and *portDestination* values

For the specific case of DG, they must specify not only their standardized classification but also additional attributes that determine where or where not they can be stowed in a given ship. Depending on the class, if they are liquid or solid state, flammable or not or the flashpoint, there are a set of rules that must be applied in order to select the best location in both the ship's area and respect to other cargo nearby. Contents on table *ALLOWED_DG_CARGO* implement this feature by including the information usually found in a document of compliance of the special requirements for a ship carrying dangerous goods, which is issued in the pursuance of the requirements of regulations of the SOLAS Convention.

4.1 (b) Layout of Generic Ships

A comprehensive description of the selected ships for the three considered types (ro-ro, ro-pax and car-carrier) can be found in [13], where the selection criteria as well as the evaluation results are also included. 3D models for these ships, as the one depicted in Figure 6, have been generated in order to provide, when necessary, a common structure reference for software components.

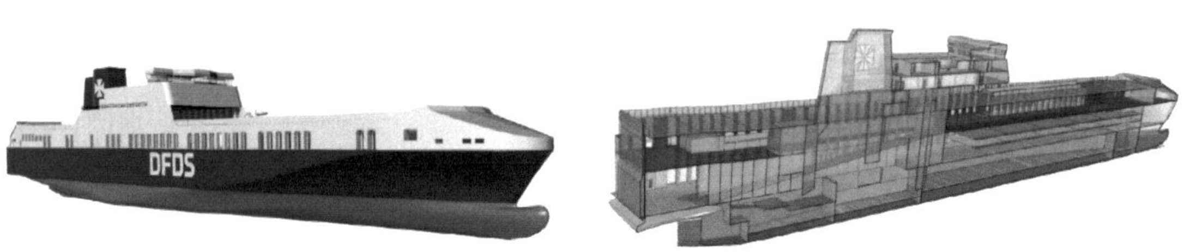

Figure 6: 3D model of the selected ro-ro cargo ship

Also, in the case of the SPT, the physical layout has been defined by means of the official drawings provided by the corresponding operator for each selected ship. These drawings are the reference of the physical layout details that populate some of the tables of the data model. The Figure 7 partially shows two data sets used by the software: the spacing between frames in order to provide accuracy for final placement of cargo units (top) or the deck type and lane definition for their available cargo space (bottom).

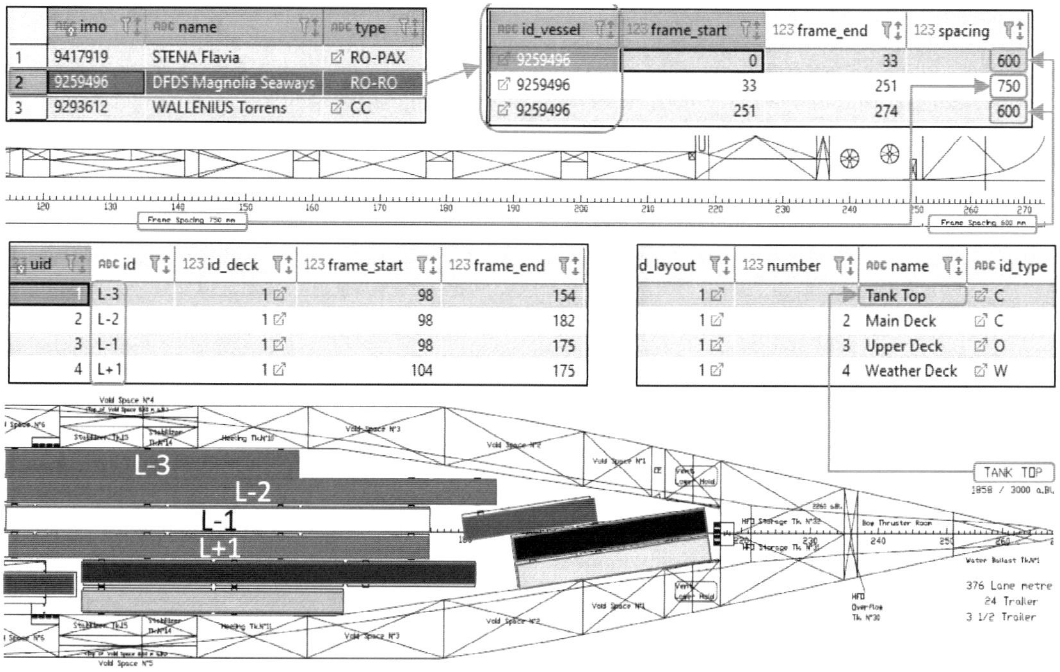

Figure 7: Example of how data model supports management of physical layout

4.1 (c) Ship elements

Some elements of the ship's infrastructure like electrical connections, LSA or openings may influence the placement of a cargo unit because safety recommendations can advise moving them away from or approaching them.

Based on the segmentation of the space previously mentioned, these elements have a scope which is approximately defined by means of the area in-between frame values of one or more lanes. Thus, as an example, when a cargo unit that must be connected during a voyage (i.e., reefer unit) is processed in the search of a suitable placement, the algorithm is expected to assign a slot which is compatible with the scope of any available electrical connection for a given deck which is compatible with the cargo type and, if necessary, in a suitable area in the case of DG as defined in the table *ALLOWED_DG_CARGO.*

4.2 CARGO DISTRIBUTION ALGORITHM WITH HAZARD MANAGEMENT

Optimization is the selection of a best (with regard to some criteria) solution from a set of different alternatives. In the context of LASH FIRE project, the optimization implemented in this algorithm consists of minimizing overall fire hazard risk which is measured by means of the risk score as previously described. The algorithm will select the "best" distribution of cargo units from a set of combinations understanding "best" as the combination that minimizes that score.

The algorithm allows multiple runs for a given ship, layout and route but using different configuration parameters (iterations or improvement objective) or different location constraints. Each execution is uniquely identified by a timestamp and eventually may contain a description.

How cargo units are expected to be stowed along the decks for a given ship (and route) is defined by means of a set of attributes:

- *Cargo unit identifier:* Unique identifier of the unit itself.
- *Deck, lane, frame_start/frame_end*: This triplet defines the exact place in the ship; the deck as stablished by the physical configuration of the ship, the lane as they are identified by the algorithm and assigned slot, which is the specific spot in the given lane (between *frame_start* and *frame_end*).
- *Id_Connection*: If the cargo unit requires to be connected during the voyage, this identifier contains the reference to the specific connection assigned by the algorithm.
- *Score*: This is the main indicator concerning the risk of a cargo unit based on the assessment using historical data. It really splits in two values: the initial score (not considering other units nearby) and final score (considering other units nearby).

Before performing any action for any use-case implemented there are a set of initialization steps that need to be completed: load configuration, load physical parameters and DG constraints, load input data and a set of pre-checks (size of units, allowed cargo types, enough capacity or quantity of electrical connections, among others).

Once all necessary checking has been done, units are processed in order to get the distribution. The two diagrams below (Figure 8) depict the main steps carried out to achieve the objective.

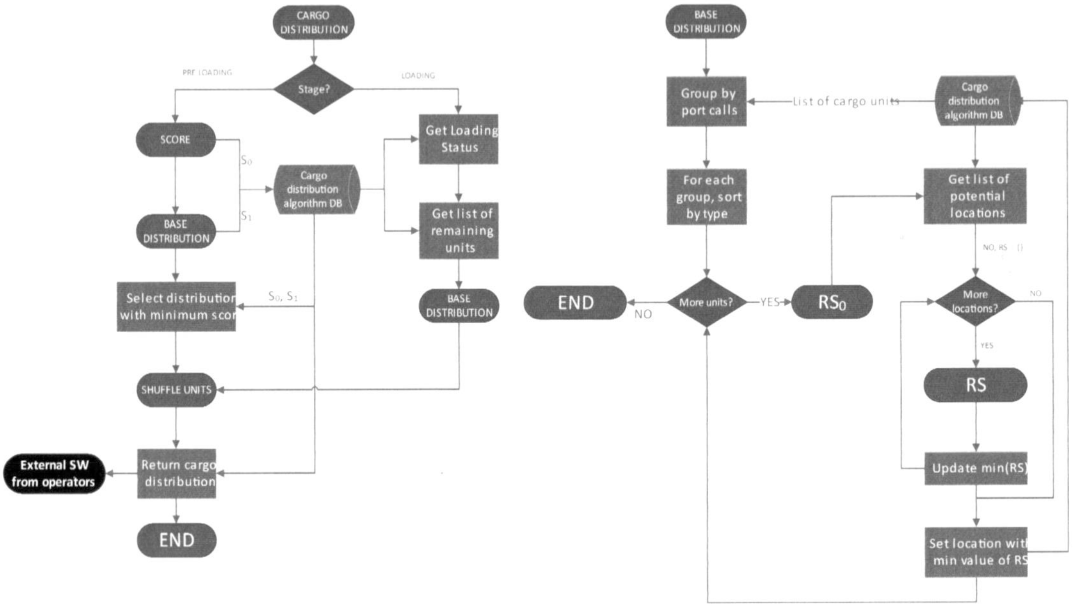

Figure 8: Workflow of the cargo distribution use-case

Distribution of cargo units can be executed before the loading process or during the loading process depending on events like alarms when inspecting the unit before entering the ship or no-shows, for example. The specific stage determines the workflow; as a general description, the use-case generates a base distribution and then, it generates combinations searching for the one with minimum risk (score) by shuffling certain units. The base distribution aims at setting an initial cargo placement where units are placed in one of the potential locations they can be stowed, selecting just any of the ones that gives the lower score. Then, the algorithm starts shuffling units (placement exchange if they can), consolidating new locations if score is reduced.

In a pre-loading stage, the base distribution is selected by choosing the one with minimum score between the suggested by the external software or a first initial one generated by the algorithm with all the units. During the loading stage, the base distribution is generated considering the loading status rather than considering an empty ship; that is, currently units loaded in the ship cannot be shuffled and the list of cargo units is not the full initial list but the remaining ones.

Finally, the new cargo distribution is sent back to the external SW from the operators.

4.3 CONTEXT OF USE AND VISUAL INTERFACE

The design envisages several profiles that are directly involved in the stowage process: shore-based stowage planner, deck crew, cargo/loading officer, truck/tractor driver, bridge officer and terminal worker/gate keeper.

The following table (Table 1) summarises the correspondence between these end users and the existing use-cases, which are listed below with a brief description of the use-cases (use-case #9 is reserved for testing purposes).

Table 1: Mapping of end users to use-cases

	#1	*#2*	*#3*	*#4*	*#5*	*#6*	*#7*	*#8*	*#10*	*#11*	*#12*	*#13*	*#14*
Shore-based stowage planner	X	X		X									X
Deck Crew			X					(X)		X			
Cargo officer/ Loading officer	X	X	X	X	X	X	X	(X)	X	X	X		X
Tug/tractor driver			X					(X)	X	X	X		
Bridge officer					X	X	X			X		X	
Terminal gate keeper								X					

- *Evaluate stowage plan fire safety (#1)*: Given a suggested stowage plan composed of a list of cargo units and their corresponding location (deck, lane and slot), a value scoring fire safety is returned.
- *Optimize stowage plan (#2)*: Takes a list of cargo units with their associated attributes: height, length or weight, DG information or cargo type, physical layout of the ship and equipment and other information about the voyage as input. Returns a cargo distribution giving the minimum value for the fire safety score.
- *Confirmation of a cargo unit location (#3)*: Once a cargo unit is loaded, its location may or may not match the suggested position. The objective is to have an updated loading status; this use case confirms the final location of the unit (deck, lane and slot) and the identifier of the electrical connection it is using, if any.
- *Disable area for loading (#4)*: If the use of specific areas of the ship become unavailable as cargo space, such information must be notified in order to be considered during cargo distributions.
- *Monitoring of loading process (#5)*: Returns the current loading status of the ship, that is, the list of already loaded units with their corresponding placement, raised alarm for each unit, if any, and additional extra information.
- *Stowage information supporting fire management (#6)*: Given a cargo unit, information which can be useful to support fire management is retrieved. It includes recommendations and constraint locations based on the cargo type (for example: Fore and Aft - Separated from LSA at least 3 metres or place on weather deck).
- *Validate distribution (#7):* Since the software operates as a plug-in, this use-case is used to consolidate a cargo distribution and avoid mismatching between involved components.
- *Get location for unit (#8):* Used to retrieve the suggested location based on the validated stowage plan.
- *Discard unit (#10):* If a cargo unit must be discarded from the suggested stowage plan, the cargo distribution algorithm must be relaunched in order to consider the new situation for the remaining units.
- *End of loading (#11):* Notifies the end of the stowage process. Full information for each cargo unit is generated, merging the last loading status and additional data from risk assessment and VHD. This information can be broadcasted/transferred to other data consumers, such as the FRMC.
- *Cargo unit feedback (#12)*: Use case that allows to add free text to a specific unit including any kind of information.
- *Fire patrol report (#13):* Generation of data needed by external systems to prepare a specific report for fire patrol purposes, containing recommendations based on the final cargo distribution.
- *No-show/rearrange cargo unit (#14):* Confirmation of a no-show for a given unit. Similar to UC#10, but the unit is tagged in a different way to preserve loading status consistency.

Figure 9 shows a mock-up of the prototype created with the collaborative browser-based interface design tool *Figma*. These mock-ups help developers to better understand the requirements of the HMI that help end users to interact with all the features implemented in the Stowage Planning Tool.

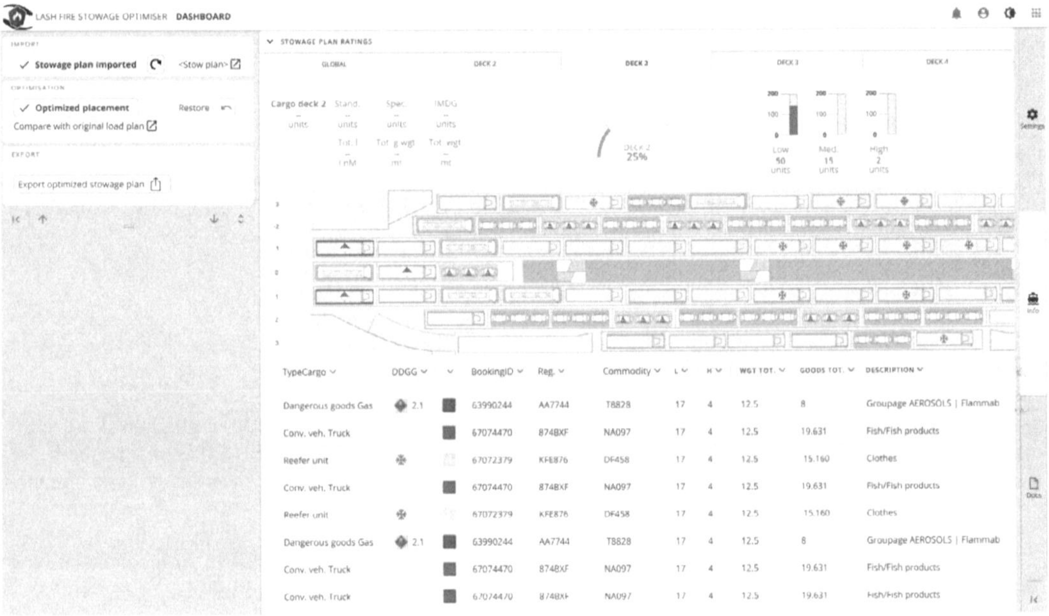

Figure 9: Mock-up of the prototype (optimization of the cargo distribution)

5. CONCLUSIONS AND NEXT STEPS

As above-mentioned, one of the main pillars of the algorithm is the risk assessment which is based on the quality and availability of fire-related incidents in the maritime sector. Nowadays, this information requires some manual tasks (mainly because the existing taxonomies are heterogeneous) before being ready to be automatically processed and, besides that, it is prone to change along the years. Thus, the first main conclusion is the need of a standardized way of reporting together with improvements on the availability by means of open-data and/or public-access repositories. This way, the software could take advantage of an automatic updating of the risk assessment if new incidents are available and cargo types increase or decrease its associated risk (as it could happen with electrical vehicles) without manual pre-processing. Standardization could also drive to easily find out which other parameters must be considered when evaluating the risk but currently discarded since it eases data processing and analysis.

The cargo distribution suggested by the algorithm will theoretically have a lower overall risk (compared to a current stowage plan where risks of cargo other than DG is not considered) by managing the cargo in a per-unit basis. That is, units are uniquely identified when they are stowed instead of grouping them by cargo type without single identification. So, in order to get the maximum benefit, cargo units should be loaded to the ship as recommended by the algorithm. However, the arrival profile of the units to the terminal is unknown and, even if it was known, capacity of the parking area in the land-side is not infinite (many small terminals do not even have pre-loading area and trucks go from the gates to the ship, often creating queues in road network that accesses the terminal). The approach to deal with this operating mode of the terminals is through the use cases #3, #10 and #14, which set the loading status as initial (non-movable) load of the ship and executes the algorithm to distribute the remaining units; the more different is the arrival profile compared to the ideal profile (units arrive in such an order that they can directly enter the ship and match the suggested distribution by the algorithm) the more different will be the final cargo distribution and therefore, the risk will not be reduced as much as it could.

When it comes to the information of a given unit, the data model includes its weight but most of times (except for DG), accurate information for this attribute is not available since the Operators book lane meters, customers are unable or unwilling to provide all relevant information and, from the naval perspective, the ship allows to proceed with approximate values of the weight since they have both passive and active systems to ensure stability during the stowage. The same goes for the contents, where units other than DG usually just provide high-level generic descriptions. This lack on detailed information prevents the development of more sophisticated tools where, for example, the results of a fire propagation simulation (where detailed information about materials and weights is essential) can be used as an additional criterion during the risk assessment in order to suggest the best cargo distribution not only from the ignition prevention perspective but also from the fire propagation point of view, if it happens.

As a continuation of the previous idea, future developments could consider the availability of accurate information about size and weight of the units in order to include early stability calculation, helping to increase safety during the stowage process.

Finally, ship types considered in the project are ro-ro, ro-pax and vehicle carrier. However, with minor modifications of the database design and adaptation of the algorithm to include specific rules, stowage of container ships could also be covered in a short-term (scalability of the software). Anyway, even if technically possible, conceptual consistency of the methodology behind the risk assessment should be checked as a compulsory initial step before further developments, especially the part related to the extension of the fire cargo hazard database.

6. ACKNOWLEDGEMENTS

The LASH FIRE project has received funding from the European Union's Horizon 2020 research and innovation programme under Grant Agreement No 814975.

The authors would like to thank the rest of the partners of the Consortium for all the interesting discussions that have taken place about the different approaches that such a tool can have, not only in the context of LASH FIRE as a demonstrator prototype but also looking beyond as a real software system which can be exploited by the operators in order to promote safer stowage.

Also, special thanks to the operators that actively participate in LASH FIRE as partners as well as the members of both MOAG and MAAG Advisory Groups for providing valuable information and arrange in-field visits during July 2021 (Valencia, Spain) and November 2021 (Ghent, Belgium and Gothenburg, Sweden).

Finally, the authors deeply appreciate the contributions by Hedvig Aminoff from Norwegian University of Science and Technology (NTNU) for providing us with Figure 9 and Vito Radolovic from FLOW Ship Design on collaboration and using data from DFDS [14] for the Figure 6.

7. REFERENCES

1. LASH FIRE, "Legislative Assessment for Safety Hazards of Fire and Innovations in Ro-ro ship Environment" official web site. https://lashfire.eu/, last accessed: July 2022.

2. LASH FIRE Public Deliverable D8.1 "Definition and parametrization of critical fire hazards, classification of cargoes, transport units, engines, fuels and vessels and identification methodologies", May 2021.

3. Wikman, J., Evegren, F., Rahm, M., Leroux, J., Breuillard, A., Kjellberg, M., Gustin, L., Efraimsson, F. (2016), "Study investigating cost effective measures for reducing the risk from fires on ro-ro passenger ships (FIRESAFE)", EMSA.

4. IMDG code: International Maritime Dangerous Goods Code. ISBN-10 9280100521

5. ISO9241-210. (2019). ISO 9241-210: Ergonomics of human-system interaction - Part 210: Human-centred design for interactive systems. In (Vol. ISO 9241-210): ISO

6. Petersen, E. S. (2010). SAM Electronics' NACOS Platinum - Reiterating Navigation Usability. European Journal of Navigation, 8(1), 14-18.

7. Correspondence Group on Casualty Analysis. (2012). Casualty statistics and investigation: Report of the Correspondence Group on Casualty Analysis. IMO, Sub-committee of Flag State Implementation.

8. GISIS, webpage: Home page. Available at: https://webaccounts.imo.org/.(accessed on 22 June 2020).

9. ForeSea, webpage: Home page. Available at: https://foresea.org/ (accessed on 22 June 2020).

10. European Marine Casualty Information Platform (2020). Available at: https://portal.emsa.europa.eu/web/emcip (accessed on 20 April 2020).

11. Marrero, A., Gasparín, P. and Rodero, F. "Main causes of cargo fire incidents in ro-ro spaces: an analysis over the last 25 year". Paper to be presented at Urban and Maritime Transport Conference 2022 (Online, 19-21 September, 2022).

12. Marrero, A. and Rodero, F. "Application of a risk analysis based on historical data to prevent cargo ignition on Ro-Ro, Ro-Pax and Car-Carrier ships". Paper to be presented at 15th International Symposium on Practical Design of Ships and Other Floating Structures PRADS 2022 (Dubrovnik, Croatia, 9-13 October, 2022)

13. LASH FIRE Public Deliverable D5.1 "Definition of generic ships", August 2021.

14. Det Forenede Dampskibs-Selskab (DFDS) web site: https://www.dfds.com/en/about

8. AUTHORS BIOGRAPHY

Francisco Rodero holds, since 2009, the position of ITS Project Manager at CENIT, the Group for Innovation in Multimodal Transport of the Transport RTD Area at CIMNE, the International Centre for Numerical Methods in Engineering, a research organization created as a partnership between the Government of Catalonia and the Technical University of Catalonia.
Bachelor of Science + Master of Science in Informatics Engineering he has been working for the last 20 years in projects where computer sciences where applied to mobility and transport areas with a strong focus in software engineering and traffic simulation. He obtained a Postgraduate in Smart Mobility: Intelligent Transport Systems from the UPC School of Professional & Executive Development in Barcelona.
Before the current position, he worked during 7 years at the INDRA group, mainly working in projects related with Satellite Navigation Systems and operations simulation for transport terminals.

África Marrero holds, since 2019, the position of Research & Development Project Manager at CENIT, the Group for Innovation in Multimodal Transport of the Transport RTD Area at CIMNE, the International Centre for Numerical Methods in Engineering, a research organization created as a partnership between the Government of Catalonia and the Technical University of Catalonia. Bachelor of Science + Master of Science in Naval and Oceanic Engineering from the Polytechnic University of Cartagena. She is finishing a Master in International Business from Universidad del Atlántico Medio.
She began her professional career as a planning and control engineer in submarine construction projects at the Navantia shipyard in Cartagena. Later, she worked as an Associate Ship Manager in Astilleros Canarios (ASTICAN Shipyard), mainly carrying out work control functions in oil rigs.

Petersen graduated as a Naval Architect (B. Sc.) in 1983, and worked as a consulting naval architect until 1996, being involved primarily in structural design and ship stability on both new-buildings and conversions. Leading an extensive series of EC-sponsored R&D projects (ATOMOS I-IV; DISC I-II, Thames), and gradually changing his focus to research, Petersen joined the Danish Maritime Institute in 1996, where he became Head of Development in 1999. Concentrating on the functionality, design and usability of marine electronics, Petersen joined Lyngsø Marine in 2002, and in 2007 he became Deputy Director of Technology in the mother company, SAM Electronics. Pursuing his interest in maritime human factors, usability and human-centered design, Petersen enrolled in Chalmers Technical University (CTH), Sweden, in 2007, where he received his Licentiate of Engineering degree in 2010, and his Doctorate in 2012, the latter in the subject of 'Engineering Usability'. After a spell (2012-2013) as Director of Product Management, SAM Electronics, and Director, Business Development in Wärtsilä SAM Electronics (2013-2015), he held a number of posts in the development organization of Wärtsilä SAM Electronics, until he became the Product Owner of Navigation as well as the Head of Strategy and Portfolio Management in 2018. Since the midsummer of 2020, Petersen holds the positions of Senior Researcher at NTNU in Trondheim, Norway, and Post Doctor at the Western University of Applied Science in Haugesund, Norway.

MAKING 3D MODEL-BASED APPROVAL A REALITY

M Sakagami and **T Otaguro,** ClassNK, Japan
T Masui and **P Lin,** NAPA Japan, Japan
Y Shimakawa and **S Hisano,** Nihon Shipyard, Japan

SUMMARY

In recent years, various classification societies and shipbuilders have conducted R&D activities on the use of 3D models created during the design phase for approval (so-called 3D Model-Based Approval, hereafter "3DMBA"). 3DMBA is an attractive concept that could bring digital transformation to the ship design process, but it also poses significant practical challenges.

In particular, there is an issue with the division of design work and information in structural design, which requires the use of different software applications for different tasks such as 3D modeling, rule calculation, direct strength calculation, and 2D drawing. This causes a situation whereby each time a design change happens, it becomes necessary to modify the data in each software application, resulting in frequent duplication of work. This is also an error-prone process, because of possible discrepancies between various types of information in each application. The introduction of 3DMBA under such circumstances would further increase the workload of shipyards, as they would have to repeat this duplication of work on each software application to complete a precise 3D model.

The authors have been conducting a project to realize an efficient 3DMBA process, and they concluded at ICCAS2019 that the key to achieving 3DMBA is to integrate the engineering process for rule and direct strength calculations into 3D CAD, and to implement viewer functions for efficient plan approval to enable shipyards to easily create 3D approval models. This paper presents the latest achievements and future prospects of the project.

NOMENCLATURE

3D CAD	3D computer-aided design software
3DMBA	3D model-based approval
3D Model	Product 3D model which includes ship-specific data produced by 3D CAD
API	Application programming interface
ClassNK	Classification society, member of IACS (International Association of Class Societies)
CSR-B&T	Common Structural Rules for Bulk Carriers and Oil Tankers
FEA	Finite element analysis
FEM	Finite element method
PSH	PrimeShip-HULL, rule calculation and FEA software provided by ClassNK
NAPA	NAPA Ltd, a software development company
NAPA Designer	State of the art 3D CAD software for ship design provided by NAPA
NSY	Nihon Shipyard Corporation, shipbuilding company
OCX	3D data format, Open Class 3D Model Exchange [5]
SPEEDS	3D data format, Smart Platform of Enhanced Engineering Data for Shipping and Shipbuilding [4]
UX	User Experience

1. INTRODUCTION

In recent years, the introduction of 3D tools into the ship design field has been increasing, driven by the growing momentum for digitalization and the enhanced capabilities of 3D CAD. Many shipyards, classification societies, and CAD vendors have initiated various activities to innovate the design process and reform the class approval process by utilizing 3D models.

Currently, the use of 3D CAD in the shipbuilding industry tends to be limited to specific purposes, however, there is potential to unlock greater benefits of new 3D technologies by reforming the entire process. Despite this, the reality is that 2D drawings are often treated as the final, "true" design information in the ship design process and the complexity of classification society's rules and regulations is increasing the designer's workload. So, introducing the concept of 3DMBA into this process merely leads to additional work and doesn't necessarily provide a meaningful solution for shipyards.

Given these circumstances, NSY, NAPA and ClassNK have been working on the development of tools and processes in order to realize effective and practical design and class approval process, with hull structural design as a first target. At ICCAS2019, the authors identified the problem of work inefficiency and data inconsistency, as caused by the division of tasks and the need to use several different engineering tools in the design process. As such, they demonstrated the potential of solving this problem by integrating design tasks and tools based on a single 3D model. [1]

The concept of 3DMBA is to utilize the 3D model created during the design process for class approval purposes. Through the study at ICCAS2019, the authors could confirm the feasibility of 3DMBA, but at the same time, found some important issues to be solved in order to develop the concept. The authors continued to work on solving problems and relevant developments for realizing 3DMBA and were able to reach an important milestone. This paper presents the achievements of the project.

2. CURRENT LANDSCAPE AROUND 3DMBA

As the title indicates, the purpose of this paper is to discuss the realization of 3DMBA. In this chapter, as a premise for the discussion, the position and the current state of 3DMBA is summarized.

2.1 MOTIVATIONS TO REALIZE 3DMBA

3DMBA, as mentioned above, is the concept of utilizing 3D models for the class approval process. While the concept itself isn't new and has been around since the introduction of 3D CAD in the shipbuilding industry, it has, in recent years, gained in popularity among various classification societies and shipyards who are increasingly interested given its connection with Digital Twins.

3DMBA is closely related to the 3D-based design process. However, it should not be addressed simply due to the passive reason of "adapting to a shipyard's design process,"; instead, it is expected to bring the following benefits:

- For classification societies:
  - It is expected that the linkage between approval systems and a 3D model will reduce the man-hours needed for, and improve the quality of, design review work. In addition to this, using a 3D model for class approval, combined with automation and other advanced IT technologies, could optimize the class approval process in ways that have not usually been possible in conventional approaches. For example, earlier timing of design review and more flexible design review by specifying target scope in consistency with shipyard's design progress, and automation of plan approval work, etc. These have potential to benefit the concerned parties, other than class societies.
- For shipyards:
  - It is hard to say that the introduction of 3DMBA can directly reduce the man-hours required for the work. On the other hand, it is expected that the work environment and tools will be streamlined to realize 3DMBA, and the work efficiency and quality will be improved because of the improved design process. Thus, the 3DMBA has the potential to bring innovation.
- For maritime industry:
  - Across the industry, there is growing interest in the concept of Digital Twins. To develop a Digital Twin, "reliable" ship model data is essential, and the 3D model that has been reviewed by the classification society for 3DMBA is considered ideal for this purpose. In this sense, it is fair to say that 3DMBA is a fundamental process for Digital Twins and other new innovations.

The authors confirmed at ICCAS2019 that 3DMBA is technically feasible, however, even if 3DMBA is implemented, it does not necessarily mean that the above benefits can be achieved. Thus, the aim of this project is not to pursue 3DMBA itself, but to explore the ways to "realize a truly effective 3DMBA" for all stakeholders. We would also like to emphasize that the goal of this project is to present new insights by utilizing the 3D model in the maritime industry.

2.2 3D TECHNOLOGY ADVANCEMENT ENABLING 3DMBA

Although 3DMBA has been seen as a dreamy concept in the past, there have been various technological advancements in recent years that have made it a more viable option. One of these developments includes advancements in cloud computing and web-based applications that enable collaboration in distributed environments; however, the primary driver must be the evolution of 3D CAD. The use of 3D models in early-stage design has become popular as a means to efficiently cope with increasingly complex classification societies' rule requirements, as safety requirements become more demanding [2]. Design work was streamlined by generating the input data needed for classification society's prescriptive rule calculations and FEM analysis from a single 3D model.

To further ensure the benefits of the 3D model utilization, a 3D design approach was adopted. Traditionally, 2D CAD, which is more familiar to designers, has been used for structural design. But dramatic improvements in UX of the software have enabled the designers to design directly on 3D CAD, which is becoming more common nowadays [3].

Improved connectivity among 3D software applications is also an important facilitator of 3DMBA. Software vendors are becoming more transparent, opening their systems' APIs to the public. In addition to this, the environment for data exchange between systems is being developed with the aim to establish industry standard 3D data formats, such as OCX and SPEEDS [4] [5]. NAPA is also developing the OCX Import/Export function, which is expected to be utilized as a means of exchanging data between companies [6].

Of course, there remain many other challenges when it comes to implementing a pragmatic 3DMBA process that offers solid benefits to all stakeholders. Either way, steady progress of relevant technologies is assisting its realization.

2.3 REMAINING KEY ISSUES

Essential technologies related to 3DMBA have been developing steadily. However, the authors believe that there are two key issues that are necessary to be solved to realize truly effective 3DMBA: "Integration" and "Visualization".

2.3 (a) "Integration" – Seamless integration between 3D CAD and strength calculation tools, keeping data consistency

While discussing effective 3DMBA, it is important to pay attention to the shipyard's design process. The ship design process has been established, refined, and operated based on the premise that 2D drawings are the final outcome of the design process for a long time. In addition to this process, recent advancements in computer technologies and the increasing complexity of structural rules have driven the use of 3D CAD in the modern design process, which makes the following possible:

- Semi-automatic modeling function of 3D CAD to enable quick initial structural design
- Drawing output function of 3D CAD to enable semi-automatic creation of structural drawings
- Data export from 3D CAD to classification society's prescriptive rule calculation software or FEM analysis software

Thus, while 3D CAD contributes to improving work efficiency, the following problems have arisen in the current situation where the roles of 2D drawings and 3D models are mixed.

- In addition to 3D CAD, several different engineering tools must be used separately, but it is difficult to learn the usage of each tool. It requires time to pre-set calculations, even though there is a certain level of interface from the 3D CAD, making overall design optimization difficult.
- The calculation model of each engineering tool has to be maintained separately. Whenever there is a design change, the change must be reflected in all calculation models via some manual operations, which causes extra man-hours and may lead to calculation errors due to failures to reflect the change.
- The final deliverable of the design work is 2D drawings, so unless there is a special reason, the 3D model will not be updated from a certain point. Hence, it becomes difficult to utilize the 3D model in subsequent processes as well as for class approval.

The author believes that one of the best solutions to these problems is to "integrate" the design process in such a way that changes to the 3D model, as a central hub for information, made by 3D CAD will reflect in the calculation models in other engineering tools in a synchronized manner. Figure 1 shows images of the current process and the ideal process that could be achieved through the "integration".

Design review during class approval is a process of comparing and matching structural drawings with rule requirements and calculation results. Although the rule calculations can be verified by classification society's rule calculation tool, the work to compare them with the structural drawings requires manual intervention. The "integration" of the 3D model and the rule calculation tools will reduce the man-hours required for the design review process as well.

These are the forms of "integration" the authors envision.

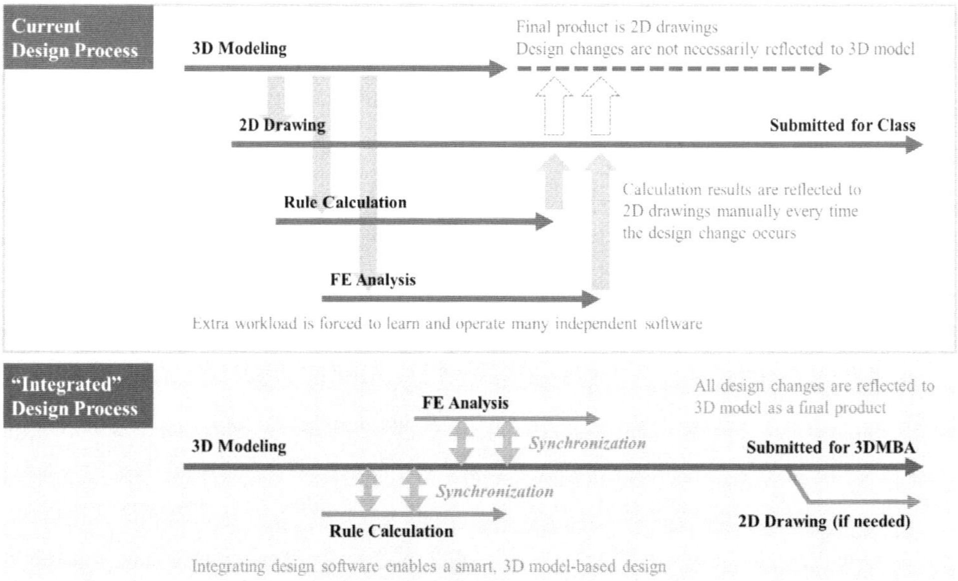

Figure 1: Current and "Integrated" Ship Design Process

2.3 (b) "Visualization" – Efficient way to visualize design information combining the advantages of 2D & 3D

Once the above issue is resolved, it is expected that the ship design and class approval process can shift from 2D-based to 3D-based. By shifting to a 3D model- a ed de ign o e , de igne i e a e o in ui i e y unde and he h ee dimensional structure to design its arrangement and reflect required scantlings in a timely manner, taking advantage of the integration with the calculation tools.

On the other hand, there is the fact that 2D drawings are better than 3D models when it comes to focusing on a particular cross-section and visually identifying the relevant design information while designing structural details. In addition, 2D drawings are better than 3D models for the purpose of presenting necessary information because 2D drawings contain well-organized and consolidated design information that should be communicated with the stakeholders. Conversely, the disadvantage of 3D models can be seen when selectively "visualizing" the intended design information, which is also considered to be one of the barriers to shifting to 3D model-based design from 2D models.

There is a similar need in the class approval process. Even if sufficient design information is included in the 3D model as attributes, it is hard to recognize these from the model's appearance. For example, to check scantling and other attribute information, the user needs to zoom in to target members, click on them, and then check the information displayed. This results in a time-consuming process and will increase the review time, diminishing the benefits of 3DMBA.

Therefore, to promote the use of 3D models in ship design and class approval, it is essential to have a 3D CAD and 3D model viewer that enables users to easily and efficiently check the design information of 3D models by combining the following advantages of 2D and 3D:

- Displaying the structures in a spatial and global way, which is an advantage of 3D
- Displaying the structures focusing on a specific subject, which is an advantage of 2D (i.e. 2D drawing-like view, annotating design information, etc.)

This is the form of "Visualization" which the authors envision.

3. ROADMAP TOWARDS PRACTICAL 3DMBA

We have reviewed the status quo around 3DMBA in the previous chapters. Based on such circumstances, this chapter discusses the roadmap to realize 3DMBA "in a truly effective manner".

Needless to say, a 3D model is essential to 3DMBA; since the 3D model will be created as a deliverable of ship design, the 3DMBA and the ship design process should be considered at the same time. As mentioned in the previous chapter, the ship design process is shifting from 2D to 3D, and in such a "transitional period," class approval is also expected to be a mixture of 2D and 3D models (Figure 2). Although both ship design and class approval would eventually shift to a 3D-based scheme, it is not realistic to aim for such a scheme all at once, ignoring the aforementioned issues. Therefore, to move the discussion forward toward the realization of 3DMBA, the optimal form of utilization of 2D drawings and a 3D model in the "transitional period" should be pursued. Then, through continued development of the systems and processes for that purpose, the scope of utilization of 3D models should be expanded and firmly established in the process of ship design and class approval.

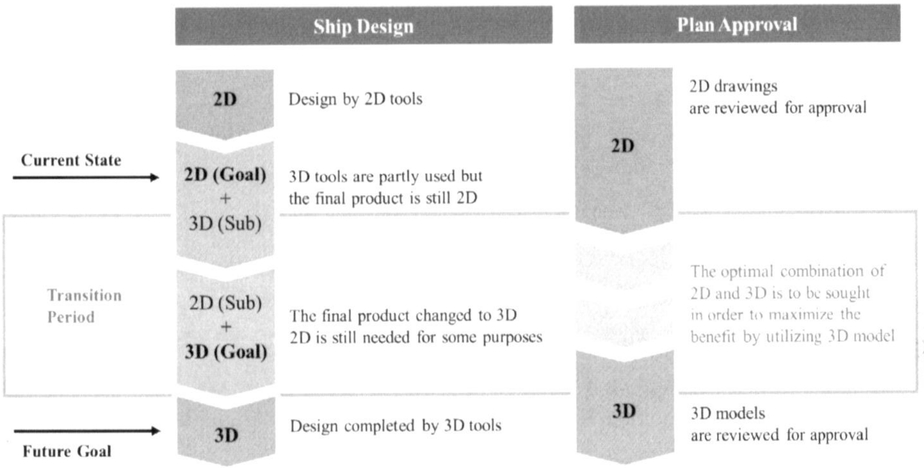

Figure 2: Future Vision of Ship Design and Plan Approval Processes

The "optimal form" in the transitional period is difficult to define uniformly, as it must change according to the peripheral technologies and conditions of the process at that time. However, we believe that the issues of "Integration" and "Visualization" raised in the previous chapter will be common development issues that need to be addressed. In other words, the authors believe that developing and implementing the systems to solve these issues first will contribute directly to the realization of the 3DMBA. Furthermore, it is expected that such attempts will provide new insights regarding the "optimal form" of the 3DMBA and the form of the realistic transitional period.

Based on this principle, the authors have engaged in the development of systems to solve the two issues and could gain certain accomplishments. The outcome of these developments is introduced in the following Chapters 4 and 5. Further discussion about the potential of 3DMBA based on this outcome is further discussed in Chapter 6.

4. SYSTEM DEVELOPMENT: INTEGRATED DESIGN SYSTEM

To realize seamless "Integration," 3D CAD and relevant engineering tools require synchronization with each other. At ICCAS2019, NSY (at that time Japan Marine United Corporation) and NAPA demonstrated a prototype of integrated systems of NAPA Designer and in-house strength calculation tools.

Based on this, the authors continued the development of integrated systems having even superior capabilities and officially released them as built-in function of PrimeShip-HULL (hereinafter "PSH") and add-on of NAPA Designer. This chapter introduces an overview of the integrated system as well as its effectiveness from the perspective of ship design and class approval.

4.1 NAPA API, A POWERFUL CUSTOMIZATION TOOL

The NAPA API provided by NAPA was used as a means of seamless integration between NAPA Designer and PSH. NAPA API consists of Napa Object Model (NOM) API and NAPA Designer Scripting API.

NOM API is a .NET API to the NAPA software. This is the high-level interface that 3rd parties are recommended to use as the basis to any integration. By using NOM API together with external development environments like Microsoft Visual Studio, users can create plugin applications which launch together with NAPA Designer.

NAPA Designer Scripting API is the object-oriented C# development tool included in NAPA Designer. It allows users to read and write NAPA model through NOM API. In addition, it offers access to the UI services of NAPA Designer including graphics, dialogues, etc. The API is intended to be used for the customization of NAPA Designer. Using this modern development environment, users can develop customized tools to run automated modeling or additional operations which are not yet supported by NAPA Designer.

4.2 INTEGRATED SYSTEM BETWEEN NAPA DESIGNER AND PSH

PSH is the structural design support software for evaluating the structural strength developed by ClassNK. It consists of a prescriptive rule calculation tool and direct strength assessment tool and has been used by many ship designers as an indispensable part of the structural design and class approval process.

The main features of PSH are:

- Comprehensive structural strength assessment functions based on ClassNK Rule and CSR-B&T
- Automation functions that contribute to the reduction of design man-hours and design support functions based on sensitivity analysis
- Comment function for easy model management
- Prescriptive rule calculation tool
    - Strength assessment for inputted cross-sectional data based on prescriptive rule requirements for longitudinal strength, local strength, strength for primary supporting members, fatigue strength, etc.
    - Transparent calculation functions allow users to check all calculation parameters and results instantly
- Direct strength assessment tool
    - Software for strength assessment based on FEA, such as cargo hold analysis and full-ship analysis, using HyperWorks provided by Altair Engineering Inc. as a platform
    - Seamless pre-post processing to assist design optimization

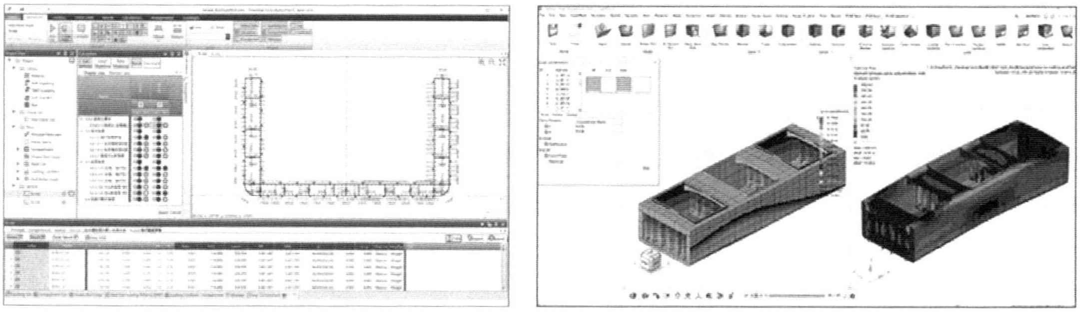

(a) Rule Calculation (b) Direct Strength Calculation

Figure 3: Functions of PSH

In the past, designers have been able to reduce man-hours at PSH, to a certain extent by importing intermediate files generated from 2D CAD or 3D CAD. The same was applied to the linkage function between NAPA Designer and PSH, which was based on an intermediate file. However, this intermediate file-based approach had a number of drawbacks, such as;

- Complex UX to use the data linkage function between NAPA Designer and PSH
- A lot of manual touch-ups due to difficulty to ensure quality level such as data lack during interfacing
- The necessity to input additional data on PSH not included in the intermediate file, which is needed every time when 3D model is changed
- The necessity to manually update the NAPA Designer model based on the determined scantling in PSH

Thus, it was very cumbersome in terms of model management and was clearly incapable of being used for 3DMBA.

Therefore, with the aim of synchronizing NAPA Designer and PSH with each other, the data linkage function was renewed using NAPA API as follows.

- Direct and seamless access to the database by utilizing the NAPA API ensures the consistency of information between the systems.
- Data needed for classification society's rule calculations (Efficiency for longitudinal strength, buckling panels, etc.), which NAPA Designer doesn't have, can now be defined on NAPA Designer with a custom plugin, and rule calculations (prescriptive/FEA) can be performed immediately after linking.
- The connectivity between the NAPA Designer model and PSH's one can be maintained always. For example, when there is some scantling update in each model, the model update can be done at once.

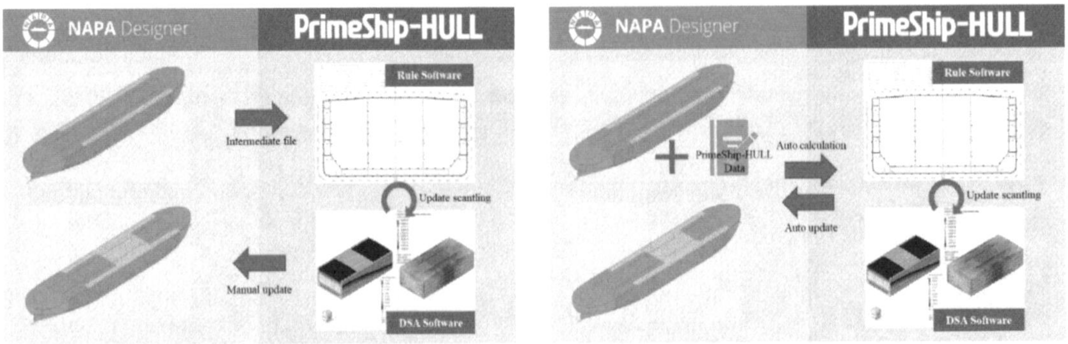

(a) Old File-based Linkage System

(b) New API-based Direct Linkage System

Figure 4: Comparison of Old and Renewed Process of Data Linkage

4.3 FUNCTIONAL TEST: SHIPYARD'S PERSPECTIVE

To validate the effectiveness of the new integrated system mentioned in the previous section with regard to the ship design process, NSY conducted a trial using sample bulk carrier and oil tanker models. An overview and verification results are described below.

4.3 (a) Prescriptive Rule Calculation Tool

In this trial, the integrated system was used to perform prescriptive rule calculations for longitudinal and local strength based on CSR-B&T.

It was verified that the new integrated system enabled to create a PSH cross-section from a NAPA Designer model within only about one minute, and immediately perform longitudinal and local strength calculations. When applied to the actual design, the following benefits can be expected.

- Significant reduction in data creation time for prescriptive rule calculations. This would make it possible to study more rational structure designs within a limited time.
- The division of work between 3D CAD and prescriptive rule calculations can be integrated, making it possible to perform prescriptive rule calculations while designing on 3D CAD.

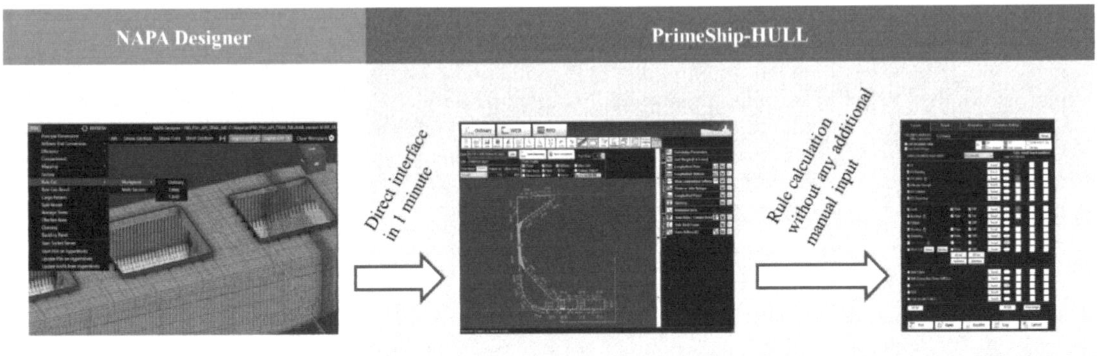

Figure 5: Trial of New Integrated System (Prescriptive Rule Calculation)

4.3 (b) Direct Strength Calculation

In this trial, the integrated system was used to perform direct strength calculations based on CSR-B&T.

It was verified that the new integrated system enabled the creation of the data needed for analysis in PSH from NAPA Designer and made it possible to run direct strength calculations within an hour. When applied to actual design, the following benefits can be expected.

- Significant time savings in FEM model creation and load/constraint setting etc. for direct strength calculations. This would make it possible to study more rational structure designs within limited time.

- The division of work between 3D CAD and direct strength calculations can be integrated, making it possible to perform direct strength calculations while designing on 3D CAD.

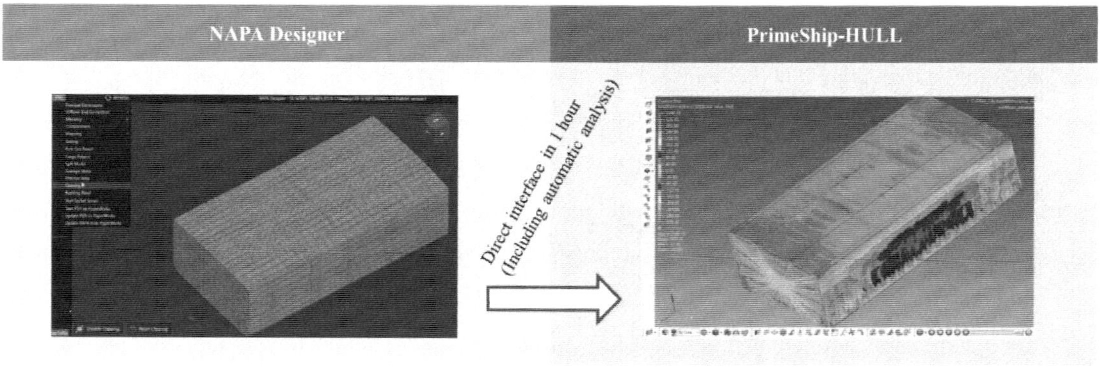

Figure 6: Trial of New Integrated System (Direct Strength Calculation)

4.4 FUNCTIONAL TEST: CLASSIFICATION SOCIETY'S PERSPECTIVE

Similar to NSY's trials in the previous section, ClassNK also validated the effectiveness of the integrated system with regard to class approval process. From the class approval's point of view, the remarkable advantages of an integrated system are as follows:

- NAPA's 3D model can now be linked to data needed for class approval, which was previously not possible to include.
- Consistency of information between NAPA's 3D model and PSH's calculation model can be ensured.

This allows the NAPA 3D model to serve as a central hub for information that can hold the necessary data alone, so that "once the NAPA 3D model is received, the rule calculation results can be verified". This means that it can potentially reduce the man-hours required to make separate calculation reports, too. In addition, previously, the consistency of the design information in structural drawings and calculation models (PSH models) had to be manually cross-checked during the design review. With this integrated system, the design information of the NAPA and PSH models can coincide, and it should reduce the man-hours required for consistency checks and further improve the quality of the review process.

It was verified through this trial that the new integrated system has sufficient functionalities to realize an effective 3DMBA process.

5. SYSTEM DEVELOPMENT: MODEL VIEWER SYSTEM

The importance of the "Visualization" related issue was discussed in Chapter 2 and solving this issue would require a sophisticated model viewer system that could replace or compensate for the traditional 2D drawings' role. The authors worked with NAPA to develop a viewer system to utilize NAPA 3D model as the central hub for information, resulting in the official release of NAPA Viewer, a new web-based application of NAPA. This chapter introduces an overview of the viewer system as well as its effectiveness from the perspective of ship design and class approval.

5.1 NAPA VIEWER

NAPA Viewer is a new web-based application developed by NAPA. It enables NAPA 3D model owners to provide all their stakeholders with secure and easy access to review, assess and comment on the models. NAPA Viewer has been developed to support 3DMBA as well as better information sharing and collaboration among all stakeholders throughout the ship design process. It provides a primary source of information for the up-to-date design information to be shared, communicated, reviewed and approved.

The main features of NAPA Viewer are:

- Web-based application for reviewing the design via a web browser
    - Application is only installed on a server, no need to install anything on the end user's device.
    - The model owner can easily control access rights.
- Visibility control for easier access to relevant information
    - Control the visibility of structures and compartments easily by arrangement trees. Users can also create an arrangement tree containing only the necessary objects (for approval).
    - Object attributes available such as plate thickness, material, stiffener profile, etc.

  - 3D Clipping allows easy control of the area to be checked.
  - Show objects in different colours according to attributes or approval status.
- 2D section views with automatic annotation
  - Annotations can be visualized automatically. This makes the 2D section views similar to traditional drawings.
  - Reviewers can check this information promptly from 2D section views.
- Commenting tool [Figure 7 (b)]
  - Enables interaction between users directly within the 3D model.
  - Comment can be mapped onto certain objects, including an option to use a saved view to quickly help to visualize the objects.
  - The tool also lets users tag others and attach documents, allowing for well-managed and targeted communication.
- Approval status management
  - Approval status can be assigned to each object individually and the model can be visualized based on the approval status. [Figure 7 (c)]

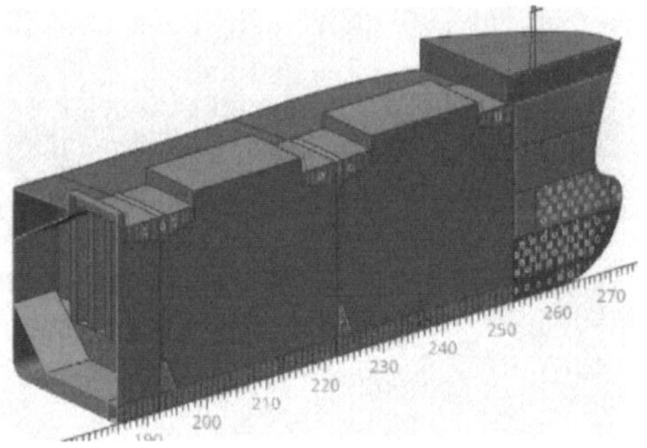

(a) Structure and compartment model in NAPA Viewer

(b) Commenting tool

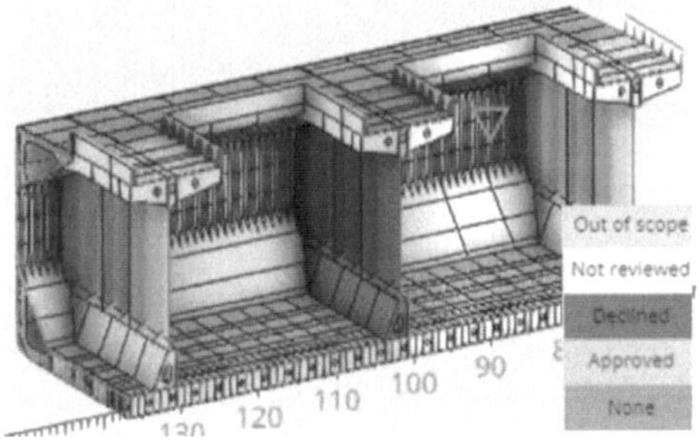

(c) 3D view with approval status

Figure 7: Features of NAPA Viewer

5.2 FUNCTIONAL TEST: SHIPYARD'S PERSPECTIVE

To validate the effectiveness of the new viewer system in ship design as described in the previous section, NSY conducted a trial of the NAPA Viewer using a sample model. An overview and the verification results are shown below.

In this trial, NAPA Viewer was used to share the 3D model among internal stakeholders, including structural design, outfitting design and production planning.

It was verified that this viewer could enable internal project stakeholders to view the up-to-date 3D model easily using any web browser such as Microsoft Edge or Google Chrome, without the need to install dedicated software in each user's computer.

With NAPA Viewer, the following benefits can be expected in terms of shipyard's design review process.

- The drawing-like 2D view can be displayed harmoniously on top of the 3D view, making it easier to visualize areas that need special attention in terms of structural strength, outfitting arrangement, and constructional workability.

- This functionality may allow a shift from 2D-based communication to 3D model-based communication with other organizations, which has been the tradition up to now.

5.3 FUNCTIONAL TEST: CLASSIFICATION SOCIETY'S VIEW

Like NSY's trial in the previous section, ClassNK also validated the effectiveness of NAPA Viewer with regard to class approval process.

In terms of design review, the important question is whether this application can deliver similar levels of quality and efficient review as the current 2D drawing-based work. The "2D View" and "Auto Annotation" features of this application are very powerful tools in this regard, as they display almost all the information that is checked during the design review process, such as plate thicknesses, materials, stiffener profiles, opening dimensions, and compartment information on any cross sections, almost automatically and easily. In addition to this, it is possible to display 3D geometry for areas where it is difficult to grasp the actual shape on top of the 2D view. It can be said that the effectiveness is even greater than current 2D drawings, depending on the subject. (Figure 8)

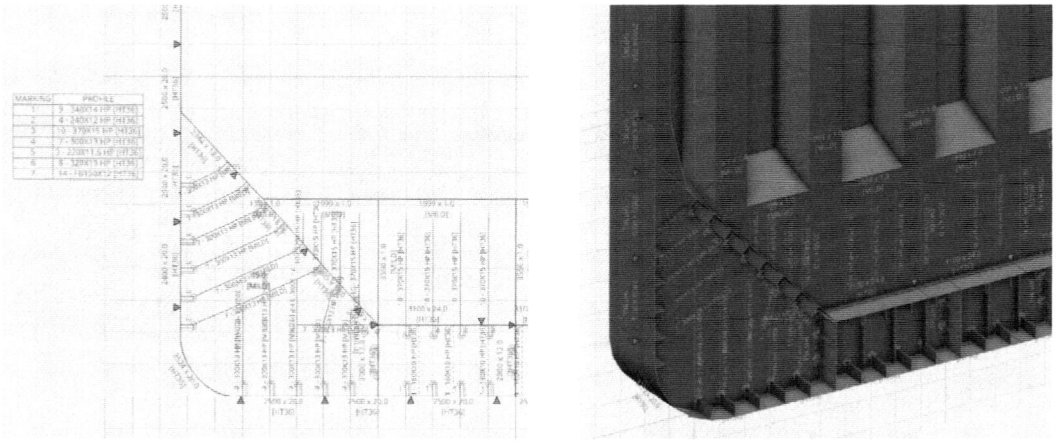

(a) 2D View with Auto Annotation (b) 3D View with Auto Annotation

Figure 8: Design Review by using NAPA Viewer

Further study is needed on how to handle design information that is not directly displayed on the viewer or not included in the 3D model; however, it can be evaluated that this application is a promising alternative to 2D drawings for design reviews.

6. CONSIDERATION AND FUTURE WORK

So far in this paper, based on the development issues summarized in Chapters 2 and 3, its development status has been introduced in Chapters 4 and 5. As discussed in Chapter 3, it is necessary to repeatedly work on the tool and process side of development to move forward with the realization of the 3DMBA. As a next step, the authors aim to conduct a pilot test using these development outcomes to refine the process and identify further development needs. This chapter evaluates the current development status and achievements and summarizes the items that will be studied in the future.

6.1 CONSIDERATION ON THE CURRENT DEVELOPMENT STATUS

As described in Chapters 4 and 5, the authors developed a new integrated system using NAPA API to cope with the "Integration" related issue and developed NAPA Viewer to cope with the "Visualization" related issue. The development of both systems has progressed very well, and it is evident that they will greatly contribute to the promotion of 3D model-based ship design and class approval.

As for class approval, we were able to indicate the possibility of realizing the design review based on the 3D model reasonably easily, with NAPA Viewer as a replacement for conventional structural drawings and API-based system integration as a replacement for strength calculation reports (Fig. 9). The next step is to incorporate these new applications into ClassNK's approval management system, which will bring us one step closer to being able to offer 3DMBA as an official service.

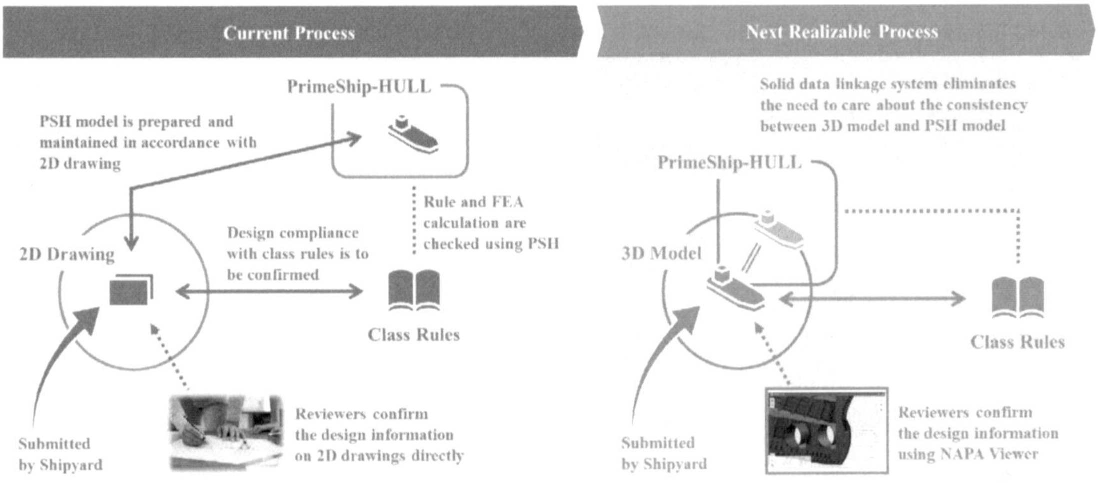

Figure 9: Possible Change of Plan Approval Process

As for ship design, the integrated system is expected to streamline design work by enabling prescriptive rule calculation and direct strength calculation to be performed in a synchronized manner with 3D CAD work. NAPA Viewer's new feature, which seamlessly displays a drawing-like 2D view on top of the 3D view, could indicate a promising possibility as a new style of "3D drawing" that will replace conventional drawings in the future. In fact, this feature would also be very useful for 3D design purposes itself. If a similar feature is available on 3D CAD, it can help designers in charge of drawing creation, who tend to prefer to design in 2D CAD, to shift to 3D-based design.

On the other hand, however, considering the entire design process, a complete transition to a 3D based model will be difficult if the 3D model cannot replace the important role of 2D drawings. One of the most important roles is the "base drawings for handing over the design information to downstream processes.

For this challenge, the following possibilities can be considered, for example:

- Option A: Semi-automatic generation of 2D drawings from 3D CAD at the end of hull structure design process. These will be used for handing over to downstream processes.
- Option B: Hand over the 3D model to the downstream processes. If another 3D CAD software is used in the downstream design, the 3D model needs to be converted to another data format.

As shown in Figure 10, NAPA has basic solutions for Option A and Option B.

For Option A, a dedicated 2D drafting tool called NAPA Drafting can be utilized to semi-automatically generate annotated drawings.

For Option B, one ideal approach would be to continue using NAPA Designer for the detailed design process, but this might require further functional improvements as it was originally targeting approval at the drawing level of 3D models. Meanwhile, it has the advanced interface capabilities to convert to the native data format of the most common production design systems. Hence, the 3D model is likely to be taken over to that system at a certain point in the detailed design process. Also, if some specialized 3D system is used to study construction procedures, like the Data Exchange function which supports more than 10 industry-standard 3D CAD formats such as STEP and JT, 3D models could be utilized. Or perhaps the OCX file export function may be useful for this kind of purpose as well.

Either way, a holistic reformation of the shipyard's design process will be required. These are important future topics to be addressed, along with further functional improvements needed to realize a "truly effective" 3DMBA.

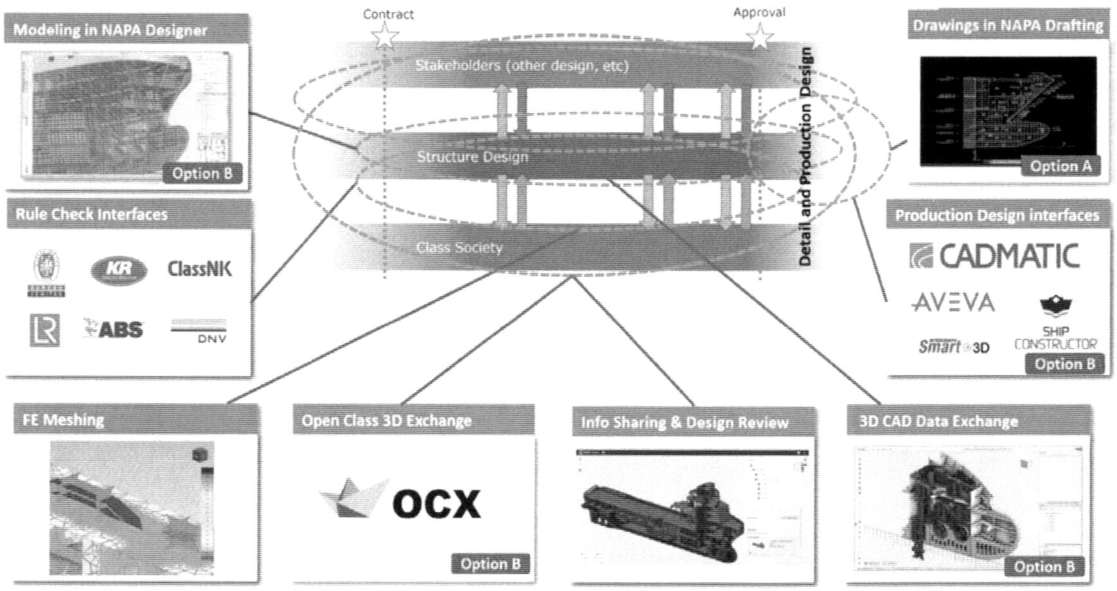

Figure 10: NAPA Offering for 3DMBA

6.2 PILOT TEST OF 3DMBA

As a next step, the authors plan to conduct a pilot test of 3DMBA with the aim of implementing the developed systems into the actual process. This pilot test aims to embody the requirements and the operational process image for the following items, for example.

- Modeling practices
- Comment management
- Revision history and approval status management
- Archiving methods

In addition to this, further considerations will be made regarding the need for functional improvements of NAPA Viewer and its integration with ClassNK's approval management system.

6.3 OPEN UP NEW POSSIBILITIES BY 3DMBA

In order to pursue the realization of a "truly effective" 3DMBA and to maximize the value it brings, the following considerations from a broader perspective shall be needed. While 3DMBA may enable new possibilities, those should require further technological and process-wise developments so that the consideration should be made in the longer term.

6.3 (a) Utilization for Class Survey

The discussion so far has mainly focused on the process during "plan approval" stage. In the subsequent stage of actual ship construction, class surveyors conduct on-site surveys by referring to the approved drawings. Therefore, when considering 3DMBA, the use of 3D models in the subsequent survey work needs to be considered as well.

Nowadays, technological developments such as web applications, cloud services, large-capacity communications, and the popularization of high-performance tablets have made it technically feasible to conduct block surveys by viewing a 3D model on a tablet. Since 3D models are useful for intuitive visualization of structural objects and those geometries, they are expected to contribute to the detection of forgotten or misplaced parts, such as brackets, collar plates, and carlings, for example. Hull blocks are assembled in various positions due to the convenience of construction, making it difficult for inexperienced surveyors to grasp the structure. 3D models may be useful for this issue as well. In addition, Augmented Reality technology may further improve survey quality by mapping the structural members of the 3D model and the structural analysis results onto the actual structure.

In addition, comments assigned through surveys could be recorded and shared through the 3D model, which would facilitate communication with shipyards and comment tracking.

6.3 (b) New Business Model

In this paper, the discussion has been mainly from the viewpoints of shipyards and classification societies. However, for 3DMBA to be viable and widely used, the value and need for 3D models must be recognized across the entire maritime industry. In this sense, the concept of Digital Twins could be encouraging for 3DMBA.

In general, Digital Twins is a concept that refers to the reproduction of real-world entities in a virtual space. In the maritime industry, it is expected to be used for ship information management throughout the ship's lifecycle and for simulations utilizing any available data. Currently, 3D models created by shipyards are rarely utilized after a ship goes into service and are rather treated as confidential information in most cases. However, there is no doubt that utilizing the 3D model will greatly contribute to the realization of Digital Twin after its service and the creation of new values through it. Therefore, it is expected that a new business model in which shipyards prepare 3D models for Digital Twin and receive rewards for the services using the models should be considered. As a result, the use of 3D models in ship design and class approval processes will be further promoted.

In this sense, although this paper discussed 3DMBA for "hull structure," in addition to this, the need for 3DMBA for ship stability, machinery, and outfitting may also grow in the future.

7. CONCLUSION

Thanks to advancements in relevant technologies, especially 3D CAD, the use of 3D models in ship design is steadily expanding, and interest in 3DMBA is growing at the same time. The 3DMBA is an attractive concept with various possibilities for classification societies, shipyards, and for the entire maritime industry, and is already technically feasible. On the other hand, several important problems have remained to realize a "truly effective" 3DMBA process in such a way that all stakeholders can enjoy the benefits.

Prime examples are the issues related to "Integration" and "Visualization" described in Chapters 2 and 3. To solve these issues, the authors worked on system development, which resulted in the official release of the "Integrated System between NAPA Designer and PSH" and the "NAPA Viewer". The authors could validate that these systems have sufficient functionalities to contribute to the improvement of the ship design and class approval processes. Although there are issues that need to be further studied in the future, the establishment of promising systems that may replace the traditional "structural drawings" and "strength calculation reports" needed for class approval is a significant achievement and has reached an important milestone.

Also, for ship design, this project helped to identify the next issues to be addressed in order to accelerate the 3D-based design process. We believe that continuous development of systems and processes for those issues will contribute to ensure a long-term competitive edge powered by 3D technologies.

The authors will continue to pilot test to implement these systems in actual processes to develop even smarter ship design and class approval processes.

8. ACKNOWLEDGEMENTS

We thank colleagues in NAPA, NSY and ClassNK who have participated in this development work and given their valuable contribution to make this paper possible.

9. REFERENCES

1. Shimakawa Y., 'Process Integration of Structural Design Promoted by 3D Model Based Approval', *ICCAS2019*, 2019.
2. Hulkkonen T., 'Process Innovation for Hull Design at Samsung Heavy Industries', *ICCAS2015*, 2015.
3. Hulkkonen T., 'Enabling a Paradigm Shift in Ship Structural Design with a 3D Approach', *ICCAS2017*, 2017.
4. Hiraki T., 'Development of Product Model Data Exchange Between Different 3D CAD Systems for Hull Designing', *ICCAS2019*, 2019.
5. Astrup O., 'Model-Based Approval – The Open Class 3D Exchange (OCX) Standard', *ICCAS2019*, 2019.
6. SON M., 'Utilization of OCX as Part of 3D Model Based Approval in Ship Design Process', *COMPIT2022*, 2022.

10. AUTHORS BIOGRAPHY

Motoki Sakagami holds the current position of Technical Engineer at Technical Solution Department, ClassNK. He is responsible for evaluation work of innovative technologies in ship industry including 3D CAD solutions such as 3D model-based approval. His previous experience includes 3 years of plan approval.

Tetsuya Otaguro holds the current position of Technical Engineer at Hull Rules Development Department, ClassNK. He is responsible for software development for hull structure design support system, including the data linkage function. He has been working in this software development team ever since joined ClassNK in 2014.

Takayoshi Masui holds the current position of Deputy Managing Director in NAPA Japan. He is currently responsible for the NAPA Steel software and business development in NAPA Group. His previous experience includes 4 years working at shipyards in Japan and 12 years in NAPA Group.

Pusheng Lin holds the current position of Technical Consultant in Napa Japan. He is mainly responsible for customer service for the NAPA Steel software. His previous experience includes 12 years working at shipyards in Japan and 5 years in NAPA Group.

Yoshiaki Shimakawa holds the current position of manager at structure design group of initial design department. He is responsible for initial design for many kinds of commercial ships and adopting 3D design at initial design stage. He joined Nihon Shipyard at the establishment of the company in 2021, after 15 years career as a structure engineer in JMU and IHIMU.

Shogo Hisano holds the current position of assistant manager at structure design group of initial design department. He is responsible for initial design for many kinds of commercial ships and adopting 3D design at initial design stage. He joined Nihon Shipyard at the establishment of the company in 2021, after 8 years career as a structure engineer in JMU.

DIGITAL PLATFORM ENABLING ROBOTIC SURVEY, REPAIR & AGILE MANUFACTURING OF SHIPS AND WATERCRAFT

F T Santo and **D Williams,** Lancaster University, UK

SUMMARY

A recent development in Friction Stir Welding (FSW) processes allow welding of steels underwater, as a result, there is a possibility to incorporate this technology to perform in-situ repair and maintenance of ships in water. The FSW requires process parameters which are chosen and monitored by experts during welding. Having such expertise on board can enhance the operation costs of the ship and make the technology expensive to adopt by small-sized European shipping industry. In this work we propose a Digital Platform (DP) that allows experts to monitor the FSW parameters and condition indicators remotely in the form of real-time visualisations. This approach potentially reduces the labour cost as the man hour of the expert can only use during the welding process. The real-time data is also stored in a central data lake which allow users to perform further analysis and determine the overall integrity of the weld.

1. INTRODUCTION

One of the main challenges faced by small to medium sized European shipping industry is to enhance the operational efficiency of the ships. The primary reason for a ship's downtime is the damages sustained in the hull during voyage. The current repair and maintenance strategy requires the ship to travel to dry dock, which may not be available in a given time frame or located closely. When a dry dock is available, the entire process of repair is not only costly but also time consuming. A survey conducted on all European maintenance stakeholder, as a part of a Horizon 2020 project [1], confirmed that their competitiveness would grow by innovations that would allow them to rapidly perform repair and maintenance of ships in water (*i.e.* without dry docking). A recent break-through in Friction Stir Welding [2] (FSW, a type of welding developed by TWI Ltd) research, has revealed that the technology can be used to weld steel structures under water. This unlocks several potential where FWS could be used in under water repair applications of ship hulls as they are also made of steel structures. FSW process creates very low distortion in the material during welding and also eliminates toxic by-product of the process as compared to arc welding technologies, this makes FSW the most suitable technique for under water repair application. The quality of the weld made through FSW is dependent on its process parameters (e.g. vertical or plunge force, torque, rotational speed and transverse speed of the tool etc.), therefore, incorrectly setting these values can result in a weld containing defects. Currently, the FSW process parameters are carefully chosen by experts in the respective field. However, having fully trained personnel on board can increase the operational cost of the ship. As a result, FSW underwater repair process may become expensive to implement by small to medium sized European shipping industry.

In this work, we developed a Digital Platform (DP) that wirelessly ingests the FSW process parameters from the welding robot and weld condition indicators, in the form of Acoustic Emission (AE) features [3], from the sensors attached to the welding structures. The data ingested in the DP is redirected it to a front-end that provides real-time visualisation through a graphical representation and saves it a Google Cloud Storage (GCS) bucket for post processing. The DP presents a wireless data-flow support, as a result, any computer dashboard registered to the DP's server can have access to its functionality. This allows the trained personal to monitor the FSW process parameters remotely, without having to be based at the ship. The real-time visualisation functionality of the DP provides users information regarding the quality of the welds (*i.e.* condition indicators) and the associated process parameters. If the quality of the welds deviates from the acceptance range, the user can re-optimise the process parameters to guide the quality within the desirable range. The state-of-the-art technology which offers some of the support is the AE data acquisition (AE-DAQ) systems which can incorporate the FSW parameters as parametric input and present them together in a computer dashboard with AE data. However, the communication between the AE-DAQ and computer dashboard is stablished through a serial bus, as a result, both the devices must remain in a close proximity. Therefore, the trained personnel will have to be on board the ship under these settings. The DP also allows the storage of data in a central data lake (GCS bucket).

The data ingestion process uses Google Pub/Sub messaging service. One stream of data ingested by Google Pub/Sub is redirected to a front-end with live visualisations functionality, whereas the other stream is processed using Apache Beam Dataflow (ABD) and Dataproc and stored in a (GCS) bucket for post analysis.

2. METHODOLOGY

FSW process parameters are initially collected by the computer controlling the robot, whereas the AE data are collected through a computer connected to an AE-DAQ. Data from these computers can then be sent to Google Cloud Platform (GCP) resources (i.e. Pub/Sub), through a python script via a Hypertext Transfer Protocol (HTTP) over the internet. An adequate authorisation and authentication are required in order to establish a connection between computers and GCP resources. In this work we used OAuth 2.0 protocol to connect with GCP resources [4]. The process begins by obtaining a credential from GCP (i.e. client ID and client secret). In the next step a token is requested by the computers to access the GCP resources, this requires a user to log in to their Google account and grant permission that the application is requesting. Following a successful authorisation, a token is generated with permission to access the required resources. Figure 1 highlights the one of the working principles of OAuth 2.0 connection protocol.

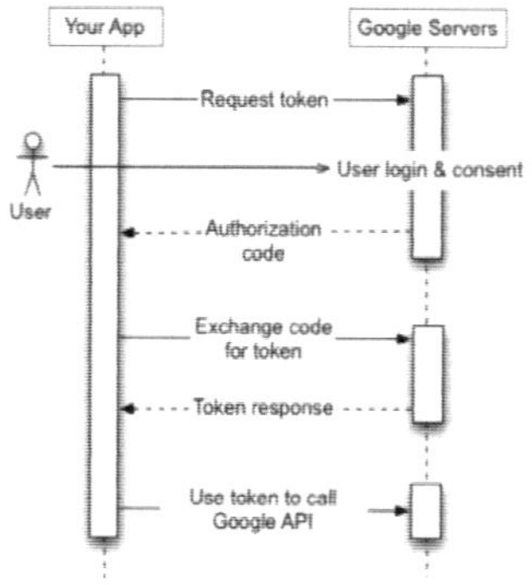

Figure 1: OAuth 2.0 connection protocol

Following a secured connection establishment, the data from computers are ingested to Pub/Sub [5], which is a real-time, auto-scalable messaging service. The main principle of Pub/Sub is based on the flow of information from publishers to subscribers by a given topic. A publisher publishes a message to a topic, which are retrieved by a number of subscribers that has a subscription to the topic. In this work we created a topic a subscription called resurgam-t and resurgam-s respectively. A pull methodology was employed to successfully retrieve the subscription messages from a topic. The sequence of operation in a pull methodology is shown in Figure 2.

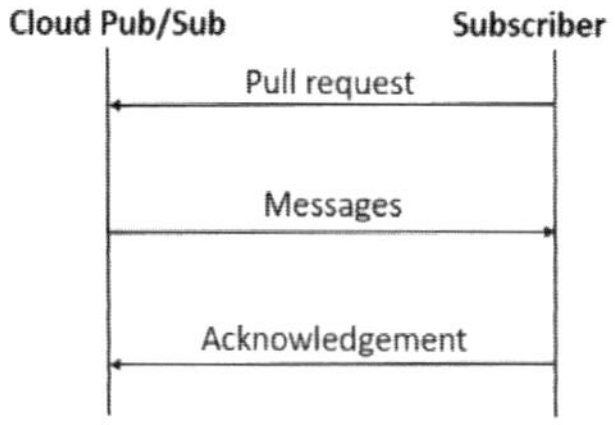

Figure 2: Pub/Sub pull subscription

The subscription data are retrieved by the DP, which uses Django as a server-side web-framework written in Python. The subscriber in DP uses OAuth 2.0 protocol to communicate/connect to the Pub/Subtopic. This was established through an appropriate service account credential file stored in the DP. The influx of real-time data from the subscriber was then rendered to a front-end written in HTML (HyperText Markup Language), CSS (Cascading Style Sheets) and JS (JavaScript). The real-time data were presented in a graphical format using Chart.js library of the JavaScript.

The massages form the topic are also retrieved in a Cloud Dataflow (CD) [6] subscribers. This is a fully managed, serverless Google cloud resource for data processing pipelines. It uses Apache Beam Framework as a programming model. The programming language used for the pipeline was Python with Apache Beam Software Development Kit (SDK). The pipeline code was executed on Dataflowrunner which uses GCP's Virtual Machines (VMs) to successfully complete the task. The CD was employed in this work to retrieve the real-time data, modify and improve their structures to the desired format and to save them in a Google Cloud Storage (GCS) as a raw file. The datasets are also divided on a per day basis, such that all the data streamed on a particular day belong to the same folder in GCS. The communication/connection between CD and GCS was stablished through OAuth 2.0 protocol using the default compute engine service account. Figure 3 below shows the pipeline for this work.

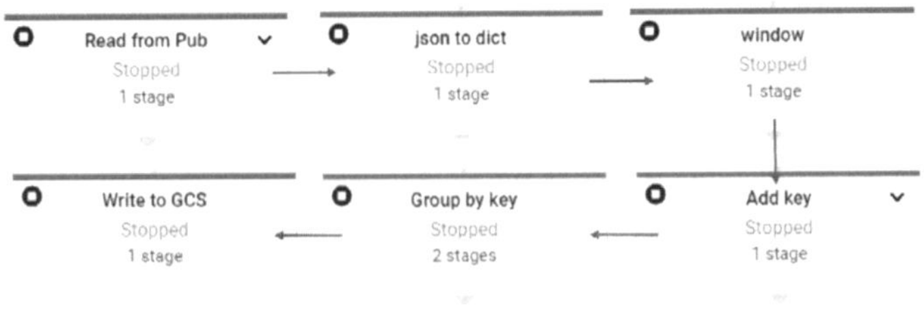

Figure 3: Dataflow pipeline execution strategy

A further structure was added to the data and represented in a graphical format at the closing of the day (*i.e.* at 23:59) prior to saving it in a different GCS bucket. This task was accomplished by creating a PySpark (an interface for Apache Spark in Python) job and adding it in a Dataproc workflow template [7]. A cluster of a master and two worker nodes comprising of n1-standard-4 (vCPUs: 4, RAM: 15 GB) were incorporated in the workflow. This enables the Dataproc to create the cluster, run the job attached to workflow and delete the cluster when the job completes. The Dataproc workflow template was further integrated with Cloud Scheduler [8] (CS) jobs in order to run it at regular interval. The communication/connection between Dataproc workflow template and CS was established through OAuth 2.0 protocol using a customised service accounts with roles that includes relevant permissions. Figure 4 shows the overall architecture of the pipeline.

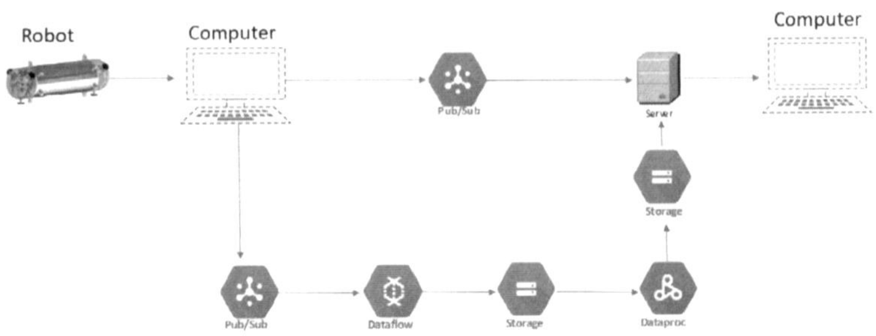

Figure 4: Overall pipeline employed

3. RESULTS AND DISCUSSION

The FSW robot that will facilitate the underwater welding of ship hulls is currently in its development phase and is expected to be completed by the end of the project [1]. However, the technology (*i.e.* FSW of steel in water) has been demonstrated through an industrial robot in a laboratory-based environment in conjunction with AE monitoring. In order to test the functionality of the DP, FSW and AE data saved from the laboratory-based tests were used to simulate the real-time data from the underwater welding bot. This was accomplished by a Python script that sends these data to Pub/Sub periodically from a local computer.

3.1 REAL-TIME VISULISATION

Figure 5 shows a dashboard of the DP that represents the real-time data ingested through Pub/Sub in a graphical format. Graph a, b and c shows the FSW parameters such as Z load or plunge force, spindle torque and spindle RPM respectively, whereas graph d shows the AE hit rate. The DP allows integration of a number of other FSW parameters (e.g. Linear speed of the spindle, X and Y load, federate etc.) and AE data analysis technique. In this paper we have shown only three FSW parameters and a simplified AE data analysis technique.

The process parameters shown in figure 5 was chosen to ultimately result in a weld with defect. This was particularly done to observe the AE response in relation to a faulty weld. Majority of the time spent in the test was to gradually increase the Z load to a required value. The onset of welding has been marked in red in all the graphs. We have observed from other tests that there is no AE activity or rate during welds without defects. In this test, there is a noticeable AE rate after the onset of welding, suggesting the presence of defect in the welds.

This functionality of the DP allows users to simultaneously monitor FSW parameters and the associated AE activity. Moreover, the monitoring can be performed remotely, as a result, the trained personnel does not have to be on board the ship.

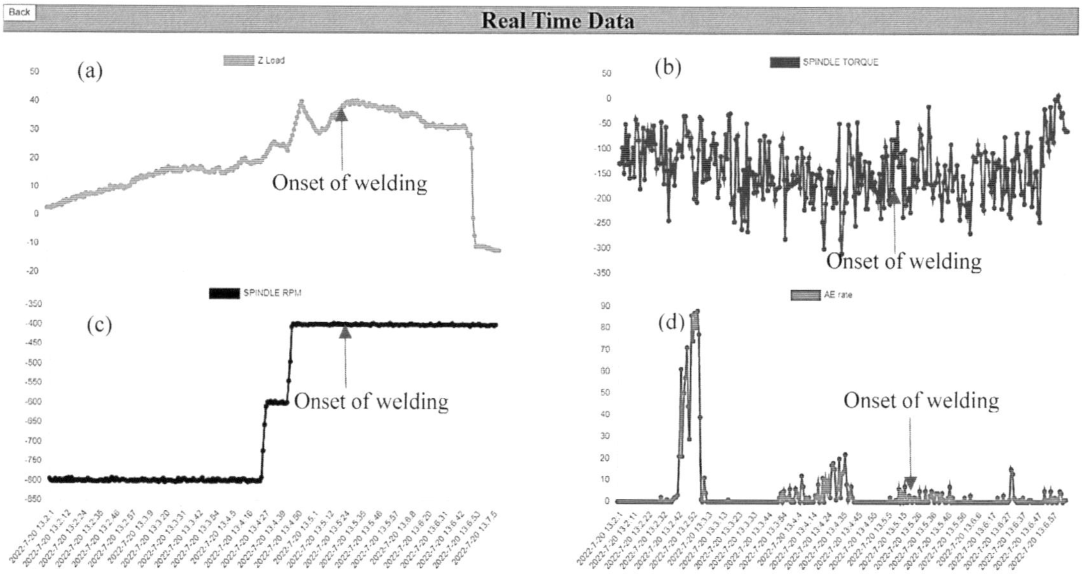

Figure 5: Real-time data visualisation Dashboard: (a) Z load or plunge force, (b) Spindle torque, (c) Spindle RPM and (d) AE hit rate.

3.2 STRUCTURE AND STORE

The real-time data from the Pub/Sub are further processed and saved in a structured and graphical format in GCS bucket. In order to test this functionality of the DP, real-time data were streamed to the Pub/Sub on three different dates (i.e. 2022/5/18, 2022/5/19 and 2022/5/23). The data immediately appears in a primary GCS bucket on each date, in a semi-structured format. Moreover, the datasets were separated by folders, respective to its individual date. Upon the availability of a new folder with an updated date at the closing of each day (*i.e.* 23:59 hour), a fully structured data along with the visualisations were extracted from the data belonging to that folder and saved in a final GCS bucket. The data in the final GCS bucket are also separated with respect to its individual date. The strategy employed to test this functionality and the folders in the primary and final GCS bucket is shown in figure 6.

The content of the final GCS bucket can then be accessed from the DP. Figure 7 shows a layout of the front-end with this functionality. The front-end allows users to input a local destination where the data will be downloaded and select the type of data, in the form of dropdown menu. As intended, the DP allows the access to both tabular data (*i.e.* in CSV format) and graphical representation (*i.e.* Analytics). The graphical representation in this case are four types of data streamed into Pub/Sub (3 FSW parameters and 1 AE data analysis technique). One of the graphical representations stored in the final GCS bucket is shown in figure 8

Figure 6: Strategy employed to structure and store the Real-time data

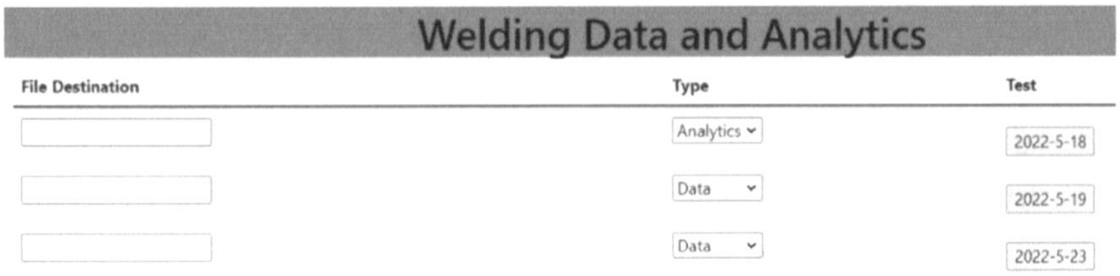

Figure 7: Front-end of the DP providing access to the data

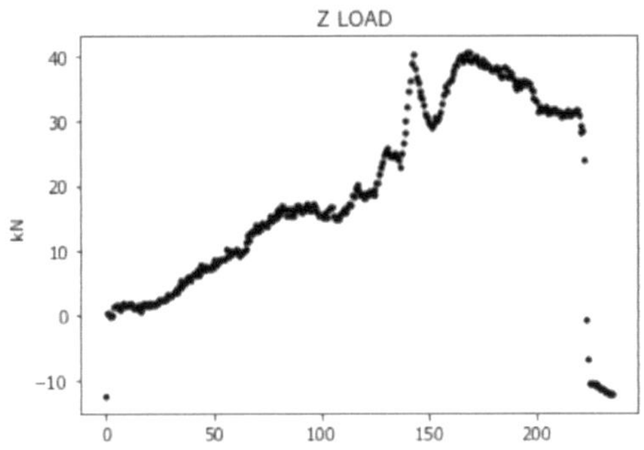

Figure 8: One of the graphical representations in the final GCS bucket

The structure and store functionality of the DP provides user opportunity to perform further analysis or post-processing of the FSW and AE data. This functionality can be particularly helpful in the final quality control, where the users may be interested to determine or investigate the quality of all the welds made during the day. The data stored in a central platform can also be used to train European Workforce in FSW processes.

4. CONCLUSION

This work incorporates the Google cloud technology and web application in a DP that facilitates the underwater FSW of ship hulls. The real-time visualisation of the DP provides data flow with very minimal latency. As a result, the proposed approach allows the experts to monitor the in-situ FSW process and condition indicators remotely without having to be based at the ship. It reduces the labour cost as the man hour of the expert are only used during welding process.

The Structure and storage support of the DP allow the expert to perform further analysis and post-processing of the FSW data. The support can also be used to identify the overall integrity of the weld.

5. REFERENCES

1. H2020, "European Commision," 6 November 2020. [Online]. Available: https://cordis.europa.eu/project/id/101007005.

2. The Welding Institute, "Friction Stir Welding of Steel," [Online]. Available: https://www.twi-global.com/what-we-do/research-and-technology/technologies/welding-joining-and-cutting/friction-welding/friction-stir-welding/materials-weldable-by-friction-stir/steel-friction-stir-welding.

3. L. Calabrese and E. Proverbio, "A Review on the Applications of Acoustic Emission Technique," Corrosion and material degradation, vol. 2, 2020.

4. Google, "Using OAuth 2.0 to Access Google APIs," [Online]. Available: https://developers.google.com/identity/protocols/oauth2.

5. Google, "Pub/Sub," [Online]. Available: https://cloud.google.com/pubsub.

6. Google, "Dataflow," [Online]. Available: https://cloud.google.com/dataflow.

7. Google, "Dataproc," [Online]. Available: https://cloud.google.com/dataproc.

8. Google, "Cloud Schedular," [Online]. Available: https://cloud.google.com/scheduler.

CONCEPTION AND IMPLEMENTATION OF A HETEROGENEOUS SHIP-SPECIFIC SENSOR SYSTEM FOR THE ACQUISITION OF RELEVANT DATA FOR AUTONOMOUS OPERATION

E Schütz, T Holzki, C Genschorek and **A Loewe** TU Berlin, Germany

SUMMARY

The major goal of this study is the development of new concepts of automated navigation to improve automation in inland navigation. For this purpose, a ship-specific sensor system has been designed and implemented to acquire time synchronous information sources. The system is composed of the following heterogeneous sensors, which were integrated within a demonstrator: RTK-GNSS receiver with antenna (high-precision corrected satellite positioning), lidar sensors (light detection and ranging), camera, inertial measuring unit, AIS receiver (automatic identification system). Measurements were realised with the demonstrator in Berlin on the inland waterway "Westhafenkanal – Charlottenburger Verbindungskanal" and subsequently the collected data were evaluated within a development environment. The presented evaluation of the test run with the created multi sensor system already provides very useful information about the interaction of different sensors and their individual advantages and disadvantages.

1. INTRODUCTION

The advancing automation of technical applications already allows the substitution of humans by machines in specific areas. This process of change can also be seen within inland shipping, so that questions arise regarding the maintenance of the security level as well as possible concerns related to the level of automation.
Automated navigation in inland waterways is of special interest because of its importance for the infrastructure. To enable autonomous operation of a ship, it is necessary to have safe, stable and redundant measurements of system-internal and system-external variables.
To guarantee a safe autonomous operation different, partly redundant sensors are merged to eliminate potential weaknesses of each sensor and to be able to balance the system. The system, hence, consists of heterogeneous sensors. The sensor system is initially installed in the laboratory on a prototype and then prepared for tests in real operation.
The following requirements apply to the measurement:
Highly accurate determination of position, location and speed

- Recording of the operating status and relevant characteristic values of the drive and energy system
- Recording of the water depth and height of the ship as reference values for power consumption
- Mobile optical traffic detection in combination with existing AIS data

The most important components of the sensor system can be found in Table 1. The sensors are managed via a central computer unit. For this purpose, we used an Ubuntu 20.04 LTS server to set up the Robot Operating System (ROS) framework [1]. The used ROS version is Noetic Ninjemys, LTS 2025. The project is based on a BSD-licensed system.
The demonstrator system is initially intended to provide information of the behaviour of the lidar sensors installed on a moving ship. For this purpose, the concept of data management as well as the data fusion of all sensor components is described and implemented into the ROS

Framework [2]. A workspace is set up in which all required packages are stored.

2. INTEGRATION PATTERNS

The communication within the sensor system as well as the integration in existing external systems requires a certain robustness of the data exchange, in order to meet the requirements. It can be assumed that connections of a network may be unreliable or slow. Thus, it can lead to delays or interruptions in the individual segments (LAN segments, switches, routers, public networks, satellite connections). Furthermore, the components of the sensor system should appeal in an individual program. This makes it is easier to expand or change a component and other modifications can be carried out more easily. The system must be able to understand different programming languages and support different operating systems.
The management of the processes is provided by an additional program, the message-oriented middleware (MOM). The communication between sender and receiver is generated via a message channel. This is a publish-subscribe channel. [see 3, p. 106] The principle is illustrated in Figure 1.

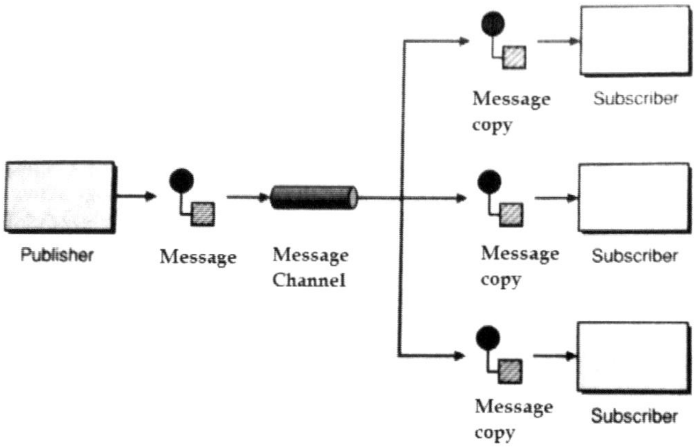

Figure 1: Publish-subscribe channel - message channel that transmits a copy of a message to each recipient (subscriber), Source: [based on 3, p. 107]

The Publishers publishes a message in the message channel. A copy of the message is then forwarded to each output channel (subscriber). This makes it possible for debugging processes to listen to a message channel without interrupting the existing connections. The possibility to listen to existing message channels can be a disadvantage for security reasons. Consequently, subscribing to a message channel should be restricted by security guidelines. [see 3, p.107f]

2. REQUIREMENTS

To demonstrate a combined use of the available data sources after an exemplary data acquisition, data fusion-based procedures are presented. The design of the demonstrator system can be divided into four functional layers:

Information sources
Describes all information fed to the system from local sensors like camera, lidar and an inertial measurement unit as well as external components such as satellites, a mobile data link and a VHF radio link.

Hardware platform
The central component of the demonstrator system is a single computer, which provides the computing capacity as well as the interfaces of all components provided. The first synchronizations are based on communication protocols between platform and sensors.

Middleware
The presented middleware is used to synchronize the individual data streams. It provides the temporal context for the individual modules, and it transfers time stamps to all messages. Within the middleware software packages can be used to add various functions to the system. This primarily involves the provision of the sensor messages in a generally freely accessible standard, as well as further complex software modules, which enables processing of the raw data.

Human-machine interaction
For a provision of higher-quality data products, resulting from data fusion and data processing of different sensor components, software packages can provide visualizations of the results. Based on the open-source structure of the presented ROS Middleware, many already established visualization programs can be used. In the elaboration, the potential is first presented in a development environment.

Table 1: Hardware overview

Components	ID	Description	Pieces
lidar	Blickfeld Cube 1/Range1	lidar based on MEMS	3
IMU	Bosch BNO055	triaxial 14-bit Accelerometer close-loop triacial 16-bit gyroscope microcontroller BSX3.0 FusionLib	1
Camera	Basler a2A2590-22gcPRO GigE camera	IMX334 ROI CMOS sensor 22fps with 5.1MP resolution	1
Objective lens	Basler Lens C125-0418-5M-P	Focal length of 4.0mm Aperture range from F1.8 - F22 5MP resolution	1
GNSS Receiver	u-blox ZED-F9P	Multi-band GNSS receiver	1
GNSS Antenna	u-blox ANN-MB-00	L1/L2 multi-band GPS, GLONASS, Galileo, BeiDou	1
AIS Receiver	WEGMATT LLC dAISy 2+	dual-channel AIS Receiver NMEA 0183 output	1
AIS Antenna	VHF Antenne SGV70PBBK	3 DB - 530MM length 6M RG58C/U COAX Kabel	1
Embedded PC	ThinkEdge SE30	Intel® Core™ i5-1145GRE Ubuntu 20.04 16 GB RAM	1
PoE Switch	Tp-link	5-Port Gigabit Desktop with 4 PoE+ Ports	1
Casing	BOX 685.011	Weatherproof plastic casing IP55 46 x 38 x 13cm	1

3. SYSTEM DESIGN

The assembly consists of a control cabinet and a base plate, which are mounted on a tripod. (see Fig. 3)
To enable the system to be expanded or adapted, the design is as modular as possible. The components presented below are designed to be independent of each other, so that the control cabinet and base plate can be adapted or exchanged as required. The base plate is used to define the geometrical alignment of the sensors. For this purpose, a CAD model was created in order to define
lidar mounts and exact drill holes for GNSS antenna and IMU. Thereby different mounting options for the lidar sensors are provided, so that the lidar's can be aligned to the on and ea , o on o h ide i h a ang e o he ongi udina axis. The arrangement of the sensor components is transferred within the ROS system into a Unified Robot Description Format (URDF). This is an XML format for the representation of a robot model. (see Fig. 1)

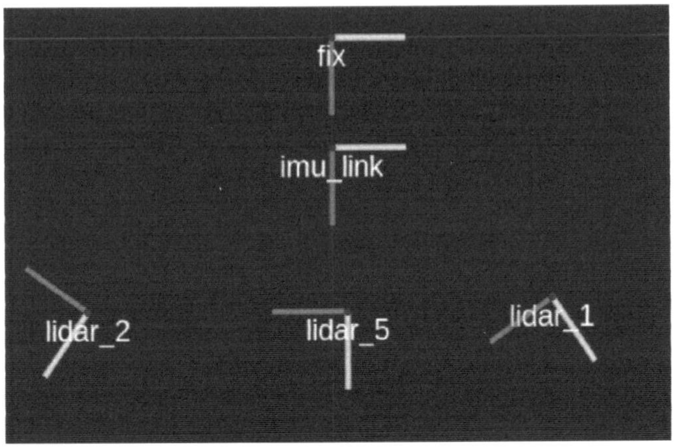

Figure 2: Sensor component alignment

3. EVALUATION OF MEASUREMENT RUNS

We performed two measurement runs in a real laboratory in the area of the Berlin Westhafen (Charlottenburger Verbindungskanal <-> Westhafen Kanal)

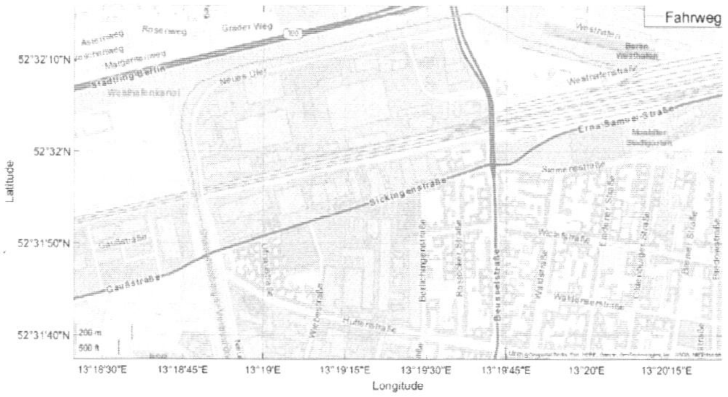

Figure 3: Track on the Ursus bulk carrier between Charlottenburger Verbindungskanal and Westhafen in Berlin, DE

For the measurement runs, the demonstrator system was set up on the bow of a bulk carrier.

Figure 4: Overall set up of the demonstrator system - sensor box on a tripod

During operation of the ship, measurements could be carried out on the outward and return voyages.
The route of the measurement run, recorded with the RTK-GNSS [4] positioning can be seen in Figure 2. A single trip covers about 2.49 km. The device passed a total of three bridges during one trip. When driving under the bridges, there are major inaccuracies in the position determination. The reason for this is, on the one hand, the intermediate disconnection of the mobile radio connection as well as the loss of some satellite connections. It became apparent from the measurement runs that if the connection is lost, it takes several minutes for the position determination to get back into the range of less than 3cm accuracy. For an exact localization at these critical points, it is advisable to use an alternative correction, for example via lidar distance measurements.
The visualization of the calculated trajectory, just by the lidar distance measurements, can be seen in Figure 4. Due to the data fusion of the three attached lidar sensors, the field of view of the sensors was large enough to detect both sides of the waterway boundary over long distances.

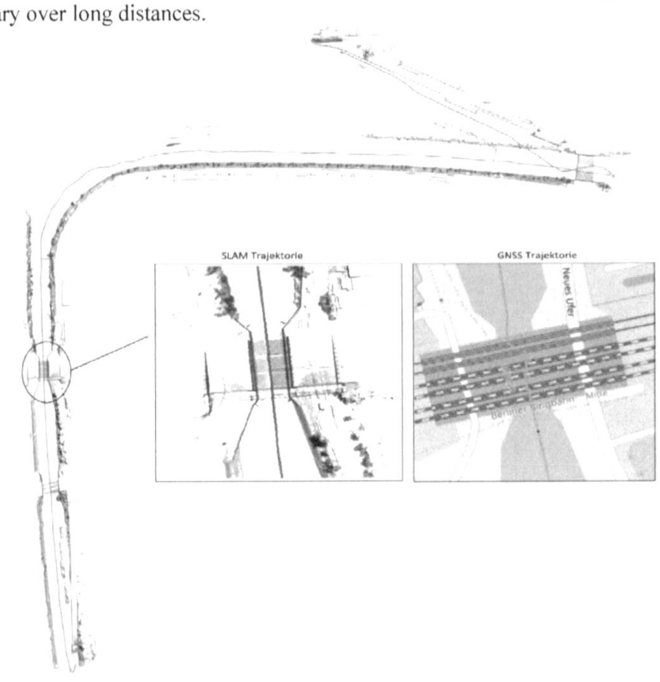

Figure 5: 2D-SLAM Cartographer - created map from lidar point clouds [5]

Figure 4 shows the calculated trajectory in comparison to the GNSS trajectory for the section under the highlighted bridge. The trajectory of the SLAM calculation for the section under consideration is a significantly better approximation to the driven trajectory.
However, there are also areas in which only few or no points at all were detected by the lidar sensors. This results in a jump in the SLAM trajectory. In order to adjust the trajectory, it is therefore necessary to include further sensor components in the calculation. For the optimization of the map seen in Figure 4, the position determination of the GNSS module is included in the global SLAM procedure (see Figure 5 & 6). The individual sensor messages can be weighted differently for the creation of the trajectory. Figure 5 shows the created map with high weighting of the GNSS position determination. In comparison to this, Figure 6 shows a higher weighting of the of the lidar distance measurement. The inclusion of the GNSS module brings a better represented trajectory in general. However, in case of disturbances of the GNSS module, a deterioration of the generated map can be seen.
For this, at the discussed bridge crossings, distortions can be seen within the map. (See highlighted area)
The marked location shows, that due to the loss of both sources (lidar and GNSS) the calculation needs more sensor messages to be able to create a reliable map.

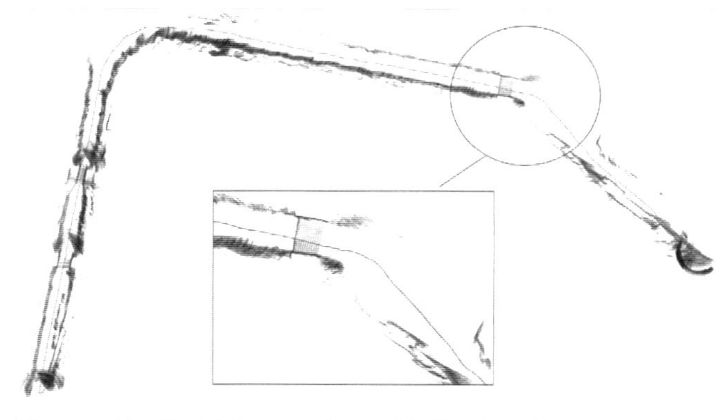

Figure 6: 2D-SLAM Cartographer - with GNSS position correction [5]

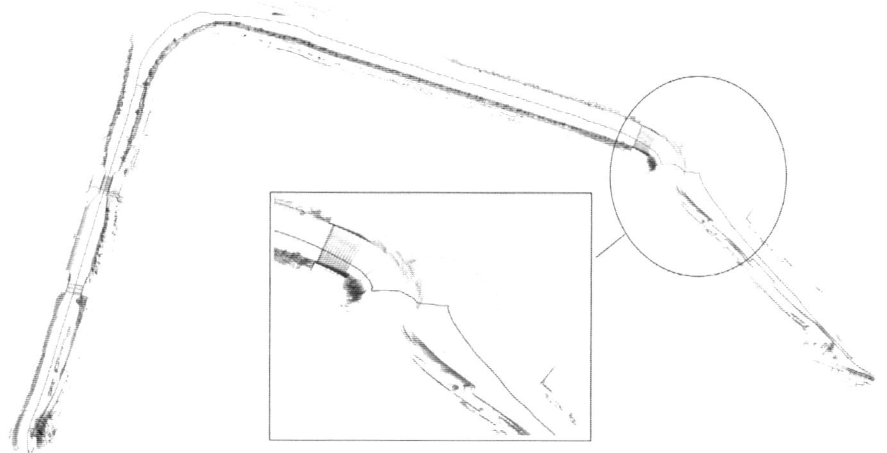

Figure 7: 2D-SLAM Cartographer - With GNSS position correction, positions of lidar points are weighted higher [5]

Thus, in both variants' errors are present within the map in both variants, because the sensor messages are not without disturbances over the whole distance.

To reduce disturbances the compass course of the IMU would be advantageous, as it ensures the orientation of the demonstrator system at error-prone locations. Since the calibration of the magnetometer for the location at the bow of the heavy lift barge was not successful, the compass heading calculations cannot be used in further analysis. Another possibility would be to extend the range of view of the distance measurements by adding more lidar sensors and thus minimize the probability of loss of information. To illustrate the redundant measurement of an event, Figure 7 shows the image of a ship passing the sensor.

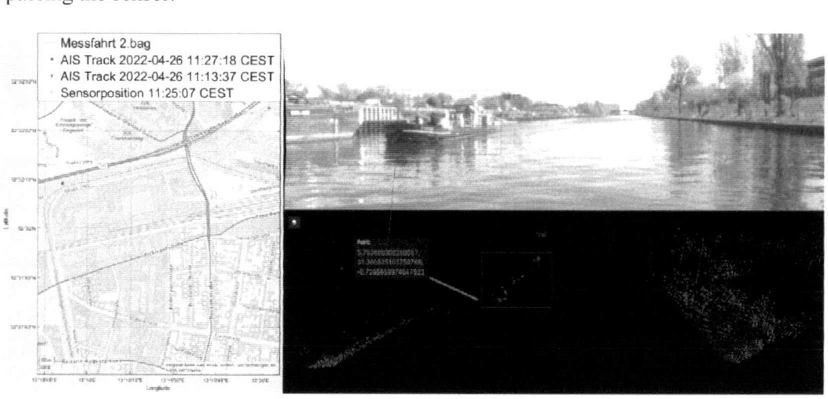

Figure 8: Redundant measurement of a passing ship

The AIS transmitter sends its current position even before the camera or lidar sensors can detect the ship. The trajectory of the passing ship was added by the online available position information for illustration.
The examination of the recordings shows that the installed camera detects the approaching vessel much earlier than the lidar sensors. This results in the sequence in which the various sensors should be able to detect the event. Thereby, the lidar sensor with the present configuration can perform reliable distance measurements of the passing ship within a radius of about 50 meters.
Figure 7 shows an example of the distance measurement of a single point on the bow of the passing ship. In order to achieve a fusion of the three data streams, the camera sensor system must first be calibrated together with the lidar sensor technology, since the lidar sensor and the camera acquire their data with respect to their own coordinate system.

4. CONCLUSIONS

The presented evaluation of the test run with the created multi sensor system already provides very useful information about the interaction of different sensors and their individual advantages and disadvantages. In a next step the designed sensor system should be integrated into an existing ship system.
In order to derive the presented system as shown in Figure 8, an exact measurement or calibration of the sensor positions is necessary. Because sensors should be mounted at the bow and stern of the ship in order to maximize the field of view, the sensors cannot be defined on a single base plate.

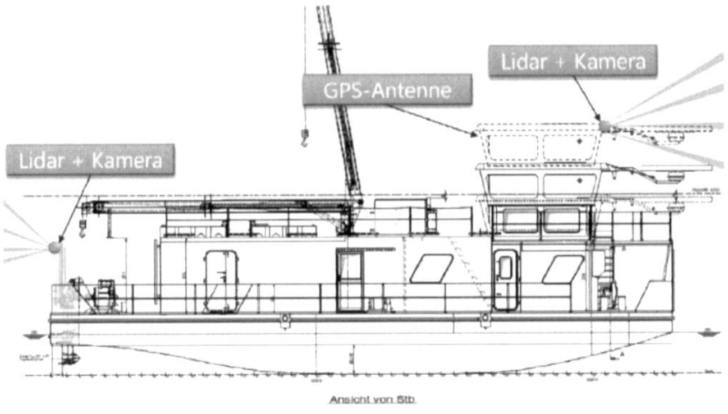

Figure 9: Mounting of the main sensor components on the tugboat ELEKTRA

This study focused on data collection above the water surface. For an extension of the sensor system, information about the topography of the basin below the water surface should be integrated. By integrating a multi-beam sonar, derivatives of the water depth profiles could be acquired. Fused with the lidar sensors above sea level, the digital map basis could be further developed. This would allow to further approximate the waterway-model to the actual real conditions. For additional measurement runs, an intermediate energy storage device is necessary to compensate unstable power supply. Further, the prevailing operating conditions of the individual hardware components must be considered in order to achieve full unrestricted functionality. Of special importance are restrictions of the components concerning thermal impacts, in order to avoid an overheating of the system. In addition, sufficient resilience to voltage changes of the power grid must be ensured in order to bridge power interruptions.
The electrical cabling within the system also needs sufficient robustness against external influences in order to ensure permanent operation. Therefore, the vibration resistance and corrosion resistance of all hardware components must be considered.

5. REFERENCES

1. Robot Operating System (ROS): The Complete Reference (Volume 3). eng. Bd. 778. Studies in Computational Intelligence. Cham: Springer International Publishing. isbn: 9783319915890.

2. Enrico Schuetz 'Konzeption und Implementierung eines heterogenen schiffsspezifischen Sensorsystems zur Erfassung relevanter Daten für einen autonomen Betrieb', Master's thesis, TU Berlin, 2022 url: https://git.tu-berlin.de/ecschuetz/ros_assembly

3. Gregor Hohpe and Bobby Woolf, 'Enterprise integration patterns : designing, building, and deploying messaging solutions', *eng. 13. print. The Addison-Wesley signature series. Boston, Mass. ; Munich u.a.: Addison-Wesley*, 2009.

4. T.Takasu, RTKLIB v 2.4.2, An Open Source Program Package for GNSS Positioning. 2018. url: https://github.com/tomojitakasu/RTKLIB

5. Wolfgang Hess a.o. 'Real-Time Loop Closure in 2D LIDAR SLAM', *IEEE International Conference on Robotics and Automation (ICRA),* 2016

6. AUTHORS BIOGRAPHY

E Schütz currently holds the position of research associate in the Department of Design and Operation of Maritime Systems at the Technical University of Berlin. He is responsible for the research project autonomSOW.

T Holzki holds the current position of R&D Engineer and Project Manager at Alberding GmbH.

C Genschorek currently holds the position of research associate in the Department of Design and Operation of Maritime Systems at the Technical University of Berlin. He is responsible for the research project ASWARM.

A Loewe currently holds the position of research associate in the Department of Design and Operation of Maritime Systems at the Technical University of Berlin. She is responsible for the research project ELEKTRA.

INFORMATION MANAGEMENT IN SHIPBUILDING PROJECTS - INFORMATION FLOW FROM 3D DESIGN TO PRODUCTION DATA

L Seppälä, CADMATIC, Finland

SUMMARY

The complexity of modern shipbuilding, alongside changes in a business environment and growing capabilities of digital solutions, challenges work processes and offer the possibility to gain efficiency by eliminating gaps in the information flows. The paper discusses the changes in modern ship design and production and the interconnections between these two phases of the shipbuilding lifecycle from the information management point of view. It offers several examples of shipyard practices and outlines the direction of the development for digital transformation in this area.

1. INTRODUCTION

The shipbuilding projects require a particular approach to the ship design process, extensive information for manufacturing, and a carefully designed information flow of data between these phases. In an ideal situation, the complete shipbuilding lifecycle solutions would support incremental creation and the use of data from one phase to another. Data refers to the 3D CAD model of a complete vessel and all of its parts: hull, structure, equipment, piping, outfitting, electrical cables, and linked engineering data from results of a simulation, electrical load calculations, as well as engineering bill of materials and production data to facilitate manufacturing, pre-outfitting, and assembly. In reality, this process in the modern shipbuilding environment is highly fragmented regarding stakeholders, work processes, and supporting IT systems. These include organization and network structures, used digital solutions systems and preferences for data formats, collaboration mode, and the scale of digitalization. Existing academic research mainly focuses on the theoretical implications of ship design as a product – highlighting its functional characteristics and partially economic profitability. The practicalities of shipbuilding process design and implications of workflows, management, and information flow design are often guarded as internal information, as they significantly impact shipyards' competitiveness.

This paper looks at information flow between two phases of the shipbuilding lifecycle – design and production - and highlights practical possibilities to gain efficiency by eliminating information flow gaps and placing data in the context.

2. 3D DESIGN IN SHIPBUILDING

The ship design stage focuses on creating the 3D model to comply with a vessel's functional requirements. Often it is assumed that CAD data includes only "a diagram or model of the final product" [1]. Typically, modern shipbuilding CAD software offers users to enrich the 3D data with production information already in the early stages of design. A safety net of settings and pre-defined values surround the designer's actions to assist with regulations of the classification society and prepare the use of design data for production. Multiple locations in the work process ensure that the design complies with the possibilities of production. For example, using settings of bending machine parameters for pipes – curves will automatically get the radius corresponding to bending machine settings, facilitating the correct space allocation for piping and providing embedded production data for later stages. Another example is using material sizes from the stock of steel plates to ensure maximum available sizes and minimization of the cutting. Embedding production data, such as templates for shell plate bending, ensures that the material needed for the templates can be visible in the procurement system, and profile cutting information is according to the automated CNC machinery format. These are just some of the thousands of automatic functions assisting ship designers in creating 3D data and setting the stage for production data output.

This type of embedded data and verification of the design model distinguishes shipbuilding-specific CAD from a general mechanical CAD, as it deals with an extreme amount of data – typically thousands of equipment and millions of piping components and steel parts, and a high degree of interconnections between the parts and topology of relations. It differentiates shipbuilding from a typical manufacturing industry approach where the CAD model produces engineering BOM, later converted to manufacturing BOM and further split according to tasks, teams, or assembly lines for production. The requirement for interconnections and scale of shipbuilding projects calls for significantly more elaborated tools for product and work breakdown structures and planning tools, and at the same time, it makes data transfer between the design and production critically important for shipbuilding.

Besides the primary functional considerations, the main trends shaping the modern ship design by digitalization and sustainability goals include the following directions:

- Increased demand for flexibility and speed in the early design stages to evaluate variants of the design and estimate the economic feasibility of different powering options

- Automation of routine design tasks and high level of embedded rules and settings, based on requirements or best practices, to assist the designer
- Interoperability of data formats to enhance concurrent execution of calculations, design, simulation, and production output

3. MODERN SHIPBUILDING PRODUCTION

The primary goal of the shipbuilding process is to manufacture the vessel according to the defined specifications and deliver it on time and without defects. Therefore the engineering and design process has to provide needed data for the production process and not only focus on fulfilling the design and functional goals. Historically, technical departments were concerned primarily with the vessel function, and production information was developed later by draftsmen in the production departments, such as the steel or piping production department [2]. Information needed from the design stage in the production process in the traditional shipbuilding process was represented by a set of drawings and listings – for ordering needed materials and for instructions on how these materials should be assembled.

This process changes in modern shipbuilding. The impact of productivity requirements, economic returns, and IT systems development change the landscape of shipbuilding production in the following directions:

- Production flow became the basis for the shipyard's layout and served as a main productivity improvement area for modernization in the late decade [3].
- Production strategy, dictated by economic reasoning, disconnected the design and production processes. It might be a "design from stock" or modular engineering process when the design is created before the production site is selected and production requirements are known. It can be an "offshoring" strategy for using a remote construction site for steel construction due to cheaper labor costs and lower requirements for quality while keeping outfitting production at the yard, like in the case of Norwegian shipyards [4]. Initiatives, such as shared production facilities in North Netherlands, require specific production data output to support high volumes of production that will be assembled in micro panels and later used in the construction of different projects and shipyards [5].
- Machinery and automation equipment of production lines - advanced robotics and computational tools applications leaped forward in the last decade resulting in more connected manufacturing and automated assembly lines in shipyard's production processes [6].

Due to these changes, production needs for data cannot be satisfied with a set of 2D documents and material take-offs from the design stage. In most cases, needed data would be adaptable output for the design stage, with a high degree of flexibility in the definition of work breakdown structure and data presentation and a possibility to take into account required machinery output. Besides design data, production needs consolidation with planning and project management to facilitate the whole process. At the methodology level, there are existing models for shipyard production simulation models [7] and the use of discrete event simulation for shipbuilding production modeling [8]. However, these discussions mostly fail to account for the 3D CAD data quality and its influence on the production process.

4. INFORMATION FLOW FROM 3D DESIGN TO PRODUCTION

Schematically, the transfer between design and production phases can be presented in a simplified Figure 1. The primary connections for the production process are with design (A), production methods (B) (such as production strategy, location, subcontracts, machinery, automation level, and similar), and planning and procurement systems (C).

While in previous decades, it was expected that connection from design to production is one way, in modern shipbuilding, these processes are often concurrent and tightly interconnected by information flows. It can be considered from a perspective of the Industry 4.0 paradigm – blurring the borders between design and manufacturing or hybridization of digital-industrial manufacturing. There are visions and expectations for shipbuilding to embrace the opportunities from digitalization [9] and the product lifecycle approach already in daily use in other industries, such as automotive or aerospace [10]. However, shipbuilding requires a specific approach and acknowledgment of the design data's role and impact.

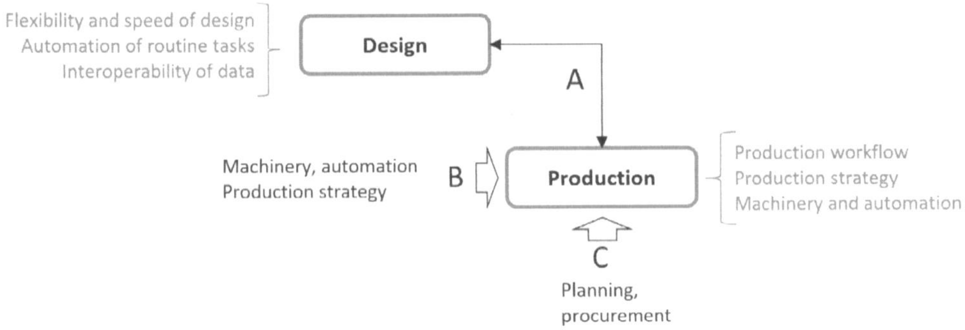

Figure 1: Simplified schematical connection of production process in shipbuilding from an information flow point of view and changes that modern shipbuilding faces.

Increased interconnection between design and production (A) is bi-directional and can be considered as design output adaptivity requirement. Ship design in detailed stages should provide flexible standards output considering machinery and automation requirements and production strategy, achieving efficiency in the production goals also consolidating data with planning and procurement. Both processes are essential and simultaneous, and therefore besides continuous improvement of production data origins in ship design, modern solutions should offer users a platform approach to consolidate data from different sources and provide data in the context of user requests.

4. EXAMPLES OF DATA FLOW GAPS MITIGATION BETWEEN ENGINEERING AND PRODUCTION

The following examples originate from shipyards' experience optimizing their work processes by implementing modern shipbuilding technology, such as Cadmatic ship design applications and the Information Management platform. While end-to-end digitalization is an attractive vision, in most cases, it is impossible to execute a significant overhaul of IT systems at once. Addressing and improving information and data flows between the existing systems provides tangible gains in productivity and a gradual increase in digitalization gains.

4.1 PLACING DATA IN THE CONTEXT OF PRODUCTION TASKS USING COLOR CODING

Access to the information does not guarantee that it can be used in time. For example, work package information might be available in a manufacturing system, but it is disconnected from the 3D model and provides an only textual status updates to production teams. Having the possibility to visualize this data in 3D and providing access to any other data in the required format, such as on a tablet or in VR/AR set, might be a game changer for the production process due to the possibility to visualize the data in digital format and thus enhance interactions. It also opens gates to industrial metaverse applications and provides applied use of 3D digital data residing in physical environments in the production process.

An advanced example from ship production practice concerns the weight of equipment information for installation teams. For the on-site teams, it is critical to have information about the weight of equipment scheduled for the installation before planning activities. Engineering data typically contain weight information, however, often, it is based on preliminary estimations without a specific modification of equipment items, and it might significantly vary from the procurement data. When equipment is ordered, the supplier provides exact information about the dry/wet weight and weight of the packaged item for every position. This information is typically available in the procurement system or ERP system, depending on what is used in the ICT landscape of the yard. This example is one of the typical challenges – information is available but not in the context of a needed task.

A solution for the challenge was found by merging available data from several sources and visualizing it for access to planning on-site activities. The first stage included the identification of the sources of data: the ERP system containing the weight information after the procurement and the 3D engineering model. Design of information flow included access to the ERP data, selection of equipment for which it is required (with weight over 25kg), and comparison criteria for engineering defined weight and data from the ERP system. Once the information flow design was clarified, it became a simple task to set up the information management tool. Results for the planning and installation included a pre-set visualization style in the information management platform with one-click coloured 3D models according to weight data. This way, on-site work can be planned to take access to heavy lifting machines and cranes into account and avoid situations where installation teams have to make unplanned hull cuts or wait for heavy equipment lifting.

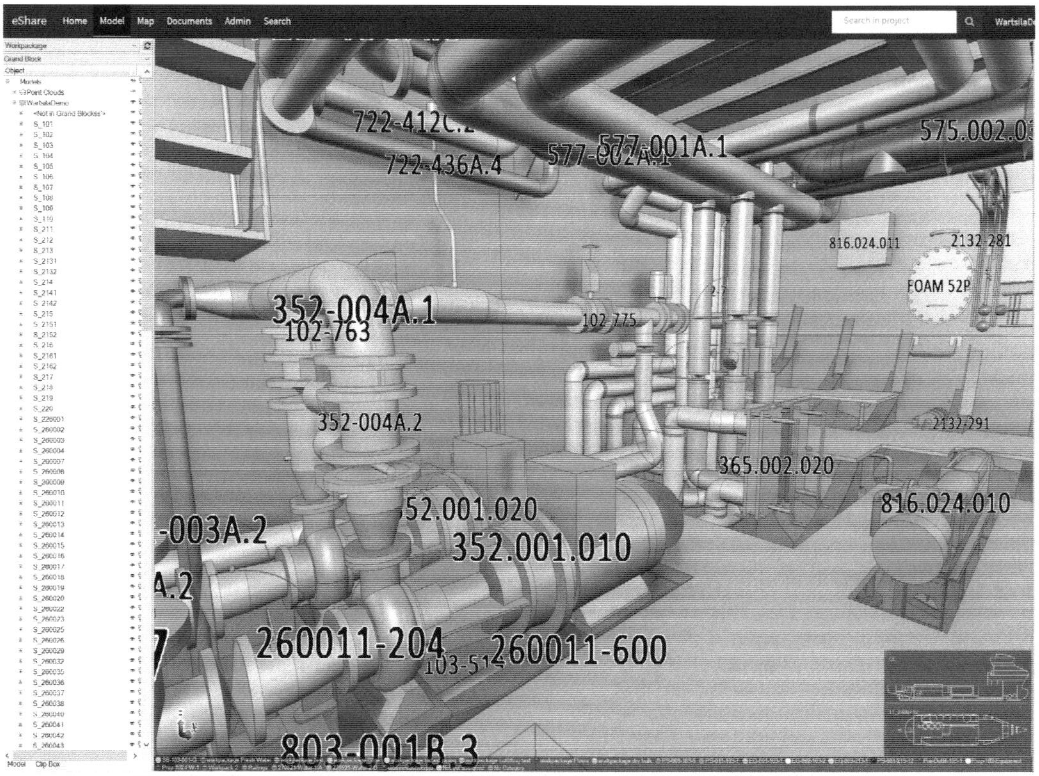

Figure 2: Example of color coding and data visualization for production in Cadmatic eShare.

Achievement is a significantly improved planning process and elimination of waiting times on-site. In this example, a simple identification of information flow and placing data in the context – all part of seizing the power of digital tools, led to measurable improvements in the work process and cost savings.

4.2 PLANNING OF INSTALLATION PACKAGES

A typical project in modern shipbuilding includes 500-1000 piping work packages for installation. The design phase for piping systems should include creating the 3D model with all related data and information for the work breakdown splitting. Depending on the workflow, the output of the design phase might be a basic 3D model, ready for a generation of production data, or a complete package of the documentation, including spool drawings, BOM, and detailed planning information.

Planning of manufacturing and installation for this process based exclusively on engineering data might result in delays and disruptions due to a lack of information about deliveries from pipe shops or suppliers. Including the data from ERP (or another type of system) provides possibilities for merging data: engineering and work break down, and delivery progress from the workshops or subcontractors and placing it in the context of the work process. Engineering and design data for piping is typically based on functional diagrams and include 3D model, component and material information, fittings and armature, joints and fittings, and additional materials, such as bolts, nuts, gaskets, insulation materials, and the like. A piping system is split into spools, and often spools from different systems are packages in a work breakdown structure for work packages installation in certain ship areas.

This example represents a case where consolidating data from engineering and installation planning using innovative technology can save significant time. It is a laborious task in case of only one project, while if a shipyard has several subcontractors delivering pipe packages and tens of simultaneous projects – it becomes essential for production planning and control.

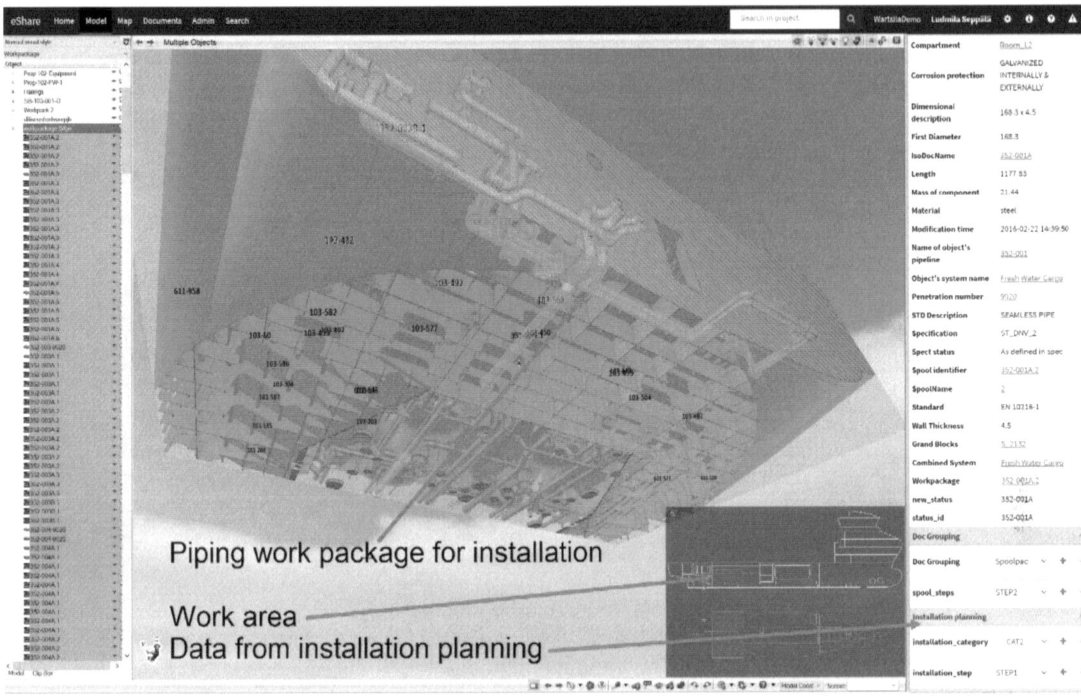

Figure 3: Example of 3D model and installation planning data in the context of work package visualization.

4.3 COMMUNICATION BETWEEN DESIGN AND PRODUCTION IN THE SHIPBUILDING NETWORK

Increasing production efficiency and the overall shipbuilding process requires more communication between the design departments and the shop floor. Providing information "on-demand" and within the context reduces the number of paper drawings and creates an interactive digital environment. It is crucial for the concurrent design and production processes when parts of the design model are released for production while yet might be affected by changes in design.

Having a "single source of truth" data can significantly increase the quality of decisions and reduce the time spent searching and verifying the information. At the same time, providing a context for production execution can increase the production quality and at least partially address the lack of skills or understanding of the overall process by production teams. Adding a link between the design model and nesting status provides a solution to track the hull building process accurately. It lets designers see in an instant what parts have been cut and welded, and the progress of building and inspection is visible, reducing the number of uncertainties and questions.

In the example presented in Figure 4, extending the usability of 3D design data and integrating it with the production tracking systems (such as nesting) enables concurrent processes and provides the needed visual context for project management.

There are many similar examples of the use of integrated information to benefit design-production communication. Powerful IT integration platforms and advanced user interfaces replace "codified" communication based on 2D static documents and open opportunities for a more interactive dynamic data exchange and use. While having an end-to-end digital thread is an important goal, eliminating the gaps while focusing on purposes is a practical strategy for digitalization in shipbuilding.

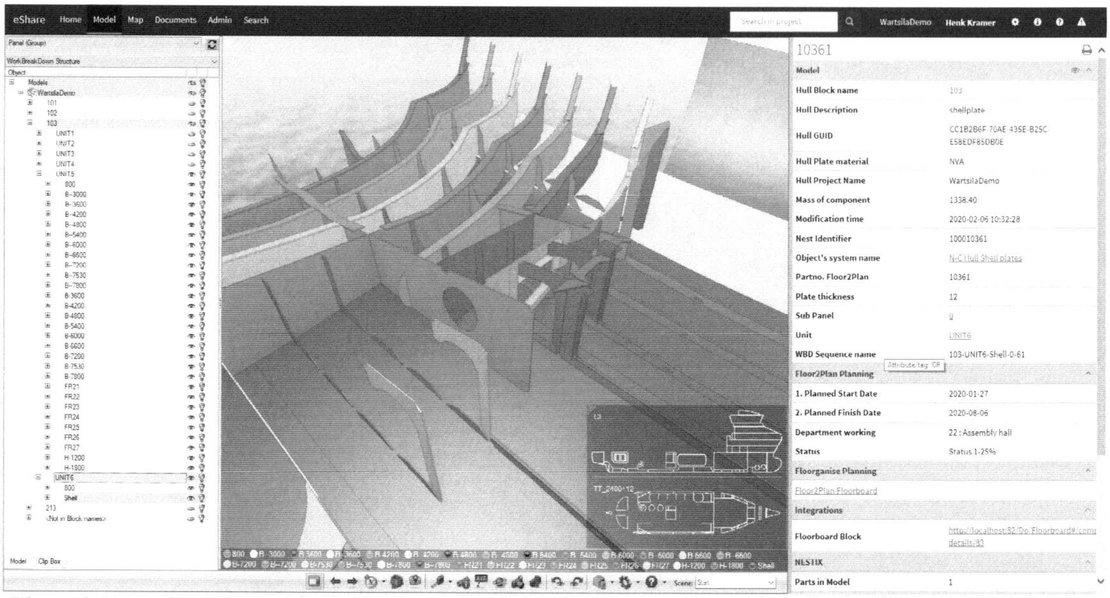

Figure 4: Example of 3D model of hull construction with colouring according to the hull material data and information about the readiness of micro panels for production from the shop floor planning.

6. INFORMATION FLOW: BENEFITS FROM DIGITALIZATION AND FLEXIBILITY FOR WORK PROCESSES

Digitalization is not just the use of digital technologies. It is a way of transforming the work processes, and new possibilities—the vision of seamless information flow and drawingless production [11] are already becoming a reality in modern shipbuilding. Besides robust functionality to support design and production phases, intelligent technology should support data integration with existing systems and allow companies to engage in a step-by-step digitalization process. Addressing the gaps in information flow grounds developing digital solutions following shipbuilders' needs. This approach embraces Industry 4.0 concepts and goes toward the Industry 5.0 concept of developing IT solutions – including sustainable goals and skills in the equation of digitalization [12]. This way, the digital solutions are not only developed in functionality according to possibilities of data handling by IT systems but follow and challenge the work patterns and serve the goals of the overall shipbuilding process.

7. CONCLUSIONS

The growth of complexity in shipbuilding projects and its distinctive differences in design and production processes from other industries call for specific solutions. Besides changes and shifts of focus in each phase, special attention is needed for the information flows between these phases. Business processes, such as offshoring production or pre-outfitting, put additional demands on IT solutions. Modern shipbuilding solutions must provide users with high-level integration possibilities and focus on innovation for use cases to support these data use and information management needs. Examples presented in this paper are based on cases where shipyards gained significant time savings and quality improvements by addressing the gaps in information flows between design and production phases and using the Cadmatic Information Management platform. These examples illustrate how digitalization transformation can be taken in steps as an evolution process, changing the way shipyards work under changes in a business environment and seizing the power IT solutions provide.

8. ACKNOWLEDGEMENTS

Screenshots for the figures in this paper are taken from the Cadmatic Information Management platform using a demo environment project courtesy of Wärtsilä Ship Design Norway AS.
Numerous discussions with Cadmatic personnel, customers, and the shipbuilding industry peers contributed to the ideas presented in the paper.

9. REFERENCES

1. Okubo, Y., Mitsuyuki, T., 'Ship Production Planning Using Shipbuilding System Modelling And Discrete Time Process Simulation', Journal of Marine Science and Engineering, 2022

2. Bruce, G., 'Shipbuilding Management', Springer, 2022.

3. Molland, A., 'The Maritime Engineering Reference Book. A Guide to Ship Design, Construction and Operation', Butterworth-Heinemann, 2008

4. Semini, M. et al, 'Offshoring Strategies in Norwegian Ship Production', Journal of Ship Production and Design, Vol. 34, No. 1, 2018

5. Conoship Inl., 'Robotization for a future-proof shipbuilding industry in the northern Netherlands', 2022

6. Mongado-Estavez, A., et al, 'Towards Automated Welding in Big Shipbuilding Assisted by Programmed Robotic Arm Using a Measuring Arm', Springer, 2017

7. Back, M-G., et al, 'A study for production simulation model generation system based on data model at a shipyard', International Journal of Naval Architecture and Ocean Engineering, Vol. 8, Iss. 5, 2016

8. Goo, B., Chung, H., Han, S., 'Layered discrete event system specification for a ship production scheduling model', Simulation Modelling Practice and Theory, Vol. 96, Elsevier, 2019

9. Munoz, J., Perez-Fernandez, R., 'Adopting Industry 4.0 Technologies in Shipbuilding Through CAD Systems', Trans RINA, Vol. 163, 2021

10. Grau, M., etc al, 'Enabling the Digital Thread for Shipbuilding and Shipping', ICCAS, 2019

11. Seppälä, L.' Drawingless Production in Digital Data-Driven Shipbuilding', COMPIT, 2019

12. European Commission, 'Industry 5.0: A Transformative Vision for Europe', European Commission, 2021

10. AUTHORS BIOGRAPHY

Ludmila Seppälä holds the position of Business Development Director, Marine Industry at CADMATIC. She is responsible for the long-term development of design, information, and data management solutions in the global shipbuilding and marine industry. Her previous experience includes design projects, production follow-up in shipyards, and numerous implementations of CAD/CAM systems in ship design companies and shipyards around the globe.

ABOUT CREATING OPEN-SOURCE 3D MODULAR VIEWER IN PYTHON

G Sikic, LINA et al, Croatia
P Prebeg, UNIZAG FSB, Croatia

SUMMARY

During the development of the ShipExplorer suit of the programs, great attention was given to building the application, modularization, and code organization. During that period, successful cooperation with UNIZAG FSB was established, resulting in a new requirement that students with no real experience in SW development create new modules that will solve some specific problems.

After the ShipExplorer project was put on hold, it was decided to start a new project with similar objectives, that will fulfil UNIZAG FSB requirements from the start, and help save and use the huge knowledge base gained during the development of ShipExplorer. As a result, project, "3D Modular Viewer" was created.

In this paper, experience designing and implementing 3DM viewer will be presented, as well as problems that had to be solved. The choice of the programming language will be discussed, together with additional constraints resulting from that decision. Design decisions and implementation details will be reviewed, and the results will be compared with the original program, ShipExplorer.

1. INTRODUCTION

During the development of the ShipExplorer[1][2] suite, there was close cooperation with UNIZAG FSB, and its development was also influenced by their requirements. Also, at the university, it was used as a primary tool for introduction of various shipbuilding concepts, as well as developing various additions designed to connect external calculation to ShipExplorer in order to prepare data and present results.

After ShipExplorer project was shut down, the need emerged to create some similar application as a replacement, and independently, a wish to save knowledge and experience gained during development of ShipExplorer from decaying and vanishing. Therefore, project D3V [3] was created.

One of the new goals was to introduce students into the development and to create a primary tool that can fulfil needs of a shipbuilding department.

2. DESIGN CONSIDERATIONS

The design of a d3v program is heavily influenced by ShipExplorer, and a starting point was its analysis. Therefore, the design of an original program was analyzed, to identify design decisions to be used in a new program. The two design characteristics that were adopted are modular design and data-centric design.

2.1 MODULAR DESIGN

The modular design means that the complete functionality of the program was divided into a set of modules that implement part of the functionality of the program. There is a core of the program that provides fundamental services, and the rest of the functionality is implemented as modules. It should be emphasized that the program should work without any module, with limited functionality. Also, design and complete source code and build system were organized to produce a series of python modules or python packages, that were loaded during a program initialization.

There are three types of modules:

- commands, general type modules that ad functionality of any kind.
- painters, specialized modules used for rendering
- io-handlers, specialized modules used for importing and exporting data.

2.1 (a) Command modules

Commands are general purpose modules that introduce new functionality of any kind. This is the preferred way to add new functionality, and after creating a new command, there are no other constraints it needs to comply with.

2.1 (b) Painter modules

Painters are specialized modules that are used to provide visual information about the loaded model. That information can be of any kind, like 3D model, selected part of the model, up to the specific legends and various 2D overlays over the 3D model. Painters must be "registered" with an OpenGL canvas widget and will be automatically called during a window refresh. There are 3 types of painters:

- background painters that "paint" the background. The default is simple, black background.
- Scene painters that paint geometry in 3D or various elements in 2D, like legends or HUD. Default is so called "basic painter" that is based on a work of one of the students of UNIZAG FSB.
- Post processing painters that are manipulating various OpenGL buffers, after the scene is drawn.

2.1 (c) IO Handler modules

IO Handlers represent a specialized type of modules that are used to import and export data to some foreign format. Another purpose of an io-handlers is to create geometry with various meta-data that correspond to the data within the foreign format.IO handlers must be provide information about which foreign formats are supported for import and export. After registration, they are called on demand to import /export data.

2.2 DATA-CENTRIC DESIGN

"Data-centric" design means that complete organisation of a program is based on a hierarchy of user defined classes, accompanied by the hierarchy of operations over those classes.

When dealing with a 3D geometry, any object is associated with some kind of "meta-data", that additionally describe the geometry. D3V system is organized the other way around, the "meta-data" are "user-data" that holds a geometry as one of the attributes. Additionally, these are "smart data", that contain a set of operations.

During a creation of a new geometry, a new type of "user data" can also be defined, with a different and extended set of operations. It should be emphasized that the complete process is controlled by the loadable module, and not by the core program.

2.3 VERSION CONTROL SYSTEM SELECTION

Source code version-control-system [4] is one of the fundamental tools in software engineering. It allows managing life cycle of every single source file, resource and literally anything the program. It also formalises team work on a software, by enforcing controlled read/write access to the managed source files. Loosely speaking, version control system can be considered to the PDM, and specific to the software development. For example, on the figure 1, comparison of two versions of the same file is shown.

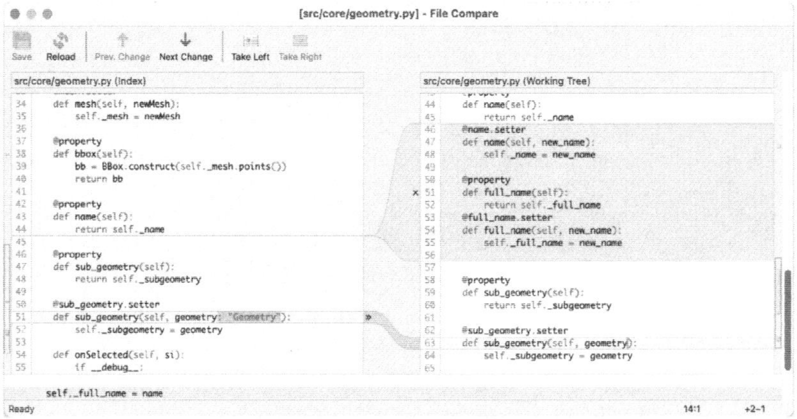

Figure 1: Comparison of two versions of the same source file.

One of the greater concerns was selection of a version control system. The development of the original system, ShipExplorer, was based on a "subversion"[5]. It is a version control system proven in the practise, and easy to use. However, git[6][7], another type of vcs is almost exclusively used. It is very capable, modern vcs, but it also has a very steep learning curve. It is also expected that someone in software development knows how to use it.

Since the one of the purposes of the program is to provide learning platform for the students, it was decided to use git as a system for sources management.

3. IMPLEMENTATION

3.1 CHOISE OF PROGRAMMING LANGUAGE

The choice of programming language was done in a context of a target developer group, i.e. students of UNIZAG FSB. Programming languages that were considered were C++ and Python. Java, JavaScript, C# were not considered.

Java, Java Script, C# et al were screened out immediately as non-suitable for this king of applications, i.e. desktop technical applications using 3D graphics, that should be extendable by beginners.

The original system ShipExplorer was build using C++ programming language. With a carefully written code, the net result was a very quick execution, with a support for multiple operating systems. However, this had an inherited problem, that the C++ is very demanding programming language, that requires quite an experience in order to achieve good results. At a time, it was not a problem because all members of a development team were proficient C++ developers. However, in a context of a beginners learning how to program C++ is no mere appealing idea.

Python is a language that gradually becomes a language of choice for technical programming. There are many "ready to use libraries" that help development. It is also used at UNIZAG FSB for "introduction to programming courses"; therefore, it was an ideal compromise and eventually it was decided to choose Python as a principal language for this project.

3.2 3RD PARTY LIBRARIES USED IN D3V

The following external libraries are used in the project:

- OpenGL[8], used for 3D model presentation,
- Qt[9], general GUI library,
- OpenMesh[10], low level library for handling triangular and quad meshes.
- NumPy[11], python libraries for numerical calculations,

3.2 (a) OpenGL

OpenGL is a well-established general-purpose library for graphical programming in 3D as well as 2D. In the original program OpenGL was also used, but much older version, V2.1. For D3V, it was decided use modern version, v4.1. The version used previously is easier to use, there is smaller amount of the boiler-plate code, but the new version is much more capable. Luckily, boiler-plate code is handled by another library Qt, so verbosity on new OpenGL version didn't create any problems.

3.2 (b) Qt library

Qt is a set of libraries for a cross-platform creation of a user interfaces. Complete visual appearance of the program is based on these libraries. It should be noted that Qt libraries solve many more problems that just User Interface. It also implements signals and slots system to share information that was adopted by the d3v and used extensively throughout the program. Also, modern OpenGL boiler-plate code is wrapped within the OpenGL related classes, making it much easier to work with.

The original system, ShipExplorer, used another library for GUI, wxWidgets[12], so this library was also considered. Although different in design, for a desktop application it provides capabilities like the Qt libraries. However, the support for the modern OpenGL is much neater in Qt, and it was chosen as a GUI library.

3.2 (c) OpenMesh library

Geometry presentation in d3v program is based on a triangulated mesh. Therefore, it was required to use supporting library that will allow mesh creation and manipulation. Unlike original system where inhouse developed library it was decided to use external library for this purpose. The main reason was the experience gained in the original program. Although the created library had a small footprint and very high performance, it turned out that much of an effort was required to maintain it. Also, maintaining it required a quite theoretical knowledge about the subject, thus limiting the number of maintainers. At the end, it was decided to use well-maintained third-party library with Python bindings, OpenMesh.

3.2 (d) NumPy

NumPy is a Python specific library numerical analysis library with a support for a multidimensional vectors and matrices. It is also used in Python course at UNIZAG FSB. Since the OpenMesh library heavily depends on NumPy, it was a chosen as a library for a numerical computation. In the early stages of the program

3.3 VISUAL APEARANCE

The visual appearance of the basic program is minimal, as shown on a following figure:

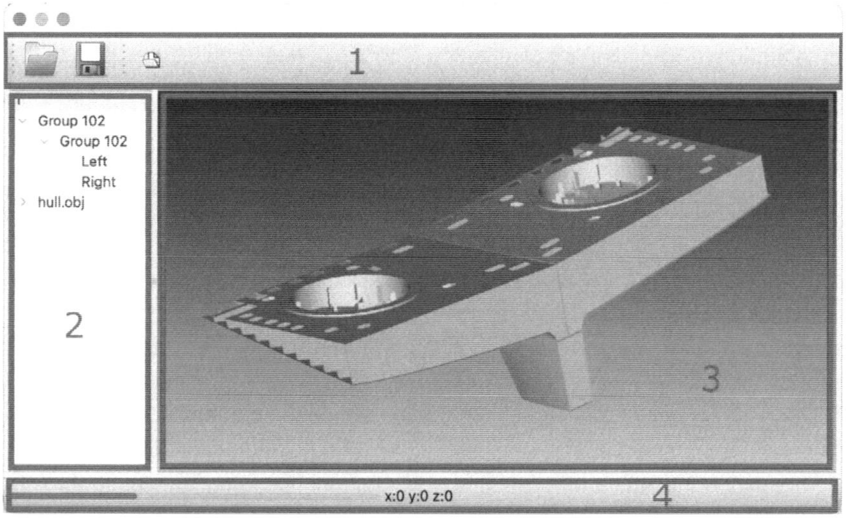

Figure 2: Four main areas of a program: Toolbar, Hierarchy tree, 3D view and status bar

There are five main areas of the gui:

- Menu bar (not visible here, since it is on MacOS, where the menu bar is detached part of the GUI)
- Tool bar (marked 1) that reflects actions from the menu bar. After initialization is done all actions found in menu bar are presented in the tool bar as well. Thus, the toolbar is a "shortcut" to the functionality contained within menu bar
- Hierarchy tree (marked 2). Presently, the hierarchy is only shown there; eventually, it will become complete hierarchy editor.
- Area for the presentation and manipulation of a 3D model (marked 3). This is the main area of the program.
- Status bar, that represents various info. Presently, it shows a progress of various operations, and coordinates under the cursor.

It must be emphasized that the visual appearance is not hard coded but controlled by the loadable modules. For example, on a following figure there is a GUI that is used for various hydrostatic calculations[13]:

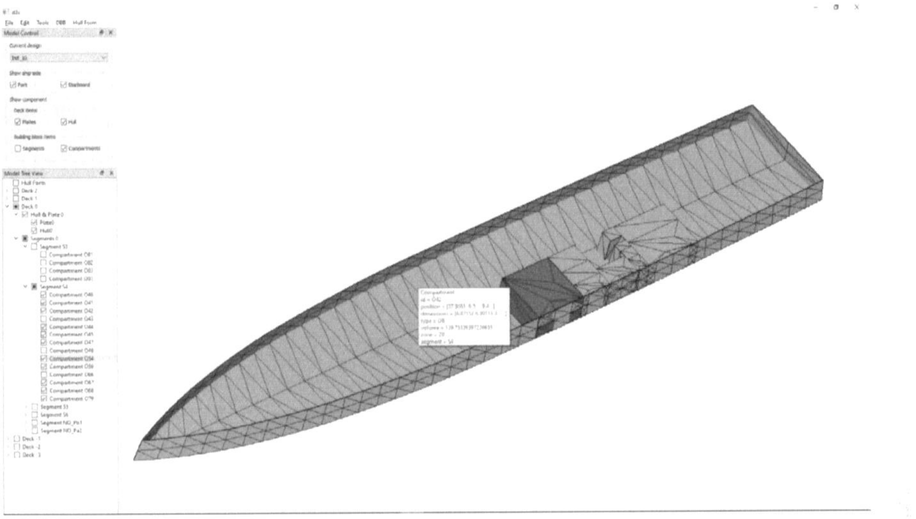

Figure 3: example of using the d3v program as a base for compartmenization and hydrostatic calculations

Choosing different painters can also have a dramatic influence on the appearance. On the following figure can be seen difference created by different painter usage: on the left-hand side there is a basic presentation with default painters, and on the right-hand side a new background painter is used, and different scene painter that uses colour code to represent different types of design data. Also, note that the new scene painter uses perspective projection, instead of orthogonal, that is used by basic scene painter.

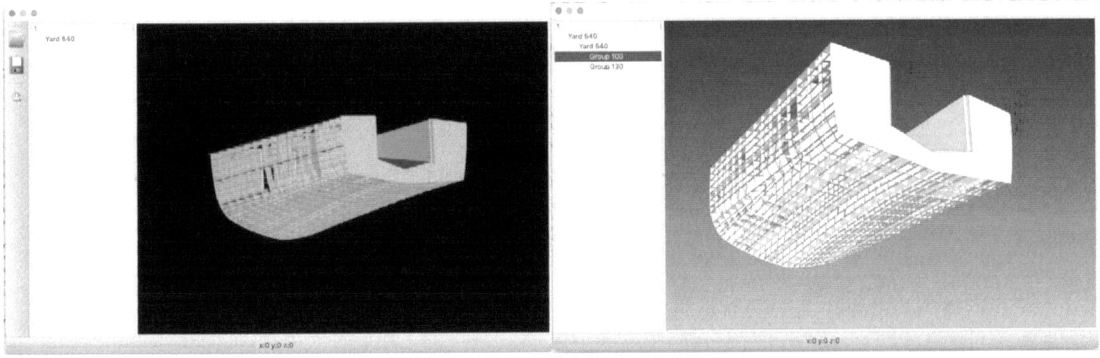

Figure 4: Difference between basic presentation versus usage of additional painters; basic painting on a left-hand side, and enhanced painting on a right-hand side.

3.4 EXTENSIBILITY IMPLEMENTATION

One of the major design requirements was to allow loading modules that were not known by the time the core program was created. Implementing it in Python would require implementing programmatic modules import instead of having "import something" in the source code. Luckily, Python programming language supports such a requirement using "importlib" package.

The process of initialization and modules loading is as follows:

- The modules hierarchy is as follows

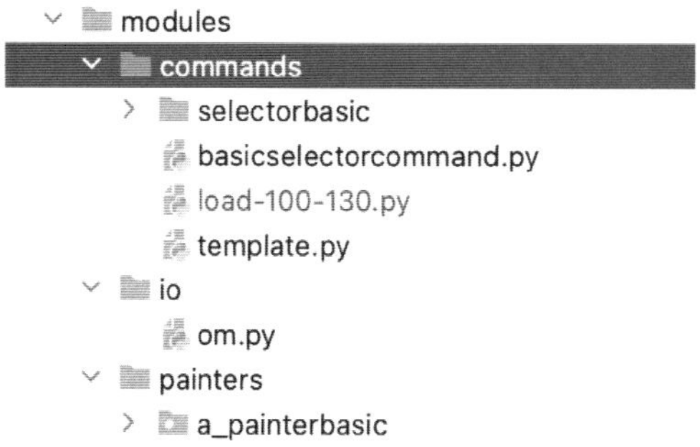

Figure 5: Organisation of a folder with modules

  - **modules** are a root modules folder. The name of the folder may be different.
  - **commands** folder is a folder where command-modules should be placed. In the figure above, there are four commands: selectorbasic, basicselectorcommand, load-100-130 and template.
  - **io** folder should contain input output handlers
  - **painters** folder contains painter modules.
- There may be multiple instances of such "modules" directories, referenced by command line or initialization file.
- All directories in all "modules" directories are appended to the PYTHONPATH variable.
- All python files or packages found within root folder (module in this example) or commands, io or painters subfolders will be loaded and parsed.

Using this approach results in very flexible extension system. Furthermore, modules that are being developed can be part that is completely unrelated to the core application.

4. PROBLEMS DURING DEVELOPMENT

Developing 3D modular viewer in Python is ongoing process and it is by no means finished. Since its design was based on another successful application, many possible problems were avoided in advance. However, there were some scenarios that were not anticipated:

- **Slow development**: being open source put the project at the end of priority list. Checking a frequency of the commits on a GitHub, it can clearly be seen that the development was not continual, but in the waves, jumps. Strictly speaking, this does not influence the net result, but is just "annoyance".
- **Code organisation**: this is a result of a different approaches in d3v development. Students that chose to participate in the project were primarily engineers and not software developers. This fact resulted in numerous situations where some problem was solved for some particular case, instead of solving it on a general level, where it should have been solved. Also, there were opposite situations, where the core of the program was unnecessary modified to solve some local problem.
- **Git and Github[14][15] adoption**: it turned out that many of the students involved in the project were not the fonds of git. Their development was not accompanied by the git commits, nor they followed git/github recommended practise. They would simply zip and download project snapshot, and continue working locally, without any git involvement. This practise created two problems: firstly, any change in core of the program was very difficult to incorporate into their work, and secondly, their work was not sufficiently acknowledged in a git repository. Temporarily, the list of authors is maintained, but it does not show amount of someone's work.
- **The lack of the documentation**: soon after starting the project, it became clear that the quality documentation about the project design is missing. Although everything is quickly explained by other means, the lack of documentation is still a handicap. This problem will be addressed as soon as possible.

5. CONCLUSIONS

The origins of D3v were presented, its design was discussed, and implementation explained. It is a work in a progress and now, the hardest part is almost over, setting the good foundation to the program. Although in the program is still in the early phase of the development, it has proven as a versatile tool that is easy to use and extend.

The next tasks will be to implement low level modules to simplify every day's tasks, extend importing and exporting from the program. Also, the 3D presentation will be upgraded, to include colours, materials and textures.

6. REFERENCES

1. i i , i i i and e o i , Development of Interactive Tool for Paint Quantity Estimation of Ship Structures. RINA ICCAS 2011 proceedings

2. R. en i , M. i ano i , . i i , . i i i , MULTI-PLATFORM COLLABORATION TOOL FOR SHIP DESIGN AND PRODUCTION, RINA ICCAS 2013 proceedings.

3. Project d3v home page at GitHub https://github.com/linaetal-fsb/d3v

4. Introduction to the version control https://en.wikipedia.org/wiki/Version_control

5. Explanation of SubVersion version control system https://en.wikipedia.org/wiki/Apache_Subversion

6. Git home page https://git-scm.com/

7. Introduction to the git on Wikipedia https://en.wikipedia.org/wiki/Git

8. OpenGL libraries home page https://opengl.org/

9. QT libraries home page https://qt.io

10. OpenMesh home page https://www.graphics.rwth-aachen.de/software/openmesh/

11. NumPy libraries home page https://numpy.org

12. WxWidgets home page https://wxwidgets.org

13. P. Prebeg, I. a i o i , . uni , . i i , . J. Erhardt, T. a o i , . e , Application of the open source program d3v for visualization and determination of the characteristics of an analytically given form, The 25th Symposium on Theory and Practice of Shipbuilding, 2022 proceedings

14. GitHub home page https://github.com

15. Introduction to GitHub https://en.wikipedia.org/wiki/GitHub

7. AUTHORS BIOGRAPHY

G is founder of "Lina et al", the software company. He was involved in 3D visualisation, OpenGL, VR, AR, mathematical modelling, software design and implementation of distributed simulations, real time visual flight simulations, design and implementation of missile control systems, terrain data visualization manipulation, and vibration calculations and measurement.

Pero Prebeg, Associate Professor at Department of Naval Architecture and Offshore Engineering, Faculty of Mechanical Engineering and Naval Architecture, University of Zagreb. Teaches courses in field of ship structural analysis and design, multicriteria design and optimization, aircraft design, aircraft graphics and engineering programming basics. He was involved in several software development projects in a field of finite element method, ship structural design and multicriteria design and optimization.

ABOUT USING THE GAME ENGINES IN SHIPBUILDING, SOFTWARE DEVELOPER′ S PERSPECTIVE

G Sikic, LINA et al, Croatia

SUMMARY

During the development of the ShipExplorer[1][2] suit of the programs, every single detail was carefully designed and implemented using low level libraries like OpenGL. The result was very sophisticated, quality program that delivered top performance with wonderful user experience. However, development of new features required theoretical and practical knowledge about many different disciplines starting from linear algebra, knowledge of OpenGL internals, and proficiency in C++ programming language. The additional problem was that every new hardware target, like mobile phones, or various HMD models had to be addressed particularly, without any code sharing.

At the same time, new class of programs started appearing, so-called "Game Engine"[3] programs. This class of programs promised maximum results with a minimum effort. Screenshots with spectacular graphics and pre-generated movies were available, showing the output of these programs. Also, "out of the box" support for all major platforms and various hardware was claimed. Taken together, it was enough to rise a question if approach of using low level libraries was a correct approach in the presence of the tools like "Unreal Engine" or "Unity".

In this paper, author's experience in working and extending Unreal Engine, will be presented. Problems using C++ programming language and internal "Blueprint" visual scripting in Unreal Engine will be analyzed. Scenario of usage as a visualization and annotation tool will be evaluated and compared to the possible alternative approaches.

1. INTRODUCTION

Last couple of years the distinction between CAD, simulations and games is becoming very blurry. Many techniques previously used in real time visual simulations are used in previously pure CAD environment. This opens a way to introduce a completely new approach to solving engineering problems.

Today's games that include simulations are very sophisticated software, that are better than the professional simulations years ago. With a progress of the computing power also progressed a software component. Many problems that previously had to be solved manually; these days can be solved automatically using existing software. Game engines are the kind of the programs that solve many of very hard problems. Looking at the visual output of such programs, and considering the time spent to create such scenes it seems like a "free launch". However, creating a scene is not enough; elements of scene need to have some specific behaviour, which usually means creating extensions to the original program, i.e. programming, either in C++ or in Blueprints.

In this article extensions of Unreal Engine[4][5] will be presented, their pros and cons, and some scenarios of Unreal Engine possible usage in shipbuilding.

2. PROGRAMMING FOR UNREAL ENGINE

There are multiple approaches to extend Unreal engine[6], C++ and Blueprints being the most known and common. It should be noted that the programming in UE might mean "programming the elements of the game" and independently, extending the capabilities of the editor used to create and configure the game. Strictly speaking, there are more ways to develop Unreal Engine, game, and editor, but C++ and Blueprints are the most common. Scripting in Python will also be presented, because it is a powerful approach to automate various tasks in Editor.

2.1 PROGRAMMING USING C++

C++ programming language is one of the fundamental ways to extend Unreal Engine. Besides Unreal Engine itself it requires a separate installation of a C++ compiler and a C++ development environment. On Windows, Visual Studio is supported, and on MacOS XCode.

After creating a C++ based UE project, all Visual studio specific files are created and controlled by Unreal Engine. From that point on, most of the development can be done in Visual Studio, using C++. However, some parts of the game still must be implemented in Blueprints. The one example of such usage would be data transfer between various user created plugins, that are mutually independent.

2.1 (a) Writing the C++ source code for the Unreal Engine

Unreal Engine has a complete very elaborate SDK to be used for a C++ programming[7]. It contains everything that might be needed, starting from various mathematical functions, smart pointers, up to the higher abstraction constructs that become a standard in C++ quite recently. This also creates a problem since UE SDK overlaps with API provided by C++ Standard Library in many areas. Unreal Engine developers' documentation mentions which parts of the standard library may be used and which parts may not. In theory this essentially means that programming for Unreal Engine requires learning completely new specific API, instead of using proven libraries. However, in practise, with a little bit of care it is possible to use all features from C++ standard libraries without any collisions with UE API, and without any problems. The only constraint that was necessary, was to use whichever approach is needed internally, and use UE approach for the parts that were visible by the Unreal Engine.

Unreal Engine SDK also enforces strict naming of the types[8]. In order to be recognized by the UE, all type names must start with strictly defined letters. For example, all classes derived from the Actor class must be prepended by letter `A`, and all enumerations must be prepended by letter 'E'. The reason for such requirements is the connection to the Unreal Engine itself.

The connection to the UE is maintained using reflection macros[9] that annotate the source code, coupled with a Unreal Hader Tool[10] application. One of the nice tricks that this requirement makes possible is to make class variables accessible from the UE Editor as parameters. For example, look at the following code snippet:

```
UCLASS(config=Input, defaultconfig, meta=(DisplayName="GRPC Properties"))
class GRPC_API UGrpcProperties final : public UDeveloperSettings
{
  GENERATED_UCLASS_BODY()
public:
  UPROPERTY(Config, EditAnywhere, Category=GRPC)
  FName ControlBoardServerAddress;

  UPROPERTY(Config, EditAnywhere, Category=GRPC)
  FName GameServerAddress;

  UPROPERTY(Config, EditAnywhere, Category=GRPC)
  int32 SimulatorId;

  UPROPERTY(Config, EditAnywhere, Category=GRPC)
  FText SimulatorName;

  UPROPERTY(Config, EditAnywhere, Category=GRPC)
  int64 SystemStateVectorTimerRate;

  UPROPERTY(Config, EditAnywhere, Category=GRPC)
  int64 SystemStateVectorTimerInitialWait;
};
```

It will result in the following GUI elements available in Project settings:

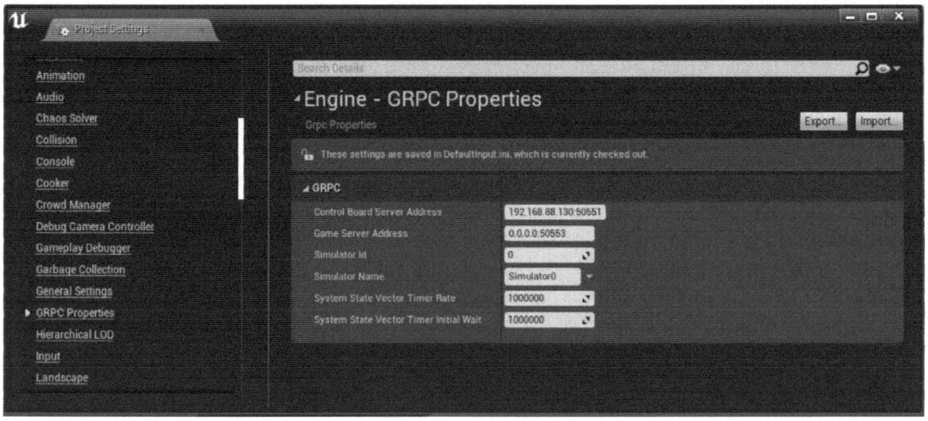

Figure 1: Part of the Project Properties GUI, that is created from the code snipped above.

Note how the names of the variables, and elements of enumeration are translated into the GUI elements. It is done by UE internally, as a part of a build process. Generally, if anything needs to be visible to the Unreal Engine, it will need to be decorated by a UObject reflection macros. In the code snippet above these macros are UCLASS and UPROPERTY.

2.1 (b) Project build system

Unreal Engine has its own in-house developed build system. Building consists of running "UnrealBuildTool" (UBT)[10]. Prior to actual compilation, Unreal Build Tool first executes "UnrealHeaderTool"[11], that parses sources and creates new generated source files. After that, dependency graph is constructed, and actual compile process started.

It should be noted that "behind the scenes" Visual Studio (or equivalent) IDE is being used as very powerful editor only, without direct control over a build process. For example, source files that will included into the build process are chosen by scanning the project folder hierarchy on the hard disk at the moment the build is initiated. The source files presented in a Visual Studio are completely irrelevant, and that information is not used for the build process.

It should also be emphasized that there are no build options in a classical sense; all build options are defined using C# sources. Every "module" in the "game" has its "<game_name>.target.cs" file that defies build options. The same applies for the plugins. Every plugin is accompanied by a "<plugin_name>.build.cs" C# source that controls the build of the plugin. It basically means that one need to know basic syntax of C# in order to change any build option, and to learn very deep internals of yet another building tool.

On the other hand, UnrealBuildTool is very versatile utility and after one becomes familiar with its requirements and shortcomings, there will be no other problems. Unfortunately, it has the same problem as the rest of Unreal Engine: the lack of proper developers' documentation. The existing documentation is perfect as a reference, when someone knows what to search for, and where, but not for an initial introduction nor learning about Unreal Engine.

2.2 PROGRAMMING USING BLUEPRINTS

Blueprints[12] are natural way of extending Unreal Engine. It is a graph-based scripting where nodes in the graph represent where the data are transformed, and the connections between the nodes represent data. In the following figure can be seen the Blueprint function that calculates the length of a vector. It is quite straightforward, but it should be stated that it is very simple function.

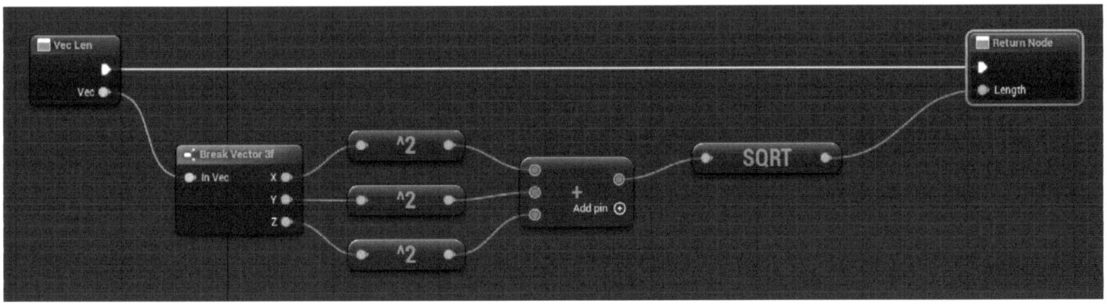

Figure 2: Blueprint function that calculates the length of a 3D vector.

Blueprints can quickly grow, as shown on a following figure; this is a snapshot of a function that fetches data from the simulation subsystem and presents is as a graph in real-time.

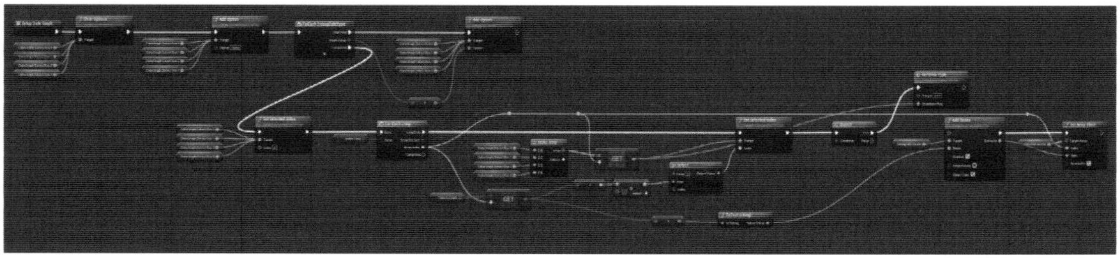

Figure 3: Blueprint function from a real world project.

The nice thing about Blueprints is that it will not allow connections between the nodes if the data flow between them iz incompatible. The problem will be spotted immediately, in the process of adding or connecting the nodes. Also, additional errors are checked for during compilation.

2.2 (a) Comparing Blueprints and C++

There are couple of reasons for Blueprints usage:

- Control flow of the program is much easier to follow in blueprints, compared to C++ implementation,
- Editing and adding new features is much quicker compared to C++
- Complete process of adding new features is faster if Blueprints are used.

Blueprints present very appealing approach to add scripting to the UE, but there are also some caveats.

- Blueprints are slower than equivalent C++ code. In their original form Blueprints can be up to 400 times slower than C++ implementation. There is also a possibility to "nativise" Blueprints, which is a process to create a C++ code from the blueprints, and then compile that code. In nativized form, execution of the Blueprints is approximately up to 2 times slower than equivalent C++ implementation. In the practise, this is usually not a problem if the usage of C++ and Blueprints is well balanced. For example, C++ code might be used in time critical parts of a simulation, and in the other areas, Blueprint implementation may be used.
- Blueprints are binary files, which makes it difficult to handle them as ordinary source files in Version Control System. For example, it is not possible to check differences between two versions of a Blueprint class using VCS tools only. This fact also makes it difficult for a more than one developer to work on one Blueprint. This problem is partly solved by plugins in UE that assume part of the functionality of a version control system.

2.2 (b) Interoperability between C++ and Blueprints

It is clear that the best approach in UE scripting is to use C++ in some areas of the simulation, usually time critical parts, and to use Blueprints it the other areas[13]. This approach opens yet another problem to solve, i.e., interoperability between C++ and Blueprints. This boils into two questions:

- how to transfer data between C++ and Blueprints,
- how to execute functionality contained in C++ from Blueprints and vice versa.

The key points here are UObject reflection macros in C++ code[14], and Unreal Header Tool (UHD). Note that the UHD is employed automatically as a part of a build process. Practically, it is enough to annotate C++ code correctly, and it will be visible from Blueprints. The same approach applies for the classes, member functions and data.

For example, the following C++ code snippet:

```
UCLASS()
class UVRMInputBlueprint : public UBlueprintFunctionLibrary
{
   GENERATED_UCLASS_BODY();

   UFUNCTION(BlueprintCallable, Category = "Input|VRM")
   static void UpdateLoopbackPedalsRPM(float deltaRPM);

   /** 1 --> constant RPM, 2 --> constant speed */
   UFUNCTION(BlueprintCallable, Category = "Input|VRM")
   static int GetLoopbackTorqueGenerationStrategy();

   UFUNCTION(BlueprintCallable, Category = "Input|VRM")
   static void UpdateLoopbackDesiredSpeed(float deltaV);
};
```

Will result in following functionality in Blueprints.

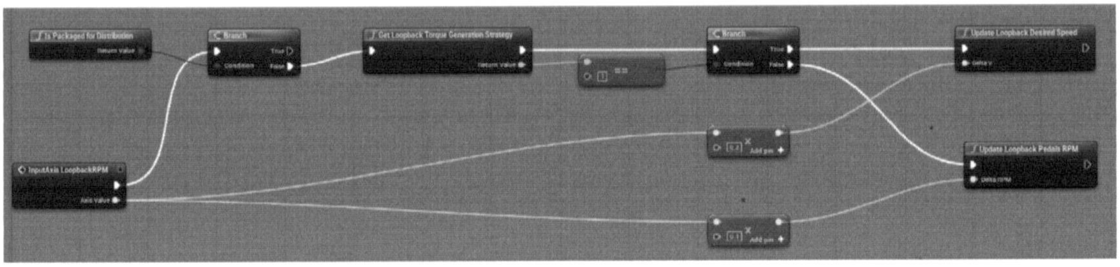

Figure 4: Usage of C++ functions in Bluperints; blue nodes and the red node are functions declared and defined in C++.

Note the functions UpdateLoopbackPedalsRPM, GetLoopbackTorqueGenerationStrategy and UpdateLoopbackDesiredSpeed. These functions are declared and defined in C++ and used s nodes in Blueprints, as shown on the figure XXX. This is accomplished by using UObject reflection macros UCLASS and UFUNCTION.

2.3 PROGRAMMING USING PYTHON

It is a little-known fact that the Unreal Engine can be programmed for using Python programming language[15], version 3. In fact, the Unreal Engine comes with Python embedded into the program, so there is no need to install Python separately. It must be noted that is a "Editor only" scripting. On the other hand, using python all elements of editor may be accessed, resulting in very powerful approach for scripting.

Unlike C++ and Blueprint scripting who's staring tightly defined, python scripts can be started numerous ways:

- from the Tools menu,
- from the console, using py command
- from the command line using `-ExecutePythonScript=script_name`
- from the command line using `-run=pythonscript -script=script_name`
- using startup scripts in project settings.

Starting python scripts from the command line is very appealing since it can work without presenting any GUI elements (i.e. headless, batch mode), that is very convenient if UE is used in every day's tasks, like importing and updating CAD model data on a daily basis.

The following code will create and animate progress bar within Unreal Editor (adopted from the UE documentation):

```
import time
import unreal                    # Import Unreal package
```

```
total_frames = 100
label = "Execiting the slow task"
with unreal.ScopedSlowTask(total_frames, label) as slow_task:
    slow_task.make_dialog(True)          # Makes the dialog visible, if it isn't already
    for i in range(total_frames):
        if slow_task.should_cancel():      # True if the user has pressed Cancel in the UI
            break
        slow_task.enter_progress_frame(1)   # Advance progress by one frame.
        time.sleep(0.1)                  #"execute" part of the slow task
```

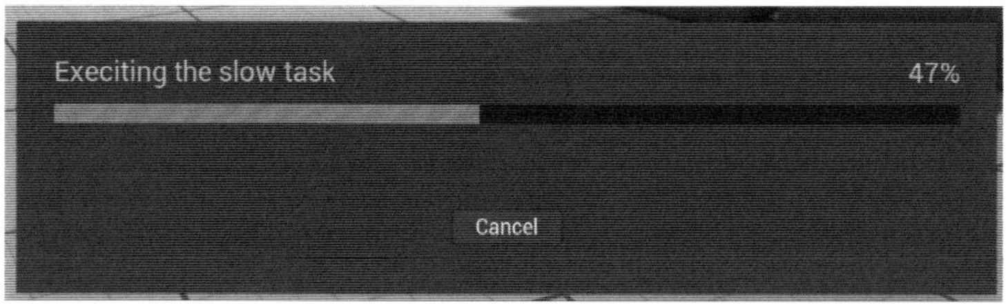

Figure 5: Result of execution the python script from above

3. USING UNREAL ENGINE IN SHIPBUILDING

Although The Unreal Engine is created for the completely different purpose, there are some scenarios where it might be used in shipbuilding. Besides creating eye catching visual presentations and animations of a future ship, one of the areas where UE might be used to the maximum is a creation of a digital twin. It can be used in daily data preparation for VR and AR examination.

All mentioned areas have one inherent problem to be solved, and it is a transfer of the data from a CAD system to the Unreal Engine. Ideally, the data transfer process should be automatic, so it can be run on a daily basis. UE supports similar requirements via its set of plugins called DataSmith[16]. It is a set of libraries designed to import complete scene from the foreign application, and not just a geometry. Also, a great attention is given to the process of reimporting the scenes, i.e., to reuse parts of the scene that has not been changes, and to minimize workload otherwise.

3.1 SOME SCENARIOS OF USING UNREAL ENGINE IN SHIPBUILDING

3.1 (a) Daily presentation of a 3Dship model.

It is hard to compete to the specialized program solutions in this area, like ShipExplorer. The quality of the presentation that UE achieves is not used, and other requirements like tight connection to some kind business-oriented software is missing.

However, if Virtual Reality presentation is needed than Unreal Engine can be used. In that case, UE is compatible with a requirement of presentation, and its support for various hardware is also used. One of the most important requirements is automatic data import. Regarding the scripting, this scenario would require usage of Blueptrints, ot Python based scripting. The following figure shows a UE-based simulation running after automatic import using DataSmith. There are no changes applied to the model after the import.

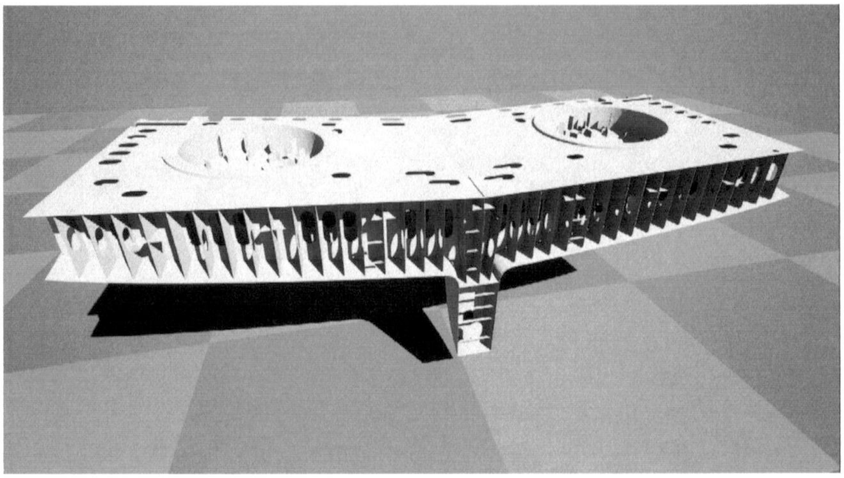

Figure 6: UE presentation of a ship's geometry after importing using DataSmith. Note that no changes were made after import.

3.1 (b) Creation of Digital Twin

Digital Twin creation would be very suitable task for Unreal Engine. Besides importing data from the CAD, this would require adding specific behaviour to the elements of imported model, and this task is identical to the implementation of a "game play". Regarding scripting, implementing new behaviour would require programming in C++ and/or Blueprint scripting. However, this implementation, if designed correctly, will need to be done only once, followed by parameter tweaking directly within Unreal Engine.

3.1 (c) Creation of a digital content

This task is similar, but simpler to the Creation of a Digital Twin. After importing geometry, there is no need for additional scripting, just additional definition of the level and recording of various movie clips.

4. CONCLUSIONS

Unreal Engine excels in data presentation, drawing quality and speed and support for literally any hardware that exists. It also makes it easy to create complete "game levels". However, to achieve really good results it required to extend original functionality of the Unreal Engine. Scripting can be done in a number of ways, with C++ and Blueprints being the most known. Both approaches has it's pro's and con's and practically, the best results are created by careful balancing of usage of C++ and Blueprints.

Scenarios of usage of the Unreal Engine in the shipbuilding environment are discussed; the main problem to be solved in this perspective is automatic and correct data import and reimport of CAD data.

5. ACKNOWLEDGEMENTS

I would like to thank my spouse for everlasting support and patience during endless hours of my playing with programming.

Also, I'd like to thank all colleagues from RINA, for their support and patience.

6. REFERENCES

1. i i , i i i and e o i , Development of Interactive Tool for Paint Quantity Estimation of Ship Structures. RINA ICCAS 2011 proceedings

2. R. en i , M. i ano i , . i i , M. i i i , MULTI-PLATFORM COLLABORATION TOOL FOR SHIP DESIGN AND PRODUCTION, RINA ICCAS 2013 proceedings.

3. Introduction to game engine programs: https://en.wikipedia.org/wiki/Game_engine

4. Unreal Engine hoem page https://www.unrealengine.com/en-US

5. Introduction to Unreal Engine: https://en.wikipedia.org/wiki/Unreal_Engine

6. Introduction to Unreal Engine scripting and programming: https://docs.unrealengine.com/5.0/en-US/unreal-engine-programming-and-scripting/

7. Introduction to UE programming using C++: https://docs.unrealengine.com/5.0/en-US/programming-with-cplusplus-in-unreal-engine

8. Unreal Engine coding standard: https://docs.unrealengine.com/5.0/en-US/epic-cplusplus-coding-standard-for-unreal-engine

9. W. Sherif, S. Whittle, Unreal Engine 4 Scripting with C++ Cookbook, Packt, 2016

10. Unreal Build Tool: https://docs.unrealengine.com/4.27/en-US/ProductionPipelines/BuildTools/UnrealBuildTool

11. Unreal Header Tool: https://docs.unrealengine.com/4.27/en-US/ProductionPipelines/BuildTools/UnrealHeaderTool

12. Introduction to the Blueprints Visual Scripting: https://docs.unrealengine.com/5.0/en-US/blueprints-visual-scripting-in-unreal-engine/

13. Balancing usage of C++ and Blueprints programming in Unreal Engine: https://docs.unrealengine.com/4.27/en-US/Resources/SampleGames/ARPG/BalancingBlueprintAndCPP/

14. Interoperability between C++ and Blueprints: https://docs.unrealengine.com/4.27/en-US/ProgrammingAndScripting/ClassCreation/CodeAndBlueprints

15. Python scripting for Unreal Engine: https://docs.unrealengine.com/5.0/en-US/scripting-the-unreal-editor-using-python/

16. Unreal Engine introduction to DataSmith importer: https://docs.unrealengine.com/5.0/en-US/datasmith-plugins-for-unreal-engine/

7. AUTHORS BIOGRAPHY

G is founder of "Lina et al", the software company. He was involved in 3D visualisation, OpenGL, VR, AR, mathematical modelling, software design and implementation of distributed simulations, real time visual flight simulations, design and implementation of missile control systems, terrain data visualization manipulation, and vibration calculations and measurement.

DYNAMIC OPTIMIZATION OF PORT OPERATIONS ONBOARD A TYPICAL RO-RO VESSEL, AIDED BY A SMART DECISION SUPPORT SYSTEM

C Singh Uppal, Indian Navy, India
S Babu PK, CUSAT, India

SUMMARY

The principal reason behind the genesis of Roll On – Roll Off vessel design was to increase the pace of loading/unloading operations by eliminating the requirement of cargo-handling equipment thus enabling the wheeled cargo to be driven to its stowage location. The practice being followed today leads to a higher dependency of port operations, associated to any vessel, on ballast water to counter the trim/list generated due to unplanned and imbalanced stowage of cargo. Though advancements in field of ship automation have led to a development of ballast water management system, still, optimization of port operations, namely loading/unloading, is missing due to operator's consideration of usage of ballast as the only natural resort. The earlier research focus at speeding up the ballasting process, whereas the need of the hour is to counter the generation of ballast, because time is a pre-requisite of all ballast operation which cannot be overlooked. The study involves the generation of an Eco-Friendly Decision Support System which counters the moment of similar cargo against each other in a three dimensionally modeled ship simulation, thus, reducing requirement of ballast. The result of the study indicates that vessel's intact statical stability could be maintained without performing ballasting operations, thus enabling cargo stowage at port to save time, as well as the surrounding marine ecosystem.
The flexible design of Decision Support System makes its application possible for various other vessel-types, while at the same time opening a plethora of opportunities to inculcate Artificial intelligence, Big-Data analysis and whereas other advancements, thus, shortening the gap between automation and optimization.

NOMENCLATURE

Ro – Ro	Roll On – Roll Off
L_{BP}	Length between the forward and aft perpendicular (m)
B_{mld}	Breadth of the vessel (excluding external plate thickness) (m)
D_{mld}	Depth of the vessel (excluding external plate thickness) (m)
T	Draft of the vessel (m)
GMt	Transverse Metacentric Height (m)
C_B	Block coefficient of vessel
C_W	Water plane area coefficient of the vessel
C_M	Midship area coefficient of the vessel
C_P	Prismatic coefficient of the vessel
LIFO	Last in First Out
MILP	Mixed Integer Linear Programming

1. INTRODUCTION

Ro-Ro vessels have built-in ramps to make the loading and unloading of the wheeled cargo much easier against using a crane. They are the true workhorses of coastal and deep-sea shipping [1] and have been under operation since the middle of the 20th century. The current loading procedures are no different than what has been used since half a century ago [2] which is highly dependent on the usage of ballast to maintain the requisite static stability. With the onset of digitalisation of maritime transport, it is quite possible that Artificial Intelligence and Big-Data aided decision support systems can provide for optimised stowage plans for Ro-Ro vessels which could eventually (post operational trials) be considered as industry standards.

1.1 BACKGROUND

Despite all the advantages of Ro-Ro transport, it has several serious disadvantages. The chief one being the time a particular vessel takes for transferring goods from one point to another. It is not the speed of ship on water which decides the operational efficiency, but the time that vessel spends on port performing loading/unloading operations, that governs it [15]. All of this is due to a huge lack of requisite operational information of vehicles being transported onboard a Ro-Ro vessel.
Yulia Liznevab [2], enumerates that majority of loading/unloading operations were being carried out solely on the basis of the physical dimensions of the vehicles(like length, breadth, height), provided at the time of registration. Not only does it pose a great risk to the structural integrity of the vessel but the sheer negligence towards gathering the vehicular weight information coupled with the principle that, " dimensionally bigger vehicle will weigh more", has transformed a weight-specific procedure into an area-specific process leading to the entire operation becoming very chaotic, un-standardised and spontaneous.

Lawrence Henesey et al. [3], further details that lack of a central decision support and data storage/management system coupled with an urge to save time leads vehicles to be concentrated on a given area (like all on port or all on aft) across various decks. This gives rise to trim, list and could even car size the vessel which is highly undesirable from statical stability point of view. All of these factors when coupled together, generate an inherent need of ballast to correct vessel's trim and heel, which is not only laborious, calculation intensive and time consuming but also impacts the environment in a harmful way as highlighted by Hu eyin et all. [4].
Furthermore Øvstebø et al.[5] and Hansen et al.[6] highlighted the need for and importance of availability of access to the cargo being unloaded at a particular port of call (Order of Unloading), in a way, that delay due to presence of cargo blocking the unloading route is minimised, thus restricting the shifting of stowed cargo and furthermore saving time at port. The cumulative time saved enables the Ro-Ro operations to become faster and generates a possibility for the vessel to complete the voyage quicker, thus increasing its commercial viability.

1.2 WORK DONE SO FAR

Ro-Ro vessels have a certain set of features that, make it stand out against other types of commercial vessels. The earliest contributions concentrated solely on solving the Ro-Ro stowage problem are given by Øvstebø et al. [5], Hansen et al. [6], Wathne [7] and Puisa [1]. They formulate a MILP model, while also demonstrating whereas other solution methods like Tabu search and genetic algorithm.
For modelling purposes, Øvstebø et al. [5] divides the decks into several logical lanes into which the vehicles are lined. Dividing the decks into lanes might be a reasonable assumption when the cargoes are very similar in size (e.g., only regular cars). Dividing the deck into lanes with the same width would give poor capacity utilization when dealing with dissimilar vehicles. Therefore, the model that this study proposes, does not rely upon this assumption. Also, it is promoting LIFO approach which is not desirable from statical stability point of view.
Hansen et al. [6], in extension to their earlier works use a heuristic approach for the storage location assignment of the incoming cargo by introducing the shifting aspect (at the time of unloading) into the model. While doing so, the study penalises (either through cost or time) the shifting of cargo rather than predicting an optimum stowage plan which prevents it. Furthermore, statical stability of the vessel takes a back seat in the study. Moreover, shifting of vehicles which is permitted in the study is not feasible from reduction of ballast point of view.
In both Hansen et al.[6] and Puisa [1], no attention was given on reducing the ballast requirement which could emerge as a necessary evil of the solutions proposed in the studies. The plethora of studies reviewed whether it is heuristic approach or MILP, all suggested towards solutions that are rigid in terms of their constraints and actions, thus leaving little scope for the developed program/solution to learn and adapt.
The answer to this lies in digitalisation which creates an extensive & controllable ecosystem very close to the real world. Thus, the goal of this study is to optimise the operation at hand by developing realistic simulations that ultimately form a smart decision support system aiming at balancing both the speed of operation and minimising the ballast requirement. A smart operation is one which uses automation and sustainable technologies to improve both its economic performance and environmental impact by utilizing such tools that include Artificial Intelligence (AI), Big-Data and Internet of Things (IOT) [3].

2. PROBLEM DESCRIPTION

2.1 MATHEMATICAL MODELLING

Just as a painter requires a canvas to work upon, this study would require a mathematical model of the decks to identify and subsequently fill the position of operable space onboard a vessel. The extent of work done on the same is quite limited, with Øvstebø et al. [5] suggesting reserving virtual lanes on deck for distinct cargoes. Unplanned stowage of vehicles into such a model will not only lead to poor deck area utilisation but will also threaten intact statical stability of the ship. Moreover Hansen et al. [6] works only on a single deck scenario by merging neighbouring cells to determine cargo stowage slots.
This study divides the individual decks into grid which are further stacked above each other to generate a 3-dimensional model. This aids easy and quick identification of operable and non-operable stowage slots as well as storing the data complementary to the slot's status, i.e., either empty or filled. The dimension of grid cell is determined based on the detailed survey of vehicles being transported by the operator, in a way, that it accommodates the maximum proportion of different types of vehicles, thus fixing the grid resolution. To address the heterogeneity in cargo dimension arising due to difference in vehicle types the study suggests using a multiple or a dividend of cells.

2.2 MINIMISING BALLAST REQUIREMENT

Though the studies of Hansen et al. [6], Puisa [1] and Øvstebø et al. [5] try to make the port operations commercially viable, the need of ballast generated through the suggested solutions to maintain intact statical stability was not addressed. This study aims to design a process which synergises both these aspects, i.e., achieving reduced usage of ballast while maintaining intact statical stability of the vessel. This is achieved by balancing the moment of similar vehicles (in terms of weights, dimension and lever) against each other in real time as discussed in later sections.

2.3 ORDER OF UNLOADING

Operational schedule demands vessel to visit multiple ports where it has to perform loading operations. Cargo slots to be unloaded at different destination ports are physically independent but logically linked [1]. Thus, the prerequisite of an optimised stowage plan is to prevent the cargo earmarked for a later destination from impending/obstructing unloading at a port arriving earlier in vessel's route.

2.4 ASSUMPTIONS

- Number of cars dealt with while performing operations is multiple and strictly non-singular
- Overall, an ideal ship is assumed which has Centre of Gravity amidships and located at the centreline
- The Ro-Ro vessel never operates under lightship condition

2.5 REASEARCH OBJECTIVES

The study is aimed :

(a) To design a geometric model of a RO-RO vessel

(b) To form a mathematical model based on constraints of transverse stability, longitudinal stability, priority of unloading, and structural strength during the loading/unloading process

(c) To create a loading plan assisted by an automated decision support program which is dynamic, self-correcting, runs in the real time and has an ability to learn & adapt

(d) To minimize the ballast requirement of the vessel while maintaining the intact statical stability

(e) To introduce a versatile approach rendering the program flexible, readily applicable on a series of Ro-Ro ships

(f) To achieve zero shifting of cargo for other ports at the time of unloading

3. DEVELOPMENT OF AUTOMATED DECISION SUPPORT SYSTEM

The decision support system guides the stowage of the cargo onboard the Ro-Ro vessel aiming chiefly at reducing the ballast requirement while maintaining statical stability and at the same time speeding up the operation. The various steps in the development process are discussed below:

3.1(a) STEP-1 : INFORMATION FROM THE OPERATOR: It is further divided into two parts:-

- System information: it includes all the data pertaining to designing the program, such as:
  - General arrangement of vessel – dimensions, details of cargo decks, ramps, lifts etc.
  - Hydrostatics parameters – Displacement, MCT, TPC, LCB, VCB at various drafts
  - Weight, LCF, VCG in the light ship condition of the vessel.
  - Most probable route of the vessel along with the information of the berthing ports.
  - Weights of vehicles planned to be transported along with their dimensions.

- Operational information: Individual Cargo dimensions and weights along with order of unloading. These are mostly collected through sensors fixed on the ramp at the time of loading.

3.1(b) STEP-2 : GENERATING MATHEMATICAL MODEL OF THE VESSEL:

- A survey of vehicle dimensions to be carried onboard in order to define the cell size of the mathematical model.
- The survey is extended to study, identify and distinguish the vehicles into categories of similar weights namely, light and heavy.

- Design of mathematical model of the vessel: -

- The process of modelling begins with identifying and focusing on the deck with the maximum possible operable area (generally observed to be the decks above WL).
- The above-mentioned deck is divided into a two-dimensional grid matrix of the cell size decided in the earlier stages.

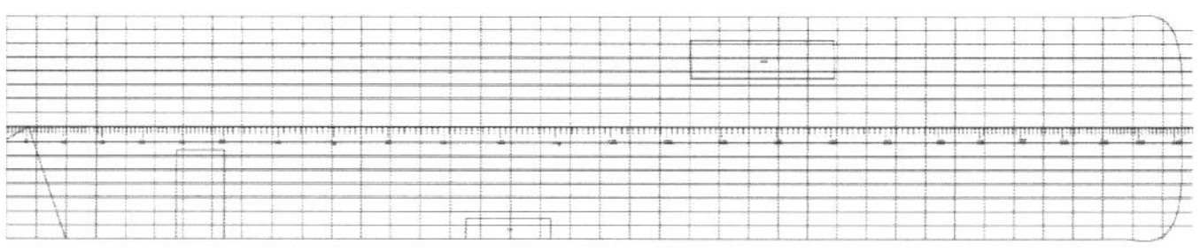

Figure 1: Division of a sample deck into grid matrix and identification of pre-occupied spaces

- Identification of all pre-occupied spaces which are not available from operational point of view (like ramps, lifts, etc.)
- Keeping the scale and the point of reference of the two-dimensional grid same for all decks, the above-mentioned steps were repeated.
- After this the decks which have been divided into two-dimensional grid, are stacked one above the other by strictly aligning the point of reference for all decks.
- The above-mentioned process models the vessel into an operable three-dimensional grid with:

  Z -Axis representing Deck number
  Y-Axis representing transverse position of the cell on grid
  X-Axis representing Longitudinal position of the cell on grid

The model thus generated is easy to perform operations and provides for quick and simple identification of the parking spaces and their availability onboard the vessel.

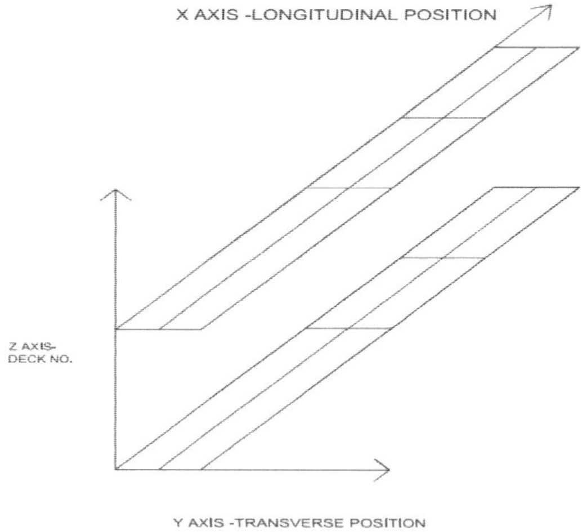

Figure 2: 3-D Coordinate system for development of mathematical model

3.1(c) STEP-3 DESIGN OF POSITION ALLOCATION PROGRAM: This is further divided into –

- Deck allocation (Z CO-ORDINATE)
  - Control conditions for allocation of deck are as follows:

- To keep G as low as possible for ease of manoeuvrability and stability
- As the vehicle moves away from the LCF it will produce a moment that could cause trim, thus minimizing the lever, i.e., the distance a particular vehicle is away from the LCF, minimizes the trim.
- Vehicle to be unloaded first is parked more accessible to the unloading points than the cargo to be unloaded later.
- Maximize the capacity of each type of vehicle under the above-mentioned constraints.

➢ The program divides the vessel's decks into an order of preference based on the weight categories of the incoming vehicles as well as the order in which they are to be unloaded.

- Deck allocation based on the order of unloading: The cargo to be unloaded first is kept closest to the most accessible deck. The most accessible deck is defined as the one, onto which the stern ramp opens up. The order of unloading is taken as an input at the time of loading itself and fed to the program as an integer, with 1 denoting unloading at first destination and 'n' denoting unloading at last destination(for a journey including 'n' stops). As this integer value increases, the vehicles are stowed farther apart from the most accessible deck in relation with their order of unloading. A sketch showing this principle for two destinations is shown in figure 3.

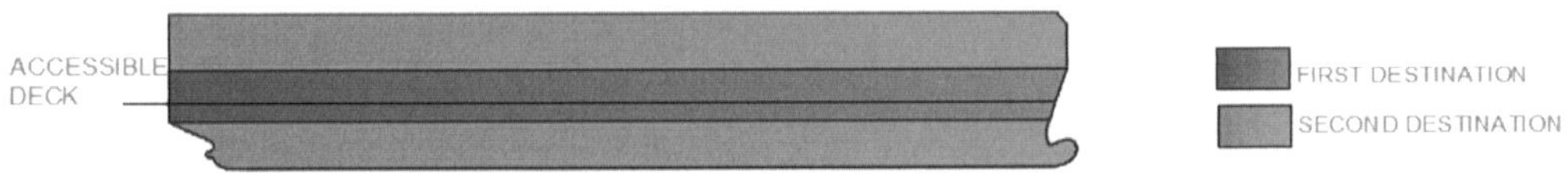

Figure 3: Division of decks based on order of unloading of vehicle for 2 destinations

- Deck allocation based on weight of incoming vehicle: Heavy vehicles occupy lower decks, and their parking density keeps on diminishing as we move up the decks, whereas light vehicles are stowed on the upper decks with their stowage preference diminishing down the decks. This will keep the center of gravity as low as possible thus fulfilling manoeuvring and stability requirements. A pictorial representation enumerating this principle is shown in figure 4, the parking density when observed from the profile view takes a Truncated Square Pyramid shape for heavier vehicles & inverted half-square pyramid for light vehicles.

Figure 4: Principle of deck allocation based on weight of the vehicles

➢ The study develops a logic by fusing the above-mentioned principles to develop an algorithm which caters for both ease of unloading and keeping the overall center of gravity as low as possible. A figure showing the stowage plan as observed from the profile view is shown in Figure 5 named 'Parking Pyramid'. Figure 5 only depicts 2 destinations which are further divided into light and heavier vehicles for ease of understanding.

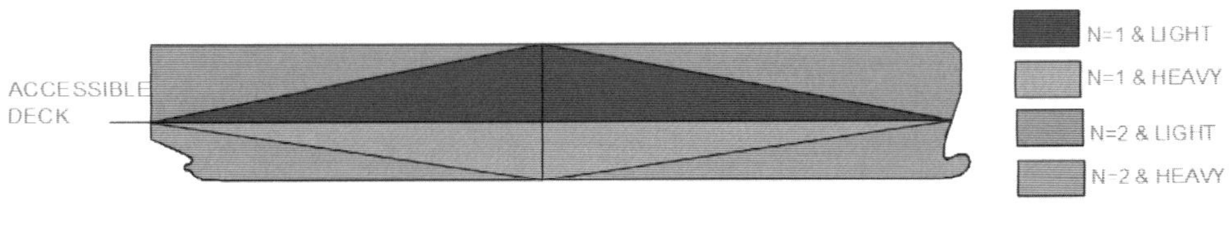

Figure 5: Parking Pyramid where N is the destination of respective unloading of cargo

- TRANSVERSE POSITION ALLOCATION (Y CO-ORDINATE)

  - Control conditions guiding allocation of transverse preference are as follows:

    - To minimize heel and list of the ship
    - To devise a stowage plan which requires minimum shifting at the time unloading

  - The program designates equal precedence to the cells equidistant from the centreline. The incoming vehicles are distributed into these cells based on:

    - Weight: Heavier vehicles are parked closer to the centreline whereas light vehicles are parked further away from the centreline.
    - Order of unloading: Vehicle that has to be unloaded first is kept closest to the driving lane.
    - Moreover, the program also identifies and keeps the driving lanes vacant until all possible spaces are occupied on all decks.
    - Since the allocation of cars based on their weight chiefly takes place at the time of deck allocation, transverse position allocation mainly designates the cells at which the vehicle has to be parked based on its order of unloading.

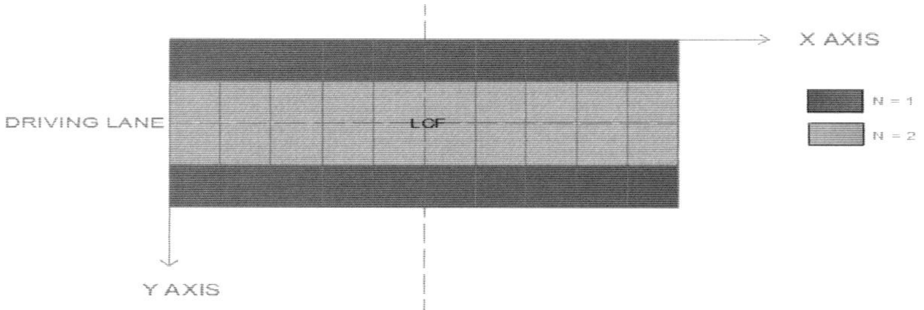

Figure 6: Order of unloading guides the lane allocation

- LONGITUDINAL POSITION ALLOCATION (X CO-ORDINATE)

  - Control conditions guiding allocation of longitudinal preference are as follows:
    - To minimize heel
    - To minimize trim.

  - As the transverse cell positions are divided in an order of preference based on the weight categories, the study uses the weight of the vehicles to balance each other on deck both, transversely and longitudinally.
  - The program begins by allocating loading position closest to the LCF to minimize trim. Once a vehicle has been parked onto a given transverse cell position, say port side of the centerline, the next vehicle of the same category will be parked on the same priority transverse position , starboard of the centerline, thus minimizing list and achieving transverse stability.
  - To counter trim the program will park the second vehicle as many units aft of the LCF, as much unit forward of the LCF the first car is placed, thus achieving longitudinal stability.
  - The logic that guides the allocation of transverse and longitudinal position onto a deck is named 'Hoping algorithm' and shown in figure 7.

❖ This is followed by development of a program which includes all these control conditions, modelling, principles and ultimately giving the user a 3-D coordinate for the stowage of incoming vehicles. The programming was carried out in C++ software to show the viability and performance of the logic developed .

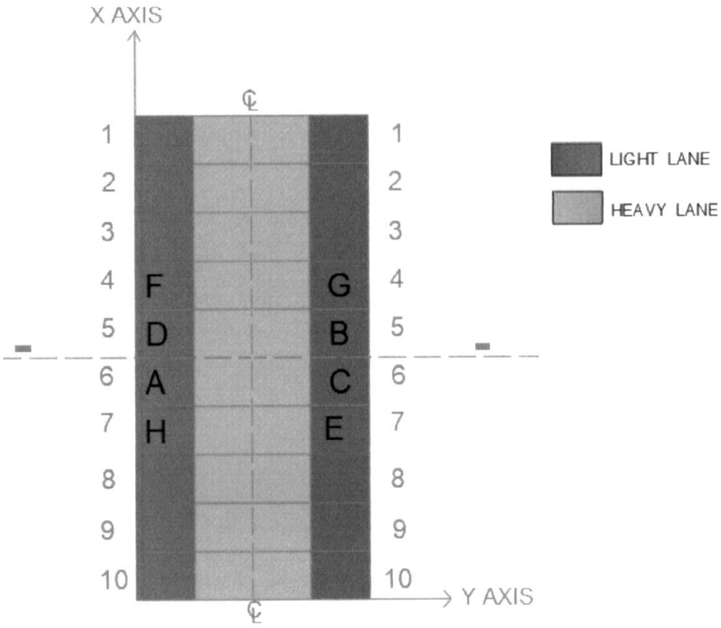

Figure 7: Position Allocation as depicted by the order of alphabets- Hoping Algorithm

4. PROOF OF WORK

Under this sub-heading, aim is to check as well as to prove the following:

- The output generated by the aforementioned Automated Decision Support System is in accordance with the expectations dictated by various control conditions and principles described in the previous section.
- To validate the result at different loading conditions and check the status of statical stability of the vessel both during and post loading operation.

4.1 CASE STUDY

- For the purpose of checking the output of the Automated Decision Support System against author's expectation as well as its validation, the preliminary design of a 6200 CEU Ro-Ro vessel consisting of 10 decks was taken. vessel information is as follows:

  - The dimensions and coefficients are:

L_{BP}	=	191 m
B_{mld}	=	32.26 m
D_{mld}	=	34.56 m
T	=	8.7 m
C_B	=	0.67
C_W	=	0.83
C_M	=	0.96
C_P	=	0.70
No. of vehicles	=	6200
No. of decks	=	11

- Along with this other relevant information were also provided such as General Arrangement, location of LCF, Hydrostatics and etc.

- Loading was carried out by conditions fixing the total number of destinations as 3. Data pertaining to individual weight and order of unloading was fed randomly into the C++ program the result of which is shown below in figure 8.

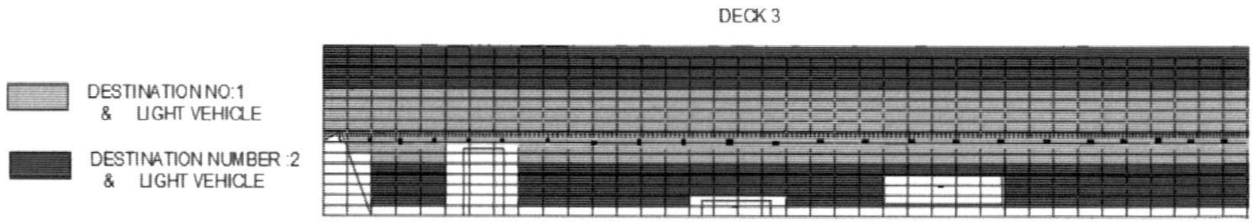

Figure 8: Loading Plan for Light Vehicles as Per The ADSS

- DISCUSSIONS:
  - Modelling: Spaces under use onboard Deck 3 like ramp , lift , etc have been identified by the program as un-occupiable and hence no vehicle has been allotted that position.
  - Deck Allocation: Since the weight category division of vehicles is light as well as their destination numbers (guiding the order of unloading) are 1 & 2, they have been correctly placed at Deck 3.
  - Row allocation: Since the driving lane is assumed to be at the centreline, it is clearly visible that unloading operation could be carried out smoothly.
  - Position allocation: The 'Hopping Algorithm' is depicted by the output of the consecutive data entries to the C++ stowage program.

- STABILITY VALIDATION
  - For checking the compliance of the vessel's intact statical stability with IMO rules, MAXSURF software was used wherein the vessel is designed and loaded for 2 load cases namely: half filled & fully filled. The results from the same are shown below:

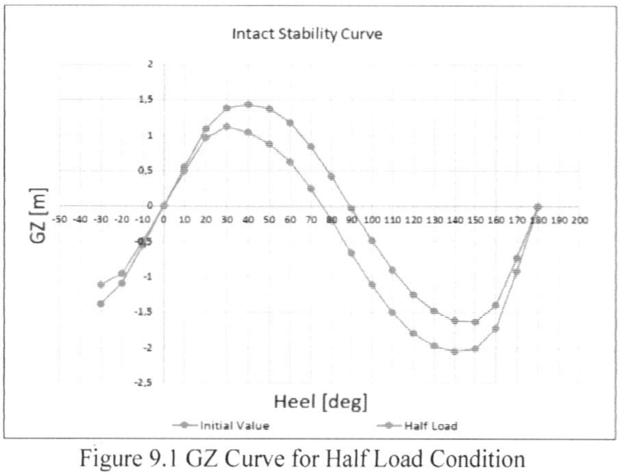

Figure 9.1 GZ Curve for Half Load Condition

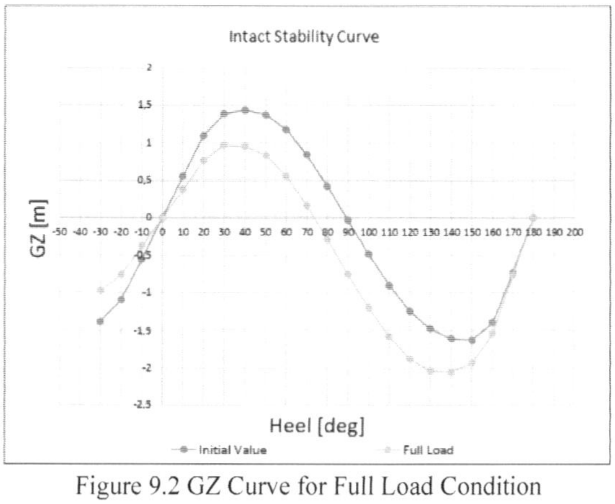

Figure 9.2 GZ Curve for Full Load Condition

SNO	CRITERIA	REQUIRED VALUE	ACTUAL VALUE	UNITS	STATUS
1	Area 0 to 30	3,15	23,75	m.deg	Pass
2	Area 0 to 40	5,15	40,16	m.deg	Pass
3	Area 30 to 40	1,71	16,41	m.deg	Pass
4	Max GZ at 30 or greater	0,20	1,79	m	Pass
5	Angle of maximum GZ	25,0	48,2	deg	Pass
6	Initial GMt	0,15	3,0	m	Pass

Table 1: Compliance of Stability Analysis Output to IMO Stability Rules

- Inference: The analysis of static stability on using MAXSURF software for different load cases produces results that comply with IMO rules and are thus good to use. Hence, it proves that the stowage plan guided by the Automated Decision Support System doesn't hinder the intact statical stability of the vessel and hence reduces the ballast requirement. Thus, the study was able to achieve its goal.

5. CONCLUSIONS

The objective of study was to optimise the port operations of Ro-Ro vessel dynamically by saving time, reducing ballast and reducing the harmful environmental impact. A decision Support System was developed which balanced the moment of similar cars against each other in a three-dimensional modelled ship simulation, thus totally eradicating the trim/list and furthermore the ballast requirement. The intact stability analysis on vessel in accordance with stowage program, showed results as per expectations. The design approach used while generating the Decision Support System is very flexible with an aim to keep the stowage program independent of both the vessel and its operational type.

6. FUTURE SCOPE

The Ro-Ro vessels have been brought under the use since the early 1960's generating a huge data-warehouse of voyage related information which could be termed as a programmers 'goldmine'. This information could be studied to make the stowage program self-learning thus rendering fixed constraint flexible.
Moreover, the reduced ballast requirements would make large ballast tanks obsolete while at the same time increasing the cargo carrying capacity of the vessel. Also, a Smart Decision Support System would rotate the cargo load in order to reduce the failure of a particular deck due to increased fatigue loads and thus increasing the longevity of vessel's life.
Perhaps the most exciting application would be the generation of a communication gateway between the firmware of a self-driven car and the developed Decision Support System. This would enable the ship operator to optimise the loading operation wherein the car drives itself to the stowage location without depending upon the availability of the driver thus cutting short an already reduced time requirement.

7. ACKNOWLEDGEMENTS

I would like to express my deep gratitude to Ms Gurpreet Kaur, Assistant Professor, Guru Tegh Bahadur Institute of Technology, GGSIPU, India, for her valuable and constructive feedback during the planning and development of this study. Her willingness to dedicate her time so generously and unconditionally has been very much appreciated.

8. REFERENCES

1. Romanas Puisa, 'Optimal stowage on Ro-Ro decks for efficiency and safety', Journal of Marine Engineering and Technology -Informa UK, 2018

2. Yulia Liznevab, et al., 'Using simulation method for improving RoPax ship loading operation', Blekinge Institute of Technology, 2021

3. Lawrence Henessey, et al., 'Improved load planning of Ro-Ro Vessels by Adopting Blockchain and Internet of Things', 22nd International Conference on Harbour, Maritime and Multimodal Logistic Modelling & Simulation-HMS, 2020

4. Adrian Parlack, et al., 'Effects of Ballast Water on Marine and Coastal Ecology', Journal of Selçuk University Natural and Applied Science, 2013

5. Brent Olav Øvstebø, et al., 'Routing and Scheduling of RoRo ships with stowage constraints', Transportation Research Part C: Emerging Technologies-Elsevier, 2011

6. Jone R. Hansen, et al., 'An adaptive large neighbourhood search heuristic for the planar storage location assignment problem: application to stowage planning for Roll-on Roll-off ships', Journal of Heuristics-Springer, 2020

7. Eivind Wathne, 'Cargo stowage planning in Ro-Ro shipping', NTNU, 2012

8. Amiadji et al., 'Development of Marine Loading and Unloading System for Ro-Ro Vessel', International Journal of Marine Engineering Innovation and Research- ITS, 2019

9. Sanghyung Park et al., 'Simulation Modelling for Automated Guided Vehicle Introduction to the Loading Process of Ro-Ro Ships', Simulation Based Maritime Design-MDPI, 2021

10. Mariusz Balaban and Tom Mastaglio, 'RoPax/RoRo: Exploring the Use of Simulation as Decision Support System', Institute of Electrical and Electronics Engineers, 2013

11. Beizhen Jia et al., 'Step-wise stowage planning of roll-on roll-off ships transporting dangerous goods', Maritime Transport Research-Elsevier, 2021

12. Jone R. Hansen et al., '2D-Packing with an Application to Stowage in Roll-on Roll-off Liner Shipping', 7th International Conference on Computational Logistics, 2016

13. Geraldo Regis Mauri et al., 'An adaptive large neighbourhood search for the discrete and continuous berth allocation problem', Computers and Operations Research-Elsevier, 2016

14. A. L. Scott, 'Damage Stability of Ro-Pax Ships with Water-on-Deck', UK Maritime and Coastguard Agency, 2019

15. 'Review of Maritime Transport', United Nations Conference on Trade and Development, 2019

9. AUTHOR'S BIOGRAPHY

Sub-Lieutenant CHARANPREET SINGH UPPAL, Naval Architect, is currently serving in the Indian Navy. He is an alumnus and holds a Bachelors of Technology in Naval Architecture and Shipbuilding from Department of Ship Technology, Cochin University of Science and Technology, India.

Dr Satheesh Babu PK, Naval Architect, currently heads the Department of Ship Technology, Cochin University of Science and Technology, India. His previous experience includes 22 years in the marine industry, including ship design, construction, repair and operations in Indian Navy as well as 7 years as an academician.

DEVELOPMENT OF A COMMUNICATION PLATFORM TO SUPPORT CONSENSUS-BUILDING AMONG STAKEHOLDERS IN SHIPPING DECARBONIZATION

K Takahashi and **Y Ushijima,** Mitsubishi Research Institute, Japan
Y Fujita and **T Saito,** Koga Software Company, Japan
H Yamato, Japan Agency for Marine-Earth Science and Technology, Japan
R Kakiuchi, Planning and Design Center for Greener Ships, Japan
S Wanaka, National Maritime Research Institute, Japan

SUMMARY

The carbon neutral shipping should be achieved no later than 2050. However, technologies for the propulsion and the fuel systems are not established, the fuel supply chain is not established and regulations for decarbonization are under discussion in IMO. Designing the carbon neutral ships in this transition phase has difficulties with these undetermined conditions. The optimum design of carbon neutral ships and the optimum time to install these to the market could be achieved with the cooperation across the sectors from shipyards, shipowners, operators, and more stakeholders including the government with clear provision of satisfactory profits or decarbonization results to each sector. Under this recognition, the simulator to provide the decarbonization data and cost estimation during this transition phase was developed together with the communication system to determine the best solution for all stakeholders. This paper shows the concept and prototyping results of the communication system.

1. BACKGROUND AND PURPOSE

In the design of the pathway to the zero emissions international shipping there is an essential difficulty. That is, (1) engine and fuel technologies are not matured, (2) no fuel supply chain is available for ammonia, hydrogen, carbon-neutral methane and so forth, and (3) the regulations for zero emissions shipping are not determined, the conditions for the pathway design are not given sufficiently. For example, a project to make some particular fleet zero emissions, many technical options like HFO-LNG-Ammonia, or HFO-LNG-Hydrogen, and so forth. Now, there is no sophisticated design and cost estimation of the system. Then, once the one option is adopted, the fuel maker and supplier which join and support this project. Solution pathway should be consistent with ESG investment program of the investors and technical requirement of IMO.

Accordingly, it is very important for the international shipping, which includes many stakeholders such as shipyards, shipowners, operators, cargo owners, refuelling companies, banks, governments and regulatory bodies, to achieve the carbon neutrality through consensus building among stakeholders. In doing so, the agreed optimal pathway should bring satisfactory benefits for each stakeholder and decarbonization effects which is accepted by the society.

In the 1980s, the Group Decision Support System (hereafter GDSS) was proposed as an applied research area of groupware for the purpose of supporting intellectual collaboration by groups [4]. The GDSS aims to realize and support the consensus building process in which participants with different positions and values can define their values. Effective support for consensus-building can be expected by externalizing each participant's subjective evaluation of the decision-making problem, identifying areas of mutual cooperation and areas of conflict, and repeating the process of persuasion or compromise focusing on the areas of conflict [5]. In the discussion toward carbon neutrality of international maritime industry, which is the subject of this study, it is effective to derive data for subjective evaluation of each sector for multiple scenarios by using simulators, and to clarify and discuss the conflicting parts of the evaluation in order to derive a solution satisfactory to each sector.

This research aims to establish a mechanism to build a consensus among stakeholders to reach a feasible solution to achieve carbon neutrality in international shipping. The mechanism of consensus building is largely composed of two systems. One is a simulator for evaluating solutions [6], and the other is a communication platform (hereafter referred to as CPF), which is a tool to support communication among stakeholders. This paper describes CPF.

2. PROPOSED METHOD

2.1 CONSENSUS BUILDING BY SPIRAL LOOP

As mentioned in the previous chapter, the zero emissions projects in the international shipping are executed based on the consensus among the stakeholders in the shipping industry. Since the pathway design is essentially based on the uncertain conditions and the executable solution may vary depending on the stakeholders participating in the particular project. Stakeholders for achieving zero emissions are typically shown in Figure 1. It suggests that a wide variety of organizations are involved. In order to achieve consensus building for such an industry structure, we propose to introduce the concept of spiral loops into the consensus building process, which has been used for design optimization studies. A conceptual diagram of the consensus building process using a spiral loop is shown in Figure 2. In the spiral loop consensus building process, rather than all stakeholders aiming to reach consensus at once, consensus building starts with shipyards, who are participants in the first loop, followed by shipowners and operators in the second loop, cargo owners in the third loop, and so on, with each stakeholder joining the discussion in turn to gradually reach consensus. Since the discussion at the next stage is based on the content agreed upon up to the previous stage, the issues at each stage will be clarified and it will be easier to discuss. If it is judged that the simulation conditions or judgment results agreed up to the previous stage need to be revised or reviewed, the loop is returned. After revising or reviewing the simulation conditions and judgment results, the consensus building process is redone again. By proceeding through the loop in this way, the consensus of all stakeholders is reached in the final loop.

In the end, as shown in Figure 1, a wide range of stakeholders will need to participate, but this paper will focus on the scope of the first loop: shipyards, the second loop: shipping companies (shipowners + operators), and the third loop: cargo owners.

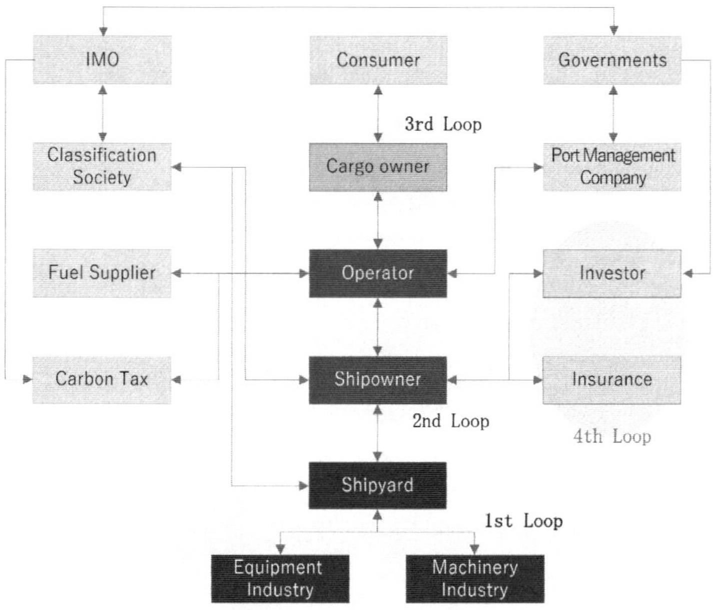

Figure 1. Stakeholders Relationship Diagram of Maritime Industry

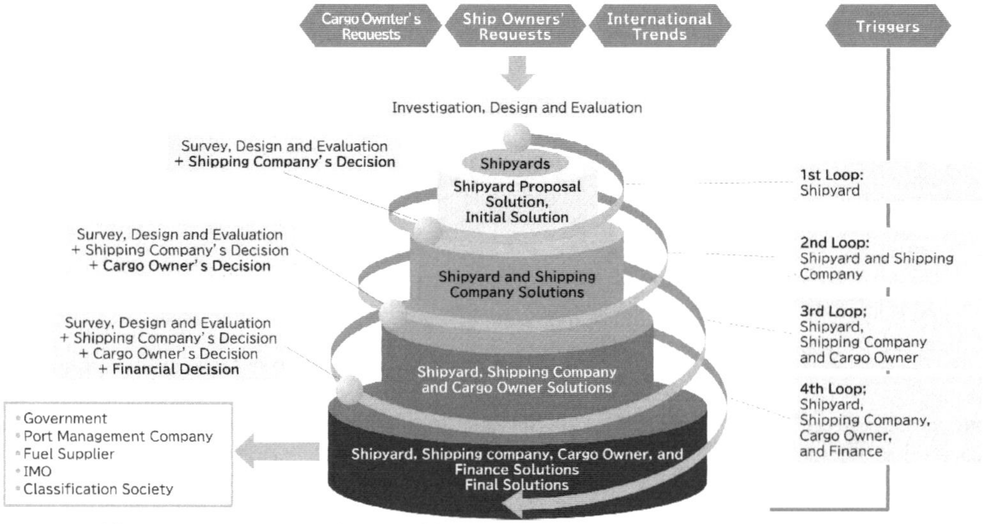

Figure 2. A Conceptual Diagram of the Consensus Building Process Using a Spiral Loop

2.2 CONSENSUS BUILDING PROCESS

In Figure 2, the spiral loop has 4 stages. First one is by the shipyard to make the first solution. and each spiral loop has inner loop to complete the solution by the shipyard. The proposal by the shipyard will be brought about to the second spiral and the criticism or additional condition to be fulfilled are given by the shipping company. Shipyard will revise design and show it to the shipping company again. Once the solution is concluded in the second loop, it will be sent to the next spiral loop. The third spiral loop is the examination by the cargo owner. The cargo owner asks the revision by their own view, then the shipyard and shipping company will revise the solution for the consensus among shipyard, shipping company and the cargo owner. This agreed solution is given to the financial company in the fourth spiral loop. The financial company will examine the solution based on the ESG criteria to make a loan to the ship owner. The consensus will be reached by this process where stakeholders appear in turn.

2.2 (a) Participants and Roles

There are three major roles for participants in spiral-loop consensus building process.

- Facilitator: Leads consensus building through a spiral loop. The facilitator sets issues to be discussed, instructs the person in charge of calculation to execute the simulation, collects the reviewer's opinions, and decides whether the response is necessary or not.
- Calculator: Runs the simulation according to the facilitator's set issues and stores the simulation results and calculation condition documents on CPF.
- Reviewer: Consists of stakeholders who are in charge of the simulation. Stakeholders in charge of each stage of the loop review the simulation results from their own perspectives. In the review, the reviewers refer to the calculation conditions as necessary, scrutinize the results and the appropriateness of the conditions, and feedback the evaluation results to the facilitator.

2.2 (b) How to Proceed with Consensus Building

The consensus building procedure within each loop is shown in Figure 3: In each loop, the discussion proceeds in four major steps.

STEP 1 Planning and Preparation: The facilitator sets an issue and announces it to all participants. The facilitator presents the simulation assumptions (input data, external environment scenario, configuration parameters) to the person in charge of the simulation, including the items to be verified in the simulation and the calculation case in which the simulation is to be performed, and instructs him or her to perform the simulation.

STEP 2 Calculation Execution: The person in charge of calculation executes the simulation using a simulator developed separately and presents the results to the reviewer.

STEP 3 Discussion and Evaluation: The reviewer closely examines and evaluates the results by referring to the simulation results and calculation conditions. The reviewer is rotated according to the stage of the loop. In this paper, it is assumed that the first loop is conducted by the shipyard, the second loop by the shipping company, and the third loop by the cargo owner. The reviewer will conduct the evaluation while communicating with the person in charge of the calculations and the participants in the previous loops by confirming and asking questions, as necessary.

STEP 4 Scrutiny and Policy Decision: When all the discussion in STEP 3 is completed, the facilitator decides whether to re-examine, proceed to the next loop, or return to the previous loop, thus the discussion of the loop is completed.

When the examination within each loop is completed, the facilitator instructs the transition of the loop, and the reviewer takes over. When this process is repeated and the final loop is completed, consensus is reached among all stakeholders on the set issue.

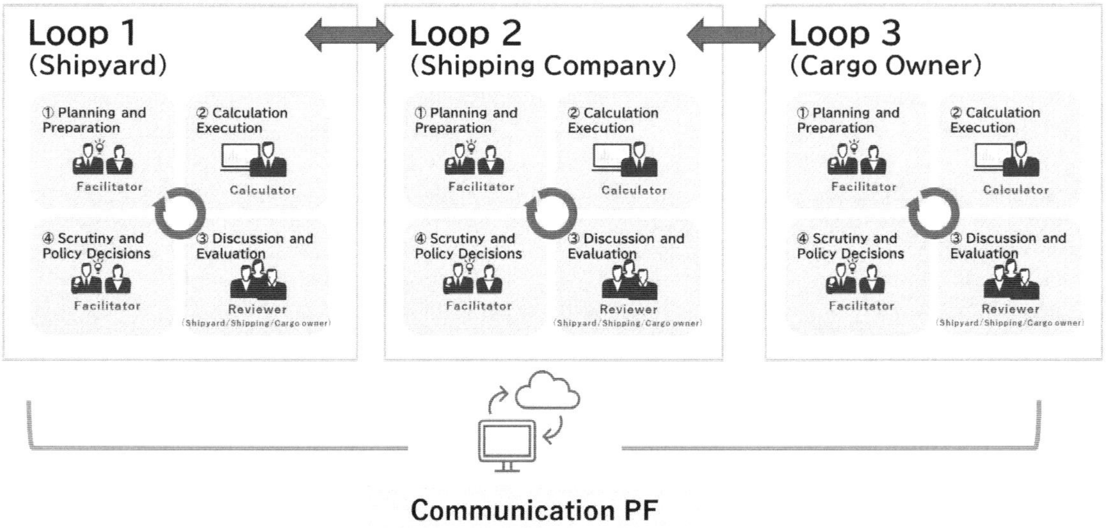

Figure 3. Consensus Building Process at Each Stage of the Loop

3. COMMUNICATION PLATFORM (CPF)

3.1 CPF OVERVIEW

We developed the CPF, a system to support the consensus building process described in the previous chapter. The CPF allows all participants (Facilitator, Calculator, Reviewer) involved in consensus building process to store and refer to information and data to be shared. It also provides a forum for discussion among participants by inputting comments. As shown in Table 1, CPF was developed on Notion, which is provided on the Internet cloud. Notion is an information sharing tool developed by Notion Labs, Inc., which allows users to easily build their own databases and screens in addition to information sharing and comment input functions [7]. Therefore, we have adopted Notion as a suitable tool for the concept study of CPF. Figure 4 shows the correspondence between the consensus building process described in the previous chapter and the main functions provided by the CPF.

3.2 FUNCTIONS PROVIDED BY CPF

This section introduces the six main functions provided by CPF.

Function 1: Entering and Referencing Setting Issues

Figure 5 (a): shows the screen shot of this function. The facilitator inputs/registers setting issues and the preconditions (input data, external environment scenario, and configuration parameters) for the simulation according to the issues. The facilitator may ask the participants to provide information/data necessary for the preparation of the preconditions, which are then referenced in the system by the person in charge of the calculations and the reviewer.

Function 2: Register and Reference Simulation Results

Figure 5 (b): shows the screen shot of this function. Simulation results are registered on CPF by the person in charge of the calculation, and the reviewer refers to them. In the study of feasible solutions to achieve zero-emissions, which is the goal of this system's consensus building, it is assumed that the simulation assumptions are complex and diverse. Therefore, it is necessary to clarify the correspondence between the simulation results managed by this function and the simulation assumptions managed by Function 1, and to manage related files and data appropriately so that reviewers can easily find necessary information

Table 1. CPF Overview

Tools of Use		Notion
System Requirements	OS	Windows / macOS
	Web Browser	Google Chrome / Firefox

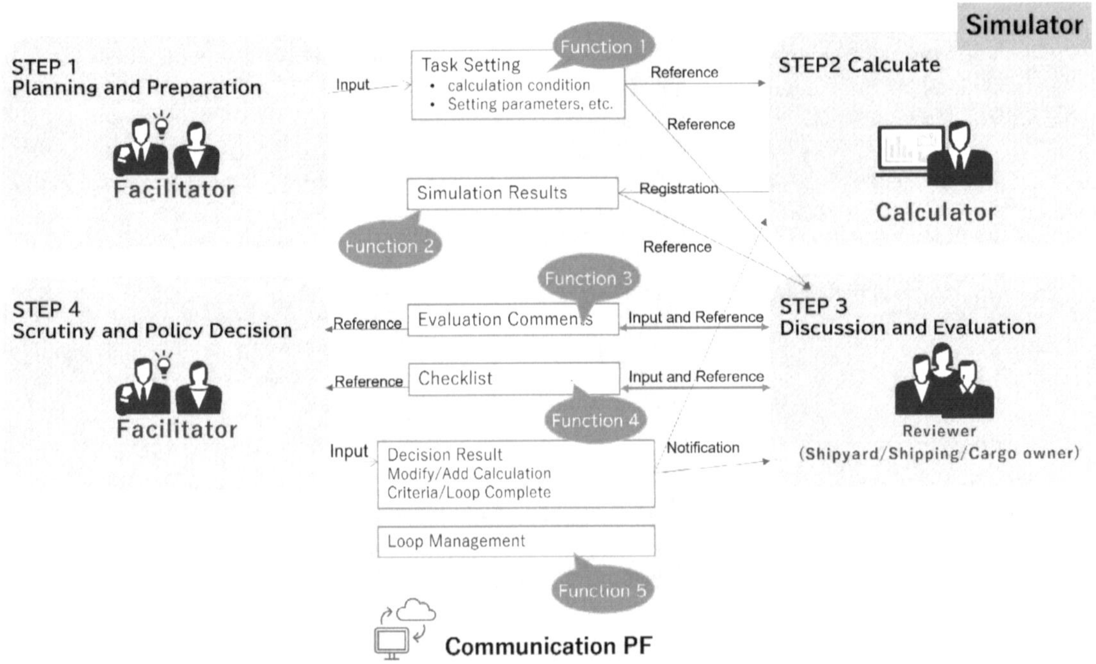

Figure 4. Consensus Building Process and Functions Provided by CPF

Function 3: Entering and Referencing Evaluation Comments

Figure 5 (c): shows a screen shot of this function. Reviewers check the simulation results and related documents on the CPF and enter their evaluation comments. Reviewers can view each other's comments, and a function for sending comments is also provided. This allows communication among reviewers and with the person in charge of the calculation on the CPF and facilitates discussion. The facilitator also checks the content of the discussion, identifies issues that need to be addressed as appropriate, and requests action from the person in charge of the calculation and other parties involved.

Function 4: Entry/Reference of Checklist

A checklist is a pre-list of viewpoints to be evaluated by the reviewer. The checklist function clarifies the issues to be discussed on CPF, and also allows the reviewer to explicitly grasp and manage the progress of the evaluation. This checklist was developed from interviews with stakeholders involved in the spiral loop. In the interviews, we presented an overview of the solution evaluation system and some simulation results, and asked questions about the simulations and use cases of this evaluation system. The evaluation viewpoints in each loop were extracted from the results of those interviews. The evaluation items to be listed in the checklist were designed so that they could be easily updated on the CPF, since they varied depending on the issue under consideration and the participants. An example of such checklist is shown in Table 2.
In addition to facilitating consensus building, the following effects can be expected by formalizing the evaluation items in this way.

- Stakeholder perspectives can be accumulated and compiled into a database.
- The perspectives of stakeholders at different stages of evaluation can be shared, leading to more efficient ship design.
- The perspectives of stakeholders at later stages can be referenced, enabling proactive responses based on their perspectives when designing a ship.

Function 5: Loop Progress Management

Figure 5 (d): shows a screen image of this function. This function does not support the consensus building process itself but is used primarily by facilitators to track the progress of discussions. This applies to the display of the current loop stage and the setting of deadlines and the reminder function as the schedule management of the reviewer.

Table 2. Check Items in the First Loop (Shipyard)

1-1	Does it prepare patterns of multiple scenarios appropriate to the problem to be solved?
1-2	Does it show a long-term ordering strategy? ① Are the results of fleet composition available? ② Are the results of product transition diagram forecasts available? ③ Are the results of GHG emissions forecasts available? ④ Are the results of freight and cost forecasts obtained? ⑤ Are the results of transportation volume forecasts obtained? ⑥ Are the results of fuel consumption forecast obtained?
1-3	Is the appropriateness of the set fares explained? ① Whether the results of fuel price forecasts are available, and if so, have they been explained? ② Has the availability of ammonia fuel been clearly indicated? ③ Have you clearly stated the concept of price differences by region of production?
1-4	Has the concept of transportation demand been explained? ① Have you presented the concept of transportation volume trends?
1-5	Has the concept of refuelling stations been clarified? ① Did you present the assumptions and rationale for fuel supply facilities? ② Concept of bunker barge, etc.
1-6	Have you clearly indicated how you view the regulations and what is reflected in the calculations? ① Have you explained the contents of the regulations, and in particular, have you considered penalties? ② Have you indicated a set timeframe for the introduction of the regulation? ③ Are the individual ship regulations designed to realize the total amount of regulation values?
1-7	Is the concept of alternative fuels (Selection of NH3, H2, etc.) described? ① Has the concept of the state of infrastructure development been presented?
1-8	Are the simulation assumptions (how realistic they are) explained? ① Concept of "free ships" ② Concept of port development in Japan ③ Whether or not supply chain emissions are taken into account, and if so, the details thereof

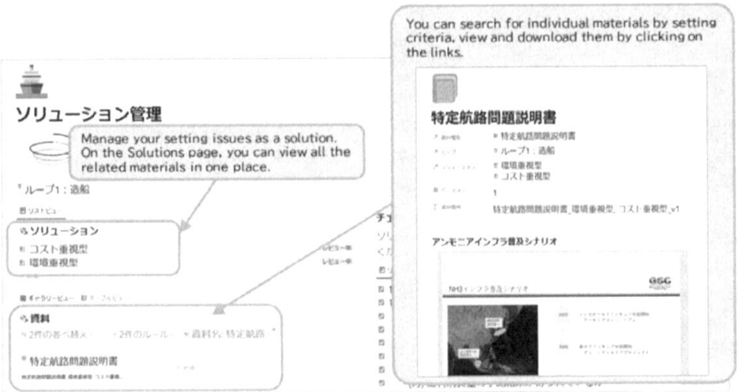

(a) Function1 Implementation Screen

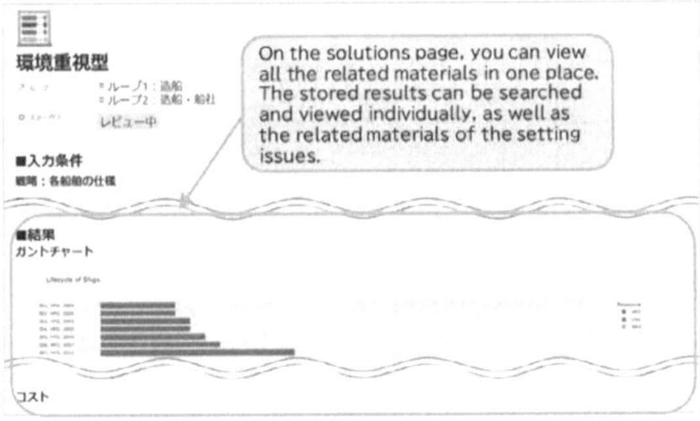

(b) Function2 Implementation Screen

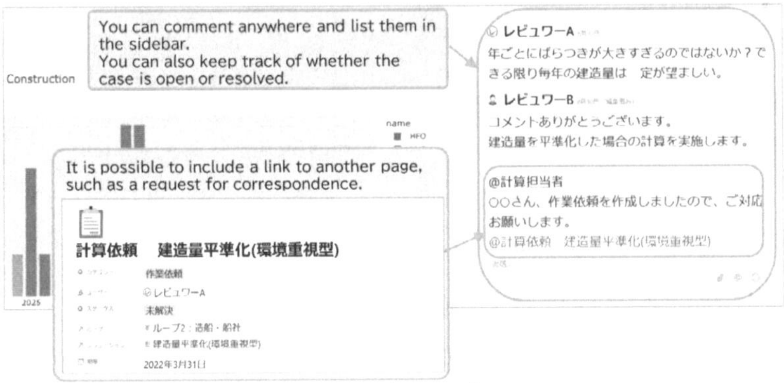

(c) Function3 Implementation Screen

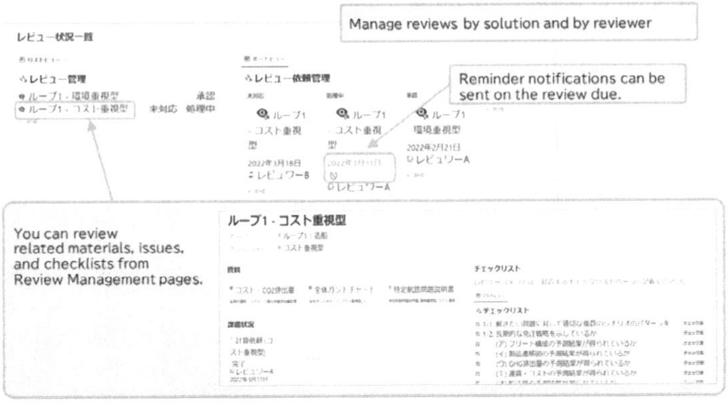

(d) Function4 Implementation Screen

Figure 5. CPF Function-Specific Implementation Screen

3.3 FLOW OF A STUDY USING CPF

An example of the flow of a study using CPF is shown in Figure 6. At the beginning of the study, the facilitator provides the shipyard with a study theme based on the results of the information collection in advance.
In the first loop, the study is first conducted according to the study theme, and Solution 1 is created. The shipyard then reviews the input information and simulator output for Solution 1. Solution 1 agreed by the shipyard in the first loop will be reviewed by the shipping company in loop 2.
In the second loop, the shipping company reviews the agreed Solution 1. As a result of the review, it is expected that changes will be required for Solution 1. For example, although the direction of the solution was agreed upon, there was a desire to level the amount of construction each year. Then, Solution 2 reflecting the request of the shipping company is newly calculated. Once again, Solution 2, which has been reviewed and agreed upon by shipyard and shipping company in the first and second loops, is subject to review by cargo owner in the third loop.
In the third loop, the cargo owner reviews the Solution 2 agreed by the shipyard and shipping company. When a request for change is made, Solution 3 reflecting the request from the cargo owner is newly calculated. Once again, the first, second, and third loops are reviewed by shipyard, shipping company, and cargo owner, and the agreed Solution 3 from each sector enters the subsequent fourth loop for implementation. In the actual business, a joint venture will be formed to realize a new ship based on Solution 3 when the third loop is completed, and negotiations will be conducted between the joint venture and financial institutions.

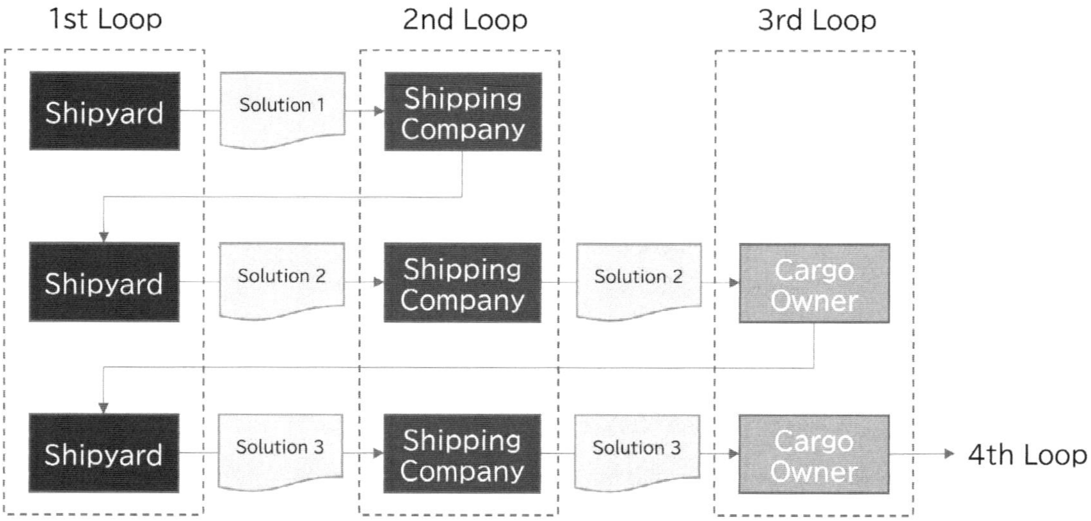

Figure 6. An Example of Flow of a Study Using CPF

4. CONCLUSIONS AND FUTURE PROSPECTS

This paper presented a proposal for a consensus-building process among stakeholders following the design spiral where the stakeholders appear in turn to reach the feasible solution and the CPF was developed as a support system for the proposed process.
The CPF developed so far is still conceptual, in the future, we plan to hold workshops attended by multiple stakeholders, including shipyards, shipping companies, cargo owners, fuel suppliers and bankers using the proposed method and the developed system. Then we will identify the issues of the proposed method and the system and improve it to make it more practical for actual business.
In addition, the proposed method and developed system will be effective not only for the maritime industry but also for complex issues requiring consensus building among multiple stakeholders. We would also like to consider applying the proposed method and the system to industries and issues other than the maritime industry.

5. ACKNOWLEDGEMENTS

This research was supported by Planning and Design Center for Greener Ships. The discussions and evaluations in the process of developing the simulation-based procedure proposed in this paper are performed in the joint research project, Solution Evaluation System project.

6. REFERENCES

1. IMO. "UN body adopts climate change strategy for shipping". 2018 https://www.imo.org/en/MediaCentre/PressBriefings/Pages/06GHGinitialstrategy.aspx. Accessed 2022.

2. S. Wanaka, K. Hiekata , Y. Horii. "Network Optimization Model for Evaluation of GHG Reduction on Ship Transportation" JASNAOE, no.31(2020): 205-212.

3. S. Wanaka, R. Kakiuchi, N. Fujita, K. Hiekata , H. Yamato. "Study on Simulation Based Evaluation of Route and Cargo Specific Project for Introducing Decarbonized Ships" ICCAS(2022).

4. G. DeSanctis , R. Gallupe. "A Foundation for the Study of Group Decision Support Systems" Management Science vol.33, no.8 (1987): 589-609.

5. N. Kato, M. Chujo , S. Kunifuji. "A Development of a Group Decision Support System Facilitating the Consensus-making Process" IPSJ vol.38, no.12 (1997): 2629-2639.

6. S. Wanaka, R. Kakiuchi, N. Fujita, K. Hiekata , H. Yamato, "Development of Simulation to Evaluate Environment, Technology, and Economics for Decarbonization of International Shipping" JASNAOE(2022).

7. Notion Labs, Inc. "Notion". 2022 https://www.notion.so/product. Accessed 2022.

7. AUTHORS BIOGRAPHY

Mr. Kazuki Takahashi holds the current position of Researcher of Mitsubishi Research Institute, Inc, Japan. He is engaged in system design and artificial intelligence at the DX Technology Division.

Dr. Yumiko Ushijima is Senior Researcher at Mitsubishi Research Institute, Inc. She specializes in DX strategy, data utilization strategy, and data-driven management at the DX Technology Division. She obtained her PhD from Keio University in the area of mechanical control.

Mr. Yoshihiro Fujita is the General Manager of Mobility Services Department of Koga Software Co., Ltd.
He participated in various industry-government-academia collaboration projects and new technology projects of the Ministry of Land, Infrastructure, Transport and Tourism. He is also the product manager for on-demand bus systems that solve mobility issues.

Mr. Takuya Saito is the Technical Section Chief of Application Development Department of Koga Software Co., Ltd.
He is an experienced front-end developer and has been involved in creating user interfaces for web and mobile applications.

Dr. Hiroyuki Yamato is currently the president of the Japan Agency for Marine-Earth Science and Technology (JAMSTEC), and Professor Emeritus of the University of Tokyo. He served as the Executive Vice President of the University of Tokyo and the President of the Japan Society of Naval Architects and Ocean Engineers. Degrees BS, MS, PhD are from the University of Tokyo in the field of Naval Architecture. He is an author of numerous papers on various fields like aircraft control systems, ship design, industrial information systems, on demand transportation systems and digital archives. He has been awarded by the government, companies and academic societies.

Mr. Ryutaro Kakiuchi is the Naval Architect and holds the Senior Researcher at Planning and Design Center for Greener Ships (GSC). He is belonging to the initial design team for Greener ships at GSC. Before this role, He worked for initial design, technical negotiation with customers, and product planning at Shin Kurushima Sanoyas Shipbuilding Co., Ltd.

Dr. Shinnosuke Wanaka holds the current position of Researcher of Knowledge and Data System Department at National Maritime Research Institute, Japan. He obtained his PhD from University of Tokyo in the area of system design and simulation. He is responsible for developing the simulation software to calculate transition of fleet to fulfil the decarbonization in the future.

International Conference on
Computer Applications in Shipbuilding
13th-15th of September 2022, Yokohama, Japan

A STUDY OF PROCESS SIMULATION BASED ON A MULTI-AGENT SYSTEM FOR SHIPBUILDING

T Taniguchi and **K Matsuo,** National Institute of Maritime, Port and Aviation Technology, Japan

SUMMARY

This paper proposes a simulation technique for the shipbuilding process based on a multi-agent system. In this paper, we propose a simulation technique for the construction process based on a multi-agent system. In this method, the worker is regarded as an agent and the virtual factory as an environment. Tasks are represented as a combination of more detailed elemental tasks. Three techniques, pathfinding with the A-star algorithm, placements of two-dimensional irregular shapes, and intersection with other equipment or occupied areas, were introduced to allow agents to take spatial constraints into account. It also simulates the manufacturing process by performing task selection through an evaluation function based on weight coefficients. The simulation results for sub-assembly and assembly process in a real shipyard confirmed that the task could be reproduced with high accuracy. It is possible to capture manufacturing features in simulation in advance, without manufacturing them.

1. INTRODUCTION

Simulation techniques are conventional methods for man-hour estimation, schedule optimization, and factory layout optimization. A lot of studies related to simulation techniques have already been conducted in this research area. In many studies, the work time is determined by dividing a total workload by its work efficiency observed on the shop floor.
On the other hand, shipbuilding work consists of more supplemental work to main work as workers operate under the constraints of three-dimensional space. Therefore, the conventional method, which consider only main works, cannot adequately respond to the present situation of shipbuilding.
To solve this problem, this paper proposes a modelling method for shipbuilding tasks that can represent supplemental works as well. The shipbuilding task is represented as a combination of elemental tasks, and each elemental task is an executable function that can consider a path in three-dimensional space. Based on this model, a multi-agent simulator for shipbuilding is developed. The simulator considers a virtual shipbuilding factory as an environment and workers as agents. By comparing the results of the simulation in a sub-assembly process with the actual work in a shipyard, it was found that the simulator can properly approximate the time of each task including supplemental works. In addition, this paper discussed the digital transformation of shipbuilding that utilizes this simulator for production planning and production control.

2. LITERATURE REVIEW

There are many studies on the simulation of manufacturing processes. Research on simulation of the construction process using discrete event simulation (DES) is very active. Krause et al. proposed a scheduling system using DES to evaluate the effect of equipment installation [1]. Shin et al. study the effect of introducing robots in small assembly processes [2], and Jeong et al. conduct a discrete-event simulation-based study on predicting the in-plant logistics time of blocks [3]. Matsuo et al. use DES to model supplemental work in a simplified manner and investigate procedures to minimize the time of supplemental work [4]. Okubo et al. have proposed a system that automatically create practical production plans using DES [5]. As described above, simulation technology has been used in shipbuilding for production planning, logistics, equipment introduction, robot introduction, and its effectiveness has been demonstrated.
Furthermore, research has been conducted on the application of multi-agent system to the manufacturing field. For example, L. Monostori et al. studied the interaction between agents and the modelling of the hierarchical structure of organizational systems and described the effectiveness of agent technology in terms of autonomy and responsiveness in the manufacturing industry [6]. Xu et al. proposed a hybrid model that combines discrete event simulation and a multi-agent system for the steelmaking process [7]. Zhang et al. proposed a multi-agent based cyber-physical system. Three types of agents have been constructed: a working machine agent, an agent for self-organization, and an agent for exception handling, and are being studied for the assembly process of machine parts [8].
On the other hand, the time for each task in these studies was calculated based on past performance data. Estimation accuracy remains an issue for newly designed products. In addition, because the industry is one in which people gather to product, spatial constraints must be considered. In this study, we propose a simulation method that applies a multi-agent system to a shipbuilding process with multiple workers. The proposed method ensures generality by representing tasks as a combination of smaller elemental tasks. It also uses pathfinding, two-dimensional packing, and interference detection to reproduce spatial constraints

3. PROCESS SIMULATION BASED ON MULTI-AGENT SYSTEM

An agent is a subject that perceives its environment and acts based on some decision mechanism. A group of these agents is called a multi-agent. In this paper, we consider the agents as workers and the environment as a virtual factory.

3.1 ENVIRONMENT

The environment in this study consists of six components as follows:

- Product: attribute information and coupling information of the parts themselves that make up the product in the completed state are defined.
- Facility: it defines the object's referent, its unique capability values, initial position and angle, and occupancy status.
- Assembly status: the information on intermediate products, components, and assembly is defined as a tree structure.
- Task: The information required for a task and the antecedent relationships among tasks are defined as a tree structure. Information about who has performed or is performing a task and with which tool is also defined as a state variable.
- Three-dimensional object: Virtual objects such as products, facilities, and tools are defined as 3D objects. The 3D coordinates and angles at representative points are state variables.
- Occupied area: In this report, we consider that not only resources are occupied, but also space is occupied. The occupied area is defined as an approximate oriented bounding box.

These are state variables that can be observed and some of which can be changed during the simulation.

3.2 MODELLING OF TASK

The tasks are represented as a combination of smaller elemental tasks so that even supplemental tasks can be considered. The elemental task is defined as a function that can be executed during the simulation and that changes the state variables of the environment defined in section 3.1. Tasks are basically expressed by connecting elemental tasks according to the procedures at the element work level, but some if-then rules are used to construct the desired procedures. The actual process should be observed in detail for constructing these tasks. Figure 1 shows an example of modelling of a placement task with a crane. The placement task is represented by a combination of elemental tasks as shown in 1~ 6. Figure 2 shows modelling example of a welding task with auto welding carriage. There is a branch along the way depending on the status of the connection, but basically it is represented by a combination of elemental work.

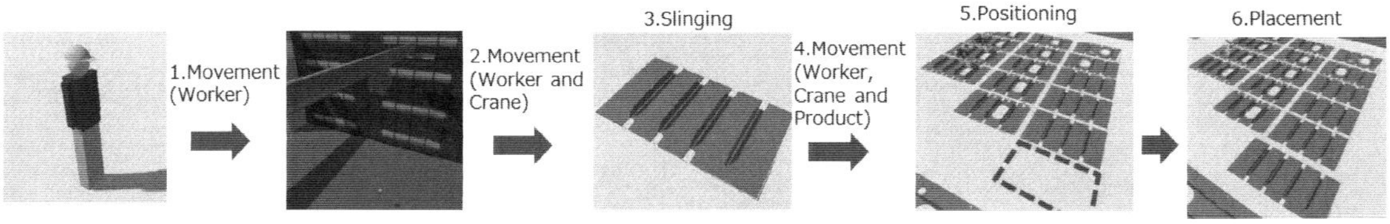

Figure 1: Modelling of a placement task.

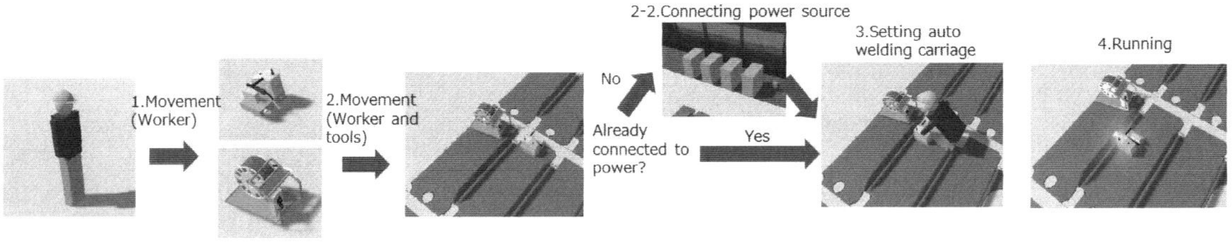

Figure 2: Modelling of a welding task with auto welding carriage.

3.3 AGENT

In this paper, the agent is defined as selecting task and determining task details, such as tools, path of movement and location. First, to determine the details of these tasks, we use three techniques as follows:

- Pathfinding with the A-star algorithm

- Placements of two-dimensional irregular shapes
- Intersection with other equipment or occupied areas

Figure 3 shows these technique images in simulation. It generates a walkable mesh, including large products, and calculates a walking path based on this mesh using the A-star algorithm. The upper right corner of Figure 3 shows the results of the calculation of the possible placement areas. The purple line shows the no-fit polygon, and the position of placement is determined using this polygon. The red boxes at the bottom of Figure 3 indicate occupied areas, and these Oriented Bounding Boxes can be used to determine interference.

Next, an evaluation function is introduced to the agent to select one task to execute from several tasks. The agent obtains the necessary parameters for evaluation from the environment and calculates the evaluation value by multiplying weight coefficients during the simulation. The value of the weight coefficient should be set so that it is close to the actual operator's judgment result.

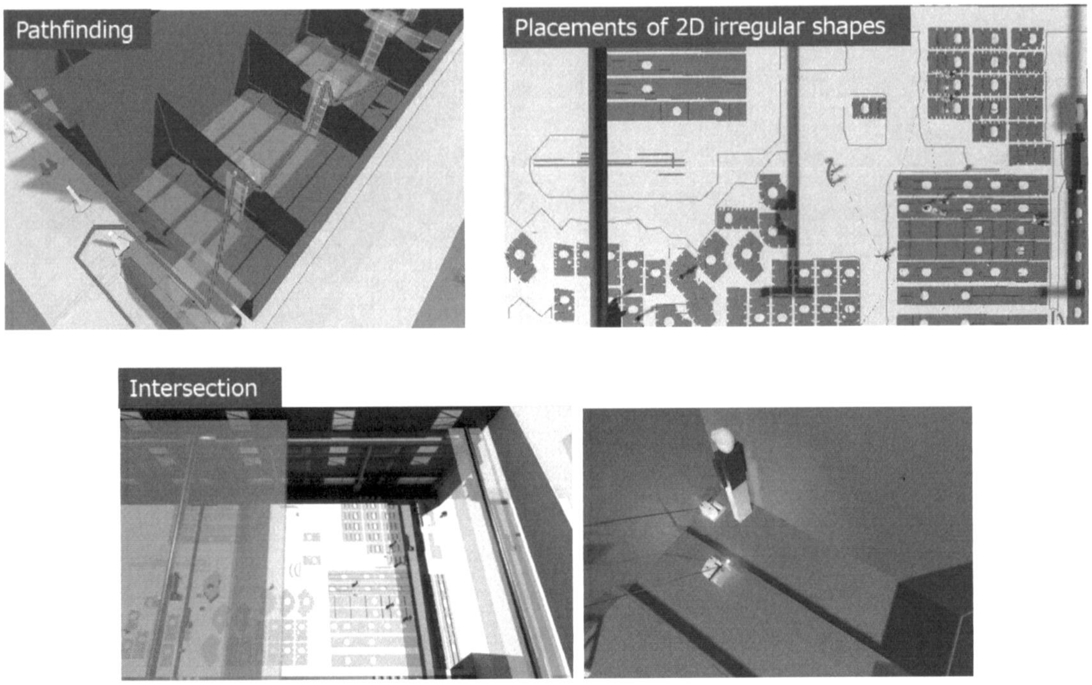

Figure 3: Three techniques for considering spatial constraints.

4. RESULTS AND DISCUSSION

4.1 RESULTS

The simulation was performed for the small and medium assembly processes. Figure 4 shows the image of simulation results of sub-assembly and assembly processes. The left side shows the actuals, and the right side indicates the simulation results. Unity was used to implement the simulation method [9], and pathfinding is conducted the tools provided by Unity Technologies [10].

Figure 5 shows the comparison of work time for a single placement task in a sub-assembly process between observation and simulation. The different colors represent the different elemental tasks described in Figure.1, and the horizontal axis shows the work time. The movement work times are calculated based on the path obtained by pathfinding, and the target position is also determined by the placement of 2D irregular shapes as described in Section 3.3. It was confirmed that the simulation results can accurately reproduce the task at the elemental work level. This figure is just one example, but similar trends have been confirmed for other tasks. It has also been observed that these work times vary with the number and type of products. Therefore, the simulator can accurately estimate reasonable working hours for various products.

Figure 6 shows the relationship between the number of trolleys used simultaneously and work speed in assembly process. The agent determines the task so that welding lines are selected that do not interfere with each other of the auto welding carriages as described in Section 3.3. For the welding process of the present product, a nonlinear relationship was revealed for the number of carriages. Saturation was observed, rather than a linear increase in work speed. The intensity of saturation depends on the size, shape, and structure of the product. Thus, the use of simulation in advance revealed trends that had previously been revealed by actual execution.

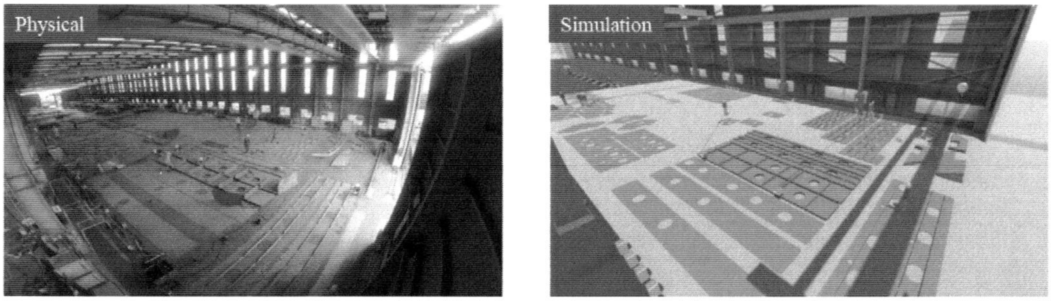

Sub-Assembly Process

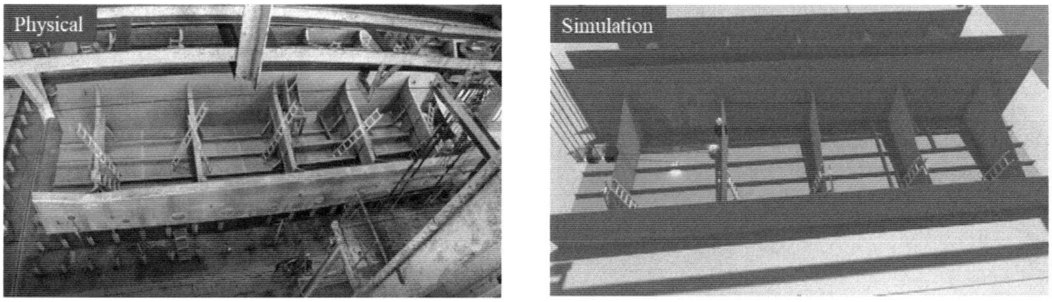

Assembly Process

Figure 4: Image of simulation results of sub-assembly and assembly processes.

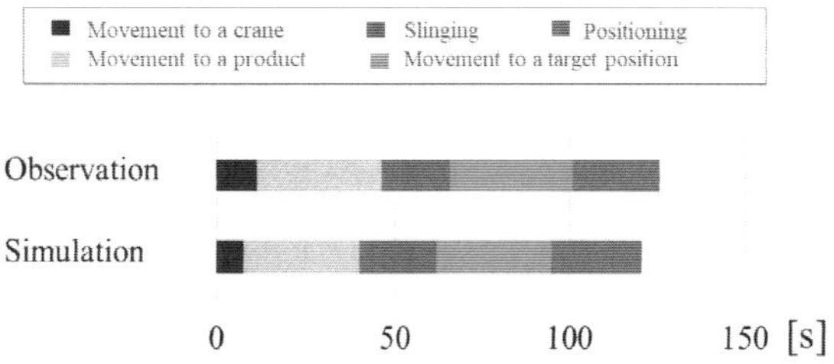

Figure 5: Comparison of work time for a single placement task in a sub-assembly process between observation and simulation.

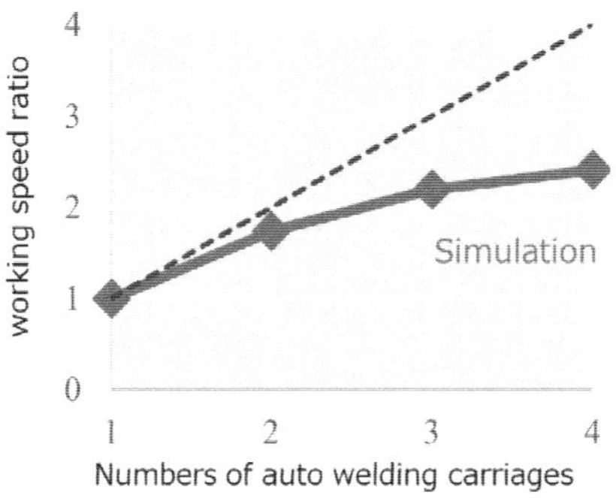

Figure 6: Relationship between the number of trolleys used simultaneously and work speed.

4.2 DISCUSSION

In this study, we evaluated the accuracy of the working time for a typical task, but, in reality, there is a part of time for conversation, short break, small rework etc., and the simulation work time itself does not necessarily equal the actual work time. Therefore, it is appropriate to consider them as standard times. The weight coefficients of the agents in the simulation were determined manually in a trial-and-error manner this time but are expected to vary depending on whether they are experienced workers or newcomers. We would like to consider future developments such as reproducing the effects of newcomers and veterans in the simulation. Furthermore, there are data linkage issues that need to be addressed in order to use this simulator in actual operations. If data equivalent to BOP can be smoothly incorporated using PLM, etc., productivity can be rationally evaluated at the design stage. If the simulation results can be utilized in the early stages of design, they will be very useful for budget planning, influencing the schedule of specification changes, and developing new vessel designs.

5. CONCLUSIONS

In this paper, we propose a simulation technique for the construction process based on a multi-agent system. In this method, the worker is regarded as an agent and the virtual factory as an environment. Tasks are represented as a combination of more detailed elemental tasks. Three techniques, pathfinding with the A-star algorithm, placements of two-dimensional irregular shapes, and intersection with other equipment or occupied areas, were introduced to allow agents to take spatial constraints into account. It also simulates the manufacturing process by performing task selection through an evaluation function based on weight coefficients. The simulation results for sub-assembly and assembly process in a real shipyard confirmed that the task could be reproduced with high accuracy. It is possible to capture manufacturing features in simulation in advance, without manufacturing them.

6. ACKNOWLEDGEMENTS

The authors would like to thank the shipyards.

7. REFERENCES

1. Krause, M., Roland, F., Steinhauer, D., and Heinemann, M. "Discrete Event Simulation: An Efficient Tool to Assist Shipyard Investment and Production Planning", *Journal of Ship Production*, Vol.20 (2004):176-182.

2. Shin, J., Lee, K., Woo, J., Kim, W., Lee, J., Kim, S., Park, J., and Yim, H., "A Modeling and Simulation of Production Process in Subassembly Lines at a Shipyard", *Journal of Ship Production*, Vol. 20 (2004): 79-83.

3. Jeong, Y., Lee, P., and Woo, J., "Shipyard Block Logistics Simulation Using Process-centric Discrete Event Simulation Method", *Journal of Ship Production and Design*, Vol. 34(2018): 168-179.

4. Matsuo, K., and Tanigawa, F. "Modelling of Work in Shipbuilding by Production Simulation", *Proceedings of Conference on Computer Applications and Information Technology in the Maritime Industries 2020*, 2020.

5. Okubo, Y., and Mitsuyuki, T. " Ship Production Planning Using Shipbuilding System Modeling and Discrete Time Process Simulation", *Journal of Marine Science and Engineering*, Vol.10, No. 176 (2022).

6. Monostori, L., Váncza, J., and Kumara, S.R.T., "Agent-Based Systems for Manufacturing", *CIRP Annals*, Vol.55 (2006): 697-720.

7. Xu, D., Lu, C., Zhou, W., and Liu, S., "Hybrid Model of Multi-agent and DEDS for Steelmaking-Continuous Casting-Hot Rolling Manufacturing Process Simulation", *The 26th Chinese Control and Decision Conference*, 2014.

8. Zhang, Y., Qian, C., Lv, J., and Liu, Y., "Agent and Cyber-Physical System Based Self-Organizing and Self-Adaptive Intelligent Shopfloor", *IEEE Transactions on Industrial Informatics*, Vol.13 (2017):737-747.

9. Unity, https://unity.com/

10. Unity Technologies, NavMeshComponent, https://github.com/Unity-Technologies/NavMeshComponents

8. AUTHORS BIOGRAPHY

Dr. Tomoyuki Taniguchi holds the current position of Chief Researcher at National Maritime Research Institute of National Institute of Maritime, Port and Aviation Technology. He is responsible for research and implementation of the utilization of computational science and technology in production and design on shipbuilding.

Dr. Kohei Matsuo holds the current position of Head of Industrial System Research Group at National Maritime Research Institute of National Institute of Maritime, Port and Aviation Technology. He is responsible for manufacturing engineering of shipbuilding. His experience includes various approaches on ICT development for shipbuilding process.

DEVELOPMENT OF A SHIPPING MARKET FORECASTING SYSTEM USING VESSEL MOVEMENT DATA AND ITS PRACTICAL APPLICATION

Y Wada and **T Majima,** National Maritime Research Institute, Japan
K Hamada and **H Hirayama**, Hiroshima University, Japan
D Watanabe, Marubeni Corporation, Japan
Y Igarashi, Kawasaki Kisen Kaisha, Ltd., Japan

SUMMARY

In the shipping industry, changes in freight rates are extremely drastic. Therefore, shipping companies needs to develop a forecasting method for freight rates to achieve sustainable growth. Recently, the details of the movement data of each ship have been obtained through the development of an automatic identification system (AIS). Therefore, we developed a practical forecasting model that combines AIS and statistical data. In our previous study, we proposed a method to develop a forecasting model for freight rates by combining AIS and statistical data using deep learning. However, there are some issues with the previous model. In this study, we analyzed the causes of deterioration in forecast accuracy from the perspective of the internal factors of the forecasting model. In addition, we developed a prediction system for demonstration tests and verified its applicability by comparing expert predictions.

1. INTRODUCTION

1.1 Background

In the shipping industry, changes in freight rates are extremely drastic. Therefore, it is important for shipping companies to develop forecasting methods for freight rates. In recent years, big data and deep learning methods have been developing rapidly and have been applied in several fields. In the shipping industry, the details of the movement data of each ship can be obtained and understood in real time by the development of an automatic identification system (AIS). Using AIS data, some studies have been conducted to support shipping companies' decision-making. In a previous study[1)], we proposed a freight rate forecasting model using AIS data and deep learning. The proposed model showed high accuracy using AIS data. However, the model is not practical, and it is insufficient to test the practical functioning of the developed forecasting model, such as real-time forecasting. The objective of this study is to develop a practical forecasting model for freight rates using a real-time forecasting system.

Since July 2002, all ships of 300 gross tons or more engaged in international navigation, all passenger ships, and all ships of 500 gross tons or more not engaged in international navigation have been obliged under the SOLAS Convention to install AIS for the purpose of avoiding ship collisions. The data transmitted by AIS include dynamic information (time (UTC), position (latitude, longitude), speed over ground, course over ground, etc.), static information (IMO number, name, captain, width, type, etc.), and navigation information (draft, destination, estimated time arrival (ETA), etc.). The transmission interval of AIS signals is 2–10 s when the vessel is navigating at 3 knots or more, and 3 min when the vessel is at anchor. In recent years, AIS signals have been collected via satellites and land-based stations, enabling the real-time monitoring of global vessel movements and the accumulation of data. Thus, AIS data covers data on global vessel movements and can be considered big data in the field of maritime transportation [2)].

1.2 Related studies

In general, analysis methods, such as trend analysis and time-series analysis, are used to understand the characteristics of the shipping market and to support decision-making in the maritime market. Manolis[3)] et al. investigated the seasonality of the freight rate for each vessel size in a dry bulk market. They found that spot freight rates for larger vessels exhibit higher seasonal fluctuations than those for smaller vessels, although differences in seasonal fluctuations between sectors were eliminated as the contract duration increased. Gu[4)] et al. used a VAR model to analyze the indicators affected by the shipping market; the VAR results showed that the Baltic Dry Index (BDI) is sensitive to FFA market movements and international oil prices.

On the other hand, vessel movement data, including AIS data, have been used in the fields of maritime market analysis and forecasting, as well as simulation of ship specification considerations in recent years. Prochazka et al.[5)] analyzed the relationship between the timing of transportation contracts, geographical characteristics, and freight rate indices by combining information on charter contracts with location information obtained from AIS data for crude oil tankers [2)]. Brancaccio et al. [6)] analyzed the variation in shipping costs by linking the AIS data of bulk carriers with fixture data and changes in shipping costs. Dimas et al.[7)] developed a system to support decision-making for ship specification planning using big data, including port calling data (generated from AIS data) and considering ship allocation, shipping cost, and GHG emissions.

In addition, in recent years, the development of machine learning methods, such as deep learning, has led to the development of technologies for predicting shipping market conditions. Sahin et al. [8] developed a model to forecast one day ahead of the BDI and demonstrated the effectiveness of the proposed model. Xiao et al. [9] used a hybrid approach integrating LSTM and ensemble learning techniques to forecast the China Coastal Bulk Coal Freight Index (CBCFI). The proposed method showed better prediction accuracy than the single ensemble learning model. Fuchigami et al. [10] applied the percolative learning method (a machine learning method) to BDI forecasting and attempted to improve its accuracy through various enhancements. Consequently, they confirmed the effectiveness of the proposed method. Kanamoto et al.[11] used AIS data and a machine learning method (multilayer neural network) to predict the Baltic Capesize Index (BCI) of a Capesize bulker. The author[1] also used AIS data and statistical data to predict bulk shipping freight rates in iron ore transportation using a multilayer neural network and showed that the addition of AIS data to the model enabled highly accurate prediction 1-2 months ahead.

Thus, research on the use of AIS data for analysis and support decision-making in the shipping market is in progress, and the effectiveness of AIS data and machine learning methods has also been confirmed. Similarly, a previous study by the authors[1] suggested that the forecasting model could improve the accuracy of shipping market forecasts by adding AIS data. However, from the standpoint of practical application, current research on the application of AIS data to shipping market forecasting is insufficient for analyzing the scope of model application, developing a real-time forecasting system, and analyzing forecast accuracy. Thus, it is necessary to verify these issues to develop a practical model and system.

2. OVERVIEW OF PREVIOUS STUDY

This section provides an overview of the previous study[1] and identifies issues.

2.1 Overview of the Prediction Model

The previous study focuses on tramp freight rates for iron ore transportation. The following are the subjects of this study:

- Target vessel: Bulk carrier
- Vessel size: 150,000 DWT - 220,000 DWT
- Target vessel: C3 (Brazil-China route)

A forecasting model for the shipping freight rate was developed based on the procedure shown in Figure 1.

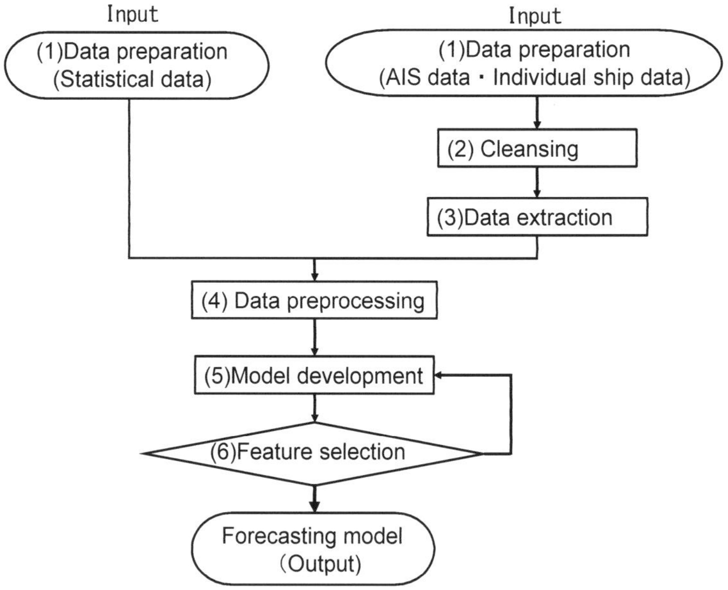

Figure 1: Overview of forecasting model development procedure.

AIS data were collected from exactEarth [12] (collected via satellite AIS) and FleetMon[13] (collected via land-based stations in coastal areas); these data were obtained from the IHI Jet Service. Statistical data such as freight rates, steel prices, iron ore prices, foreign exchange rates, and bunker prices were obtained from Clarkson[14], Investing.com[15], TRAMP Data Service [16], and other sources. Individual vessel data were obtained from Sea-web ships [17].

The AIS data were cleansed, extracted, pre-processed, and organized prior to model development.

2.2 Application of deep learning

The forecasting model is developed using the data processed in (1)-(4) shown in Figure 1. Figure 2 shows the structure of deep learning. A deep neural network (DNN), which is a simple multilayered version of a neural network, was used in the study. D denotes the set of variables (original series) of the statistical data and vessel movement data. In this study, the input data for D (original series) is the past 60 days from the current value. For example, if the number of vessels in surrounding Western Australia (extracted from AIS data), the number of vessels surrounding East Asia (extracted from AIS data), iron ore prices, and steel prices are input variables. In that case, D becomes a data set of four variables. Additionally, the input data o di e ence series) of the change from the previous day is the past 59 days from the current value. The output is the amount of change (difference) in freight rates one month (30 days) and two months (60 days) ahead.

The parameters used to develop the prediction model are listed in Table 1. A linear function was applied to the input and output layers, and the ReLU function was applied to the intermediate layers. Dropout was incorporated into all intermediate layers, and the dropout rate was set to 0.2. Batch normalization was also incorporated into all the layers. In addition, the hold-out method was used to develop the model, which was divided into a learning dataset and validation dataset, as shown in Table 1. The forecast dataset was used to analyze the prediction accuracy of the model during its unlearning period.

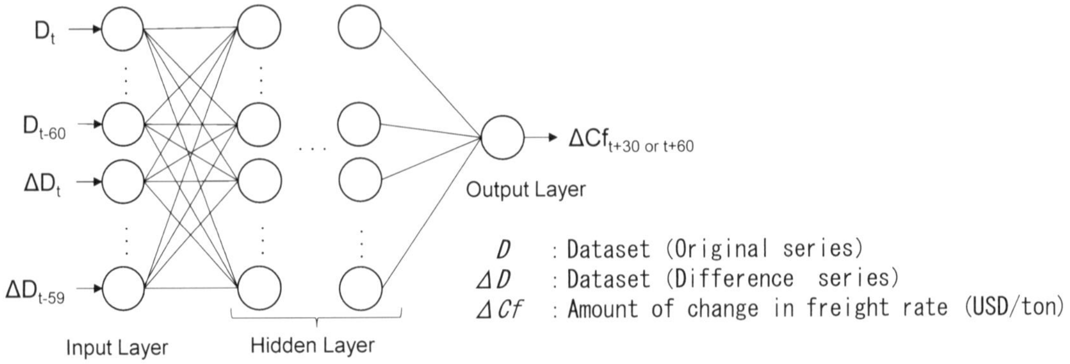

Figure 2: Structure of deep learning.

Table 1: Parameters for model development.

Hyperparameters of deep learning	
Normalization method	Min-Max normalization
Middle layers	9
Loss Function	Mean squared error (MSE)
Optimizer	Adam
Batch size	10
Epoch number	2000
Dataset distribution	
Learning dataset	990 days (October 28, 2014 - July 13, 2017)
Validation dataset	320 days (July 14, 2017 - May 29, 2018)
Forecast dataset	216 days (May 30, 2018 - December 31,2018)

In addition, feature selection was performed to improve the accuracy of the model. The influence of the input variables on the model was evaluated, and the model's input variables were selected. The following procedure was used to select the input variables:

A) The influence of the input variables used in the model development was evaluated using sensitivity analysis. If better results than the current status can be obtained by removing the input variables from the validation dataset of the model, the input variables are removed.

B) Perform the process in A) for all input variables and add a new variable if a better result than the current one cannot be obtained.

C) Re-develop the model with the new input variables.

Steps A)–C) are repeated until the accuracy of the model is no longer improved and the optimal configuration of the input variables is examined.

2.3 Simulation Results

To investigate the importance of AIS data and the effects of feature selection, three case simulations were conducted. The simulation span was set from May 30, 2018, to December 31, 2018. The accuracy of the forecasting model was tested using data from an untrained period (the forecast dataset in Table 1), which was not used to develop the model. The model was evaluated using the mean squared error (MSE) and mean directional accuracy (MDA)[18], which represents the degree of agreement between the forecast results and the direction of fluctuation of actual values.

Case 1: A model in which the input data were selected by combining statistical and AIS data.
Case 2: A model in which input data were selected based only on statistical data.
Case 3: A model in which statistical data and AIS data were combined. Feature selection was not performed.

Figure 3 shows the results of C3 (between Brazil and China) forecasting one-month (30 days) and two-month (60 days) ahead. Table 2 presents the prediction accuracy of each case model. Therefore, the accuracy of one-month-ahead and two-month-ahead forecasting in Case 1 was achieved at over 80% in MDA when feature selection was executed. In Case 1, the MSE also becomes minimized compared with Cases 2 and 3. However, MDA and MSE become worse, and it is difficult to follow the actual results in Cases 2 and 3.

Based on these results, the forecasting accuracy was improved by combining AIS data with statistical data. In addition, a model with high accuracy can be developed using feature selection.

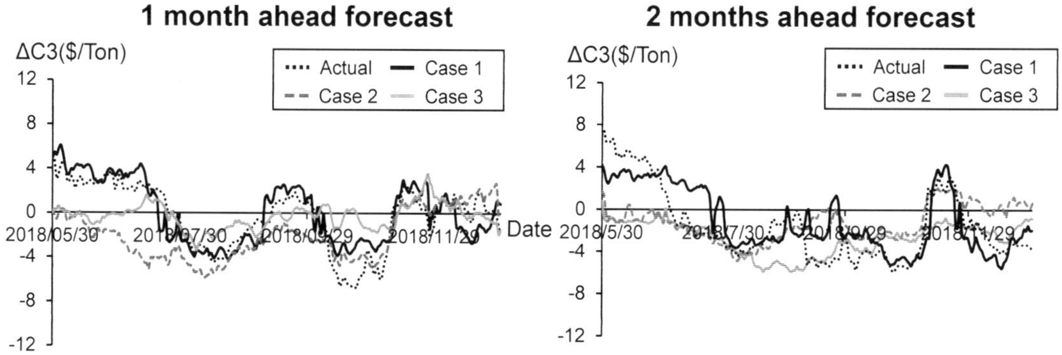

Figure 3: Simulation results (May 30, 2018 – December 31, 2018).

Table 2: Model accuracy (May 30, 2018 – December 31, 2018).

Target	Simulation Case	MDA	MSE
1 month ahead forecast	**Case 1**	**82.4%**	**2.59**
	Case 2	60.2%	9.07
	Case 3	67.6%	6.45
2 months ahead forecast	**Case 1**	**83.8%**	**4.89**
	Case 2	70.8%	9.50
	Case 3	76.9%	9.21

2.4 Issues of the Previous Study

As previously mentioned, in our previous study, we developed a forecasting model using AIS data and deep learning and demonstrated the high forecasting accuracy of the model and the effectiveness of AIS data in forecasting freight rates. However, the following issues must be addressed for practical applications:

- The reliability of the forecast is not sufficiently ensured, because the predictable range of the forecast model is not clear.
- The accuracy of forecasting models compared with forecasts by experts is not fully understood.

- It is not possible to verify the feasibility of the forecasting model in actual operation.

Based on the aforementioned issues, this study analyzes the internal factors of the forecasting models to determine the causes of the deterioration in forecasting accuracy. We will also develop a forecasting system and confirm the superiority of the forecasting model by comparing the experts' predictions.

3. BASIC CONCEPT

The subjects of this research are as follows:
Target vessels: Capesize bulk carriers (150,000 DWT – 220,000 DWT),
Target: Bulk carrier freight rates for iron ore transport - Target routes: BCI C5 (Western Australia-China freight rates) and BCI C3 (Brazil-China freight rates).

To address the issues raised in the previous studies described in Section 2.4, this study examines the following points:

- Applicability of the model

The proposed forecasting model does not sufficiently ensure the reliability of the prediction because there are periods when the prediction accuracy deteriorates and the causes have not been identified. Therefore, we examined the robustness of the forecasting model to extrapolated data inputs, which are generally considered difficult to predict, and analyzed the range of possible predictions. In general, deep learning methods are difficult to apply to data outside the range of the training data[19]. In this study, we analyzed the prediction accuracy of the extrapolation model. Additionally, we devised a method for handling cases in which the accuracy deteriorates and proposed a method for developing a robust shipping market forecasting model with a certain level of accuracy under all conditions.

In this study, the analysis was conducted one month ahead of the freight rates of C5 (between Australia and China). In the model development, we used 14 variables, including statistical data and vessel movement data extracted from AIS data, which were applied for feature selection. The model was developed using 1694 input variables. The forecasting model was developed with a training period of 990 days (from November 27, 2014, to August 12, 2017) and a validation period of 320 days (from August 13, 2017, to June 28, 2018). The test period was 947 days (June 29, 2018, to January 30, 2021).

- Demonstration test

Previous studies have only confirmed the accuracy of the models using archived data. Therefore, verification of the applicability of real-time forecasting is insufficient. In this study, we verified the accuracy of the forecasting model in real time by comparing it with the experts' predictions for a shipping company. To conduct a demonstration test, we developed a system for real-time forecasting of shipping market conditions. We also verified the feasibility of the forecasting model in an actual operation.

An overview of the demonstration test is presented in Table 3. This demonstration test compares the model's predictions with those of experts to gain an understanding of the model's forecasting ability. The average value of the forecasts by the five experts was used to evaluate the accuracy of the developed model. Owing to the nature of the data used, there is a time lag of approximately 2 days between data updates. The forecasting interval of the forecasting model was 30 days ahead, and the human forecasts were 28 days ahead to account for the effect of data lag.

Table 3: Demonstration test overview

	Method	Prediction target	Verification Period
Case 1	Deep learning model (the proposed model in this study)	1 month (30 days) ahead of freight rate for iron ore transportation from Brazil to China C3	April 6, 2021 – September 28, 2021
Case 2	Average forecasting result of five experts		

4. ANALYSIS OF THE MODEL'S SCOPE OF APPLICATION

4.1 Analysis Result

Figure 4 (a) shows the forecasting results using the model, and graph (b) shows the results of analyzing the extrapolation of the input variables. The first axis (stacked graph) shows the total number of extrapolations, and the second axis (line graph) shows the number of extrapolated items. The accuracy of the forecasting results cannot reflect the actual results for periods 1 and 2. These were common periods in which both the total number of extrapolations and the number of extrapolated items increased. Thus, we found that forecasting models have difficulty forecasting when conditions change significantly from the past shipping market.

Additionally, the model accuracy worsens when the total number of extrapolations exceeds 25 and the number of extrapolated items exceeds three. In this study, we defined "in alert" when the total number of extrapolations and the number of extrapolated items exceed the limits. The other case defines "in normal." Table 4 presents the accuracy of the model. The MDA shows 65.5% and MSE shows 3.09 in total period. when in the "alert" case, the MDA shows 59.3% and MSE shows 5.03. On the other hand, when in the "normal" case, the MDA shows 67.0% and MSE shows 2.61. The accuracy becomes worse compared with the "in normal" case when in the "alert" case.

Based on this analysis, we consider the improvement method of the model.

4.2 Improvement of the Model

Based on the analysis results in Figure 4, we conducted the following measures when the "in alert" case is occurred.

A) Redevelop the model by adding the most recent data as training data. After that, the model is replaced to new redevelop model.

B) Switch to FFA (forward freight agreement) price if the alert still occurs after taking countermeasures A).

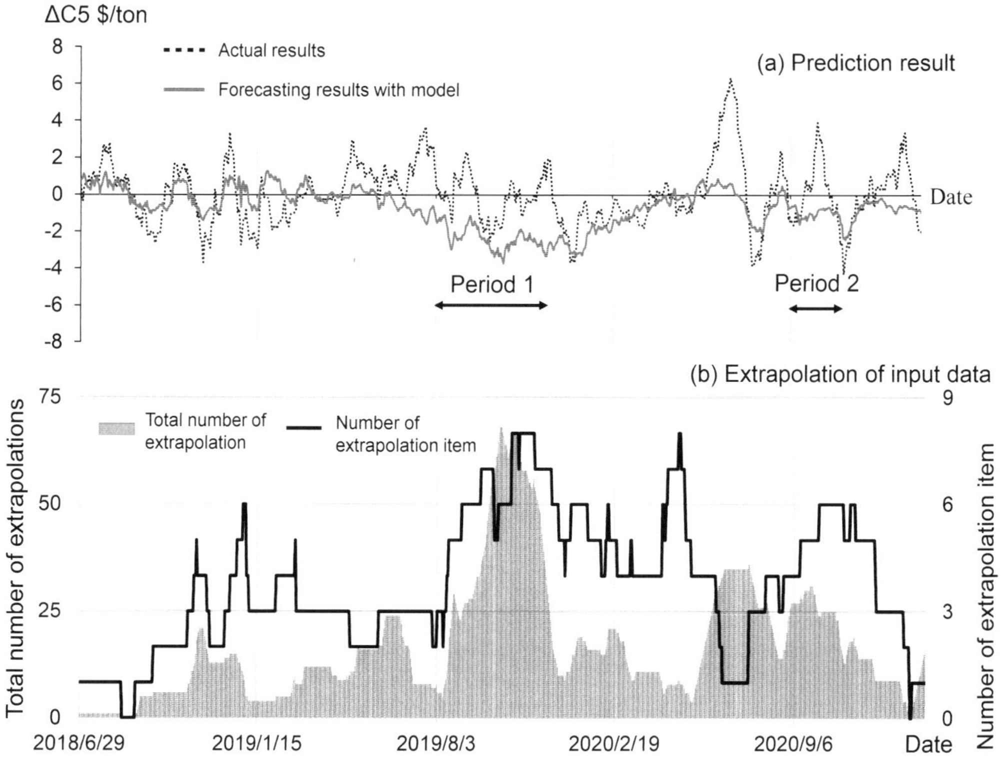

Figure 4: Simulation results and extrapolation analysis (June 29, 2018- January 30, 2021).

Table 4. Model accuracy (June 29, 2018- January 30, 2021).

	Total		In alert		In normal	
	MDA	MSE	MDA	MSE	MDA	MSE
Forecasting model	65.5%	3.09	59.3%	5.03	67.0%	2.61

4.3 Application of proposed methods

Figure 5 shows the time series of the forecast results. As shown in Figure 5, the introduction of FFA price has resulted in a significant change in the results from August 2019 to January 2020. The results also show a significant change from July 2020 to January 2021 owing to the redevelopment of the model. Table 5 shows the forecasting results using the aforementioned measures. As shown in Table 5, MDA shows 70.5% and MSE shows 2.29 in total period using the improved method. It can be observed that the accuracy of both MDA and MSE is improved compared to the results when the alerts are not countermeasure.

The results show that the model maintaining the accuracy can be developed even when extrapolations occur.

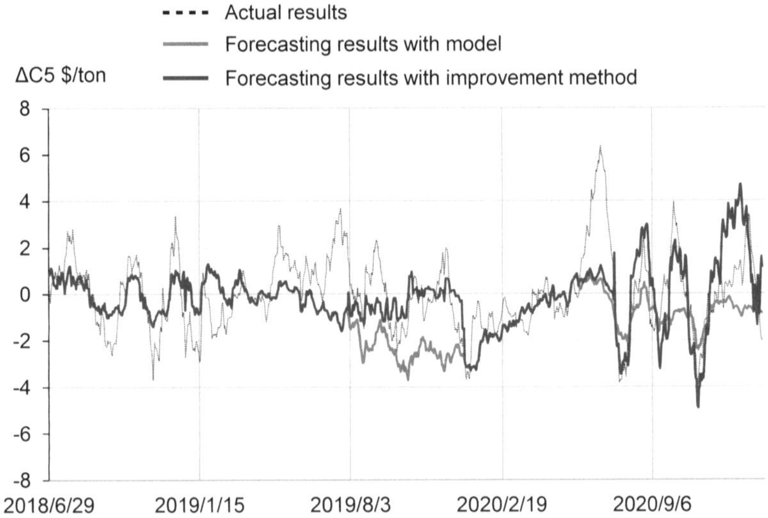

Figure 5: Simulation results (June 29, 2018- January 30, 2021).

Table 5: Model accuracy (Forecasting model vs Improvement method).

Method	MDA	MSE
Forecasting model	65.5%	3.09
Improvement method	70.5%	2.29

5. DEMONSTRATION TEST

5.1 Development of Maritime Market Forecasting System

A real-time forecasting system for freight rates was developed to verify the actual operation. Figure 6 shows an overview of the proposed system. The system consists of three functions: (1) storing the latest data in a database, (2) generating data to be input into a forecast model and running the forecast model, and (3) sharing the values calculated by the forecast model and information on vessel movements in each sea area with system users on the Web. The results of the forecasts can be viewed on the web browser, and the forecast values one month ahead (original series and difference series), two months ahead (original series and difference series), and the actual results (original series and difference series) can be checked (Figure 7).

The system was implemented in a Python environment using Microsoft Azure[20)], Internet Information Services (IIS) [21)], wfastCGI [22)], and Flask [23)]. This enabled to access the forecast results externally via the web. Owing to the data distribution of the data source, delays of several days occasionally occurred in updating the results. However, no major delays occurred in the delivery of data, and it was confirmed that the system could be used in actual operations without stopping in the middle of the project.

5.2 Compared with Model Prediction and Experts' Prediction

Figure 8 and Table 6 show the results of predicting the freight rates between Brazil and China one month ahead. From April 6, 2021, to June 2, 2021, the ver. 1 model's accuracy deteriorated significantly because the price of steel, which is one of the input variables, soared to an all-time high. It can be observed that experts were also unable to predict this steep rise in freight rates.

Therefore, ver. 2 models were redeveloped by adding the period when the prices soared as training data, and forecasts were created from June 3, 2021. Therefore, as compared to the ver. 1 model, the ver. 2 model significantly improved accuracy and obtained a result 14.2% better than the practitioner's forecast for MDA.

These results indicate that when extrapolations occur, it is possible to redevelop the model to achieve a higher forecasting accuracy than that of experts.

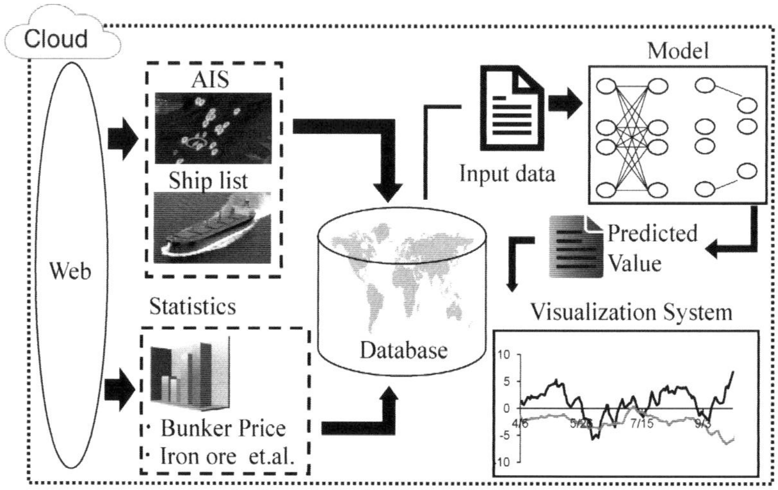

Figure 6: Overview of real-time forecasting system.

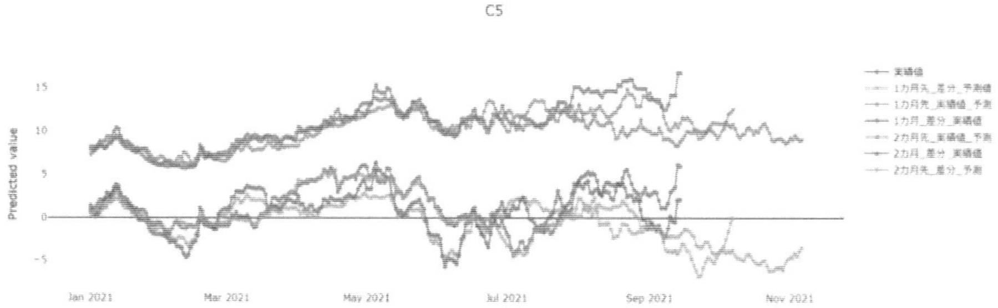

Figure 7: Example of visualization results.

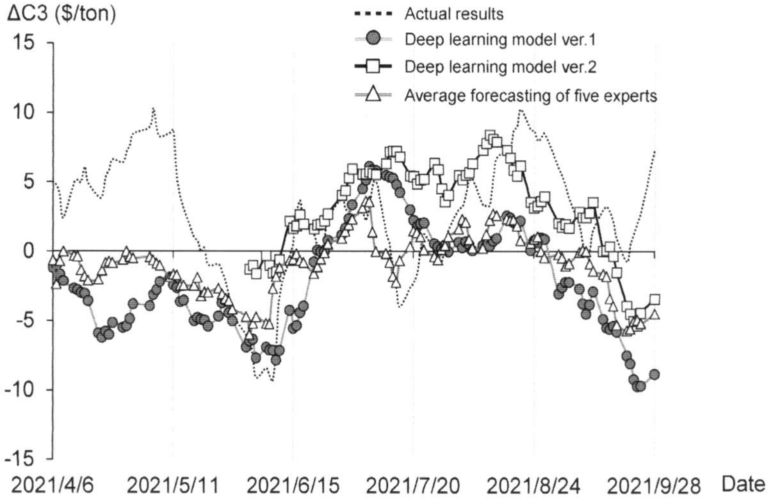

Figure 8: Simulation results (April 6, 2021 – September 28, 2021).

Table 6: Model accuracy (April 6, 2021 – September 28, 2021).

	April 6, 2021 – June 2, 2021		June 3, 2021 - September 28, 2021	
Method	MDA	MSE	MDA	MSE
Deep learning model ver.1	21.1%	74.2	59.8%	32.3
Deep learning model ver.2		-	75.6%	21.3
Average forecasting of five experts	28.9%	34.2	61.4%	16.4

6. CONCLUSIONS

In this study, we identified the scope of application of the model and conducted demonstration test in order to develop a practical forecasting model of freight rate. The findings of this study are summarized as follows;

- The impact of extrapolation of input variables on forecast accuracy was quantitatively analyzed. It was found that accuracy deteriorated significantly when the number of extrapolations exceeded 25 and the number of items exceeded 3.
- When extrapolation increases, the model can be redeveloped by learning the previous data. If extrapolation continues to increase after the model is redeveloped and does not improve the model accuracy, the forecasting accuracy can be improved and maintained by switching to FFA.
- For verification of actual operation of the model, we developed a system to forecast freight rates in real-time. The system was verified to be operational without any interruptions.
- A demonstration test was conducted to compare the accuracy of the forecasting model with experts. Although the initial version of the forecasting model was low accuracy by the effects of many extrapolations that occurred during the demonstration test period, the redeveloped model, which had been trained up to the immediately preceding period, achieved an accuracy that exceeded that of the experts.

In this study, tramp freight rates for iron ore transportation were analyzed. Future research should focus on other vessel sizes and types.

7. ACKNOWLEDGMENTS

This work was supported in part by JSPS KAKENHI Grant Numbers JP21K04501, JP20H00286, and JP20H02371.

8. REFERENCES

1. Wada, Y., Hamada, K., 'Dry Bulk Freight Rate Forecasting based on Maritime Big Data Using Deep Learning', *Kanrin : bulletin of the Japan Society of Naval Architects and Ocean Engineers 97: 46-50*, 2021. (in Japanese)

2. Wada, Y., Shibasaki, R., Kosaka, H., Watanabe, D., Ito, H., Tsubota, K., Aratani, T., Izumiyama, T., Iwasa, R., Literature Review and Future Prospects on Vessel Movement Data Usage including AIS □ Focusing on Logistics, Shipping, Shipbuilding, Environment, and Cruise Fields□ □ *Journal of the Japan Society of Naval Architects and Ocean Engineers 34:123-138*, 2021. (in Japanese)

3. Manolis, K., Amir, A., 'Seasonality patterns in dry bulk shipping spot and time charter freight rates', *Transportation Research Part E: Logistics and Transportation Review 37(6):443-467*, 2001.

4. Gu,Y., Dong,X., Chen, Z., 'The relation between the international and China shipping markets', *Research in Transportation Business & Management 34*, 2020.

5. Prochazka, V., Adland, R., Wolff, F.C., 'Contracting decisions in the crude oil transportation market: Evidence from fixtures matched with AIS data', *Transportation Research Part A: Policy and Practice 130:37–53*, 2019.

6. Brancaccio, G., Kalouptsidi, M., Papageorgiou, T., 'Geography, transportation, and endogenous trade costs', *Econometrica 88(2): 657-691*, 2020.

7. Muzhoffar, D. A. F., Hamada, K., Wada, Y., Miyake, Y., Kawamura, S. Basic Ship-Planning Support System Using Big Data in Maritime Logistics for Simulating Demand Generation. Journal of Marine Science and Engineering 10(2):186, 2022..

8. ahin, , gen, , n e , , n, , 'Forecasting the Baltic Dry Index by using an artificial neural network approach', *Turkish Journal of Electrical Engineering & Computer Sciences 26.3:1673-1684*, 2018.

9. Xiao, W., Xu, C., Liu, H., Liu, X., 'A Hybrid LSTM-Based Ensemble Learning Approach for China Coastal Bulk Coal Freight Index Prediction', *Journal of Advanced Transportation 2021*, 2021.

10. Fuchikami, J., Nagao, T., 'Proposal of Percolative Learning Method for the Baltic Dry Index Forecasting', *Journal of the Japan Society of Naval Architects and Ocean Engineers 33: 199-207*, 2021.

11. Kanamoto, K., Wada, Y., Shibasaki, R., 'Predicting a dry bulk freight index by deep learning with global vessel movement data', *Advances in Transdisciplinary Engineering 10: 105−114*, 2019.

12. exactEarth: https://spire.com/maritime/?utm_campaign=maritime_2022_exactearth_redirect&utm_source=exactearth&utm_medium=website&utm_content=homepage (Accessed on 12 Apr 2022).

13. FleetMon: https://www.fleetmon.com/ (Accessed on 12 Apr 2022).

14. Clarkson: https://www.clarksons.com (Accessed on 12 Apr 2022).

15. Investing.com: https://jp.investing.com/ (Accessed on 12 Apr 2022).

16. TRAMP Data Service: http://www.tramp.co.jp/ (Accessed on 12 Apr 2022).

17. Sea-web ships: https://ihsmarkit.com/products/sea-web-vessel-search.html (Accessed on 12 Apr 2022).

18. Bergmeir, C., Costantini, M., and Benitez, J.M., 'On the usefulness of cross-validation for directional forecast Evaluation', *Computational Statistics and Data Analysis 76:132-143*, 2014.

19. Reichstein, M., Camps-Valls, G., Stevens, B., Jung, M., Denzler, J., Carvalhais, N., 'Deep learning and process understanding for data-driven Earth system science', *Nature 566:195–204*, 2019.

20. Microsoft Azure: https://azure.microsoft.com/ja-jp/get-started/azure-portal/ (Accessed on 05 Jul 2022).

21. Internet Information Services: https://docs.microsoft.com/ja-jp/security-updates/planningandimplementationguide/19870063 (Accessed on 05 Jul 2022).

22. WfastCGI: https://pypi.org/project/wfastcgi/ (Accessed on 05 Jul 2022).

23. Flask: https://flask.palletsprojects.com/en/2.1.x/ (Accessed on 05 Jul 2022).

9. AUTHORS BIOGRAPHY

Yujiro Wada holds the current position of chief researcher at the Knowledge and Data System Department of the National Maritime Research Institute. His previous experience includes maritime economics analysis, modelling, and system development.

Kunihiro Hamada is currently a professor at the Graduate School of Advanced Science and Engineering, Hiroshima University. His previous experience includes research on design support systems for shipbuilding.

Hiroki Hirayama was a Master's student at the Graduate School of Advanced Science and Engineering, Hiroshima University, at the time of this research.

Takahiro Majima holds the current position of director at the Knowledge and Data System Department, National Maritime Research Institute. His previous experience includes research on shipping optimization.

Daisuke Watanabe holds the current position of data scientist at Marubeni Corporation. His previous experience includes maritime market modelling.

Yusuke Igarashi holds the current position of project manager at Kawasaki Kisen Kaisha, Ltd. His previous experience includes the management and introduction of big data systems.

STUDY ON SIMULATION BASED EVALUATION OF ROUTE AND CARGO SPECIFIC PROJECT FOR INTRODUCING DECARBONIZED SHIPS

S Wanaka, National Maritime Research Institute, Japan
R Kakiuchi, Planning and Design Center for Greener Ships, Japan
N Fujita, Mitsubishi Research Institute, Japan
K Hiekata, University of Tokyo, Japan
H Yamato, Japan Agency for Marin-Earth Science and Technology, Japan

SUMMARY

In recent years, International Maritime Organization (IMO) adopted an initial strategy for GHG reduction from ships. Cargo owners, shipping companies, and ship owners have started to consider how to respond to the strategy. The introduction of decarbonized ships is a future direction to fulfil the IMO's strategy. However, for each route and cargo, the situation is different, and it is necessary to consider the introduction plan of decarbonized ships that is appropriate for each.
This paper proposes an evaluation method for the introduction project of decarbonized ships. The method assumes specific route and cargo, and various scenarios of decarbonization technologies, environmental regulations, fuel prices, and infrastructure deployment for alternative fuels, and evaluates costs such as OPEX, CAPEX, and amount of GHG emission based on a simulation of fleet transition aiming to the zero-emissions target time. The method provides a quantitative evaluation of the introduction project to the stakeholders and supports their decision-making. By changing the inputted parameters, sensitivity analysis to consider future uncertainty can be conducted based on the method.
As a case study, we assume several types of alternative fuelled ships, and alternative fuel-ready ships, and evaluate the environmental performance and costs of introducing them to iron ore transportation between Australia and Japan. Scenarios of infrastructure deployment for alternative fuels is assumed as well, and by comparison of the cases, the usefulness of the fuels and technologies in each case is discussed.

1. INTRODUCTION

In response to growing global demand to strengthen efforts to prevent global warming, leading international cargo owners and shipping companies have announced to achieve zero emissions by 2050. In addition, to evaluate the reduction of GHG emissions from international shipping by financial institutions, such as those represented by the Poseidon Principles [1], are emerging. Against these developments, the IMO is about to consider net zero by 2050 as a readjust of its initial strategy [2] on GHG reduction. With the fuel efficiency performance rating system for existing vessels that are scheduled to be introduced in 2023, there is a possibility that international shipping will be required to respond to zero carbon including existing ships in the latter half of the 2030s.
On the other hand, about the realization of zero emissions in international shipping, factors which are impactful for zero emissions are still uncertain, such as GHG emission regulations, frameworks for related economic measures, technological development such as corresponding propulsion plants, and the supply chain and price of zero-carbon and carbon-neutral fuels [3].
For this reason, various scenarios are considered to achieve zero emissions in international shipping, and in these circumstances, it is meaningful for related shipowners, ship operators, and financial institutions to conduct economic and environmental simulations for various scenarios and to select appropriate responses and countermeasures. Especially for planning an actual decarbonization project, each project has a different target cargo, route, and duration to achieve decarbonization. Therefore, an evaluation method for the project with a specific route and cargo is meaningful to support the shipowners', operators', and financial institution's decision-making and planning.
The objective of this study is to propose an evaluation method for the introduction project of decarbonized ships. The method assumes specific routes and cargo, and scenarios of decarbonization technologies, environmental regulations, fuel prices, and infrastructure deployment for alternative fuels, and evaluates costs such as OPEX, CAPEX, and amount of CO_2 emissions based on a simulation of fleet transition. Based on the outputs, it is examined how the fuel composition ratio of fleets engaged in the target route will change toward zero emissions. Furthermore, it is possible to evaluate how the transition of the fleet composition changes depending on the infrastructure development circumstances in the target route. In this paper, some cases are calculated by the simulation, and it demonstrates the usefulness of the proposed method of project-based evaluation for the decarbonization of shipping.

2. SIMULATION METHOD

2.1 OVERVIEW

The authors have proposed a global scale model to analyze regulations effectiveness and shipbuilding demand from the macro perspective [4], but this paper takes a more micro perspective and aims to develop a simulation method for project evaluation, focusing on specific routes and cargoes. Table 1 shows an overview of the global scale model and the project-based model.

Table 1: Summary of Global Scale and Project-based Model

	Global Scale Model [4]	Project-based Model
Objective	Macro scale analysis of regulations and shipbuilding demands	Project evaluation of decarbonization of fleet
Outline	The simulation is conducted for ships operated all over the world to analyze the macro scale behaviour in changing regulations and/or ship's performance newly built. The main target of the simulation is to support decision-making for the regulations or to determine the roadmap to achieve decarbonization from international shipping.	The simulation is conducted for a case in which the route, cargo, and ships are assumed specifically. It evaluates the impact of infrastructure development on fleet transition, and the cost. The main target is for the evaluation of environmental or economic performance when planning a specific project.
Target	Ships operated all over the world. About 10,300 bulk carriers are covered in the study [4].	Ships operated in the specific route. In this study, as an example, 23 Capesize bulk carriers engaged in the transport of iron ore between Japan and Australia are focused.

2.2 PROJECT-BASED SIMULATION FOR DECARBONIZATION OF SHIPPING

Figure 1 shows an overview of the simulation method of this study, which consists of 3 processes: operation, scrapping, and shipbuilding. The base of the method's concept is the same as the global scale model, but the detail of the model in the simulation is different. Each model is described as follows.

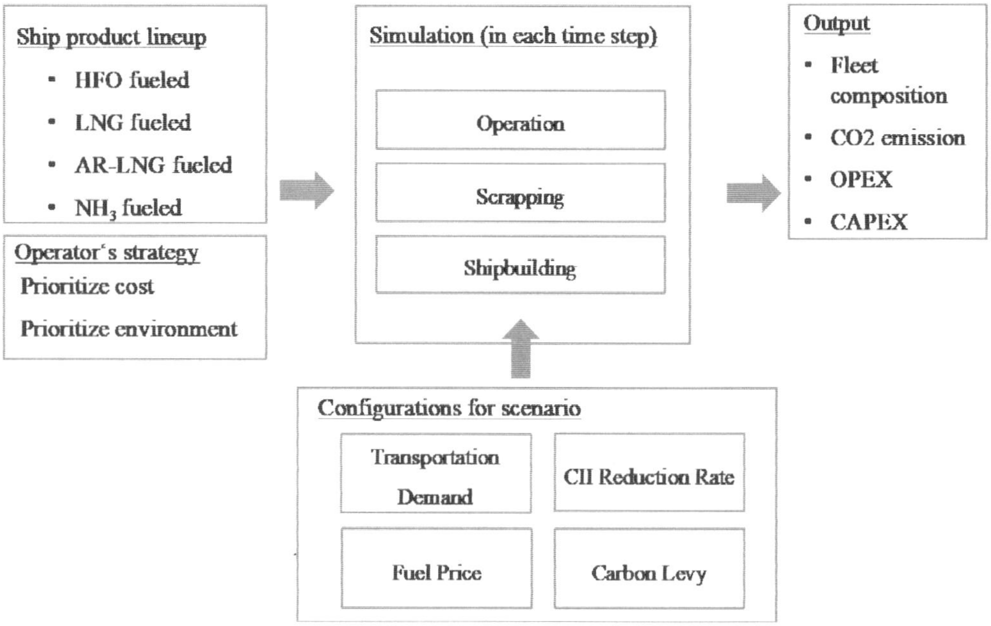

Figure 1: Overview of the simulation's input and output

For the operation model, we solve optimization problems such as equations (1) - (4) to obtain the number of roundtrips x_i for each ship i in the fleet. $Transportwork$ is the annual transport demand (tons), dwt_i is the DWT of vessel i, V is the ship speed in knots, and $distance$ is the voyage distance required for one round trip (miles), W_{ME} is the main engine output (kW), SFC_{ME} is the fuel consumption rate of the main engine (g/kWh), W_{AUX} is the auxiliary engine output (kW), and SFC_{AUX} is the fuel consumption rate of the auxiliary engine, $Price_{fuel}$ is the fuel price (USD/g), and $Cost_{op,i}$ is the cost of one day of operation, max_round is the maximum number of round trips calculated from the number of days at sea. The mathematical programming library PulP [5] was used to solve the optimization problem. In the model, differences in bunkering port and fuel price of each port are expressed by configuration files for $distance_i$, and $Price_{fuel}$.

$$\min \sum_i Cost_i \cdot x_i \tag{1}$$

$$Transportwork - \sum_i dwt_i \cdot x_i \leq 0 \tag{2}$$

$$Cost_i = \frac{distance_i}{V_i}\left((W_{ME}SFC_{ME} + W_{AUX}SFC_{AUX}) \cdot Price_{fuel} + \frac{Cost_{op,i}}{24}\right) \tag{3}$$

$$x_i \leq \max\_round \tag{4}$$

The speed V_i is pre-defined as input, but operational speed drop can be occurred by CII rating evaluation [6]. CO_2 emission is calculated in each time step, and if the CII rate is D for 3 years or E, the ship speed is reduced until the CII rate becomes C. Equations (1) – (4) assume operators prioritize their cost effectiveness to decide their operation. However, the assumption depends on the operator, and it can be considered some operators prioritize environmental performance more than cost effectiveness. Therefore, this simulation can input the operator's strategy: prioritizing cost or prioritizing environment. If the prioritizing environment is selected, the fleet is operated to minimize the amount of CO_2 emission by changing the objective function. The strategies affect the shipbuilding process as well.

The scrapping model is very simple and based on ships' age. Around 20 years are assumed as the lifetime, and ships that reached that age are scrapped. For the shipbuilding model, the number of ships newly built is determined for fulfilling transportation capacity equals the transportation demand. After the number of ships to be built is determined, it is needed to select the type of fuel for new ships to be built. In this study, heavy fuel oil (HFO), LNG, ammonia-ready LNG (AR-LNG), and ammonia (NH_3) can be selected. In the step, the operator's strategy affects the decision-making. If the strategy is prioritizing cost, the vessel with the lowest lifecycle cost is selected, and if the strategy is prioritizing environment, the vessel with the lowest CO_2 emissions over its lifecycle is selected. The lifecycle cost is calculated by Equation (5). ASY is the assumed years of operation, $FC_{i,t}$ is the fuel cost, $OC_{i,t}$ is the operating cost, $CC_{i,t}$ is the carbon charge, SC_i is the ship price (USD), IR is the assumed annual interest rate, RP is the assumed payback period (12 years).

$$AC_i = \frac{1}{\sum_{t=1}^{ASY} dwt \cdot D_t}(CAPEX_i + \sum_{t=1}^{ASY}(FC_{i,t} + OC_{i,t} + CC_{i,t})) \tag{5}$$

By the simulation, the transition of fleet composition, amount of CO_2 emission, and cost including OPEX, and CAPEX is calculated. In this study as an operation cost (OPEX), only fuel consumption cost is considered. Figure 2 shows the UI image for the simulation output. For the explanation, A ~ E is assigned in the figure. The Gantt chart A expresses when the vessels in the fleet are built and scrapped, and which fuel is utilized for the operation. Each row represents a ship in the fleet, the bar is the lifespan for the ship. The color represents the fuel used, blue is HFO, red is LNG, and green is NH_3. CO_2 emission is shown in line chart B, and the cost is shown in chart C. In the cost chart, blue is OPEX, red is CAPEX, purple is carbon levy, and green is the modification cost of AR-LNG at the modification from LNG fuelled to NH_3 fuelled. Chart D represents percentages of fuel used in the fleet, and Chart E represents the number of ships newly built.

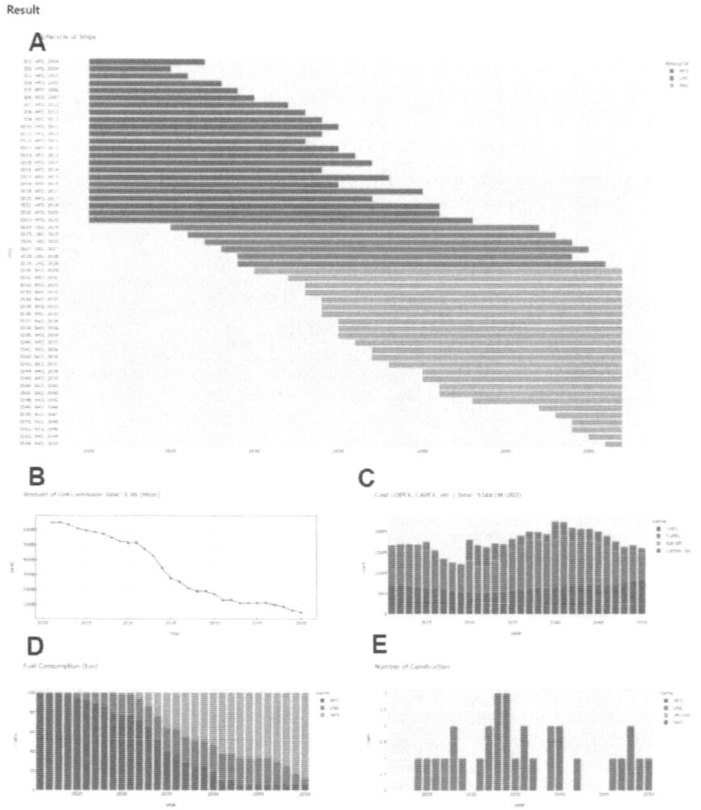

Figure 2: Output and UI example of the simulation

3. CASE STUDY

3.1 OVERVIEW

This chapter assumes the transport of iron ore between Japan and Australia as a specific route, and simulations are conducted for Capesize bulk carriers engaged in this route. Table 2 summarizes the conditions for iron ore transportation between Japan and Australia assumed in this case study. The target vessels are 200,000 DWT class Capesize bulk carriers. The loading port is Port Hedland, one of the leading iron ore export ports in Western Australia. The iron ore carriers are assumed to be destined for Japanese steel mills, and the case study did not take into account the differences in range and equipment limitations of different ports in Japan. The initial fleet of 23 vessels is assumed to be engaged on the route. The transportation demand for iron ore on the route is assumed to be 35 million tons per year, with no increase or decrease in demand during the entire period of the simulation. The NH_3 fuel technology is assumed to become widely available after 2025, and NH_3-fueled vessels can be introduced after 2025. Bunkering ports are assumed to be Japan as the landing port and Singapore as the intermediate port, details of those are shown in 3.2(a).

Table 2: General Setting for Case Study

Items	Contents
Target fleet	200kDWT class cape size bulk carrier
Target route	West Australia – Japan
Cargo	Iron ore
Loading port	Port Hedland
Unloading port	Japan
Initial fleet	23 ships
Transportation demand	35 M ton / year
Introduction of NH_3	After 2025
Bunkering port	Port for unloading (Japan) or intermediate port (Singapore)

3.2 CASE SETTING AND INPUT DATA

3.2 (a) Route and Bunkering Ports

The detailed route in this case study is shown in Figure 3. It is assumed that the route of the voyage at ballast condition varies depending on the location where bunkering is performed. The route from the loading port to the unloading port is assumed to be 4,000(SM), regardless of the bunkering location. On the other hand, the voyage at ballast condition is assumed to be 4,000(SM) when bunkering in Japan, as in the case of outbound voyages, but 5,000(SM) when bunkering in Singapore. HFO and LNG are assumed to be bunkering in Japan for the entire simulation period. Bunkering locations change at different times of the year for operations by NH3. In this case study, the bunkering of NH_3 takes place in Singapore from 2025 to 2035 and in Japan after 2035. This is an assumption that the development of infrastructure for NH_3 bunkering proceeds in Singapore, and as NH_3 becomes common fuel for shipping, the development of infrastructure in Japan follows. For Japan-Australia routes, bunkering in Singapore would be operationally disadvantageous due to the longer route so once the NH_3 supply infrastructure in Japan is developed, bunkering of NH_3 is conducted in Japan without calling Singapore.

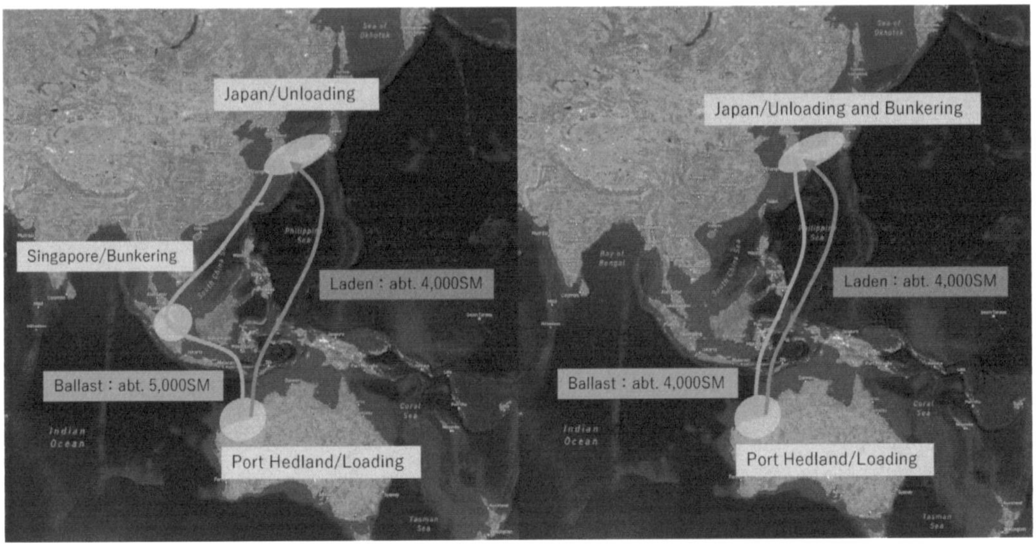

Figure 3: Route and bunkering port setting (Left: bunkering at Singapore, right: bunkering at Japan)

3.2 (b) Regulations

The reduction rates in the CII evaluation are set as shown in Figure 4. In this case study, it is assumed that the reduction rate is 1.5% per year until 2030, with 2018 as the base year, and then linearly reduced to zero in 2050. This CII reduction rates definition is still under discussion in IMO, and Figure 4 is not the latest one. The carbon charge is set at USD 200 per CO2-ton starting in 2030 and USD 500 per CO2-ton starting in 2040, as shown in Figure 5.

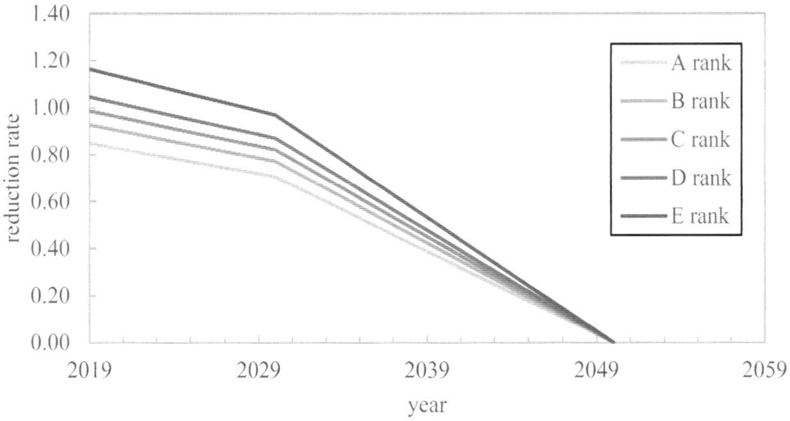

Figure 4: Setting of CII Reduction Rate for the Case Study

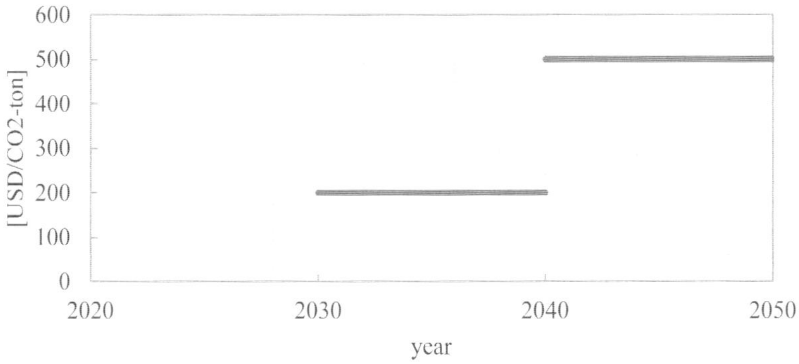

Figure 5: Setting of Carbon Levy

3.2 (c) Fuel Prices

The fuel price per GJ (USD) used in the calculation is set as shown in Figure 6. These figures are also studied in various industry sectors. The values in Figure 6 are selected as examples for the simulation. The ammonia fuel is blue ammonia produced from LNG with carbon capture and storage. NH_3's prices are calculated assuming that the fuel is produced at the natural gas production site (Middle East) and transported to the supply points in Japan and Singapore respectively. The regional difference in the fuels is driven by the difference in transportation costs only. The bunkering costs are not considered. The prices of HFO and LNG do not take into account regional price differences and bunkering costs.

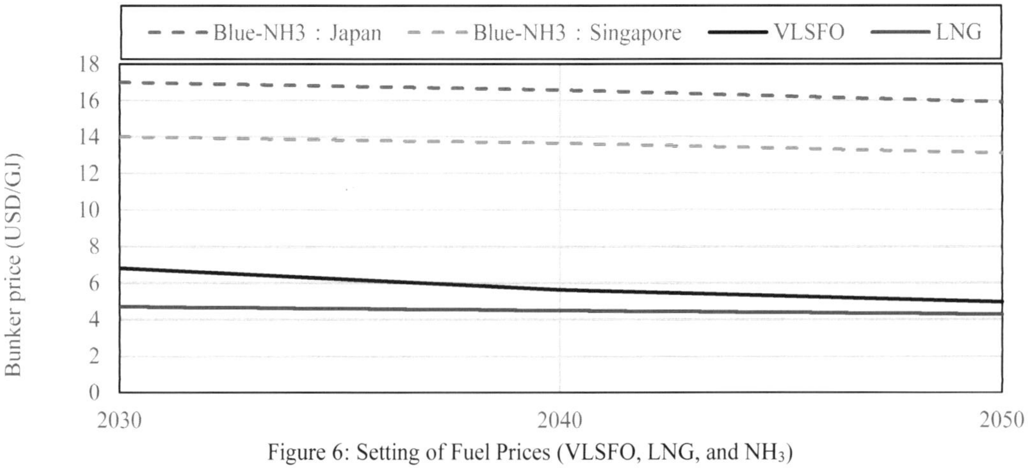

Figure 6: Setting of Fuel Prices (VLSFO, LNG, and NH_3)

3.2 (d) Initial Fleet and Product Line-ups for Ships Newly built

Based on AIS data for vessels engaged in iron ore transportation between Japan and Australia, the year-of-build distribution of the 23 vessels in the initial fleet is set as shown in Figure 7. The performance and main engine output of the initial fleet are based on average data for vessels in each year of construction, taking into account the requirements in each phase of the EEDI.

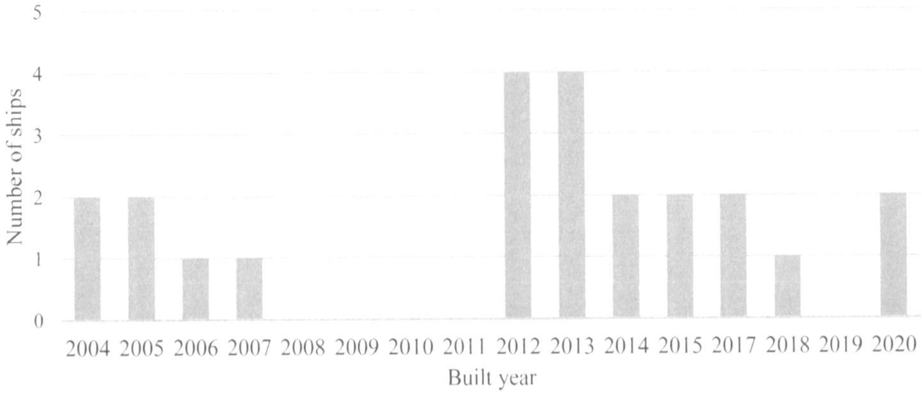

Figure 7: Assumed distribution of construction years for the 23 vessels in the initial fleet

For product line-ups which are the input of the simulation, 4 types of vessels are assumed; HFO fuelled, LNG fuelled, AR-LNG fuelled, and NH_3 fuelled ships. The performance and ship's price of each fuel type concept ship were established based on the IMO 4th GHG Study [7] and the author's experience. Especially, the performance of AR-LNG-fuelled and NH3-fueled vessels are designed based on LNG-fuelled vessels. The estimated costs of the ships are shown in Figure 8, where between the AR-LNG-fuelled and NH_3-fueled vessels, the modification cost is about 18% of the ship's price, and the price of NH_3 fuelled vessel increases by about 2% from the LNG-fuelled vessel's price.

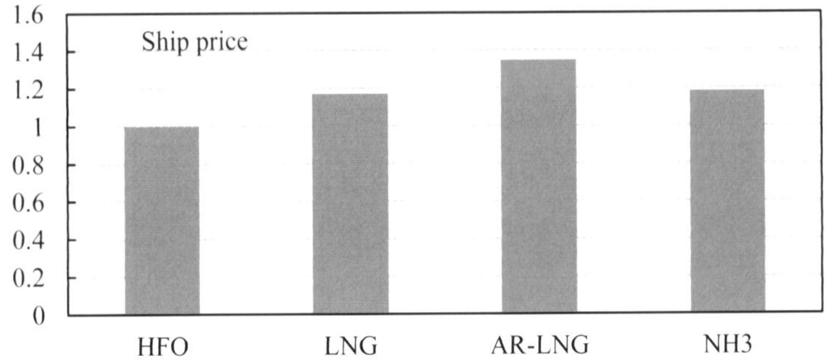

Figure 8: Assumed cost of each product concept

Each product concept has available timing and propulsion performance as shown in Figure 9. HFO-fuelled ships are built with performance equivalent to EEDI Phase 2 until 2024 and EEDI Phase 3 after 2025. Based on the performance of HFO-fuelled vessels, LNG-fuelled vessels were assumed to have a 10% reduction in CO_2 emissions by conversion to LNG fuel. AR-LNG-fuelled vessels are assumed to have the same propulsive performance as LNG-fuelled vessels, and when the CII rating is D for 3 consecutive years or E for a single year, the vessel's fuel switches from LNG fuel to NH_3 fuel. NH_3-fueled vessels can be introduced after 2025 as mentioned above, and the performance of the vessels themselves is assumed to be equivalent to EEDI Phase 3 for HFO-fuelled vessels. Since pilot fuel is not considered, NH3-fueled vessels are assumed to be zero emissions. Methane slip and N_2O are also not considered.

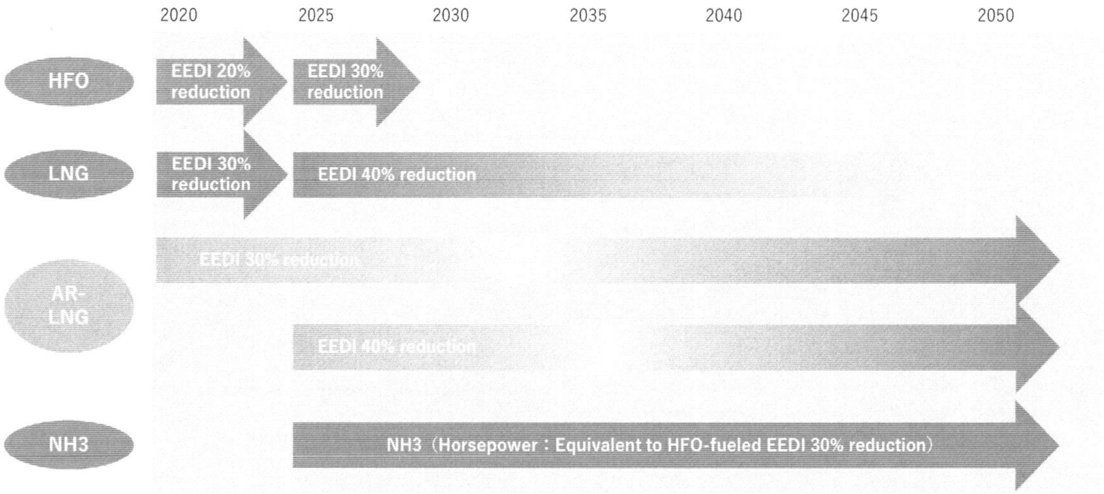

Figure 9: Available timing of each product and the performance improvement

3.2 (e) Summary of the Cases

The objective of this case study is to compare the simulation results in changing the bunkering port and the operator's strategy and demonstrate how to utilize the simulation to discuss and evaluate the decarbonization project. The list of cases in this case study is shown in Table 3. In case 1, all bunkering of fuels is conducted in Japan, while in cases 2 and 3, HFO and LNG are bunkered in Japan, NH_3 is bunkered in Singapore from 2025 to 2035, and in Japan after 2035. The operator's strategies for case 1 and case 2 are prioritizing cost, while case 3 is prioritizing environment; comparison and evaluation can be made between cases 1 and 2 based on differences in bunkering locations and between cases 2 and 3 based on differences in operator's strategies.

Table 3: Case lists of the Case Study

Case No.	Case 1	Case 2	Case 3
Case title	Base case	Singapore bunkering	Prioritizing environment
Bunkering port	Japan NH_3 can be introduced after 2025	HFO, LNG□ Japan NH3□ 2025-2035 Singapore 2035- Japan	HFO, LNG□ Japan NH3□ 2025-2035 Singapore 2035- Japan
Operator's strategy	Prioritize cost	Prioritize cost	Prioritize environment

3.3 RESULT

The results of the simulations for cases 1~3 is shown in Figure 10 ~ 18 expressing lifetime and fuel used for each ship, time series data of CO_2 emission (M ton), and cost (M USD). The horizontal axis of each graph is a time series (year). The vertical axis of the Gantt chart shows the ID of the vessel. The color in the charts is explained in 2.2

Case 1 assumes that all fuels, whether HFO, LNG, or NH_3, can be bunkered in Japan. The calculation results show that the fuel type concept of the fleet is transitioning from HFO to AR-LNG to NH3 as shown in Figure 10. By 2043, all ships are NH3 fuelled, achieving zero emissions without waiting for 2050. The total CO_2 emissions and costs until 2050 are 8.92 (M tons) and 5,927 (M USD).

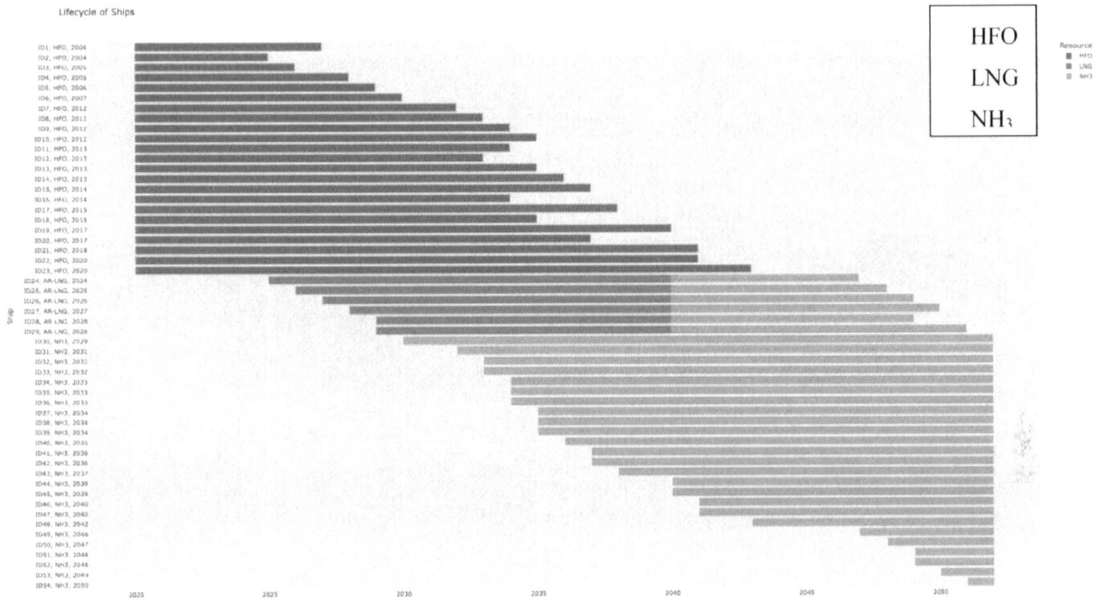

Figure 10: Lifetime and fuel used for vessels in the fleet (Case 1)

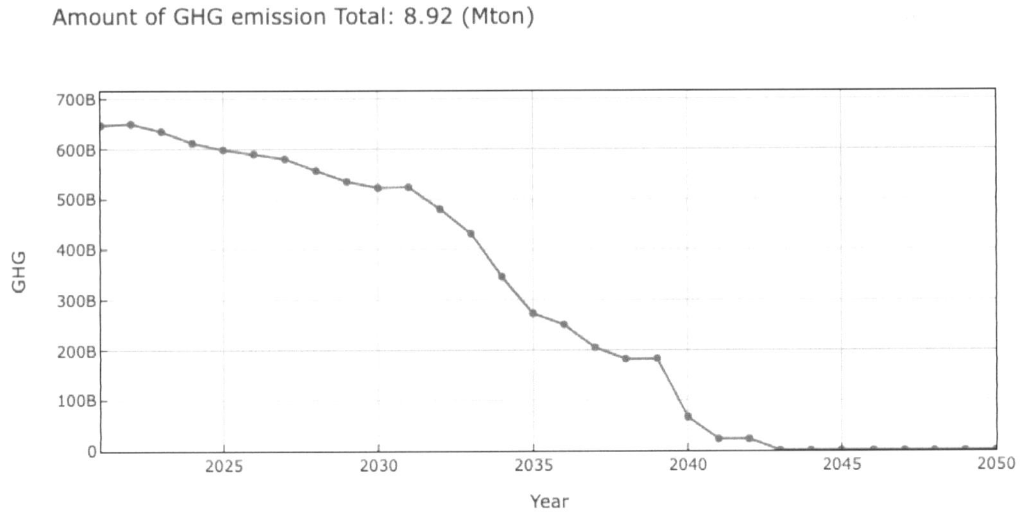

Figure 11: Time series data of CO_2 emission during the simulation (Case 1)

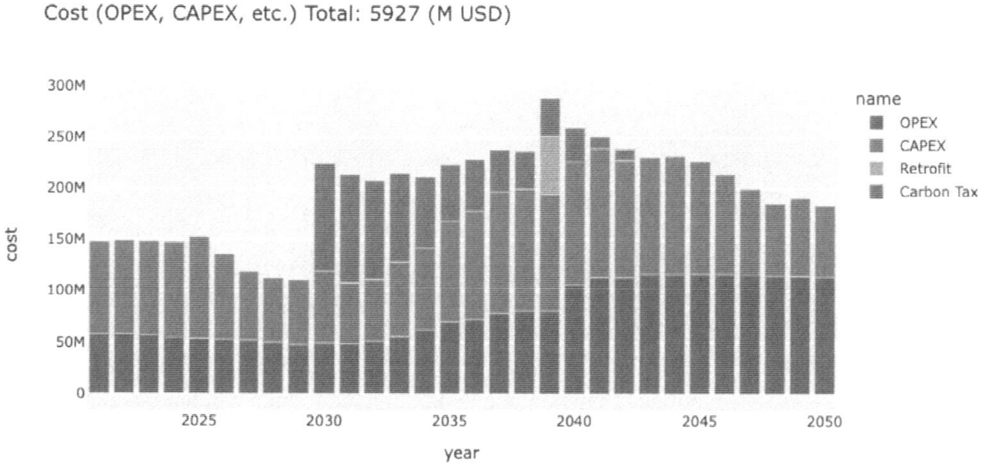

Figure 12: Time series data of cost including OPEX, CAPEX, retrofit cost and carbon tax (Case 1)

In case 2, HFO and LNG are assumed to be bunkering in Japan, and the scenario is that NH_3 fuel is bunkered in Singapore from 2025 to 2035, followed by NH_3 is bunkered in Japan after 2035 as infrastructure is developed separately in various regions. Compared to case 1, the introduction of NH_3 bunkering is accelerated by 1 year, and the total CO_2 emissions and costs are reduced to 8.6 (M tons) and 5,913 (M USD). This is because the introduction of NH_3-fueled vessels earlier than case 1 has a total cost advantage when higher carbon charges and conversion costs of AR-LNG-fuelled vessels are considered, although the vessels have a longer distance by calling at Singapore.

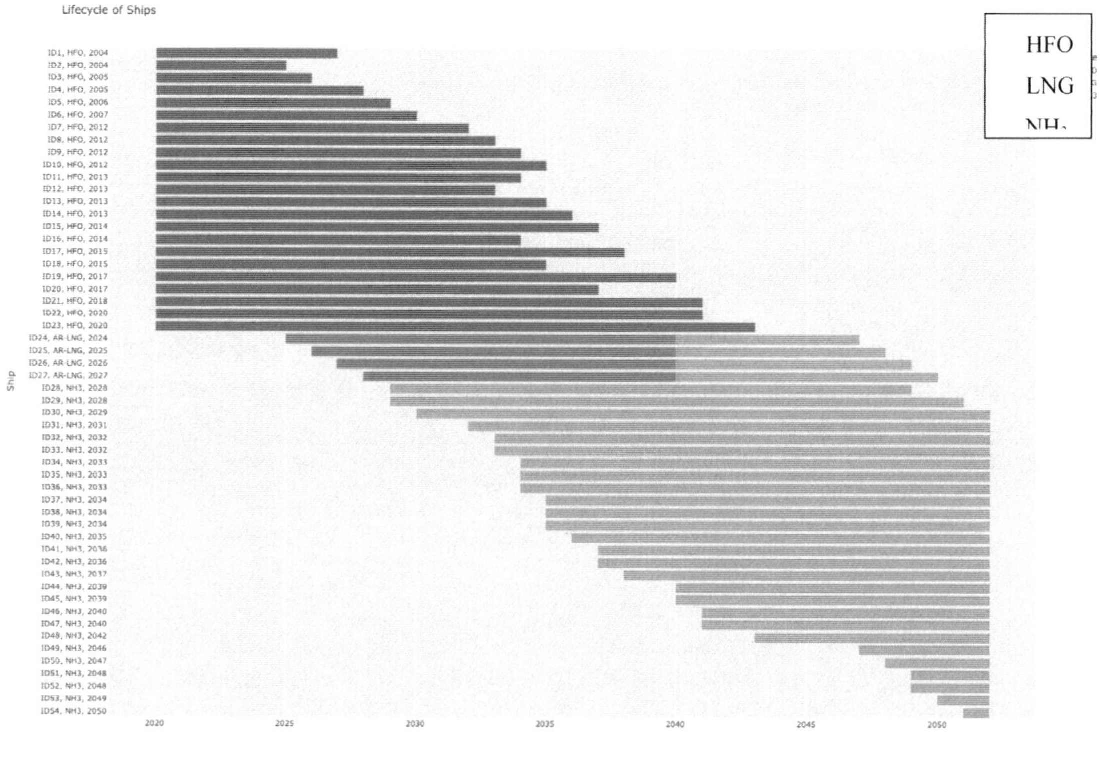

Figure 13: Lifetime and fuel used for vessels in the fleet (Case 2)

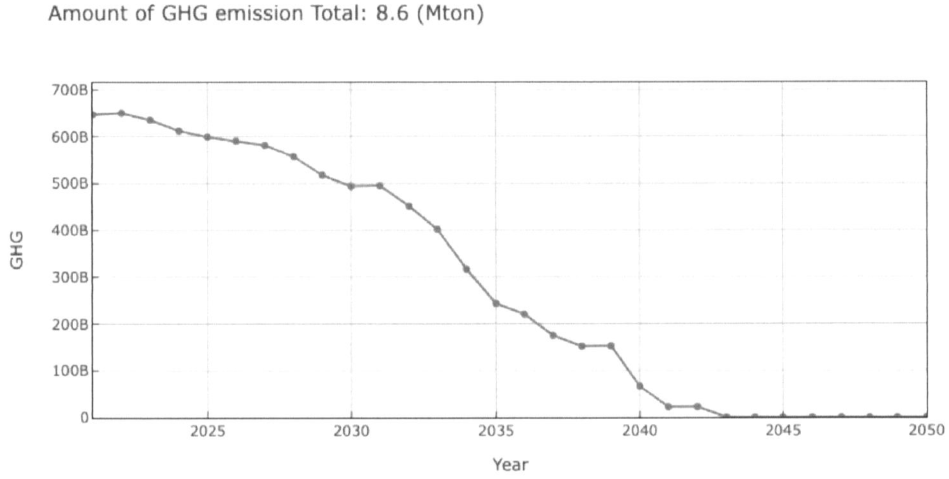

Figure 14: Time series data of CO_2 emission during the simulation (Case 2)

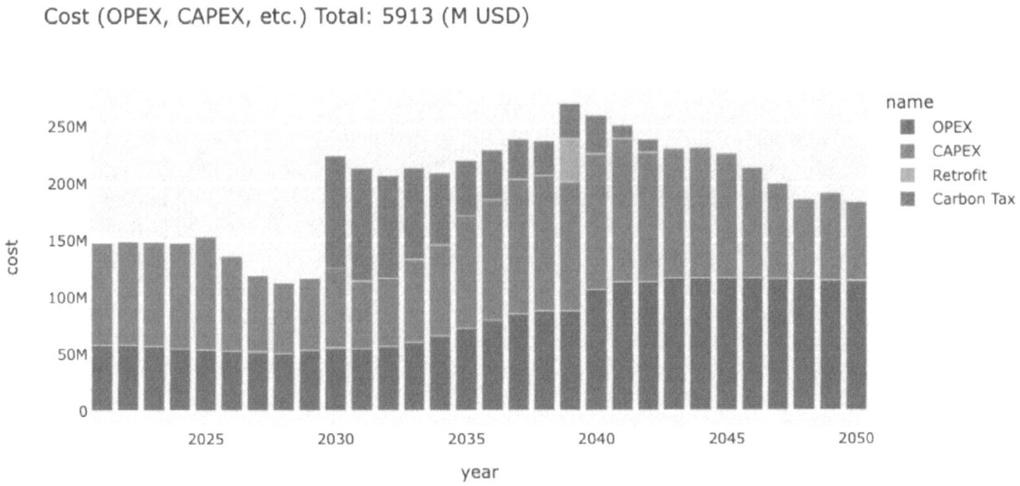

Figure 15: Time series data of cost including OPEX, CAPEX, retrofit cost and carbon tax (Case 2)

Figures 16 ~ 18 show the results of case 3, in which the ship operator's strategy in case 2 was changed to the prioritizing environment. In the case of prioritizing the environment, NH_3-fueled vessels, which have zero emissions, are basically selected. The results also show that NH_3-fueled vessels are built after 2025 when NH_3 can be introduced, and that only vessels built in 2024 are AR-LNG-fuelled vessels. The total CO_2 emissions are 7.72 (M ton), about 8% lower than the case 2, due to the prioritizing environmental performance.

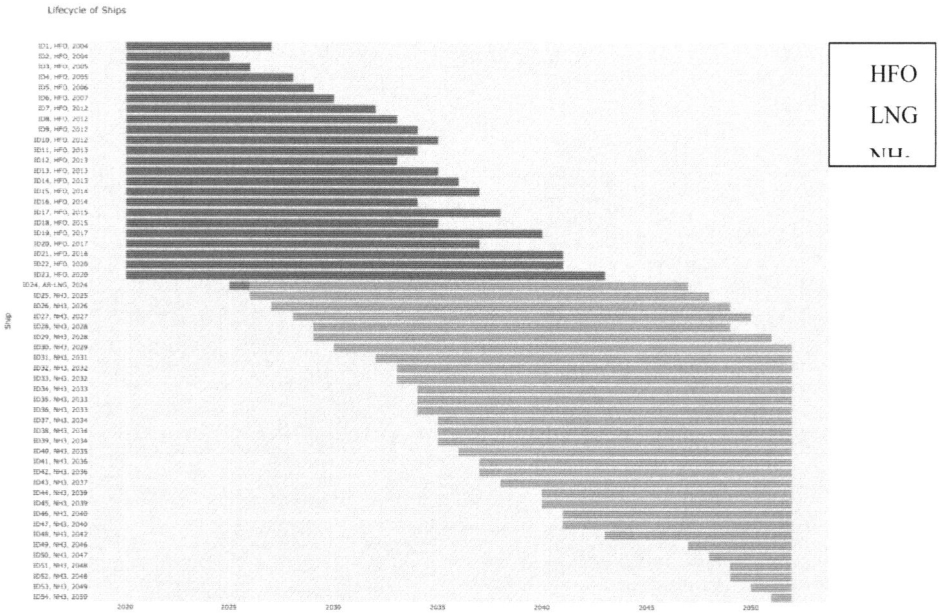

Figure 16: Lifetime and fuel used for vessels in the fleet (Case 3)

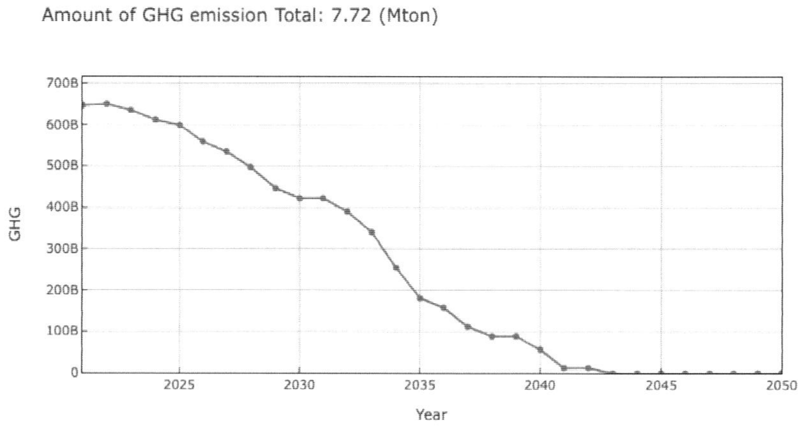

Figure 17: Time series data of CO_2 emission during the simulation (Case 3)

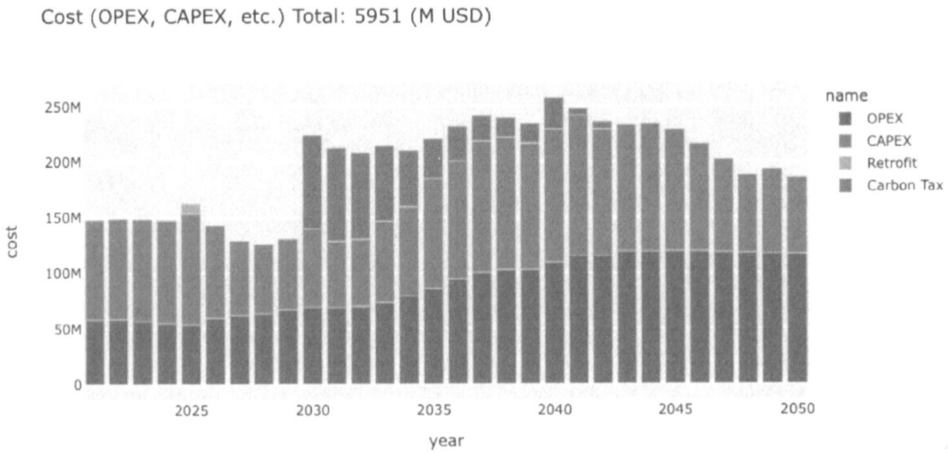

Figure 18: Time series data of cost including OPEX, CAPEX, retrofit cost and carbon tax (Case 3)

The calculation results of total cost and total CO_2 emissions for cases 1-3 up to 2050 are summarized in Table 4. As shown in the result, the simulation in this study enables us to compare and evaluate the effects of different NH_3 fuel supply locations, and differences in operator's strategy between prioritizing cost and prioritizing environment in terms of economic and environmental performance. These results vary depending on the input parameters' values and assumptions. This simulation procedure, therefore, can support the discussions and evaluations for various kinds of future fleet project planning, by properly changing these inputs and assumptions.

Table 4: Summary of the Case Study Result

Case No.	Case1	Case2	Case3
Bunkering port	Japan	2025-2035 Singapore 2035- Japan	2025-2035 Singapore 2035- Japan
Operator's strategy	Prioritize cost	Prioritize cost	Prioritize environment
Cost (M USD)	5,927	5,913	5,951
CO_2 emission (Mt)	8.92	8.6	7.72

4. CONCLUSIONS

This paper proposed the evaluation method for the introduction project of decarbonized ships. The method assumes specific routes and cargo, and scenarios of decarbonization technologies, environmental regulations, fuel prices, and infrastructure deployment for alternative fuels, and evaluates costs such as OPEX, CAPEX, and amount of CO_2 emission based on the simulation of fleet transition. Some cases were calculated by the simulation, and it demonstrates the usefulness of the proposed method of project-based evaluation for decarbonization of shipping in comparing the difference between bunkering ports and ship operator's strategy quantitatively.

As in the introduction, there are many future uncertainties regarding the decarbonization of shipping. To utilize the simulation and support decision-making for a specific project under the uncertainties, it is important to conduct many cases based on various scenarios and explore better directions for the uncertain situation. Moreover, the decision-making is across many stakeholders not only ship operators but also ship owners, shipbuilding companies, and financial institutions. To summarize the cases and build consensus among the stakeholders, the software called communication platform (CPF) is developing [8], which is for knowledge management and facilitating communication among stakeholders. As a future direction, we are planning to evaluate and design actually by the simulation with such software.

5. ACKNOWLEDGEMENTS

This research was supported by JSPS KAKENHI Grant Numbers JP20K14967 and Planning and Design Center for Greener Ships. The discussions and evaluations in the process of developing the simulation-based procedure proposed in this paper are performed in the joint research project, the Solution Evaluation System project.

6. REFERENCES

1. Barreiro, J., Zaragoza, S., & Diaz-Casas, V.: "Review of ship energy efficiency." *Ocean Engineering*, 257, (2022): 111594.

2. Faber, J, et al: Fourth IMO GHG Study, International Maritime Organization, London, UK, 2020.

3. Xuan, H., Liu, Q., Wang, L., & Yang, L.: "Decision-Making on the Selection of Clean Energy Technology for Green Ships Based on the Rough Set and TOPSIS Method." *Journal of Marine Science and Engineering,* 10(5), (2022): 579.

4. IMO: 2021 Guidelines on the reference lines for use with operational carbon intensity indicators (CII reference lines guidelines, G2), Resolution MEPC.337(76), 2021.

5. Wanaka, S., Kakiuchi, R., Fujita, N., Hiekata, K., and Yamato, H. "Simulation Method of Fleet Transition Based on Technology, Economics, and Regulation Scenario for Decarbonization of Shipping", International Conference on Computer Application in Shipbuilding, 2022.

6. Optimization with PuLP, https://coin-or.github.io/pulp/ Accessed 2022.

7. IMO: Initial IMO Strategy on Reduction of GHG emissions from ships, Resolution MEPC.304(72), 2018.

8. Takahashi, K., Ushijima, Y., Fujita, Y., Saito, T., Yamato, H., Kakiuchi, R., and Wanaka, S. "Development of a communication platform to support consensus-building among stakeholders in shipping decarbonization", International Conference on Computer Application in Shipbuilding, 2022.

7. AUTHORS BIOGRAPHY

Dr. Shinnosuke Wanaka holds the current position of Researcher in Knowledge and Data System Department at the National Maritime Research Institute, Japan. He obtained his Ph.D. from the University of Tokyo in the area of system design and simulation. He is responsible for developing the simulation software to calculate the transition of fleet to fulfil the decarbonization in the future.

Mr. Ryutaro Kakiuchi is the Naval Architect and holds the Senior Researcher at Planning and Design Center for Greener Ships (GSC). He is belonging to the initial design team for Greener ships at GSC. Before this role, He worked on initial design, technical negotiation with customers, and product planning at Shin Kurushima Sanoyas Shipbuilding Co., Ltd.

Mr. Naoya Fujita is a Senior Researcher at Mitsubishi Research Institute, Inc. He researches and consults on maritime matters, including the next-generation marine vehicle. Before his current responsibilities, he has experience assessing the risks of energy infrastructure facilities against natural disasters such as tsunamis and floods.

Dr. Kazuo Hiekata holds the current position of Professor at Graduate School of Frontier Sciences, The University of Tokyo. He received B.S., M.S. and Ph.D. degrees from The University of Tokyo in 1998, 2000 and 2008 respectively. After his graduation, he joined IBM Japan Ltd., as a Software Engineer. In 2004, he joined the School of Engineering, the University of Tokyo as an Assistant Professor.

Dr. Hiroyuki Yamato is currently the president of the Japan Agency for Marine-Earth Science and Technology (JAMSTEC), and Professor Emeritus of the University of Tokyo. He served as the Executive Vice President of the University of Tokyo and the President of the Japan Society of Naval Architects and Ocean Engineers. Degrees BS, MS, PhD are from the University of Tokyo in the field of Naval Architecture.

SIMULATION METHOD OF FLEET TRANSITION BASED ON TECHNOLOGY, ECONOMICS, AND REGULATION SCENARIO FOR DECARBONIZATION OF SHIPPING

S Wanaka, National Maritime Research Institute, Japan
R Kakiuchi, Planning and Design Center for Greener Ships, Japan
N Fujita, Mitsubishi Research Institute, Japan
K Hiekata, University of Tokyo, Japan
H Yamato, Japan Agency for Marin-Earth Science and Technology, Japan

SUMMARY

International Maritime Organization (IMO) adopted an initial strategy for GHG reduction from ships, and the industry has started to consider how to respond to the strategy. To strategically advance decarbonization, it is necessary to develop tools to discuss how to implement the IMO's strategy quantitatively.
This paper proposes a method to simulate fleet transition toward decarbonization based on multiple scenarios of technology, economics, and regulation. The simulation model includes not only technical models such as energy consumption from ships and the GHG emissions but also social models of regulation for operation and operator's decision-making. By inputting future assumptions of ship's performance, fuel prices, regulation, and transportation demand, the simulation projects a transition of fleet and calculates OPEX, CAPEX, and amount of GHG emissions during the transition. The simulation enables the stakeholders of decarbonization of shipping to analyse and compare future possible strategies of the fleet transition quantitatively.

1. INTRODUCTION

The International Maritime Organization (IMO) is discussing targets and regulations for decarbonizing international shipping [1][2]. Initially, a goal of a 50% reduction in GHG emissions from ships by 2050 was proposed, but in recent years there have been discussions about a goal that would achieve zero emissions by 2050.
To achieve the target, the maritime sector is considering how to reduce GHG emissions in terms of fuels, policies, and other technologies [3]. Alternative fuels are the most common solution, and various kinds of zero-emission ships are studied such as ammonia [4] [5] [6], fuel-cell [7] [8], biofuel [9], and so on. As for the policies, regulation for the fuels and market-based measures (MBM) are discussed [10] [11]. A carbon capture and storage system [12][13] is considered as a feasible technology to reduce the GHG from shipping before the maturity of alternative fuels. It is important to select and combine appropriate options from many possible candidates and proceed with decarbonization strategically all over the sector.
On the other hand, there are many uncertainties [14], such as technical issues including which alternative fuels to use, regulations, fuel prices, and the development of alternative fuel supply infrastructure [15], and decision-making and consensus building among various stakeholders are difficult due to the diverse their interests involved [16].
To strategically promote decarbonization, it is necessary to develop a tool to quantitatively discuss the decarbonization of the current fleet.
The objective of this study is to develop a simulator that can calculate the transition of fleet composition until international shipping achieves zero emissions for each scenario, including technology, transportation demand, regulations, and so on. The developed simulator evaluates the changes in fleet composition, costs, and CO_2 emissions for each scenario, and supports decision-making and consensus-building among stakeholders in achieving zero emissions. As a case study, several scenarios are calculated, and it demonstrates the usefulness of the simulator.

2. SIMULATION METHOD OF FLEET TRANSITION BASED ON TECHNOLOGY, ECONOMICS, AND REGULATION SCENARIO

An overview of the simulation developed in this study is shown in Figure 1. The developed simulation takes as input the product lineups which are introduced to the market in the future, slow steaming and scrapping scenarios, and external environmental scenarios such as transportation demand forecasts, regulatory trends, and fuel prices. Based on the input data, the system simulates future operations, shipbuilding, and scrapping, and outputs time series of fleet composition, CO_2 emissions, OPEX, CAPEX, and required freight rates that represent the cost per ton-mile required for transportation. Details of each input data and simulation are described below.

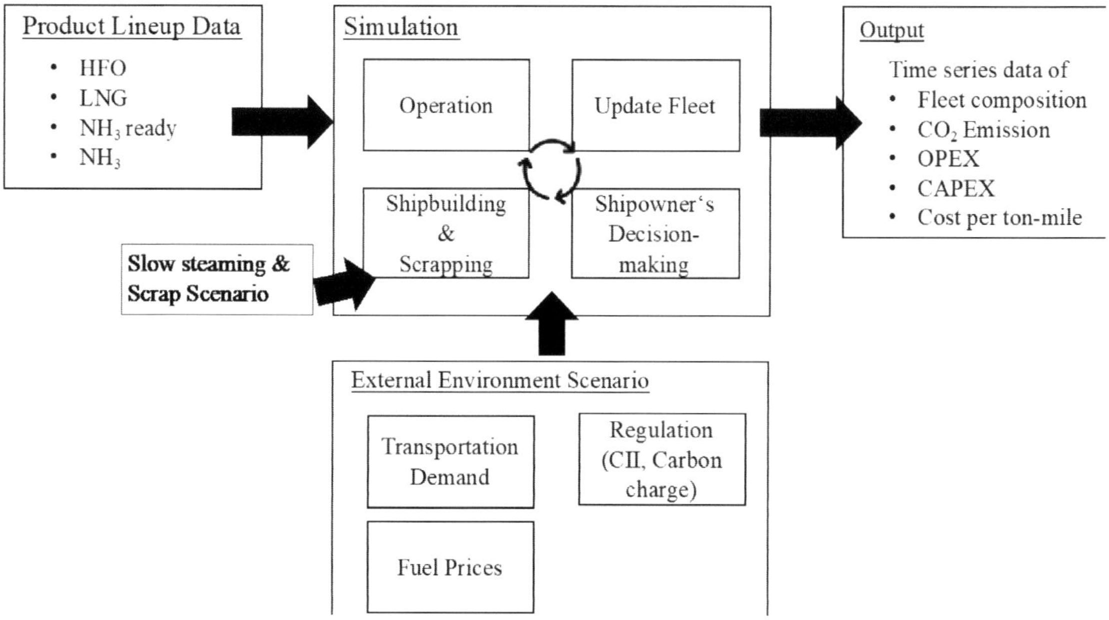

Figure 1: Overview of the Simulation

2.1 PRODUCT LINEUP DATA

A shipyard's product lineup is defined by the following data: fuel type, size (DWT), cargo amount (DWT), design speed (knot), main engine performance (g/kWh), auxiliary engine performance (g/kWh), main engine output (kW), auxiliary engine output (kW), construction cost (M$), modification cost (M$), and the start of deployment period. In this study, heavy fuel oil (HFO), liquefied natural gas (LNG), ammonia-ready LNG (AR-LNG), and ammonia (NH_3) are considered as fuel types. AR-LNG fuelled ship is a ship that is initially operated as an LNG-fuelled ship but can be converted to an NH_3-fueled ship with the modification cost.

2.2 Slow steaming and scrapping scenario

The slow steaming and scrapping scenario is input data as scenarios for determining the amount of slow steaming and scrapping in the simulation. Various environmental indexes and regulations are discussed in the IMO, there are multiple scenarios of the implementation and industries' response to the regulation. For example, a carbon intensity indicator (CII) is an index to evaluate vessels' environmental performance, and the regulation to utilize the index is discussed. As for the response to the regulation, multiple scenarios can be considered: either indicator will affect the scrap, or the indicator will only affect the operation and not the scrap. In this simulation, two types of input are possible: slow steaming and scrap based on CII evaluation and slow steaming based on CII and scrap based on ship age. How the slow steaming and scrapping are simulated based on each scenario is explained in 2.4.

2.3 External Environment Scenario

The external environment scenarios are defined as time series data of transportation demand forecasts, CII reduction rates, carbon price, and fuel price. More detailed examples are presented in the case study.

2.4 Simulation Details

The development simulation consists of the following steps: (a). calculation of costs and CO_2 emissions using an operation model, (b). calculation of scrapping and shipbuilding volume, (c). determination of ships' concepts newly built using an operator's decision-making model, and (d). updating the fleet database.

2.4 (a) Calculation of costs and CO_2 Emissions Using an Operational Model

The calculation flow of the operation model is based on the bottom-up emissions calculation method in the IMO 4th GHG Study [18]. In the fleet database, the fleet is defined by the following data: ship ID, size type, fuel type, size (DWT), cargo amount (DWT), ship speed (knot), the performance of main and auxiliary engines (g/kWh), the output of main and auxiliary engines (kW), operating cost (USD), transported distance (miles) and the built year. From the database, calculate the fuel consumption FOC_i(g/ton-mile) for each ship ID i using Equation (1). Then, calculate the fuel consumption per ton-mile $\overline{FOC}$(g/ton-mile) for the fleet using equation (2). By multiplying $\overline{FOC}$ by the transport demand for the year (ton-mile), the fuel consumption can be calculated, and the CO_2 emissions (ton) and fuel cost (M$) are calculated accordingly. i is the ship ID, t is the time (year), $\overline{dwt_\iota}$ is the amount of cargo carried (DWT), $\dot{W}_{ME_\iota}$ is the main engine output (kW), SFC_{ME_i} is the fuel consumption rate of the main engine (g/kWh), $\dot{W}_{AE|BO_\iota}$ means auxiliary engine output (kW), $SFC_{base\,i}$ means fuel consumption rate of the auxiliary engine (g/kWh), V_i means ship speed (knot), $\eta_{a,t}$ means aging rate, η_w means weather effect factor, η_f means fouling effect, D_i means distance sailed (miles), and n_i means number of ships.

$$FOC_i = \frac{\eta_{a,\,t}\left(\dot{W}_{ME_\iota} \cdot SFC_{ME_i} + \dot{W}_{AE|BO_\iota} \cdot SFC_{base_i}\right)}{\overline{dwt_\iota} \cdot V_i \cdot \eta_w \cdot \eta_f} \quad (1)$$

$$\overline{FOC} = \frac{\sum_i (FOC_i \cdot \overline{dwt_\iota} \cdot D_i \cdot n_i)}{\sum_i \left(\overline{dwt_\iota} \cdot D_i \cdot n_i\right)} \quad (2)$$

The aging rate was set to 0.5%, and the weather and fouling effect coefficients were set to those of the IMO 4th GHG Study [17].

2.4 (b) Calculation of Scrapping and Shipbuilding Volume

This step varies depending on the input slow steaming and scrapping scenario. In the case of slow steaming and scrap based on CII, two types of scrap volume, scrap volume calculated based on historical data and scrap based on CII evaluation, are considered to determine the scrap volume. As for the scrap volume based on historical data, the amount of scrapping is calculated using the regression equation of transport demand, shipping capacity, and scrapping amount, referring to the study by Wada et al. [18]. As for the scrap volume based on CII, CII is calculated for each ship, and the ship whose CII rate becomes D for 3 years or E is counted as a ship to be scrapped. Scrap volume based on historical data and CII, and the ships are scrapped to fulfil the larger volume in order of annual efficiency ratio (AER) which is CO_2 emissions per ton-mile. The CII evaluation formula is shown in Equation (3) [19]. $CII_{i,A-D,t0+\Delta t}$ means the CII evaluation value of the ship i at the simulation's time step $t0 + \Delta t$. 4 values from A-D are calculated and used for ranking, for example, rate A is assigned if the value is smaller than CII_A, rate B is assigned if the value is smaller than CII_B, and so on. In the case of bulk carriers, $a = 4745, c = 0.622, k_A = 0.86, k_B = 0.94, k_C = 1.06, k_D = 1.18$. RR means the reduction rate, and it means that the criteria become more stringent as time goes by.

$$CII_{i,A-D,t0+\Delta t} = (1 - RR \cdot \Delta t)) \cdot k_{A-D} \cdot a \cdot \overline{dwt_\iota}^{-c} \quad (3)$$

In the slow steaming based on the CII evaluation and scrap based on ship age scenario, a ship whose CII evaluation has been D for three years or E is decelerated so that the evaluation recovers to C. Specifically, this is obtained by solving an equation in which $\dot{W}_{ME_\iota}$ in Equation (1) is used as a variable and FOC_i, which is obtained from the CII standard value, is substituted. The cubic law between a ship's speed and output is assumed to hold. 30% of the maximum power is assumed to be the lower limit of slow steaming, and if slow steaming beyond that limit becomes necessary, the ship continues to operate at the lowest speed. The amount of scrapping is calculated based on the relationship between ship age and remaining ratio as shown in Figure 2, and ships are assumed to be scrapped in order of AER to satisfy the amount of scrapping.

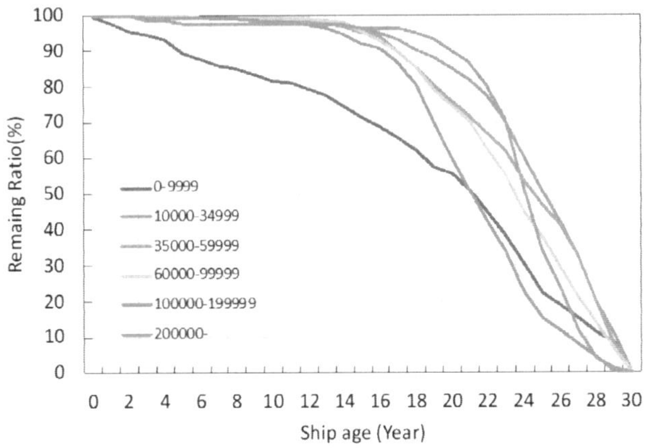

Figure 2. Relationship between ship age and remaining ratio

The amount of shipbuilding is determined based on the obtained scarp volume. The transport capacity of vessels newly built, BT_t, is calculated using Equation (4). Where T_t is the transportation demand (ton-mile), VT_t is the transportation capacity of the fleet (ton-mile), and ST_t is the transportation capacity of the scrapped vessel. The obtained capacity of new vessels is allocated by size. The distribution ratio is determined based on the historical data which is shown in Table 1.

$$BT_t = T_{t+1} - VT_t + ST_t \tag{4}$$

Table 1: Distribution Ratios of Each Size for Scrapping and Newbuilding

Size (DWT)	0-9999	10000-34999	35000-59999	60000-99999	100000-199999	200000-
Scrapping	0.006	0.08	0.152	0.203	0.345	0.214
Newbuilding	0.003	0.073	0.083	0.410	0.123	0.308

2.4 (C) Determination of the concept of vessels newly built by ship owner's decision-making model
This step determines which fuel types of vessels will be built to fulfil the number of new vessels determined in step 2.4(b). This step models the shipowner's decision to select the vessels to be purchased each year. Life cycle cost for each ship lineup considering carbon charges and regulations is calculated and the ship with the lowest cost per ton-mile is selected as the vessel to be built. The cost calculations are shown in Equations (5) and (6). i is the ID for each product lineup, ASY is the assumed years of operation, $FC_{i,t}$ is the fuel cost, $OC_{i,t}$ is the operating cost, $CC_{i,t}$ is the carbon charge, SC_i is the ship's price (USD), IR is the assumed annual interest rate, RP is the assumed payback period (12 years). ASY is basically 20 years, and only when slow steaming and scrap based on CII scenario is applied, it is calculated as the duration up to the vessel's CII rate becomes D for three years or E. Vessels with shorter operational years have a larger CAPEX per year, making them less likely to be selected.

$$AC_i = \frac{1}{\sum_{t=1}^{ASY} dwt \cdot D_t}\left(CAPEX_i + \sum_{t=1}^{ASY}(FC_{i,t} + OC_{i,t} + CC_{i,t})\right) \tag{5}$$

$$CAPEX_i = \sum_{y=1}^{RP} \frac{SC_i}{RP} \cdot (1 + IR)^y \tag{6}$$

2.4 (D) Update fleet database
Steps 2.4(B) and 2.4(C) determine the CII rating, operating speed, scrapped vessels, and newly built vessels. The fleet database is updated with that information, and the simulation returns to step 2.4 (A). This is repeated until t reaches a predetermined value, and the time series of CO_2 emissions and costs are calculated as a result.

3. CASE STUDY

3.1 OVERVIEW

In this chapter, we compare the changes in fleet composition transitions and CO_2 emissions of bulk carriers operating worldwide under different slow steaming and scrap scenarios. By comparing time series data on CO_2 emissions in each of the cases of slow steaming and scrap scenarios shown in 2.2, the advantages and disadvantages of different regulations and their implementation are analysed. Moreover, the future demand for AR-LNG fuelled ships is analyzed by comparing the case where AR-LNG is included and not included as a part of the product lineup. The simulation period starts in 2019 and calculations are performed until zero emissions are achieved. All the fuels are supplied as well as HFO and all the ship lineups are available throughout the period the simulation calculates.

3.2 INPUT DATA

Table 2: Simulation Setup for Case Study

	Case 1	Case 2	Case 3
Transportation demand	RCP 1.9 OECD		
Slow steaming scenario	CII based	CII based	CII based
Scrapping scenario	CII based	Ship age based	Ship age based
Carbon charge	N/A	200USD/CO_2-ton (2030~) 500USD/CO_2-ton (2040~)	200 USD/CO_2-ton (2030~) 500 USD/CO_2-ton (2040~)
AR-LNG in lineups	No	No	Yes

The calculation conditions of the cases are summarized in Table 2. The RCP1.9 OECD scenario presented in the IMO 4th GHG Study [18] is adopted as the transport demand scenario. In the scenario, the transportation demand for coal is decreased, but the total transportation demand for bulk carrier increase year by year. The reduction rate in the CII rating is shown in Figure 3. The fuel price per GJ (USD) is set as shown in Figure 4, based on a survey by the authors. The carbon charge was set at 200 USD/CO_2-ton from 2030 and 500 USD/CO_2-ton from 2040 only in the scenarios of slow steaming based on the CII and scrapping based on ship age.

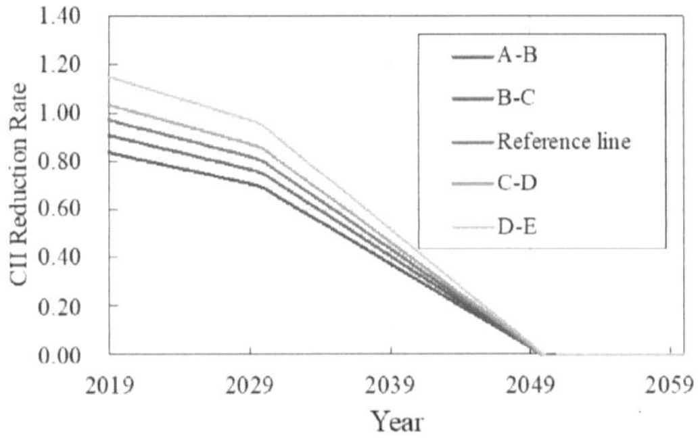

Figure 3. CII Reduction Rate Setting for the Case Study

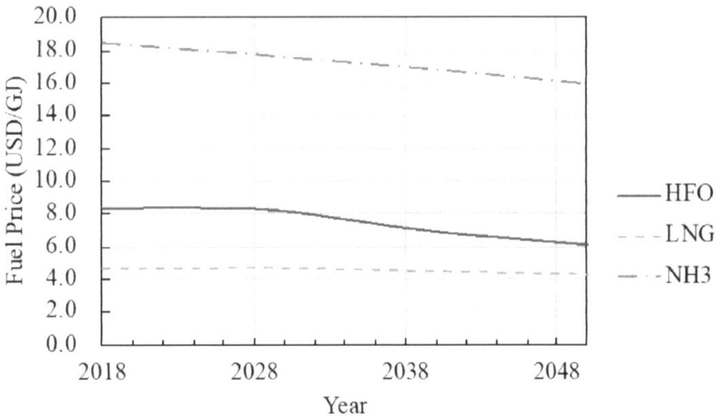

Figure 4: Fuel Prices of HFO, LNG, NH_3

The initial database for the fleet was set up as a database of 10279 vessels, based on the IHS Seaweb data (2018) [20]. To calculate the CO_2 emissions and CII rating in the development simulation, the AERs for each ship is required, and the data was created assuming that the AERs are distributed as shown in Table 3.

Table 3: Assumption of CII Rating Distribution for Each Size (%)

	A	B	C	D	E
0-9999	37.6	17.6	25.9	7.1	11.8
100000-34999	12.6	17.1	27.6	21.6	21.1
35000-59999	10.9	12.6	26.1	24.7	25.6
60000-99999	22.6	25.1	32.6	13.2	6.5
100000-199999	7.41	14.5	35.5	25.9	16.7
200000-	6.80	14.5	34.2	20.1	24.4

For the product lineup data, 24 data were generated for six size categories (0-9999, 10000-34999, 35000-59999, 60000-9999, 100000-199999, 200000-(DWT)) and four fuel types (HFO, LNG, AR-LNG, NH_3). The values of performance and ship's price were developed based on the IMO 4th GHG Study and interviews with shipbuilding companies, and the values for AR-LNG and NH_3 were set based on LNG-fuelled vessels, estimating a decrease in loading capacity due to an increase in fuel tank capacity, a 13% - 18% retrofit cost relative to ship's price, and a 2% increase in ship's price itself.

3.3 RESULT

The results of the 2 types of slow steaming and scrap scenarios are shown in Figures 5-8: slow steaming and scrapping based on CII, and slow steaming based on CII and scrapping based on ship age. Each shows the change in fleet composition, CO_2 emissions, and scrap volume. In the case of the slow steaming and scrapping based on the CII scenario (Case 1), zero emissions are achieved in 2042. The introduction of NH_3 fuel started in 2029. By offsetting the increase in transportation demand through slow steaming, CO_2 emissions change gradually at the early stage of simulation and rapidly decrease after NH_3 becomes popular fuel around 2038, and decarbonization is achieved. Figure 5 shows that when scrapping based on CII rating is introduced, the number of ships scrapped is highly increased. Especially, there is a threshold in which all the HFO and LNG fuelled ships cannot be operated around 2038, and a sudden increase in demand for scrapping occurs. In practice, there is a limit to the amount that can be scrapped and built, and such behaviour may not be acceptable for the shipbuilding market.

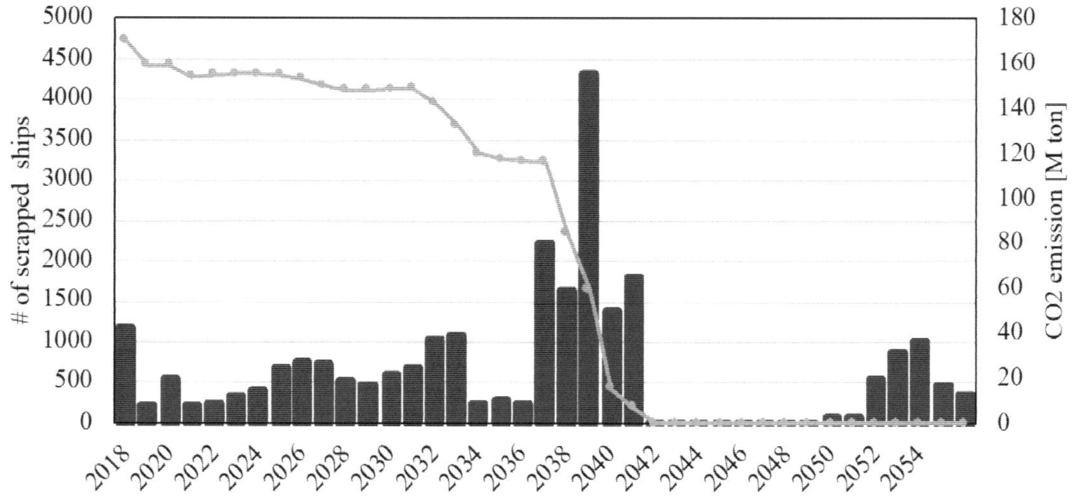

Figure 5: CO_2 emission and number of scrapped ships (Case 1: slow steaming and scrapping based on CII)

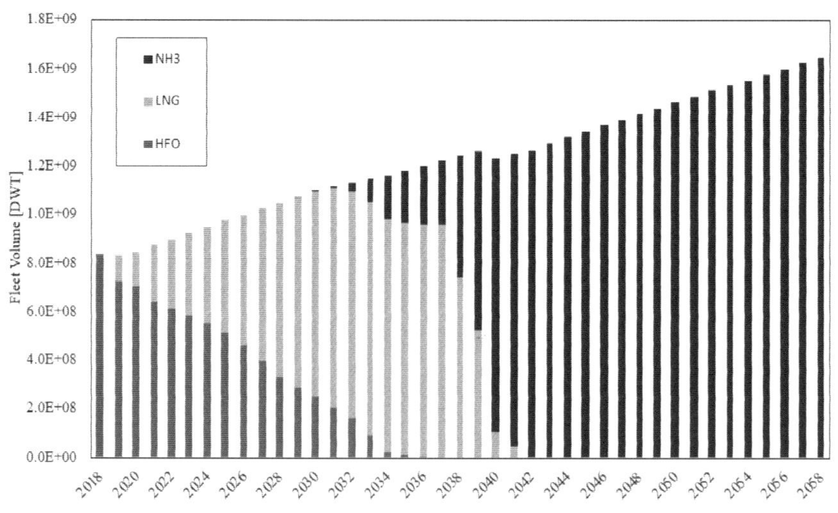

Figure 6: Time series change of fleet composition (Case 1: slow steaming and scrapping based on CII)

With slow steaming based on CII and scrapping based on ship age (Case 2), decarbonization of shipping is not achieved by 2050, but achieved in 2062; NH_3 fuel is introduced in 2029 when carbon charges begin, but because scrapping is based on ships' age, older ships using fossil fuels are not replaced and zero emission is not achieved by 2050. Before the introduction of NH_3 in 2029, CO_2 emissions are almost constant by offsetting the increase in transportation demand through slow steaming and the use of LNG fuel. From there, CO_2 emissions will decrease monotonically, reaching zero emissions by 2062, when all fossil fuel ships will be replaced. Figure 7 shows the number of scrapped ships and CO_2 emissions during the simulation. Figure 8 shows the time series data of fleet composition. Compared to case 1, the amount of scrapping is smaller and the transition to the decarbonized fleet is achieved gradually.

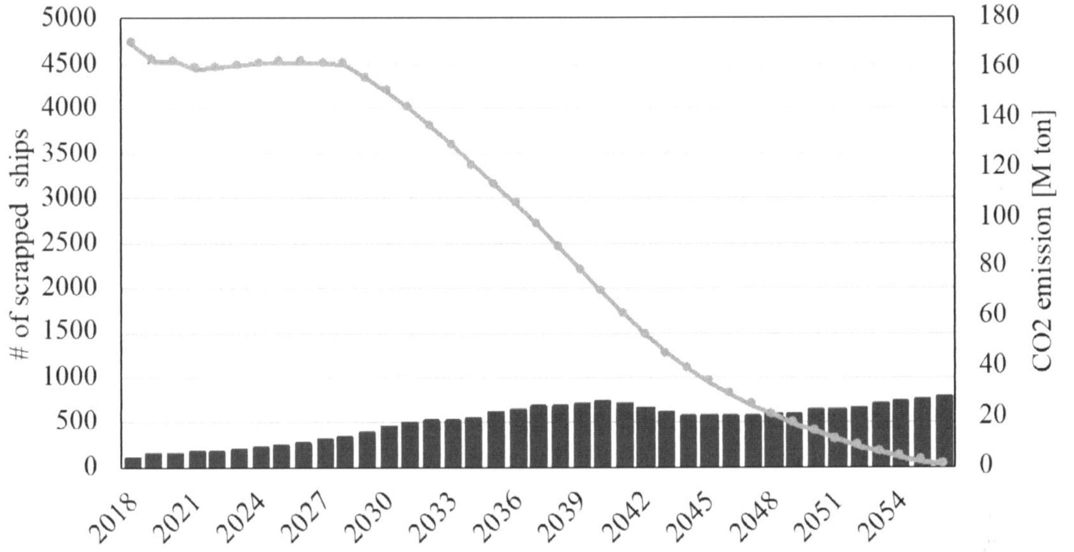

Figure 7: CO_2 emission and number of scrapped ships (Case 2: slow steaming based on CII and scrapping based on ship age)

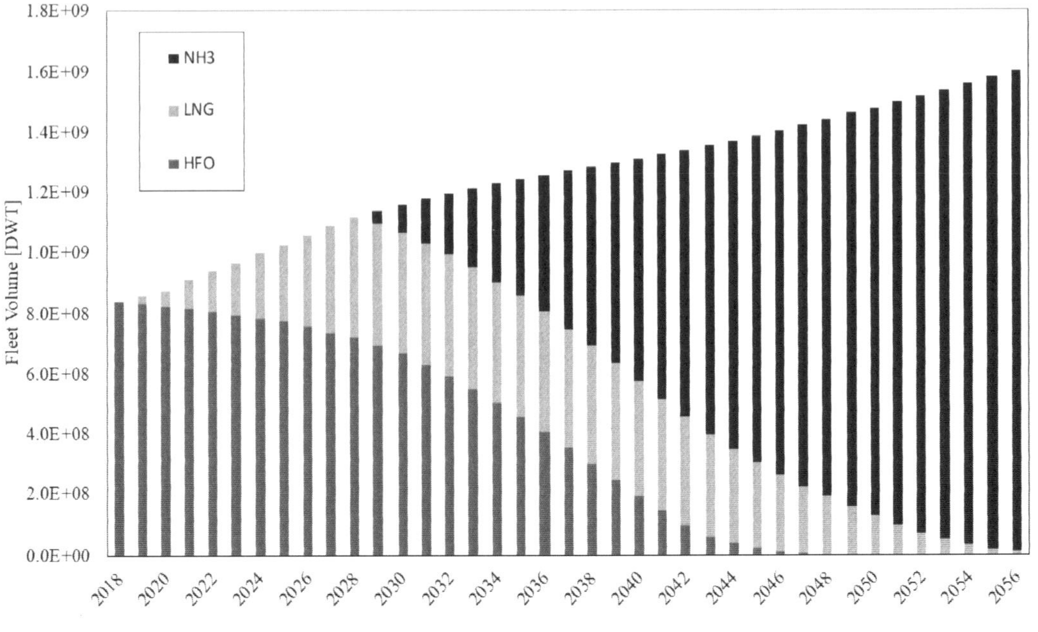

Figure 8: Time series change of fleet composition (Case 2: slow steaming based on CII and scrapping based on ship age)

By comparing the scenarios, the advantages, and disadvantages of each are discussed. While zero emissions by 2050 can certainly be achieved with a scrapping operation based on CII, which is the performance of individual vessels, a sudden switch to alternative fuels at a certain point will cause wild fluctuations in the demand for shipbuilding and scrapping. If a carbon charge is used to encourage fuel conversion, but scrapping is based on ship age, as usual, it is not guaranteed that zero emissions will be achieved by 2050, but the effect of promoting the introduction of alternative fuel ships at an early stage can be seen in the gradual conversion process.

Figures 9 and 10 show the result of case 3, what happened if AR-LNG ships are introduced into the shipbuilding market. AR-LNG-fuelled vessels are selected as a product to bridge the gap between LNG-fuelled and NH_3-fueled vessels. This is because the concept of operating LNG-fuelled vessels initially and switching to NH_3-fueled vessels as regulations dictate is cost-effective in situations where LNG-fuelled vessels do not have sufficient service life but NH_3 fuel is more expensive.

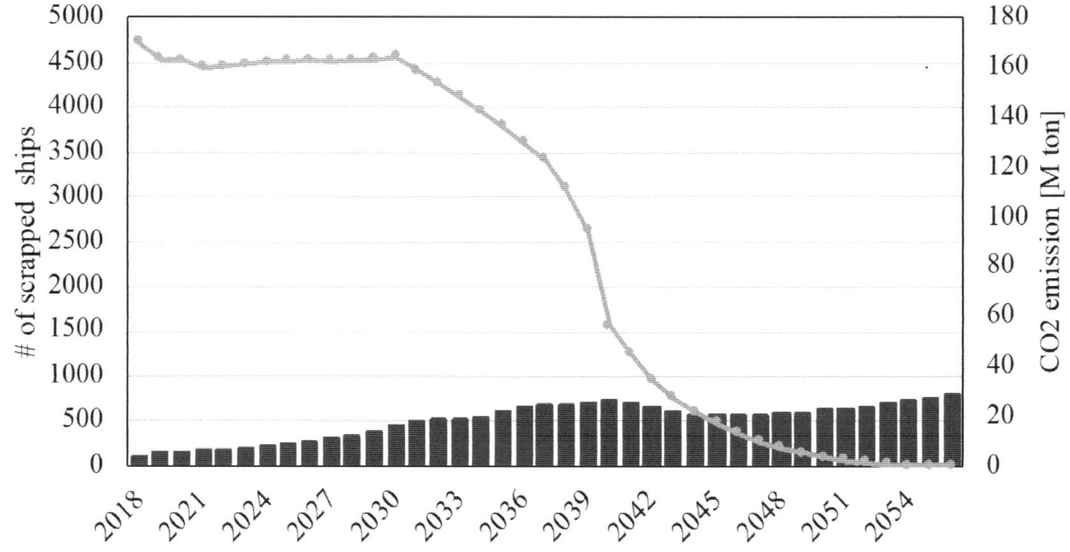

Figure 9: Time series change of fleet composition (Case 3: AR-LNG applied)

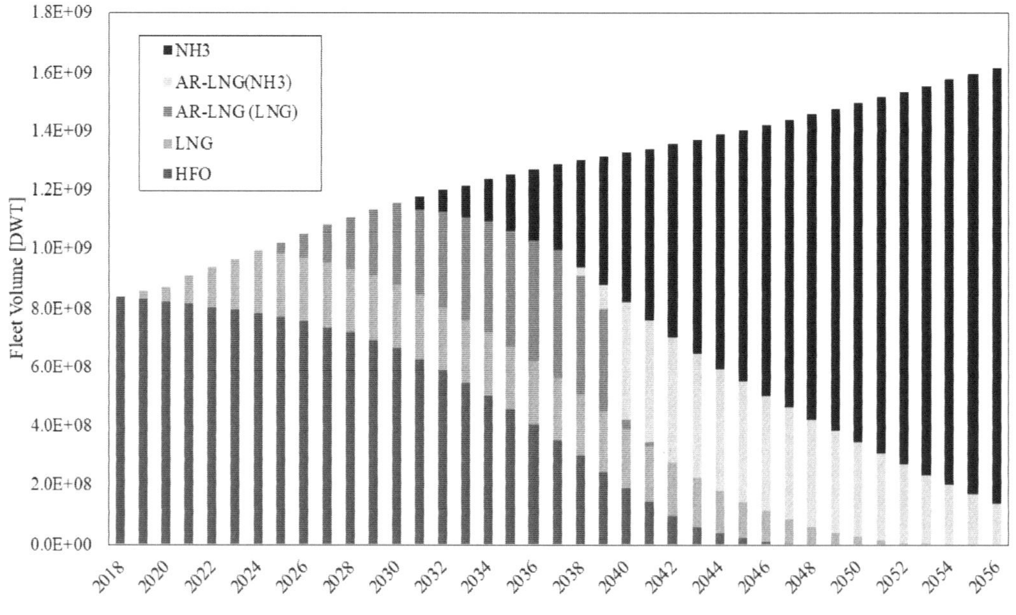

Figure 10: Number of Scrapped Ships (Case 3: AR-LNG applied)

4. DISCUSSION

The case study demonstrates that the advantages and disadvantages of each of the slow steaming and scrapping scenarios can be analysed through the simulation results. Such an analysis is very useful in making decisions about future operations and regulations. From the findings obtained in this case, it can be considered that it would be effective to operate and regulate by a combination of scenarios, such as gradually introducing NH_3 fuel by introducing a carbon charge in the initial phase and proceeding with CII-based scrapping in the final phase.

Figures 11 and 12 show the results of CO_2 emissions when such an operation and regulation are introduced. Until 2044, the transition of the fleet is promoted by carbon charge, and between 2045 – 2050, ships with D or E rates by CII evaluation are scrapped equally. Zero emissions by 2050 can be achieved by gradually introducing NH_3 fuel through the introduction of a carbon charge and switching to the removal of NH_3 fuel based on CII evaluation at the end of the process. Thus, the simulation can be used to study future operations and regulations.

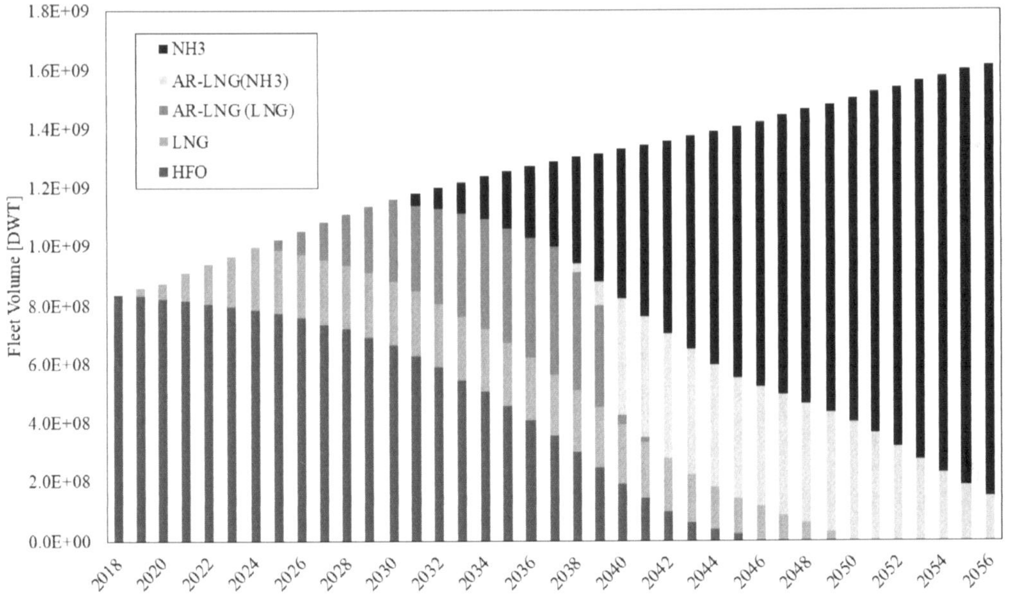

Figure 11. Time series change of fleet composition (Combination of the slow steaming and scrapping scenario)

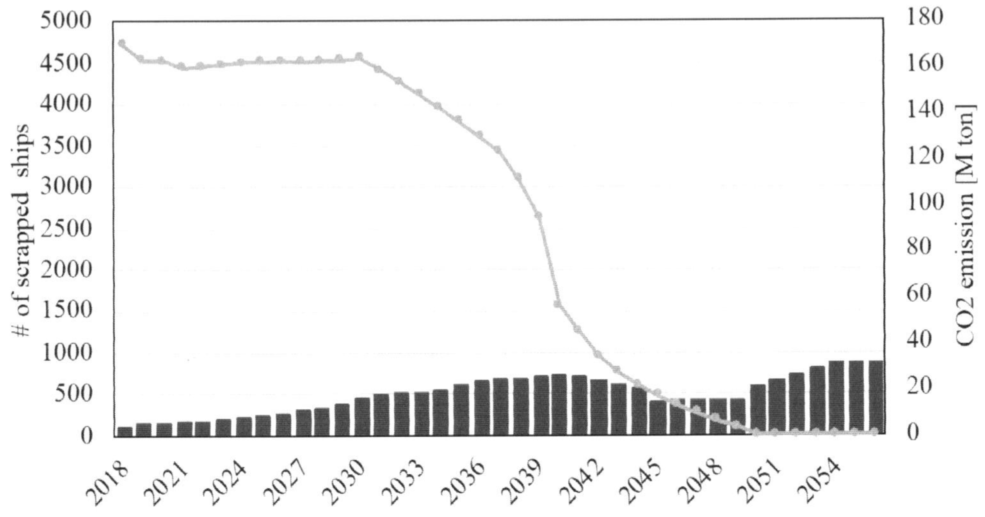

Figure 12. Time series change of CO_2 emission (Combination of the slow steaming and scrapping scenario)

Of course, the results of this study were calculated using one of many possible scenarios as input, and future work is needed to determine what kind of input would be appropriate. Especially, regulations such as CII and carbon charge, fuel prices, and performance of ships using alternative fuels should be considered by a more detailed survey.

From another point of view, it is possible to discuss the demand for ships fuelled by greener fuel. The result of the case study shows that there is some demand for AR-LNG ships in the future. To achieve decarbonization, it is not required to introduce AR-LNG ships. However, the simulation calculated the cost for each ship newly introduce and determined that AR-LNG ships are the most cost-effective ships for some duration. In the case of slow steaming based on CII and scrapping based on ship age, 4122 ships are constructed as AR-LNG ships. Whether this demand is considered large or not is an individual discussion for shipyards, but the simulations developed can be used to estimate future demand for ships, which is useful for shipyards when considering their product strategies.

Finally, some limitations of this study are discussed. First, the accuracy of the initial fleet data input to the simulation should be considered. To project the future transition of fleet composition, it is important to capture the current fleet situation including ship age distribution, environmental performance, and so on. By providing more accurate initial fleet data, it is expected that more accurate and detailed discussion becomes possible based on the simulation. Taking other alternative fuels other than NH_3 into account is another future work for the simulation. In this study, the fuel used for ship operation is determined by the cost. However, there are many possible fuels such as biofuels and H_2, and which and how much the fuel is utilized are determined by not only the cost but also their availability. Considering a larger variety of fuels and their availability is an important aspect to improve our simulation. Also, the capacity of the shipbuilding and scrapping is missing in the simulation as well. In the simulation, the amount of shipbuilding and scrapping is determined without the constraint of the capacity. However, there should be an upper limit on shipbuilding and scrapping capacity, and it should be a constraint in the actual market. Especially, in a situation where regulation strongly affects the scrapping, the amount of shipbuilding and scrapping can vary intensively so it is important to consider the capacity limitation of shipbuilding and scrapping.

5. CONCLUSION

In this study, we developed a simulation that can calculate the transition of fleet composition until international shipping achieves zero emissions for each scenario, including technology, transportation demand, and regulations. The developed simulator evaluates the changes in fleet composition, costs, and CO_2 emissions for each scenario, and supports decision-making and consensus-building among stakeholders in achieving zero emissions. As a case study, several slow steaming and scrapping scenarios were considered and compared. The simulation enables us to analyse the advantages and disadvantages of those scenarios and to propose and evaluate the combined scenario that promotes conversion based on carbon charges in the early stage and scrapping based on individual ship performance in the last stage. Also, by inputting a new concept of a ship such as AR-LNG as lineup data, the simulation calculated how much demand happens on the assumption of fuel prices, regulation, and transportation demand. The results demonstrate that the simulation is useful to study and discuss the future scenario toward the decarbonization of shipping quantitatively. To utilize the simulation for supporting consensus-building and decision-making about the direction of future decarbonization among the stakeholders, there are some steps to achieve. Especially, it is necessary to manage and facilitate the communication based on the simulation among the stakeholders and converge their opinions in one direction. Takahashi et al. [21] propose a platform to facilitate consensus-building, and combining the simulation and platform is future work.

6. ACKNOWLEDGEMENTS

This research was supported by JSPS KAKENHI Grant Numbers JP20K14967 and Planning and Design Center for Greener Ships. The discussions and evaluations in the process of developing the simulation-based procedure proposed in the paper are performed in the joint research project called Solution Evaluation System Project.

7. REFERENCES

1. IMO, "Initial IMO Strategy on Reduction of GHG emissions from ships", *Resolution MEPC*, 304(72) 2018.

2. Joung, T. H., Kang, S. G., Lee, J. K., and Ahn, J., "The IMO initial strategy for reducing Greenhouse Gas (GHG) emissions, and its follow-up actions towards 2050", *Journal of International Maritime Safety, Environmental Affairs, and Shipping*, Vol. 4, No. 1 (2020): 1-7.

3. Balcombe, P., Brierley, J., Lewis, C., Skatvedt, L., Speirs, J., Hawkes, A., and Staffell, I., "How to decarbonise international shipping: Options for fuels, technologies and policies" *Energy conversion and management*, 182, (2019) :72-88.

4. Kim, K., Roh, G., Kim, W., and Chun, K., "A preliminary study on an alternative ship propulsion system fueled by ammonia: Environmental and economic assessments." *Journal of Marine Science and Engineering*, 8(3), 183 (2020).

5. Seo, Y., and Han, S. "Economic Evaluation of an Ammonia-Fueled Ammonia Carrier Depending on Methods of Ammonia Fuel Storage." *Energies*, 14(24), 8326 (2021).

6. Hansson, J., Brynolf, S., Fridell, E., and Lehtveer, M. "The potential role of ammonia as marine fuel—based on energy systems modeling and multi-criteria decision analysis." *Sustainability*, 12(8), 3265 (2020).

7. Choi, C. H., Yu, S., Han, I. S., Kho, B. K., Kang, D. G., Lee, H. Y., Seo, M., Kong J., Kim, G., Ahn, J., Park, S., Jang D., Lee J. H., and Kim, M. "Development and demonstration of PEM fuel-cell-battery hybrid system for propulsion of tourist boat." *International Journal of Hydrogen Energy*, 41(5), (2016): 3591-3599.

8. Shakeri, N., Zadeh, M., and Nielsen, J. B. "Hydrogen fuel cells for ship electric propulsion: Moving toward greener ships." *IEEE Electrification Magazine*, 8(2), (2020): 27-43.

9. Kesieme, U., Pazouki, K., Murphy, A., and Chrysanthou, A. "Biofuel as an alternative shipping fuel: technological, environmental and economic assessment." *Sustainable Energy & Fuels*, 3(4) (2019): 899-909.

10. Lagouvardou, S., Psaraftis, H. N., and Zis, T. "A literature survey on market-based measures for the decarbonization of shipping." *Sustainability*, 12(10), (2020): 3953.

11. Psaraftis, H. N., Zis, T., and Lagouvardou, S. "A comparative evaluation of market based measures for shipping decarbonization." Maritime Transport Research, 2, (2021): 100019.

12. Fang, S., Xu, Y., Li, Z., Ding, Z., Liu, L., and Wang, H. "Optimal sizing of shipboard carbon capture system for maritime greenhouse emission control." *IEEE Transactions on Industry Applications*, 55(6), (2019): 5543-5553.

13. Fang, S., Xu, Y., and Li, Z. "Joint generation and demand-side management for shipboard carbon capture and storage system." IEEE/IAS 55th Industrial and Commercial Power Systems Technical Conference (I&CPS). (2019): 1-8

14. Yuan, J., Ng, S. H., and Sou, W. S. "Uncertainty quantification of CO_2 emission reduction for maritime shipping." *Energy Policy*, 88, (2016): 113-130.

15. Hu, H., Yuan, J., and Nian, V.: "Development of a multi-objective decision-making method to evaluate correlated decarbonization measures under uncertainty–The example of international shipping." *Transport Policy*, Vol. 82, 2019: 148-157

16. Halim, R. A., Kirstein, L., Merk, O., and Martinez, L. M. "Decarbonization pathways for international maritime transport: A model-based policy impact assessment." *Sustainability*, Vol. 10, No. 7, (2018): 2243

17. Faber, J, et al "Fourth IMO GHG Study" International Maritime Organization, London, UK, 2020.

18. Wada, Y., Hamada, K., Hirata, N., Seki, K., and Yamada, S. "A system dynamics model for shipbuilding demand forecasting." *Journal of Marine Science and Technology*, Vol. 23, No. 2, (2018): 236-252

19. IMO: 2021 Guidelines on the reference lines for use with operational carbon intensity indicators (CII reference lines guidelines, G2), Resolution MEPC.337(76), 2021.

20. Maritime IHS, Sea-web ships https://maritime.ihs.com/Account2/Index, [Accessed 25 February 2022].

21. Takahashi, K., Ushijima, Y., Fujita, Y., Saito, T., Yamato, H., Kakiuchi, R., and Wanaka, S. "Development of a communication platform to support consensus-building among stakeholders in shipping decarbonization", International Conference on Computer Application in Shipbuilding, 2022.

8. AUTHORS BIOGRAPHY

Dr. Shinnosuke Wanaka holds the current position of Researcher in the Knowledge and Data System Department at the National Maritime Research Institute, Japan. He obtained his Ph.D. from the University of Tokyo in the area of system design and simulation. He is responsible for developing the simulation software to calculate the transition of the fleet to fulfill the decarbonization in the future.

Ryutaro Kakiuchi is the Naval Architect and holds the Senior Researcher at Planning and Design Center for Greener Ships (GSC). He is belonging to the initial design team for Greener ships at GSC. Before this role, He worked on an initial design, technical negotiation with customers, and product planning at Shin Kurushima Sanoyas Shipbuilding Co., Ltd.

Naoki Fujita is a Senior Researcher at Mitsubishi Research Institute, Inc. He researches and consults on maritime matters, including the next-generation marine vehicle. Before his current responsibilities, he has experience assessing the risks of energy infrastructure facilities against natural disasters such as tsunamis and floods.

Dr. Kazuo Hiekata holds the current position of Professor at the Graduate School of Frontier Sciences, The University of Tokyo. He received B.S., M.S., and Ph.D. degrees from The University of Tokyo in 1998, 2000, and 2008 respectively. After his graduation, he joined IBM Japan Ltd., as a Software Engineer. In 2004, he joined the School of Engineering, The University of Tokyo as an Assistant Professor.

Dr. Hiroyuki Yamato is currently the president of the Japan Agency for Marine-Earth Science and Technology (JAMSTEC) and Professor Emeritus of the University of Tokyo. He served as the Executive Vice President of the University of Tokyo and the President of the Japan Society of Naval Architects and Ocean Engineers. Degrees BS, MS, Ph.D. are from the University of Tokyo in the field of Naval Architecture.

International Conference on
Computer Applications in Shipbuilding
13th-15th of September 2022, Yokohama, Japan

***** Formatting Issue - Best Available Paper/Graphic *****

AI/ML APPLICATIONS FOR SHIP DESIGN

J Hodges, M P Wheeler, and **M Belhocine** Siemens Digital Industry Software, USA
J Henry, Monolith AI, UK

SUMMARY

This work uses Simcenter STAR-CCM+ to generate CFD data for determining the powering characteristics of a vessel and Simcenter ROM Builder and Monolith for using that data in no-code machine learning pipelines to allow hull designers to understand how design variants affect vessel performance instantly. Firstly, in terms of capabilities, the machine learning models were able to predict several 0D performance metrics (torque, total resistance, powering, and propulsion metrics) with respect to 12 independent geometrical parameters involved in the ship design. These predictions were then supplemented with additional machine learning predictions of the spatial flow field as well to provide designers with more detailed understanding of the hydrodynamic aspects of the vessel. Regarding the spatial fields, there are two main capabilities the machine learning models provided: predict results of spatial fields in the form of local contours (loads on the hull, free surface deformation, and in-flow into the propeller), and thereby also an optimization pipeline which can morph the geometry according to the desired cost function for any combination of these machine learning predictions.

NOMENCLATURE

AI	Artificial intelligence
CAE	Computer-aided engineering
CC	Pearson coefficient
CFD	Computational fluid dynamics
CPU	Central processing unit
(°)	Degrees
G_i	Geometric parameters, specified by CFD practitioner from their design expertise
ML	Machine learning
N	Sample count
NN	Neural network
NRSME	Normalized Root Mean Square Error
P	Pressure (kg·m ^{1}·s)
R^2	A coefficient, between 0 and 1, measuring goodness of fit of a statistical model
ROM	Reduced order model
$\tau_{i,j,k}$	Three individual wall shear stress components (kg·m ^{1}·s)
VAE	Variational autoencoder
Z_i	Latent parameter, generated by the encoder in the VAE-based workflow

1. INTRODUCTION

Recent advancements in machine learning and artificial intelligence have been applied to practical ship design problems. A well-known hull was simulated using computational fluid dynamics (CFD). Then, a small database of CFD runs was created by randomly picking design variants to analyse. Reduced order models (ROM) and artificial intelligence techniques have been applied to the generated data sets. Global scalar values such as resistance, trim, sinkage, and shaft power can be instantly predicted when the ship changes geometrically or updated to the ship operating conditions of interest. In addition, field variables such as in-flow to the propeller, generated wakes, local pressure and shear stresses on the hull can also be predicted giving ship designers instant insight to design variants. The background of these models is discussed, and then an evaluation of productivity and accuracy is investigated using a practical example.

2. LITERATURE REVIEW

People have been designing ships for centuries. Ever since towing tanks appeared over 150 years ago ship designers i a i y e ied on he e e e i en a a i i ie o de e o hi Several well-known hull shapes have been validated in towing tanks for the purpose of sharing data, such as KVLCC2 [1], DTMB [2], and inland water way vessels [3].

However, these tests can be both expensive and time consuming to perform. In today's world, simulation tools can be e e aged o eed u he de ign o e While they may be faster than traditional tests, Mizze et al. [4] suggested that

*** Formatting Issue - Best Available Paper/Graphic ***

the use of computational fluid dynamics tools can provide accurate answers but are still difficult to use and not practical in a de ign en i on en Whee e ho a ed a enchmarking of several well-known hulls using computational fluid dynamics and employed speed up techniques to obtain results in a reasonable amount of time.

However, in design environments, it is often not feasible to run high fidelity CFD simulations every time something hange As such, the use of ROMs and ode a e o in e e ing he e ode a o o in an edi ion o results when things change.

The advent of machine learning in the late 2010's has paved the way for many impressive applications of such technologies into the CFD landscape for several years now. Some seminal works as early examples focused on turbulence modelling [6] and, of equal importance, the concept of constructing a machine learning pipeline that has physically interpretable attributes and results [7,8]. More recently, there are many examples of machine learning applied to technical solver development and alternatives [9], industrial applications of machine learning pipelines for designers [10], and much more. This work stands to showcase the end-user advantages in an industrial design environment, as opposed to more technical developments per se. However, the algorithms used in the machine learning models for the present study are quite advanced and state-of-the-art for tackling a core challenge that is somewhat unique to that of the CFD space: modelling the large point clouds of scalar fields generated from sweeps of simulations.

3. RESEARCH METHODOLOGIES

The CFD models employed in this work follows the same procedure discussed y Whee e [5]. The computational fluid dynamics code, Simcenter STAR-CCM+ version 2021.2 was used. This is a well-established and well-known code in the marine industry. In this work, a segregated solver with incompressible fluids was used. The free surface was used using the Volume of Fluid (VOF) with high resolution interface capturing scheme (HRIC) model. The hull motions in trim and heave were solved for in every simulation. The propellers were modelled in every case using the virtual disk model.

The hull form used in this study in the MV Regal. The Regal is an existing ship that has been analysed in the Lloyd's Register CFD workshop [11] and is also being investigated in the ongoing JORES workshop [12]. An image of the Regal is shown below in Figure 1 in greyscale, with two different sample bow geometries generated with the parameterized CAD model. The ship was modelled in full scale, and the ship has fixed main particular throughout the studies executed as shown in the figure. Whee e a ida ed CF i u a ion i h hi hu

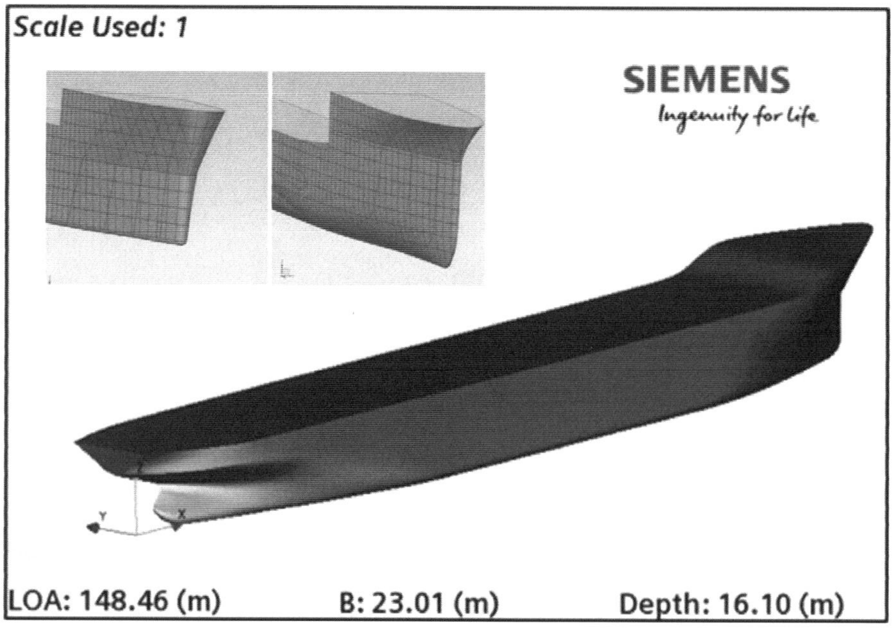

Figure 1: Image of the MV Regal with two different bow geometries shown as a sample (top-left)

The hull form was then parameterized in Siemens NX CAD software. The model was parametrized with 9 independent variables that can control the rear stern tube shape, the bow shape, the bilge radius, the deck height of the fore and aft decks, placement of knuckles, and the relative length of the fore, aft, and mid sections of the vessel. Two sample geometries are shown in Figure 1.

From these CFD models, several results of various forms are extracted and used in a pipeline towards creating a machine learning model. Since there are several predictions being made from various machine learning models, it is convenient to easily summarize all such data packages in Table 1. Two different tools were used for creating the machine learning models. Simcenter ROM Builder was used for the creation of surrogates capturing the 0D data (integral metrics). For results with more spatial variation and complicated patterns in the data, Monolith was used. This would also include the last row in Table 1 that signifies training a machine learning model on CAD (.stl) data for learning the geometrical representation of the design space.

Table 1: CFD results of interest which are used in machine learning model generation. Results from Simcenter ROM Builder were used where possible, whereby Monolith was used elsewhere.

Type	Name	Scalar
Integral Metric	Single metrics signifying overall performance	Total resistance [kN] Propeller Torque [N m] Propeller Rotation Rate (RPM) Shaft Power [W]
Planar	Scalar field in propeller plane	Wa e a ion
Planar	Scalar field on water free surface	Displacement Height
3D Result	Scalar field on hull surface	$\tau_{i,j,k}$, Pressure
3D Geometry	Entire hull geometry	n/a

A single batch of simulations was generated which was thereby the basis (origin) for where the data in Table 1 was extracted. There were 128 simulations ran in total to comprise the database for training, which were driven by a Latin Hypercube Sampling. For a defined parameter space, in this case the geometric variables, the Latin Hypercube Sampling does a sequence of near-random sampling for the geometric parameters. Often, it is a good approach for generating a machine learning database on a limited number of samples. The simulations took between 1-2 hours using 96 CPUs for each run. 10 simulations were run concurrently, and the entire data set was generated over the span of 24 hours. All simulations were simulated at 10 knots (Froude number =0.136) with a prescribed draft of 5.24 m. Hull roughness was assumed to be smooth.

When generating machine learning models to predict the integral performance metrics, the result was that a surrogate or reduced order model, can be created to be used in coupling with other simulation tools. Several file formats are possible to export the model: Simcenter Amesim, FMU 2.0, ONNX, or S-Function. Initially, the CFD data was input into Simcenter ROM Builder and divided with a 80/20 split for training and testing, respectively, meaning 80% of the data was used to train the model and the rest of the data set was used to validate the model's predictions. The imported data was a standard tabular form in a .csv file which summarized the database of simulations (columns: inputs and outputs, row: different simulations). To provide the best data possible, a 'wizard' type feature was used to evaluate many different models, similar to auto-ml functionality. Neural networks, Response Surface Models, and Gaussian Processes were included in this evaluation. The best model to fit the data was a Neural Network, whereby the ideal hyperparameters were 3 dense layers of 37-18-10 sizes and a ReLu activation function. On a standard laptop in serial, 50 epochs of training took 19 seconds and resulted in 97.8% and 89% accuracy for training and testing, respectively.

Next, the details will be covered for the models/process to model the wake fraction in the propeller plane with Monolith. The local (cell) scalars were accessed and visualized from within Simcenter STAR-CCM+ CFD simulations in a plane just in front of the propeller and orthogonal to the nominal streamwise direction of the mean local flow-field. Each design resulted in one data file (.csv) that all shared the same vantage point when plotting wake fraction in the aforementioned plane. This would be the input data for the machine learning model to reproduce the propeller plane scalar result. The .csv files are accompanied by another tabular file which contains the independent parameters and results for each case to summarize the full set of design simulations completed (i.e. a single table with 128 rows and each column is an input variable). The point-clouds of wake fraction, with spatial coordinates, are combined with for each respective design in the summary of integral metrics (inputs and outputs). This pre-processing of the data enables a machine learning model to relate the input values to the associated point clouds of wake fraction data. In other words, the CFD practitioner can specify nine values to describe the geometric input and a machine learning model can predict the entire scalar field as the result.

*** Formatting Issue - Best Available Paper/Graphic ***

Next, the details will be covered for the models/process to model the predictions of the free surface displacement height, or generated wakes, in Monolith. Much like the propeller plane pipeline, the data is extracted from each simulation to be passed to the machine learning model for training. However, in this case the data is of tabular form when extracted from the simulation as a .csv file. This machine learning model is tasked to predict the local values of the free surface displacement in the field for various geometric configurations. As such, the position coordinates, local surface height, and geometric variables need to be included for each design to pass the machine learning model. These pieces of information for each case are stored as separate columns in separate .csv files, whereby each cell is designated a separate row. Later, these tables are used to generate three-dimensional point clouds of the position coordinates and field data, then concatenated and fed to the machine learning model for training as one dataset. Monolith's Surface Field model is a perfect machine learning model to pick for this task, as this model uses AI models that are geared towards three-dimensional volume fields of engineering scalars. They are convolutional models, which means they scan the 3D data files as a whole and identify patterns in how 3D topology affects 3D fields, rather than making point-wise predictions. This is very powerful, as the point cloud nature of spatial data fields in engineering simulation is adverse to the traditional methods in machine learning because popular models for addressing point-wise predictions cannot scale to large enough sizes to be feasible. Tabular data still best benefit primarily from tree-based methods [13], such as the first row in Table 1. Convolutional approaches would be an attractive consideration, however their limitation is that they are often in two-dimensional space and are quite limited to small interrogation windows (e.g.as in medical applications, where 1024x1024 would be considered a large size).

Lastly, the models and methods based on the three-dimensional spatial data will be covered next. The scalar fields, in this case the pressure and the individual components of the wall shear stress, are predicted on the hull wall in a pipeline that is nearly identical to that of the above displacement height scalar on the free surface. Whe e adi iona i e neu a networks would break down in scale when going into three-dimensional space, Monolith's Surface Field model continues to be a viable choice for such a dataset.

The final row for predictions in Table 1 signifies that the machine learning model can learn from the CAD (.stl files) for the various designs and become able to recreate the full geometric design space with a unique parameterization (different from the parameterization that was done when setting up the CAD/CFD designs). Variational autoencoders (VAEs) are the specific machine learning model that is used to learn the surface meshes, which are a specific type of neural network. Despite that a parametrization exists for the geometry already in the CAD, such is not specifically needed by the VAE pipeline, as it only requires surface tessellations to subsequently learn the geometry. The autoencoders are comprised of several blocks, as illustrated in Figure 2. In the software, the user has the freedom to specify how many latent parameters to have. One benefit of creating a latent space directly is not only new ways to create designs apart from traditional methods, but also to create smaller sets of parameters can then be executed in a DoE or optimization sweep. Otherwise, it is not uncommon to have large sets of parameters which prohibit a uniform or random sampling of the space. For example, a 20-parameter set to describe a geometry for which you specify 3 values each yields 8,000 cases to be simulated, whereas a latent space of 7 parameters would only yield 343 simulations. More details on the VAEs in Monolith can be found in one of their recent publications which was the sole focus of the work [14].

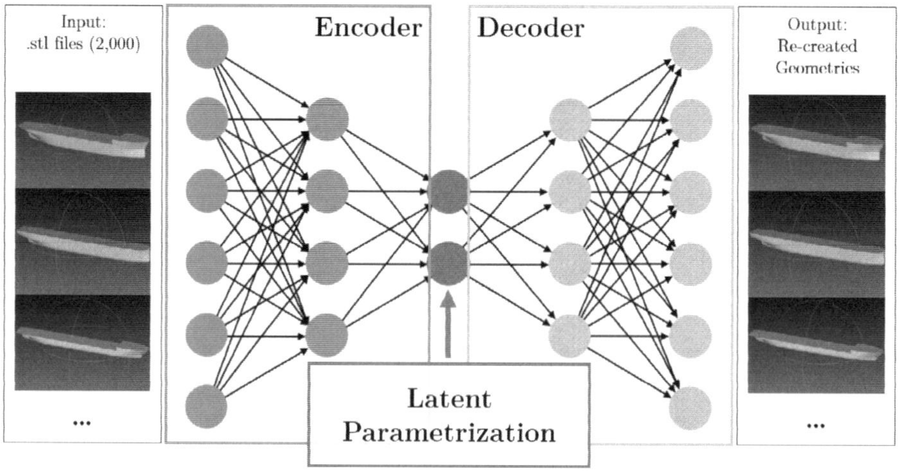

Figure 2: Variational Autoencoder (VAE) structure diagram. The .stl files associated to different designs are used as input the encoder. The encoder creates a latent parameterization of these geometries, specifically the coordinates of the mesh points, which can be used to accurately re-create geometries at will by the decoder.

*** Formatting Issue - Best Available Paper/Graphic ***

From the variety of machine learning predictions produced in this work and summarized in Table 1, the end capability is a utility to the CAE/CFD engineer which can expedite the design task by combining simulation and machine learning techniques. Figure 3 summarizes the different workflows possible, from the point of data ingestion to final prediction, and how they can be used. The final step in the red block provides real-time prediction via surrogates that can be used for rapid design space exploration and the possibility of predicting high-fidelity results (e.g. three-dimensional data fields) from lower fidelity simulations (focused on 0D metrics alone). These ML models can also be consumed in optimization workflows in place of simulation, which is indicated by the green box.

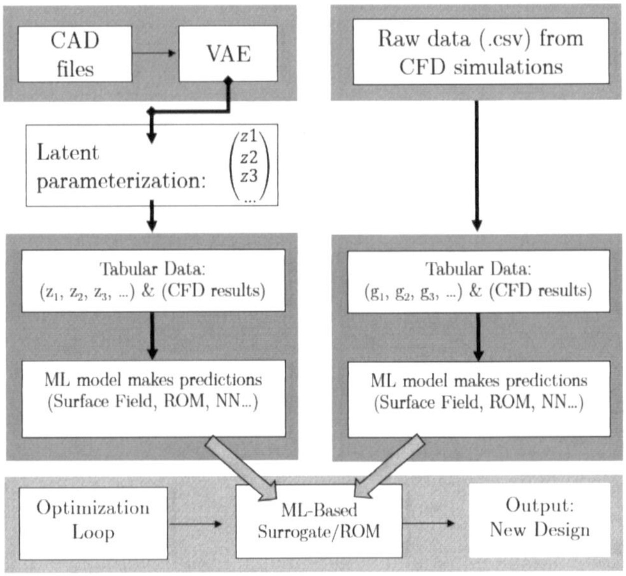

Figure 3: Concept of different workflows available. Blue: the starting point where data is available (CAD files and/or CFD results). Red: machine learning models can be generated for various types of results of interest, whereby rapid predictions can be made with trained surrogates. Green: optional further usage of trained ML models to replace simulation in optimization routine.

4. RESULTS AND DISCUSSION

4.1 INTEGRAL METRICS

The first set of results to discuss will be that of the integral performance metrics, which are single scalar values. The total number of CFD simulations available for the machine learning pipeline is quite limited, but this is intentional, as it reflects a normal scenario where the CFD practitioner would prefer running as few of simulations as possible due to their resource cost and time investment needed. As such, any further removal of training samples will tend to result in trained models which perform worse in validation, since now the trained models are predicting outside of the parameter ranges they trained in.

As mentioned, the original CFD data set was generated using a Latin Hypercube Sampling technique was used populate designs representative of the parameter space. As such, Figure 4 displays a histogram of the sampling on each of the geometric parameters (signified as g) for the training (blue) and testing (orange) dataset. Because the training and testing data is split by a random process in an 80/20 split, respectively, it is observed that most of the parameter values occur in both the training and testing data. However, the relative count for each geometric value (the bar height) is not similar between the training and testing datasets. This is assuring that the tuneable coefficients, which in the case of neural networks are the weights and biases, are not being skewed towards specific ranges for the geometric parameters which are used in the test evaluations.

***** Formatting Issue - Best Available Paper/Graphic *****

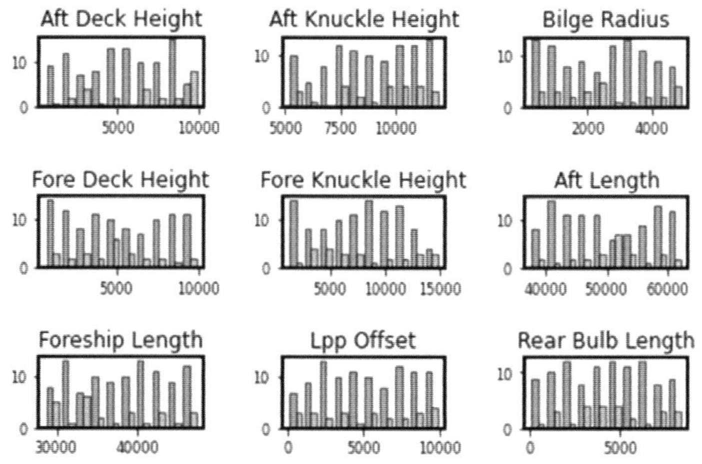

Figure 4: Histograms for each geometric parameter used as inputs to the machine learning models. Blue: training data, orange: testing data.

For the testing dataset, plots of the predicted versus actual data are shown in Figure 5. The ideal scenario where machine learning model predicts the actual values is a linear line, shown in blue. These plots can be anchored by singular error metrics, which are tabulated in Table 2. The fidelity index is defined as one minus the normalized root mean square error multiplied by the Pearson coefficient, as seen in Equation 1.

Equation 1: Fidelity index

$$Fidelity\ Index\ [\%] = |100 \cdot (1 - NRSME) \cdot (Pearson\ Coefficient)|,\ NRSME = \frac{\sqrt{\frac{1}{N}\sum_1^N (y - y_P)^2}}{\max(y) - \min(y)}$$

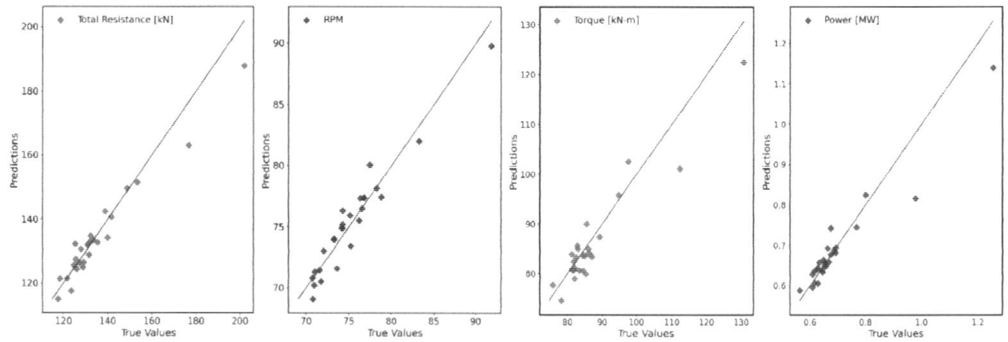

Figure 5: Actual values versus the predicted values by the machine learning model on the test dataset

During training, the training and testing losses were monitored so that overfitting would be identifiable and avoided. After only 300 training steps, both loss curves flattened to their long-term values. An adaptive learning rate was used since the magnitude of loss would vary over the duration of training and therefore it is much faster for the user to receive a converged model if the rate of change of the weights and biases are also allowed to vary adaptively throughout the optimization process. As a result, training was very quick and took less than 30 seconds to complete. The data was shuffled during training and fed in batches of 16. Upon testing linear, ReLu, and tanh activation functions, the tanh activation function produced the best results. These are popular choices for activation functions in a neural network and all perform reasonably well, whereby activation functions scale the output value from individual neurons according to the function selected. There were several architectures which proved successful for this dataset, whether it as simple as a single hidden layer with 14 neurons, two hidden layers of 26 and 11 neurons, or three hidden layers with 35-16-11 layers. These hy e a a e e e e dedu ed y he Wi a d un iona i y, hi h i e ui a en o an u o un ion ha ou u recommender architecture and hyperparameter sets that perform the best. Neural networks were superior for this dataset and selected in lieu of the Gaussian Processes and Response Surface Model choices which were also tested.

Table 2: Accuracy quantification (fidelity index) for the machine learning predictions for each of the results and the datasets.

Result Metric	Dataset	Fidelity Index Value
Total resistance [kN]	Training	96.6%
	Test	92.0%
Torque [N-m]	Training	98.5%
	Testing	88.7%
RPM	Training	92.2%
	Testing	90.7%
Power [HP]	Training	100%
	Testing	92.8%

4.2 SCALAR FIELDS IN THE PROPELLAR PLANE

From the same group of 128 CFD simulations, results are extracted from the propeller plane to build machine learning models. The input variables for this result prediction are identical to the nine geometrical parameters for the 0D results prediction, except that also the two spatial coordinates are included as independent variables. The result of interest is the local wake fraction, which is the velocity magnitude normalized by the vessel speed, such that a machine learning model could provide flow-field predictions in real-time. Training and testing were only done on 50% of the cases, whereby the remaining 50% was excluded from the training process entirely and used for validation testing thereafter.

The Surface Field model in Monolith was used and trained for 250 epochs, as that was sufficient to reach more than two orders of magnitude reduction in the training and validation loss without experiencing overfitting. The model only took 101 seconds to train and used 51,000 rows of data (which equates to spatial cells in the simulation). This number is the result of approximately 850 cells in the plane extracted from each simulation from 60 CFD simulations. The data for each simulation is stored in .csv files after extraction from each simulation and brought uploaded into the Monolith platform. No data transformation of any kind is done before training, with exception to splitting for training and post-training validation testing.

Figure 6 below displays such results for the wake fraction in the propeller plane on three validation cases, as well as the associated error in the form of a percentage. The values of wake fraction vary between zero and one, with error on the order of 10% (note small wake fraction values can yield large error percentages). The same nine parameters shown in Figure 4 are used as inputs to the machine learning model for predicting the wake fraction fields. Some of which are located far from the propeller plane, and therefore might have an indirect and lesser impact on the result wake fraction result Wi h this application-specific knowledge, a 'ranking' on which geometric parameters would have the most influence could be formed with fewer parameters, and therefore gradually expanded when training individual machine learning models. This would be a systematic way to reduce the prediction error, however in this paper the priority was to maintain the same geometric parameters for a variety of results predictions. The benefit of this approach is to judge how robust the machine learning models can be on the same simulations, but with different data, without the expert knowledge for tuning in each prediction.

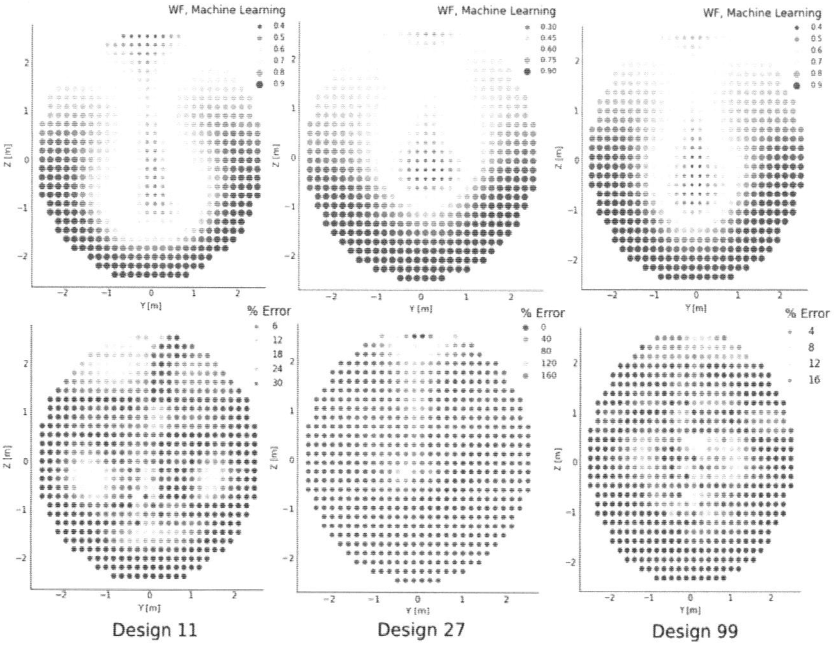

Figure 6: Wa e a ion e e o for several randomly selected designs which were excluded from the training process. **Top row**: machine learning prediction. **Bottom row**: percentage error compared to ground truth (CFD) data.

4.3 C F W F F C

The next results to be discussed are that of the free surface prediction. These results can be obtained as a scalar in the form of the z-coordinate as the displacement of the water surface. These wave patterns are the steady state results from a transient simulation. As in Figure 4, the same nine independent geometric parameters are used as inputs also to this machine learning prediction task. As in the propeller case, in addition to these nine, the spatial coordinates are also included such that the spatial distributions can be learned. The training/testing/validation split was for this use case was 70/20/10, respectively. Monolith's Surface Field model was used, since the convolutional feature is key to reduce the inherently large size of the point clouds from many different simulations. When o a ing o a anda d neu a ne o , the predictions had pervasive amounts of noise and were unable to learn the global and generic patterns. It should be noted that the predictions with the Surface Field model are only reasonably accurate close to the hull at the bow and stern, whereby the results were often poor in the far field. As a note for possible future work, this could likely be improved if the input variables were changed to that of a latent encoding provided by an Autoencoder (see section 4.5 for more details).

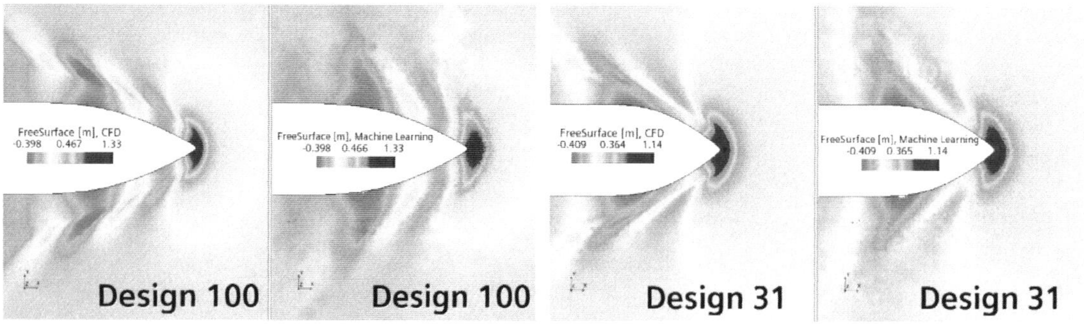

Figure 7. Wa e ee u a e o a i on e een g ound u h CF i u a ion and machine learning predictions

These results can be seen in Figure 7 for two specific designs within the validation set, which were omitted from training entirely and thus ensure a blind evaluation of the prepared model. The machine learning results capture a big difference in the case results; the steepness in the deflection angle pink coloured area representing the lower water levels. In this case,

***** Formatting Issue - Best Available Paper/Graphic *****

the number of training steps is an important hyperparameter. Experiments were done by training for various amounts of steps, and at/beyond 10,000 training steps was critical for good results.

4.4 SCALAR FIELDS ON THE HULL SURFACE

The next results to be discussed are that of the local results on the hull surface. This includes all components of the wall shear stress, with the *i* direction plotted in Figure 8 and the piezometric pressure as shown in Figure 9. The motivation is that it would be desirable to have a machine learning model that can accurately predict the drag not only as an integral performance metric for the entire design, but also in the form of detailed (and local) shear stress and pressure constituent components that can help aid in design decisions and provide a fundamental understanding of the vessel hydrodynamics.

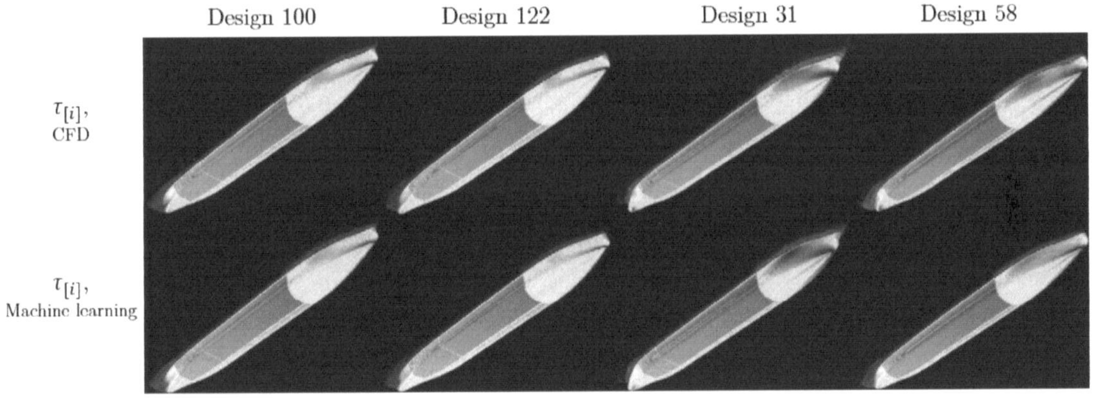

Figure 8: Wa hea e i o onen on hu no-slip) wall. The machine learning predictions represent cases not used in training

In Figure 8, several plots are shown for designs that were in the validation testing set (indicating they were not used in the training process but held aside for blind testing of the trained model). Designs 122 and 58 have noticeably different form than in Design 100 and 31, and such is evident in the wall shear stress contours. The machine learning model does a reasonable job at picking up on these salient different flow patterns, owing to the separation patterns in the flow-field, in the front and end of the hull. The geometric parameters used as inputs to the machine learning model are the same nine from Figure 4. Hence, these parameters strongly dictate the separation patterns and therefore are great candidate variables for correlating to the wall shear stress fields. If a broader set of parameters were used, such as operational conditions, then the wall shear stress predictions from the machine learning models could become more error prone.

Next, in Figure 9, similar local contours on the bow are shown for the piezometric pressure; defined as the pressure after removing the hydrostatic component. Plotting the piezometric pressure should on the hull surface should elucidate where the accelerating and stagnation zones are. Downstream of the stagnation point, the piezometric pressure gradients are noticeably different between the two cases pictured. The machine learning model captures this trend, as well as the differing thicknesses of the low-pressure region (pink colour, in 'V' shape). These patterns are highly sensitive to the local geometry, and thus requires the machine learning model to accurately relate the difference in the geometries to specific scalar distributions.

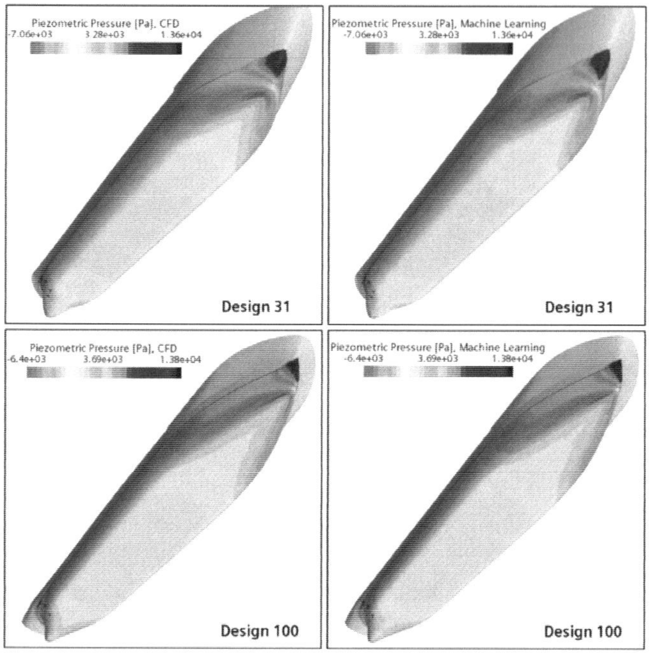

Figure 9: Piezometric pressure on the bow, for CFD and machine learning predictions two validation cases (excluded from model training)

4.5 LEARNING THE HULL GEOMETRY

The final component of the study to discuss, which pertains to the left column in Figure 3, is in regard to a machine learning pipeline to learn the physical geometry in Monolith. The first step, as input, is to generate and provide 2,000 individual geometries in the form of .stl files. These geometries are created in NX, which took about ten minutes to create through automated generation on a standard workstation. In Monolith, these geometries are converted from three-dimensional surface tessellations to voxelised meshes for training (which took about 20 minutes to complete). Next, the geometry is 'learned' from the voxels by a variational autoencoder model, which is a specific type of neural network. The combination of a large number of geometries, and the requirement to train the machine learning model for a substantial number of training steps, led to a total training time of 9.5 hours. This was done on a standard CPU cluster and only needs to be completed once. Furthermore, using a GPU cluster would have significantly reduced the compute time. Once trained, the model can then decode, or reproduce, three-dimensional geometries upon user request in real-time. The Variational Autoencoder (VAE) learns the geometric representation and exposes a set of latent parameters the user can adjust to create different geometries. The number of latent parameters can freely be changed to any reasonable value (approximately 4 or more), whereby in this study ten latent parameters were created. Figure 10 shows some randomly selected cases, among the full set of 2000, for the VAE to re-create (decode) and then compare to the 'real' CAD. The corresponding loss curve for the training is also shown in Figure 10. Both the convergence behaviour and re-created geometries reflect that the VAE has accurately learned the geometry-space for which the CFD simulations were ran in, by its own latent parameterization.

Figure 10: (Left) Nine randomly selected geometries generated by the autoencoder workflow; Blue: original (.stl), Red: autoencoder recreation through decoding. (Right) Convergence plot of the loss over the training steps for the autoencoder learning the geometry from the voxels.

*** Formatting Issue - Best Available Paper/Graphic ***

One immediate value is the benefit of a new, and yet still accurate, representation of a geometric design space through a latent space that is condensed to fewer parameters. For example, if there are 20 geometric parameters originally, a uniform sampling for a CFD DoE would yield 8,000 cases if each parameter had only three values. However, if a latent space of ten parameters were used instead, that would result in only 1,000 cases with again three values for each parameter. A more powerful capability yet is illustrated in the green box of Figure 3; using these machine learnings model in an optimization study Wi hou on ide ing he -based portion, already there is the opportunity to predict results from a machine learning surrogate in real-time and therefore significantly reduce the time for such an optimization study. Robustness, reliability, and CAD sensitivity studies are other exercises were thousands of design points need to be solved, which in turn are an ideal application for surrogates that provide results in real-time. When further considering the VAE's ability to learn the geometry, then inverse design methodologies are possible; the required results (integral metrics and/or spatial fields, as in this study) can be stipulated upfront and the optimization study instead solves for the geometry which satisfies the criteria/cost function. This rapid and comprehensive understanding that the VAEs provide will be explored in subsequent work in this context of an optimization cycle.

7. CONCLUSIONS

This work has showcased ways to quickly assess true powering requirements for ships operating at full scale in a design environment. A modest set (128 cases) of simulation data was generated over the course of a day by running randomly generated design variants. ROM and AI models were applied to the simulation database and tested to see how well powering can be predicted on a vessel that may undergo design iterations. Predicting global scalar values such as resistance and shaft power were remarkably good. The study also investigated looking at how well these models predicted the local field variables; wake fraction at the propeller, the generated wake, and the shear and pressure loads on the hull were able to be investigated which provided a solid understanding of the flow characteristics. The beauty of these models is a designer can see how well a new design variant will be affected in a matter of seconds with high confidence in leu of having to wait several hours to understand a design variants operational performance. The last part of the study showed we can reduce the number of independent geometric variables to help aid in an optimization task using latent parameterization, which will be concluded in future work.

8. ACKNOWLEDGEMENTS

The authors would like to thank Felix Muehloff for providing the parameterized hull form used in this study. Also, many thanks to Dr. Dejan Radosavljevic for his contributions and encouragement to work on this project. The authors would like to thank the JORES organizing committee and participating members for providing valuable insight, hull geometries, and validation data that helped validate our numerical simulations and allows for full scale sea trail information to be showcased in a public forum.

9. REFERENCES

1. Stern, F., K. Agdraup, S. Y. Kim, A. C. Hochbaum, K. P. Rhee, F. H. H. A. Quadvlieg, P. Perdon, T. Hino, R. Broglia, and J. Gorski. "Experience from SIMMAN 2008—the first workshop on verification and validation of ship maneuvering simulation methods." *Journal of Ship Research* 55, no. 02 (2011): 135-147.

2. Gui, La, J. Longo, and F. Stern. "Towing tank PIV measurement system, data and uncertainty assessment for DTMB Model 5512." *Experiments in Fluids* 31, no. 3 (2001): 336-346.

3. Karczewski, Artur, and Janusz Kozak. "Comparison of selected parametric methods for prediction of inland waterways ship hull resistance in towing tank test." *Polish Maritime Research* (2018).

4. Mizzi, Kurt, Paula Kellett, Yigit K. Demirel, Richard Martin, and Osman Turan. "HPC and CFD in the marine industry: past, present and future." (2015): 92-97.

5. Wheeler, Miles Patrick. "Speeding up Hydrodynamic Simulations with a Parametric, Templated Approach." SNAME Maritime Convention. OnePetro, 2021.

6. Ling, Julia, Reese Jones, and Jeremy Templeton. "Machine learning strategies for systems with invariance properties." *Journal of Computational Physics* 318 (2016): 22-35.

7. Wang, Jian-Xun, Jin- ong Wu, and Heng Xiao hy i -informed machine learning approach for reconstructing Reynolds stress modeling discrepancies based on DNS data." *Physical Review Fluids* 2.3 (2017): 034603.

***** Formatting Issue - Best Available Paper/Graphic *****

8. Milani, Pedro M., Julia Ling, and John K. Eaton. "Physical interpretation of machine learning models applied to film cooling flows." *Journal of Turbomachinery* 141.1 (2019): 011004.

9. Raissi, Maziar, Paris Perdikaris, and George E. Karniadakis. "Physics-informed neural networks: A deep learning framework for solving forward and inverse problems involving nonlinear partial differential equations." *Journal of Computational physics* 378 (2019): 686-707.

10. Jaffar, Farwa, et al. "Prediction of drag force on vehicles in a platoon configuration using machine learning." *IEEE Access* 8 (2020): 201823-201834.

11. Lloyds Register. "Proceedings". 2017 http://info.lr.org/l/12702/2017-02-20/3m372v/12702/156863/Proceedings.zip. Accessed 2020.

12. JoRes "Joint Research Project". 2022 https://jores.net/. Accessed 2022.

13. Grinsztajn, o, doua d ya on, and a a o uau Why do ee-based models still outperform deep learning on tabular data?." *arXiv preprint arXiv:2207.08815* (2022).

14. Pongetti, Julie, et al. "Using Autoencoders and Output Consolidation to Improve Machine Learning Models for Turbomachinery Applications." Turbo Expo: Power for Land, Sea, and Air. Vol. 84935. American Society of Mechanical Engineers, 2021.

10. AUTHORS BIOGRAPHY

Dr. Justin Hodges is the AI/ML technical specialist in product management for the Simcenter products at Siemens Digital Industry Software. His duties are dynamic for each product and season, but ultimately are to provide technical ownership and guidance in the development, release, and usage of the matrix of simulation and test products in the Simcenter portfolio.

Dr. Miles Wheeler holds the current position of Senior Marine Applications Specialist at Siemens Digital Industry Software. He is responsible for ensuring that Simcenter STAR-CCM+ is meeting the needs of the marine community. He o ained hi h o Wa hing on a e ni e i y in computational hydrodynamics. Before this role, Miles worked as a computational fluid dynamics consultant in the marine and automotive sectors.

Dr. Mohamed Belhocine is a Product Developer for Simcenter ROM Builder Siemens Digital Industry Software. After a PhD in simulation and controls of large electrical networks, Mohamed developed control strategies for electrical systems and aircraft controls before joining Siemens as a control expert. He is now part of ROM group which takes care of identification and model reduction topics for Simcenter simulation and test products.

Dr. Joel Henry is a Principal Engineer at Monolith AI and has worked with the majority of Monolith's clients to create lasting value for them. He has won the Imperial College Award for the best PhD thesis and has previously worked on improving test and simulation methods in the aerospace industry.

AUTOMATIC DATA EXTRACTION AND UNFOLDING FOR SHIP HULL PLATES ON CATIA

Yu Fu[2], **Wei Wei**[3], **Wuyi Xie**[3], **Yong Hu**[1, 2] and **Chaoyan Huang**[1, 2],
1 Key Laboratory of High Performance Ship Technology of Ministry of Education, Wuhan University of Technology, P.R. China
2 School of Naval Architecture, Ocean and Energy Power Engineering, Wuhan University of Technology, P.R. China
3 Wuchang Shipbuilding Industry Group Co., LTD, P.R. China.

SUMMARY

In view of the current low efficiency of ship production and the shortcomings of various software in the unfolding of ship hull plates, in this study, a method that automatic data extraction and unfolding of ship hull plates is proposed. This method has carried out by the add-on development of the Component Application Architecture (CAA) on CATIA 3DEXPERIENCE platform (3DE platform). The sequence of the method is as the following: Firstly, the model is switched into the editable mode. Then, the file tree is traversed for judging the type of parts and identifying the geometric features, and the intersection lines between surfaces and the structural lines are generated. Thirdly, the neutral layer of parts are extracted and the process information (e.g. margin, compensation and bevel) are added. Finally, the information extracted are integrated into an exchange file for unfolding or processing. Different methods are proposed for flat plate, curved plate and combined plate. As for the double-curved hull plate which cannot be directly unfolded, an approximate unfolding method of dividing the plate into strips and small triangles and then splicing them together is proposed according to the forming method of plate. The test result shows that this method can reliably extract the plate parts and unfold the curved plate.

KEYWORDS: Ship Hull Plates; Unfolding of Curved Plate; Add-on Development of CATIA; CAA; 3DE Platform

1. INTRODUCTION

A ship is a complex marine engineering structure. Ship hull plates are fabricated by plate forming process before they are welded together for assembly. However, the production process of hull plates is difficult because they are generally complex in shape and curvature. In addition, the hull plate should be unfolded from three-dimension (3-D) model to two-dimension (2-D) plane before nesting and cutting. Therefore, the unfolding of hull plates is an important procedure in the process of ship production design.

In the traditional process of shipbuilding, it has been practically carried out by the method of manual calculation with several models and sample boxes in inefficient. The curved plate is processed by the line heating, which will cause the shrinkage and deformation of the plate parts. The amount of shrinkage and deformation is difficult to calculate accurately, so more margins are usually added to the curved plate, and then the excess margin is cut off after the processing is completed and its shape is checked, which not only causes waste of materials, but also consumes a lot of labor costs.

Digital shipbuilding are gradually adopted in modern shipbuilding, and how to unfold the curved plate on has become a new problem. Currently, several software are used for ship production design, such as HD-SHM, TRIBON, AVEVA Marine (AM) and CATIA. Ship hull plates can be unfolded in HD-SHM of Hudong Zhonghua Shipbuilding Co.LTD. It adopts the traditional manual geometric approach, and the plates are divided into diamond-shaped plates, fan-shaped plates and other types, which is complicated in operation and low in accuracy. TRIBON software lacks interface with common databases. The quality of AM software's drawing depends on the operator. The output standards of drawing information are inconsistent, so the quality of drawing are not so well [1]. The unfolding function in CATIA software is a geometrical unfolding, and the result cannot be used for ship production.

Although CATIA software lacks the function of unfolding the hull plates, its generative shape design (GSD) has the characteristics of surface stitching in high accuracy and freely switching of multiple modes. In addition, users can carry out a variety of add-on development on CATIA software, and file transfer is convenient. Accordingly, in this study, automatic data extraction and unfolding for ship hull plates has been carried out by the add-on development on CATIA software.

2. INTRODUCTION OF CATIA SOFTWARE AND CLASSIFICATION OF HULL PLATE PARTS

2.1 INTRODUCTION OF CATIA SOFTWARE

CATIA of Dassault system is a software that integrates CAD/CAM/CAE/PDM, which is widely used in aviation, aerospace, automotive, marine and electronic industries. In shipbuilding, CATIA software is mainly used for ship outfitting, piping layout and assembly design, but CATIA lacks functions in the ship hull production design. Especially, CATIA has few functions for unfolding of ship hull plates.

CATIA provides users with four add-on development tools, including User Defined Features, Knowledge Engineering, Automation Application Interface, and Component Application Architecture (CAA). Among them, CAA is a customized development based on components, which is the combination and extension of its component objects. The Component Object Model (COM) and Object Linking and Embedding (OLE) are adopted, which standardizes the program design of CAA easily [2].
The aim of the study is to realize the unfolding function of the hull plates. The specific process is as follows: firstly, the information of various plate parts are extracted, and the process information of parts are added interactively.Then, all the information are integrated into a data set, and finally the data set are used for unfolding the hull plates. It can be seen that the program is complex and multiple interfaces are used. Most importantly, it is hoped that a standardized application program can be formed. Hence, CAA add-on development of CATIA software is chosen in this study.

2.2 CLASSIFICATION OF HULL PLATE PARTS

Ships are products of many varieties and small batches. The hull structure is complex and diverse with a large number of parts, and there are less hull plate parts with the same surface shape. Hence, it is necessary to classify the plate parts on the hull plate parts in order to adopt different methods of information extraction and unfolding of ship hull plates to improve the efficiency of the program. Considering the above reasons, in this study, the ship hull plate parts are divided into flat plate, curved plate and combined plate.
(1) Flat plate: The flat plate in the hull structure includes deck plate, longitudinal and transverse bulkhead plate, inner bottom plate, watertight bulkhead, bottom plate at midship, floor plate, longitudinal deck, inner side plate, inner bottom sloping plate, platform plate, bulwark plate, etc. Since the complexity of the hull structure, the flat plates are generally different in shape with curved edges and openings.
(2) Curved plate: The curved plate in the hull structure includes most of the hull shell plate, such as the hull shell plate at the bow and stern, the bilge plate, etc. Curved plates are fabricated by plate forming process. They are generally formed by flame heat or mechanical tools such as roller and press. Moreover, flame heat source is essentially used in the case of forming for doubly curved shape. The principle of thermoplastic forming has adopted by the flame heat processing, which will cause the shrinkage of the plates. Therefore, the surface area of the plate is reduced after the flame heat processing. Surface unfolding is the inverse process of surface forming. In the unfolded results, the surface area of the unfolded plane should be larger than the actual surface area in the 3D model, thereby offsetting the reduction in the surface area of the surface due to the shrinkage of flame heat processing.
(3) Combined plate: The combined plate in the hull structure mainly refers to the folded plate, which generally contains both flat plate and curved plate, including hatch coaming, cargo hatchway end coaming, folded bracket and other complex combined plates. This kind of plate is complex and needs to be carried out in some special ways.

3. DATA EXTRACTION OF SHIP PLATE PARTS ON 3DE PLATFORM

3.1 PROCEDURE OF AUTOMATIC DATA EXTRACTION METHOD

In order to realize the unfolding function for hull plate parts, the 3D hull model should be firstly identified as well as the relevant data information of parts should be obtained on the CATIA software.
Therefore, an automatic data extraction method for hull plate parts is proposed in this study. The sequence of the procedure is as the following: traverse the file tree, judge the type of parts, identify the geometric features, extract the neutral layer of the intersecting parts, generate the temporary structure line, add the process information manually, extract the data information of parts and create an XML file for each parts.
When a plate is bent under a force, the area of the tensioned side increases and the area of the compressed side decreases due to the thickness of plate. The neutral layer is the surface whose area is equal to that before bending, as shown in Figure 1. Hence, the neutral layer is used as the extraction result of a plate, which can reduce the accuracy influence of the plate thickness on the unfolding of the curved plate. It has been practically carried out by half of the plate thickness since it is a very difficult task to determine the neutral layer of a plate.

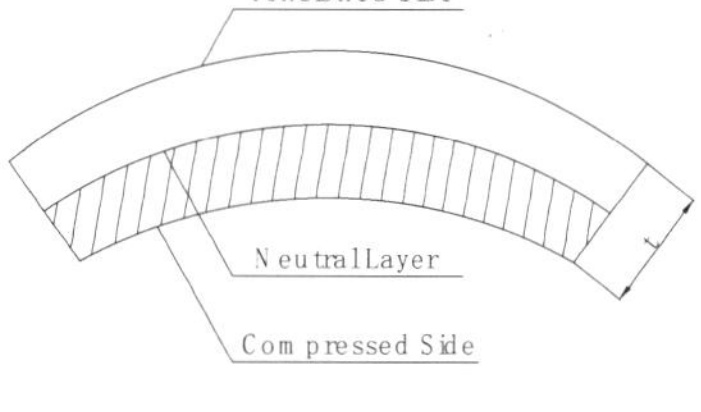

Figure 1. Neutral Layar

The automatic data extraction method of hull plate parts is proposed in this study, and the details are as follows.
(1) Traversing the file tree and judging the type of parts: The entity of the target part is obtained by traversing the file tree, and the set of faces and boundaries in topological level is obtained by the topological relationship. In addition, the type of parts can be judged through the interface named CATIMeasurablePlane, whose return value of 1 indicates that the plate is curved plate, and return value of 0 indicates that the plate is flat plate. And the curved plate and the folded plate can be distinguished by the interface named IsTypeOf, one with more than two faces is a folded plate.
(2) Identifying the geometric features: Geometric feature recognition is based on the Boundary Representation (B-rep) method, which is a way of representing shapes using limits. An entity is represented as a set of connected surface elements, that is, the boundary between entity and non-entity. A B-rep model is consisted by two parts: topology (points, edges, faces, entities, loops, etc.) and geometry (surfaces, curves, and points, etc). The geometric information and topology information of the part can be obtained by geometric feature recognition.
(3) Generating the temporary structure line: The structural theory line of the ship is one of the effective measures to ensure the assembly accuracy. It is an indispensable construction basis for lofting, processing, assembly. Accordingly, it is also considered to display the structural lines on the plate. The process is as follows: firstly, the datum plane of the target part and adjacent parts are extracted, and then the intersection line of the two datum planes are generated through the surface intersection algorithm. The rib plane, waterplane, and longitudinal section are used as references in this process.
It should be pointed out that the datum plane is a quadrilateral enveloping the actual plane obtained by the program, as shown in Figure 2. It aims to uniformly deal with non-quadrilateral hull plates, such as triangular plates, pentagonal plates and polygonal plates, so as to improve the operation efficiency of the program.

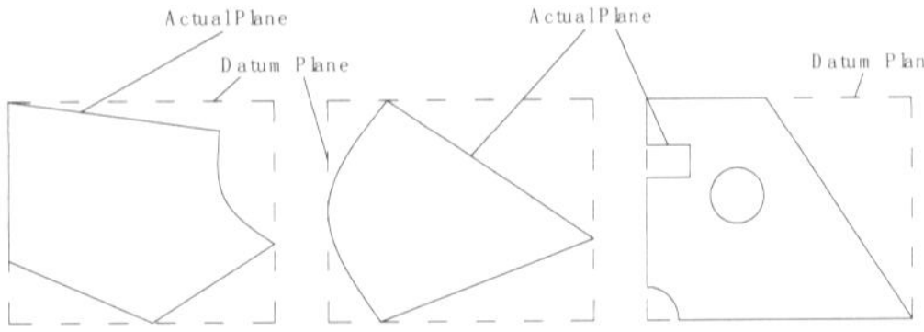

Figure 2. Datum Plane and Actual Plane of Various Plates

(4)Adding the process information: In this study, human-computer interaction is used to add the process information, e.g. margin, compensation, bevel. The boundary location of the plate can be automatically determined by the program. Margin refers to adding a certain amount on several sides of the plate during lofting. Since the size of the hull plate will be affected by the lofting, assembly, welding and other processes. The amount of compensation is considered since the shrinkage deformation of hull parts during welding. The welding bevel is a groove with geometric shape (e.g. X, Y, V).
(5) Extracting the part information and create an XML file: The geometric information, attribute information and process information of plate are integrated into a data set. The geometric and non-geometric information of the plate are matched. Finally, the data of each plate is written into an XML file for subsequent unfolding or processing.

3.2 DATA EXTRACTION METHOD FOR VARIOUS PLATES

In this study, different data extraction methods are applied to different types of plates, which facilitates the unfolding of plate as well as improves the running efficiency of the program.
(1) Flat plate: Since the flat plate can be nested without unfolding, it is only necessary to obtain the neutral layer, contours and openings of the flat part during extraction. The contours are divided into straight edge and curved edge.
(2) Curved plate: Since most of the curved plates of the ship hull parts are double-curved shape, in this study, the curved plate is firstly divided into a series of $n \times n$ spatial grid points, and then extracted, where $n = 32, 64, 128$ or 256. A non-uniform rational B-spline (NURBS) curve is then fitted through this series of grid points. This method is beneficial for subdivision surfaces and unfolding the curved plate.
(3) Combined plate: A combined plate generally consists of both flat plate and curved plate. The combined plate of the hull plate can be roughly simplified into two cases as shown in Figure 3: the left folded plate is actually one plate in the files tree; the right folded plate is actually three plates in the files tree. Therefore, the three plates are extracted separately, and then all the data is combined into one XML file.

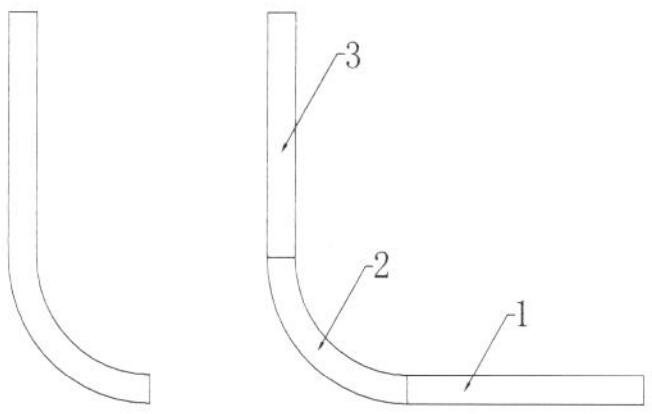

Figure 3. Combined Plate

The extraction of combined plate is different from that of single flat plate and curved plate. Due to the complexity of combined plate in the hull structure, some more complex plates, such as the plates at the base room, may need to be extracted manually.

4. UNFOLDING OF SHIP HULL PLATES ON 3DE PLATFORM

4.1 INTRODUCTION OF UNFOLDING THE SHIP HULL PLATES

Unfolding of hull plate is the process of finding out the real shape of a spatial surface that cannot be represented on the projection drawing. The purpose of unfolding the hull plate is to draw sketch and make templates. Since the accuracy of unfolding directly affects the accuracy of hull lofting, so it is necessary to seek a better method to unfold a plate.

The curved plate in the hull structure is divided into a developable surface and a mathematically non-developable surface. Surfaces whose Gaussian curvatures are zero at every point are described as developable surfaces. Outfitting parts, most of the longitudinal plates and the hull plates in the parallel medium structure are described as developable surfaces. These plates can be accurately obtained the real shape through the geometric approach, so it is only required to obtain their actual shape when unfolding. Ship hull plates and parts with complex curvature in engine room, fore and aft peak tank are described as non-developable surfaces. These plates cannot be accurately obtained the real shape through the geometric approach, so it is necessary to unfold them by some approximate approaches.

The unfolding of the non-developable surfaces has been studied by many researchers. The triangular plane mesh method was proposed by Parida [3] et al. The way of obtaining a flat pattern is to triangulate the surface into approximate planar triangles, and then to unfold them into a plane. The problem that occurs is that the resulting shapes can have somewhat irregular boundaries. A method to transform curved surfaces using the finite element method has already been proposed by Shimada [4] et al. However, the domain decomposition algorithm it uses is not suitable for deal with complex surfaces. Bennis [5] et al proposed a new general flattening algorithm by defining the notion of geodesic curvature. This approach however causes accumulated errors and overlaps and gaps between grids. A method of approximating complex surfaces with ruled surfaces was proposed by XiPing [6]. The non-ruled-complex-surfaces are subdivided into several regions and each region is approximated with a ruled surface. A 3-D forming method for saddle shape was carried out by geometrical approach [7]. By dividing 3-D surface into several patches, the surface was approximated to a set of plane patches. Thus geometrical parameters such as shrinkage and bending angle can be achieved.

A method of mathematical unfolding for 3-D curvature hull plate, small curve triangle method, was presented by XiaoXiong [8] et al. Based on this study, the effect of plate thickness and forming method in the process of curved plate splicing is considered by ChenXingbu [9].

4.2 UNFOLDING OF VARIOUS TYPES OF PLATES

According to the classification of hull plate parts in Section 2.2, flat plates do not need to be unfolded. In this study, the unfolding of curved plates and combined plates is discussion.

(1)Curved plate: The method of small curve triangles [8] is used in this study. As described in Section 3.2, the surface is divided into a series of n × n spatial grid points, and the NURBS method is used to fit the surface for calculating the curvature and the intersection of the surface. The sequence of the unfolding for curved plate is as the following:

1) Determine a datum line as the basis for dividing the plate into strips;
2) Dividing the plate into strips;
3) Dividing the strips into triangles;
4) Unfolding each triangle and splice them into strips;
5) Splicing all the strips into a plate.

The spliced plate is the result of unfolding, as shown in Figure 4.

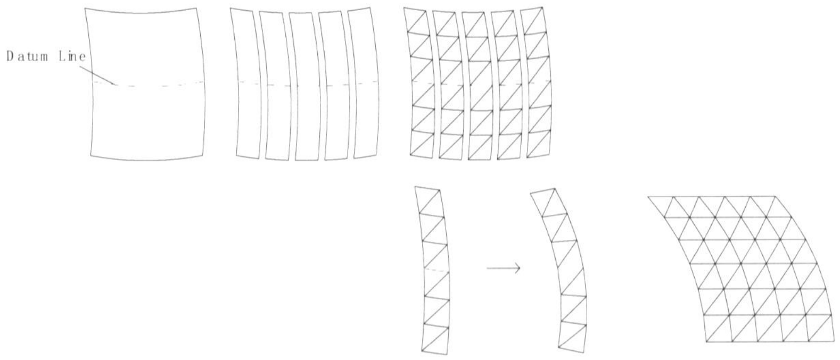

Figure 4. Schematic of Unfolding the Curved Plate

(2) Combined plate: As shown in Figure 5, the folded plate is consisted of three plates, which are two flat plate (Plate a and Plate b) and a curved plate. The sequence of the unfolding for folded plate is as the following:

1) Determining a datum plane (Plate a);

2) Finding the common edge of the plate adjacent to the datum plane, and calculating the midpoint of the common edge and the center of the datum plane;

3) Determining a direction vector according to the two points;

4) Calculating the length of the curved edge of the curved plate according to the extracted data area of the curved plate and the length of the common edge;

5) Rotating and translating the plate b and then splice it to the datum plane according to the determined direction vector;

6) Combining all the plates into one plate as the result of unfolding the folded plate.

The schematic of unfolding the folded plate is shown in Figure 5.

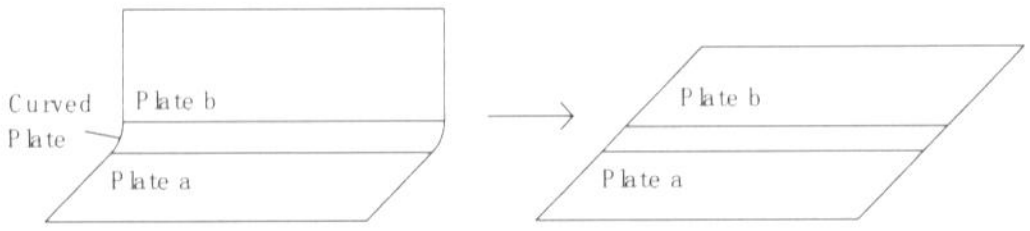

Figure 5. Schematic of Unfolding the Folded Plate

5. PROGRAM IMPLEMENTATION

The specific program implementation process of the above functions on CATIA 3DE platform is as follows:

(1) Compiling the program, clicking the Run Time Prompt button, and typing cnext in the command box to open the 3DE platform. The toolbar of interface in 3DE platform is shown in Figure 6, which includes four buttons: EditMode, SurfaceIntersect, EdgeInfor and Unfold.

(2) Clicking the EditMode command to switch the model into editable mode. As shown in Figure 6, it is a combined plate.

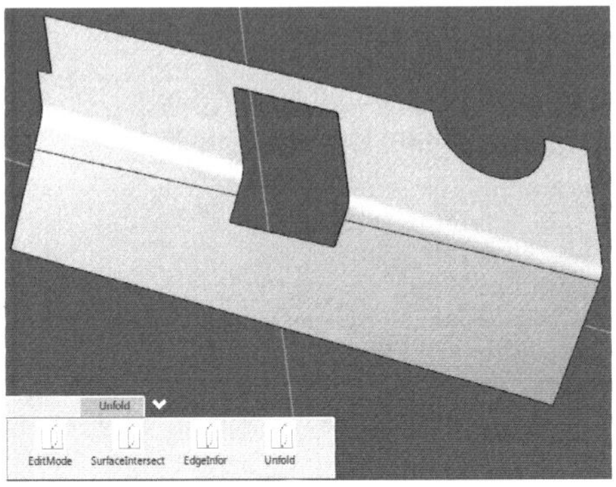

Figure 6. A Combined Plate in 3DE Platform

(3)Clicking the SurfaceIntersect command to find intersections between surfaces and generate structural lines. The structural lines on flat plate, curved plate and combined plate are shown in Figure 7.

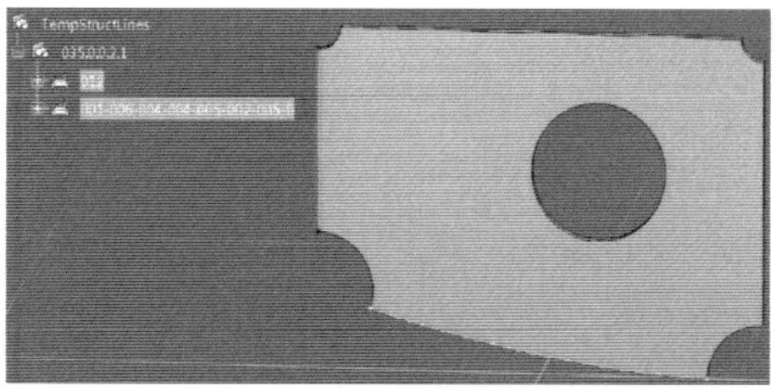

(a) Structural Lines on Flat Plate

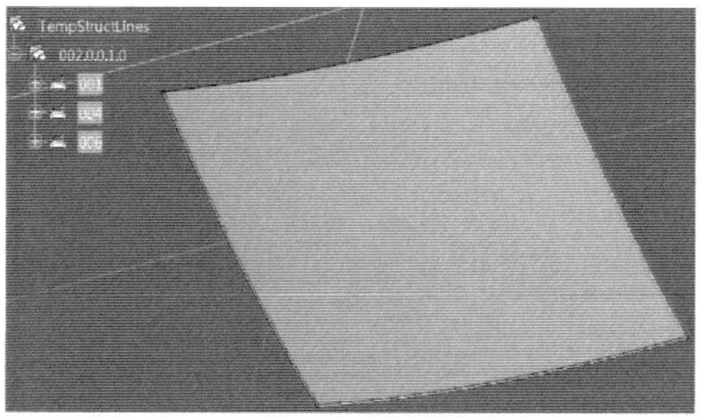

(b) Structural Lines on Curved Plate

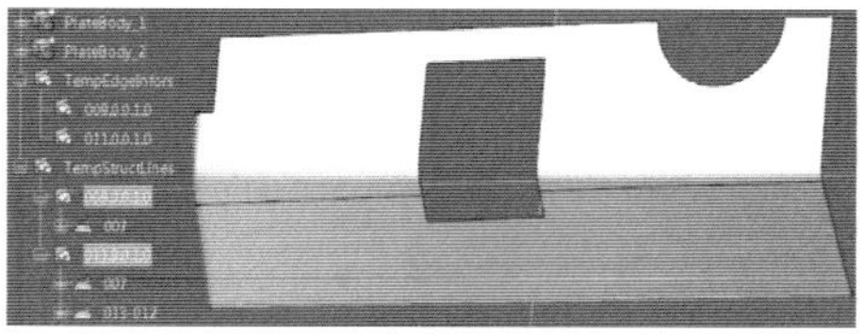

(c) Structural Lines on Combined Plate

Figure 7. Structural Lines

(4) Clicking the EdgeInfor command to add the process information of edges, which includes location, margin, compensation and bevel. Figure 8 shows that the process information of edges of various plates.
(5) Clicking the Unfold command to extract and unfold plates after setting the welding shrinkage of the rib and the number of surface mesh points. Figure 9 shows that the unfolding result of plates in AutoCAD.
The visualization of the extracted XML file is shown in Figure 10, and the extracted XML file is shown in Figure 11.

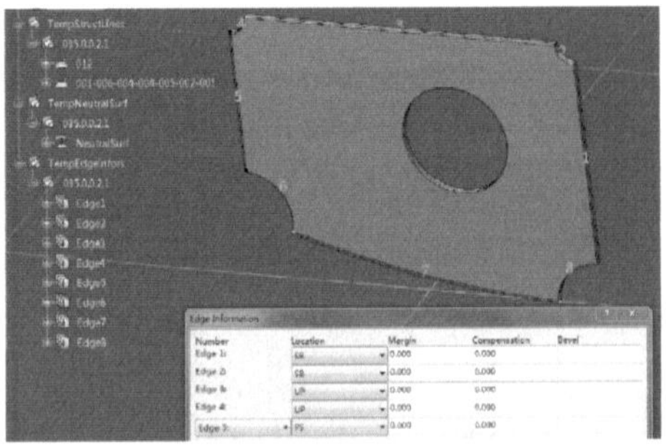

(a) Flat Plate

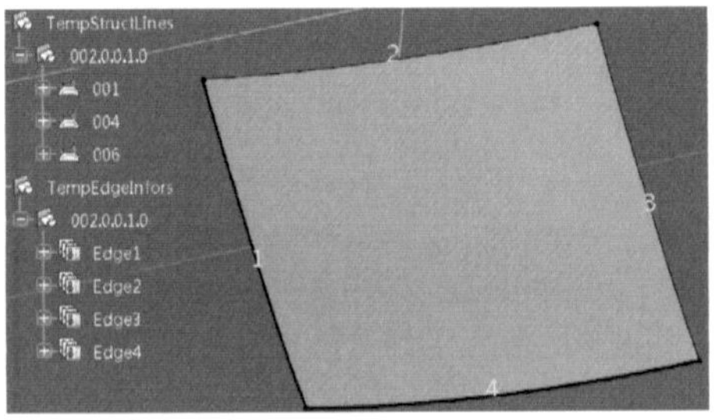

(b) Curved Plate

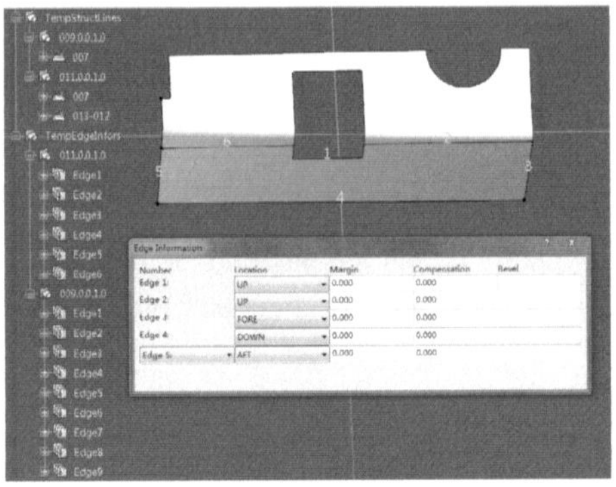

(c) Combined Plate

Figure 8. The Process Information of Edges of Various Plates

(a) Curved Plate

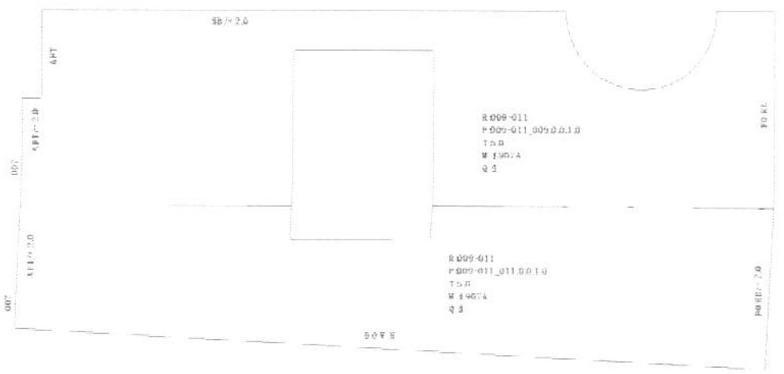

(b) Combined Plate

Figure 9. The Unfolding Result of Curved Plate and Combined Plate in AutoCAD

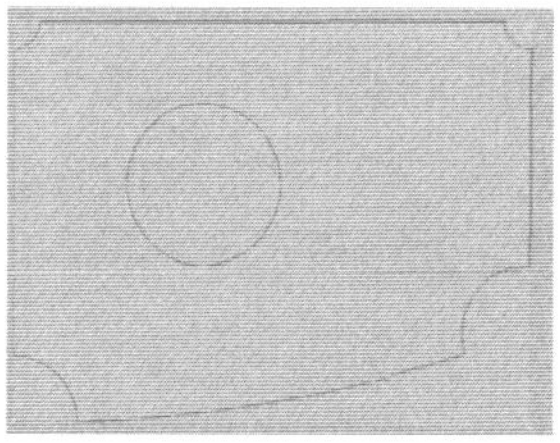

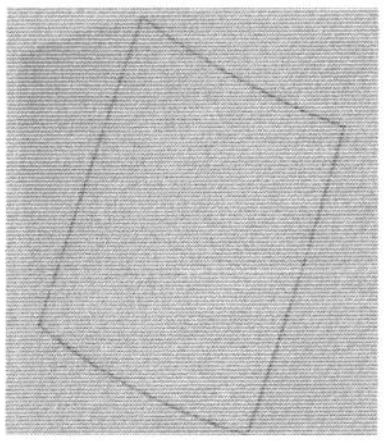

(a) Flat Plate (b) The Curved Plate

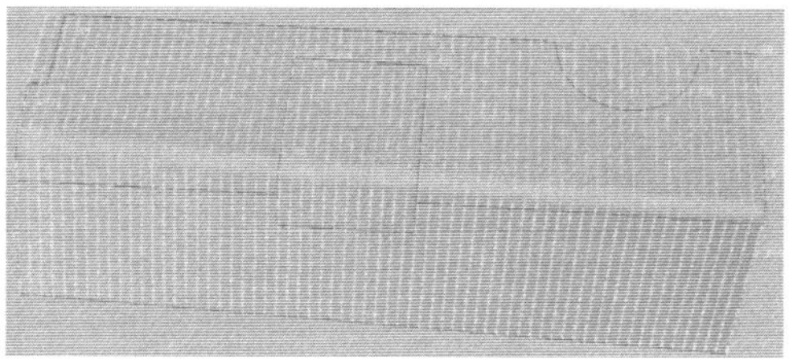

(c) The Combined Plate

Figure 10. The Extracting Result of Plates


```
<?xml version="1.0" encoding="UTF-8"?>
- <ExportShellInfo>
    <PlateInfo>
      <BlockName Value="035"/>
      <PartName Value="035_035.0.0.2.1"/>
      <Thickness Value="8"/>
      <FrameWeldingShrinkage Value="1"/>
      <PartNum Value="1"/>
    </PlateInfo>
    <DesignData/>
    <ProcessData>
    - <BoundaryData>
        <Edge EdgeNo="1" Location="" Margin="" Bevel="">
          <Point X="-1496.000" Y="-400.000" Z="1.992"/>
          <Point X="-1496.000" Y="-400.000" Z="300.000"/>
        </Edge>
        ......
      </BoundaryData>
    - <Reality_BoundaryData>
      - <Edge EdgeNo="1" Location="SB" Margin="2.000000" Bevel="Y">
          <Point X="-1496.000" Y="-2.500" Z="50.818"/>
          <Point X="-1496.000" Y="-2.500" Z="280.000"/>
          ......
        </Edge>
      </Reality_BoundaryData>
    - <SurfData>
        <Group GroupNo="1">
          <Point X="-1496.000" Y="-400.000" Z="1.992"/>
          <Point X="-1496.000" Y="-2.500" Z="1.992"/>
        </Group>
        ......
      </SurfData>
    </ProcessData>
</ExportShellInfo>
```


Figure 11. The Extracted XML File

6. CONCLUSIONS

In view of the low efficiency of ship production design and the shortcomings of TRIBON, AM, HD-SHM, CATIA and other systems in the unfolding of ship hull plates, in this study, automatic data extraction and unfolding of ship hull plates has carried out by the add-on development on CATIA software.

However, some complex hull plates still need to be treated manually. In addition, the efficiency of extraction and unfolding depends on the method of hull modeling. During the test, it was found that the staff used different ways to establish the 3-

D model of the ship hull, which would affect the extraction of part information to a certain extent. For unfolding complex shape hull plates, the accuracy of results should be further investigated.

7. REFERENCES

1. YUAN LI, MINGTAO CHEN, 'Application of Intelligen-t Drawing Software for Hull Production Design', *GSI S HIPBUILDING TECHNOLOGY, 41(03):43-44+30*, 2021.
2. DAICHUN LIANG, WEIMIN ZHANG, LIJIANG SUI, 'Brie-f Analysis on Secondary Development of CATIA Based on CAA', *Aeronautical Manufacturing Tec-hnology*, 2012(10):65-68, 2012.
3. LAXMI PARIDA, S.P, 'Constraint-satisfying planar development of complex surfaces', *Computer-Aided Design*, 25(4).225-232. 1993.
4. SHIMADA T, TADA Y, 'Approximate transformation of an arbitrary curved surface into a plane using d-namic programming', *Computer-Aided Design*, 23(2):153-159, 1991.
5. BENNIS C, VÉZIEN J M, IGLÉSIAS G, 'Piecewise surface flattening for non-distorted texture mapping', *Conference on Computer Graphics and Interactive Techniques*, ACM, 237-246, 1991.
6. PING XI, 'Geometric approach of 3D surface development', *Chinese Journal of Computers*, 20(4):315~322, 1997.
7. SEONG W J, JEON Y C, NA S J, 'Ship-hull plate forming of saddle shape by geometrical approach', *Journal of Materials Processing Technology*,213(11):1885-1893, 2013.
8. XIONG XIAO, Chaoyan Huang, Yong Hu, Ping Yuan, 'Study on the Method of Mathematical Expansion of 3-D Curvature Hull Plate', *SHIP&OCEAN ENGINEERING*, 42(01):74-77, 2013.
9. XINGBU CHEN, 'Research on Development Algorithm for Double Curvature Hull Plate and Development of Software', *Wuhan University of Technology*, 2019.

8. AUTHORS BIOGRAPHY

Yu Fu is a master degree candidate at Wuhan University of Technology. His research interest is advanced shipbuilding technology.

Wei Wei holds the current position of engineer at Wuchang Shipbuilding Industry Group Co., LTD.

Wuyi Xie holds the current position of manager at Wuchang Shipbuilding Industry Group Co., LTD.

Hu Yong holds the current position of professor at Wuhan University of Technology. He is responsible for unfolding algorithm. His previous experience includes precision forming technology of ship parts, such as profile, pipe, hull plate.

Chaoyan Huang is a Ph.D. degree candidate at Wuhan University of Technology. His research interest is advanced shipbuilding technology.

THE ROYAL INSTITUTION
OF NAVAL ARCHITECTS

CONFERENCE PROCEEDINGS

THE ROYAL INSTITUTION OF NAVAL ARCHITECTS

8-9 Northumberland Street, London WC2N 5DA

Tel: +44 (0)207 235 4622 Fax: +44 (0)207 259 5912

events@rina.org.uk

www.rina.org.uk

Stay connected with our social platforms

The Royal Institution of Naval Architects
8-9 Northumberland Street
London, WC2N 5DA
United Kingdom

ISBN 978-1-7138-8296-1

CURRAN ASSOCIATES INC.
proceedings.com

Workshop on Induction of Linguistic Structure 2012

Held at the 2012 Conference of the North American Chapter of the Association for Computational Linguistics: Human Language Technologies (NAACL-HLT 2012)

Montreal, Quebec, Canada
3-8 June 2012

ISBN: 978-1-62276-547-8